3rd China-Romania Science and Technology Seminar (CRSTS 2018)

IOP Conference Series: Materials Science and Engineering
Volume 399

Brasov, Romania
24 - 27 April 2018

Editors:

Ioan Vasile Abrudan
Tiemao Shi
Simona Lache
Yuhou Wu
Radu Muntean
Gheorghe Oancea

ISBN: 978-1-5108-6992-9
ISSN: 1757-8981

Printed from e-media with permission by:

Curran Associates, Inc.
57 Morehouse Lane
Red Hook, NY 12571

Some format issues inherent in the e-media version may also appear in this print version.

Printed by Curran Associates, Inc. (2018)

For permission requests, please contact the Institute of Physics at the address below.

Institute of Physics
Dirac House, Temple Back
Bristol BS1 6BE UK

Phone: 44 1 17 929 7481
Fax: 44 1 17 920 0979

techtracking@iop.org

Additional copies of this publication are available from:

Curran Associates, Inc.
57 Morehouse Lane
Red Hook, NY 12571 USA
Phone: 845-758-0400
Fax: 845-758-2633
Email: curran@proceedings.com
Web: www.proceedings.com

Table of contents

Volume 399

3rd China-Romania Science and Technology Seminar (CRSTS 2018) 24–27 April 2018, Brasov, Romania

Accepted papers received: 25 July 2018 Published online: 3 October 2018

Preface

Papers

The 3rd China - Romania Science and Technology Seminar CRSTS 2018

24 - 27 April 2018, Brasov, Romania

Selected, peer reviewed papers from the 3rd China - Romania Science and Technology Seminar (CRSTS 2018)

Edited by:

Ioan Vasile **ABRUDAN**
Transilvania University of Brasov, Romania

Tiemao **SHI**
Shenyang Jianzhu University, China

Simona **LACHE**
Transilvania University of Brasov, Romania

Yuhou **WU**
Shenyang Jianzhu University, China

Radu **MUNTEAN**
Transilvania University of Brasov, Romania

Gheorghe **OANCEA**
Transilvania University of Brasov, Romania

IOP Conf. Series: Materials Science and Engineering **399** (2018) 011001 doi:10.1088/1757-899X/399/1/011001

FOREWORD

After the establishment of the People's Republic of China, Romania was among the first countries in the world to have developed diplomatic relations with China (October 5th, 1949) and, in the following decades, the political, economic and cultural relations between the two countries have continuously evolved, being based on friendship, confidence and reciprocal support. Thousands of cooperation projects in almost all sectors of industry, agriculture, education, science, culture etc. were implemented for the benefit of China and Romania. Whilst the post-1989 changes in the Romanian society partially affected the intensity of this fruitful cooperation, more and more Romanian decision-makers and representatives of the business sector, the academia, as well as of the scientific and cultural life have recently understood that reviving the cooperation with China should be a priority. And not only because of the tremendous economic and technological development of China nowadays, but also because of the long-lasting friendship and trust between the two countries.

In recent years, Transilvania University of Brasov has been at the forefront of the academic and scientific cooperation between Romania and China. The establishment of the Confucius Institute in 2012, in partnership with Shenyang Jianzhu University, represented the impulse for a new stage of intensive cooperation between the two universities and also between the two countries. Hundreds of staff and student mobility exchanges, the participation in scientific events in China and Romania, the establishment of new cooperation partnerships in Beijing, Tianjin, Hunan province etc., the establishment of a Chinese language Bachelor's study programme, the implementation of an international joint laboratory project and a "111" project, and the active involvement in the "16+1 cooperation" in the field of forestry and education are only a few examples of this excellent cooperation in recent years. Our university is fully committed to the further development of the academic and scientific cooperation between Romania and China, and we are convinced that the approaching of the 70th anniversary of the People's Republic of China and 100 years of modern Romania, will represent the beginning of a new stage of intensive cooperation between the two countries for the benefit of our peoples.

In this context, Transilvania University was honoured to organize in partnership with Shenyang Jianzhu University and to host the 3rd China-Romania Science and Technology Seminar – CRSTS 2018 in Brasov in April 2018. More than 100 scientists for China and Romania, as well as from Germany, United States of America, Poland, Spain and Sri Lanka submitted 119 papers to the scientific board, out of which 90 accepted papers (52 from Romania, 34 from China, and one from Germania, USA, Spain and Sri Lanka) were presented during the Seminar. After a rigorous peer-review process (which included for each article at least two reviewers from the scientific committee), 62 papers were finally accepted for publication in the CRSTS 2018 proceedings volume. We would like to address special thanks to all those who were involved in the successful organisation of the 3rd China-Romania Science and Technology Seminar – CRSTS 2018, as well as to our distinguished publisher.

Prof. Dr. Ioan Vasile Abrudan
Rector of Transilvania University of Brasov, Romania

IOP Conf. Series: Materials Science and Engineering **399** (2018) 011001 doi:10.1088/1757-899X/399/1/011001

The 3rd China - Romania Science and Technology Seminar CRSTS 2018

■ **General Co-Chairs:**
Dong Yukuan, Party Secretary of Shenyang Jianzhu University
Mircea Tierean, President of the Senate of Transilvania University of Brasov

■ **Seminar Co-Chairs:**
Shi Tiemao, Rector of Shenyang Jianzhu University
Ioan Vasile Abrudan, Rector of Transilvania University of Brasov

■ **Academic Committee Co-Chairs:**
Wu Yuhou, Shenyang Jianzhu University
Teofil Mihailescu, Transilvania University of Brasov

Committee members

China

Prof. dr. Shi Tiemao,
Shenyang Jianzhu University

Prof. dr. Zhuang Zhuo,
Tsinghua University
Prof. Chen Ruisan,
Shenyang Jianzhu University
Prof. dr. Wang Youwei,
Chinese Society For Urban Studies
Prof. dr. Feng Guohui,
Shenyang Jianzhu University
Prof. dr. Zhang Ke,
Shenyang Jianzhu University

Prof. dr. Xu Ying,
Shenyang Jianzhu University
Prof. dr. Jiao Anliang,
China Construction Seventh Group Co. Ltd.
Prof. dr. Luo Shiwei,
China Construction First Group Co. Ltd.
Prof. dr. Liu Zijin,
China Academy of Building Research
Prof. dr. Lu Feng,
Shenyang Jianzhu University
Prof. dr. Li Hongnan,
Shenyang Jianzhu University
Prof. dr. Zhao Weijian,
Shenyang Jianzhu University

Romania

Prof. dr. Polidor Bratu,
Institute for Research on Equipment and Technologies in Construction Bucharest
Conf. dr. Luminita Barote,
Transilvania University of Brasov
Prof. dr. Ramona Clinciu,
Transilvania University of Brasov
Prof. dr. Mircea Viorel Dragoi,
Transilvania University of Brasov
Conf. dr. Nicolae Eftimie,
Transilvania University of Brasov
Prof. dr. Dan Georgescu,
Technical University of Civil Engineering Bucharest
Prof. dr. Simona Lache,
Transilvania University of Brasov
Conf. dr. Cristin Olimpiu Morariu,
Transilvania University of Brasov

Prof. dr. Daniela Lucia Manea,
Technical University of Cluj-Napoca
Prof. dr. Gheorghe Oancea,
Transilvania University of Brasov
Conf. dr. Cristian Pisarciuc,
Transilvania University of Brasov
Prof. dr. Ioan Calin Rosca,
Transilvania University of Brasov
Prof. dr. Dan Deaconu Sotropa,
Technical University "Gheorghe Asachi" of Iasi

IOP Conf. Series: Materials Science and Engineering 399 (2018) 011001 doi:10.1088/1757-899X/399/1/011001

Prof. dr. Li Songhua,
Shenyang Jianzhu University
Prof. dr. Zhang Lixiu,
Shenyang Jianzhu University
Prof. dr. Sun Hong,
Shenyang Jianzhu University

Prof. dr. Horatiu Teodorescu Draghicescu,
Transilvania University of Brasov
Prof. dr. Ioan Tuns,
Transilvania University of Brasov
Prof. dr. Sorin Vlase,
Transilvania University of Brasov
Prof. dr. Raul Zaharia,
Politehnica University of Timisoara

■ Organizing Committee Co-Chairs

Zhang Ke, Shenyang Jianzhu University
Simona Lache, Transilvania University of Brasov

Committee members

China	Romania
Prof. dr. Li Songhua	Prof. dr. Camelia Cerbu
Prof. Sun Yuanyuan	Lect. dr. Alexandru Catalin Filip
Associate Prof. dr. Zhou Peng	Lect. dr. Teofil Florin Galalatanu
Associate Prof. dr. Shi Huaitao	Asist. dr. Maria Violeta Guiman
Associate Prof. dr. Fan Liting	Lect. dr. Corina Silvia Micu
Associate Prof. dr. Wang He	Lect. dr. Radu Muntean
Associate Prof. dr. Yuan Zhe	Asist. dr. Violeta Mihaela Munteanu
Associate Prof. dr. Wei Chunyu	Associate Prof. dr. Luminita Aurica Parv
Associate Prof. dr. Wang Zhan	Lect. dr. Costel Plescan
Lect. dr. Bai Xiaotian	Asist. dr. Dorin Radu
Asist. Chen Zongsheng	Lect. dr. Marian Nicolae Velea
Asist. Yu Mingfu	Lect. dr. Marian Sebastian Zaharia

IOP Conf. Series: Materials Science and Engineering **399** (2018) 011002 doi:10.1088/1757-899X/399/1/011002

Peer review statement

All papers published in this volume of *IOP Conference Series: Materials Science and Engineering* have been peer reviewed through processes administered by the proceedings Editors. Reviews were conducted by expert referees to the professional and scientific standards expected of a proceedings journal published by IOP Publishing.

Green manufacturing by using organic cooling-lubrication fluids

B Arsene[1], G Pasca Pascariu[1*], F A Sarbu[1], M Barbu[1] and G Calefariu[1]

[1] Department of Engineering and Industrial Management, Transilvania University of Brasov, Brasov, 500174, Romania

[*] E-mail: arsene.bogdan@unitbv.ro

Abstract. The global competition and trends force the manufacturing industry players to permanently innovate, reduce costs and become more environmentally friendly, thus the pressure for eco-friendly, sustainable and more reliable products is rising. The environmental concerns have begun to be increasingly included in sustainable development strategies of the companies since the effects of worldwide pollution have become more and more visible. In manufacturing industry, in machining units, the cutting fluids are an important source of expenses and they can also be a threat for environment and employee's health. This paper presents a study regarding the possibility of replacing the traditional flood cooling in hard turning with minimal quantity lubrication using vegetable biodegradable oils since the scientific community has brought many contributions to the phenomenon of replacing grinding with hard turning. The capability of hard turning to replace grinding in many industrial applications is proved and thus the elimination of harmful coolant and transition to biodegradable cutting fluids seems to be the way of the future in machining.

1. Introduction

The environmental concerns have begun to be increasingly included in sustainable development strategies of the companies. Nowadays, machining processes constantly meet cost pressure and high quality expectations [1] , thus to minimize the environmental hazards due to industries a new approach in manufacturing is necessary i.e. green manufacturing, which is a key element under the overarching umbrella of sustainability [2, 3]. According to Maruthi&Rashmi "*Green Manufacturing is a philosophy rather than a standard or a process*" which aims *"to minimize waste and pollution through product and process design"* having as main goal the sustainability [4].

In the effort to achieve and implement sustainable manufacturing, the companies understand that sustainability is an important strategy for business environment i.e. saving resources (energy, raw material etc.), alignment with legal regulations and customer demands, and obtaining competitive advantage, which may lead to innovation in manufacturing [5]. Thus, the transition to green manufacturing is not only helpful for environment, it is helpful also for business [2].

Green manufacturing refers to methods and techniques to reduce the costs and environmental impact [6], to create safe and clean work-places [1], to reduce harmful emissions, to minimize or eliminate wasteful resources consumption and to recycle as much as possible [2, 4]. In manufacturing, the machining process is a major resource consumer, thus, the concerns and future research in green manufacturing regarding machining processes shall be focused on the main elements which lead to green and sustainable machining e.g. energy-efficient machine-tool, recyclable cutting tools and biodegradable cutting fluids, to achieve green production and eco-innovation, as they are presented in figure 1.

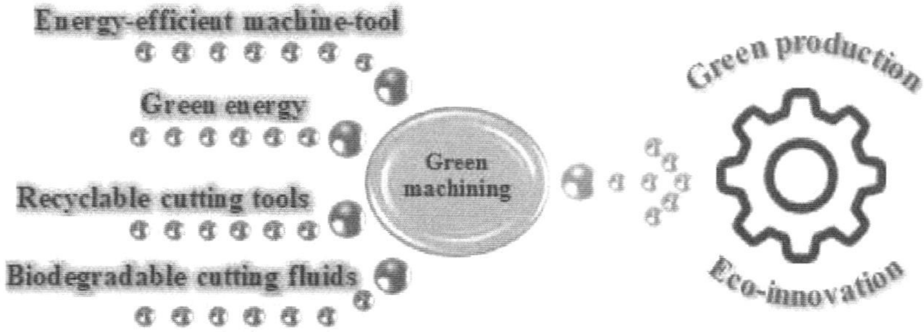

Figure 1. Green resources for green & sustainable machining.

For many years, manufactures are using different green machining strategies to respond to high pressures for sustainability, which include process time minimizing, optimizing the energy consumption and different actions to save oil, water, raw material and energy [7].

Green machining is an approach which doesn't use traditional cooling system, since the coolants have some unhealthy and dangerous chemical substances e.g. biocide to control the microbial growth [8]. The traditional method used in machining for cooling, i.e. flood cooling, has some important disadvantages e.g. it is harmful both for environment and workers due to chemical content of MWF such as chlorine, sulfur and phosphorus which are harmful and cause environmental pollution [9], it can cause serious ecological damage, it requires proper infrastructure (e.g. for storing, pumping, transport, reclamation, filtration and waste-water treatment) [1, 6], it demands continuous monitoring and treatment of the coolant to avoid fungal and bacterial growth [6] and it represents over 10% of the machining cost [1,2,6,10]. Sivarajan&Padmanabhan [1] reported that 16 – 20 % of the manufacturing costs in metal cutting industries are represented by CF and lubricants, while Xiao&Zhang [9] stated that 15 –30% of the machining cost, for a manufacturer from the automotive industry, consists in the use of grinding fluid. On the other hand, Tai et.al. [6] claimed that within powertrain operations, the cost of MWF is about 10-17% of the total manufacturing cost and Elmunafi et.al.[10] indicated that the cost for coolant use and handling can be about 17% of the overall production costs. These issues led to new approaches in machining e.g. nearly dry machining (NDM), cryogenic machining and dry machining. All these approaches are in agreement with the green machining requirements. The nearly dry machining methods consist in minimal quantity lubrication (MQL) and minimal fluid application (MFA). An overview of approaches for cooling in machining, i.e. in turning, is presented in figure 2 [11, 12].

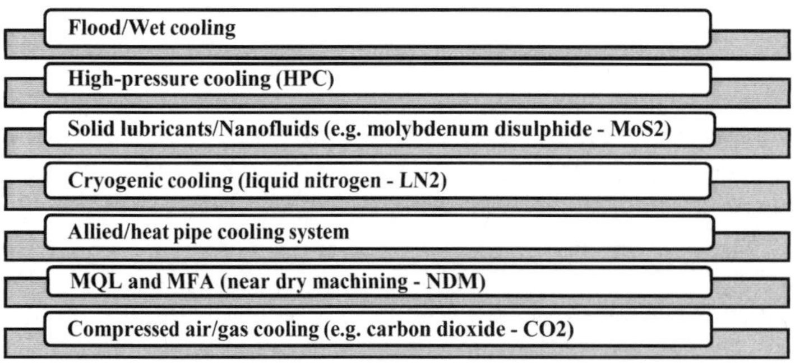

Figure 2. Cooling approaches in turning [11, 12].

2. Traditional cooling vs. MQL

In many machining units, the conventional coolant is still considered an essential factor in machining, "*a necessary evil*"[1], since it helps reduce friction, cutting temperature, thermal expansion of workpiece, fixture, and machine, remove chips [1, 6] and prevent adhesion between tool and chip, with impact in obtaining a stable machining process, a high-quality surface and a longer tool life [1]. Nevertheless, the elimination of harmful coolant, i.e. transition to biodegradable and environmental friendly CF and the wide implementation of MQL is the future in machining.

According to [13] the MQL technique was proposed a decade ago *"as a mean for addressing the issues of environmental intrusiveness and occupational hazard associated with airborne cutting fluid particles"*. MQL is an ecological alternative to traditional cooling systems (e.g. flooding the work-piece) [2, 6] used in machining to acquire cooling and lubrication during processes, which consists in spraying directly into the cutting zone through nozzles of a small amount of oil or metal working fluid (MWF) based on compressed air [2, 6], at a low rate of 10 – 200 mL/h [6, 10, 14]. This method has shown in the last 10-15 years numerous advantages in comparison with flood cooling, e.g. significant savings in CF, energy and equipment costs, it consumes less energy, it offers a high flexibility to relocate the machines, it reduces ecological hazards, it provides a healthier and cleaner work-place, it reduces greenhouse gas emission, and ensures better quality of products and higher productivity [2, 6], with effects in cost optimization and competitiveness increase.

One of the first applications of MQL in hard turning was designed by Ko et.al. [15] that developed an air-oil mix cooling system to provide cooling and lubrication effects concurrently when hard turning AISI 52100 (62-64 HRC) steel. They have used several types of inserts e.g. coated tungsten carbides inserts, cermet-based inserts and CBN inserts. First, they have found that the coolant should be ejected on the rake and clearance faces, since the higher temperature was located in these zones rather than in other zones. The highest temperature was observed on the rake face. Second, they have found that this system provides better results in terms of Ra roughness, radial force and tool wear comparative to dry, flood and air-jet cooling when turning hard material with TIN coated carbide inserts due to the simultaneous reduction of temperature and friction by the mist coolant. Obviously, the carbide inserted tool life was shorter than cermet and CBN insets, but the most important conclusion of the study is that the proposed method i.e. air-oil mix cooling is superior and more suitable than the rest of methods.

This technique was studied in machining by various authors on different types of oils. The most environment friendly oils used in experiments are vegetable oils e.g. castor oil [16, 17], coconut oil [18-23], sunflower oil [16, 22, 24-27], soya bean oil [22, 27-29], palm oil [19, 30-32] and canola oil [16, 20, 21, 25, 26]. The outcomes have shown improvements in terms of tool wear, tool life, cutting and feed forces and surface roughness (Ra) compared to dry machining or flood cooling and according to [22] the performance of the vegetable oil which is superior to mineral-based oils. Vegetable oils have the status of environmental friendliness regarding biodegradability, resource renewability and cutting process efficiency [10], they are less toxic and decrease the waste treatment cost due to their great biodegradability [25] and according to [33] they are considered "*as viable alternative to petroleum-based fluids*" due to the following reasons:

- the nature and structure of the molecules create a strong and dense homogeneous lubricating film with high capacity to absorb pressure [33, 34];
- the vegetable oil lubricating film is intrinsically strong and lubricious and improves quality of the products, it reduces frictions and reduces heat generation [25, 33, 34];
- the high boiling point and major molecular weight ensure less losses from vaporization and misting [33];
- vegetable oils have a high flash point, thus the risk of fire hazards is lower and of smoke formation is reduced [33- 35];
- vegetable oils are environmentally friendly, they do not cause health issues [33] and they have high biodegradability [35];
- vegetable oils are compatibility with additives [35];

- vegetable oils have a high viscosity index which provides a stable lubricity [34, 35];

Another technique named Minimal Fluid Application (MFA) or Minimal Cutting Fluid Application (MFCA) was developed by Varadarajan et.al [36]. The proposed solution contains in a fuel pump coupled to an infinitely variable electric drive which generates and sends high velocity pulsing jet of cutting fluid on tool-part interface. The main advantage of this system is that it facilitates the independent variation of the injection pressure, frequency of pulsing and the rate of application of cutting fluid [36-38]. Also, the system can deliver fluid at a flow rate of 0.5 ml/min (i.e. 30 ml/h) through six nozzles simultaneously [36]. Some pieces of research [38, 39] have shown that MCFA technique provides better results than MQL in terms of surface roughness and cutting temperature at soft turning of steel with 45 HRC hardness.

3. Current researches in MQL machining on vegetable oils

3.1. Soft machining on vegetable oils as cooling-lubrication fluids

In soft machining there are several papers which aim at the use of vegetable oil as cooling-lubrication fluid. Belluco & De Chiffre [40] studied the drilling AISI 316L austenitic stainless with high speed steel (HSS) having as CF a mineral oil-based and a mixture based on rapeseed oil with an amount of sulfur and phosphor containing additives. The experiments indicated that the vegetable-based oils produced better results than mineral oil (MO) regarding tool life (177% increase) and thrust force (7% reduction). On the other hand, Rahim & Sasahara [30]describe the drilling of titanium alloy Ti–6Al–4V with an indexable carbide drill. The palm oil and synthetic ester was selected as CF. MQL on palm oil produced lower thrust force, torque, workpiece temperatures, flank wear and corner wear compared with MQL on synthetic ester and flood cooling. Similar results were obtained in another study of the same authors [31] when drilling Inconel 718. Kuram et.al. [24] presented the end milling of AISI 304 on vegetable CF as sunflower and canola oils with 8% lubricating additive (sulphur based additive) and commercial semi-synthetic CF. The outcomes have shown better results regarding specific energy (J/mm3) and surface roughness for canola oils. From the experiments of Pereira et. al. [16] who studied the end milling of heat resistant super-alloy (HRSA) Inconel 718 using different types of vegetable lubricating fluids e.g. sunflower oils, castor oil, canola oil and a recycled oil, we know that the highest tool life was achieved by using canola oil and recycled oil and the lower friction coefficient was obtained when on castor oil.

Regarding the turning of materials, Elmunafi et.al [17] studied the turning of AISI 420 stainless steel (47-48 HRC) on castor oil at a flow rate of 50 mL/h and 5 bar air pressure, to observe the influence of cutting parameters on tool life, surface roughness and cutting forces. Their observations revealed that vegetable oil is superior to dry hard turning since the tool life was much higher due to reduction in friction and temperature in cutting zone, and the surface roughness and cutting forces enhanced slightly. Raj et. al. [28] conducted experiments on turning (MFA) of AISI 4340 (45 HRC) using soya bean and MO as lubrication fluid in order to identify the dynamic of cutting force. The cooling parameters were set on three levels and for oil pressure, oil quantity and frequency of pulsating the values were 50 bar, 75 bar and 100 bar, 3ml/min, 6ml/min and 9ml/min, respectively 250, 500 și 750 pulses/min. The result have shown that the use of MFA technique on soya bean oil provided the minimum value of cutting force compared to flood cooling (soya bean or MO) and dry cutting , regarding MFA, the highest oil rate and pressure provided the lowest cutting force. Xavior & Adithan [18] evaluated the influence of cutting fluids on tool wear and surface roughness, using coconut oil, soluble oil and straight cutting oil, in turning of AISI 304. In general, coconut oil was found to be a better cutting fluid than the conventional mineral oils and according to ANOVA it is the second influencing factor after feed rate on surface roughness. Some authors dealt with turning of AISI 1040 (32 HRC) using conventional CF and vegetable oil based nano cutting fluids, at a flow rate of 10 mL/min [19-21]. Vamsi Krishna et.al. [19] studied the machining on coconut oil and synthetic oil prepared with solid nanoboric acid particles (50 nm) in different percentage (0.25%, 0.5% and 1%). Thermal conductivity increased and specific heat decreased with percentage increase in nanoboric acid in base oil. The coconut oil showed better performance compared to synthetic oil in terms of cutting temperatures, tool flank wear and surface roughness and in all cases the palm oil with 0.5% suspensions was superior. On the other hand, Padmini et.al. [20, 21] investigated the machining on canola, sesame and coconut oil prepared by dispersing nano-suspensions of molybdenum disulphide (MoS2) with the same nano-particle inclusion (NPI-%) i.e. 0.25, 0.5 and 1. The coconut oil with 0.5% MoS2 leads to lowest cutting force, lowest cutting temperatures, lowest tool wear and best surface roughness compared to all other lubricating conditions.

Cetin et.al. [25] and Ozcelik et.al. [26] investigated the turning of AISI 304L (32 HRC) stainless steel on two types of vegetable oil (canola and sunflower), a commercial mineral based CF and a commercial semi-synthetic CF. The vegetable oils were prepared with 8% and 12% extreme pressure (EP) chemical additives, whose functions refer to corrosion prevention, pH regulation, anti-foaming and improving the flash point. In [25] polysorbate (Tween 85 andTween 20) was used as additive and in [26] sulfur based EP additive (Polartech XP 9018) was used. From [25] we know that higher rate (12%) of EP additives provide better performance in terms of forces (cutting and feed) and affects negatively the surface roughness due to chemical interaction with the machined surface. Anyway, the vegetable oils are superior to mineral and semi-synthetic CF. Ozcelik et.al. [26] indicated that canola oil with 8% EP additives gave the best performance regarding surface roughness, feed forces and tool wears. Ghuge & Mahalle [22] investigated the soft turning of AISI 4130 dry, flood and MQL (50mL/h) using 5 types of oils, one mineral and four vegetable (soya beam, coconut, sunflower and groundnut). The cutting temperature was reduced by 18% in case of MQL compared with dry and 5% compared to flood. Regarding the oils, soya beam oil provides better results than the others in terms of cutting forces and roughness. In another study, [27], the same authors evaluated the cutting force and power consumption at turning of the same steel, using soya beam, sunflower and MO. As in the previous study, soya beam oil provided the best result i.e. 9% reduction in cutting force and power consumed compared with MO and an average of 7 % compared to sunflower oil.

A summary of the parts considered in the research, performed in soft machining by replacing flood cooling with vegetable oils is presented in table 1.

Table 1. Parts considered in research, performed in soft machining using vegetable oils.

Machining operation	Type of oil	Materials	References
Drilling	• Palm and rapeseed /canola	AISI 316L, Ti–6Al–4V, Inconel 718	[30,31,40]
Milling	• Sunflower, castor and canola	Inconel 718, AISI 304	[16,24]
Turning	• Castor, soya bean, coconut, canola, palm, sesame, sunflower, groundnut	AISI 420, AISI 4340, AISI 304, AISI 1040, AISI 304L, AISI 431	[17-22, 25-27, 28, 32]

3.2. Hard machining on vegetable oils as cooling-lubrication fluids

Since the scientific community has brought many contributions to the phenomenon of replacing grinding with hard turning and the capability is proved, the elimination of flood cooling and the transition to MQL or MFA on vegetable oil is the next step for hard turning. Even if, the literature presents lots of papers regarding soft machining (e.g. soft turning) on vegetable oils as cooling-lubrication fluids, there is a lack of papers regarding hard turning of materials with hardness higher than 55 HRC.

Chinchanikaret.al. [23] studied the hard turning of AISI 52100 (60-62 HRC) under dry condition and on two different types of coolant, a milky emulsion (composed of water and soluble cutting oil) and vegetable oil (coconut oil). The outcomes have shown that at high speed rate (over 160 m/min) and high depth of cut (over 0.,35 mm) the coconut oil has provided better surface finish in terms of roughness compared to the other two. This result indicates that using vegetable oil is the most suitable way in order to increase the productivity. Even if the tool wear and cutting force were not considered, the perspective for coconut oil as coolant is real. Gajrani et.al. [41] presented the hard turning of AISI H13 (56 HRC) on a mineral oil-based and a vegetable-based eco-friendly bio-cutting fluid. The techniques chosen were flood cooling and MQL at an air pressure of 0.5 MPa (5 bar) and oil flow rate of 35 mL/h. Compared to flood cooling, MQL produced better results in terms of feed and cutting forces, surface roughness and friction coefficient and regarding MQL CF, the biodegradable fluid performed better than mineral oil-based fluid taking into account the same features and it has shown higher thermal conductivity and better ability to penetrate the chip-tool interface.

Gunjal & Patil [29] conducted experiments on hard turning of AISI 4340 with coated carbide inserts under MQL and dry conditions, on three types of vegetable based oil, namely canola, soya bean and coconut, at a flow rate of 50mL/h and 5 bar pressure. The best results in terms of tool life, tool wear and surface roughness were obtained by the machining on canola oil. Last but not least, Mia et.al. [14] investigated the hard turning of AISI 1060 with hardness 56-60 HRC using olive oil and coated carbide inserts, under MQL at a flow rate of 150 ml/h. Their study focused on the influence of cutting parameters on surface finish characteristics e.g. average surface roughness (Ra), root mean square surface roughness (Rq) and maximum height of surface roughness (Rz) and tool wear e.g. average principal flank wear (VB) and auxiliary principal flank wear (VS). The outcomes indicated that the lowest level of roughness was obtained with the highest cutting parameters, which means that olive oil may be a viable alternative to increase the productivity and reduce costs. Even if the experiments have not indicated the difference between flood or dry hard turning and MQL on olive oil, this study may be the starting point for future research with regard to MQL hard turning on olive oil, due to superior lubrication property.

A summary of the parts considered in the research, performed in hard turning of hardened steels (> 55 HRC) using vegetable oils is presented in table 2:

Table 2. Parts considered in research, performed in hard turning using vegetable oils.

Reference	Material	Hardness (HRC)	Tool material	Lubricant	Discussion
[14]	C60 (AISI 1060)	55-60	• Coated carbide inserts	• olive oil	• olive oil was used in order to identity the effect of cutting parameters on tool wear and surface roughness
[23]	100Cr6 (AISI 52100)	60-62	• PVD[a] coated TiSiN[b]-TiAlN[c] carbide insert	• coconut oil	• coconut oil provided better surface roughness
[29]	40NiCrMo6 (AISI 4340)	52-54	• PVD AlTiN[d] coated carbide inserts	• coconut oil • canola oil • soyabean	• canola oil performed better than soya bean and coconut oils in terms of tool life, tool wear and surface roughness
[41]	X40CrMoV5 (AISI H13)	56	• TiN[e] coated carbide insert	• vegetable-based eco-friendly bio-CF	• biodegradable CF performed better than mineral oil-based in terms of feed and cutting forces, surface roughness, friction coefficient and thermal conductivity

[a],Physical Vapor Deposition;
[b],TiSiN - Titanium Silicon Nitride;
[c],Titanium Aluminium Nitride;
[d],Aluminium Titanium Nitride;
[e],Titanium Nitride.

As it can be seen, there is a lack of research regarding MQL hard turning on vegetable oil. Whether the outcomes of future work are similar or better than the result from soft machining e.g. reduction in cutting force, thus the perspective of HT forecost reduction or productivity increase seems to be real.

From the fundamentals of machining we know that the cutting power (Pc - kW) depends largely on the cutting speed (Vc) and on the main cutting force (Fc). Thus, the energy consumed (kWh) depends mainly on these two parameters. As well, the tool wear (flank and crater) is influenced by Vc and Fc. In this way, the perspective for HT on vegetable oil as cooling-lubrication fluid is presented in table 3.

Table 3. Perspectives of HT on vegetable oil as BCF using MQL technique.

»»Reduction in Fc&ConstantVc--->Decrease of energyconsumption i.e. lowerenergy cost, lower tool wear, higher tool lifeandlower toolscost
»»Reduction in Fc&Increase of Vc --->Constant energy consumption&increase of productivity

4. Conclusions and further work

The transition to eco-friendly CF is necessary to ensure a sustainable and greener manufacturing and machining, more safety and cleaner workplaces, in order to enhance the competitiveness and to gain the technology competitive advantage.

The vegetable oils seem to be a real, viable and cheap option to replace traditional flood cooling in many industrial application, since the outcomes proved that this types of cooling-lubricating fluids yield better performance in terms of roughness, cutting force, power consumption, tool wear and specific energy. Even if, the literature presents lots of paper regarding soft machining (e.g. soft turning, drilling and milling) on vegetable oils as cooling-lubrication fluids, there is a lack of papers regarding hard turning of materials with hardness higher than 55 HRC, in order to replace grinding in an eco-efficient way. Thus, further research shall be focused in this direction, using ceramic and CBN inserts with wider geometry and different types of vegetable oils, applying MQL, in order to investigate the surface characteristics (e.g. micro-hardness, white layers, residual stress and surface roughness), tool wear, power consumption and process costs.

References:

[1] Sivarajan S and Padmanabhan R 2014 Green machining and forming by the use of surface coated tools *Procedia Eng.* **97** 15
[2] Paul I D, Bhole G P and Chaudhari J R 2014 A Review on Green Manufacturing: It's Important, Methodology and its Application *Procedia Mater. Sci.* **6** 1644
[3] Singh A, Philip D and Ramkumar J 2015 Quantifying green manufacturability of a unit production process using simulation *Procedia CIRP* **29** 257
[4] Dilip Maruthi G and Rashmi R 2015 Green Manufacturing: It's Tools and Techniques that can be implemented in Manufacturing Sectors *Mater. Today Proc.* **2** 3350
[5] Dornfeld D A 2014 Moving towards green and sustainable manufacturing *Int. J. Precis. Eng. Manuf. - Green Technol.* **1** 63
[6] Tai B L, Stephenson D A, Furness R J and Shih A J 2014 Minimum quantity lubrication (MQL) in automotive powertrain machining *Procedia CIRP* **14** 523
[7] Helu M, Behmann B, Meier H, Dornfeld D, Lanza G and Schulze V 2012 Impact of green machining strategies on achieved surface quality *CIRP Ann. - Manuf. Technol.* **61** 55
[8] Rahim E A, Ibrahim M R, Rahim A A, Aziz S and Mohid Z 2015 Experimental investigation of minimum quantity lubrication (MQL) as a sustainable cooling technique *Procedia CIRP* **26** 351
[9] Xiao K Q and Zhang L C 2006 The effect of compressed cold air and vegetable oil on the subsurface residual stress of ground tool steel *J. Mater. Process. Technol.* **178** 9
[10] Elmunafi M H S, Mohd Yusof N and Kurniawan D 2015 Effect of cutting speed and feed in

turning hardened stainless steel using coated carbide cutting tool under minimum quantity lubrication using castor oil *Adv. Mech. Eng.***7** 168781401560066

[11] Sharma V S, Dogra M and Suri N M 2009 Cooling techniques for improved productivity in turning *Int. J. Mach. Tools Manuf.***49** 435

[12] Liew P J, Shaaroni A, Sidik N A C and Yan J 2017 An overview of current status of cutting fluids and cooling techniques of turning hard steel *Int. J. Heat Mass Transf.* **114** 380

[13] Boubekri N and Shaikh V 2014 Minimum Quantity Lubrication (MQL) in Machining: Benefits and Drawbacks *J. Ind. Intell. Inf.* **3** 205

[14] Mia M, Dey P R, Hossain M S, Arafat M T, Asaduzzaman M, Shoriat Ullah M and Tareq Zobaer S M 2018 Taguchi S/N based optimization of machining parameters for surface roughness, tool wear and material removal rate in hard turning under MQL cutting condition *Measurement* **122** 380

[15] Ko T J, Kim H S and Chung B G 1999 Air-oil cooling method for turning of hardened material *Int. J. Adv. Manuf. Technol.* **15** 470

[16] Pereira O, Martín-Alfonso J E, Rodríguez A, Calleja A, Fernández-Valdivielso A and López de Lacalle L N 2017 Sustainability analysis of lubricant oils for minimum quantity lubrication based on their tribo-rheological performance *J. Clean. Prod.***164** 1419

[17] Elmunafi M H S, Kurniawan D and Noordin M Y 2015 Use of castor oil as cutting fluid in machining of hardened stainless steel with minimum quantity of lubricant *Procedia CIRP* **26** 408

[18] Xavior M A and Adithan M 2009 Determining the influence of cutting fluids on tool wear and surface roughness during turning of AISI 304 austenitic stainless steel *J. Mater. Process. Technol.***209** 900

[19] Vamsi Krishna P, Srikant R R and Nageswara Rao D 2010 Experimental investigation on the performance of nanoboric acid suspensions in SAE-40 and coconut oil during turning of AISI 1040 steel *Int. J. Mach. Tools Manuf.***50** 911

[20] Padmini R, Vamsi Krishna P and Krishna Mohana Rao G 2016 Effectiveness of vegetable oil based nanofluids as potential cutting fluids in turning AISI 1040 steel *Tribol. Int.***94** 490

[21] Padmini R, Vamsi Krishna P, Krishna Mohana Rao G and Rukmini Srikant R 2016 Performance evaluation of vegetable oil based nano cutting fluids in machining using grey relational analysis-A step towards sustainable manufacturing *J. Clean. Prod.***172** 2862

[22] Ghuge N C and Mahalle A M 2016 Performance Evaluation of Vegetable Oil during Turning of AISI4130 in Terms of Temperature and Surface Roughness Performance Evaluation of Vegetable Oil during Turning of AISI4130 in Terms of Temperature and Surface Roughness *IJESC* **6** 5879-

[23] Chinchanikar S, Salve A V, Netake P, More A, Kendre S and Kumar R 2014 Comparative Evaluations of Surface Roughness During Hard Turning under Dry and with Water-based and Vegetable Oil-based Cutting Fluids *Procedia Mater. Sci.***5** 1966

[24] Kuram E, Ozcelik B, Bayramoglu M, Demirbas E and Simsek B T 2013 Optimization of cutting fluids and cutting parameters during end milling by using D-optimal design of experiments *J. Clean. Prod.***42** 159

[25] Cetin M H, Ozcelik B, Kuram E and Demirbas E 2011 Evaluation of vegetable based cutting fluids with extreme pressure and cutting parameters in turning of AISI 304L by Taguchi method *J. Clean. Prod.***19** 2049

[26] Ozcelik B, Kuram E, Huseyin Cetin M and Demirbas E 2011 Experimental investigations of vegetable based cutting fluids with extreme pressure during turning of AISI 304L *Tribol. Int.***44** 1864

[27] Ghuge N C and Mahalle A M 2017 Response Surface Modeling for Cutting Force and Power Consumption during Turning using Vegetable Oils *JET* **7** 75

[28] Raj A, Leo Dev Wins K, Dev Anand M and Varadarajan A S 2016 Performance Evaluation of Vegetable Oil based Cutting Fluid during Hard Turning of AISI 4340 Steel with Minimal

Cutting Fluid Application *Indian J. Sci. Technol.* **9** 0

[29] Gunjal S U and Patil N G 2018 Experimental Investigations into Turning of Hardened AISI 4340 Steel using Vegetable based Cutting Fluids under Minimum Quantity Lubrication *Procedia Manuf.* **20** 18

[30] Rahim E A and Sasahara H 2011 A study of the effect of palm oil as MQL lubricant on high speed drilling of titanium alloys *Tribol. Int.* **44** 309

[31] Rahim E and Sasahara H 2017 Performance of palm oil as a biobased machining lubricant when drilling inconel 718 *MATEC Web Conf.* **101** 80

[32] Mahadi M A, Choudhury I A, Azuddin M and Nukman Y 2017 Use of Boric Acid Powder Aided Vegetable Oil Lubricant in Turning AISI 431 Steel *Procedia Eng.* **184** 128

[33] Khan M M A and Dhar N R 2006 Performance evaluation of minimum quantity lubrication by vegetable oil in terms of cutting force, cutting zone temperature, tool wear, job dimension and surface finish in turning AISI-1060 steel *J. Zhejiang Univ. A* **7** 1790

[34] Kuram E, Ozcelik B and Demirbas E 2013 *Green Manufacturing Processes and Systems* (Berlin: Springer-Verlag Berlin Heidelberg) Ed Paulo Davim J

[35] Shashidhara Y M and Jayaram S R 2010 Vegetable oils as a potential cutting fluid-An evolution *Tribol. Int.* **43** 1073

[36] Varadarajan A S, Philip P K and Ramamoorthy B 2002 Investigations on hard turning with minimal cutting fluid application (HTMF) and its comparison with dry and wet turning *Int. J. Mach. Tools Manuf.* **42** 193

[37] Kumar C R V and Ramamoorthy B 2007 Performance of coated tools during hard turning under minimum fluid application *J. Mater. Process. Technol.* **185** 210

[38] Raj A, Wins K L D and Varadarajan A S 2016 Comparison of surface roughness and chip characteristics obtained under different modes of lubrication during hard turning of AISI H13 tool work steel. *IOP Conf. Ser. Mater. Sci. Eng.* **149**

[39] Raj A, Leo Dev Wins K, Dev Anand M and Varadarajan A S 2016 Performance Evaluation of Vegetable Oil based Cutting Fluid during Hard Turning of AISI 4340 Steel with Minimal Cutting Fluid Application *Indian J. Sci. Technol.* **9** 0

[40] Belluco W and De Chiffre L 2004 Performance evaluation of vegetable-based oils in drilling austenitic stainless steel *J. Mater. Process. Technol.* **148** 171

[41] Gajrani K K, Ram D and Ravi Sankar M 2017 Biodegradation and hard machining performance comparison of eco-friendly cutting fluid and mineral oil using flood cooling and minimum quantity cutting fluid techniques *J. Clean. Prod.* **165** 1420

IOP Conf. Series: Materials Science and Engineering **399** (2018) 012002 doi:10.1088/1757-899X/399/1/012002

Impact of axial loads on the sound directivity of angular contact ceramic ball bearings

X T Bai[1, 2*], Z Yuan[1, 2] and H P Yan[1]

[1] School of Mechanical Engineering, Shenyang Jianzhu University, Shenyang, Liaoning, 110168, China
[2] International Joint Laboratory on Modern Construction Engineering Equipment, Shenyang, Liaoning, 110168, China

[*] E-mail: acetyws@163.com

Abstract. This paper mainly focuses on the sound directivity of the angular contact ceramic ball bearing applied to the ceramic motorized spindle. The nonlinear dynamic model is proposed with consideration of axial load, nonlinear hydrodynamic forces and interaction between balls and cage. Then the sub-source decomposition method is introduced for the establishment of an acoustic radiation model. Sound pressure levels along the circumferential direction are used as evaluation indexes of sound directivity, and the sound directivity with axial loads at different rotation speed are analyzed. Results indicate that the sound directivity tends to be obvious with the increase of rotation speed, and the axial loads have significant influence on the deviation angles of directivity performance. This work provides insights into monitoring the acoustic performance of ceramic bearings, and can guide further research and development.

1. Introduction

As the key components of mechanical systems, bearings are required to have high precision and reliability [1]. Ceramic ball bearings take ceramic as the material of rolling elements and has been proved of higher turning accuracy and thermal stability over traditional steel ball bearings. Therefore, the ceramic ball bearings are widely applied in research fields such as aeronautics and navigation. However, with the increase of rotation speed, the excessive radiation noise affects the entire acoustic performance, and hinders the further improvement of working speed. The radiation noise results from the friction and impact between the components of the bearing, and has great impact on the surrounding acoustic environment [2]. Sound directivity is an important evaluation index of acoustic performance and varies greatly with the working parameters. Researches on the sound directivity can help get the accurate sound radiation performance and lead to the noise reduction.

Ceramic ball bearings applied on mechanical systems are mainly angular contact bearings, which work under axial and radial loads. Axial loads are widely used to eliminate deflections due to external loads and prevent skidding. However, excessive axial loads may lead to large deformations on the rolling elements and raceways, and bring about extra friction and noise [3]. The deformations on the steel raceways and ceramic balls are different because of different stiffness and rigidity. The influence of axial loads on the sound directivity is complex, hence the theoretical analysis of the impact of axial loads on the sound directivity of ceramic bearings is quite rare as far as the author's knowledge [4-6].

Many researchers have conducted researches on the dynamic model of angular contact bearings. In 1995, Chapman J J proposed an alternative theoretical model of angular contact ball bearing, and validated the results by an experimental investigation [7]. As one of the following influential studies, Cao Y Z and Altintas Y presents a model of spindle bearing and machine tool system, and used Hertzian contact theory to calculate the contact force and displacement of bearing. The centrifugal force and gyroscopic moment were also taken into consideration to get the dynamic response of the bearing-spindle system [8]. It has been proved that the vibration of rolling elements and rings are the main sources of noise radiation at low rotation speed, and the noise radiated by the cage is relatively slight because of the flexibility of the material. Wang Y L focused on the ball skidding and proposed an analytical model to investigate the contact balls and raceways, cage and lubricant. It was pointed out that the internal load influenced the skidding behavior significantly and appropriate loads should be determined to reduce skidding [9]. Zhao C J and Yu X K analyzed the relationship between the gyroscopic torque and friction coefficient between rolling element and groove and built a quasi-statics model of axial load for calculation. The variation of the outer contact angle with axial force was obtained and the ratio of contact forces with the increase of axial forces was also studied to get the influence situation of axial load on the dynamic characteristics [10]. Kerst S proposed a semi-analytical dynamic model addressing the flexibility of the outer ring. The results showed that the flexibility of outer ring had a significant effect on the deformation of the raceway, thus affecting the dynamic characteristics. The model was proved to be of low computation cost and high accuracy [11]. Gunduz A made an investigation of the vibration response of angular contact ball bearings with respect to bearing preload, and proved that the axial preload affects the dynamic characteristics by changing the diagonal and off-diagonal elements of the stiffness matrix [12]. However, with the increase of rotation speed, the oil whirl tends to be obvious and the cage runs more irregularly. The cage has to be taken as a sub-source at high rotation speed accordingly. Deng S E conducted several studies on the impact of lubricant traction coefficient on cage's dynamic characteristics in high-speed angular contact bearing and the stability of cage under various working conditions were assessed by the slip ratio of cage [13-15]. Cao H R, Niu L K *et al* also stressed in their review that the cage dynamic response was easily influenced with the change of working conditions and the contact between cage and other elements could vary greatly [16].

As are displayed above, the researchers have conducted detailed research on the nonlinear dynamic response of the elements in angular contact bearing, but refused to go a step further to study the acoustic characteristics. It has been figured out in previous work that the acoustic performance of the bearing can be obtained by the sub-source decomposition method and the acoustic model has been proved of consistency with the experiment result. As one of the key acoustic characteristics, the sound directivity shows the circumferential distribution of the radiation noise and is affected by the working parameters which may bring about nonlinear and unstable factors. The studies on the trends of directivity with the working conditions or structural parameters are quite rare to the authors' knowledge. In this paper, the nonlinear dynamic model of ceramic bearing is proposed and the sub-source decomposition method is introduced for the calculation of radiation noise. The variation of sound directivity with axial loads and rotation speed are analysed and discussed afterwards.

2. Ceramic ball bearing model

In this paper, the outer ring is fixed to simulate the actual working condition and the inner ring rotates under stable speed. Assuming that the mass centers of the components coincide with the geometrical centers and the balls are of the same diameters. Here we neglect the impact of surface roughness of the raceways and the dynamic model of the ceramic ball bearing is shown in figure 1.

Multiple coordinate systems are applied in the analysis of ceramic ball bearing as shown in figure 1. Coordinate system $\{O;X,Y,Z\}$ is fixed, and coordinate systems $\{O_i;X_i,Y_i,Z_i\}$, $\{O_c;X_c,Y_c,Z_c\}$ are used to describe the dynamic response of inner ring and cage, respectively. O_{ir} and O_{or} show the center of curvatures of inner and outer raceway. Q_{ij} and Q_{oj} show the contact forces given by the inner ring and outer ring, and Q_{cj} shows the impact force between the jth ball and the cage. Q_{cxj}, Q_{cyj} and Q_{czj} are decomposition components along axes O_cX_c, O_cY_c and O_cZ_c. α_{ij} and α_{oj} show the contact angles between the *j*th ball and inner/outer raceway, respectively. F_{Rij} shows the frictional forces between the jth ball and inner ring and F_{Roj} shows that between the *j*th ball and outer ring. Subscripts ξ and η show the forces in plane *YOZ* and *XOZ*, respectively. $P_{R\eta j}$ and $P_{R\xi j}$ are frictional forces between the *j*th ball and cage. F_a is the axial load, and F_c shows the hydrodynamic force acting on the cage, with F_{cy} and F_{cz} the decomposition components along axes O_cY_c and O_cZ_c, respectively. $T_{\xi ij}$ is the traction force contact surface between the *j*th ball and inner raceway. ϕ_j and φ_j are phase angles of the *j*th ball in coordinate systems $O_cY_cZ_c$ and $O_iY_iZ_i$. m_j is the mass of the *j*th ball. When the bearing is running, and axial load F_a is applied on the inner ring, the contact forces bring about deformations on the elements.

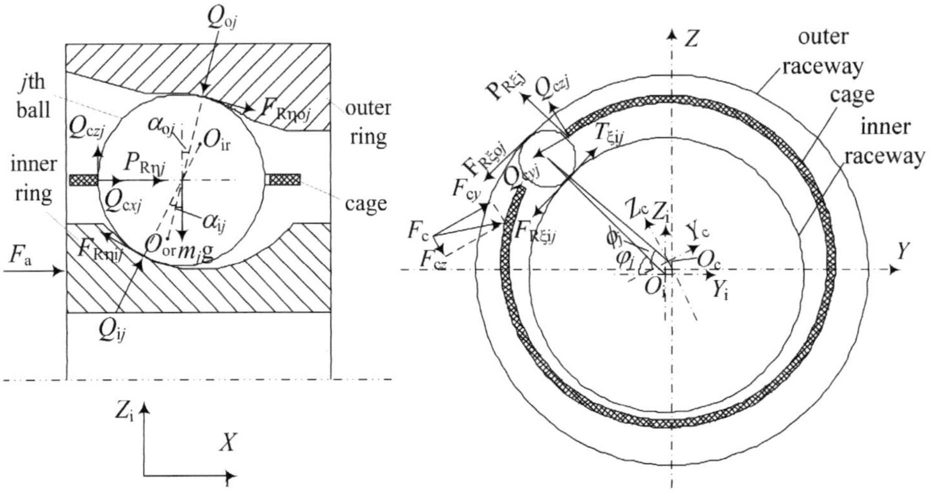

Figure 1. Dynamic model of ceramic ball bearing

Assuming that the deformations of the balls are negligible compared with the deformations of the rings owing to the great stiffness of the ceramic balls, and the contact forces can be expressed as:

$$\begin{cases} Q_{oj} = K_{oj}\cdot\delta_{oj}^{1.5} \\ Q_{ij} = K_{ij}\cdot\delta_{ij}^{1.5} \end{cases} \tag{1}$$

where K_{oj} and K_{ij} are the contact stiffness between the *j*th ball and the outer/inner raceway, respectively, and δ_{oj} and δ_{ij} show the deformation of the corresponding raceway. The inner ring is in contact with the balls and the dynamic differential equations can be described as:

$$F_a - \sum_{j=1}^{N}\left(Q_{ij}\sin\alpha_{ij} - F_{R\eta ij}\cos\alpha_{ij}\right) = m_i\ddot{x}_i \tag{2}$$

$$F_y + \sum_{j=1}^{N}\left[\left(Q_{ij}\cos\alpha_{ij} + F_{R\eta ij}\sin\alpha_{ij}\right)\cos\varphi_j + \left(T_{\xi ij} - F_{R\xi ij}\right)\sin\varphi_j\right] = m_i\ddot{y}_i \tag{3}$$

$$F_z - \sum_{j=1}^{N}\left[\left(Q_{ij}\cos\alpha_{ij} + F_{R\eta ij}\sin\alpha_{ij}\right)\sin\varphi_j - \left(T_{\xi ij} - F_{R\xi ij}\right)\cos\varphi_j\right] = m_i\ddot{z}_i \tag{4}$$

$$M_y + \sum_{j=1}^{N}\left[r_{ij}\left(Q_{ij}\sin\alpha_{ij} - F_{R\eta ij}\cos\alpha_{ij}\right)\sin\varphi_j + \frac{D_W}{2} f_i T_{\xi ij}\sin\alpha_{ij}\cos\varphi_j\right] = I_{iy}\dot{\omega}_{iy} - \left(I_{iz} - I_{ix}\right)\omega_{iz}\omega_{ix} \tag{5}$$

$$M_z + \sum_{j=1}^{N}\left[r_{ij}\left(Q_{ij}\sin\alpha_{ij} - F_{R\eta ij}\cos\alpha_{ij}\right)\cos\varphi_j - \frac{D_W}{2} f_i T_{\xi ij}\sin\alpha_{ij}\sin\varphi_j\right] = I_{iz}\dot{\omega}_{iz} - \left(I_{ix} - I_{iy}\right)\omega_{ix}\omega_{iy} \tag{6}$$

where: m_i shows the mass of inner ring, $\ddot{x}_i$, $\ddot{y}_i$, $\ddot{z}_i$ are the accelerations along axes O_iX_i, O_iY_i and O_iZ_i, respectively. F_y, F_z, M_y, M_z are external loads and D_W is the ball diameter. f_i is the inner ring raceway curvature radius coefficient. I_{ix}, I_{iy}, I_{iz} are the moments of inertia of inner ring, and ω_{ix}, ω_{iy}, ω_{iz} are the angular velocities. r_{ij} is the rolling radius and can be expressed as:

$$r_{ij} = 0.5d_m - 0.5D_W g_i \cos\alpha_{ij} \tag{7}$$

where: d_m is the pitch diameter of the bearing, and g_i is the inner ring raceway curvature radius coefficient.

The contact between balls and cage lead to impact and friction noise and the dynamic equations of the cage can be shown as:

$$\sum_{j=1}^{N}\left(P_{R\eta j} + Q_{cxj}\right) = m_c\ddot{x}_c \tag{8}$$

$$\sum_{j=1}^{N}\left[F_{cy} - \left(Q_{cyj} - P_{R\xi j}\cos\phi_j\right)\right] = m_c\ddot{y}_c \tag{9}$$

$$\sum_{j=1}^{N}\left(Q_{czj} + P_{R\xi j}\sin\phi_j - F_{cz}\right) = m_c\ddot{z}_c \tag{10}$$

$$\sum_{j=1}^{N}\left(P_{R\xi j}\cdot\frac{D_W}{2} - \sqrt{Q_{cyj}^2 + Q_{czj}^2}\cdot\frac{d_m}{2}\right) + M_{cx} = I_{cx}\dot{\omega}_{cx} - \left(I_{cy} - I_{cz}\right)\omega_{cy}\omega_{cz} \tag{11}$$

$$\sum_{j=1}^{N}\left(P_{R\eta j} + Q_{cxj}\right)\cdot\frac{d_m}{2}\sin\phi_j = I_{cy}\dot{\omega}_{cy} - \left(I_{cz} - I_{cx}\right)\omega_{cz}\omega_{cx} \tag{12}$$

$$\sum_{j=1}^{N}\left(P_{R\eta j} + Q_{cxj}\right)\cdot\frac{d_m}{2}\cos\phi_j = I_{cz}\dot{\omega}_{cz} - \left(I_{cx} - I_{cy}\right)\omega_{cx}\omega_{cy} \tag{13}$$

where m_c is the mass of the cage, $\ddot{x}_c$, $\ddot{y}_c$, $\ddot{z}_c$ are the accelerations of the cage along axes O_cX_c, O_cY_c and O_cZ_c, respectively. M_{cx} is the external load, and I_{cx}, I_{cy}, I_{cz} are moments of inertia of the cage. ω_{cx}, ω_{cy}, ω_{cz} are angular velocities of the cage, and $\dot{\omega}_{cx}$, $\dot{\omega}_{cy}$ $\dot{\omega}_{cz}$ are angular accelerations.

The balls are in contact with the cage, inner ring and outer ring, and are the main sources of the radiation noise. The dynamic differential equations of the jth ball can be expressed as:

$$F_{bx} + F_{R\eta\eta j}\cos\alpha_{oj} - F_{R\eta\eta i}\cos\alpha_{ij} + Q_{ij}\sin\alpha_{ij} - Q_{oj}\sin\alpha_{oj} + Q_{cxj} + P_{R\eta j} = m_b\ddot{x}_{bj} \tag{14}$$

$$F_{by} + m_b g\cos\varphi_j + Q_{czj}\cos\phi_j + Q_{cyj}\sin\phi_j + F_{R\xi\xi i} - T_{\xi ij} - F_{R\xi\xi j} = m_b\ddot{y}_{bj} \tag{15}$$

$$F_{bz} + Q_{ij}\cos\alpha_{ij} - Q_{oj}\cos\alpha_{oj} + F_{R\eta\eta i}\sin\alpha_{ij} - F_{R\eta\eta j}\sin\alpha_{oj} - m_b g\sin\varphi_j - P_{R\xi j} = m_b\ddot{z}_{bj} \tag{16}$$

$$\left(T_{\xi ij} - F_{R\xi\xi j} - F_{R\xi\xi i} + P_{R\xi j}\right)\cdot\frac{D_W}{2} = I_b\omega_{bxj} + J_x\dot{\omega}_{xj} \tag{17}$$

$$\left(F_{R\eta\eta i} + F_{R\eta\eta j} + P_{R\eta j}\right)\cdot\frac{D_W}{2} = I_b\omega_{byj} + J_y\dot{\omega}_{yj} + I_b\omega_{bzj}\dot{\theta}_{bj} \tag{18}$$

$$P_{R\eta j}\cdot\frac{D_W}{2} = I_b\omega_{bzj} - I_b\omega_{yj}\dot{\theta}_{bj} + J_z\omega_{zj} \tag{19}$$

where m_b is the mass of the jth ball, $\ddot{x}_{bj}$, $\ddot{y}_{bj}$, $\ddot{z}_{bj}$ are the accelerations of the ball, and F_{bx}, F_{by}, F_{bz} are the hydrodynamic forces acting on the ball in different directions. $\dot{\omega}_{xj}$, $\dot{\omega}_{yj}$, $\dot{\omega}_{zj}$ are angular accelerations of x_{bj}, y_{bj} and z_{bj}, respectively. ω_{bxj}, ω_{byj}, ω_{bzj} are angular velocities of the jth ball in coordinate $\{O; X, Y, Z\}$, $\dot{\theta}_{bj}$ is the orbit speed of the jth ball in coordinate $\{O; X, Y, Z\}$.

3. Numerical simulations

Assuming that the bearing works with perfect lubrication and ignore the interference during rotation. The rotation speed is set from 15000r min^{-1} to 30000r min^{-1}, and bearing works under stable rotation speed without fluctuation. The axial loads are set to vary from 1000N to 3000N. The cage is made of phenolic resin, and the balls are made of silicon nitride ceramics. The bearing has the same structural parameters with SCHAEFFLER HCB7009-C-T-P4S, and the parameters are shown in table 1.

Table1. Major parameters of the bearing.

Item	Value
Bearing outside diameter(mm)	75
Bearing width(mm)	16
Initial contact angle(degree)	15
Cage outside diameter(mm)	65
Pocket diameter(mm)	10
Cage width(mm)	14
Ball diameter(mm)	9.5
Ball number	15
Inner ring bore diameter(mm)	45

The inner ring, the cage and the balls can be regarded as sub-sources of the radiation noise and the radiation noise can be obtained by a sub-source decomposition method derived from previous work in ref. [2]. The bearing runs with a horizontal axis, and the radial loads are ignored.

4. Results and discussion

4.1. Sound directivity performance with different axial loads

The field points are distributed in a circle on a plane 10mm from the YOZ plane and the diameter of the circle is set to be 200mm. The calculation step is set as 6 degree and therefore there are 60 field points on the circumference. Here we take the 12 o'clock direction as 0 degree, and the other points arranged clockwise. The bearing is set to run at 30000 r min^{-1} in clockwise direction, and the axial loads are 1000N, 1500N, 2000N, 2500N and 3000N. The sound pressure levels (SPL) at the field points with different axial loads are shown in figure 2.

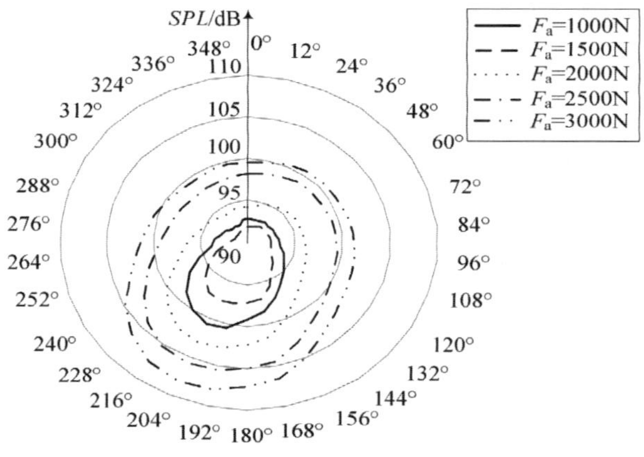

Figure 2. Sound pressure levels with different axial loads.

Here we define SPL_{max} and SPL_{min} as the maximum and minimum sound pressure level along the circumference, and ϕ_m as the directivity angle where it arrives at SPL_{max}, then the variation trends of SPL_{max} and ϕ_m are shown in figure 3.

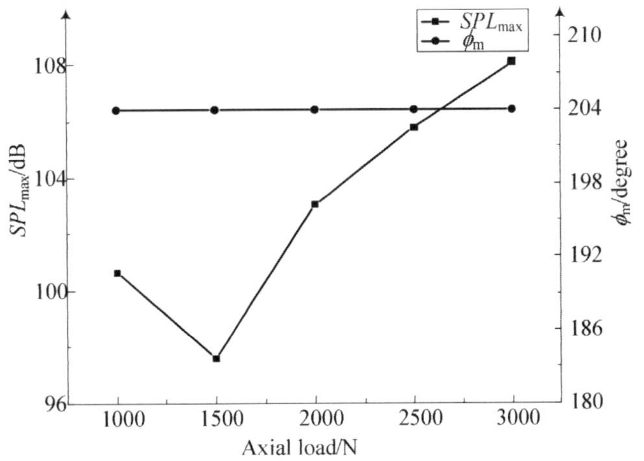

Figure 3. Variation trends of SPL_{max} and ϕ_m at ω=30000r min^{-1}.

In figure 2, the solid line, dashed line, dotted line, dot-dashed line and double dot-dashed line show the circumferential distributions of radiation noise with F_a=1000N, 1500N, 2000N,2500N and 3000N, respectively. In figure 3, the line with squares shows the variation trends of SPL_{max}, while the line with circles shows the variation trends of ϕ_m. It can be seen that the radiation noise at a fixed axial load varies little from 300° to 84°, but shows an obvious directivity in the lower semicircle at around 204°. The overall radiation noise shows an uptrend with the increase of axial load except a decline at F_a=1500N, which indicates that F_a=1500N is more applicable to the rotation speed of ω= 30000 r min^{-1}.

When the axial load continues to increase, SPL_{max} tends to increase and ϕ_m stays still which indicates that the friction and impact raise with the axial load, but the place where the largest deformation took place remains unchanged. Furthermore, the *SPL* curve tends to be more smooth which indicates that the friction and impact at the directivity angle increase more obviously and thus show a clearer directivity.

4.2. Sound directivity performance at different rotation speed

The circumferential distributions vary with the axial load, and the effect of rotation speed also needs to be taken into consideration. F_a=1500N has already been proved to be the optimized value at ω=30000 r min^{-1} in the former study, and therefore here we take F_a=1500N as a fixed axial load and the rotation speed is set as a variable. The arrangement of field points is the same with that in section *4.1* and the sound pressure levels at ω=15000 r min^{-1}, 20000 r min^{-1}, 25000 r min^{-1}, 30000 r min^{-1} are shown in figure 4 and the variation trends of SPL_{max} and ϕ_m are shown in figure 5.

In figure 4, the solid line, dashed line, dotted line and dot-dashed line show the sound pressure distribution at ω=15000 r min^{-1}, 20000 r min^{-1}, 25000 r min^{-1} and 30000 r min^{-1}, respectively. In figure 5, the lines with rectangles and circles show the variation of SPL_{max} and ϕ_m, respectively. It can be seen from figure 4 that the radiation noise has an overall rise with the increase of rotation speed and the gap between SPL_{max} and SPL_{min} also has a great disparity with the increase of rotation speed. The *SPL* curve turns from circumferential average to showing obvious directivity which is due to the aggravation of friction and impact. Similar with that in figure 2, the curves in figure 4 vary little in circumference from 300° to 84° and the changed amplitudes between curves are also not very large in this section. In the lower semicircles, the curves show clearer directivity with the increase of rotation speed, and there seem to be small differences between the directivity angles. According to figure5, ϕ_m appear at right below the bearing, but slightly skewed to the direction of rotation and the skewed angle increases with the increase of rotation speed (from 0° to 24°). This is mainly caused by the deviation of load zone. As is shown in figure 1, there are eccentricities between O, O_i and O_c. The centrifugal forces brought by the eccentricities grow rapidly with the increase of rotation speed and the angles of resultant forces are changed accordingly, thus leading to the deviation of directivity angles.

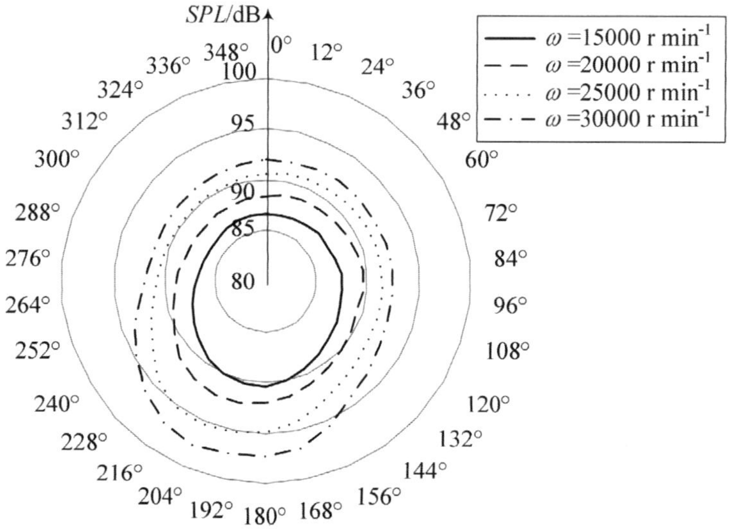

Figure 4. Sound pressure levels at different rotation speed.

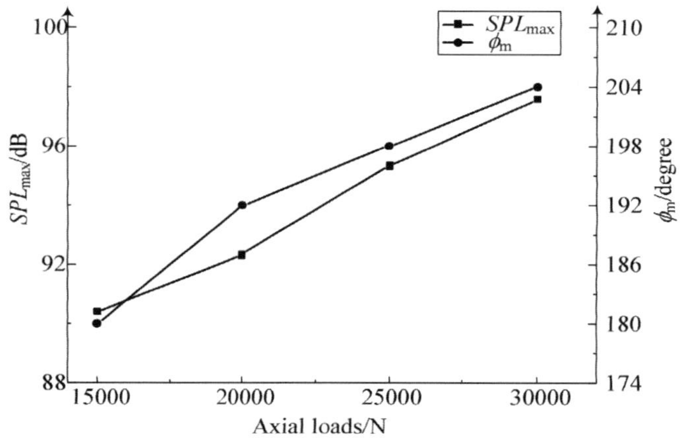

Figure 5. Variation trends of SPL_{max} and ϕ_m at F_a=1500N.

Compared with figure 2 and figure 3, it can be seen that the rotation speed acts differently on the directivity of the radiation noise. The *SPL* curves at ω=30000r min^{-1} have almost the same tendencies, and the shapes tend to be smooth with the increase of axial load, while the curves with F_a=1500N have almost the same smoothness with different directivity angles. As the nonlinear dynamic model has universal applicability, the findings can be extended to the series of ceramic ball angular contact bearings for the further analysis of sound directivity. As is illustrated in section 2, an obvious sound directivity will lead to fault or even breakdown in the directivity angle owing to severe wear. It can be concluded that the overall radiation noise raises with the rotation speed at any fixed axial load, however, a specific axial load at a fixed rotation speed can be obtained to minimize the radiation noise and make the directivity less obvious.

5. Conclusion

In this paper, the dynamic model of the ceramic ball bearing is established, and the sound directivity of ceramic ball angular contact bearings with different axial loads and at different rotation speed are analysed. In terms of *SPL* amplitudes, it is found that the radiation noise varies greatly with the change of axial load and rotation speed. The *SPL* distribution tends to be stable and even in the upper semicircles and shows obvious directivity in the lower ones. When the axial load increases at a given rotation speed, the directivity generally turns obvious, but the directivity angle stays unchanged. Similarly, when the rotation speed rises with a fixed axial load, the directivity becomes clearer with the directivity angle skewed to a large angle towards the running direction. The radiation noise of ceramic ball bearings increases with the rotation speed and an optimized axial load can be selected at each rotation speed to reduce the radiation noise and make the directivity less obvious. The investigation provides detailed information of the acoustic performance of ceramic ball bearings, thus putting forward the optimization direction of further development.

Acknowledgement
This work was supported by Doctor Start-up Foundation of Liaoning Province [grant number 20170520147] and Postdoctoral Fund Project of Shenyang Jianzhu University [grant number SJZUBSH201707].

References

[1] Cao H R, Niu L K, Xi S T and Chen X F 2018 Mechanical model development of rolling bearing- motor systems: A review *Mech. Syst. Signal. Pr.* **102** 37
[2] Bai X T, Wu Y H, Zhang K, Chen C Z and Yan H P 2017 Radiation noise of the bearing applied to the motorized spindle based on the sub-source decomposition method *J. Sound. Vib.* **410** 35
[3] Xu T, Xu G H, Zhang Q, Hua C, Tan H H, Zhang S C and Luo A L 2013 A preload analytical method for ball bearings utilising bearing skidding criterion *Tribol. Int.* **67** 44
[4] Niu L K, Cao H R and Xiong X Y 2017 Dynamic modelling and vibration response simulations of angular contact balls with ball defects considering the three- dimensional motion of balls *Tribol. Int.* **109** 26
[5] Bizarre L, Nonato F and Cavalca K L 2018 Formulation of five degrees of freedom ball bearing model accounting for the nonlinear stiffness and damping of elastohydro- dynamic point contacts *Mech. Mach. Theory.* **124** 179
[6] Kumar S, Goyal D and Dhami S S 2018 Statistical and frequency analysis of acoustical signals for condition monitoring of ball bearing *Mater. Today. Proc.* **5** 5186
[7] Chapman J J 1995 Angular Contact Ball Bearing Dynamics, an Experimental and Theoretical Investigation *Tribol. Ser.* **30** 435
[8] Cao Y Z and Altinas Y 2007 Modeling of spindle-bearing and machine tool systems for virtual simulation of milling operations *Int. J. Mach. Tool. Manu.* **47** 1342
[9] Wang Y L, Wang W Z, Zhang S G and Zhao Z Q 2015 Investigation of skidding in angular contact ball bearings under high speed *Tribol. Int.* **92** 404
[10] Zhao C J, Yu X K, Huang Q X, Ge S D and Gao X 2015 Analysis on the load characteristics and coefficient of friction of angular contact ball bearing at high speed *Tribol. Int.* **87** 50
[11] Kerst S, Shyrokau B and Holweg E 2018 A semi-analytical bearing model considering outer race flexibility for model based bearing load monitoring *Mech. Syst. Signal. Pr.* **104** 384
[12] Gunduz A, Dreyer J T and Singh R 2012 Effect of bearing preloads on the modal characteristics of a shaft-bearing assembly:Experiments on double row angular contact ball bearings *Mech. Syst. Signal. Pr.* **31** 176

[13] Deng S E, Li X L, Wang J G and Teng H F 2011 Frictional torque characteristic of angular contact ball bearings *J. Mech. Eng.* **47** 114 (In Chinese)
[14] Zhang W H, Deng S E, Chen G D and Cui Y C 2016 Study on the impact of roller convexity excursion of high-speed cylindrical roller bearing on roller's dynamic characteristics *Mech. Mach. Theory.* **103** 21
[15] Zhang W H, Deng S E, Chen G D and Cui Y C 2017 Impact of lubricant traction coefficient on cage's dynamic characteristics in high-speed angular contact ball bearing *Chinese. J. Aeronaut.* **30** 827
[16] Wang H, Han Q K and Zhou D N 2017 Nonlinear dynamic modelling of rotor system supported by angular contact ball bearings *Mech. Syst. Signal. Pr.* **85** 16

IOP Conf. Series: Materials Science and Engineering **399** (2018) 012003 doi:10.1088/1757-899X/399/1/012003

Surface analysis of SAE 3310 carburized steel in environment saturated with carbon nanofibers

S A Barabas[1*], A Florescu[1] and F Sarbu[1]

[1] Department of Engineering and Industrial Management, Transilvania University of Brasov

[*] E-mail: sorin.barabas@unitbv.ro

Abstract. The carburizing process is described by the reaction between environment atmosphere and metal surface followed by carbon transfer in a metal matrix. The main issue of heat treatment consist in ensuring the carburizing controlled atmosphere with a superior carbon potential level than metal surface of steel, in the carburizing furnace at a temperature that ensures process development. Obtaining carbon nanofibers relatively easily led to a new carburizing treatment, in the atmosphere full of carbon nanofibers. The use of nanomaterials for carburizing requires less time and best diffusion. The most common area of carbon concentration and residual austenite presence, the edges and corners are sensible reduced, the hardness is higher than the hardness obtained in regular carburizing treatment. Experiments made on samples of steel SAE 3310 have demonstrated the beneficial role of the presence of carbon nanofibers in carburizing atmosphere. It has developed the conclusion that the carburized layer properties depend on the amount of carbon nanofibers, during carburizing.

1. Introduction

The use of nanomaterials in carburizing heat treatment is a less researched option, presenting a high potential for innovation. The possibilities presented by carbon nanotubes to be captured and embedded in the steel matrix were investigated in heat treatment furnace. The morphology of single walled nanotubes is characterized by the properties of the chirality vector of graphene, which limited the configuration of nanotubes to armchair or zig-zag forms. The researches confirmed the thermal stability of SWCNT (single walled carbon nanotubes) in vacuum up to 2800^0 C and in the air up to750^0C [1].

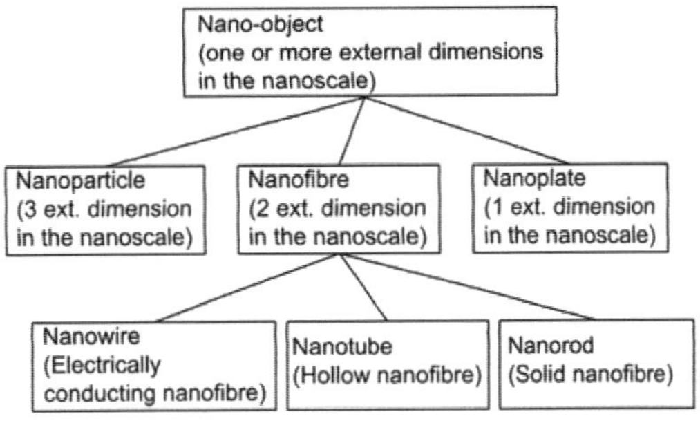

Figure 1. The position of nanotubes in nanomaterials group.

In the carburizing process, multi walled carbon nanotubes from commerce were used, with pronounced variability (length, diameter of fibers, number of walls and chirality), mixed in emulsion. Nanotubes are hollow fibers with 2 dimensions at nanoscale, with cylindrical nanostructure of carbon allotrope (figure 1).The interaction between metal matrix and CNT (carbon nanotubes) is possible to achieve by several methods: laser surface melting in presence of carbon nanotubes [2], electrophoretic deposition [3], powder metallurgy [4]. These methods are synthesized in figure 2.

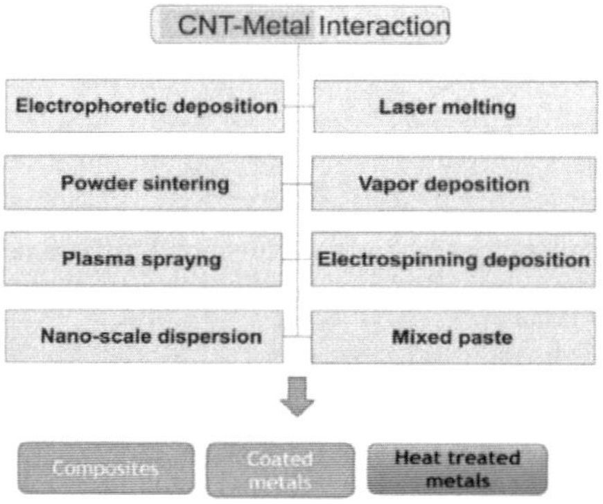

Figure 2. Interaction between CNT (carbon nanotubes) and metal matrix and its results.

The results of the interaction are composites, coated metals and heat treated metals. The carburizing process consists in mass transfer on the surface of the metallic products of carbon atoms, which can be adsorbed by the solid. The process is the same as a solid carburization in graphite, with the difference that the graphite is replaced by an emulsion impregnated with MWCNT (multi walled carbon nanotubes).

2. Theoretical consideration

Mass transfer from the carburizing environment on the metallic surface occurs due to the difference between the average concentration of carbon in the environment and the metal surface [5, 6]. Diffusion occurs in metallic material due to its appearance in the superficial layer, a surplus of carbon through mechanisms of adsorption [7, 8]. Nanotubes from the metal surface will diffuse into the surface layer which results in the formation of a carburized layer. Some atoms of metal will be absorbed in a carbon nanotubes structure. The carburized layer formation mechanism is presented in figure 3.

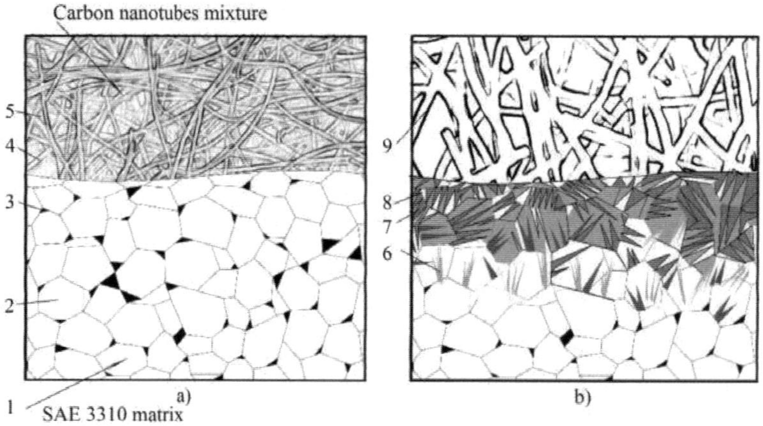

Figure 3. Carburized layer mechanism a) at starting time b) absorption of carbon atoms (1-steel structure; 2-ferrite; 3-pearlite; 4-carbon nanotubes; 5-emulsion; 6-cementite; 7- martensite; 8-austenite; 9-carbon nanotubes).

3. Experimental work

For the production of samples, the study material will be steel used for bearings components, SAE 3310. The samples will be coated with a sodium silicate emulsion. The samples were heated at 980^{0}C, the holding time is 2 hours. Carbon diffusion is made by cooling at 600^{0}C. A different quantity of carbon nanotubes and graphite are mixed in an emulsion with sodium silicate (Na_2O_3Si) -20%, isopropyl alcohol (C_3H_8O) -30%, phenol formaldehyde resin -30% (multi walled carbon nanotubes powder acquired in commerce with the following parameters: purity > 97%, average outside diameter >55 nm, average inside diameter 8 nm, length 10-30 μm, color black, density ~2.1 g/cm^3). The carburizing process was made according to the diagram presented in figure 4.

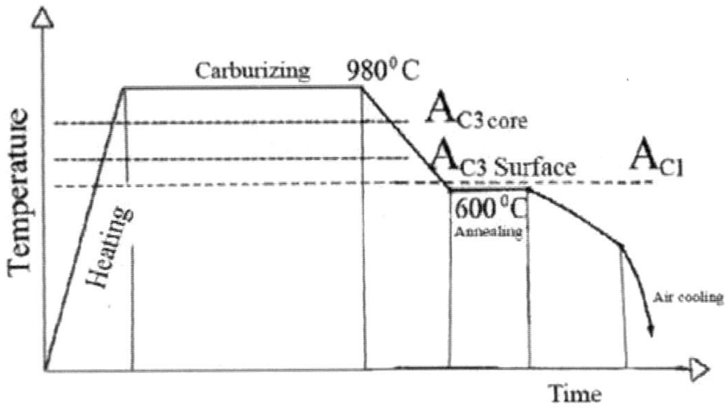

Figure 4. Diagram of carburizing process [6].

In order to establish an optimum process, time and temperature, some restrictions are made. The restrictions consist in control carbides formation and retained austenite formation. These drawbacks can be avoided by an optimum carbon potential. The microscopic investigation showed the presence of retained austenite on the surface of carburized samples, the presence of martensite and carbides in the layer and the presence of sorbite in the core (figure 5).

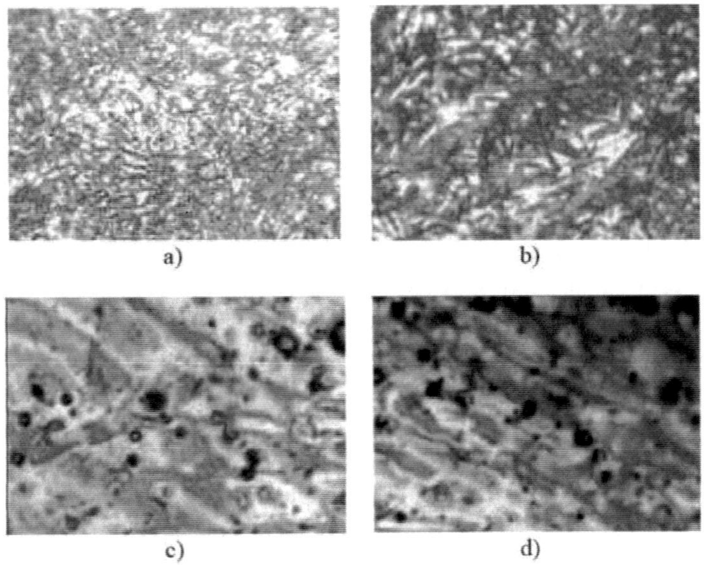

Figure 5. Microstructures of carburized samples a) core structure, b) Carburized layer structure (0.5 mm), c) Carburized layer structure (0.2 mm), d) Carburized layer structure (at surface).

Figure 5 reveals the microstructure obtained after 2 hours of treatment, depending on the depth of carburizing: in core - fine sorbate with hardness 36 HRC (attack: nital 2%, magnification 100x); at 0.5 mm - hardening martensite plus retained austenite plus fine carbides with hardness 58 HRC (attack: nital 2%, magnification 400x); at 0.2 mm, - retained austenite plus fine carbides plus traces of hardening martensite with hardness 52 HRC (attack: nital 2%, magnification 800x); at surface, carburizing: 2 hours plus diffusion: 30 min - hardening martensite plus retained austenite plus fine carbides with hardness 61 HRC (attack: nital 2%, magnification 800x). The formation of a quantity of residual austenite is almost inevitable in microstructures of carburized layers containing high levels of carbon, which, in quantities of more than 50%, cause a significant decrease in the level of hardness and reduced fatigue. The main cause of excess residual austenite is the existence of higher carbon content in the surface layer. The most common areas of the surface carbon concentration are edges and corners of the piece. Another consequence of the high carbon content is revealed by the formation of massive carbides, which are formed at the limit of austenite grains and may have different morphologies in interdependence with the degree of alloying of the carburized steel. In table 1, table 2 and table 3 the hardness of the samples is shown after 2 hours of carburizing and 30 min for diffusion treatment, with different emulsions.

Table 1. Value of hardness at surface of samples.

	Hardness [HRC]				
Samples / Type of carburization	In controlled atmosphere	Emulsion 10% graphite	Emulsion 5% graphite + 5% CNT	Emulsion 5% CNT	Emulsion 10%CNT
1	60	57	59	60	61
2	59	54	60	59	62
3	59	56	61	59	61
4	58	59	60	58	62
5	60	58	60	59	60
Average	59.2	56.8	60.0	59.0	61.2

1-graphite 10%, 2- controlled atmosphere, 3- 5% CNT, 4- 5% CNT+5 % graphite, 5- 10% CNT

Table 2. Value of hardness at 0.5 mm of samples.

	Hardness [HRC]				
Samples / Type of carburization	In controlled atmosphere	Emulsion 10% graphite	Emulsion 5% graphite +5% CNT	Emulsion 5% CNT	Emulsion 10% CNT
1	49	44	49	49	51
2	48	41	51	49	52
3	48	46	51	48	51
4	47	45	49	48	50
5	49	47	49	47	50
Average	48.2	44.6	49.8	48.2	51.8

1-graphite 10%, 2- controlled atmosphere, 3- 5% CNT, 4- 5% CNT+5 % graphite, 5- 10% CNT

Table 3. Value of hardness at 1 mm of samples.

	Hardness [HRC]				
Samples / Type of carburization	In controlled atmosphere	Emulsion 10% graphite	Emulsion 5% graphite +5% CNT	Emulsion 5% CNT	Emulsion 10% CNT
1	39	36	39	40	41
2	38	37	41	41	43
3	37	35	41	40	41
4	38	36	42	39	40
5	39	34	41	39	42
Average	38.2	33.6	40.8	39.8	41.4

1-graphite 10%, 2- controlled atmosphere, 3- 5% CNT, 4- 5% CNT+5 % graphite, 5- 10% CNT

4. Conclusions

The experiments made it possible to plot the graphic in figure 6.

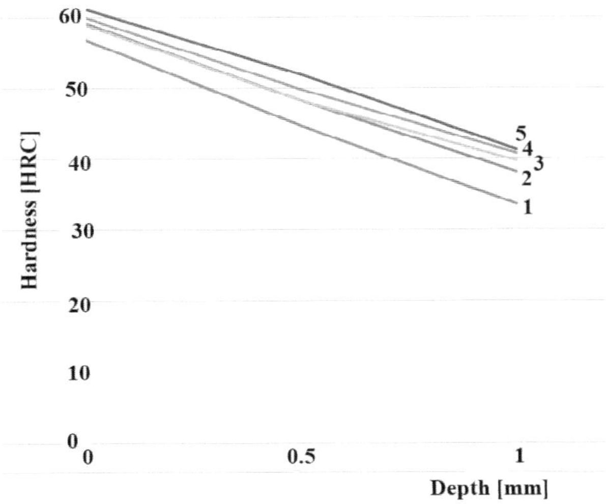

Figure 6. Influence of carburizing environment to the hardness and depth.

The curves of figure 6 represent the results of the experiments shown in table 1, table 2 and table 3: 1- pack carburizing with graphite 10%; 2 – carburizing in controlled atmosphere; 3 – pack carburizing with 5%CNT; 4 – pack carburizing with 5% graphite+5% CNT; 5 – pack carburizing with 10%CNT. Hardness increase with amount of carbon nanotubes. In the proposed treatment with carbon nanotubes a decrease of residual austenite due to a better diffusion of carbon in the layer is expected, which realizes redistribution of carbon atoms in the layer. The consumed energy will decrease because of low time carburizing and due to simplicity of carburizing installations towards carburizing in controlled gaseous environment.

5. References

[1] Thostenson E T, Li C and Chou T W 2005 *Comp. Sci. Technol.* **65** 491
[2] Yao J, Zhang Q, Gao M and Zhang W 2008 *Appl. Surf. Sci.* **254** 7092
[3] Zhang S, Wang Y, Han X, Cai Y and Xua S *Progr. Nat. Sci.: Mat. Int.* in press
[4] Bunakov N A, Kozlov D V, Golovanov V N, Klimov E S, Grebchuk E E, Efimov M S and Kostishko B B 2016 *Res. Phys.* **6** 231
[5] Karabelchtchikova O, Maniruzzaman M and Sisson R D 2006 *Proc. MS&T, Cincinnati* 375-386
[6] Barabas S and Fota A 2013 *Rec. J.* **14** 212
[7] Hikata A, Okubo S, Utsunomiya R, Higashi Y and Matsuba T 2014 *Sei Tech. Rev.* **9**
[8] Afolabi A S, Hassan A S and Daramola M O 2015 *Proc. W. Cong. Eng. Comp. Sci.* San Francisco

IOP Conf. Series: Materials Science and Engineering **399** (2018) 012004 doi:10.1088/1757-899X/399/1/012004

Optimization of the hydro abrasive jet cutting process of thick steel sheets

F B Barabas[1*] and T Deaconescu[1]

[1] Department of Engineering and Industrial Management, Transilvania University of Brasov

[*] E-mail: bogdan.fb@gmail.com

Abstract. In this paper is studied the issue of thick steel sheets cutting on abrasive waterjet machine, due to the abrupt curvature of the jet, caused by energy loss at high thicknesses. Depending on the hardness of the processed material, incomplete cutting occurs and it is evident at a thickness of sheets larger than 30-50 mm. Achieving a cutting with different configuration regimes implies on most of the equipment, turning off the machine, a new configuration and restarting the process of abrasive waterjet cutting which leads to the appearance of traces and marks on cut surface. Choosing the parameters in such a way that cutting is achieved (automatic configuration), leads to obtaining an uncut zone at the exit of the material. This uncut area can be eliminated by an oversized configuration of input parameters, or just reduced by a randomly selected configuration. Both cases lead to large economic losses. The correct choice of machining regime for abrasive cutting of thick sheets, depending on the thickness and the hardness of the material is the aim of these studies and proposed calculation method can be implemented.

1. Introduction

Hydro abrasive jet processing, due to the advantages it possesses, has been a major development in recent years. The possibilities of the technological process have led to intense research in the field and to a diversification manifested in particular by an exponential growth of equipment manufacturers and an increase in the type of processed materials. The technological hydro abrasive process allowed the production of work pieces with thermally unaffected surfaces, with high precision and association with CAD programs has made it possible to obtain variants, simple and easy, at a reasonable price. The parameters of the technological process are configured according to the capabilities of the equipment used, the possibilities offered by the producer as well as the processed material. In hydro abrasive jet cutting of thick steel sheets, the specific elements of the machine (the water pressure p, made with a pump, the diameter d of the orifice to create a jet, type of mixing chamber, the length and diameter of the mixing tube) as well as the selected parameters (distance to the part and abrasive type) are chosen depending on the possibilities.

Configuration is limited by the capabilities available. The only continuous variable element is the traverse speed V_t, who, in condition of maintaining certain input parameters, depends on the thickness of the material h and its machinability M. Traverse speed is set so that the minimum condition to cutting the material. In the case of additional quality and precision requirements, this may be modified in the sense of lowering it, until the desired roughness is achieved. From the exit of the mixing tube, to the

IOP Conf. Series: Materials Science and Engineering **399** (2018) 012004 doi:10.1088/1757-899X/399/1/012004

part, due to the very small distance between the tube and the part, as well as due to the high velocity of the jet V_j, we can consider the trace of a particle, perfectly perpendicular to the part, no matter how big they are traverse speed V_t. With the contact between the jet component and the surface of the piece, t_s velocity begins to decrease due to the loss of kinetic energy occurring at the collision of the jet particles with the material of the part [1]. If the cutting is bigger and the machinability is lower, the speed V_t decreases after a curve with accelerated slope, the angle of this slope can be seen on striations appearing on the surface of the part (figure 1).

Figure 1. Hydro abrasive cutting of thick steel part.

As shown in figure 1, when cutting the thick sheets, an unprocessed area with length L appears, due to the energy losses of the jet caused by speed decrease, and due to the tendency of the particles to migrate to the area where they do not encounter resistance (area already processed). At the exit of the jet from the material, in the upper part, trajectory of the particles, without any resistance returns to rectilinear shape, thus favors the occurrence of uncut areas. The present paper studies the influence of the traverse speed and the machinability of the material on the unprocessed area, for steel sheets, with a thickness of 50 mm, 60 mm and 70 mm. The processed materials was SAE 3310, SAE 8620 and 316L.

2. Theoretical consideration

The jet curvature occurs due to deflection of particles caused by kinetic energy loss [2, 3]. The impact of a waterjet on steel causes minor material removal, this is determined by the kinetic energy of the abrasive particles. In the zone I (figure 1), particles have a sufficient level of kinetic energy, so the resulted surface is free of striation and the unprocessed area does not occur. In the zone II, due to the kinetic energy loss, the cutting power of particles, decrease and to the end of cut, occurs an unprocessed area. This phenomenon is illustrated in figure 1 and figure 2 and must be looked at a process of cutting in waves. The first wave consists in particles with the lowest energy, the next waves having more energy. This, leaves trace marks on the surface, generally called striations disposed at a variable angle which depends on the jet's traverse speed V_t, on the jet's velocity V_j and on the material's machinability M. This angle leads to a gap between the jet entry point and the exit point, noted with L. As the cutting depth increases, the gap becomes bigger and the unprocessed area occurs [4, 5].

$$V_t = \frac{R_m}{d_n h}, \tag{1}$$

where: V_t is traverse speed of abrasive jet [mm/s]; R_m is the rate of removal unit volume of material [mm^3/s]; d_n is the nozzle bore diameter [mm]; h is jet cutting depth in the chosen material [mm].

$$R_m = \frac{1}{2} I_m V_j^2 M, \tag{2}$$

where: I_m is impulse given to the abrasive mixture in unit time [g m/s]; V_j is abrasive jet velocity at the exit of the nozzle [m/s]; M_s is specific machinability of material [6, 7].

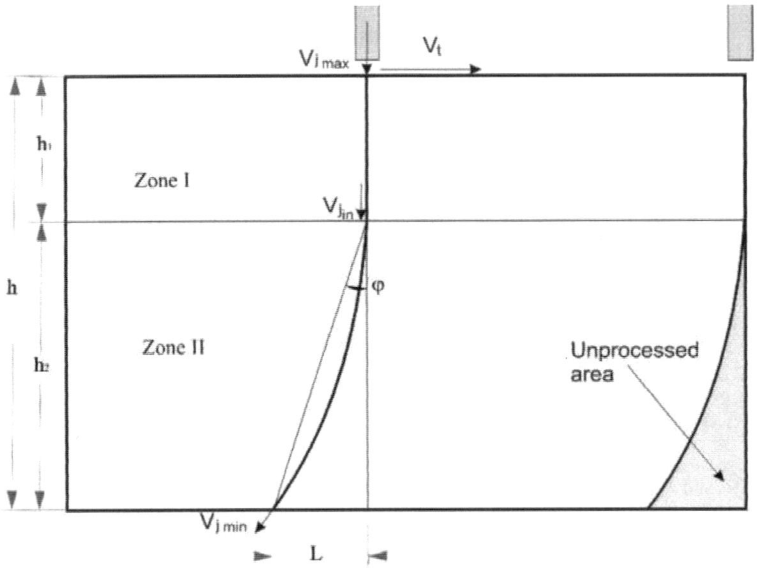

Figure 2. Formation of unprocessed area.

$$V_j = \mu \xi \sqrt{\frac{2p}{\rho_a}}, \tag{3}$$

where: μ is acceleration coefficient of the orifice ring for high speed; ξ is a coefficient that takes into account the characteristics of abrasive, grinding, friability, inhomogenities; p is the pressure of the liquid in the mixing tube [MPa]; ρ_a is the density of the abrasive fluid [g/mm^3].

According Eq. (1) and Eq. (2), material influence on traverse speed is given by the thickness and specific machinability M_s.

$$V_t = \frac{\frac{1}{2} I_m V_j^2 M}{d_n h} \tag{4}$$

The condition for eliminating the unprocessed area is:

$$V_{jin} = V_{jmin} = \sqrt{\frac{2V_t d_n h}{I_m M}}, \tag{5}$$

where: V_{jin} is speed of abrasive jet [mm/s] at final of zone I; V_{jmin} is jet speed at output of processed part.

3. Experimental work

For experiments were chosen SAE 3310, SAE 8620 and 316L alloy steel with thickness h = 30 mm, 50mm and 70 mm, The materials have the following properties: SAE 3310 - Hardness Brinell, HBW = 240; machinability, M = 78; modulus of elasticity, E = 200 GPa, SAE 8620 - HBW = 150; M = 90; E =190 GPa and 316L - HBW = 210; M = 82.5; E =210 GPa

Setting the AWJ machine, Maxiem 1530 from Omax Corporation, has been executed with water pressure provided by the pump to the maximum value p = 350 MPa (figure 3).Total length of cut is l=100 mm. Higher virtual thicknesses were chosen to change the traverse speed and this has been established after the timing of the run of total length (figure 4).

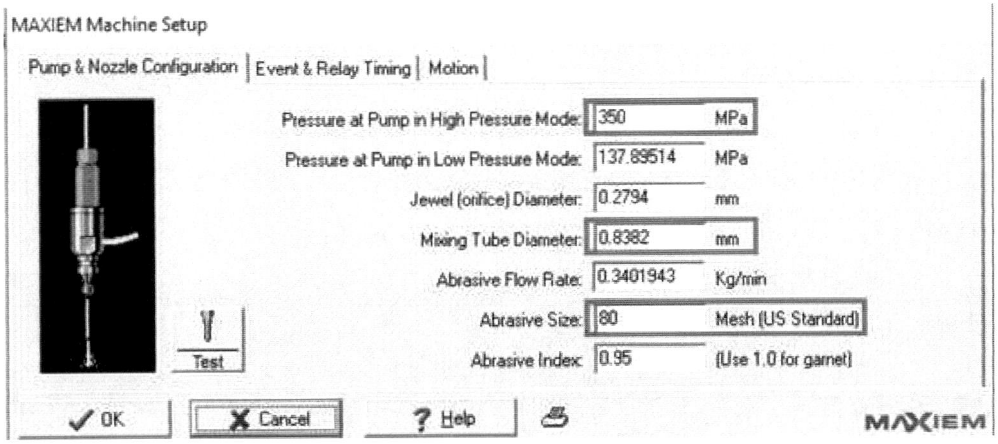

Figure 3. Settings of AWJ machine, Maxiem 1530.

The method of changing the traverse speed is shown in figure 4.

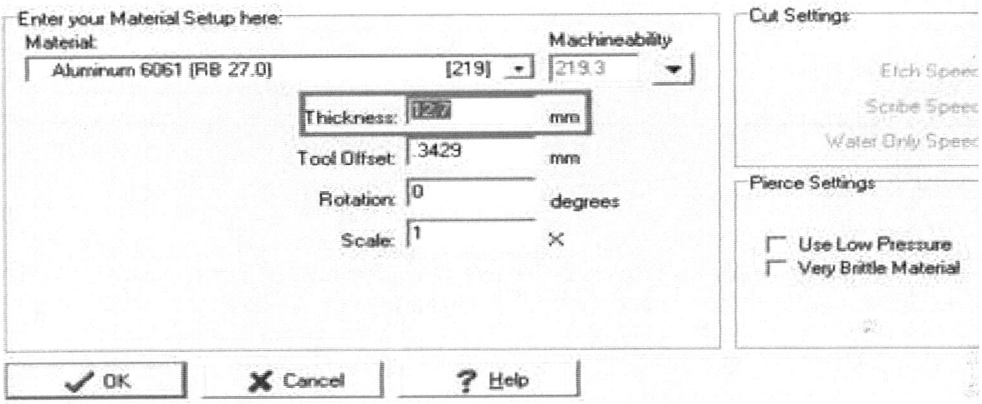

Figure 4. Changing the traverse speed by choosing virtual thickness.

The measurements of length L of unprocessed area were performed using a digital caliper and an optical comparator measurement machine 321 GL. The media of results of measurement are shown in table 1, table 2 and table 3.

Table 1. Value of length L of unprocessed area at steel sheets (30 mm).

h = 30 mm									
Type of steel	Machinability	Traverse speed V_t [mm/min]				Length of uncut area L [mm]			
		1	2	3	4	1	2	3	4
SAE 3310	78	40	30	20	10	0.5	0.2	0	0
SAE 8620	90	60	50	40	30	0.3	0	0	0
316 L	82.5	50	40	30	20	0.4	0.1	0	0

Table 2. Value of length L of unprocessed area at steel sheets (50 mm).

h = 50 mm									
Type of steel	Machinability	Traverse speed V_t [mm/min]				Length of uncut area L [mm]			
		1	2	3	4	1	2	3	4
SAE 3310	78	30	20	10	5	1.6	1.0	0.7	0.1
SAE 8620	90	40	30	20	10	1.4	0.9	0.5	0
316 L	82.5	35	25	20	8	1.4	1.0	0.6	0.1

Table 3. Value of length L of unprocessed area at steel sheets (70 mm).

h = 70 mm									
Type of steel	Machinability	Traverse speed V_t [mm/min]				Length of uncut area L [mm]			
		1	2	3	4	1	2	3	4
SAE 3310	78	15	10	6	3	2.5	1.9	1.2	0.5
SAE 8620	90	20	15	10	5	2.3	2.0	1.0	0.3
316 L	82.5	20	12	8	4	2.4	1.8	1.2	0.4

4. Conclusions

The wide use of cutting jet do occur in the cutting process, a growing number of materials such as multilayer materials or welded materials. This research opens the door to extensive research in this area, in order to optimize cutting abrasive water jet. As shown in table 1, 2 and 3, the growth of thickness of materials leads to the appearance of uncut area with length L. With decrease of traverse speed, L tend to decrease.

Chart for the SAE 3310, for traverse speed V_t=10 mm/min and V_t=20 mm/min is shown in figure 5.

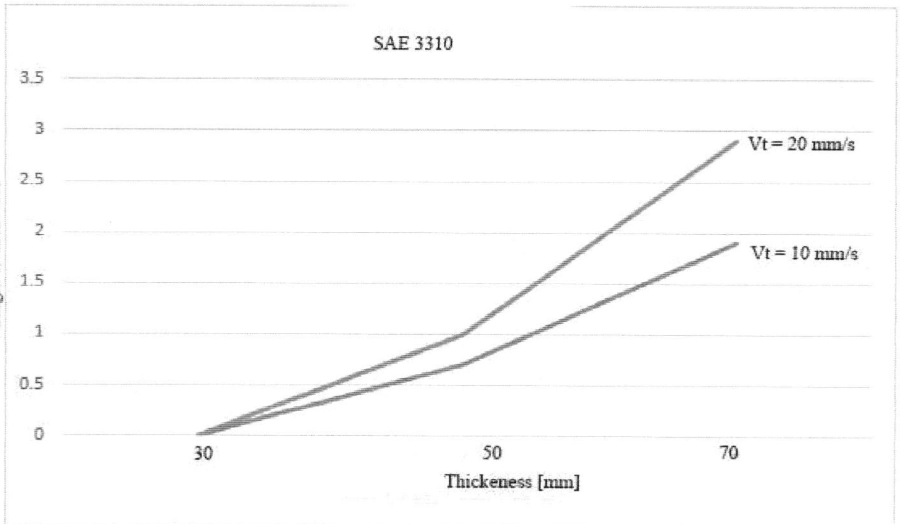

Figure 5. Diagram of length of uncut zone function of thickness of material for SAE 3310.

It can be seen up to the thicknesses of sheets h = 30 mm, uncut area does not occur. When the thickness to h = 50 mm, an uncut area with L comprised between 0.5 mm and 1 mm, appears causing an issue when cutting plates of steel. When the thickness increases over 50 mm, the uncut area grows strongly.

Chart for the SAE 8620, for traverse speed V_t=20 mm/min and V_t=30 mm/min is shown in figure 6.

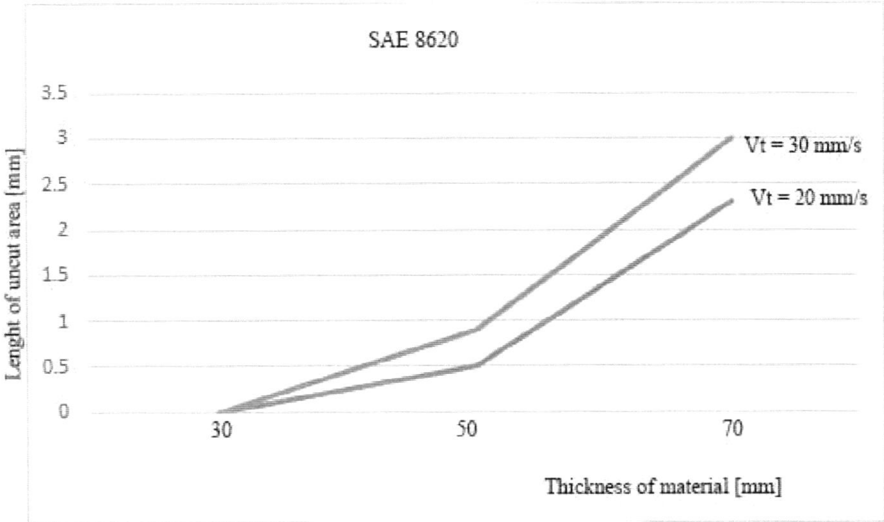

Figure 6. Diagram of length of uncut zone function of material thickness for SAE 8620.

For SAE 8620, with hardness much less than the hardness of SAE 3310 steel, when the thickness of materials grows to 50 mm the uncut area is lower than to the SAE 3310.

Chart for the SAE 3310, for traverse speed V_t=8 mm/min and V_t=20 mm/min is shown in figure 7.

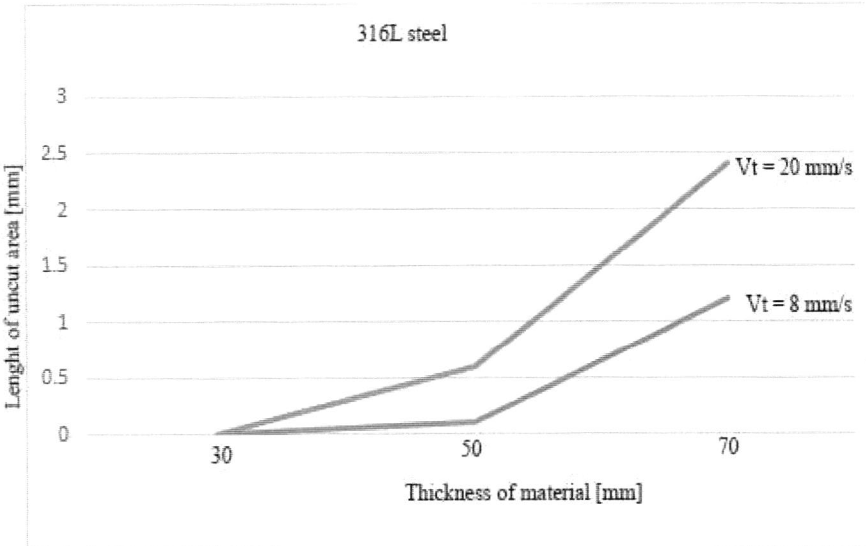

Figure 7. Diagram of length of uncut zone function of material thickness for SAE 316L.

The main difference between the three materials is machinability. It can be concluded that the length of the uncut area depends on three important factors: material thickness, traverse speed and material machinability (the machinability, at the same values of traverse speed, at bigger values leads to shrinking the uncut area). Both, the thickness of the material and its machinability are properties of the work piece that can not be changed. One possibility to fix this problem is to work with 2 traverse speeds, on the last part, lowering the traverse speed to low values.

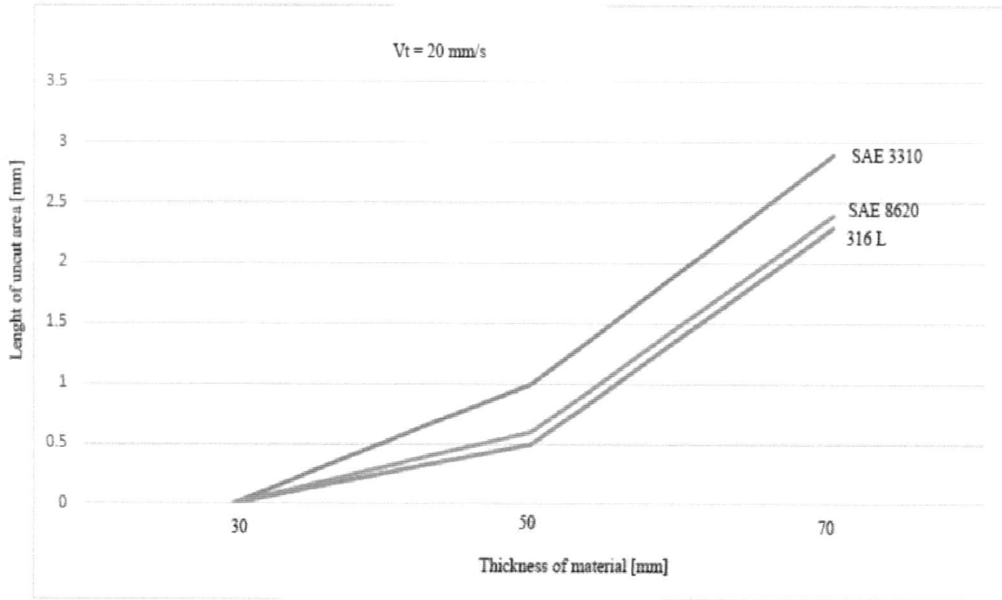

Figure 8. Diagram of length of uncut zone function of material thickness for SAE 3310, SAE 8620 and 316L steel at traverse speed $V_t = 20$ mm/s.

In figure 8 it is shown the increase of uncut area for different materials at the same speed. All graphics revealed the rise of uncut zone with the increase of thickness of materials and with the increase of traverse speed. The length of uncut zone may be controlled with the optimization of values of traverse speed. Optimization results in increased dimensional precision and increased surface quality.

References

[1] Ahmed D H, Naser J and Deam R T, 2016 *J. Mat. Proc. Technol.* **232** 116–130
[2] Schwartzentruber J, Spel J K and Papini M, 2017 *Int. J. Mach. Tools Manuf.* **12** 21–17
[3] Hlavac R M 2009 *J. Mat. Proc. Technol.* **209** 4154–4161
[4] Barabas B and Fota A 2015 *Int. Conf. Innov. Technol.* Dubrovnik Croatia.
[5] Jafar R H M and all 2015 *J. Manuf. Proc.* **17** 127–140
[6] Hlavac R M and all 2018 *Prec. Eng.* in press
[7] Miao X and all 2018 *Int. J. Adv. Manuf. Technol.* in press Springer Verlag London

Study on the reduction of the general / overall torsion on multi – story, rectangular, reinforced concrete structures

M F Botiș[1*], C Cerbu[2] and H Shi[3]

[1] Department of Civil Engineering, Faculty of Civil Engineering, Transilvania University of Brasov, B-dul Eroilor No. 29, 500036, Romania
[2] Department of Mechanical Engineering, Faculty of Mechanical Engineering, Transilvania University of Brasov, B-dul Eroilor No. 29, 500036, Romania
[3] College of Mechanical Engineering, Shenyang Jianzhu University China

[*] E-mail: mbotis@unitbv.ro

Abstract: This paper researches the reduction of the general / overall torsion of multi – story, rectangular, reinforced concrete structures. The general torsion is caused by the distance (eccentricity) between the centre of stiffness and the centre of mass of the structure. During the seismic shaking of the structural systems, the inertia force acts through the centre of mass, while the resistive force acts through the centre of stiffness. Under this coupled lateral - torsional motion, structural members located along the perimeter of the buildings develop increased deformations and stresses due to buildings' twisting, which results in a higher risk of collapse (brittle, non - ductile failure, damage). This leads to the torsional behaviour of buildings, which is one of the most frequent sources of structural damage and failure during strong ground motions. This paper investigates the influence of the cross sectional properties of the structural components / members, to reduce the above mentioned effect as much as possible. The reduction of the eccentricity can be obtained by modifying cross sectional properties of some rectangular reinforced concrete walls, such as the moment of inertia (length and width of the reinforced concrete walls), which are asymmetrically arranged relative to the central reinforced concrete core. The position of these rectangular reinforced concrete walls is known. Finding the optimal height and width of these walls can be done by using a MatLab function, which minimizes the distance between the centre of stiffness and the centre of mass. The obtained results are then verified in a structural software (ETABS 2016), using modal analysis, or the mode – superposition method, to determine the mode shapes of the structure.

1. Introduction

Buildings are mainly classified into regular and irregular. Asymmetric - plan buildings, namely buildings with in - plan asymmetric mass and strength distributions, are systems characterized by a coupled torsional - translational seismic response.

Asymmetric building structures are almost unavoidable in modern construction, due to various types of functional and architectural requirements. Even in symmetric structures, the asymmetric position of the structural components (non - symmetric distributions of mass and stiffness) tends to produce an effective asymmetric structure. Such an asymmetry, even if it is small, can produce a torsional response, coupled with a translational response.

The presence of structural irregularities has an adverse effect on the seismic response of the structure. Structural irregularities can be broadly classified as plan irregularities and vertical irregularities. In the present study, the effect of plan irregularity on the seismic response of a structure is studied [1].

1.1. Plan Irregularity

The condition of being non - uniform in the plan of a structure is called plan irregularity. These can be characterized by five different types such as torsional, reentrant corners, diaphragms discontinuity, out of plane offset and non – parallel system for plan irregularity [1].

1.2. Vertical Irregularity

Structures having significant physical discontinuities in a vertical configuration or in their lateral force resisting systems are termed as vertically irregular structure. The vertical irregularities in structures are Stiffness irregularity, Mass irregularity, Vertical geometric irregularity, Discontinuity in capacity [1].

1.3. Reinforced concrete (R.C.) cores

Reinforced concrete (R.C.) cores are used in many residential multi - story buildings as the primary seismic force resisting system. To minimize the effect of the structural system on the architectural layout, shear walls often act as stairway and elevator shafts in many R.C. buildings. Seismic damage surveys and analyses conducted on modes of failure of building structures during past severe earthquakes concluded that the most vulnerable building structures are those which are asymmetric in nature [2].

The lateral - torsional coupling due to the eccentricity between the centre of mass (C.M.) and the centre of stiffness (C.S.) in asymmetric building structures (in their floor plans or along their height) generates torsional vibration even under purely translational ground shaking, with significant swaying and twisting.

During the seismic shaking of the structural systems, the inertia force acts through the centre of mass, while the resistive force acts through the centre of stiffness, as shown in figure 1 [3].

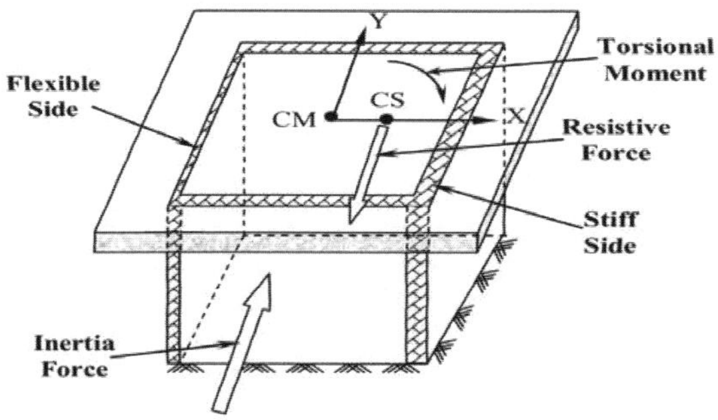

Figure 1. Generation of torsional moment in asymmetric structures during seismic excitation.

The natural eccentricity is generally defined as the distance between the centre of mass (C.M.) and the centre of stiffness (C.S.) for a considered floor, while accidental eccentricity generally accounts for factors such as the rotational component of ground motion about the vertical axis, the difference between computed and actual values of the mass and stiffness, and an unfavourable distribution of live load mass.

1.4. Types of eccentricities

The natural eccentricity is generally defined as the distance between the centre of mass (C.M.) and the centre of stiffness (C.S.) for a considered floor, while accidental eccentricity generally accounts for factors such as the rotational component of ground motion about the vertical axis, the

IOP Conf. Series: Materials Science and Engineering **399** (2018) 012005 doi:10.1088/1757-899X/399/1/012005

difference between computed and actual values of the mass and stiffness and an unfavourable distribution of live load mass.

1.5. Centre of mass
The centre of mass is a position defined as the average position of all the parts of the system, weighted according to their masses. The distribution of mass is balanced around the centre of mass and the average of the weighted position coordinates of the distributed mass defines its coordinates.

During an earthquake, acceleration-induced inertia forces will be generated at each floor level and they will act at a point where the mass of the entire story may be assumed to be concentrated. In a building having a symmetrical distribution of mass, the positions of the centres of floor masses will not differ from floor to floor. However, irregular mass distribution over the height of a building may result in variation in the centres of masses at various floors [1].

1.6. Centre of stiffness
The centre of stiffness is a point at a particular story, as the location of application of lateral load at that point will not produce rotation of that story. This definition is valid when the slab is modelled with an infinite in - plane floor behaviour. An infinite in - plane floor behaviour causes all of its constrained joints to move together as a planar diaphragm that is rigid against membrane deformation. As a function of structural properties, the centre of stiffness is independent of loading [1].

2. Building details
The structural analysis of a six story R.C. building structure has been done with the help of ETABS 2016 [4] structural software. The building is assumed as commercial complex. A linear static analysis has been performed. The regular grid plan of the structure is shown in figure 2. The structure is assumed to be located in Vrancea. The building contains a highly "accentuated" plan irregularity.

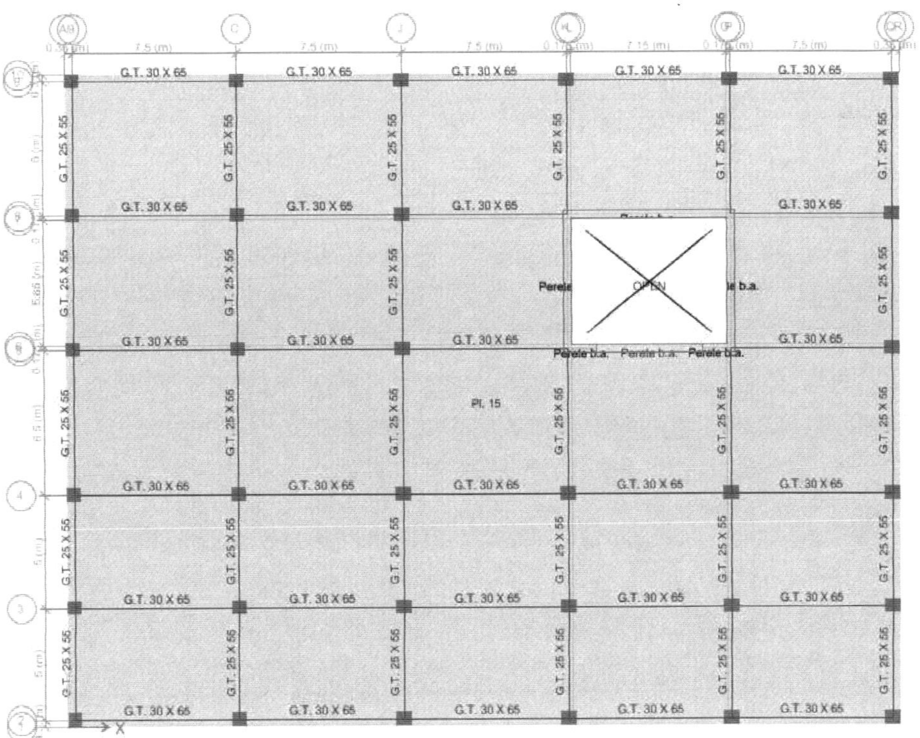

Figure 2. Building plan "before".

IOP Conf. Series: Materials Science and Engineering **399** (2018) 012005 doi:10.1088/1757-899X/399/1/012005

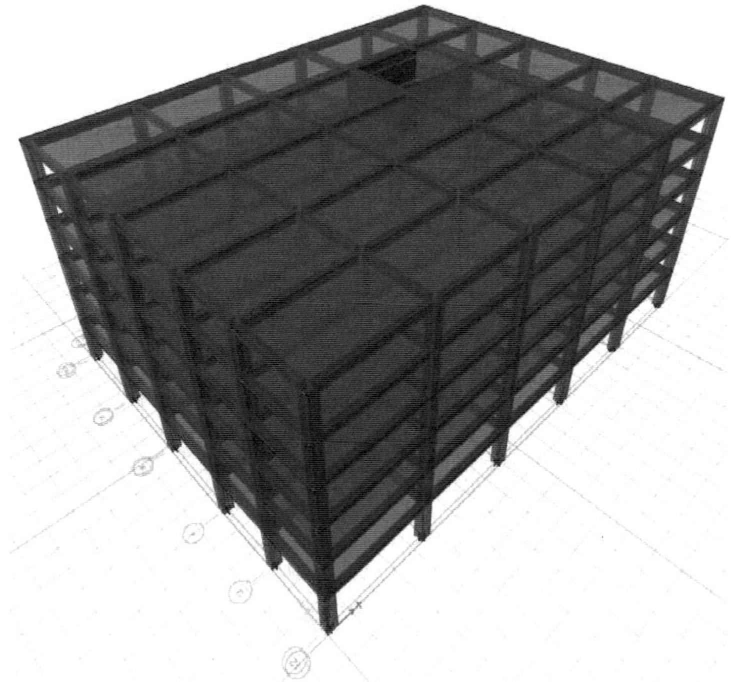

Figure 3. Isometric view of the structure "before".

2.1. Structural data

Table 1. Structural data.

Number of stories	6
Ground story height	3.50 [m]
Intermediate story height	3.00 [m]
Total number of columns	32
Slab thickness	15 [cm]
Core wall thickness	35 [cm]
Outer wall thickness	W_1 = 60 [cm] W_2 = 70 [cm]
Beam size	B_1 = 30 X 65 [cm] B_2 = 25 X 55 [cm]
Column size	C_1 = 70 X 60 [cm]
Grade of concrete	C 20/25
Grade of steel	BST 500S
Density of concrete	25 [kN/m^3]
Super dead load	1.3 [kN/m^2]
Live load	3.0 [kN/m^2]
Roof load	2.0 [kN/m^2]

IOP Conf. Series: Materials Science and Engineering **399** (2018) 012005 doi:10.1088/1757-899X/399/1/012005

Super dead loads are basically superimposed dead loads, which are applied on a structure; the load of any finished, partitioning, cladding, false ceiling are all super dead loads.

The live load is the load superimposed by the use or occupancy of the building, not including the environmental loads, such as wind load, rain load, earthquake load or dead load.

Table 2. Earthquake data [5].

Location, Country	Vrancea, Romania
a_g	3.924 [m/s^2]
T_c	1.0 [s]
T_B	0.20 [s]
T_D	3.00 [s]
Importance factor	1.00
Framing type	D.C.H.
Behaviour factor	6.75
Base shear coefficient, c	0.10
ν	0.45
Damping	5 [%]

3. Analysis

The building is modelled in ETABS 2016; an infinite in-plane floor behaviour is assigned at every story level, as shown in figure 4. Supports are assigned as fixed supports, neglecting soil - structure interaction. A linear static analysis was performed for earthquake cases: X Dir.; X Dir. ± Eccentricity; Y Dir.; Y Dir. ± Eccentricity.

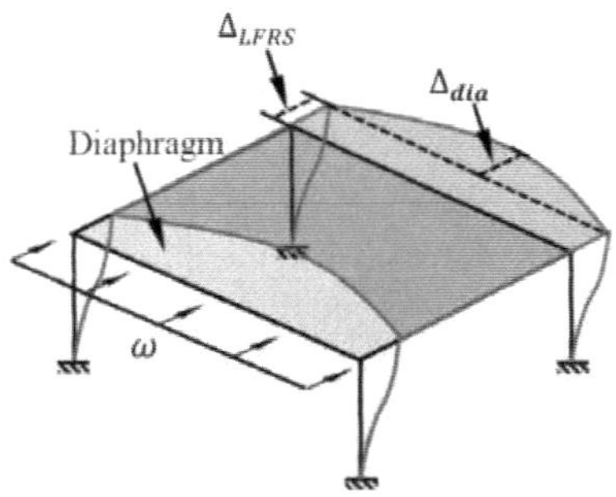

Figure 4. Prescribed diaphragm classification [6].

$$\Delta_{dia} \leq 0{,}50 \cdot \Delta_{L.F.R.S.} \rightarrow \text{Rigid diaphragm} \quad (1)$$

$$0{,}50 \cdot \Delta_{L.F.R.S.} < 0{,}50 \cdot \Delta_{dia} \leq 2{,}0 \cdot \Delta_{L.F.R.S.} \rightarrow \text{Stiff diaphragm} \quad (2)$$

$$\Delta_{dia} > 2{,}0 \cdot \Delta_{L.F.R.S.} \rightarrow \text{Flexible diaphragm} \quad (3)$$

An infinite in - plane floor behaviour is considered for this study. In this case, its midpoint displacement, under lateral load, is less than half the average displacement of the lateral resisting system. The rigid diaphragm distributes the horizontal forces to the lateral resisting elements, in direct proportion to their relative stiffness. It is based on the assumption that the diaphragm does not deform itself and will cause each vertical element to deflect the same amount. Rigid diaphragms capable of transferring torsional and shear deflections and forces are also based on the assumption that the diaphragm and shear walls undergo rigid body rotation, and this produces additional shear forces in the shear wall.

3.1. Analysis results

Table 3. Modal participating mass ratios "before".

Case	Mode	Period [sec]	UX	UY	RZ	Sum UX	Sum UY	SumRZ
Modal	1	0.508	0.2387	0.2083	0.3861	0.2387	0.2083	0.3861
Modal	2	0.337	0.2822	0.4757	0.0089	0.5209	0.684	0.395
Modal	3	0.255	0.2546	0.074	0.4586	0.7755	0.758	0.8535
Modal	4	0.159	0.0535	0.0387	0.0242	0.829	0.7966	0.8777
Modal	5	0.095	0.0642	0.0959	0.0007	0.8932	0.8925	0.8784

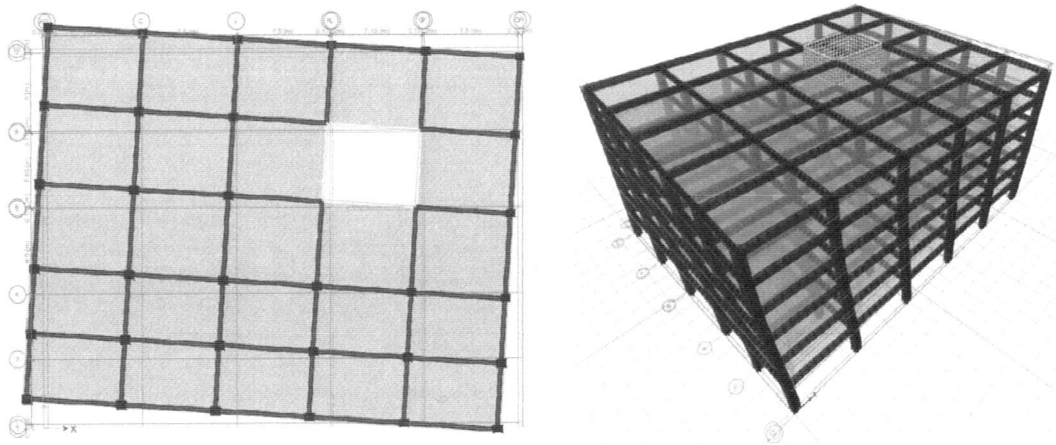

Figure 5. Mode shape 1 – $T_1 = 0.508$ s.

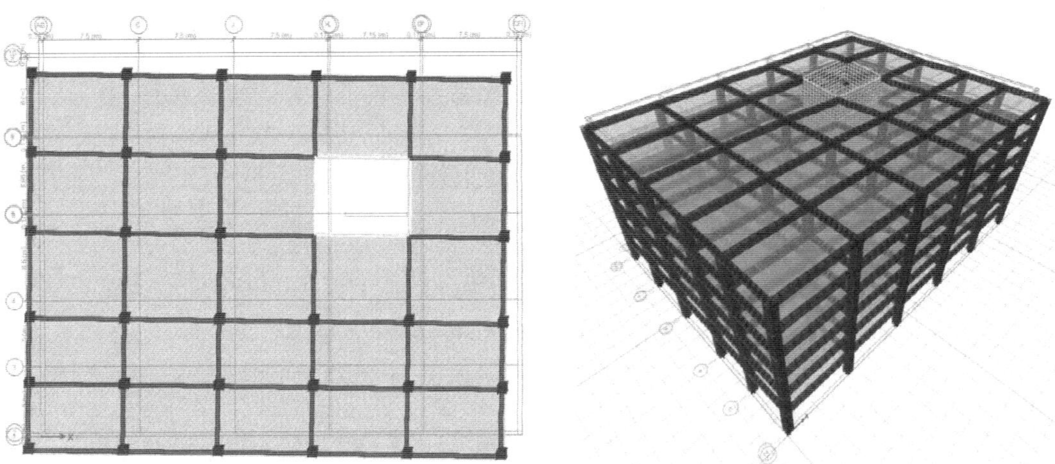

Figure 6. Mode shape 2 – $T_2 = 0.337$ s.

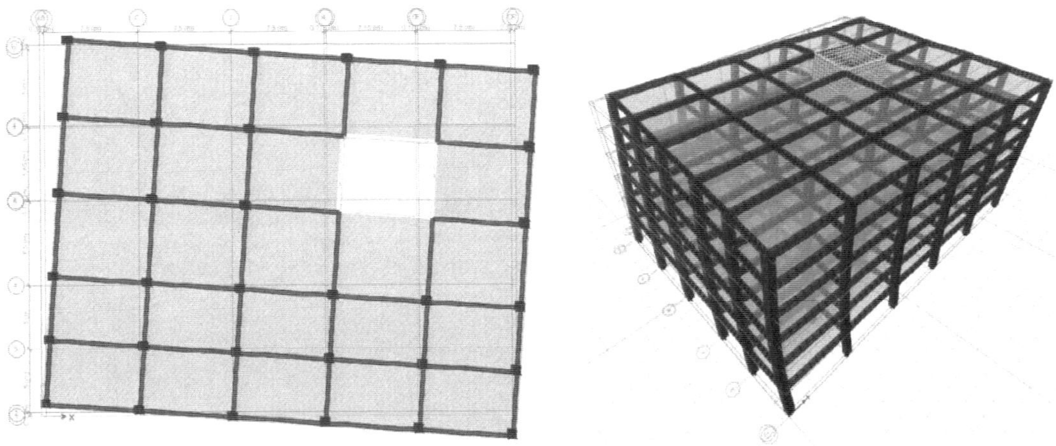

Figure 7. Mode shape 3 – T_3 = 0.255 s.

The coupled lateral - torsional motion appears already in the first mode shape, the structure is subjected to a predominant torsional motion (table 3), after that a lateral coupled motion can be observed in the following mode shapes (figure 5, figure 6 and figure 7).

4. Solution

As expected, in the first three mode shapes a coupled lateral - torsional motion appears. To avoid this torsional behaviour, two R.C. structural walls are placed in different positions, with different sectional properties (length and width of the reinforced concrete walls), as shown in figure 6 and 7. The position of these rectangular reinforced concrete walls is known.

To optimize the length of these elements, the Nelder - Mead algorithm or downhill simplex method or amoeba method is used, which is a commonly applied numerical method, used to find the minimum or maximum of an objective function in a multidimensional space. It is applied to nonlinear optimization problems, for which derivatives may not be known.

For the given system of equations:

$$\begin{cases} f_1(x_1,x_2,x_3,x_4,\dots,x_n)=0 \\ f_2(x_1,x_2,x_3,x_4,\dots,x_n)=0 \\ \dots\dots\dots\dots\dots\dots\dots\dots\dots\dots\dots\dots\dots\dots \\ f_{n-1}(x_1,x_2,x_3,x_4,\dots,x_n)=0 \\ f_n(x_1,x_2,x_3,x_4,\dots,x_n)=0 \end{cases} \tag{4}$$

the following function is constructed:

$$M_{(X)} = f_1^2(x_1,x_2,\dots,x_n,x_{n-1})^2+\dots+f_n^2(x_1,x_2,\dots,x_n,x_{n-1})^2=\sum_{i=1}^{n}\left(f_i(x_1,x_2,\dots,x_n,x_{n-1})\right)^2 \tag{5}$$

Thus the system of equations becomes:

$$\begin{cases} x_{C.M.}-x_{C.S.}=0 \\ y_{C.M.}-y_{C.S.}=0 \end{cases} \tag{6}$$

where:

$x_{C.M.}, y_{C.M.}$; $x_{C.S.}, y_{C.S.}$ are the coordinates of the centre of the mass, respectively of the stiffness of the studied structure.

Observation:

During this preliminary phase, the additional accidental eccentricity, according to P100 / 1 – 2013 [5], was not taken into account:

$$e_{a_i} = \pm 0.05 \cdot L_i \tag{7}$$

where:

e_{a_i} is the accidental eccentricity of the mass, at level i, relative to the calculated position of the centre of the masses, applied to the same direction at all levels.

L_i is the dimension of the floor, perpendicular to the direction of the seismic action.

$$\begin{cases} \dfrac{\sum_{i=1}^{n} A_i \cdot x_i}{\sum_{i=1}^{n} A_i} - \dfrac{\sum_{i=1}^{n} I_{x_i} \cdot x_i}{\sum_{i=1}^{n} I_{x_i}} = 0 \\ \dfrac{\sum_{i=1}^{n} A_i \cdot y_i}{\sum_{i=1}^{n} A_i} - \dfrac{\sum_{i=1}^{n} I_{y_i} \cdot y_i}{\sum_{i=1}^{n} I_{y_i}} = 0 \end{cases} \tag{8}$$

$$\begin{cases} f_1(L_1, L_2) = 21.2491 - \left(\dfrac{\left(\frac{0.6^3 \cdot L_1}{12} \cdot 15.35\right) + \left(\frac{0.7 \cdot L_2^3}{12} \cdot 0.35\right) + (41.39219 \cdot 26.60) + 7.12530}{\left(\frac{0{,}6^3 \cdot L_1}{12}\right) + \left(\frac{0{,}7 \cdot L_2^3}{12}\right) + 41{,}39219 + 0{,}37800} \right) = 0 \\ f_2(L_1, L_2) = 15.9379 - \left(\dfrac{\left(\frac{0.6 \cdot L_1^3}{12} \cdot 0.30\right) + \left(\frac{0.7^3 \cdot L_2}{12} \cdot 5.30\right) + (80.54802 \cdot 20.4569) + 7.22015}{\left(\frac{0.6 \cdot L_1^3}{12}\right) + \left(\frac{0.7^3 \cdot L_2}{12}\right) + 80.54802 + 0.51450} \right) = 0 \end{cases} \tag{9}$$

where:

L_1, L_2 are the optimal length of the reinforced concrete (R.C.) walls.

$$M(L_1, L_2) = \sum_{i=1}^{2} f_i^2(L_1, L_2) = f_1^2(L_1, L_2) + f_2^2(L_1, L_2) \tag{10}$$

Using the MatLab [3] function fminsearch ('func', [1 1]), with the initial solution vector $X_0 = [1\ 1]$ the following solutions are obtained:

$$L_1 = 7.731\ \text{m} \cong 7.75\ \text{m};\ L_2 = 5.649\ \text{m} \cong 5.65\ \text{m} \tag{11}$$

To verify the results, another numerical method was used (genetic algorithm). The genetic algorithm (G.A.) is a metaheuristic method inspired by the process of natural selection that belongs to the larger class of evolutionary algorithms (E.A.). Genetic algorithms are commonly used to generate high-quality solutions to optimization and search problems by relying on bio-inspired operators such as mutation, crossover and selection.

The obtained results are:

$$L_1 = 7.736\ \text{m} \cong 7.75\ \text{m};\ L_2 = 5.647\ \text{m} \cong 5.65\ \text{m} \tag{12}$$

After finding the optimal length of the reinforced concrete walls, a linear static analysis was performed again to see the differences between the two models (with and without reinforced concrete walls).

IOP Conf. Series: Materials Science and Engineering **399** (2018) 012005 doi:10.1088/1757-899X/399/1/012005

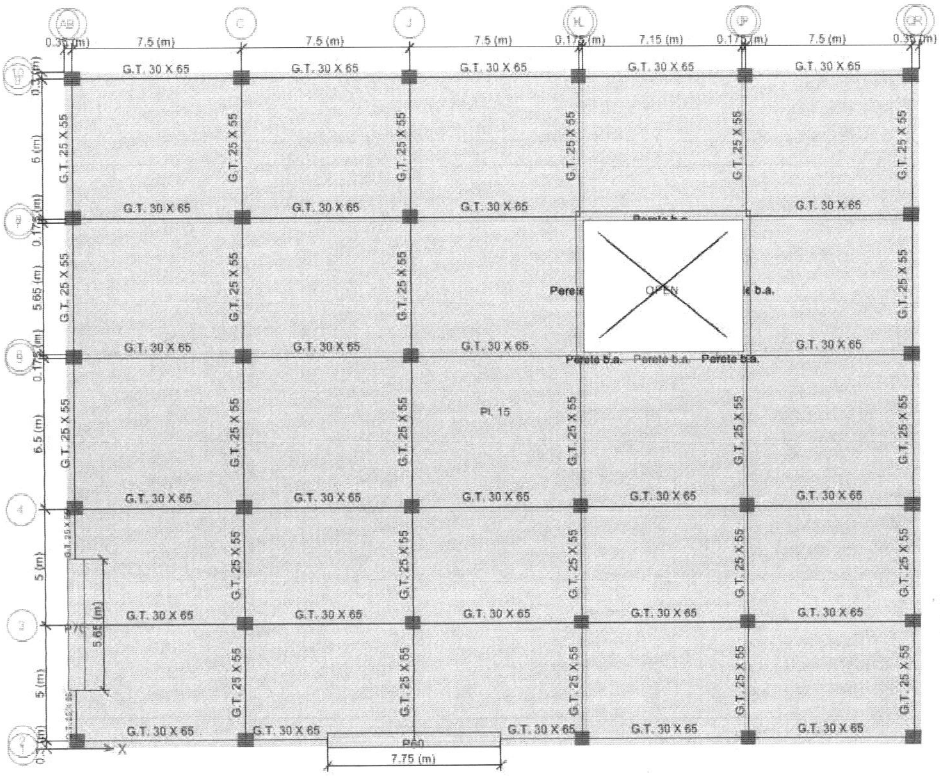

Figure 8. Building plan "after".

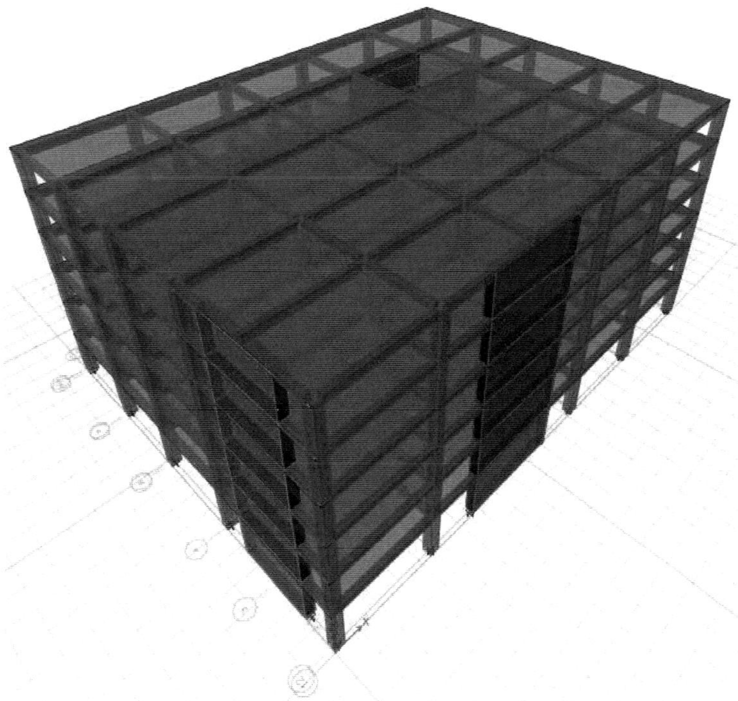

Figure 9. Isometric view of the structure "after".

Table 4. Modal participating mass ratios "after".

Case	Mode	Period [sec]	UX	UY	RZ	Sum UX	Sum UY	Sum RZ
Modal	1	0.304	0.0001	0.7422	0.0003	0.0001	0.7422	0.0003
Modal	2	0.270	0.7419	0.00003687	0.0109	0.742	0.7423	0.0112
Modal	3	0.228	0.0171	0.00003025	0.751	0.7591	0.7423	0.7622
Modal	4	0.079	0.0042	0.1755	0.0039	0.7633	0.9177	0.7661
Modal	5	0.075	0.1536	0.0076	0.0053	0.9168	0.9254	0.7714

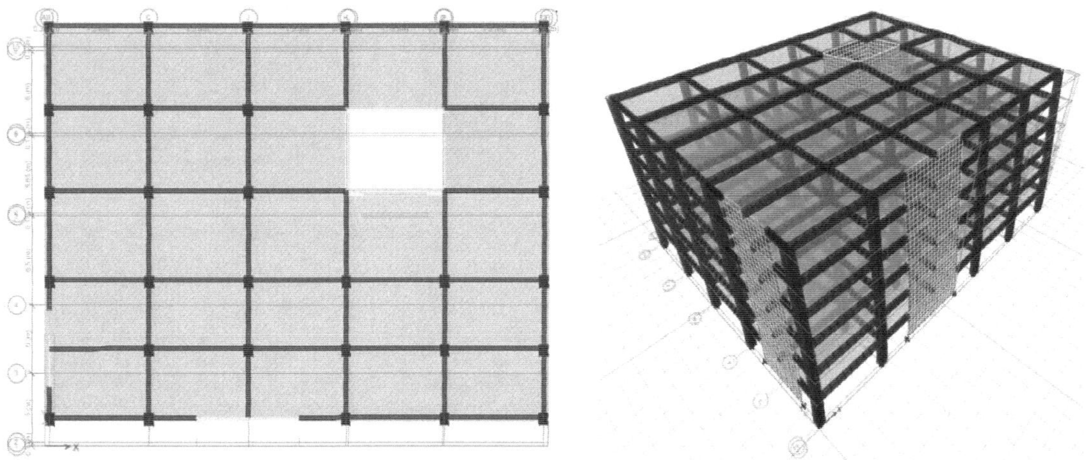

Figure 10. Mode shape 1 – $T_1 = 0.304$ s.

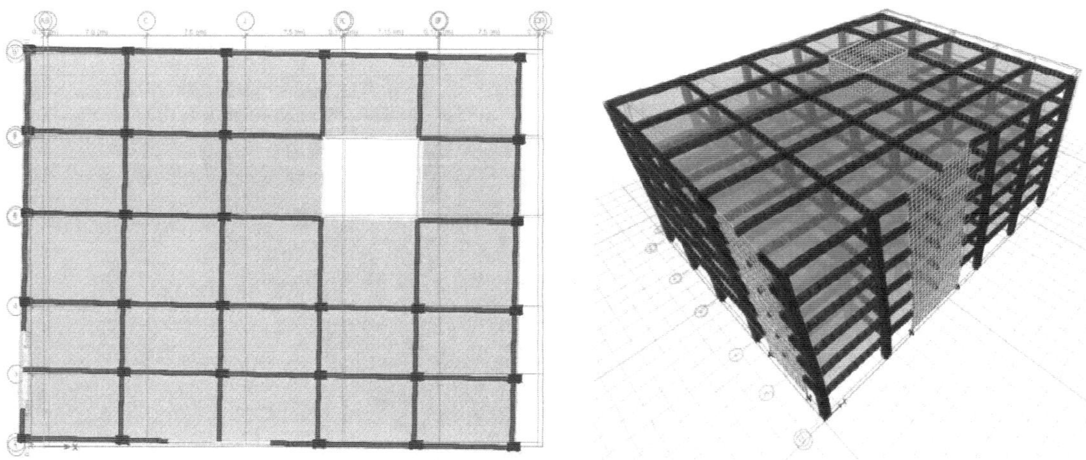

Figure 11. Mode shape 2 – $T_2 = 0.270$ s.

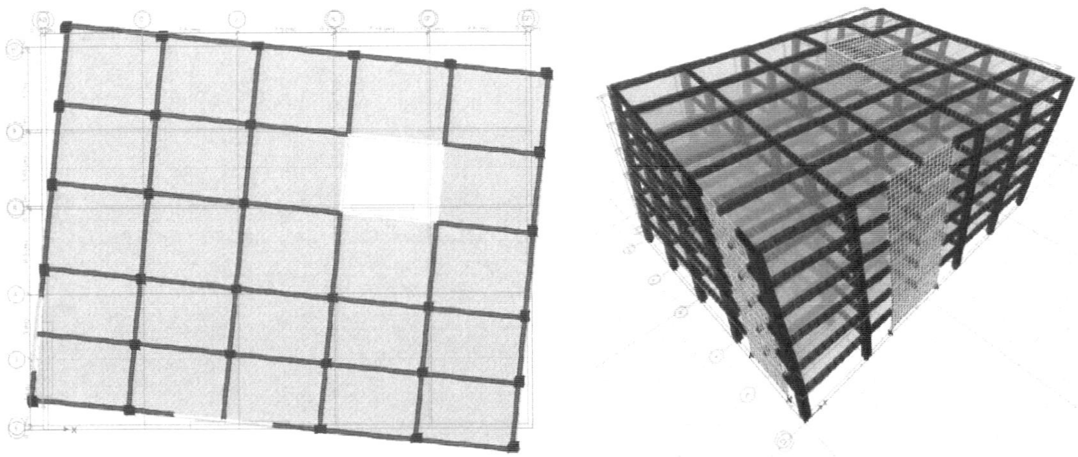

Figure 12. Mode shape 3 – T_3 = 0.228 s.

It is clearly visible now that, after optimizing the R.C. walls, the structure has a smooth orthogonal motion in the first two mode shapes (figure 10, figure 11) and the coupled lateral - torsional motion has been reduced to the minimum (table 5). The torsional motion appears only in the third mode shape (figure 12).

5. Conclusions

From the above mentioned results it can be seen that now the structure has a proper mode shape in the first three modes, thus preventing the coupled lateral - torsional motion, increased deformations and stresses due to buildings' twisting.

Asymmetric buildings are no longer a problem to create / design, due to this method, applied with numerical methods, as the distance between the centre of mass (C.M.) and the centre of stiffness (C.S.) can be reduced to 0.

For the structure, a much more favourable behaviour can be obtained, such as post-yield behaviour and energy dissipation when subjected to actual earthquake ground motions, but with the following important observation:

Such a building / structure must have only an elastic structural behaviour, because, in a postelastic response, if the structural walls are damaged, the whole building will get a strong torsion behaviour (and a possible structural collapse) is expected, due to the additional mass of these R.C. walls, which will not work as structural members. Hence, supplementary safety measures for the seismic structural behaviour must be applied if a post-elastic design is considered.

References

[1] Naresh Kumar B G, Punith N, Bhyrav Raj B and Arpitha T P 2017 Assessment of Location of Center of Mass and Center of Stiffness for Different Setback Buildings *International Journal of Modern Engineering Research* (IJMER)

[2] Pelletier K and Léger P 2017 Nonlinear seismic modelling of reinforced concrete cores including torsion *Engineering Structures*

[3] Maske S G and Pajgade P S 2013 Torsional Behavior of Asymmetrical Buildings *International Journal of Modern Engineering Research* (IJMER)

[4] Computers and Structures, Inc. (C.S.I.), ETABS 2016 Software, V16.2; 2017

[5] P 100-1 / 2013

[6] Srisangeerthanan S, Javad Hashemi M, Rajeev P, Gad E and Fernando S 2018 Numerical study on the effects of diaphragm stiffness and strength on the seismic response of multi-story modular buildings *Engineering Structures*

Ecological criteria considering the connections between the construction, building-site and the spatial planning development

C Cazacu[1*] and G C Chiţonu[1]

[1] Faculty of Civil Engineering, Transilvania University of Brasov, Turnului st. no.5, Brasov, Romania

[*] E-mail: cazacu.christiana@unitbv.ro

Abstract. The integration purpose of the ecological criteria in the projecting process is preceded by an intrinsic connection between the existential building space and global environment. The starting point is given by the evaluation of the balance amongst the functional resources of both systems, natural and anthropic. Thus, the ecological criteria open a path for the project leaders to identify and to evaluate the renewable resources potential, as alternative sources, and also their impact on the natural environment. In regard of the anthropic geo-system, following the increase of the population density in the urban areas, or of the depopulation of the rural areas due to the migration of the population to the metropolitan areas, the ecological answer is necessary to be concluded with the sustainable spatial planning. This study aims to develop and compare new categories of factors involved in spatial planning projects, having the goal to mitigate the irreversible damages of the natural environment.

1. Introduction

One of the greatest challenges of our time is undoubtedly related to the changes to the environment, the effects of which we are currently witnessing. Although the causes of these effects are being debated, substantial sections of the research communities are focusing and working together on reducing the multiple factors which are generating the global warming with its climate consequences for the entire planet. For the industrial countries (at the very least), one of the key ways to deal with negative man-made climate effects lies in the development of building environments focusing on reduction of carbon emissions. The urban environment can most easily be upgraded by changing the current approach, and creating a sustainable spatial development, especially for the big cities and metropolitan areas.

A correct attitude in the integrated ecological planning of buildings is the process starting from the urban planning.

The main purpose on this level has a prevention character, avoiding some measures, which once having been taken, even if they are supported by the community progress, can create irreversible damages [1].

Decoding the concept of sustainable development requires a holistic analysis. The impact of the research group's results on planning development has changed the importance of promoting the progress to socio-human implications which has been determined in the ecological context [2].

2. Basic requirements in the ecological planning

The aim of this study is to clarify the potential impact of the ecological environment and eco-system concepts, and how they can help to produce a balance to sustainable development in the building and urban design fields.

In the field of integrated architectural research, an important role in the development of sustainable concepts has been studied at University of Architecture and Urbanism "Ion Mincu"-Bucharest, one of the promoters being Prof. dr. arch. Cristina Victoria Ochinciuc. In her book, "Sustainable development concepts in architecture. Integrated planning", a synthesis of the important ways of integrating the ecological criteria in planning and implementing the urban development is shown by the following six criteria [2]:

- the connection between the environment and the resources (natural and human);
- managing and implementing the development projects;
- major problems of the operational ways in the projects frames;
- socio-cultural factors;
- environment protection with regard to preserving the population health,
- long term estimations[2].

The first ecological criterion refers mainly to the characterisation of the local, natural and anthropic geo-system (economics resources), geological, hydrological and seismological studies and analysing the microclimate, the landscape perception with regard to spatial typologies and perspectives. The predictable changes of the environment factors, including the social and sanitary issues, are also mentioned.

The second criterion is about the importance of including environment protection measures and population health protection, cchoosing measures for soil protection, controlling forestry, erosions, landslides, other secondary pollution effects; evaluating the side-effects on the rivers: rate of flow, pollution of the water; aadopting protection measures on the ecosystem: fauna, vegetation; ddevelopment possibilities as functionality, landscape perception, environment protection.

The third criterion includes projects in reducing the negative impact on the environment, studying the ways and the systems of treating the waste water, gas emissions, the ground-water, recycling the solid-waste, rreducing the negative impacts on the well-being of the local population, avoiding increasing migration, teaching the active population in getting involved in the process of detecting and combating the unpredictable harmful effects, as well as in detecting olfactory and sound pollution, transporting and depositing the solid waste, explosion risks, accidental presence of toxic substances [3].

The fourth domain is enhancing the evaluation of the consequences of the demographic rate and the population density in relation to the social and cultural behaviour in the local communities, and, in the case of the population migration, making sure sufficient supplies, accommodation, transporting solutions are available. This criterion also refers to estimating the needs regarding buildings and functions: housing, educational system, administrative and socio-cultural programs and having perspectives for upgrading the programs, and connecting them with the development of the society.

The fifth criterion links the natural environment with population health by evaluating the consequences in case of behaviour changes in food-production or culinary habits, controlling the water quality, controlling wastes to prevent health damages to the population; evaluating the evolution of the life-hope and the morbidity (on age groups and different pathologies) following the chemical and physical mutations of some elements or factors in the natural geo-system which can lead to increasing undesirable species or causing diseases; adopting measures for preventing contamination with pathological factors in case of population migration to other, less populated areas.

The last ecological criterion described long terms procedures: Evaluating the effects on the microclimate in case of the felling of trees and other changes due to the excessive utilisation of lands [4]; elaborate solutions for further landscaping to prevent the ecological disasters which can accidentally appear through development planning; evaluating the possible consequences in case of further extension of some functions over the anthropic geo-system; natural ventilation of the neighbourhood, planting surfaces having protection role against the wind, or the excessive sunlight in the summer; Considering water surfaces which can improve the climate by changing the ambient temperature can contribute to decreasing energy consumption [5] .

It is thought that an urban planning thinking can lead to a 50% reduction in the needed heat of a building, when in hard climate conditions buildings have received an optimal insulation.

By knowing the balance between the functional resources of both natural and anthropic geo-systems, the ecological urban planning allows the researchers to identify and evaluate the usage potential of some regenerating resources, such as alternative energy solutions: sun, wind, geo-thermal heat, bio-masse [6] .

Many countries have developed their own regulations in the fields of sustainable energy, ecological and environmental development, but some of them are specific at national level, while some of them can be implemented at international level.

In a European context, it is desirable to establish common standards, norms and criteria for these mentioned scientific fields, so that one can more easily and effectively share best practices and increase the speed of implementation of these concepts. These norms and standards and the extent to which, in terms of local specific implementation, it is reasonable or sensible to implement them across Europe and beyond will naturally vary, but there are concepts that can easily be shared across countries and regions, and one should seek to get them unified and commonly agreed upon.

3. Ecological buildings versus intelligent buildings

The word sustainable has the roots in the Latin "subtenir", which means *hold up* or *support*. A community must be supported by its own population, in present and future. Some areas having a good combination of physical, cultural, spiritual characteristics inspire the people to take care of their community. These places have the best chances where the ecological concepts can be developed.

The concepts with regard to "intelligent buildings", high-tech or eco-tech (ecological high-tech) belong to the same field with ecological architecture and sustainable development. The access to advanced technology permitted many advantages, such as using the buildings with high energy savings, automatic control, intelligent structures [3]. On the other hand, the materials used in building this type and the maximum performance could be in contradiction with the recycle rate or reutilisation. There are differences in on-site integration for an ecological building, as compared to a high-tech one.

Another point of view reveals the fact that the intelligent architecture is always ecological, but the ecological architecture is not always intelligent.

In an intelligent building, the interior and exterior spaces and the comfort parameters are balanced and they are easy to control with electronic systems. A new architectural methodology is developing, having as purpose to keep the balance in relation to interior comfort, as an answer to excess solar heating, exterior temperature or wind effects. Also, the analysis must take account of the bio-electro-magnetism or local ionization radiations or those which are emanated by some building materials (even if they are recycled or natural, from the ground in case of underground constructions) [7].

An intelligent building is characterized by 4 types of interactions [6]:

- the interaction between the annual energy consumption, the optimal interior ecological climate and the thermal and visual proprieties of the transparent façades;
- the interaction between the annual energy consumption, the optimal interior ecological climate and the cardinal orientation (N, S, E, V);
- the interaction between the annual energy consumption, the optimal interior ecological climate and the controlled or uncontrolled indoor air exchange;
- the interaction between solar system (active, passive and combined) and the energy consumption level of the building.

The intelligent building is dynamic, having all the parameters controlled by electronic systems. Due to this process, the balance of the indoor space and the exterior space is constant.

The ecological solutions are based on environment protection and they promote energy and materials savings, recycling, proving the progress of sustainable development. The intelligent solutions solve two problems: mitigating the environment changes and increasing the utilisation capacity. Recent projects prove the possibility of evaluating ecological buildings as a new type of architecture. The new building techniques, the energy concepts and facades systems bring innovative ideas [8].

Through intelligent planning, it is possible to create buildings with an optimal living and working environment. There are 3 ways to accomplish this result: technical, ecological and reprogramming.

The technical approach focusses on improving the facades and has been launched by introducing the climate facades.

The ecological approach is related to analysing the whole building. The building becomes a system which is exchanging energy with the environment, avoiding the necessity of conventional fuels.

An example of reference, both in terms of observance of the ecological architecture as well as related to sustainable development or the generation of energy from unconventional sources, is the traditional vernacular architecture and the technical equipment used in the past [9].

All the materials used were fully recyclable, without exception, most of them brought from the immediate vicinity of the site where they were found. Examples of building materials, as clay, hemp, lime, in addition to building green roofs and wooden houses show the connection between the past and the present with regard of the ecological concepts [10].

The most useful approach is performing an analysis before the conventional planning process, with the focus on the efficiency of the buildings. This concept can be called reprogramming.

This new concept of architecture has brought the concept of energy self-sufficient buildings like Zero-energy houses and flats, or plus-homes. If these buildings prove that a living space can be created without using polluting energy sources, then this concept should be considered on a larger scale. The planning process herein can be defined in several ways:

- the building as an organism which can function more efficiently (the building and the climate installation make up a whole);
- the building as a system which can be organized more efficiently, having connections with the environment;
- the building as a system which provides better effects on the well-being of its occupants.

The main question is whether a narrow ecological definition is desirable or whether a broader definition containing also social, cultural and spiritual dimensions is preferable. The narrow definition is seen in many both official and grass roots contexts and may make it easier at this time to get funding. The need for framework for education and self-audits as measures of how far we have come and where we go underscores this [5].

4. Conclusions. Architectural answers

Ecological criteria included in the development of urban planning lead to an enhancement of the architectural program. By evaluating the natural and human existing resources, architecture can provide new answers to the needs of our eco systems. To do this, we have to define the functional potential of the natural and anthropised geo-system and settle the development perspectives.

It is necessary to make sure that the spatial planning development leads to good living conditions and optimizes the needs of the construction. An economic analysis has to consider the costs for the main utilization, and the post-utilization of the future buildings.

The planning team will prevent the damages which can be created to the local communities and natural geo-system by the new construction. They will solve the accesses, preserve the landscape, creating transitory spaces, sanitary protection, solving the techno-utility challenges, and facilitating waste recycling. The focus and balance of the environmental efforts depend on the local conditions, including resources, policies and individual actions and the community's unique characteristics. The concept of sustainable community has been applied to various aspects, such as urban spreading, redevelopment of built-up areas, economic development and growth, ecosystem management, agriculture, biodiversity, green buildings, water management and pollution prevention.

The results provided by involving specialists from different fields (social, cultural, economic, health) establish a balance to the ecosystem in the short term and also for the future development of the areas.

References

[1] Ioanid V 1991 *Urbanism şi mediu* [Urbanism and environment] (Bucureşti: Editura Tehnica)
[2] Ochinciuc C V 2002 *Conceptul dezvoltării durabile în arhitectură. Proiectarea integrată.* [The sustainable development concept in architecture. The integrated planning] (Bucureşti: Editura Univesitara "Ion Mincu") pp 145-198
[3] Voica M 2008 *Arhitectura ecologică – tradiţional şi tehnologie contemporană. Dezvoltare durabilă şi management ecologic.* [The ecological architecture-traditional and contemporary technology. Suistanable development and ecological management] (Bucureşti: Ph.D. thesis "Ion Mincu" University of Architecture and Urban Planning) pp 120-122
[4] Smith P F 2005 *Architecture in a climate of change* (Oxford: Elsevier/Architectural press) pp 21-34
[5] McDonnough W and Braungart M 2002 *Cradle to Cradle* (New York: North Point Press)
[6] Ray V 1992 *De l'écologie à l'environnement.Environnement et cadre de vie.*[From the ecology to the environment. The environment is the life frame] (Bucharest: L. Harmattan. L' Institute francais)
[7] Cazacu C E 2015 Study on environmentally friendly constructions. *Bulletin of the Transilvania University of Braşov. Series 1 special issue no.1* **8(57)** 21
[8] Muntean R and Cazacu C E 2011 Using PET (polyethylene trephtalate) waste for buildings. *Analele Universităţii din Oradea JAES issue 3* **1(14)** 73
[9] Chiţonu G C 2010 *Sustainable Development reflected in the values of Traditional Arhitecture. Scientific Session Constructions and Instalations CIB* (Brasov: Editura Universităţii Transilvania)
[10] Cazacu C E Muntean R Galatanu T and Taus D 2016 Hemp-lime technology. *Bulletin of the Transilvania University of Braşov. Series 1 special issue no.1* **9(58)** 19

IOP Conf. Series: Materials Science and Engineering **399** (2018) 012007 doi:10.1088/1757-899X/399/1/012007

The use of Digital Image Correlation in determining the mechanical properties of materials

C Cerbu[1*], D Xu[2], H Wang[2] and I C Roşca[1]

[1] Department of Mechanical Engineering, Faculty of Mechanical Engineering, Transilvania University of Braşov, B-dul Eroilor No. 29, 500036, Romania
[2] Tianjin Key Laboratory of Refrigeration Technology, Tianjin University of Commerce, Guangrong Road No. 409, Beichen District, Tianjin 300134, PR China

[*] E-mail: cerbu@unitbv.ro

Abstract. The main purpose of this paper is to study the use of Digital Image Correlation in order to determine the Poisson's ratio in the case of aluminium alloy and composite materials under tensile tests. The satisfactory results obtained in the case of specimens made of aluminium alloy prove that Digital Image Correlation is able to satisfy the requirements in order to determine the Poisson's ratio. The testing procedure was carried out for two kinds of composite materials: (1) jute / epoxy composite material; (2) jute / glass / epoxy composite material. During the tests, a digital camera was used to capture the images of the tensile specimen during the tensile test. In order to measure both the longitudinal strain in direction of the tensile force and the transverse strain, the Digital Image Correlation method is used in the post-processing of all images recorded for each specimen. The data of transverse strain related to the longitudinal strain were graphically shown. The initial portion of curve corresponding to the linear portion of the stress-strain curve, was approximated by a linear regression function for each specimen tested. The slope of the fitting line represents Poisson's ratio. The Poisson's ratio, corresponding to the reinforcing plane with fabric for Jute / epoxy composite, is 35.5% greater than the corresponding to Jute / glass / epoxy composite.

1. Introduction

Due to several distinct advantages such as low weight, high ratio between strength and weight, and high stiffness-weight ratio, composite materials have been widely used in industry in recent years, even in the aerospace field [1-3]. Thus it is essential to propose a reliable and efficient method to characterize the mechanical properties of the composite materials. The Poisson's ratio of composite materials is an indispensable parameter in the finite element analysis, but the studies concerning the determination of the Poisson's ratio are rarely reported [4].

During tests, extensometers and strain gauges are generally used in order to measure the displacements or strains of specimens and then, in order to determine the mechanical properties [5]. However, what they can provide is just an average value of the displacement or strain over the gauge length at every moment in the experiment [6, 7]. The experimental information regarding the deformation field of composite materials is too limited because the mechanical behaviour of composite materials is more complicated in comparison with that of isotropic materials [7].

The Digital Image Correlation (DIC), a powerful optical testing technique which emerged in recent years, has been widely used to observe the stress-strain behaviour [8, 9], residual stress [10] and crack propagation [11] . Compared with traditional methods, its prominent advantage lies in the non-contact and full-field measurement [12, 13]. By means of the DIC, some researchers have successfully investigated the strain fields of composite materials both in macro- and micro-scales [14]. Among these studies, the reinforcements involve carbon fibre [15], glass fibre [16], woven synthetic [17] and natural fibers [18]. It is worth noting that most of the existing studies focus on the description of strain distribution or failure evolution.

On the other hand, over the past years, there has been a great interest in the use of natural fibres (jute, flax, hemp fibres) to reinforce the composite materials used as building materials for ecological buildings.

Taking into account the above-mentioned aspects, this paper aims to determine and compare Poisson's ratio of two kinds of composite materials reinforced with vegetable fibers by means of the DIC technique. The composite materials tested are the following: Jute / epoxy composite material; Jute / glass / epoxy composite material whose core layers are reinforced with jute fabric, while the outer layers are reinforced with glass fabric. For this purpose, the strain fields of specimens involving longitudinal strain ε_l and transverse strain ε_t were calculated by Vic-2D software. The values of Poisson's ratio can be finally achieved by fitting the data of the transverse strain ε_t related to the longitudinal strain ε_l corresponding to the elastic portion of the stress-strain $(\sigma - \varepsilon)$ curve.

2. Materials. Testing method

2.1. Materials

The materials tested were: the EN AW-6060-T6 aluminium alloy; jute / epoxy composite material; jute / glass / epoxy composite material. The aluminium alloy was tested to determine the Poisson's ratio in order to validate the experimental technique that consists in combining of the tensile test with the Digital Image Correlation method.

In accordance with the European Standard EN 573 – 3 / 2010 [19], the EN AW 6060 aluminium alloy belongs to the series of 6000 of the aluminium alloys AlMgSi (aluminium – magnesium – silicon). The chemical composition of the aluminium alloy EN AW 6060 is: 0.3÷0.6% Si; 0.1÷0.3% Fe; 0.10% Cu; 0.35÷0.60 % Mg; 0.05 % Cr; 0.15% Zn; 0.10 % Ti; 0.05 % other metallic components so as the sum does not exceed 0.15 %; the difference is covered by the aluminium [19]. According to [20], the EN AW 6060 aluminium alloy being in T6 heat treatment condition is encoded EN AW-6060-T6.

Aluminium tensile specimens are manufactured according to EN ISO 6892-1: 2002 [21]. Some dimensions of the tensile specimen are: total length – $l = 150\,mm$; active length $l_0 = 60\,mm$; width $b = 10\,mm$ of the active length; width $B = 10\,mm$ of end part that is clamped in the tensile machine.

The composite materials involved in this paper are made of Epolam epoxy resin. One kind of composite material contains four layers reinforced only with jute woven fabric. The other composite material contains also four layers: two core layers reinforced with flax woven fabric; one bottom outer layer (shell) reinforced with E-glass woven fabric; one top outer layer (shell) reinforced also with E-glass fabric. E-glass and jute woven fabrics have the same kind of yarn on both warp and weft directions. The densities corresponding to jute and glass woven fabrics are $400\,g/m^2$ and $200\,g/m^2$, respectively.

The Epolam epoxy resin whose physical and chemical characteristics are shown in table 1, is used to manufacture the composite materials [22]. To initiate and to accelerate the polymerisation process, the hardener agent was mixed with the epoxy resin before the impregnation of the fabrics. The mechanical characteristics of the epoxy resin with hardener are shown in table 2 [22].

One panel was manufactured for each kind of composite material, whose dimensions were 500x500 mm^2. The jute and glass fabrics used to reinforce the layers have the same orientation in all layers. A lower forming pressure was used to manufacture each panel by using hand lay-up technology. The

IOP Conf. Series: Materials Science and Engineering **399** (2018) 012007 doi:10.1088/1757-899X/399/1/012007

composite panels were kept for one week at room temperature $\left(\approx 20^{\circ} C\right)$ before cutting of the tensile specimens.

For both composite materials, the tensile specimens were cut so as their length should be parallel with the same direction of the reinforcing fabrics. It was not necessary to cut two specimen sets (one set corresponding to warp direction of jute fabric and another set corresponding to weft direction) because both jute and glass fabrics used as reinforcement are made of the same yarn in warp direction like in weft direction. The shape and dimensions of the tensile specimens made of composite materials were in accordance with standard [23].

Five tensile specimens were tested in case of each material involved in this paper.

Table 1. Physical and chemical characteristics of the Epolam epoxy resin, in liquid state [22].

Characteristic	Value	Unit of measure	Method
Density, 25 °C	1.15	g/cm^3	ISO 1675: 1985
Viscosity, 25 °C	1550	mPa·s	Brookfield LVT
Mixture ratio with hardener agent	32 (weight ratio) 38 (volume ratio)	%	-
Gel-time, at 23 °C (100 g resin + 32 g hardener)	2.5	Hours	-
Manipulation time (100 g resin + 32 g hardener)	60	Minutes	-
Glass transition temperature	80	°C	ISO 11359: 2002°C9188

Table 2. Mechanical characteristics of the Epolam epoxy resin (with hardener) without reinforcing [22].

Characteristic	Value	Unit of measure	Method
Tensile stress in tension	70	MPa	ISO 527: 1993
Flexural stress	120	MPa	ISO 178: 2001
Modulus of elasticity E	3100	MPa	ISO 178 :2001
Impact strength - Charpy (unnotch specimen)	40	kJ/m^2	ISO 179
Elongation in tensile test	5	%	ISO 527: 1993
Toughness	83	Shore D15	ISO 868: 2003

2.2. Testing method

Herein, the testing method consists in combining the tensile test with the digital image correlation method in order to determine the Poisson's ratio (transverse contraction coefficient), in addition to the other tensile properties corresponding to the composite materials described above. Firstly, the technique is used in the case of the aluminium alloy, just to validate the experimental procedure.

The Poisson ratio ν is another elastic characteristic of a material and it is computed by using the following relationship (1):

$$\nu = \left|\frac{\varepsilon_t}{\varepsilon_l}\right|, \tag{1}$$

where: ε_l represents the strain in the longitudinal direction of the tensile specimen that coincides with the direction of the tensile force; ε_t is the strain of the tensile specimen in the transverse direction that is the direction perpendicular to the tensile force.

The composite material reinforced with bidirectional woven fabrics is an orthotropic material with respect to the material coordinate system denoted with 123, whose 1 and 2 axes coincide with the directions of the fabrics. In this context, in the case of the composite materials which are the focus of this paper, one determines the Poisson's ratio ν_{12} corresponding to the reinforcing plane with fibers because the plane of the tensile specimens is parallel to the reinforcing fabrics and all tensile specimens were cut so as their length be parallel to the same direction of the reinforcement fabrics.

A photo of the set-up of the experimental test is shown in the figure 1.

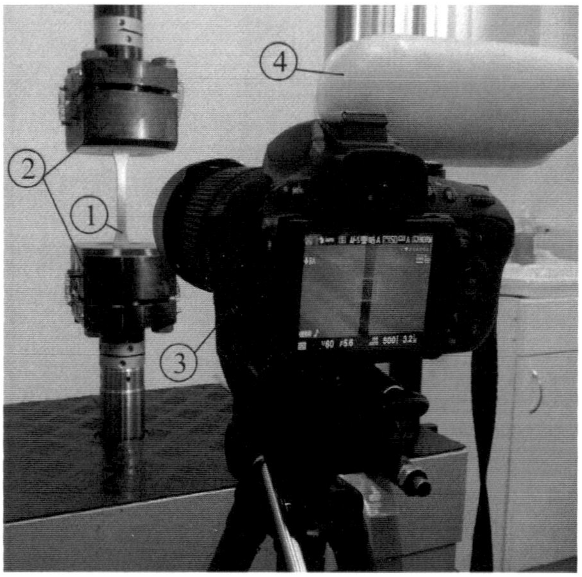

Figure 1. Set-up of the tensile test combined with DIC method: 1 – Tensile specimen; 2 - Fixing device of the tensile machine; 3 – Photo camera; 4 – Light source

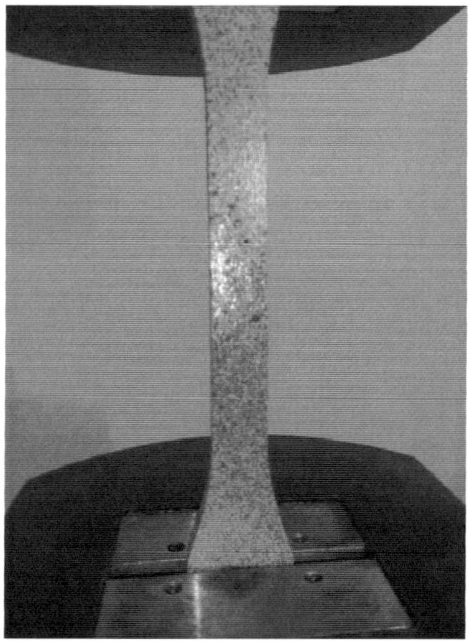

Figure 2. Tensile specimen clamped in tensile machine

Tensile tests were conducted on the universal digitally-controlled testing machine type LFV 50-HM, 980 (Walter & Bai, Switzerland). The universal machine whose maximum force is 200 kN is equipped with devices for tensile, compressive and bending tests. The software of the machine permits the recording, with a high acquisition frequency, of the following data: tensile force F, elongation Δl and the time t. The speed of loading was 1 mm/min. in tensile tests. The data of the elastic portion of the tensile stress-strain $(\sigma - \varepsilon)$ curve were approximated by using a linear function whose slope represents the modulus of elasticity E (Young's modulus). The curve of the longitudinal strain ε_l related to the time t is also plotted in case of each specimen in order to correlate it with the data obtained by using the digital image correlation method.

In order to measure both the longitudinal strain ε_l and the transverse strain ε_t, the VIC-2D system is used in the digital image correlation technique. This system consists of the computer software VIC-2D and the digital camera denoted by 3 in figure 1. In fact, the VIC-2D system may be used to provide two dimensional maps of any planar specimen mechanically loaded in the plane of the specimen.

The digital image correlation technique consists in capturing a set of photos of the specimen during the deformation by using the digital camera. Then, the digital photos recorded are analysed in order to create a contour plot of the strains. Before the test, it is necessary to prepare the specimen by spraying a random dot pattern (speckled pattern) on the surface of the specimen that is photographed. Figure 2 shows the speckled pattern on the tensile specimen clamped in the tensile machine. Before the test begins, a reference photo is taken. During the mechanical test, the camera takes other photos. The VIC-2D software analyses the displacements expressed in pixels by comparing the photos captured during deformation with the reference photo and, finally, correlate them to show the plot contour of the strains. The displacement plots may also be obtained after calibration by using a calibration plate.

A DSLR Nikon D5100 camera was used whose maximum resolution is 4928x3264 pixels and whose lens 18-55 mm f/3.5-5.6 permits the sequential shooting. The photo camera was set to capture one photo per second for 200 photos in case of each tensile test. Then, all photos were post-processed by using the VIC-2D software in order to obtain the graph of the transverse strain ε_t related to the longitudinal strain ε_l in the case of each tested specimen (figure 3).

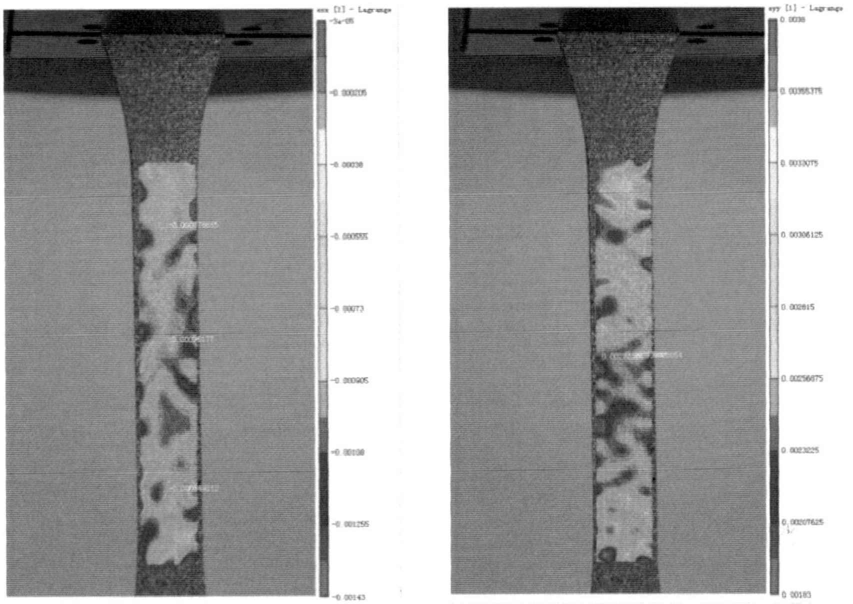

a. b.

Figure 3. Contour plot of the strains by using VIC 2009 software:
a. Transverse strain $\varepsilon_t\left(\varepsilon_{xx}\right)$; b. Longitudinal strain $\varepsilon_l\left(\varepsilon_{yy}\right)$.

For each tensile specimen tested, the $\varepsilon_t - \varepsilon_l$ curve was correlated with the stress-strain $(\sigma - \varepsilon)$ curve and longitudinal strain – time $(\varepsilon_l - t)$ curve in order to determine the Poisson's ratio. In other words, the data of the $\varepsilon_t - \varepsilon_l$ curve corresponding to the elastic portion of the $\sigma - \varepsilon$ curve, were approximated with a linear function whose slope represents the Poisson's ratio ν in accordance with the relationship (1).

3. Results

Firstly, the medium stress-strain curve recorded in the case of the EN AW-6060-T6 aluminium alloy is shown in the figure 4, and the average values of the tensile properties are shown in the table 3.

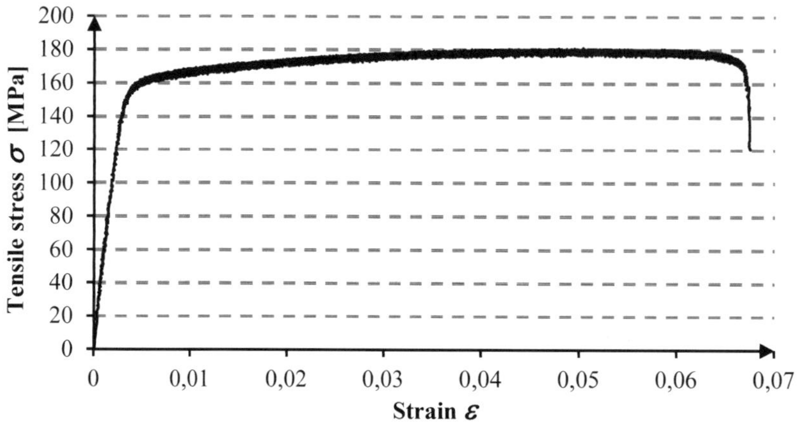

Figure 4. Stress-strain $(\sigma - \varepsilon)$ curve recorded in tensile test for EN AW-6060-T6 aluminium alloy.

Table 3. Tensile properties and Poisson's ratio ν determined for EN AW-6060-T6 aluminium.

	Young's modulus E (MPa)	Stress σ_e at the elastic limit	Max. stress σ_{max} (MPa)	Strain ε_e at the elastic limit	Strain ε la F_{max}	Poisson's ratio ν
Average value	52780	144	182	0.002784	0.047019	0.353
Stdev	2393	28	12	0.0004	0.0052	0.010

The Poisson's ratio $\nu = 0.353$ of the aluminium alloy is the average value and it is approximately equal to the value given in specialty literature [24]. Two $\varepsilon_t - \varepsilon_l$ recorded in the case of two tensile aluminium specimens are plotted in figure 5. In table 3, one may also remarks a good result concerning the standard deviation value (Stdev) corresponding to Poisson's ratio ν in the case of the EN AW-6060-T6 aluminium alloy.

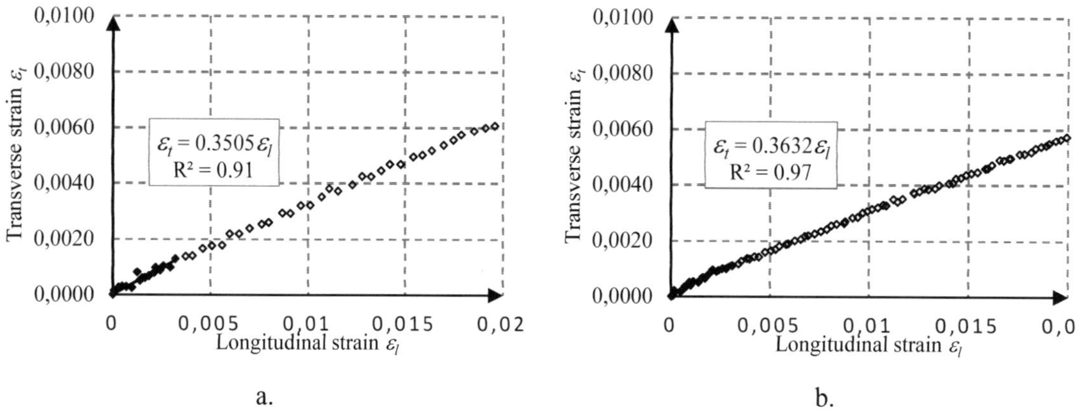

Figure 5. Transverse strain related ε_t to longitudinal strain ε_l in case of EN AW-6060-T6 aluminium alloy: a. Specimen 1; b. Specimen 3.

Figure 6 shows the tensile stress-strain $(\sigma - \varepsilon)$ curves recorded for the two kinds of composite materials tested. The tensile properties corresponding to each tested specimen made of composite material are shown in table 4, including the average values and Stdev values of these properties.

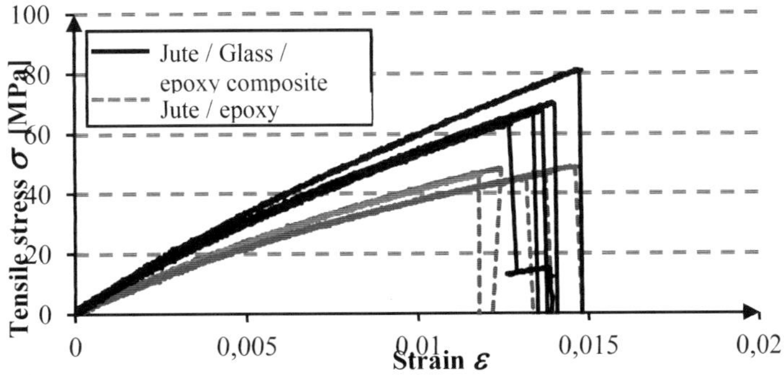

Figure 6. Stress-strain $(\sigma - \varepsilon)$curves recorded in tensile test in case of the composite materials tested.

Table 4. Tensile properties of the composite materials tested.

Specimen	Jute / epoxy			Jute / glass / epoxy		
	Young's modulus E (MPa)	Max. stress σ_{max} (MPa)	Strain ε la F_{max}	Young's modulus E (MPa)	Max. stress σ_{max} (MPa)	Strain ε la F_{max}
1	5250.3	48.36	0.0123	6913.2	67.86	0.0137
2	5122.4	46.25	0.0117	6634.0	63.19	0.0126
3	4583.0	45.23	0.0131	6831.0	70.30	0.0139
4	4707.5	48.96	0.0145	6555.6	68.68	0.0135
5	5086.3	46.48	0.0137	7146.6	81.17	0.0146
Average value	4950	47	0.0131	6816	70	0.0137
Stdev	288	1.6	0.0011	234	6.7	0.0007

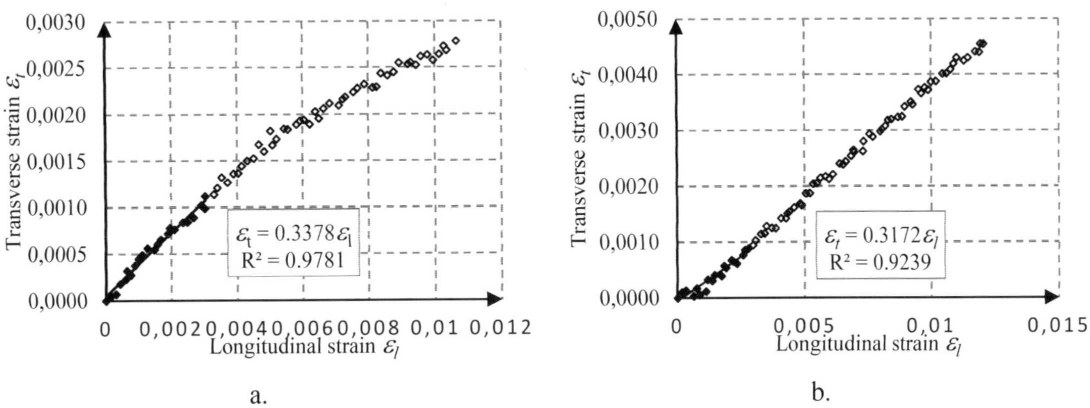

Figure 7. Transverse strain related ε_t to longitudinal strain ε_l in case of the Jute / epoxy composite: a. Specimen 2; b. Specimen 3.

Regarding the curves of the transverse strain ε_t related to the longitudinal strain ε_l, this paper shows only two curves corresponding to two kinds of composite material tested: figure 7 for Jute / epoxy composite material; figure 8 for Jute / glass / epoxy composite material. The Poisson's ratio ν_{12} (figure 7 and 8) was computed by taking into account the data of $\varepsilon_t - \varepsilon_l$ curve corresponding to $\varepsilon_l = 0 \div 0.03$ in accordance to the linear portion of the stress-strain $(\sigma - \varepsilon)$ curve (figure 6).

The values of the Poisson's ratio ν_{12} computed for all specimens made of composite materials including the average values and Stdev are shown in table 5.

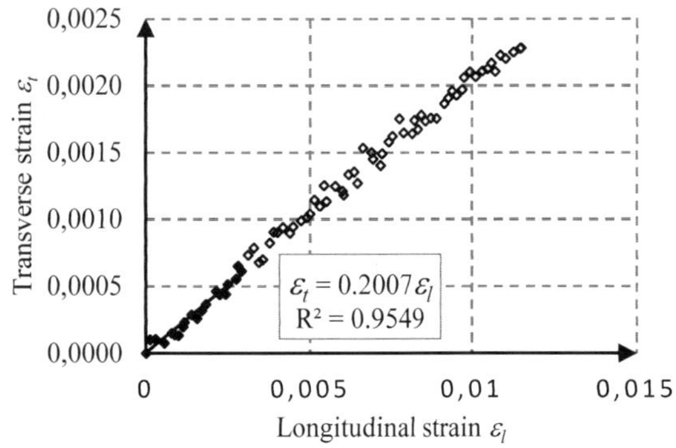

a.

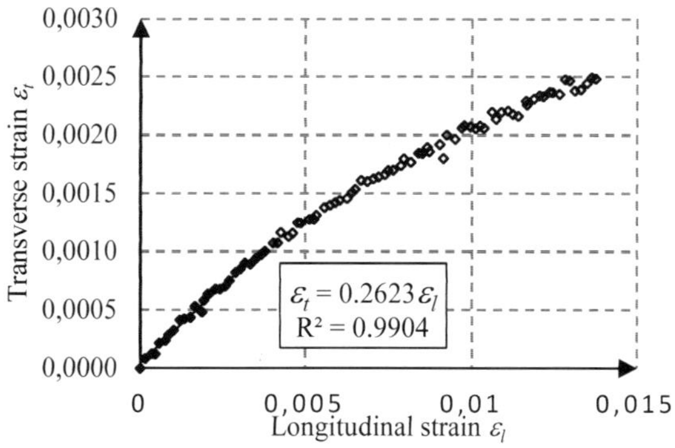

b.

Figure 8. Transverse strain related ε_t to longitudinal strain ε_l in case of the Jute / glass / epoxy composite material: a. Specimen 1; b. Specimen 5.

Table 5. Poisson's ratio ν_{12} computed for the composite materials tested.

No. of tensile specimen	Jute / epoxy	Jute / glass / epoxy
Specimen 1	0.3149	0.2007
Specimen 2	0.3378	0.2639
Specimen 3	0.3172	0.2011
Specimen 4	0.3816	0.2593
Specimen 5	0.2579	0.2623
Average value	0.3219	0.2375
Stdev	0.0447	0.0334

One is to note the higher degree of scattering of the results for the Poisson's ratio ν_{12} in the case of the Jute/glass composite material in comparison with the Poisson's ratio ν_{12} corresponding to Jute / glass / epoxy composite material. The reason is the poorer homogeneity corresponding to the Jute / epoxy composite material at microscopic level in comparison with the homogeneity corresponding to the Jute / glass / epoxy composite material. The outer layers of the Jute / glass / epoxy hybrid composite material are reinforced with glass fabric whose density is 200 g/m^2 that is equal to half of the density of 400 g/m^2 corresponding to the jute fabrics. In other words, the yarns of the glass fabric are thinner than the yarns of the jute fabric, which ensures a better homogeneity at the surface of the tensile specimens made of Jute / glass / epoxy hybrid composite material.

The Poisson's ratio ν_{12} corresponding to the reinforcing plane of the composite is in fact an equivalent Poisson's ratio of the laminated composite material because it corresponds to a fictitious orthotropic material whose mechanical behaviour is similar to the behaviour of the laminated composite material.

4. Conclusions

The experimental technique presented in this paper is very useful to determine the Poisson's ratio of the materials taking into account that this elastic property is required to characterize the material of any mechanical structure analysed by using the finite element analysis. This work also extends the application of the DIC method especially in testing composite materials, and can provide valuable data for finite element simulation.

The Poisson's ratio ν_{12} corresponding to Jute / epoxy composite material is 35.5% greater in comparison with Poisson's ratio ν_{12} corresponding to Jute / glass / epoxy composite material. Moreover, Poisson's ratio of the Jute / glass / epoxy composite material is greater than Poisson's ratio corresponding to the composite material reinforced only with glass fabric that is $\nu_{12} = 0.15$ in accordance with the results reported in previous work by the authors [4].

The above remarks lead to the conclusion that the reinforcing with jute fabric of all layers of the polymeric composite materials involves the increase of the *Poisson*'s ratio comparatively to the *Poisson*'s ratio of the hybrid composite material whose outer layers are reinforced with glass fibers.

As the yarns of the glass fabric are thinner than the yarns of the jute fabric, the above findings could be caused by this aspect. The density of the fabrics used to reinforce the outer layers of the polymeric composite materials could have an effect on the Poisson's ratio ν_{12} corresponding to the reinforcing plane.

The article reports elastic properties of composite materials reinforced with natural fibers (jute fabric) which has been a large area of interest in recent years, taking into account their ecological advantages. These kinds of composite materials are used in the field of the automotive industry (the interior of the doors) and in the construction of the ecological buildings (thermal or sound insulation panels, interior design panels etc.).

Acknowledgments

The authors hereby acknowledge the structural founds project PRO-DD (POS-CCE, O.2.2.1., ID 123, SMIS 2637, ctr. No 11/2009) and Transilvania University of Brasov for providing the infrastructure used in this work.

References

[1] Boufaida Z, Boisse J, André S and Farge L 2017 Mesoscopic strain field analysis in a woven composite using a spectral solver and 3D-DIC measurements *Composite Structures* **160** 604-612

[2] Perrier A, Touchard F, Chocinski-Arnault L and Mellier D 2016 Mechanical behaviour analysis of the interface in single hemp yarn composites: DIC measurements and FEM calculations *Polymer Testing* **52** 1-8

[3] Cerbu C and Curtu I 2011 Mechanical characterization of the glass fibres / rubber / resin composite material *Materiale Plastice* **48** (1) 93-97

[4] Cerbu C, Curtu I, Constantinescu D M and Miron M C 2011 Aspects concerning the transversal contraction in the case of some composite materials reinforced with glass fabric *Materiale Plastice* **48** (4) 341-345

[5] Nguyen V-T, Kwon S-J and Kwon O-H 2017 Mechanical properties identification of sheet metals by 2D-digital image correlation method *Procedia Engineering* **184** 381-389

[6] Wang Y H, Jiang J H, Wanintrudal C, Du C, Zhou D, Smith L M and Yang L X 2010 Whole field sheet-metal tensile test using digital image correlation *Experimental Techniques* **34** (2) 54–59

[7] Bruno L 2018 Mechanical characterization of composite materials by optical techniques: A review *Optics & Lasers in Engineering* **104** 192-203

[8] Wang H 2013 Marker Identification Technique for Deformation Measurement *Advances in Mechanical Engineering* **2013** Article ID 246318, 8 pages.

[9] Wang H and Kang Y 2002 Improved digital speckle correlation method and its application in fracture analysis of metallic foil *Optical Engineering* **41**(11) 436-45

[10] Lord J D, Penn D and Whitehead P 2008 The Application of Digital Image Correlation for Measuring Residual Stress by Incremental Hole Drilling *Applied Mechanics and Materials* **13-14** 65-73

[11] Abanto-Bueno J and Lambros J 2002 Investigation of crack growth in functionally graded materials using digital image correlation *Engineering Fracture Mechanics* **69**(14–16) 1695-1711

[12] Pan B, Qian K, Xie H and Asundi A 2009 Two-dimensional digital image correlation for in-plane displacement and strain measurement: A review *Measurement Science & Technology* **20**(6) 152-154

[13] Wang H and Hou Z 2010 Application of Genetic Algorithms in a surface deformation measurement technique *2010 Sixth International Conference on Natural Computation* 2291-2295 doi: 10.1109/ICNC.2010.5584215

[14] Mehdikhani M, Aravand M, Sabuncuoglu B, Callens M G, Lomov S V and Gorbatikh L 2016 Full-field strain measurements at the micro-scale in fiber-reinforced composites using digital image correlation *Composite Structures* **140** 192-201

[15] Brynk T, Molak R M, Janiszewska M and Pakiela Z 2012 Digital Image Correlation measurements as a tool of composites deformation description *Computational Materials Science* **64** 157-161

[16] Cascelli C 2014 Application of digital image correlation to glass fibre-reinforced composites under transverse tensile loading *Master of Science Thesis Delft University of Technology*

[17] Nicoletto G, Anzelotti G and Enrica R 2009 Mesoscopic strain fields in woven composites: Experiments vs. finite element modeling *Optics & Lasers in Engineering* **47**(3-4) 352-359

[18] Perrier A, de Vasconcellos D S, Touchard F, Chocinski-Arnault L and Mellier D 2015 Full-field Measurement at the Weave Scale in Hemp/Epoxy Composite Using Digital Image Correlation Polymers & Polymer Composites Shropshire 23(9) 589-599

[19] EN 573-3: 2010. Aluminium and aluminium alloys. Chemical composition and form of wrought products. Part 3: Chemical composition and form of products. Romanian Standard version, ASRO, 2010

[20] EN 573-5: 2008. Aluminium and aluminium alloys. Chemical composition and form of wrought products. Part 5: Codification of standardized wrought products. Romanian Standard version, ASRO, 2008

[21] EN ISO 6892-1: 2010 (or EN ISO 6892-1: 2002). Metallic materials. Tensile testing. Part 1: Method of test at room temperature

[22] Data sheet of the Epolam 2015 epoxy resin. Axson Technologies (U.S.A.), 2006

[23] EN ISO 527-4:1997. Plastics – Determination of tensile properties. Part 4: Test conditions for isotropic and orthotropic fiber-reinforced plastic composites

[24] https://www.makeitfrom.com/material-properties/6060-T6-Aluminum (accessed in April 2018)

IOP Conf. Series: Materials Science and Engineering **399** (2018) 012008 doi:10.1088/1757-899X/399/1/012008

Determining the predictive reliability of the water meters using the failure rate estimation. Case study - single jet cold water meter

M R Clinciu[1*] and R Clinciu[1]

[1] Department of Manufacturing Engineering, Transilvania University of Brasov, Romania

[*] E-mail: clinciumr@unitbv.ro

Abstract. The paper presents a method of estimating the failure rate by applying the extrapolation method, for the case of a single jet cold water meter. The predictive reliability of a system (water meter) is the reliability expressed by the reliability indicators resulting from forecast calculations based on the reliability of the component elements. In order to analyze the reliability of the considered water meter, it is divided into component elements or blocks of elements, which are functional or constructive units, whose reliability indicators are determined in operation. The water meter must be seen as a complex product, the structure of which includes a number of elements forming a system comprising a series of elements connected in one whole, that fulfills a specific technical function. The logical connection of the elements and the component blocks gives the logical reliability scheme with the subsystems and structural elements of the pressure measuring means. Finally, the logic mechanical reliability scheme of the water meter is determined, this being analyzed as a functional complex system, expressing the link between the reliability of the component elements and the reliability of the system as a whole.

1. Introduction

At present, considerable interest is granted to the conservation of the quality of the environment, the rational utilization of all natural resources, including water resources, identification and understanding of the causes of deterioration in the quality of the natural environment structure and components in order to prevent irrational use, waste of the resources. In the category of rational use of natural resources, which has been and still is a major concern, both nationally and internationally, fall the water and especially drinking water. Concerns for the rational use, thus eliminating the losses of drinking water, are priority objectives.

Since the quality of any process depends on the value of the measurement information and due to the necessity of a higher and higher precision, at present, there is an increasing interest in metrological reliability [1-5].

2. Case study - single jet cold water meter

Depending on the structural scheme of a water meter (figure 1), one can construct the functional block diagram of the mechanical means of measuring volumes [6].

The reliability of a water meter (mechanical device for measuring volumes) cannot be treated as the reliability of a product, which consists of an element [7]. A water meter must be regarded as a complex structure which includes a number of elements forming a system comprising a set of elements connected in a whole, which performs a specific function technique. These elements included in the composition of a water meter form: sub input unit - water meter body, unit of measurement - rotor (turbine), transmission unit (magnetic coupling), and integration and indicating subassembly - mechanical integrator.

A water meter consists of [6]:

- brass meter body provided with inlet channels for admission/exhaust of water. The exhaust channel is provided with a filter;
- turbine - measuring device operating under water pressure;
- magnetic coupling, which ensures the transmission of motion from the measurement unit to the integration unit;
- Integrating unit consisting of eight cylindrical drums and a dial with pointer and a graduated scale with division value of 0.05 dm^3. The mechanism is equipped with a device for removing condensation from the rolls.

The fluid flow through the meter produces the movement of the rotor, whose speed is proportional to the circulating water flow through the device. Rotational motion is demultiplicated and transmitted to the integration and indicating units. The transmission of motion is achieved by magnetic coupling since the mechanism is of a dry type.

In case of a single jet water meter, the body is made in such a way that the water jet hits the turbine tangentially, in a single point.

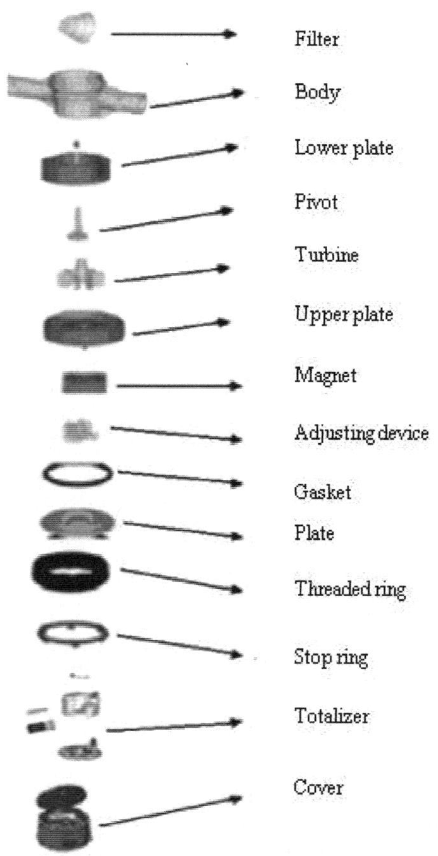

Figure 1. The single jet cold water meter – components.

IOP Conf. Series: Materials Science and Engineering **399** (2018) 012008 doi:10.1088/1757-899X/399/1/012008

3. Logical scheme of reliability for a cold single jet water meter

In order to analyze the reliability of the systems, they are divided into component elements or blocks of elements, which are functional or constructive units, whose reliability indicators are determined in exploitation. This decomposition is not unique, depending on how the system reliability is analyzed, the information that is available about the system components or their reliability parameters. By logically connecting the components and blocks, the logical reliability scheme with the subsystems and structural elements of the pressure measurement means is obtained [6].

Finally, the logical mechanical reliability scheme of the pressure measuring instrument is determined, analyzed as a functional complex system, expressing the link between the reliability of the component elements and the reliability of the system as a whole.

The logical scheme of reliability of a mechanically metered volume measurement means, which is representatively adopted, is shown in figure 2, and the integration and indicating subassembly unit, which is part of this scheme, is shown in figure 3.

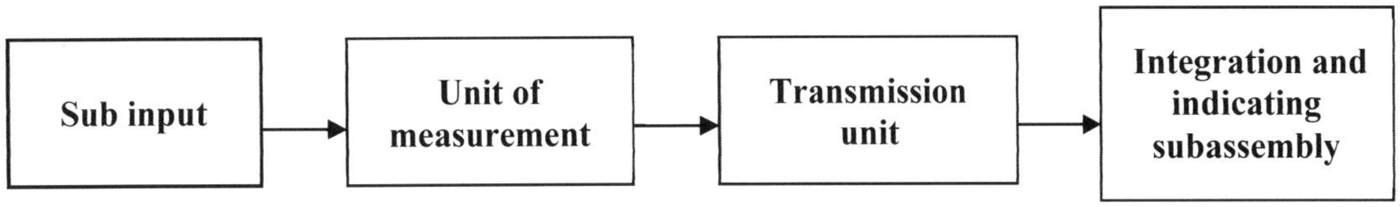

Figure 2. Logical scheme of reliability of a single jet cold water meter.

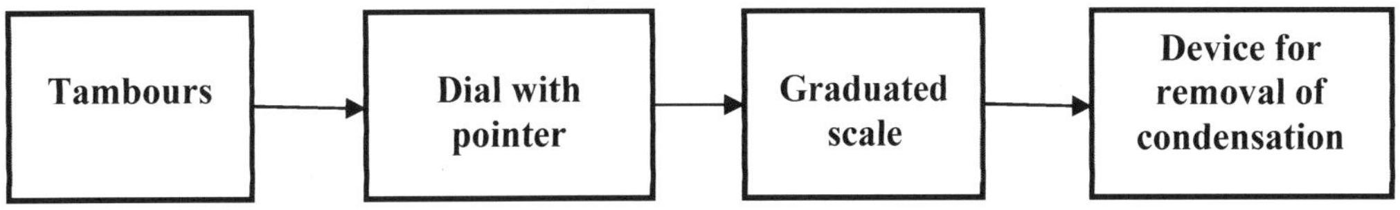

Figure 3. Integration and indicating subassembly unit for a single jet cold water meter.

4. Estimation of failure rate by the extrapolation method

In the logical scheme of reliability of the chosen model (figure 2), the following issues are taken into account: each element of the system has a random operating time and can be found in two states: operating or non-operating; serially related items have only one way of communicating. The block, module, or subsystem is a group of elements within the system, typically having the same connection mode that works correctly if it correctly transfers the input-to-output information [6].

The components of the system can be characterized by the probability of operation R(t) or the probability of failure Fi(t) = 1 - R(t). The system is characterized, in its turn, by the probability of operation R(t) or the probability of malfunction Fs(t) = 1 - R(t). The unequivocal structural conveys the relationship between R(t), or Fs(t) and Ri(t) or Fi(t). Finding the analytical expression of this relation, the mathematical model or numerical values of Rs (t) or Fs (t) for given Ri (t) or Fi (t), represent the fundamental problem of the system reliability [7-9]. In conclusion, the purpose of the structural reliability analysis is to determine the functional relationship between the reliability function R(t) of the system and the reliability functions Ri of the component elements:

$$R_s = \psi (R_1, R_2, \ldots, R_n) \tag{1}$$

The logical model introduces essential simplifications to the functional model, where the analysis is based on knowing the probability densities associated with component parameters.

The first simplification introduced in the logical model is that the reliability functions Ri that characterize the system components are assumed to be known according to the actual failure criteria imposed by the component parameter system; thus, the logic-based analysis does not require a prior calculation of the individual reliability functions, starting from the probability densities of the parameters.

The second simplification is that the structure of the system is described by the Boolean function (1), which is not particularly complex.

$$R_s(t) = \prod_{i=1}^{n} R_i(t) \tag{2}$$

Since the probability Rs(t) is the product of exponential functions (2), its expression will also be an exponential function:

$$R_s(t) = \prod_{i=1}^{n} e^{-\lambda_i t} = e^{-\sum_{i=1}^{n} \lambda_i t} = e^{-\lambda_s t} \tag{3}$$

where:

$$\lambda_s = \sum_{i=1}^{n} \lambda_i \tag{4}$$

The reliability of a system is noticed to be lower as its complexity is greater.

In order to estimate the system failure rate λ_s of a representatively adopted water meter (figure 1), whose logic scheme is shown in figure 2, based on the data from the literature, one can consider table 1, which contains the values of the fault intensities λ_i of the component elements of the analyzed water meter.

Table 1. Values of failure intensity for component parts of water meters [7, 9].

Current number	Element name	Values of exponential distribution parameters E(x; λ) $\lambda \cdot 10^{-6}$/h		
		Minimum value	Medium value	Maximum value
1.	Filter	0.045	0.3	0.8
2.	Body	0.0005	0.0125	0.041
3.	Lower plate	0.015	0.145	0.175
4.	Pivot	0.08	0.2	0.35
5.	Turbine	0.5	0.68	1.039
6.	Upper plate	0.015	0.175	0.700
7.	Magnet	0.24	0.600	0.930
8.	Adjusting device	0.7	2.14	5.54
9.	Gasket	0.01	0.02	0.03
10.	Plate	0.02	0.5	1
11.	Threaded ring	0.250	0.700	1.120
12.	Stop ring	0.01	0.02	0.03
13.	Totalizer	0.011	0.02	1.96
14.	Cover	0.002	0.006	0.01

Thus, based on the numerical values in table 1, using the relation (4), the value of the failure intensity of the analyzed system is determined as follows [7, 8]:

$$\lambda_s = \sum_{i=1}^{9} \lambda_i = \lambda_1 + \lambda_2 + ... + \lambda_{14} = 3.834 \cdot 10^{-6} \quad (5)$$

The times to failure for the components of a water meter have exponential distributions [6, 9].

5. Conclusions

The paper presents a method for determining the predictive reliability of a single jet cold water meter using the failure rate estimation.

The analysis of the predictive reliability represents a basic tool in product design, in order to predict their performance in terms of reliability and to optimize the constructive design of the product in order to ensure a certain level of reliability. The importance of the predictive design reliability analysis is argued by the fact that the reliability of the products is determined by 40-80% of the quality of the project. Predictive reliability indicators provide only an indicative value for the performance of the product. The calculation of predictive reliability is made by taking into account the specificity and particularities of each system or each concrete use. For this it is necessary to know the relationships between the system parameters and the parameters of the elements.

The predictive reliability analysis addresses the maturity period of the product in which the defects are random, at a relatively constant and stable rate. This is the reason why most of the predictive reliability analyzes are based on the statistical model of the exponential distribution. The predictive analysis is based on previous experience on the testing or operation of identical or similar components under identical or similar conditions to the equipment under consideration.

The operational reliability of the product is generally lower than the predictive or experimental reliability. An important cause of the differences between predictive and operational reliability is given by the conditions of use of that product. This is why, in calculating predictive reliability, these factors have to be taken into account by means of environmental coefficients.

As further work, it is intended to apply this study for different types of water meters and to compare the results obtained.

6. References

[1] Kundzewicz Z and Kindler J 1995 *IAHS Publ.* **231** 217-224
[2] Arregui F J, Pardo M A, Parra J C and Soriano J 2007 *IWA Water Loss Conf.* Sept 2007 Bucharest
[3] Arregui F J, Martinez B, Soriano J and Parra J C 2009 *IWA Water Loss Conf.* Apr 2007 Cape Town
[4] Toteva P, Slavov S and Vasileva D 2017 *Matec Web Conf.* **94** 04012
[5] Vasilevskyi O M 2014 *Int. J.Metrol. Qual.Eng.***5** 403-412
[6] Clinciu M R 2009 *Research on the Reliability of the Water Meters - PhD Thesis* (Brasov: Transilvania University of Brasov)
[7] Hohan I 1982 *Technology and system reliability* (Bucuresti: Didactical and Pedagogical Publishing House)
[8] Martinescu I and Popescu I 1995 *Reliability* (Brasov: Gryphon Publishing House)
[9] Martinescu I and Popescu I 2002 *Analysis of system reliability and security* (Brasov: Transilvania University of Brasov)

IOP Conf. Series: Materials Science and Engineering **399** (2018) 012009 doi:10.1088/1757-899X/399/1/012009

Study on the metrological reliability of the water meters

R Clinciu[1*] and M R Clinciu[1]

[1] Department of Manufacturing Engineering, Transilvania University of Brasov, Romania

* E-mail: r.clinciu@unitbv.ro

Abstract. The paper presents the results of an experimental study conducted on specific measuring devices - water meters of a certain precision class. The aims of the study are to determine the calibration life of the measuring devices as a function of the metrological reliability and, at the same time, to emphasize the parameter deviation as an important research tool for the quality assurance in metrology. This experimental research may be added to former researches performed on different types of measuring devices. Parameter deviation represents an important research tool for the estimation of the calibration life of the measuring devices. It helps to predict the moment in time when the probability density function for the measurement errors exceeds the tolerance interval by a certain percentage, corresponding to the established level of significance. This moment corresponds to the stage in which the considered measuring device needs to be checked and, if necessary, calibrated. The experimental study conducted is aimed at determining the calibration life of water meters of the same type and of a certain precision class. The research performed on the considered water meters aims to establish which the appropriate moment for performing the calibration is for a specific set of measuring devices, moment which can be estimated by using the parameter deviation functions, these being obtained by means of regression analysis.

1. Introduction

A measuring device is characterized by a specific tolerance interval. In case the measuring device is correctly adjusted, the measurement errors are placed inside the tolerance interval. In time, by using the measuring device, the measurement errors might start to grow such as, at a certain moment, to exceed the tolerance interval. If the probability density function (pdf) for the measurement errors exceeds the tolerance interval by a certain percentage, it is considered that a metrological disturbance has appeared [1]. Metrological disturbances are difficult to discover at the moment they are produced, they can be discovered only during the periodic checking of the measuring device or at encountering the negative consequences of the use of such a measuring device, so that it becomes very important to predict, with a certain precision, the moment when the measuring device starts to measure with greater errors than the tolerated ones [1]. At present, since the quality of any process depends on the measurement information, a remarkable interest is given to the metrological reliability [2, 3].

Parameter deviation represents an important research tool for the estimation of the calibration life of the measuring devices. It helps to predict the moment in time when the probability density function for the measurement errors exceeds the tolerance interval by a certain percentage, corresponding to the established level of significance. This moment corresponds to the stage in which the considered measuring device needs to be checked and, if necessary, calibrated [1, 4]. This experimental research may be added to former researches performed on different types of measuring devices [5-8].

The paper presents the results obtained by performing an experimental study on specific measuring devices - water meters of precision class B, a medium precision for the water meters. The experimental data is obtained in laboratory conditions and it is represented by the measuring errors obtained by measuring the same volume of water by a single person; experimental data is grouped into samples produced at a certain interval of time.

An appropriate algorithm, based on the parameter deviation functions, is used for estimating the calibration life of the water meters considered. The estimated calibration life, obtained as a function of the metrological reliability, is then compared to the requirements set in the appropriate standards, such as to enable the conclusions towards the use of the water meters analysed, from the calibration life point of view.

2. Algorithm for estimating the calibration life of water meters

General metrology uses especially the normal distribution, although positive and asymmetrical distributions, such as Weibull distribution, are frequently met in the analysis of the metrological reliability [9].

2.1. Notations used in the algorithm

The algorithm developed for the estimation of the calibration life of the measuring devices uses the following notations: α - the level of significance; m - the mean; σ - the standard deviation; n - the volume of the sample; k - the number of samples; x_{ji} - the experimental data, which is assumed to be normally distributed, according to the normal distribution function $N(m, \sigma, t)$ ($j = 1,..., k$, $i = 1,..., n$) having both parameters m and σ variable; m_0 - the initial value of the mean; a - the variation coefficient of the mean; σ_0 - the initial value of the standard deviation; b - the variation coefficient of the standard deviation; T_ε - the tolerance interval for variable ε, as a measurement error; ti - the initial moment; dt - the period of time at which samples are collected (it is considered that dt is constant); t_r - the calibration time [4-6].

The case of a normal distribution $N(m_0, \sigma_0, t)$ is considered. In time, due to the internal parameter deviation, the parameters of the distribution become m and σ and the normal distribution becomes $N(m, \sigma, t)$. It is considered that the mean m and the standard deviation σ may have a linear or an exponential variation, according to the equations:

$$m = m_0 + a \cdot t \quad (1)$$

$$\sigma = \sigma_0 + b \cdot t \quad (2)$$

$$m = m_0 \cdot \exp(a \cdot t) \quad (3)$$

$$\sigma = \sigma_0 \cdot \exp(b \cdot t). \quad (4)$$

2.2. Presentation of the algorithm

The algorithm for estimating the calibration life of the measuring devices as a function of metrological reliability includes the following steps [4]:

- collecting experimental data, represented by measurement errors;
- performing data outlier tests- Grubbs test [10];
- graphical representation (probability plot), distribution assumption – normal distribution [10, 11];
- parameter estimation for the normal distribution [10];
- goodness-of-fit tests; it is considered a general goodness-of-fit test -Kolmogorov-Smirnov and a normality goodness-of-fit test -Lilliefors. The optimum distribution is chosen by comparing the values obtained for the statistics of the tests considered with the critical values of the tests: d_j - the value of the statistics of the Kolmogorov-Smirnov goodness-of-fit test is to be compared to $d_{n, \alpha}$ - the critical value of the statistics of the Kolmogorov-Smirnov goodness-of-fit test; L_j - the value of the statistics of

the Lilliefors goodness-of-fit test is to be compared to $L_{n,\alpha}$ - the critical value of the statistics of the Lilliefors goodness-of-fit test [9-11];

- parameter deviation functions are determined by using regression analysis and the correlation coefficients for the linear variation, as well as for the exponential variation [10, 11]; the type of the parameter deviation function, linear or exponential, for parameters m and σ, is indicated by the highest value of the two calculated correlation coefficients for each parameter;
- determining the calibration life, by using the parameter deviation functions and the method described in [4]. The period for the metrological control of water meters is set to seven years, as stated in standards [12].

3. Experimental study

The experimental study is conducted on water meters of the same type, of precision class B, a medium precision for the water meters, connected in series arrangement, to a water source. The water meters considered are of a different type than those considered in a former study [8]. At the time the experimental study was conducted, all water meters were in their specified working period, they had been used in the installations in the apartments in a block of flats, for a specific period of time, 36 months, and were still in their working period according to the standards prescriptions [12].

The considered level of significance is $\alpha = 0.05$, the volume of a sample is $n = 18$, the number of samples $k = 6$, initial moment $ti = 0$ and the interval between the moments at which samples are collected $dt = 1$.

The experimental data, x_{ji}, is represented by the measuring errors, obtained by measuring a certain volume of water, 0.01 m^3, the measuring errors being obtained as the difference between the indications of each water meter used and the true value of the volume of water (0.01 m^3). Data is grouped in six samples produced at a certain period of time (one month), by measuring the same volume of water with the group of water meters, in laboratory conditions. Table 1 presents the measuring errors x_{ji}, obtained by measuring the same volume of water with the water meters considered; they are all located within the appropriate tolerance interval $T_\varepsilon = 0.0004$ m^3. The tolerated maximum measuring errors are ±2% for cold water (if water temperature <= 30°C) [12], such as for the considered case, the tolerance interval is $T_\varepsilon = 0.0004$ m^3.

Table 1. Measuring errors.

Sample 1	Sample 2	Sample 3	Sample 4	Sample 5	Sample 6
-0.00007	0.00007	0.00005	0.00007	0.00017	0.00017
-0.00011	-0.00011	-0.00010	-0.00010	0.00011	0.00011
-0.00006	-0.00006	-0.00006	-0.00006	-0.00006	-0.00006
-0.0002	-0.0002	-0.0002	-0.0002	-0.0002	-0.0002
-0.00011	-0.00011	-0.00011	-0.00011	-0.00011	-0.00011
-0.00004	-0.00004	-0.00004	-0.00004	0.00004	0.00004
-0.00016	-0.00016	0.00016	0.00016	0.00019	0.00019
0.00008	0.00008	0.00008	0.00008	0.00008	0.00008
-0.00001	-0.00001	-0.00001	-0.00001	-0.00001	-0.00001
-0.00011	-0.00011	-0.00011	-0.00011	0.00011	0.00011
-0.00002	-0.00002	0.00002	0.00002	-0.00002	-0.00002
0.00002	0.00002	0.00002	0.00002	0.00002	0.00002
-0.00007	-0.00007	0.00007	0.00007	-0.00007	-0.00007
-0.00012	-0.00012	0.00012	0.00012	-0.00012	-0.00012
-0.00003	-0.00003	0.00003	0.00003	-0.00003	-0.00004
-0.00002	-0.00002	0.00002	0.00002	0.00002	-0.00002
0.00002	0.00002	0.00002	0.00002	0.00002	0.00003
0.00001	0.00001	-0.00001	0.00001	0.00002	0.00012

The results of the application of the data outlier test, the Grubbs test, are presented in figures 1 to 6. The conclusions of the applications of this test to the six samples considered are that there is no outlier at 5% level of significance in any of the samples considered.

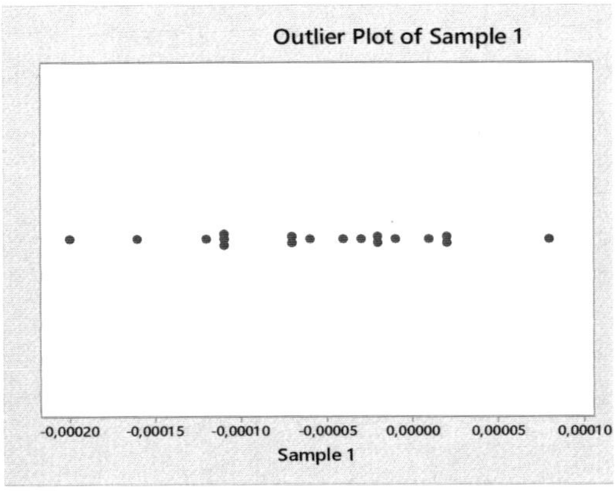

Figure 1. Outlier plot of sample 1.

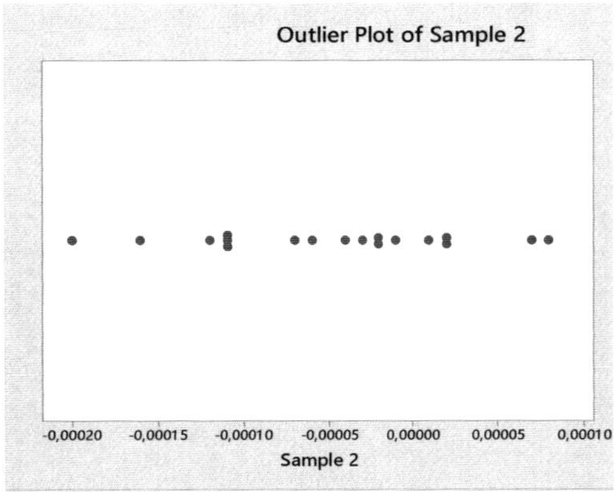

Figure 2. Outlier plot of sample 2.

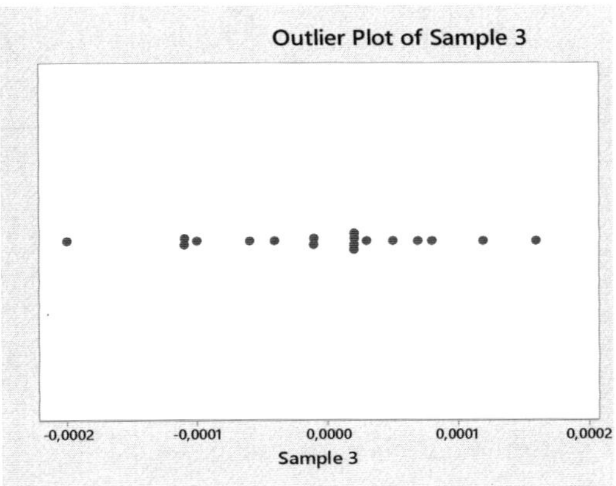

Figure 3. Outlier plot of sample 3.

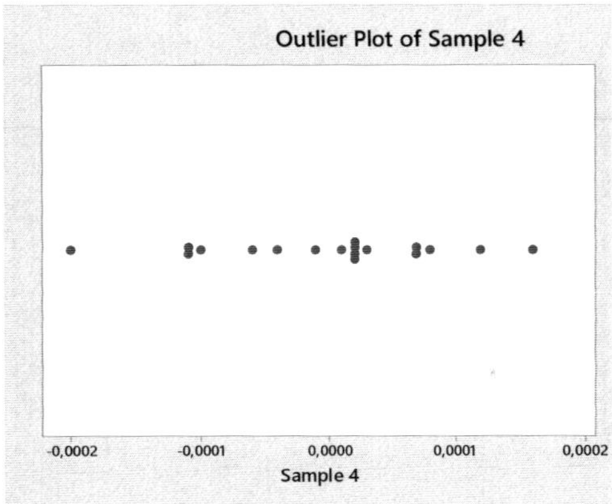

Figure 4. Outlier plot of sample 4.

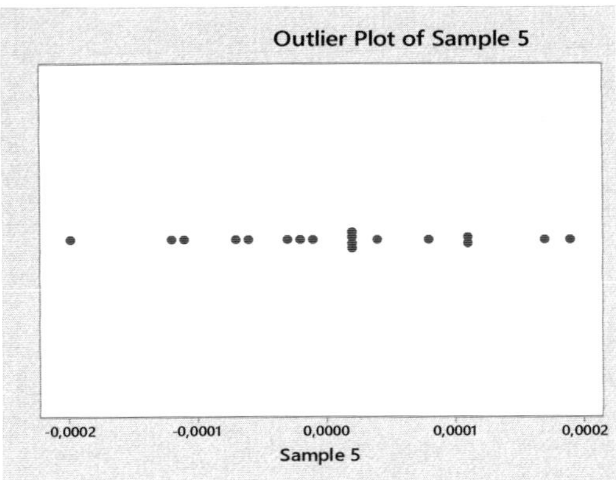

Figure 5. Outlier plot of sample 5.

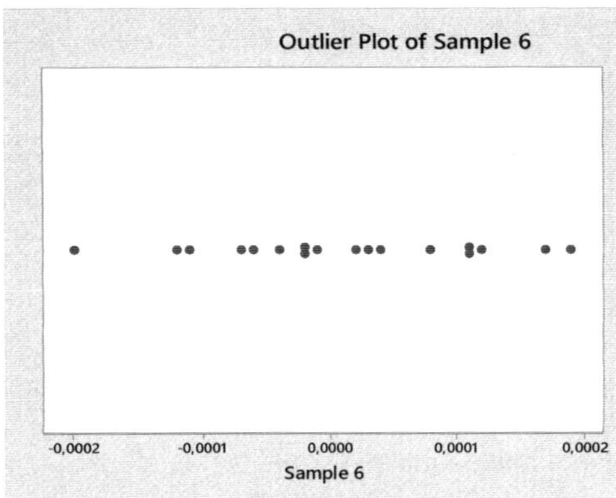

Figure 6. Outlier plot of sample 6.

IOP Conf. Series: Materials Science and Engineering **399** (2018) 012009 doi:10.1088/1757-899X/399/1/012009

The probability plots of the six samples considered are presented in figures 7 to 12. By analyzing the probability plots, one can assume that the appropriate distribution is the normal distribution and this is to be tested by using the goodness-of-fit tests.

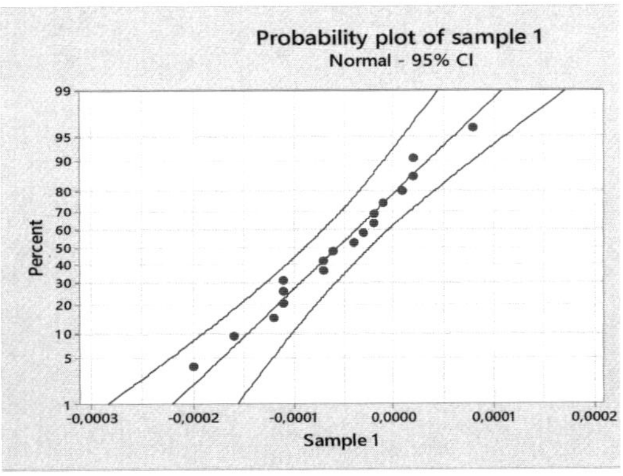

Figure 7. Probability plot of sample 1.

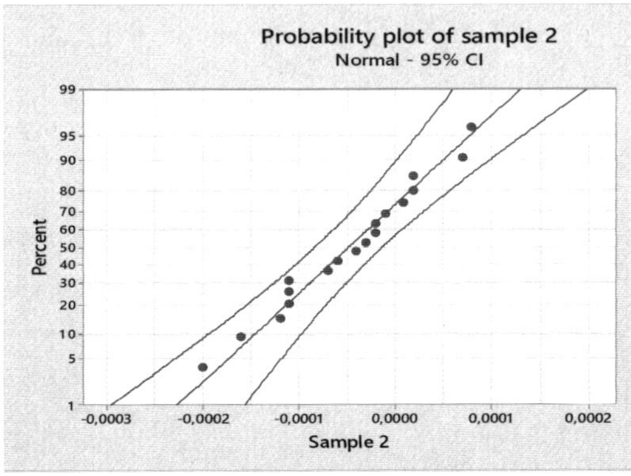

Figure 8. Probability plot of sample 2.

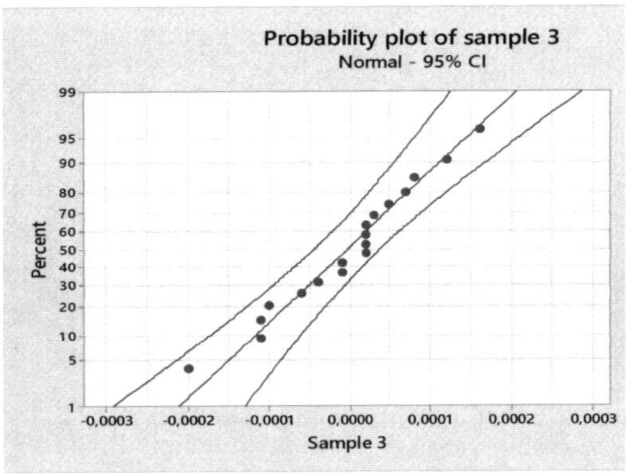

Figure 9. Probability plot of sample 3.

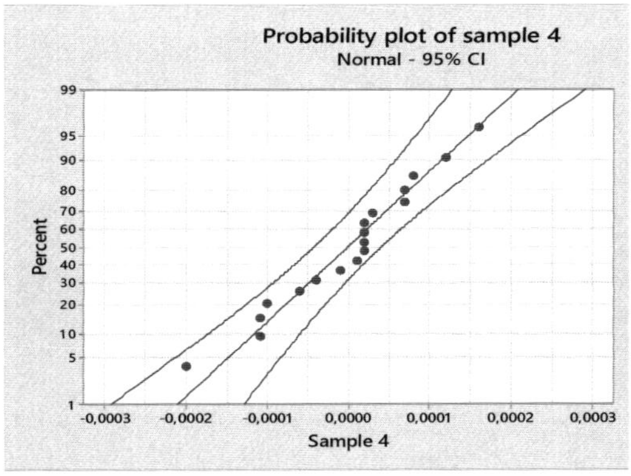

Figure 10. Probability plot of sample 4.

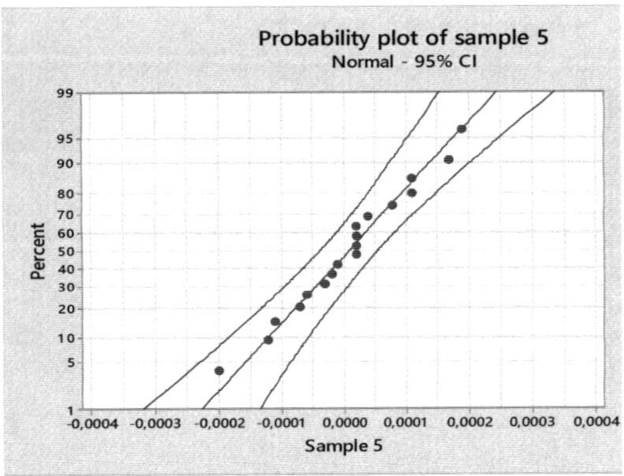

Figure 11. Probability plot of sample 5.

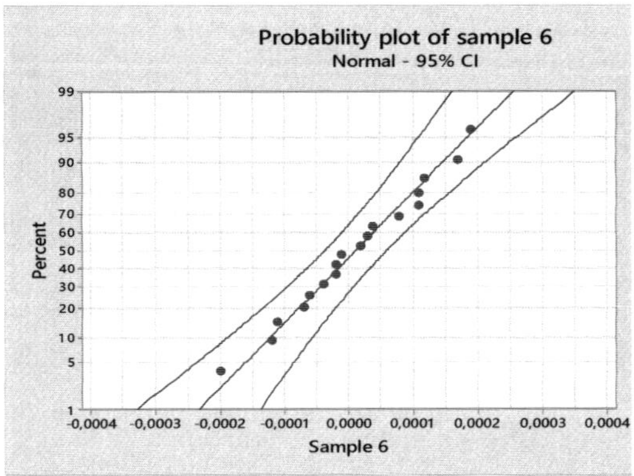

Figure 12. Probability plot of sample 6.

The critical value of the statistic of the Kolmogorov-Smirnov goodness-of-fit test for the considered sample size is $d_{n,\alpha} = 0.309$ and the critical value of the statistic of the Lilliefors goodness-of-fit test for the considered sample size is $L_{n,\alpha} = 0.200$ [9-11].

Table 2 presents the values obtained for the statistics of both Kolmogorov-Smirnov and Lilliefors goodness-of-fit tests, for each sample. These values are according to both conditions: $d_j \leq d_{n,\alpha} = 0.309$, for Kolmogorov-Smirnov goodness-of-fit test and respectively $L_j \leq L_{n,\alpha} = 0.200$ for Lilliefors goodness-of-fit and this indicates normal distributions [9-11].

Table 2. Values of the statistics of the goodness-of-fit tests.

	Sample 1	Sample 2	Sample 3	Sample 4	Sample 5	Sample 6
d_j	0.1129	0.1255	0.1562	0.1577	0.1229	0.1017
L_j	0.1129	0.1255	0.1006	0.1021	0.1229	0.0837

The values of the estimated parameters m and σ for each sample are given in table 3.

Table 3. Values of the estimated parameters of the samples.

	Sample 1	Sample 2	Sample 3	Sample 4	Sample 5	Sample 6
m	-0.000056	-0.000048	-0.00003	-0.00001	0.00009	0.00012
σ	0.000068	0.000074	0.000087	0.000088	0.000098	0.000102

By using the values of the means and standard deviations of the samples, presented in table 3, one can obtain the parameter deviation functions, by means of regression analysis [10, 11]. The correlation coefficients for linear variation of the parameters are $r_{m\text{-}l} = 0.9350$ and $r_{\sigma\text{-}l} = 0.9831$ and the correlation coefficients for exponential variation of the parameters are $r_{m\text{-}exp} = 0.8184$ and $r_{\sigma\text{-}exp} = 0.9774$; these values indicate a linear variation for both parameters, m and σ, according to the equations presented below:

$$m = -0.00012 + 0.000038 \cdot t. \tag{5}$$

$$\sigma = 0.000062 + 0.000007 \cdot t. \tag{6}$$

The calibration life for the considered case is calculated by means of parameter deviation functions presented in equations (5) and (6), according to the method described in [4]. The value which is obtained for the calibration time for the case considered is $t_r = 38.768$ time units (months), value which is less than the remaining period for the use of the water meters (water meters had been used for 36 months before the experiment), according to the pre-set period for the metrological control of the water meters, which is set to 7 years/84 months in the appropriate standards [12].

The estimated calibration life in this case is close to the pre-set period for the metrological control of the water meters, which proves the fact that the type of water meters analysed in this research is better than other types analysed [8]. Still, the estimated calibration life is less than the remaining period for the use of the water meters, till the pre-set period for the metrological control. This means that the use of the respective water meters, more than the period indicated by the value t_r, might produce errors in metering.

4. Conclusions

The method for determining the calibration life based on parameter deviation functions can be used for estimating the right moment for performing the metrological control/calibration for a specific set of measuring devices. This moment might be later than the pre-set one, and in this case financial costs with the metrological control are diminished, or earlier than the pre-set calibration time, and in this case greater errors than the tolerated ones are prevented. As further work, it is necessary to perform a study on the metrological reliability of different types of water meters which are being used at present such as to provide information on the calibration life of the types of water meters considered.

By estimating the calibration life of the measuring devices as a function of metrological reliability, based on parameter deviation functions, it might be possible to test whether, for a certain type of measuring devices, the values obtained for the calibration life t_r are according to the requirements set in the appropriate standards. As future research, it is intended to consider other types of metrological devices, to test whether the values obtained for the calibration life meet the requirements set in the appropriate standards.

References

[1] Wiener U and Veres F 1972 *Qual. Prod. Metr. J.* **5** 265
[2] Toteva P, Slavov S and Vasileva D 2017 *Matec Web Conf.* **94** 04012
[3] Arregui F J, Pardo M A, Parra J C and Soriano J 2007 *IWA Water Loss Conf.* Sept 2007 Bucharest
[4] Clinciu R 2000 *Bulletin.of Transilvania Univ. of Brașov* **7(42)** 65
[5] Clinciu R 2013 *Applied Mech. Mat.* **371** 695
[6] Clinciu R 2014 *Applied Mech. Mat.* **657** 694
[7] Clinciu R 2014 *Academic J. of Manuf. Eng.* **12/1** 76
[8] Clinciu M R and Clinciu R 2017 *Matec Web Conf.* **94** 04003
[9] Vodă V G 1979 *Applied metrology* **XXVI(1)** 11
[10] Martinescu I and Popescu I 1995 *Reliability* (Brasov: Gryphon)
[11] Kececioglu D 1991 *Reliability Engineering Handbook* (N J: Prentice Hall, Englewood Cliffs)
[12] NML 003-05 2005 *Water meters* (Bucuresti: Monitorul Oficial al Romaniei)

Soil-structure interaction and motion incoherency effects on a concrete bridge with deep foundations

M Conțiu[1*], D M Ghiocel[2] and D Crețu[3]

[1] Faculty of Civil Engineering, Transilvania University of Brasov, Romania
[2] Ghiocel Predictive Technologies, Inc., Rochester, New York, USA
[3] Dan Ghiocel Research Center, Technical University of Civil Engineering Bucharest, Romania

[*] E-mail: mircea.contiu@unitbv.ro

Abstract. The paper presents a seismic SSI study that has been conducted for a 242 m long concrete bridge in Brașov, that has deep foundations on drilled piles. The structure is typical for Romania and highly used throughout the country. The seismic soil-structure interaction were performed using the state-of-the-art ACS SASSI software. The soil-structure interaction analyses were performed for both coherent (synchronous) and incoherent (non-synchronous) seismic inputs. The main focus of the study is to evaluate the effects of soil – structure interaction and of the seismic motion incoherency on the bridge dynamic response. The paper includes comparisons of results obtained using the Eurocode 8 analysis procedures and the state-of-the-art seismic soil-structure interaction analysis using ACS SASSI. The final conclusions indicate the limitations of the current bridge design soil-structure interaction modelling and the simplified motion spatial variability modelling in Eurocode 8. Furthermore, the future directions of the research study are presented.

1. Introduction

Because most types of civil engineering structures are in direct contact with the surrounding soil, their behaviour is influenced by the soil properties. This fact is true for both static and dynamic loads (wind, earthquakes, etc.).The influence may be so strong for earthquakes that it could completely change the behaviour of the structure: in some cases, the overall response may be favourable, reducing the forces on the structures (and the execution costs), but may also be unfavourable, which can lead to disasters.

Observations and studies on recent earthquakes have highlighted the need to include SSI (soil-structure interaction) effects in the structural design, like the 1995 Kobe earthquake which severely damaged bridges along the Hanshin Expressway [1].

The study aims to draw a comparison between the typical seismic bridge design methodology according to Eurocode 8 and the state-of-the-art seismic SSI analysis methods, developed for nuclear structures, which include sophisticated SSI models and load modelling procedures, using measured data to estimate the incoherent behaviour of the seismic motion.

2. Description of the bridge structure and the local soil conditions

The structure being studied is a newly designed and built 242m long concrete bridge in Brașov, Romania. The bridge superstructure is composed of precast prestressed beams connected by a concrete slab and with cast-in-place concrete diaphragms on the supports. The piers are composed of 2 circular columns with a diameter of 1.50 m, which are connected through a pier cap. The pier height varies between 6.50 – 9.00 m. All foundations consist of 6 drilled piles 17.00 m long connected by a pile cap [2].

The geotechnical study revealed a mixture of layers of gravel and sand up to a depth of about 20-25m [2].

Figure 1. The studied concrete bridge in Brasov, Romania.

3. Analysis models and SSI methodology

Two analyses will be carried out: a typical design analysis according to Eurocode 8 and a state-of-the-art sophisticated SSI analysis.

For the bridge structure itself, the same structural model composed of beam and shell elements will be used. The differences between the analyses lie in the SSI methodology and the way the seismic load is applied.

For the typical design analysis, the soil – pile interaction is considered through the widely accepted "beam on elastic foundation" (Winkler model) approach, where the stiffness of the springs varies linearly with depth (figure 2. a.). The seismic loading is introduced as uniform acceleration, acting on the whole structure, through three linear time-history analyses: one for each orthogonal direction. The dynamic earth pressure on the abutments is modelled according to the Mononobe-Okabe theory as linear varying pressures. The spatial variability of the seismic action is taken into account through static differential displacements at each foundation, according to the simplified model from EC 8.

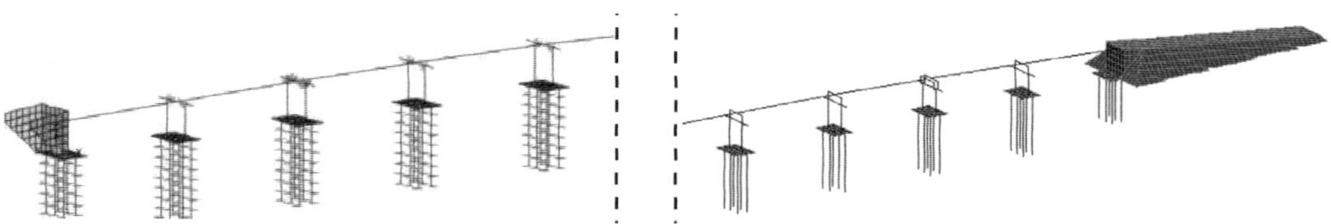

a. Typical design model b. Sophisticated SSI model

Figure 2. Bridge models used in the study.

The second sophisticated analysis uses the modern SSI (soil-structure interaction) methodology currently implemented in the ACS SASSI software for the design of nuclear structures in the US and all over the globe (figure 2. b.). The site is modelled as a series of horizontal layers overlying a uniform half space at considerable depth; the soil materials properties are assumed to be visco-elastic. The analysis uses a complex frequency domain solution, so all systems have to be linearized [3, 4]. Seed and Idriss (1970) offer an equivalent linear procedure for the approximation of the non-linear hysteretic behaviour of the soil [5].All points on the piles and pile cap are considered as interaction points between the structure and the soil column model used for the site analysis. The backfill is modeled with solid elements having the material properties of the type of soil used (gravel). Of course, the nodes at the base of the backfill are also interaction points.

The usual wave field combination is used: SV-waves for the bridge longitudinal direction (X), SH-waves for the bridge transverse direction (Y) and P-waves for the vertical direction (Z). The seismic waves propagate through the soil layers until they reach the structure; thus, the local soil conditions can have a strong effect on the seismic waves, before they reach the structure.

The analysis is carried out for both coherent and incoherent seismic input. The incoherency or the spatial variability of the ground motion is defined by the one-dimensional coherence function, assuming a homogenous and isotropic Gaussian stochastic field. Abrahamson (1993, 2005, 2006, 2007) has created various statistical databases of seismological information recorded in many dense arrays, and generated empirical plane-waves incoherency models for multiple soil conditions and foundation types, through the formulation on the coherency function [6, 7]. This study uses the 2005 Abrahamson plane-wave coherency model, along with introducing a wave passage effect for an apparent wave velocity of 1300 m/s along the bridge. The ACS SASSI software was used to compute 15 incoherent simulations on top of the coherent analysis at the beginning.

A uniform soil profile with average characteristics between gravel and dense sand is used in both models. Both models have been subjected to the same input accelerations, which generated from the site elastic response spectrum (figure 3).

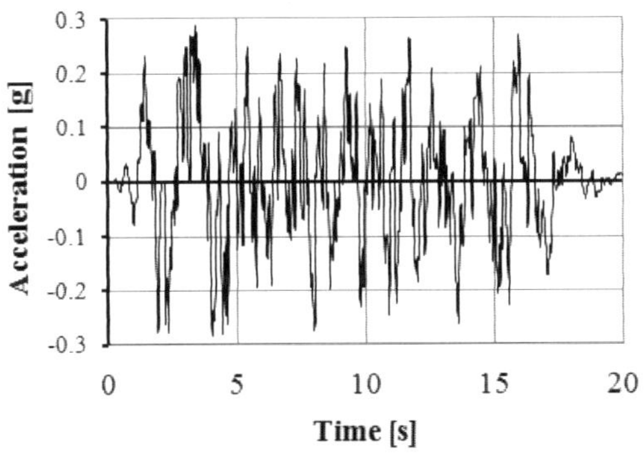

Artificial acceleration time-history

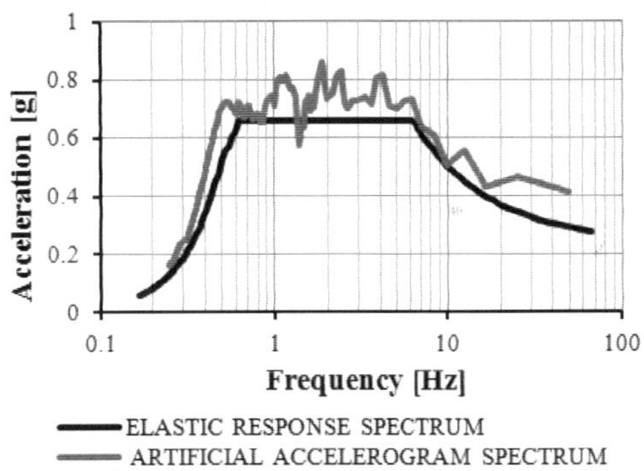

Spectrum comparison

Figure 3. Seismic input [2].

4. Results and comparison

The results that will be compared are the structural stresses of the pier columns and a corresponding pile. Two piers are studied: a side pier (next to the abutment), which is the shortest and stiffest (about 5.90 m) and the mid pier (the one in the middle of the structure), which is the longest and most flexible one (about 8.0 m).

The following notations will be used throughout this section: both DESIGN and SPA-VAR refer to the typical design model; the first one shows the results from the time-history analysis and the second one from the static differential displacement loading according to the simplified procedure in the Eurocode 8. As for the sophisticated SSI model, the coherent analysis is abbreviated COH, the 15 incoherent simulations are named INCOH_RUNS; INCOH_MEAN is the average response computed from the incoherent runs. X, Y and Z refer to the direction of loading, meaning along the bridge, transverse to the bridge direction and vertical.

"M", "V" and "N" are the bending moments, shear and axial forces; the local beam element directions are 1 – axial and vertical, 2 – along the bridge and 3 – transverse to the bridge axis. The results are presented in a graph with stresses along the horizontal axis and elevation on the vertical axis; the values above 0 elevation represent one of the pier columns and below 0 a corner pile from the corresponding foundation. All stresses have been computed in the analysis, but only the ones which offer an interesting insight about the goal of the paper will be presented.

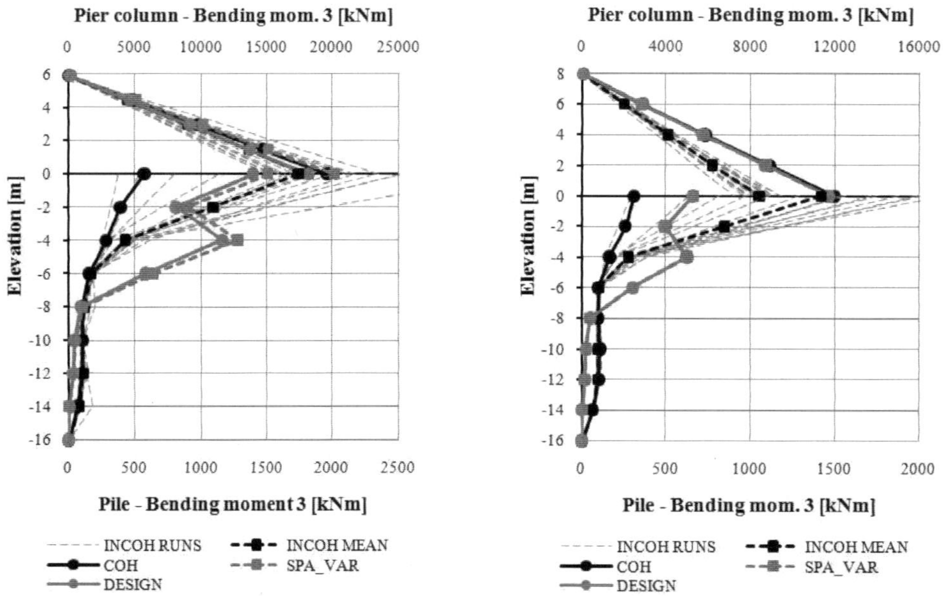

Figure 4. Bending moment 3 for earthquake loading in X direction (left=side pier, right=mid pier) [2].

Figure 4 presents the bending moments generated by seismic loading in the X direction (along the bridge axis). There is a notable difference in the shape of the bending moment diagram for the piles: the classic design model yields a large value at a depth of about 2-3 times the pile diameter, while the SSI model shows an exponential growth to a maximum value at the pile cap level.

The SPA_VAR cases which represent the results on the design model, including the simplified spatial variability loading, underestimate the bending moment in the piles by comparison to the mean incoherent values INCOH_MEAN. An increase in the pile bending moment of 25% for the side pier and 100% for the mid pier is observed.

As for the pier columns, all analyses give approximately the same bending moment diagram for the side pier. The mid pier shows an interesting effect: while the coherent analysis and the design model show similar results, the incoherent analysis reduces the stresses with about 30% from the other analyses. This especially happens to flexible elements, like the mid pier, because, due to incoherency, the ground and the structure don't move together reaching big displacements like they do in a coherent analysis; now the ground movement is chaotic, with waves coming from all directions that cancel each other out.

Figure 5 presents the bending moments due to earthquake loading in the transverse direction to the bridge axis. The bridge response in the transverse direction is much stiffer, due to development of the foundations along this axis and the fact that the pier columns and pier cap act together as a frame.

The incoherent analysis generates an increase in the pile bending moments from the design values of about 55% for the very stiff side pier. The design values are spot on for the piles of the more flexible mid pier.

The incoherent analysis offers another interesting effect for the pier columns. The design model and the coherent analysis give similar results and exhibit a classic moment diagram for a laterally loaded embedded frame, with maximum values at the base and the top and values close to 0 at the mid of the column. The side pier doesn't show this classic behaviour for the incoherent ground motion; the value at the middle of the pier is very large, while the values at the base and top of the pier are smaller than the ones from the design model / coherent analysis. This can generate real issues for the pier column, as the reinforcement in the middle of the element is often reduced in typical design and detailing practice. For the mid pier, both the coherent and incoherent analyses produce a reduction of column bending moments from the design model of about 30%.

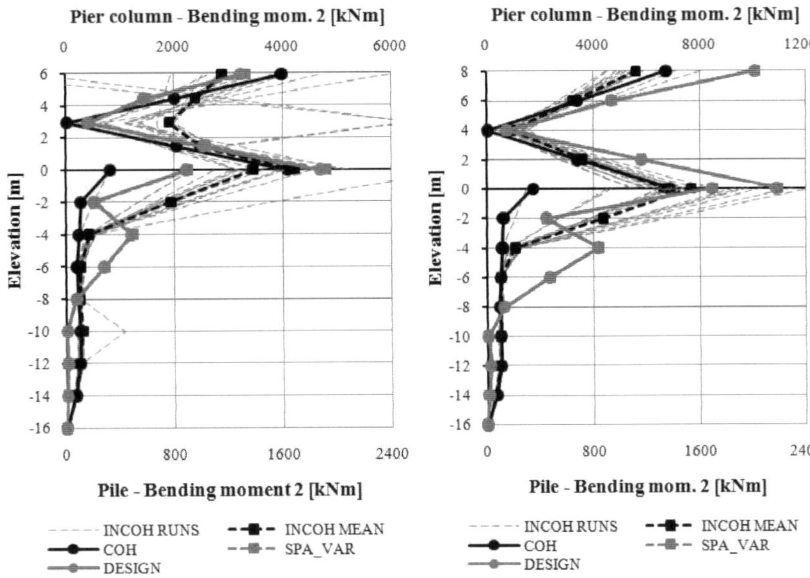

Figure 5. Bending moment 2 for earthquake loading in Y direction (left=side pier, right=mid pier) [2].

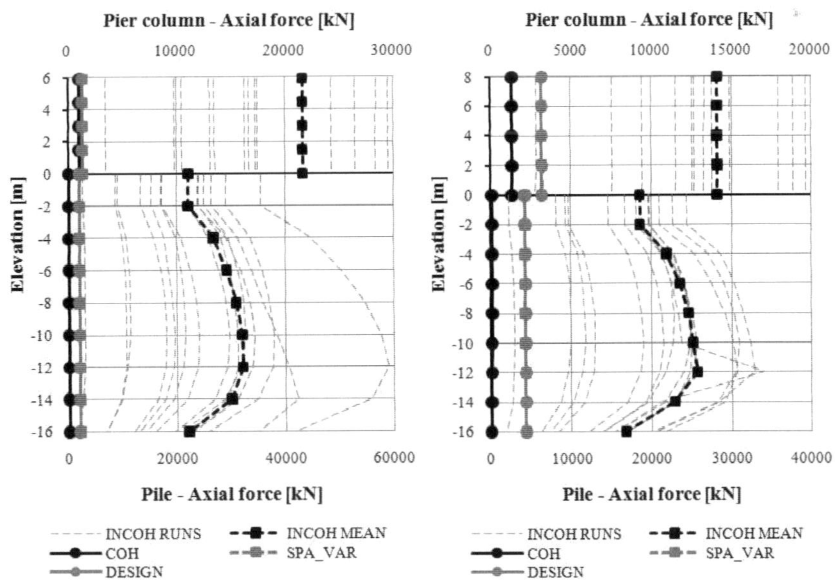

Figure 6. Axial force for earthquake loading in Y direction (left=side pier, right=mid pier) [2].

The incoherent analysis shows a very large increase in the axial forces for both piles and pier columns (figure 6). The reason for this may be the fact that the incoherent vertical ground movement is rigidly transmitted to the piles as axial displacements through the interaction nodes; of course, the piles themselves behave extremely stiff to these axial displacements and generate very high axial forces, as those exhibited in the analysis results. This is an issue that is being treated in the next phase of the study.

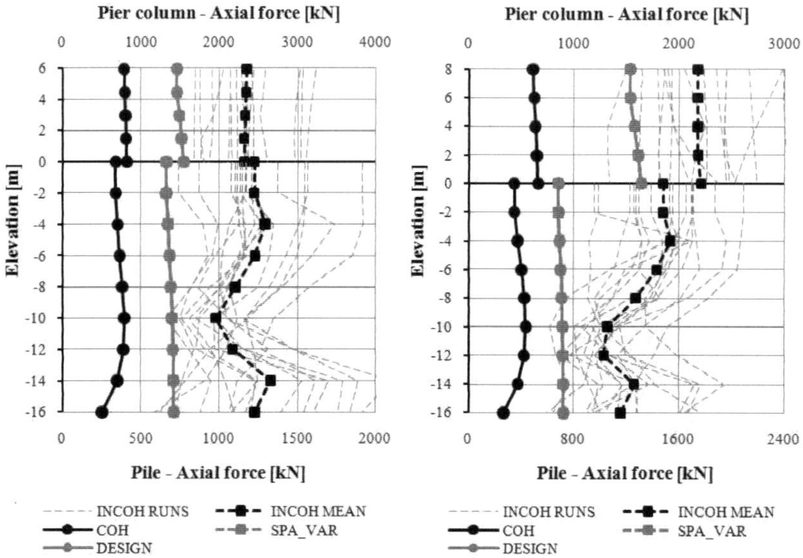

Figure 7. Axial force for earthquake loading in Z direction (left=side pier, right=mid pier) [2].

The axial forces due to earthquake loading in the vertical direction are plotted in figure 7. The incoherent analysis shows an increase in the axial forces of about 50-80%. Still, the increase isn't as big as the one from the transverse seismic action described before; this may mean that the extreme axial forces from before have something to do with the very strong rocking motion of the bridge in the transverse direction. This is supported by the fact that the increase of the axial forces from the incoherent ground motion in the longitudinal direction (X) are similar to the one exhibited here and do not show the extreme values from the Y direction.

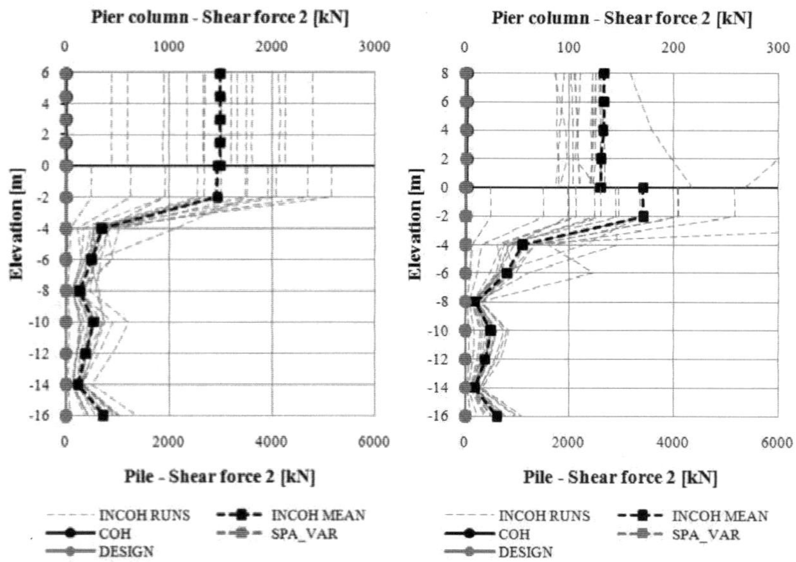

Figure 8. Shear Force 2 for earthquake loading in Z direction (left=side pier, right=mid pier) [2].

A behaviour that is completely ignored in the design model and the coherent analysis is that the vertical seismic action produces shear forces and bending moments in the structural elements. This is attributed to the fact that the vertically propagating seismic waves also generate horizontal waves through the reflection / refraction that occurs at the interface between soil layers. As figure 8 points out, the response is much stronger for stiffer elements (side pier), than that of the more flexible ones.

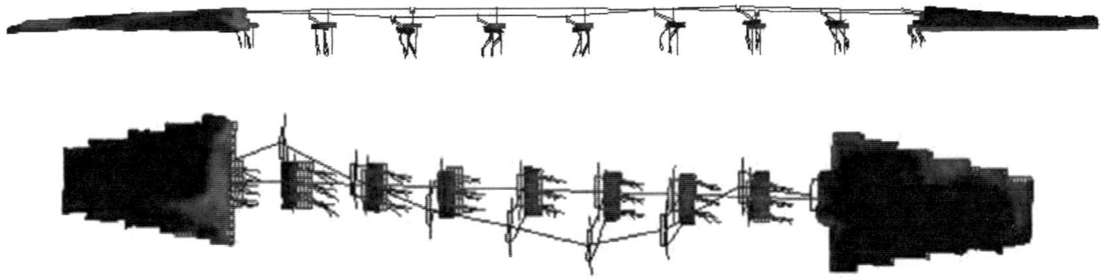

Figure 9. Instant incoherent accelerations of bridge structure in elevation and planview (deformed shape shown in blue, undeformed shape in red) [2].

Figure 9 shows two acceleration plots of the SSI model from the incoherent analyses. The bridge is not subjected to a constant acceleration in the same direction, but to accelerations that seem random both in amplitude and direction. The shape of the plots seems "broken", especially for the piles and the bridge deck; these elements are subjected to variable accelerations along their length.

In the above described study, differences in behaviour have been observed between the typical design model and the sophisticated SSI model that need to be taken into account in the design stage of the bridge. The issue of the extremely high axial forces still needs to be resolved. Therefore, the study focuses now on improving the foundation model. The results obtained so far are presented next.

5. Improving the foundation model

As the complete bridge model is too big for a complex analysis on the foundations, the study will focus, at first, on a single pier with the corresponding deck weight attached. After a satisfactory solution has been obtained, the complete bridge analysis will be recomputed.

A phenomenon that all studies on the matter acknowledge is the degrading of the soil in the immediate vicinity of the foundation, both due to the execution of the foundations and the hysteretic ground motion during an earthquake. This effect cannot be included in the present beam model. Furthermore, the wave propagation effect in the area near the beams will be distorted. Thus, the decision has been made to model the foundation and the near-field soil with solid elements.

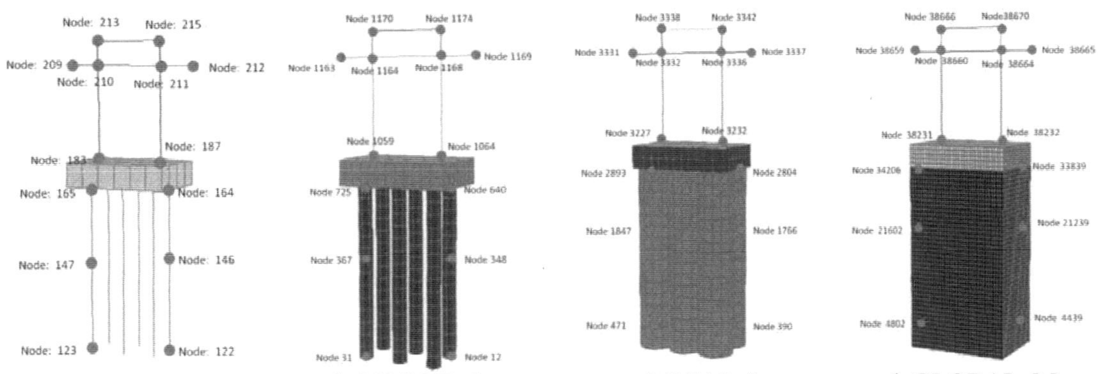

a. BEAMS b. NOSOIL_L c. LOCAL_L d. GLOBAL_M

Figure 10. Foundation models.

IOP Conf. Series: Materials Science and Engineering **399** (2018) 012010 doi:10.1088/1757-899X/399/1/012010

For this stage, the pier and its corresponding foundation will be modeled in 4 methods:

- BEAMS – the piles are modeled as beam elements, similar to the global bridge model
- NOSOIL – both the piles and the pile cap are modeled with solid elements
- LOCAL – in addition to the NOSOIL model, a soil column is modeled with solid elements around each pile; this is the same procedure as described by Novak (2003)
- GLOBAL – a soil box is modeled with solid elements around the foundation, and all piles are embedded into it

Each of the models with solid elements (NOSOIL, LOCAL, GLOBAL) has a couple of variants, depending on mesh size (large L and medium M) and the substructuring methodology (flexible volume FV and flexible interface FIT). Figure 10 contains a picture of each model.

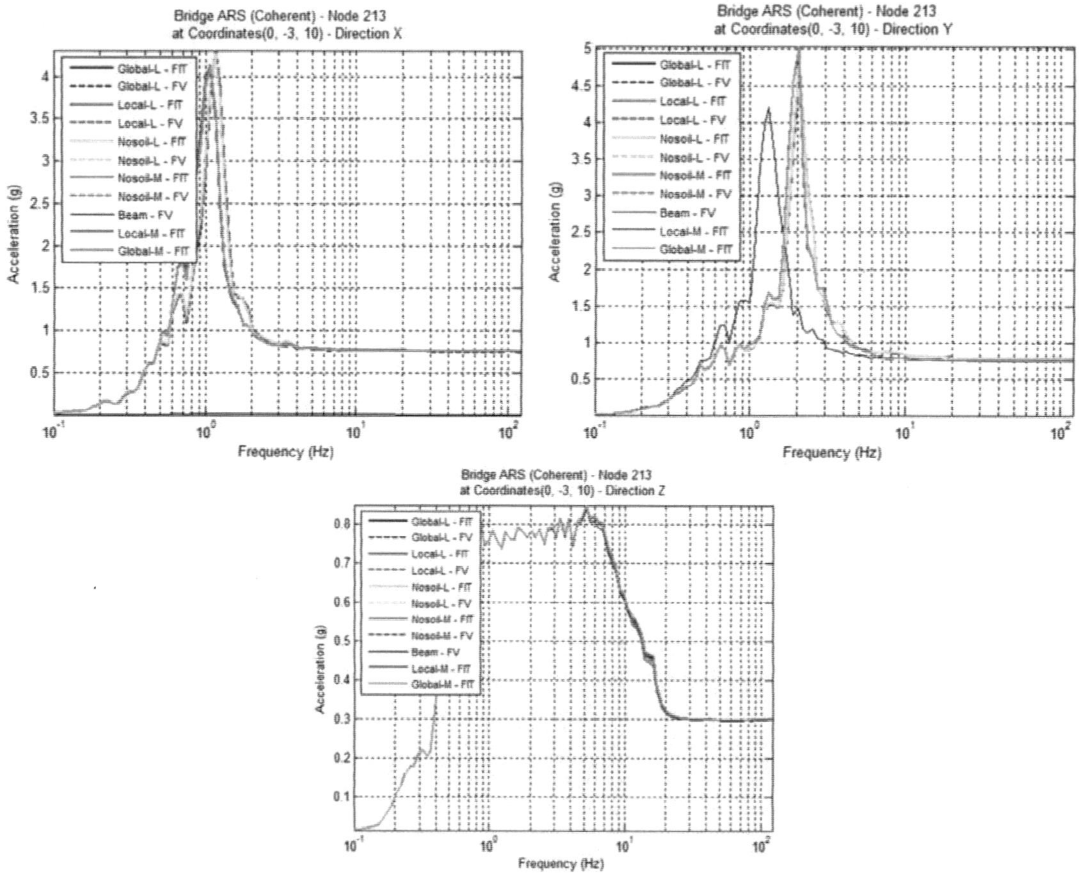

Figure 11. Foundation models acceleration response spectrum.

Figure 11 contains the acceleration response spectra at deck level (the top of the pier). The use of the solid near-field soil elements does not have almost any effect; that was to be expected, as their influence will be truly felt when nonlinearity is added to their behaviour.

The BEAMS model underestimates the stiffness of the model, as can be seen in the acceleration response spectrum in Y direction. The differences in behaviour of the models with solid elements are very small and may be neglected. The larger mesh size (L) will be preferred for the next stages of analysis to the more refined mesh (S) due to reduction in model size and analysis duration it provides. The good correlation between the flexible volume FV and flexible interface FIT substructuring methods, validates to use of the FIT method for the rest of the study, which, again, reduces the model size and analysis times considerably.

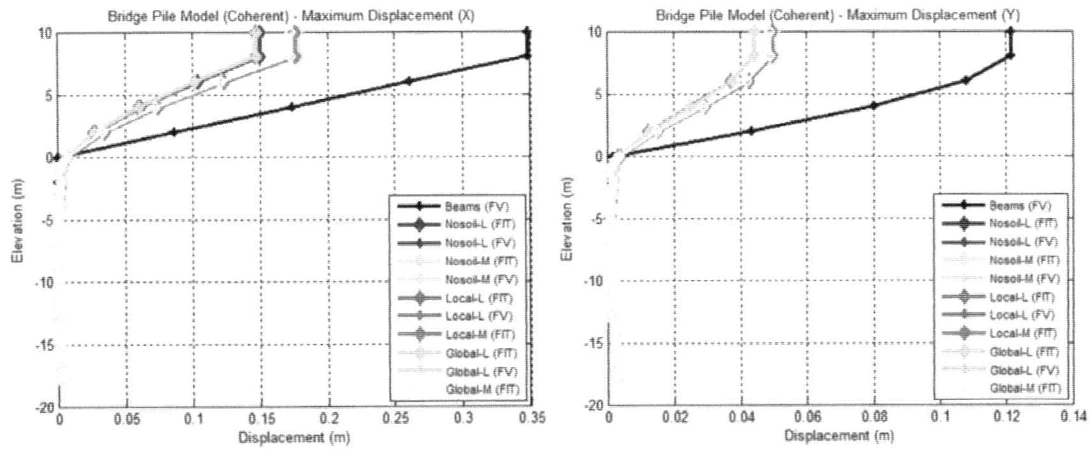

Figure 12. Foundation models maximum horizontal displacements.

The BEAMS model generates very large displacements and will be disregarded in the following analyses (figure 12). The other models yield similar displacements with a total variation of less than 10%.

Having decided that the models with solid elements will be used further, the next step is to add the nonlinear behaviour of the near field soil to the analysis. For this, the same Seed and Idriss (1970) linearized model will be used.

For the nonlinear analysis, the foundation model needed to be changed, because the pile cap was in direct contact with the soil layers from the site analysis. This would ensure a rigid connection between the structure and the site model and the near-field soil elements would not be loaded at their true capacity. Thus, a solid element "belt" with soil mechanical properties and nonlinear behaviour has been added around the pile cap. Figure 13 shows the complete analysis model to the left and the two main components to the right: the structural concrete model and the near-field soil elements.

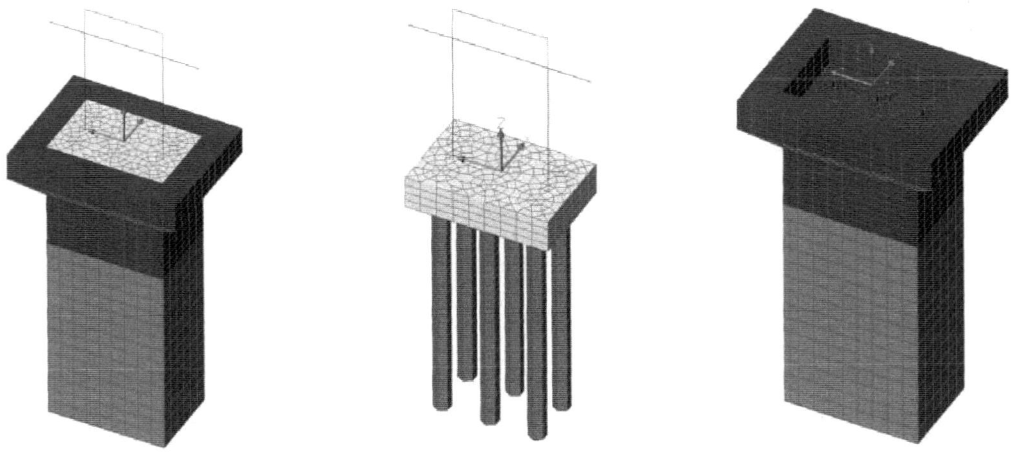

a. The complete model b. Structural concrete elements c. Near-field soil

Figure 13. Foundation model for nonlinear analysis.

Figure 14 presents the acceleration response spectrum at deck level (top of the pier) for the three orthogonal directions. Both coherent / incoherent and linear / nonlinear analyses are presented. The incoherent graph represents the mean values of the 15 incoherent simulations that were studied.

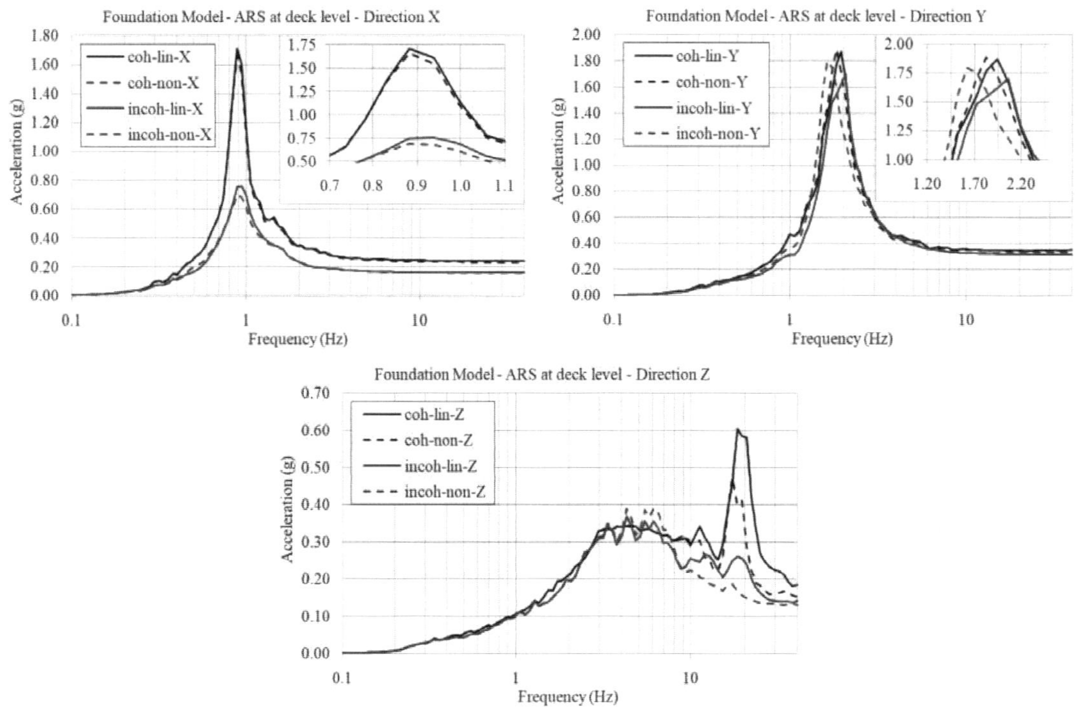

Figure 14. Foundation model acceleration response spectrum for coherent / incoherent and linear / nonlinear analysis.

For the X direction (along the bridge axis), the incoherency reduces the seismic response of the bridge to less than 50%, having an important positive effect on the structure. There is no frequency shift between the peaks, which means that the seismic waves scatter in all directions and partially cancel each other out. The nonlinear behaviour provides an additional reduction of around 5% which can be attributed to damping in the soil, because almost no frequency shift can be observed.

In the transverse direction (Y), the incoherency in the linear analysis generates a reduction of about 5%. The nonlinear analysis exhibits a frequency shift of 0.1 Hz for the coherent analysis and 0.5 Hz for the incoherent analysis. Also an amplification of the spectral acceleration of about 6% can be observed in the incoherent analysis between the nonlinear and linear results; this phenomena leads to an increase in the forces acting on the structure.

The vertical direction (Z) shows a strong decrease of spectral acceleration at the 11 Hz peak due to incoherency effects.

Figure 15 presents pile and pier column element forces corresponding to the seismic loads acting in longitudinal (X) and transverse direction (Y).The following conclusions can be drawn for the foundation model:

- for the piles, the incoherency effect generates a considerable increase in structural response in comparison to the coherent SSI analyses; for the pier columns the effect is the opposite, as incoherency reduces the element forces.
- the nonlinearity of the near-field soil doesn't have an important effect on the structural forces in the pier columns; for the piles, an interesting effect can be observed: the axial forces are reduced, while the transverse forces (shear forces that further generate bending) increase by more than 25%;
- the typical design procedure will generate higher structural forces, with the exception of the pile bending moment due to earthquake loading in the longitudinal direction (X), where the design clearly underestimates the structural response; for this structure, a more economical

design could be achieved for the pier columns and the length of the piles by considering the sophisticated SSI model.

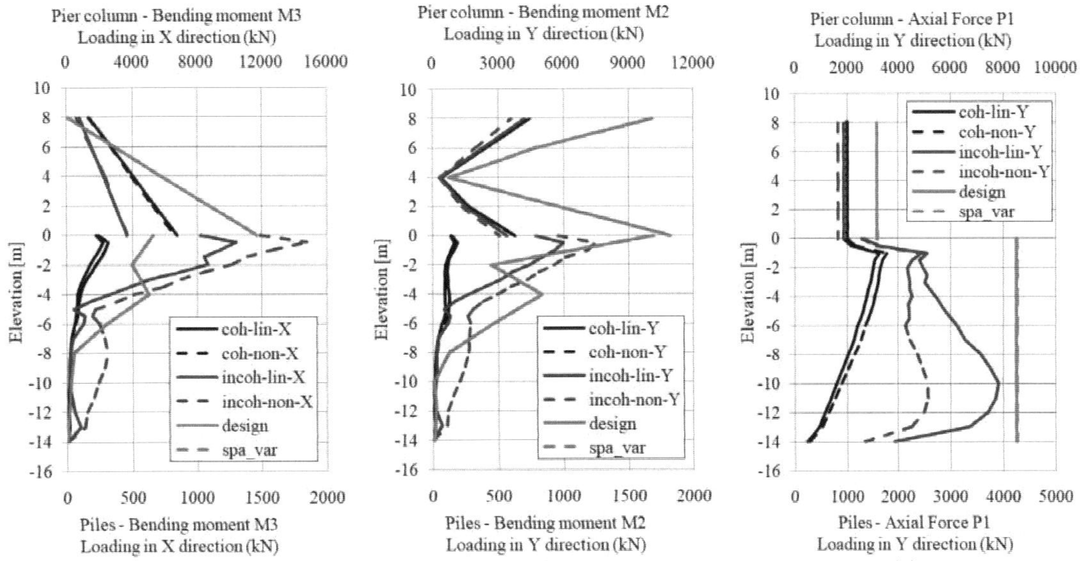

(a) (b) (c)

Figure 15. Bending moment M3 for earthquake loading in X direction / Bending moment M2 and axial force P1 for earthquake loading in Y direction (coherent / incoherent and linear / nonlinear analysis, design and simplified spatial variation model).

6. Conclusions and future directions

The scope of this paper was to compare the results of a state-of-the-art seismic SSI analysis of a bridge structure to the current design procedure according to Eurocode 8.Along the course of the study, a methodology has been developed for modelling the bridge structure and foundations in order to obtain adequate results from the SSI analysis.

The overall conclusion is that the Eurocode 8 design can be considered acceptable for small structures where SSI effects and motion incoherency do not play such a large role. However, when dealing with large structures, the simplified spatial variation model given in Eurocode 8 will influence the structure very little. Furthermore, this simplification completely ignores the large shear forces that the piles are subjected to, due to incoherent differential displacements in the soil.

Several interesting effects of motion incoherency on bridge structures have been highlighted throughout the paper. Based on the investigated case study results, no clear answers can be given to whether the design model overestimates or underestimates the structural response. The only thing that can be clearly stated is that the current design practice does not capture all the effects displayed by the state-of-the-art SSI analysis. Incoherent SSI analyses should be mandatory for the seismic design of large bridges and are still recommended for smaller structures.

As a future direction of the study, the complete bridge SSI model will be analysed, taking into account the nonlinear behaviour of soil in direct contact or in the immediate vicinity of the foundations. Also, an improved model for the superstructure will be implemented, that offers a better representation of the spatial distribution of stiffness and mass.

References

[1] Mylonakis G, Gazetas G, Nikolaou S and Michaelides O 2000 The role of soil on the collapse of 18 piers of the Hanshin expressway in the Kobe earthquake *12th World Conference on Earthquake Engineering*, Auckland, New Zealand, 1074

[2] Contiu M, Ghiocel D M and Cretu D 2014 Incoherent soil-structure interaction (SSI) effects for a 242 m long concrete bridge founded on deep piles *Proceedings of 2ECCES : EAEE Session,* Istanbul, Turkey

[3] Ghiocel D M 2007 *Stochastic and Deterministic Approaches for Incoherent Seismic SSI Analysis as Implemented in ACS SASSI*, Appendix C, Electric Power Research Institute, Palo Alto, CA and US Department of Energy, Germantown, D, EPRI Report TR-1015111, USA

[4] Ghiocel D M 2009 *ACS SASSI NQA Version 2.3.0 - An Advanced Computational Software for 3D Dynamic Analysis Including Soil Structure Interaction*, User Manuals, Technical Reports, GP Technologies, Inc., Rochester, New York, USAA

[5] Seed H B, Idriss I M 1970 *Soil Moduli and Damping Factors for Dynamic Response Analysis*, Earthquake Engineering Research Center, Berkeley California, Report EERC 70-10, USA

[6] Abrahamson N 2006*Spatial Coherency for Soil-Structure Interaction*, Electric Power Research Institute, Palo Alto, CA and US Department of Energy, Germantown, MD, EPRI TR-1014101, USA

[7] Abrahamson N 2007 *Effects of Seismic Motion Incoherency Effects*, Electric Power Research Institute, Palo Alto, CA and US Department of Energy, Germantown, MD, EPRI TR-1015110, USA

Ecological approach for assessing drill quality. A case study

R Daicu[1*]**, V Dițu**[1] **and G Oancea**[1]

[1] Transilvania University of Brasov, Department of Manufacturing Engineering, Mihai Viteazu No.5, Brasov, Romania

[*] E-mail: raluca.daicu@unitbv.ro

Abstract. The quality of drills is classically appreciated by measuring the effective cutting time until the appearance of the admissible wear on the land surface. The experimental tries lead to material consumption and also to resulted chips which, in most cases, are cast as a pollution element. The proposed method for assessing quality is based on measuring the electrical current at cutting and substantially reduces the quantity of resulted chips. The case study was developed for HSS $\phi 8$ drills and shows the advantages of the method and its ecological impact. The experiments were performed using 1C45 steel workpiece material.

1. Introduction

Making clean technologies has become a necessity to keep the environment clean. It is well known that a clean technology presents the following characteristics [1]:

- not to pollute the environment of outside and inside the organization;
- a very small amount of waste resulting in the machining processes;
- to have less material and energy consumption;
- the water consumption should be as low as possible;
- not to pollute the environment by noise.

The comparative analysis of various techniques and technologies from an ecological point of view has become a necessity in terms of protecting the environment in which we live. The ecological approach must be developed in all areas of human activity. A lot of solutions in this way are provided by the surrounding nature itself. Pollution elements should not be treated with other elements of man-made pollution. The reduction of pollution in a certain area has as a first approach the reduction of the quantity of polluting material resulting from various processes.

As machining is a basic method of obtaining different parts, any element resulting from the cutting process must be analysed, reduced the amount in which it is produced, or even eliminated. At present, mankind concentrates on large quantities of polluting materials and neglects small amounts resulting from processes. An example in this direction, which represents the objective of the paper, is the polluting elements resulting from the cutting process which takes place in short time in the case of the experiments carried out for quality appreciation of a cutting tool.

A profound analysis of pollutants in a process cannot be done without a systemic approach of the process.

The case study refers to finding a less polluting method for quality assessing of a batch of cutting tools. The classical test method consists in cutting with a batch of cutting tools until the appearance of admissible wear on the land surface, and the effective cutting time represents the tool life. From the

IOP Conf. Series: Materials Science and Engineering **399** (2018) 012011 doi:10.1088/1757-899X/399/1/012011

cutting process developed during the tests pollutants result. The systemic approach of the cutting process is presented in figure 1 [2]. Analysing this figure, the pollutants resulted from the machining process can be easily identified. Also, according to [2], the cutting process is a primary system (it cannot be decomposed into physical subsystems) with a degenerative reaction loop.

The elements that can pollute during the cutting process are found both at the entrance into the system and at the exit. Figure 2 illustrates the elements with pollution potential from the cutting process.

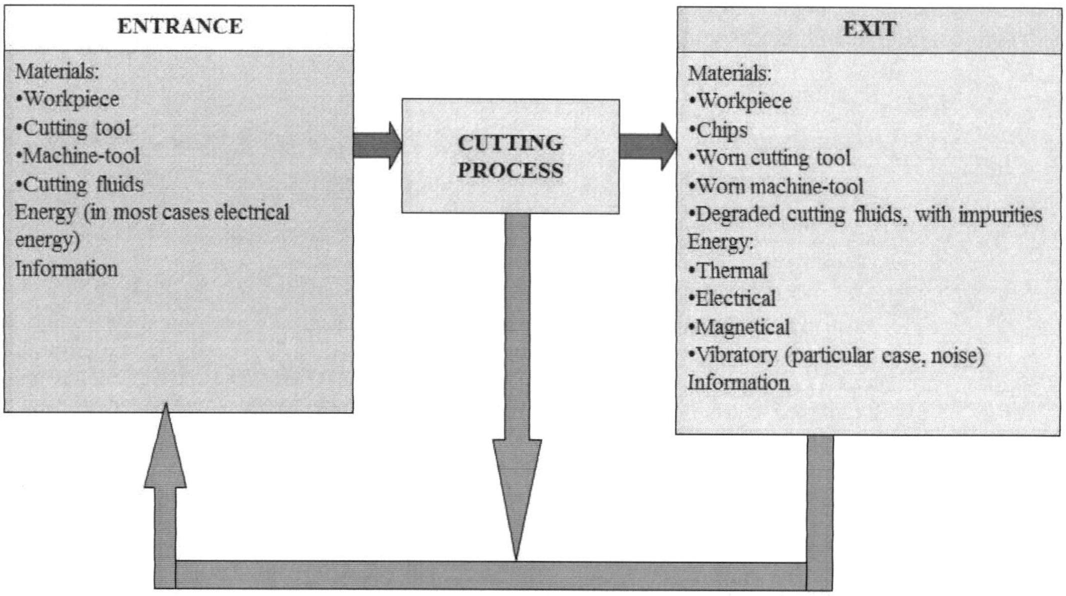

Figure 1. Cutting process as a system.

POLLUTANTS FROM CUTTING PROCESS	
Pollutant	Placement in cutting process
1. Unclean workpieces	1. Entrance
2. Machine-tool with oil/cutting fluids leaks	2. Entrance
3. Cutting fluids	3. Entrance
4. Chips	4. Exit
5. Worn cutting tools	5. Exit
6. Degraded cutting fluids with impurities	6. Exit
7. Vibrations (noises)	7. Exit

Figure 2. Identification of pollutants from the cutting process.

This paper proposes a method for assessing the drills quality, which, as compared to the classical method (life tests), has the advantage of a very low pollution.

The method proposes the measurement of the electrical current at cutting [3-7] for the quality assessment of drills, instead of life tests. As compared to the classical method, due to the short cutting time (20 seconds on average) the quantity of resulted chips is very low.

In the approached case study 2 batches of drills are compared.

2. The quality analysis of two drills batches, from two different producers, using the electrical current at cutting

The paper continues the approach from another paper [8], by comparing two HSS-Co steel drills batches with the diameter of 8 mm. The batches of 10 drills each were purchased from two different producers.

For quality assessing, the electrical current at cutting was measured, by following the next steps:

- measuring the electrical current at cutting for each drill using the experimental stand presented in figure 3;
- microscopically visualisation of drill edges to obtain additional information regarding the connection between the electrical current at drilling and the state of the cutting edge.
- processing the experimental data and concluding.

The experiments were done on 1C45 material with the following chemical composition: C = 0.49%, Si = 0.19%, Mn = 0.77%, P = 0.027%, S = 0.003%. The 1C45 steel tensile strength is 806MPa (80.6 daN/mm^2) and the hardness is 224 HB.

Figure 3. Experimental stand for measuring electrical current at drilling.

The experimental stand is composed of: 1 – mandrel for fixing the drill, 2 – collector brushes, 3a- instrument for measuring the electrical current voltage with data acquisition, 3b – PC with software for data acquisition and 4 – gripping device for multiple drilling.

The obtained values for the first batch of drills, from one producer, are presented in table 1, and for the second batch, from another producer, in table 2. The same cutting parameters as in the paper [8] were used for additional comparisons.

Based on tables 1 and 2, table 3 was conceived, which presents, comparatively, the number of drills that have the same value of the electrical current at cutting and the minimum and maximum values for each batch.

Table 1. Values for electrical current at cutting "U_0" at drilling 1C45 with HSS-Co drills, for batch I.

Drill no. from batch I	Drill diameter [mm]	Spindle speed [rev/min]	Feed [mm/rev]	Voltage "U_0" of electrical current at cutting[mV]
1	8	560	0.25	1.3
2	8	560	0.25	1.4
3	8	560	0.25	1.3
4	8	560	0.25	1.1
5	8	560	0.25	1.4
6	8	560	0.25	1.1
7	8	560	0.25	1.2
8	8	560	0.25	1.6
9	8	560	0.25	1.3
10	8	560	0.25	1.1

Table 2. Values for electrical current at cutting "U_0" at drilling 1C45 with HSS-Co drills, for batch II.

Drill no. from batch II	Drill diameter [mm]	Spindle speed [rev/min]	Feed [mm/rev]	Voltage "U_0" of electrical current at cutting[mV]
1	8	560	0.25	1.0
2	8	560	0.25	1.2
3	8	560	0.25	1.2
4	8	560	0.25	1.0
5	8	560	0.25	1.2
6	8	560	0.25	0.9
7	8	560	0.25	1.0
8	8	560	0.25	1.0
9	8	560	0.25	1.1
10	8	560	0.25	1.1

Table 3. Number of drills with the same value for electrical current at cutting for batch I and II.

No. crt.	Cutting electrical current voltage U_0 [mV]	No. of drills from batch I	No. of drills from batch II
1	0.9	0	1
2	1.0	0	4
3	1.1	3	2
4	1.2	1	3
5	1.3	3	0
6	1.4	2	0
7	1.5	0	0
8	1.6	1	0
	Total	10	10

The results from table 3 are graphically presented in figure 4.

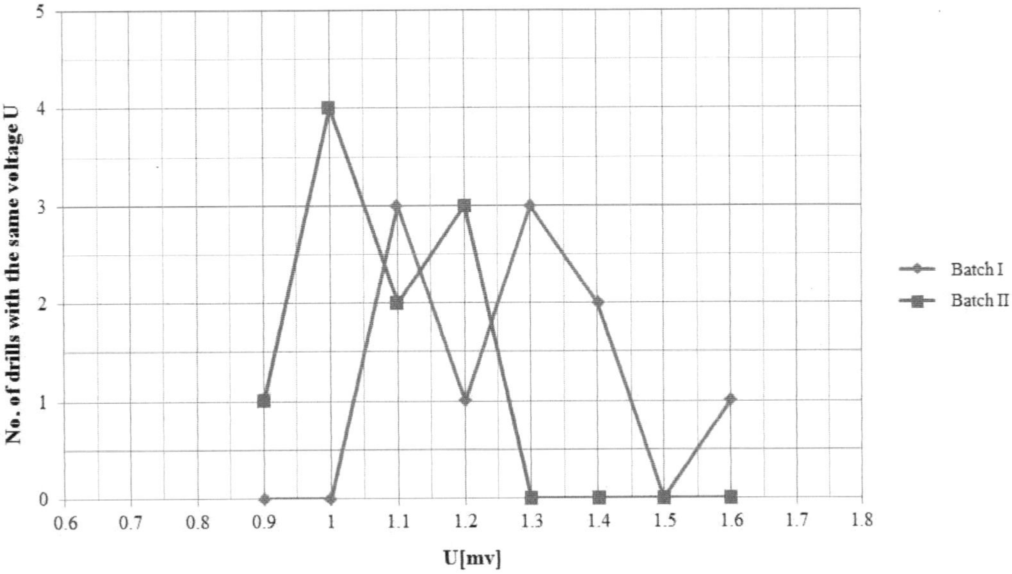

Figure 4. No. of drills with the same value for electrical current for batches I and II.

Analysing figure 4, it can be observed that the drills from batch II have a lower voltage during the cutting process, meaning that the temperature is lower and the tool life will be longer.

Also, the variation interval of the electrical current voltage is shorter, meaning that the technological process for drills manufacturing is better handled by the second producer.

The microscopically images from figure 5 show that the qualitative differences between the two batches are given by the shape of chisel edge adjustment and a better roughness on the land surface of the drill from the batch II.

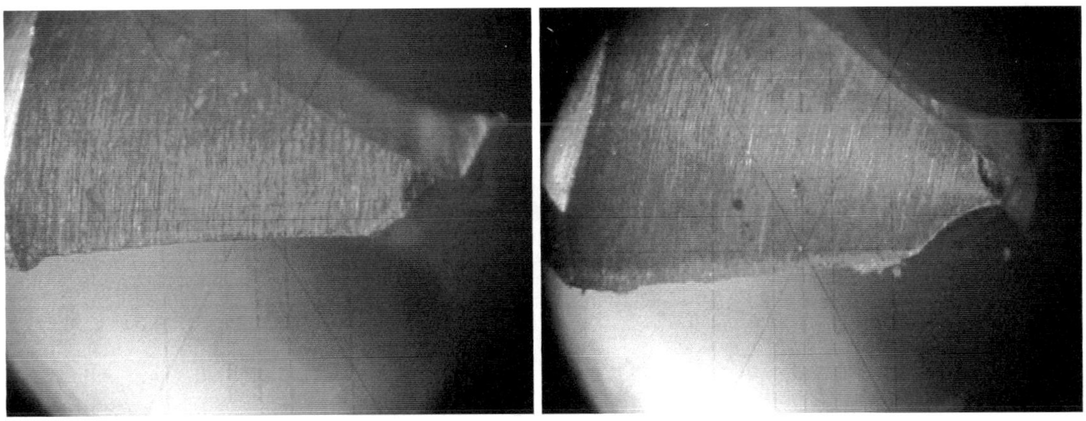

Figure 5. Images with a drill from the batch I (left) and batch II (right).

3. Determination of the amount of chips obtained from the experiments

In the classic method, the experiments would have been carried out until the admissible wear, the effective cutting time being the drill life. According to [9] the tool life for the used cutting parameters would be 25 minutes. The total length of the machined hole is:

$$L = T \cdot s \cdot f \quad [mm] = 25 \cdot 560 \cdot 0.25 = 3500 \; mm, \tag{1}$$

where:

- T [min] is the effective processing time with a drill;
- s [rev/min] is the spindle speed;
- f [mm/rev] is the cutting feed.

The volume of the removed material for a drill would be:

$$V_1 = L{\cdot}f{\cdot}d = 3500{\cdot}0.25{\cdot}4 = 3500\ mm^3, \tag{2}$$

where d [mm] is the cutting depth.

For the two batches, the total quantity of removed material would be:

$$V_{1,\ total} = 3500{\cdot}20 = 70000\ mm^3. \tag{3}$$

For the method used by the authors, the volume of removed material is:

- $L_2 = 0.33{\cdot}560{\cdot}0.25 = 46$ mm;
- $V_2 = 46{\cdot}0.25{\cdot}4 = 46$ mm^3;
- $V_{2,\ total} = 46{\cdot}20 = 920$ mm^3.

It can be observed that the quantity of chips is reduced by 98.7%, being in the same proportion with the removed material.

4. Conclusion

From the case study presented in this paper, the following conclusions can be drawn:

- electrical current at cutting is a simple and sensitive method for quality assessing of a cutting edge;
- the drills from batch II are better from qualitative point of view than drills from batch I produced by a different company;
- the quality difference between the two batches is given by the land surface roughness (the drills from batch II present a smaller roughness) and also by the quality of the chisel edge adjustment (the drills from batch II present a more accentuated adjustment of the chisel edge);
- using this method, reduction of chips pollution is very significant and, in the same time, is also reduced the consumption of material along with the electricity consumption.

References

[1] Nedelcu A and Ardeleanu D 2007 *Ecological technologies, environment management, and safety in industry* (Braşov: Transilvania University of Braşov) p 9
[2] Diţu V 2008 *Basis of cutting metals* (Bucharest: Matrix Rom) p 177
[3] Diţu V 2008 *Academic Journal of Manufacturing Engineering* **1** 63
[4] Murata M, Kurokawa S and Ohnishi O 2013 *CSME* **79** 2546
[5] Murata M, Kurokawa S, Ohnishi O, Uneda M and Doi T 2012 *CSME* **6** 958
[6] Tanaka R, Hosokawa A, Yamada K and Ueda T 2004 *Proceedings of the Seventh International Conference on Progress of Machining Technology* 786
[7] Hirota H, Murata M, Kojima A and Yanagihara K 1998 *JSPE* **64** 1191
[8] Diţu V, Oancea G and Daicu R 2017 *MATEC Web Confonferences* **94** 02005
[9] Picoş C and Pruteanu O 1992 *Design of technologies for mechanical processing by cutting* vol 2 (Chişinău: Chişinău) p 23

IOP Conf. Series: Materials Science and Engineering **399** (2018) 012012 doi:10.1088/1757-899X/399/1/012012

Log cottage – a case study

M Dancs[1*] and G Onul[2]

[1] Technical University of Cluj-Napoca, 28 Memorandumului Street, Cluj, Romania
[2] Transylvania Forestry College, 1 Garii Street, Nasaud, Romania

[*] E-mail: dancs_madalina@yahoo.com

Abstract. Wood is considered to be one of the oldest building materials, alongside stone. Although wood is characterized by a lower durability, due to multiple factors, we can say that it has many uses, especially in rural areas, in the proximity of woods. An important reason for using wood as a building material was its lower processing cost. However, we cannot omit the fact that, in today's Romania, due to changes in the forestry legislation, a "wood crises" has been generated, which led to higher costs in purchasing, processing and transport. Over time, wood has proved its qualities as an excellent building material, due to its physical and mechanical properties. It is a good thermic and phonic isolator, it is light and resistant, especially if it has been treated against fire, woodworms and fungi. This paper presents generalities regarding wood as a building material, updated and specific data about a log cottage, highlighting its advantages and disadvantages, its building and production costs. The case study could represent a small guidance for investors, giving, at the same time, an overview of the current availability of the total volume of wood in Romania.

1. Introduction

The world wood plan and superior wood products have exceeded the period of their use as roof and auxiliary structures, being increasingly present in the construction of important buildings, such as the ones with large openings (e.g. sports halls, warehouses and industrial halls), and not only. Building holiday cottages in secluded or hard-to-reach areas in Romania has led many investors, architects and builders to prefer wood as the main building material, both for economic and aesthetic reasons. The beauty of the seemingly wooden construction elements, the architectural and structural success, the safety in operation and, last but not least, the low price have attracted positive reviews from many users.

Having multiple qualities, wood is a living material, with cosy, pleasant, physical and mechanical properties. Wooden constructions have proved good behaviour and safety over time, especially if the wood is properly treated against biological agents (insects, fungi) and fire [1].

2. The present situation of the volume of wood in Romania

In Romania, there is the question of the real existence of the quantity of wood that is harvested for production. The National Forestry Institute (NFI) statistical data show that less than half of the growth of forests of the national forest fund is harvested.

According to the definitions that are used in our country, the forest vegetation consists of the national forest fund and the forest vegetation outside the national forest fund (two categories).

Romania has a total forest area of 7.046 million hectares, out of which the land actually covered with trees is 6.9 million hectares. The largest forest area is found in Transylvania, where 37% of the total land area is covered by forest vegetation, according to data from the first cycle of the National Forestry Inventory (NFI) [2].

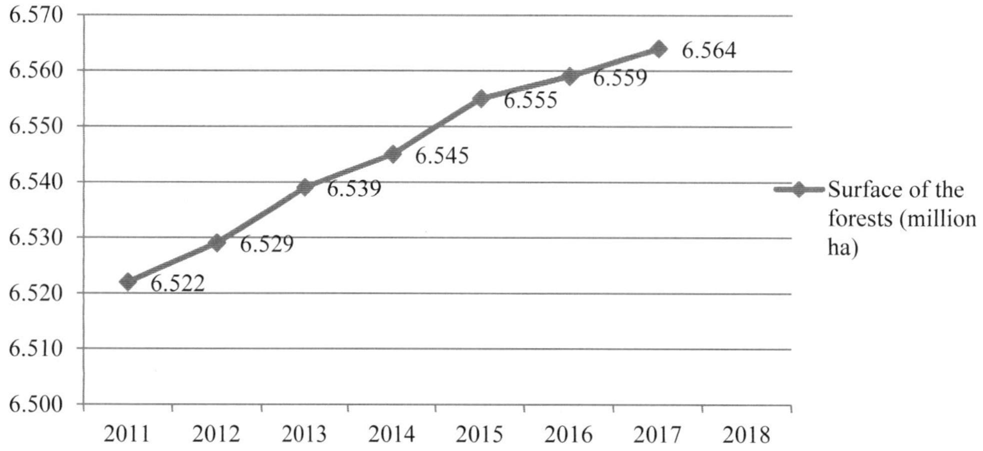

Figure 1. Evolution of the forest fund between 2011 and 2018.

Beech is the main forest species in our country. It occupies 31% of the area covered by the forest. It is followed by softwood species (spruce, fir, pine etc.) with a share of 26%, various hardwood species (hornbeam, acacia, ash, maple etc.) with 20% share and cvercinee or oak species (pedunculate oak etc.) with a share of 16% in the forest area. On the last place there is the group of various softwood species (linden, poplar, willow etc.) which occupies 7% of the forest area [1].

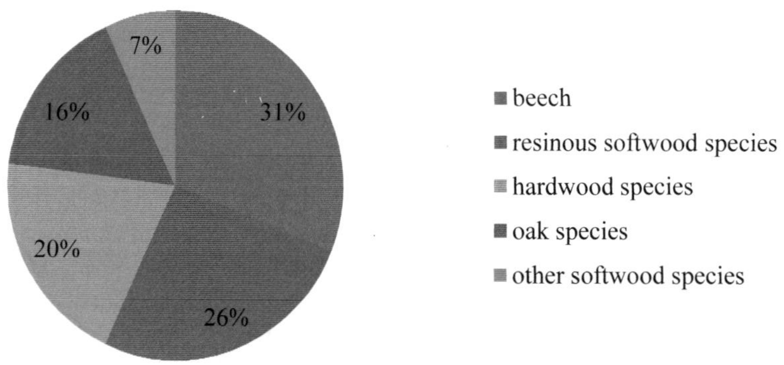

Figure 2. Tree species that can be found in Romanian forests.

Depending on the topography, 65% of the total mountainous region in Romania is covered with forest. For the hill region, the forests represent 27 % of the area and, for the plain region, forests occupy 5% of the area.

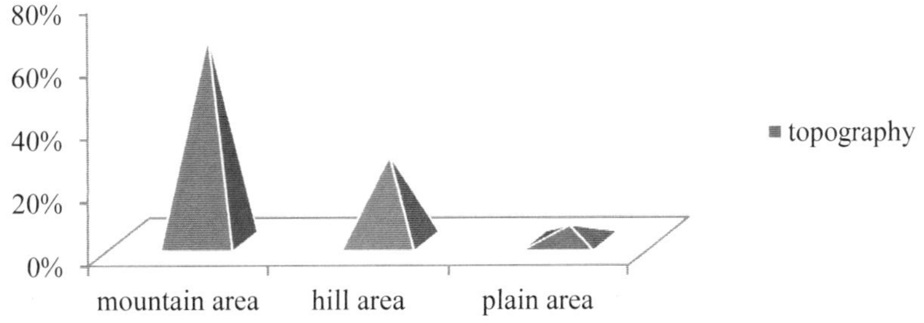

Figure 3. Forest area depending on the topography.

NFI data also reveal that, according to the age class distribution, the largest area of the forest (21%) is in the IInd age class (21- 40 years old), followed by the IVth age class (61-80 years old) and the IIIrd one (41-60 years old) with 19% and 18% of the forest area. Summing up, the forest area of the three age classes represents over 58% of the total forest area.

Nearby age categories are known as age classes. Depending on the source of the trees and the treatments applied, the arboreta can be grouped into 5, 10 or 20 years old. The age classes are expressed in hectares, their summation representing the actual wooded area of the forest in question. However, this area does not always correspond to the total area under regeneration: the difference is non-regenerated land or land affected by the forest administration.

Figure 4. Age classes of arboreta [3].

Out of the total state forests, 60% are in the Ist functional group (having special protection features) and 40% are in the Ind functional group (having both protection and production features).

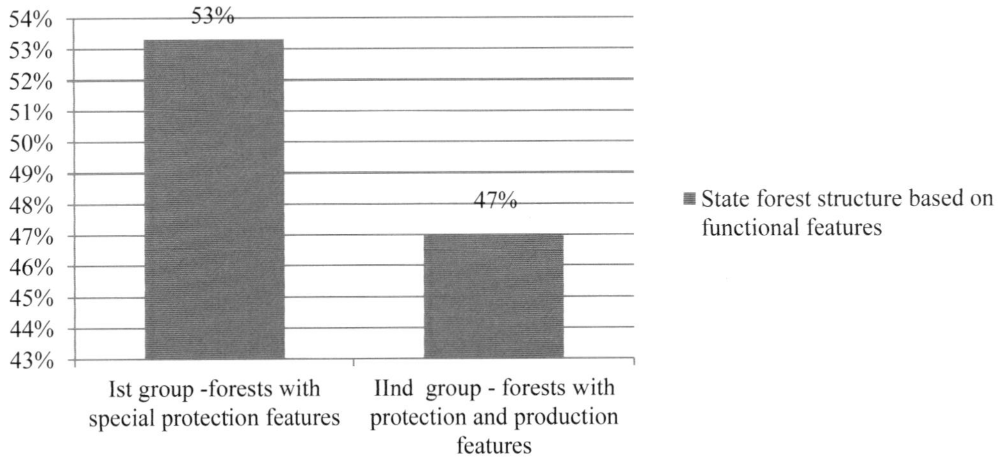

Figure 5. State forest structure based on functional features.

At present, Romania harvests around 18 million m^3 of wood material annually. Considering the annual wood exploitation rhythm, we can say that it is influenced by natural calamities, which temporarily affect certain areas, as well as the periods of restriction provided by the legal regulations. As a result, the most important volumes of wood are exploited within March and April, as well as between September and November.

Considering the wood harvesting species, the largest exploitation is of beech species (33%) and, close to it, the resinous species (31%), as presented in figure 6 [4].

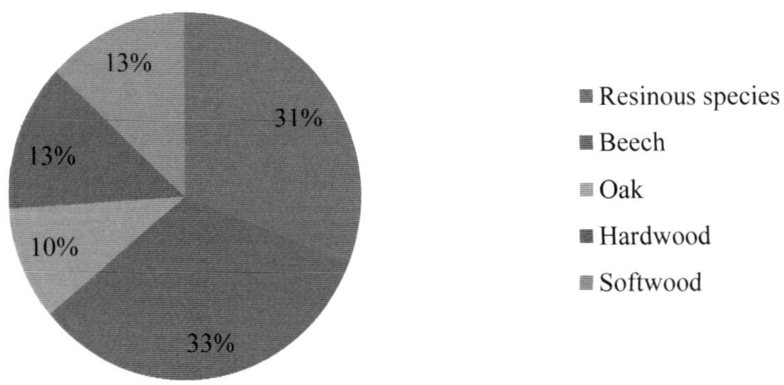

Figure 6. Situation of the wood species exploited per year.

As a result of the legal measures taken in 2015 in order to reduce illegal deforestation and illegal tree harvesting, it can be said that, at the moment, we have enough wood to support the development of sustainable wooden constructions. Probably the most important measure taken was the foundation of the Forest Guards, institution with legal personality established under the Emergency Ordinance no. 32/2015, subordinated to the Ministry of the Environment, Waters and Forests. The Forest Guards provide forestry and hunting inspections and controls of the central public authority responsible for forestry, regardless the form of ownership, of the national forest fund and the forest vegetation outside it. Also, the Forest Radar was a sustainable measure that lead to a significant reduction of illegal harvesting, due to the easy process of checking the legal status of a tree transport and reporting it to the responsible authorities in case of illegal proof.

Concerns about the regeneration of arboreta on forest funds from which wood has been harvested as a result of the application of main product cuttings, for the afforestation of land without forest vegetation, which have not had other uses attributed to forestry, as well as for the ecological reconstruction of the lands affected by different forms of degradation, have also constituted priority objectives in the last years. The harvesting of forests based on the age cycle classes insures a sustainable and, at the same time, healthy exploitation of wood.

3. Building a wooden cottage

3.1. The Advantages and Disadvantages of Wooden Buildings

This chapter intends to emphasize the main advantages and disadvantages of wood as construction material. As any construction material, it has positive and negative aspects that one must take into consideration before deciding if it is the suitable material for a future house or cottage.

3.1.1. The Advantages of Wooden Buildings. One of the main advantages of wood as constrution material would be its apparently low density compared to relatively high resistance. Compared to the density of the other main building materials (masonry, reinforced concrete, steel etc.), it can be seen that wood is lighter and the ratio between strength and density has the value comparable with the one of steel, both in compression and stretching.

Another advantage is due to the low weight of wood that makes all constructions made of this material exhibit favorable behavior in seismic action, or be placed more easily on difficult foundation terrains. It also requires less consumption of materials in foundation structures. There is a well-known seismic experiment based on wood constructions on the famous pagoda style in Japan, which have been proved to resist to an earthquake of more than 9.0 magnitude.

Also, the easy machining and shaping of wood, both in the plant and on the site, with the possibility of building wood construction in any season without requiring special measures is another important advantage. The execution speed is high, by eliminating wet works specific to reinforced concrete or masonry. The commissioning of timber constructions is possible immediately after the finishing works. The easy shaping also leads to the possibility of making special shapes and gauges that are difficult or even impossible to achieve with other building materials. There are wooden constructions in the form of arches or domes with openings reaching 100 m.

Regarding the thermal properties, we can say that wood is a favorable construction material. Compared to steel, concrete and even brick, the wood has a much lower coefficient of thermal conductivity, which justifies its use as a proper insulation material with good efficiency. Wood provides thermal resistance when passing a heat flow through it, 300-400 times larger than steel and 7-10 times larger than concrete. The linear thermal expansion coefficient along the reduced fiber makes it unnecessary to have thermal expansion joints in wooden constructions and to exhibit a good fire resistance behavior. For resinous wood, for example, the coefficient is 2-3 times lower than the coefficient of thermal expansion of steel and of reinforced concrete [5].

If we take into consideration the economical aspect while building a house or a cottage, the high durability of wooden constructions, which are in an optimum operating system, in terms of environmental conditions could be a plus in choosing wood as construction material. The maintenance costs are normal, except for the exterior polish that requires periodic maintenance (laquer at 7, 8 years). Interventions on wood elements, for consolidation or restoration, are easily made on the spot.

Although it may seem unbelievable at a first glance, wood also has a relatively good behavior in terms of fire resistance. It is a fuel material, but it behaves well in terms of structural resistance to fire since the massive elements are consumed relatively slowly at a rate of 0.5 - 0.7 mm/minute, which implies a decrease in the transversal section of 1 cm on each face, within a quarter of an hour. In the meantime, the fire temperature can reach 700-800° C. On the other hand, the strength and stiffness of wood inside the charred section remain practically unchanged [5].

The possibility of re-using wood in making other building elements or using it for energy production triggers reduced waste. Also, the outstanding architectural features and cosiness of wood make it be used not only as a structural material, but also as a finishing or apparent material, with special aesthetic effects. There is the possibility to associate wood with steel or concrete, conducive to the formation of efficient mixed structures.

3.1.2. The Disadvantages of Wooden Buildings. Of course that, as any other construction material, wood has its negative parts too, mostly due to the fact that it is a natural product of organic origin, having a non-homogeneous and anisotropic structure. An important disadvantage is the very high variability of characteristics between various species and within the same species, due to very diverse variables. There is also a high variation of mechanical and physical characteristics in different directions than the fiber's direction. Due to the non-homogeneity of the wood structure, the resistance is different along the trunk of the wood and on the cross-section, its variation being between 10 and 40% [5].

Another disadvantage is the high influence of humidity on the physical-mechanical characteristics, on the dimensions and the durability of the wood. For example, moisture variation from 5 to 15% leads to some wood species dropping compression strength by almost 2 times. Increasing humidity also favours the biological degradation of wood, in particular due to the action of fungi, and creates health problems for the people living in the building [5].

The limited range of wood material, both in terms of cross-sectional dimensions and lengths is also a disadvantage. The use of elements in the form of beams or pillars, with large cross-sections (usually over 20 cm) or great lenght (over 5-6 m) often leads to high prices. This disadvantage can be eliminated by using composite elements or elements made of glued battens.

Natural defects in wood (defects in shape and structure, cracks etc.), defects caused by fungi, insects or chemicals, as well as the effects of decontraction and swelling phenomena are important drawbacks in choosing wood as construction material. It is vital to apply a proper treatament against natural attacks on wood when deciding to use it as construction material for a house or a cottage.

3.2. Description of the construction

3.2.1. Functional description. The aim of this paper is to present a case-study on how to make a wooden holiday cottage. At a volumetric level, the concept consists of using the land surface as efficiently as possible. From a functional point of view, the ground floor is the "living space" where the dining room, the kitchen and two service bathrooms are located, and, in the attic, there is the "sleeping space", where there are bedrooms with bathrooms. Having in mind that the cottage could serve as a guest house, the intention was to provide a high level of comfort, so every bedroom has its own bathroom, various in size.

The designed construction has ground floor and attic, with a built area of: Ba = 89.44 m^2 and a usable area of: Ua = 159.1 m^2, of which the ground floor has 66.17 m^2, the attic, designed as first floor, has 69.0 m^2, the terrace has 7.0 m^2, and the balcony has 6.93 m^2. The wood cottage is built of calibrated roundwood with a diameter of 20 cm, and for the construction 75.0 m^3 of finished product are necessary - calibrated wood, equivalent of 94.0 m^3 of raw product - raw wood.

Figure 7. Calibrated log (roundwood) cottage.

The ground floor consists of a living room, a kitchen, two service bathrooms, a lobby, a pantry, the staircase and a covered terrace, as shown below.

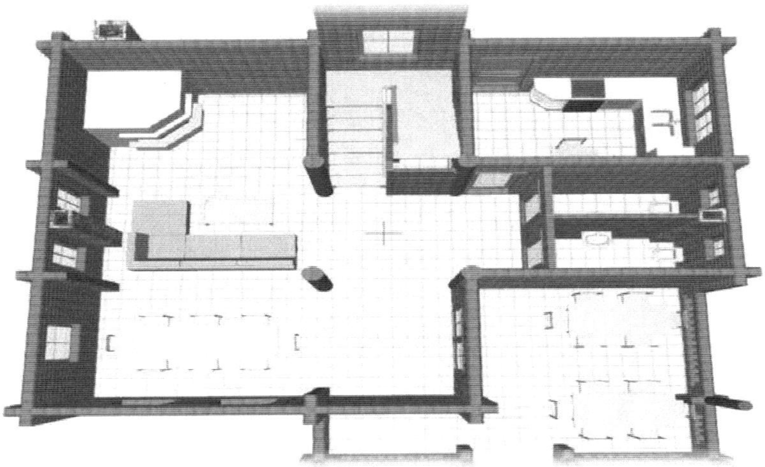

Figure 8. Ground floor of the wooden cottage.

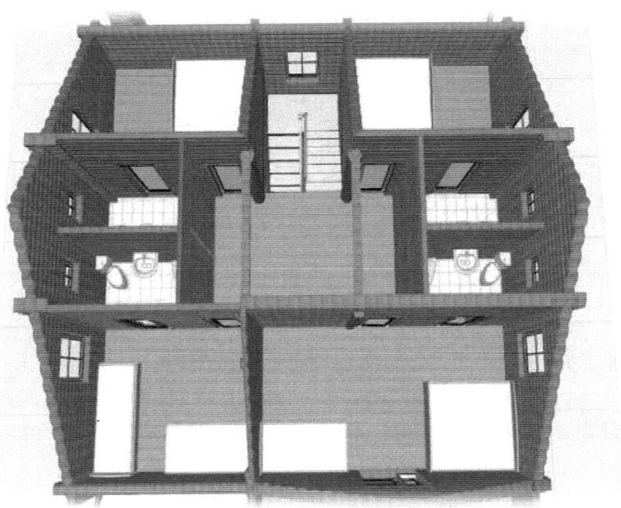

Figure 9. Attic of the wooden cottage.

The access in the building is made through the terrace, entering the living room. The first floor, designed as a completely used attic, has four bedrooms, each having a small bathroom, a lobby and the staircase and a covered balcony, as shown in figure 9.

3.2.2. Description of the construction solution. The foundation solution designed for this log cottage consists of rigid continuous foundations with plain concrete soles, class C8/10, with monolithic reinforced concrete elevations and C16/20 reinforced concrete soles and polypropylene fibers. For a better behavior over time, especially in the outdoor terrace area and patio access steps, the concrete slab was additionally reinforced with polypropylene fibers dispersed throughout the whole volume of concrete [6 - 8].

The resistence structure is made up of calibrated roundwood walls, with a diameter of 20 cm. The floor plan over the ground floor is made of square wooden beams that discharge onto the resistence walls. The roof is designed to have a wooden structure with metallic tile cover. The wood used to make the structure of the roof is fir of C24m class and the tiles have been chosen to have a proper colour to fit the natural environment.

The inner compartments are made of calibrated roundwood walls, having a diameter of 20 cm and 10x10 cm trimmed wood. For the bathrooms, specific waterproofing is planned to be used, then ceramic tiles will be attached on the walls and floors. The exterior walls will be treated with special solutions, so as to avoid insects and fungi attacks. On the ground floor, the ceiling will be left in apparent solution. In the attic, there will be a false ceiling made of larch wainscoting. Due to the natural aspect that is intented to be kept in the building and on the outside area surrounding the cottage, natural material and colours were planned to be used, so as to give a pleasant and comfortable view, a cosy atmosphere and a good rapport between price and quality.

The inner and outer carpentry will be made of solid wood, dark brown colour, for a perfect integration with the roundwooded walls. For the external finishes, socket and exterior walls, good quality treatament solutions and a natural laquer will be used to keep the natural color of the roundwood walls. The roof will also be in the same color with the carpentry, a dark brown colour that will give elegance to the entire building but, at the same time, it will make it fit perfectly in the natural environment. Due to the fact that the attic will be fully used for sleeping spaces, the roof system will ensure a good insulation, both against cold and humidity. It will consist of metallic tiles cover, slats, waterproof foil, rafters of 10x15cm, between which fireproof mineral wool of 15 cm width is placed. On the roof, the chimney will also be colorly integrated in the whole assembly of the building.

Concerning the outdoor landscaping, walkways were designed around the cottage, made of prefabricated slabs, which will be placed on a sand bed with 5% slopes. A covered terrace having the floor made of natural stone and green spaces that will be a pleasant scenery for the viewer are also intended to be part of the entire ensemble of the cottage.

4. Conclusions

The estimated average price of such a log cottage in Romania, including the foundation as described above and finishings, could be about 560 Euro/m^2, considering local building and human resources. If we take into account a neccessity of 54.0 m^3 of concrete for the foundation, at the price of 100 euros/m^3, plus 90.0 m^2 of calibrated spruce roundwood, for which the price varies between 40 and 130 euros/m^3, depending on the age class and quality, the tiles and insulation for the roof surface and the interior and exterior facilities, the price will be more than 50,000 euros. It it important to know that the prices mentioned above are the final prices, for the finite materials, including harvesting, calibration, transport, and any other additional costs that may be involved between the harvesting place and the location of the construction. The same is valid for the concrete price, which includes all the costs of production and transport to the final location, and, also, for all the materials used during the building process of the roundwood cottage.

As a conclusion, if one asks whether wood can be considered a sustainable construction material, the answer is certainly YES! According to IFN data, the volume of wood material harvested annually in

Romania is slightly increasing, the measures to keep the balance between the age classes of the harvesting forests is very properly guarded by the responsibles of the forest administrations. More and more people choose wood for their cottages and even for their houses, both as raw material or prefabricated components. In the near future the green buildings and the eco buildings could be considered to have a strong wood basis, hoping to reach energetical independence too, through their facilities [9]. So, wood is a sustainable building material and there is no danger of loosing our forests if we use it.

References

[1] Andreica H A and Beridean A D 2007 *Structuri din lemn* (Cluj –Napoca) Editura U.T Press
[2] http://www.mmediu.ro/categorie/starea-padurilor/209
[3] www.silvanews.ro/silvicultura/amenajari/clasele-de-varsta/
[4] http://www.mmediu.ro/categorie/inventarul-forestier-national/185
[5] http://www.arhiconoradea.ro/info%20studenti/note%20de%20curs/fekete%20luminita /curs%20lemn.pdf
[6] Mustea A, Manea D L, Orban Y A, Jumate E, Muntean R 2017 Obtaining physical properties of a composite plaster mortar through a CAE analysis *Bulletin of the Transilvania University of Brasov* **10** (59) Special Issue No. 1 p. 135-140
[7] Muntean R and Muntean G 2015 The concordance between theoretical and experimental model for formworks made of dispersely reinforced concrete with polypropylene fibers *International Scientific Conference CIBv (Braşov)* p. 67-74
[8] Muntean R and Muntean G 2008 Proprietatile betoanelor armate cu fibre din polipropilena (Properties of reinforced concrete with polypropylene fibers) *Proc. Int. Scientific Conf.* - CIB p. 235-240
[9] Rinkesh Kukreja 2016 What is a green building? *Green & Sustenable Business*

Virtual Reality for Real Estate – a case study

B A Deaky[1*] and A L Parv[1]

[1] Department of Manufacturing Engineering, Faculty of Technological Engineering and Industrial Management, Transilvania University of Brasov, Romania

[*] E-mail: bogdan-alexandru.deaky@unitbv.ro

Abstract. This paper presents a particular VR implementation done for a real estate developer. This implementation focuses on increasing immersion and is most suitable for properties that are to be built. It is the latest development of the VR4RE (Virtual Reality for Real Estate) project, which aims at saving time and money for both real estate sellers and buyers by employing modern technologies. VR4RE is one of the innovative projects developed by Bluemind Software. This paper also summarizes the history of in-house technological attempts at creating appropriate presentation tools.

1. Introduction

Sutherland presented the key concepts of immersion in a virtual world in "The Ultimate Display" [1] in 1965. The paper also introduced the challenge to make that virtual world look, act, sound and feel real.

Brooks defines a VR experience as an experience in which a person is effectively immersed in a virtual world. The virtual world is responsive by means of a control of the viewpoint [2]. Immersive virtual reality is defined by Andries van Dam et al. as technology that gives the user the psychophysical experience of being surrounded by a virtual environment. This experience is created with a combination of hardware, software, and interaction devices to establish a human rather centric point of view, as opposed to a computer-determined point of view [3].

The authors do benefit from existing research done in our university regarding haptics [4] and from several years of experience in developing VR applications of Bluemind Software. VR technology is currently entering the Real Estate market (it is one of its many application areas) worldwide through various companies which develop applications. As [5] also concluded, with the continuous reduction of costs it does gain a bigger and bigger role to play in the real estate market.

VR4RE (Virtual Reality for Real Estate) is one of the innovative ideas developed at Bluemind Software and it is one of the research directions in which the authors are involved.

Out of the other innovative projects of Bluemind Software (of which a few were created in collaboration with universities), we mention 3DCar (the first interactive 3D car configurator), ASIC Client (the first Android client application to access MIT's iLabs Remote Laboratory Management System – with Charintia University of Applied Sciences and MIT) and Lab Configurator (a simple and secure method to use mobile devices to configure embedded devices which run online laboratories – with UPORTO). The recent Museum MR project (created in collaboration with the Casa Muresenilor museum of Brasov and Trasilvania University of Brasov) combines AR and VR to create a mixed reality solution which helps museums add a wow facture to their exhibitions and exhibit more in the existing physical space.

VR4RE is focused on helping real-estate owners by teleporting visitors from anywhere (in VR) directly to the property that is for sale or rent and helps visitors save time and money by visiting a broad range of properties from only the most compelling ones in person.

This paper presents a brief presentation of implementations done so far and focuses on particular VR implementation done for a real estate developer. The focus of this implementation is the better immersion of the visitor by using recent hardware.

2. The history of in-house VR attempts

The starting idea for VR4RE emerged during discussions with a real estate developer in 2008. After having seen 3DCar, the real estate developer asked if the same technology could be used to present real estate. After working on a simple prototype, the 2008 real estate crisis, along with the fact that the tools to develop solutions were much more expensive, interfered with creating a finished solution. The first good looking prototype came out during a Startup Weekend event in Brasov, Romania, in 2013.

It was an apartment visualization application used of real time 3D model rendering for PC. The application included interactive doors. The first finished application was developed to run in a web-page (in Chrome and, with the help of add-ons, also in other browsers).

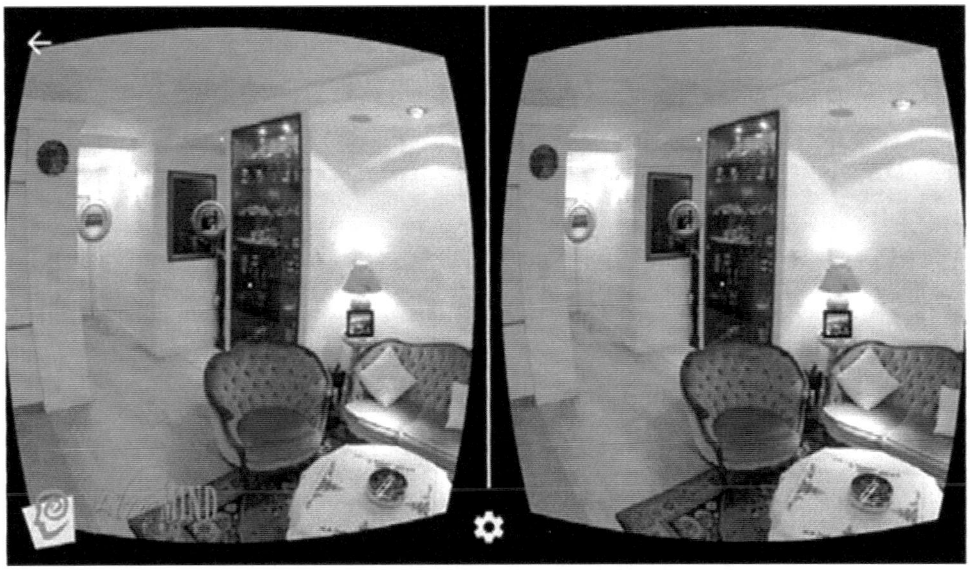

Figure 1. Online VR apartment tour demo in the browser using 360 degrees images (on smartphone).

TimeWalk [6] will be presented in more detail in a subsequent paper. It is important to know that, among others, TimeWalk involves creating a virtual copy of the world to enable virtual tourism.

The apartment prototype (at that time called 3DGate) was included in TimeWalk's proof of concept during the next couple of years, thus enabling VR with Oculus Rift for the first time. The inclusion also facilitated the first window view on the surroundings of the apartment. In this case, the apartment was placed in one of the skyscrapers of Times Square, New York.

In 2015, we started experimenting with creating mobile apps for smartphones used together with Google Cardboard devices. The results were good when using a better Google Cardboard, such as the Dive's one.

We started experimenting with these simpler devices because our existing HMDs (head-mounted displays) were great but just too expensive for possible users. Involving Cardboard devices (including Gear VR users) helps us reach many more people because of its lower hardware cost.

Also in 2015, we integrated a new apartment and made it accessible for smartphones with Google Cardboard. For the first time, we added exterior views using 360-degree photos, in a successful attempt to replicate the results achieved during the integration in Timewalk.

In early 2016, we developed our first project for a real estate developer. It included a whole level of the in-construction residence in Brasov, Romania. The project was conducted with the assistance of a real-estate-agency.

The result consists of an Android application for Google Cardboard and a Windows application, which can work both with and without Oculus Rift.

Because of the bigger and more complex scene and the limits of mobile devices, it required the development of better optimization capabilities.

The feedback that we got with the assistance of our partners showed that our 3D VR technology is suitable for presenting in-construction and expensive real estate.

We have also learned that, for wide adoption, less expensive solutions are needed. We realized that creating VR tours using 360-degree photos makes them accessible to common apartment owners and the technology suits existing real estate property.

In 2017, we created an in-house 360 degrees VR technology to provide the cheaper alternative:

- it can be deployed online as a webpage to make it easily accessible;
- it can be switched to VR mode and it works with smartphones together with Google Cardboard, Samsung Gear and other similar devices. Additionally, it can be used even without any VR headsets or goggles, directly on the PC or smartphone.

Figure 1 presents an implementation done using JavaScript VR solutions.

We continued to improve this technology but also continued to focus on the more expensive but also more immersive solutions. The most recent implementation is presented in the next chapter.

3. Case study

The most recent implementation consists of creating a VR application, in which visitors can visit a one-bedroom apartment. Its layout can be seen in figure 2. It consists of a hallway from which a living room, a bedroom, a kitchen and a bathroom can be entered. Both the living room and the bedroom have balconies.

Figure 2. Apartment floorplan (top view of the scene).

The apartment had to be an accurate replica of a real one found in an in-construction block of flats owned by a well-known real estate developer located in Brasov, Romania.

For this to happen, the first step was to find out the date related to the apartment (footprint, sizes, layout etc.). It was decided that there was need to involve the architect who designs the apartments for the real estate developer. This was done with the hope that we can get not only accurate data, but also a clear floorplan and maybe even 3D models.

The ideal envisioned workflow was:

- get 3D models from the architect;
- import 3D models into the VR application development environment;
- set up lighting;
- integrate VR;
- build test versions and test/fix;
- build the final application.

This workflow was followed and we managed to get 3D models from the architect. ArhiCAD was used to design the apartments and, with some efforts, we managed to obtain the models in a format which can be used to import them in the application development environment.

A scene was setup in the development environment and the 3D models were imported. Lighting was set up. However, during these steps, the first problems were noticed: the 3D models contained too many polygons to be able to use them for real-time renderings (in VR, we need to be able to render the model many times per second not only from one but from two simultaneous viewpoints corresponding to the two eyes of the viewer). Many visible mesh-related errors were also found.

A long period of trial and error followed while we experimented with several ways of decimating and recreating the 3D models.

Expectations of the real estate developer and of the architect related to keeping the ArchiCAD models used during design were a supplementary hindrance.

In the end, in order to be able to deliver a working result with a good quality, the application development workflow included an additional and time-consuming step, where a part of the 3D models was recreated from scratch using 3D modelling software and the other part of the 3D models was kept, but only after undergoing decimation procedures executed with various software tools. The testing and fixing step also took considerably longer than anticipated.

HTC Vive (SteamVR) was used for the VR implementation as it is the most immersive recent consumer-level VR hardware available). It creates the best immersion by allowing the user to move and walk freely in a pre-setup space. The two controllers allow to track and represent movements of the hands as well.

One of the controllers was programmed to be used so as to enable the visitor of the apartment to teleport himself in the apartment when the real-world available space is smaller than the size of the apartment (which is usually the case) (see figure 3). The other controller is programmed to offer a menu with various options in VR.

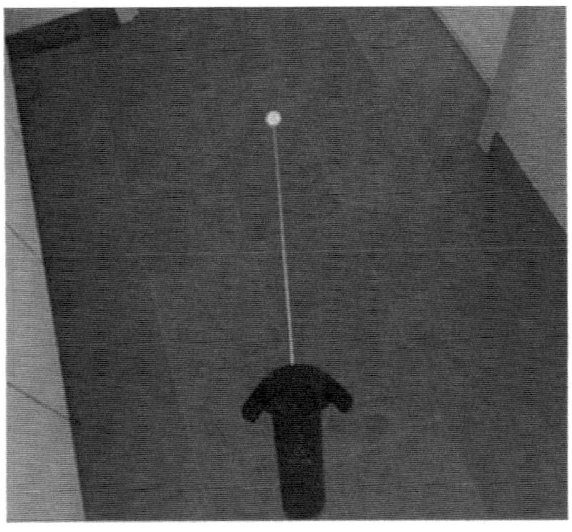

Figure 3. Controller used to teleport (VR view).

The following images display various views in the apartment. Figure 4 shows the living room, while the user is initiating the teleport to the hallway.

Figure 4. Living room.

Figure 5. Bedroom.

Figure 5 displays the bedroom. It can be noticed that the exterior views using 360-degree photos mentioned in [7] have been used also in this project.

Figure 6 shows a view from the balcony of the bedroom.

Figure 6. Bedroom balcony.

4. Conclusions and further research

This paper presented a summary of the evolution of the VR4RE research direction and its latest developments. For more details and images regarding the history of real estate VR application development at Bluemind Software please see [7].

This paper shows that efforts are being and will keep being directed towards both of the research directions and market segments identified in [7]. It focused on 3D VR used to achieve a much more complete immersion with haptics (the Vive controllers do feature very basic haptic feedback) and positional tracking. The solution is ideal for properties to be built.

Acknowledgment
The authors hereby thank Bluemind Software for the permission to publish this information.

References

[1] Sutherland I E 1965 The ultimate display *Proceedings of IFIPS Congress* (New York City, NY, May 1965), 2, 506
[2] Van Dam A 2002 Immersive VR for Scientific Visualization: A Progress Report *IEEE Computer Graphics and Applications* **20** 26
[3] Brooks F P 1999 What's Real About Virtual Reality *IEEE Computer Graphics and Applications* **19** 16
[4] Butnaru T 2011 An approach for teaching mechanisms using haptic systems *The 6th International Conference on Virtual Learning ICVL* 30
[5] Wang K and Hu Z 2009 Application Study of Virtual Reality in Real Estate Industry, *Proceedings of the 3rd WSEAS International Conference on Computer Engineering and Applications (CEA'09)* 247
[6] TimeWalk Inc., TimeWalk, http://timewalk.tw/
[7] Deaky B A and Parv L 2017 Virtual Reality for Real Estate. Its evolution in Bluemind Software *Book Series IEEE-Experiment at International Conference, Proceedings of 2017 4th Experiment@International Conference (EXP.AT'17)* 83

A review of STEP-NC compliant CNC systems and possibilities of closed loop manufacturing

M Dharmawardhana[1*], G Oancea[2] and A Ratnaweera[1]

[1] Faculty of Engineering, University of Peradeniya, Sri Lanka.
[2] Department of Manufacturing Engineering, Transilvania University of Brasov, Romania.

[*] E-mail: mahanamad@gmail.com

Abstract. A new programming interface, ISO14649, named STEP-NC, has been introduced recently to overcome the drawbacks of the conventional CNC system. Enhanced machining flexibility, interoperability, and adaptability are the key features of this new programming strategy. STEP-NC technology is based on object-oriented concepts and considers machining processes rather than tool motions as in ISO 6983 for machine tool operation. As such, high level machining features and machining parameters are transferred to the CNC controller, which are eventually used for generating axis movements and other operations. The bi-directional data communication facility of this new strategy enables last minute shop floor level modifications of machining operations at the controller level and establishes a communication pathway to feed major modifications, if required, back to the CAD level as well. Accordingly, the new programming interface allows seamless integration in the CAD/CAM/CNC chain and paves the way for Closed Loop Manufacturing facilities. Most CNC controller vendors have not released STEP-NC compliant controllers yet. Therefore, researchers are developing Open Architecture Control systems to execute STEP-NC and operate CNC machines. This paper reviews research and development of STEP-NC controllers in the last decade and the capabilities of Closed Loop Manufacturing with STEP-NC based systems.

1. Introduction

The manufacturing industry plays a vital role in the global economy. Product manufacturing for diverse customer requirements is one of the challenges in manufacturing today. Computerised Numerical Control (CNC) machines have played an important role in manufacturing precise, accurate, and high quality products from the inception of technology. On the other hand, CNC machines are important to get high productivity rates, uniformity of product, reduced rejection, reduced tooling costs, and less operator involvement [1]. G and M codes have been used to command CNC machines since the introduction of CNC machines in 1950's. In 1960's, the Electronic Industry Association (EIA) developed a standard named RS274D to program CNC machines. Later in 1982, this was adopted by ISO and a new standard was introduced, called ISO6983. In this method, preparatory functions and miscellaneous commands, defined in ISO 6983, were used to move the cutting tool with respect to the axes. In the last 60 years, the same programming format has been used to program CNC machines, remaining unchanged. This low level programming method describes only the elementary actions and tool movements. The CAD-CAM-CNC numerical chain broke with this format and

IOP Conf. Series: Materials Science and Engineering **399** (2018) 012014 doi:10.1088/1757-899X/399/1/012014

prevented the feedback from the shop floor [2]. On the other hand, sustainable manufacturing with environmental conscious and energy efficient manufacturing systems have become the modern trend in manufacturing. As a result of the rapid advancement of computer based technologies, feature based CAD/CAPP/CAM systems with highly advanced capabilities and powerful CNC controllers were introduced in the manufacturing industry. However, a weak interconnection between CAD/CAPP/CAM systems and CNC controllers, and hence the lack of provisions for closed-loop control of machining processes, have remained a major deficiency in ISO 6983 standard. Different CNC controller vendors have added new supplement commands to their own controller to improve some deficient features of ISO6983. This created further incompatibility in the part programs generated for different CNC machines [3].

Because of these problems, the requirement to develop a standard to describe the product data throughout the life cycle arose, which is independent from the controller and the computer system. In order to address the above requirements, a new standard called Standard for the Exchange of Product Model Data (STEP), ISO 10303, was introduced to exchange CAD data between CAx systems. Later, STEP compliant data interface was developed to exchange the product data between CAD/CAM systems and CNC controllers. This standard is known as STEP Data Model for Computerized Numerical Controllers (STEP-NC) and defined under ISO14649 [1]. As opposed to ISO6983, STEP-NC provides feature based programming with an object oriented data model, information on the features to be machined, tool types, operations to be performed, and sequence of operations [3].

Since existing commercial controllers could not understand STEP-NC codes, interpreters were developed during the early stages to convert STEP-NC code to ISO 6983 format, which could be uploaded directly to CNC machines. However, this type of controllers did not support the full capabilities of STEP-NC format. As such, interpreted type CNC controllers were introduced as the next step, where STEP-NC files can directly be uploaded to a controller, and the tool path and axis commands are generated within the controller. Recently, STEP-NC based intelligent controllers were developed to perform more optimized NC operations with self-adaptive capabilities.

2. ISO 14649, STEP-NC architecture

The application of STEP methods to NC machines is defined as STEP-NC [4]. STEP-NC, ISO 14649 is a set of standards, which are defined under the subtitle of "Data model for computerized numerical controllers". It has a number of parts as listed below [4]:

ISO 14649 - Part 1: Overview and fundamental principles;
ISO 14649 - Part 10: General process data;
ISO 14649 - Part 11: Process data for milling;
ISO 14649 - Part 12: Process data for turning;
ISO 14649 - Part 111: Tools for milling;
ISO 14649 - Part 121: Tools for turning [4, 5].

2.1. STEP-NC program organization

The structure of STEP-NC data is mainly divided into two sections. The first section of the part program is the header section marked by the keyword "HEADER". In this section, some general information and comments such as author, date, organization etc. are included. The second main section is the data section and it starts with the keyword "DATA". The data section is divided into three main parts, namely Workplan and executables, Manufacturing features, and Geometry description [6].

The starting point for executing a part program is the PROJECT entity in the DATA section. The PROJECT describes the main *"Workplan"* which contains executables (manufacturing tasks or commands) and details of Workpiece to be machined. There can be three types of executables, namely *"Workingstep"*, Program structure, and NC function. *"Workingstep"* details a single manufacturing operation using one cutting tool. The next executable is program structure, and it can be either a *"Workplan"* or an execution flow statement. The last type of executable is NC function, and it may include setting of the workpiece coordinate system, security plane, and other auxiliary commands [6]. Figure 1 shows the overall structure of the STEP-NC data model.

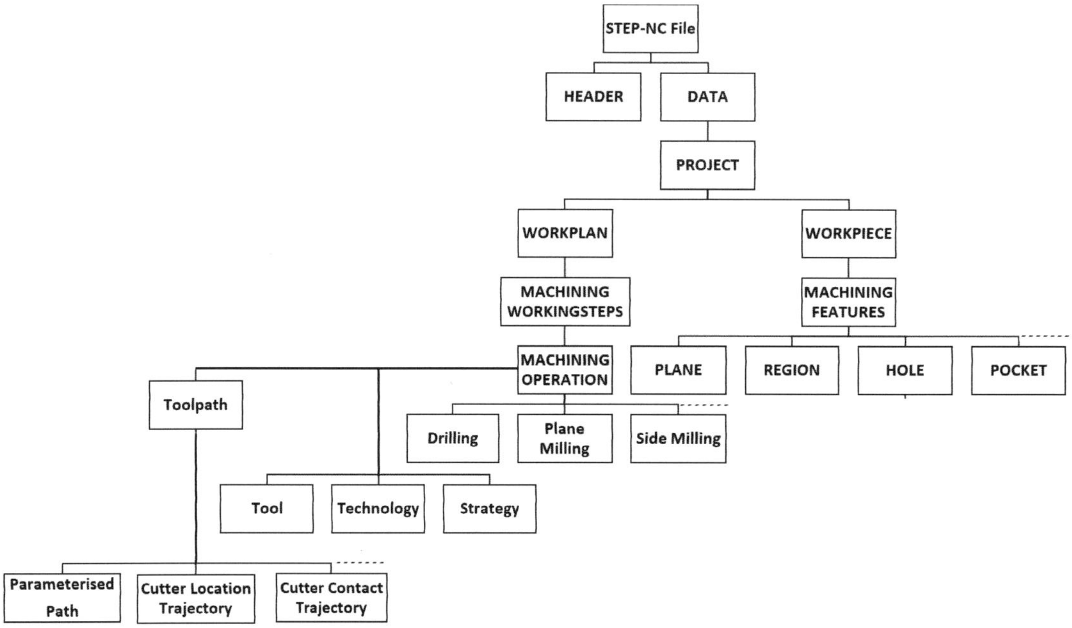

Figure 1. Structure of the STEP-NC data model [7].

3. STEP-NC based controllers

According to the research carried out in the last decade or so, STEP-NC controllers can be classified into four categories, according to their functions [2, 8].

3.1. In-direct STEP-NC controller

This category uses the existing legacy of CNC controllers and G/M codes to program toolpaths. An interpreter is added to the front end of the system to translate STEP-NC code to ISO6984 format. As such, this is called In-direct STEP-NC controller and can easily be implemented in the existing CNC machine tools [2]. Several prototypes, interpreters, and front end applications were developed to be used with existing CNCs [9-11]. However, the innovative features available in STEP-NC format cannot be effectively implemented with this type of controller, and hence the major benefits cannot be achieved [2].

3.2. Interpreted STEP-NC controller

The second category of controllers, Interpreted STEP-NC controller, does not require G codes anymore. The STEP-NC file can be uploaded directly to the machine controller for processing. File interpretation, feature based data extraction, toolpath generation, and axis drive command generation are totally integrated in the machine controller [1].

Most of the early developments in STEP-NC controllers, including the Interpreted type controllers made in different countries and different research groups, have been reviewed by Xu et al. [12].

Because of the closed nature of the commercial CNC machines, the user cannot upgrade existing CNC machines with new hardware and software from different manufacturers and vendors. Most manufacturers are of the view that conventional CNCs are powerful enough to handle machining tasks. Therefore, CNC and CAM vendors are not keen on promoting STEP-NC for future development

[13]. The Open Architecture Control (OAC) technology provided a solution for this problem and most researchers tend to use OAC for their implementation [14].

3.3. Intelligent STEP-NC controller

CNC controllers of this type, identified in the third category, are equipped with the features in the second category, but are also capable of performing NC tasks intelligently and autonomously. Accordingly, real time optimization of machining parameters and toolpaths can be performed with the help of online process data. Automatic collision free toolpath generation, automatic cutting condition selections, status monitoring and recovery, and machining status and feedback results may be incorporated as intelligent functions of the controller [15].

3.4. Collaborative Intelligent STEP-NC controller

The fourth category is STEP-NC enabled machining, which supports web-based, distributed, and collaborative manufacturing with all the facilities available in the second and third categories. This is the ultimate goal of the STEP-NC compliant CNC controllers [3]. A web based platform provides an information protocol for manufacturing operations which take place around the globe via the internet. The final requirement of the various stakeholders of CAD, CAPP, CAM, and CNC is to achieve a manufacturing solution with the seamless DA-BA-SA (Design-Anywhere, Build-Anywhere, Support-Anywhere) facility [16]. The fourth category of STEP-NC compliant controllers fulfils this requirement.

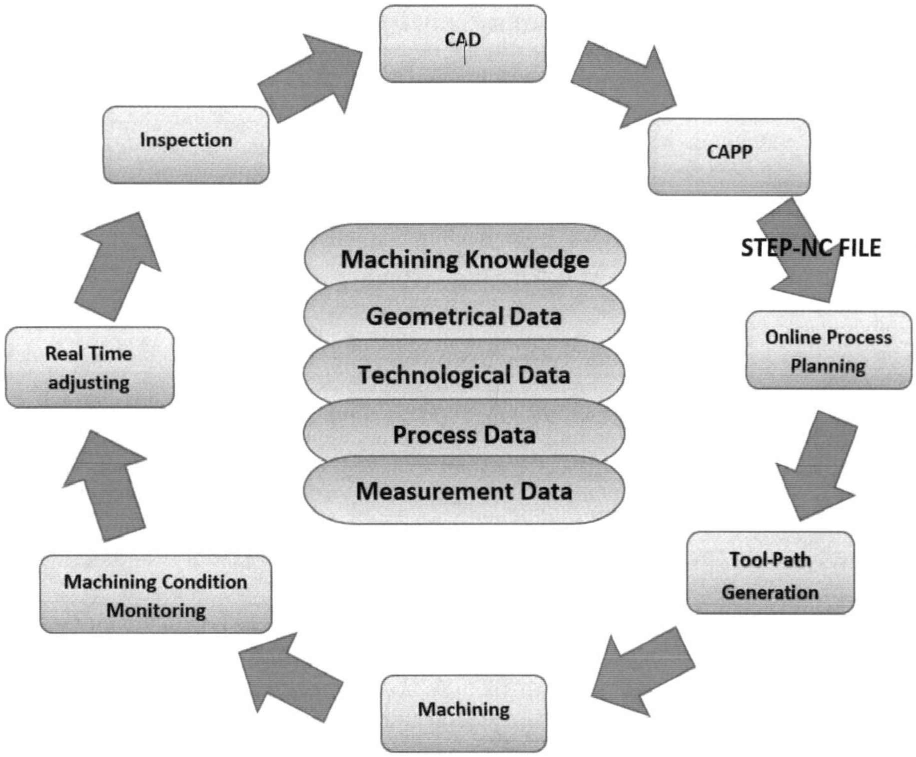

Figure 2. Closed-loop and self-learning STEP-NC machining System [16].

4. Development of STEP-NC compliant controllers with closed loop capability

The Research Division of Numeric Control Technology in the Harbin Institute of Technology proposed an indirect type open CNC controller named HITCNC [17]. This was further developed into an adaptive controller by Hu et al. by effectively utilizing the high level information in the STEP-NC file, the machining process condition data, and the inspection results [18]. Online and real time machining process control was developed on a 5 axis CNC machine. The STEP-NC interpreter adjusts the technological parameters according to the machining knowledge, the machine tool status, and the cutting tool condition. The feedback sensors provide the necessary data on the machining process conditions to estimate the optimal technological parameters using an adaptive control algorithm. After completing a "*Machining Workingstep*", it can be verified by online inspection for the required tolerances [18, 19].

Manufacturing knowledge obtained through the feedback from the machining to the product design and production engineering phases was demonstrated by Danjou et al. [20]. In this method, machining data is extracted from a CNC machine and validated in order to effectively use the results. The extracted knowledge is stored in a Manufacturing Process Management (MPM) system to provide a guideline to generate similar programs in the future. The CATIA environment is used for CAD/CAM and the NC Simul Machine simulator is used instead of a real CNC machine to simulate toolpath. The proposal is to use a parser to translate the generated G code program to STEP-NC [20-22].

This concept was developed by Wosnik et al. [23] using the drive signals feedback to the CAPP system for process monitoring and optimization. Algorithms have been developed to process and exchange drive signal data in the application of open digital servo drives in machine tools. Wosnik et al. proposed a method to compare previous force profiles in similar machining conditions with the feedback data and update process force calculation models and machining parameters [23].

A STEP-NC enabled machine condition monitoring system was developed by Ridwan and Xu [24]. It consists of three sub-systems. The first sub-system is to generate optimum machining parameters for time critical machining operations and quality critical machining operations. A simulator has been developed to verify the optimization algorithms, the real-time process control, and the monitoring algorithm. The second sub-system is the adaptive execution of the Canonical Machining Command program with fuzzy feed rate optimization. This will help to keep a constant load within the machine tool's capability and reduce chatter amplitude. The last sub-system is the knowledge based evaluation system which utilises accurate, informative, and updated machining know-how to achieve automated and intelligent machining operations [24].

The STEP-NC compliant intelligent CNC controller was developed with the RCM3700 embedded microcontroller to drive a 2D plasma cutter. The program can be directly uploaded to the controller since there is a STEP-NC interpreter. Next, the interpreted data is fed to the High level controller which evaluates its feasibility of machining and decides the machining operation to execute. A machining inspection facility and web based monitoring and execution facility has been developed by the researchers [25, 26].

5. Development of STEP-NC compliant controllers CLM with inspection

Inspection is an important element in CLM. It can be used to get precise measurements and monitor the performance of the machine tool. One of the parts in the Data model of ISO14649, Part 16, describes touch probing operations on a CNC machine tool or on a CMM. The interface between the programming system and the CNC controller is defined for inspection with this data model. Zhao et al. proposed a framework of STEP-NC enabled, real time CLM system, which uses on-machine touch probes. The machining and inspection information is saved in a STEP-NC file by using CAPP software according to the design requirements. Apart from the "*MachiningWorkingsteps*" the file contains "*InspectionWorkingsteps*", which should be carried out before machining, in between "*Machining Workingsteps*" and after machining. The inspection results are analyzed, compared with design requirements feedback to the STEP-NC data model if necessary, to generate or modify machining operations. Zhao et al. have reported a case study according to Example 1 given in ISO14649-11 [27] by adding some inspection points and generated an intermediate file to demonstrate the correctness of the data model [28].

In manufacturing, human workers' problem solving capabilities, dexterity, and cognitive capabilities are very important. These will help to improve flexibility, adaptability, and reliability of the system. Zhao et al. proposed a consolidated data model and a system structure for process planning, inspection and feedback with cognitive manufacturing capabilities. Pre-processed STEP part 21 file, with machining features and tolerance, is the input for the system. The developed algorithm traces the tolerance related features together with geometrical data and decides what geometry is to be measured and the tolerance limit. The machining working steps are rearranged according to the chosen machine accuracy and the tightness of the tolerance, and newly generated "*InspectionWorkingsteps*" are inserted. The machining and inspection commands generated according to the new process plan are passed to the CNC machine for execution. After each inspection operation, the measured data is analyzed and updated online for subsequent machining operations [29].

A similar study was conducted by Ali et al., who have developed a STEP-Compliant framework for the generation of inspection files for discrete components [30].

Breacher et al [30] demonstrated a closed loop process chain, which includes the generation and execution of the STEP-NC program and feedback. This paper describes Part 16 of ISO14649, Data for touch probing based inspection, which uses integration of the inspection tasks. The STEP-NC data file is converted to a suitable format to extract geometrical and inspection data. The data is reorganized and exported into a STEP AP203 format while sorting according to the inspection criteria. The final inspection data is fed to the CMM and the measured information is fed back to complete the process chain and further modifications if required [29].

Traceability is one of the many important aspects in the manufacturing process. By recording manufacturing information, such as machines, tools, raw materials, employees, time, and storage conditions, etc. it is possible to trace back the origin of product defect in the process [32, 33].

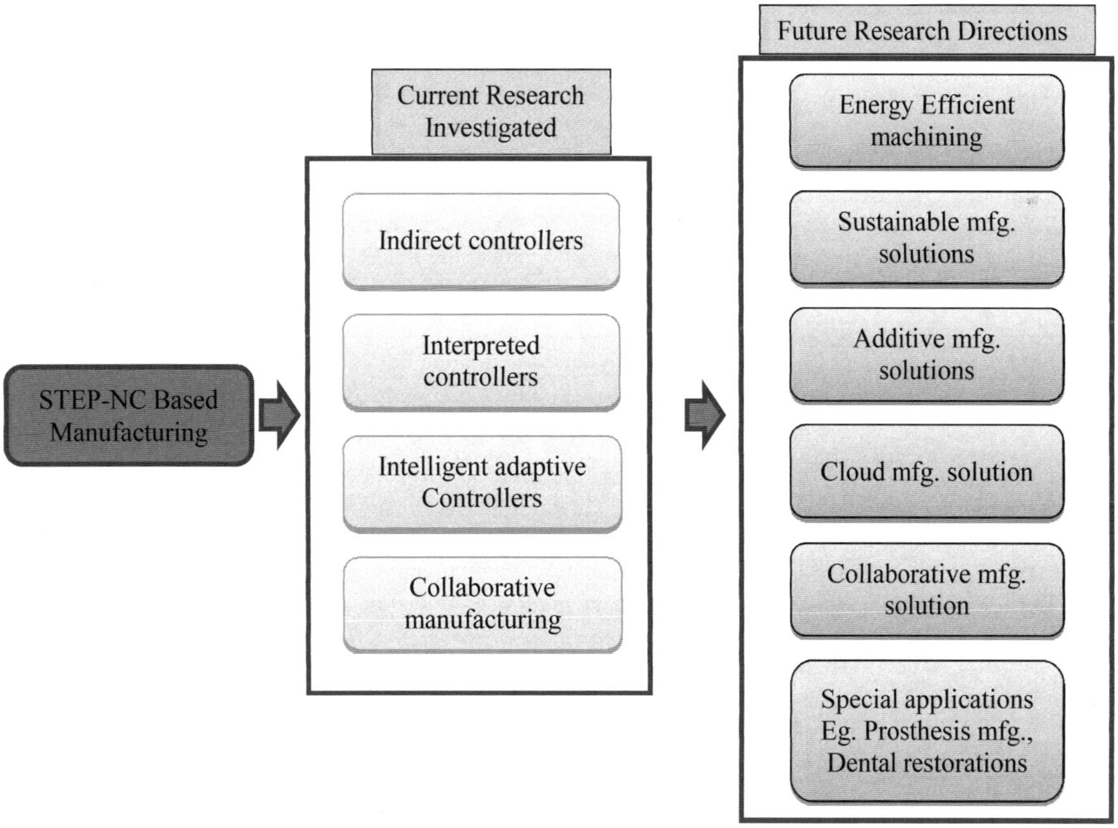

Figure 3. Current and future research directions.

Although there are few traceability functions available in the STEP-NC standard, it does not define any monitoring or traceability capability. Campos and Miguez [34] proposed a new set of traceability commands which could be incorporated into the STEP-NC code. The new traceability functions are time, tool position, sensor data, process start/stop, position deviation, operator, and tool data, controller events, and alarms. NC-Explorer and STEP-NC software developed by STEPTools, Inc. is used as core tools to process STEP-NC files and programming the interfaces [34].

6. Future research directions

Nowadays, long-term sustainable strategies rather than traditional short-term financial considerations are implemented in the manufacturing industry, to ensure competitiveness [35]. Therefore, different paradigms such as flexible manufacturing, lean manufacturing, sustainable manufacturing, and cloud manufacturing solutions have been introduced into the manufacturing environment. Energy consumption in the machining domain is one of the most interested in building a sustainable machining future [36]. The STEP-NC machining technologies have introduced a good platform to enable energy efficient manufacturing environment in CNC machining. This will be one of the main research areas to be addressed in machining. Currently, there is a great amount of research published on energy efficient machining with STEP-NC, but there are more topics to be researched to develop industrial level intelligent controller. In the future, energy efficient, sustainable, intelligent manufacturing solutions with the STEP-NC technology will enhance the different application areas, such as additive manufacturing, cloud manufacturing, and medical and dental applications etc. Figure 3 shows some future research possibilities in STEP-NC manufacturing.

7. Conclusion

The ultimate goal of STEP-NC enabled machining is to develop a vendor-neutral, tool-neutral, and controller-neutral architecture for intelligent CNC machining which supports Web-based, distributed, and collaborative manufacturing [37, 11]. This paper provides details on the development of the STEP-NC programming interface, architecture, and a literature review on STEP-NC based research carried out around the world. Special emphasis is placed on closed loop manufacturing systems and inspection data feedback systems. All this research can be categorised as the development of intelligent manufacturing systems for STEP-NC compliant machining and inspection systems. This will contribute to shaping up the future advanced, green, intelligent, autonomous, and collaborative CNC controller designs.

References

[1] Yusri Y, Kassim N D and Tan N Z Z 2011 *Int. J. Com. Inter. Manuf.* **24** 126
[2] Rauch M, Laguionie R, Hascoet J and Suh S 2012 *Robo. Com. Inter. Manuf.* **28** 375
[3] Cha J, Suh S, Hascoet J and Stroud I 2014 *J. Intell. Manuf.* **27** 959
[4] Hardwick M et al. 2013 *Int. J. Adv. Manuf. Technol.* **68** 1023
[5] Kramer t and Xu X 2009*Advanced* Design *and Manufacturing Based on STEP* ed Xu X, Nee A Y C (London: Springer) p 16
[6] ISO 14649–1 2003 *Industrial automation systems and integration - Physical device control - Data model for computerized numerical controllers Part 1: Overview and fundamental principles*
[7] Latif K, Yusof Y B and Bux Q 2016 *New Open CNC Cell Controller Based on STEP-NC and G Code*(Germany: Lambert Academic Publishing) p 29
[8] Xiao W, Zheng L, Huan J and Lei p 2015 *Robo. Com. Inter. Manuf.* **31** 1-10
[9] Gutierrez M E and Alvares A J 2013 *22nd Inter. Congress of Mech. Eng.* Nov 3-7, 2013 Brazil
[10] Newman S T, Allen R D and Rosso R S U 2003 *Int. J. Com. Inter. Manuf.* **16** 590
[11] Wang H, Xu X and Tedford J D 2007 *Int. J. Prod. Res.* **45** 3809
[12] Xu X W , Wang H, Mao J, Newman S T, Kramer T R, Proctor F M and Michaloski J L 2005 *Int.J. Prod. Res.***43** 3703

[13] Latif K, Yusof Y 2016 *Proc. CIRP* **40** 230
[14] Minhat M, Vyatkin V, Xua X, Wong S and Al-Bayaa Z 2008 *Robo. Com. Inter. Manuf.* **25** 560
[15] Othman M A, Minhat M and Jamaludin Z 2017 *IOP Conf. Series: Mat. Sci. and Eng.* **257** 012048
[16] Suh S, Kang S, Chung D and Stroud I Theory and Design of CNC Systems 2008 (London: Springer)
[17] T Liu, Wang Y and Fu H 2006 *Proce. 6th World Cong. Intell, Cont. Auto.*June 21-23, 2006 China
[18] Hu P, Han Z, Fu H and Han D 2016 *Int. J. Adv. Manuf. Technol.* **83** 1361
[19] Po H, Hongya F and Dedong H 2014 *Int. Conf. on Adv. Intell. Mechat. (AIM)* July 8-11, 2014 France
[20] Danjou C, Duigou J and Eynard B 2016 *Int. J. Com. Inter. Manuf.* **30** 995
[21] Danjou C, Duigou J and Eynard B 2015 *CIRP CMS* **41**852
[22] Dajou C, DUigou and Eynard B 2017 *Int. J. Interact. Des. Manuf.***11** 233
[23] Wosnik M, Kramer C, Selig A and Klemm P 2007 *Int. J. Com. Inter. Manuf.***19** 559
[24] Ridwan F and Xu X 2013 *Robo. Com. Inter. Manuf.* **25** 12
[25] Calabrese F and Celentano 2007 *Emerg. Tech. and Fac. Auto.*Sept. 25-28, 2007
[26] Hongya H P, Zhenyu H and Dedong H 2014 *Int. Conf. onAdv. Intell. Mechat. (AIM)* July 8-11, 2014 France
[27] ISO14649-112004 Industrial automation systems and integration - Physical device control - Data model for computerized numerical controllers Part 11: Process data for milling
[28] Zhao F, Xu X and Xie S 2008 *Robo. Com. Inter. Manuf.* **24** 200
[29] Zhao Y F and Xu X 2010 *Adv. Eng. Informatics* **24** 269
[30] Ali L, Newman S T and Petzing J N 2005 *Proc. Institu. Mech. Eng. Part B: Eng. Manuf.* **219** 557
[31] Brecher C, Vitr M and Wolf J 2007 *Int. J. Com. Inter. Manuf.***19** 570
[32] Rizzi A and Zamboni R 1999 *Logistics Info. Manag.* **12** 367
[33] Drop K V 2002 1999 *Logistics Info. Manag.* **15** 24
[34] Campos J G and Miguez L R 2011 *Copm. In Indus.* **62** 311
[35] Peng T and Xu X 2014 *Int. J. Adv. Manuf. Technol.* **72** 1389
[36] Peng T, Xu X and Wang L 2014 *Jour. Mauf. Sys.* **33** 196
[37] Xu X and Newman S T 2006 *Comp. in Ind.* **57** 141

IOP Conf. Series: Materials Science and Engineering **399** (2018) 012015 doi:10.1088/1757-899X/399/1/012015

Vertical wind turbine with self-limitation system of speed

M V Drăgoi[1*], D B Vrinceanu[1], V M Stamate[1] and N F Cofaru[2]

[1] Transilvania University of Brasov, Manufacturing Engineering Department, Str. Mihai Viteazul nr. 5, 500174, Brasov, Romania
[2] "Lucian Blaga" University Sibiu, Department of Management and Industrial Engineering, 10, Victoriei Bd, Sibiu, 550024, Romania

[*] E-mail: dragoi.m@unitbv.ro

Abstract. The proper work of vertical wind turbine Savonius typed may be negatively influenced by very high wind speed. Usually the wind turbine speed is limited by means of a brake. The turbine speed is limited, but the spindle of the turbine is still subject of increased torque. The paper proposes a new, innovative technical solution that does not counteract increasing the turbine speed above a limit value, but prevents it. The new concept is based on dividing the blade of the turbine into several parallel vertical strips, properly displayed, as to form the shape of a conventional blade. Each strip can rotate against its own vertical axis, to change its orientation in vertical plane. This brings the benefit that when the turbine speed tends to overcome a certain value (dangerous for the turbine) the strips orientation changes, and the surface of the blade becomes a non-continuous one. The area of the blade that faces to the wind reduces, and prevents increasing the turbine speed. When wind speed slows down, the strips are guided back to their initial position by means of some springs. Reorienting the blades is based on centrifugal force generated of turbine rotation, so, it can be concluded that this forms a self-limitation system of turbine speed.

1. Introduction

Wind turbines are green devices which produce electrical energy. They are quite various in design, but some types got a special development due to their advantages. One of them is Savonius typed wind turbine. Savonius turbines are vertical ones, and they are widely used for some specific features. The most important advantages of this type of turbines are, according to [1]: they are compact, can start by themselves even at low wind speed, have a low noise level, and are cheap and simple to assembly. Savonius turbines are suitable both for urban zones, and for insulated areas, like mountains, for instance.

Meanwhile, Savonius typed turbines have a drawback, their power coefficient is relatively low [1]. Both advantages and disadvantages make Savonius turbines an attractive research subject for scientists and engineers.

A survey of scientific literature reveals a large interest of scientist to study Savonius typed turbines. On ISI WoS can be found more than 200 papers dedicated to the subject. Because the relatively low power coefficient of these turbines, most of researchers focused on ways to increase it. Some samples of researches in the area are mentioned bellow.

In [2] a study related to the influence of number of blades on the performance of Savonius turbine is presented. The conclusion is that three is the optimum number of blades. In [3] a study on optimal

IOP Conf. Series: Materials Science and Engineering **399** (2018) 012015 doi:10.1088/1757-899X/399/1/012015

orientation of the blades in order to deflect the air flow to increase the power coefficient is presented. It was found that the performance of turbine can be increased in this way up to 30%.

A large category of research papers refer to numerical simulation and analytical studies on Savonius turbines [3-9]

Many articles approach the optimization of blades shape. In [1] is described how genetic algorithms are used to find an optimum shape of the blade. The result is that using the proposed optimized blades, an increase of 33% of power coefficient can be achieved. In [10] an inverse method to optimize the blade shape is presented. That is, the whole optimization process is carried on taking into account some targets to be obtained, i. e. a certain torque and power. The new found shape has been validated by experimental research, and the conclusion was the area of the blade can be reduced with up to 9%, compared to a classic shape, without affecting the power or torque of the turbine.

A small amount of academic papers study the ways to control the limit the turbine speed in strong wind [3, 11]. Usually, the solution proposed for high speed of wind is the usage of a brake to *stop* turbine from rotating.

2. The problem

Authors consider that using a brake to stop the turbine from rotating while wind becomes too strong is a non-effective approach, because of the following reasons:

- even if the turbine rotation stops, its spindle is still subject to big stress (torsion);
- an additional system is needed to detect when the wind speed slowed down enough, as the turbine may be restarted safely;
- this approach counteracts an unwanted event that already happened (over-speeding the turbine), instead of preventing it;
- stopping the turbine causes a loss of produced energy.

The present work proposes a new approach to the problem: prevent the over-speeding of turbine, instead of fighting against it.

3. The new concept of Savonius typed turbine

It is considered that it is much better to prevent an unwanted event as over-speeding the turbine, than fighting against it when it has already been produced or is going to occur. It is known that the speed of the turbine, depends directly (among other factors) on the area of the blades. Based on these statements a new concept of turbine was developed.

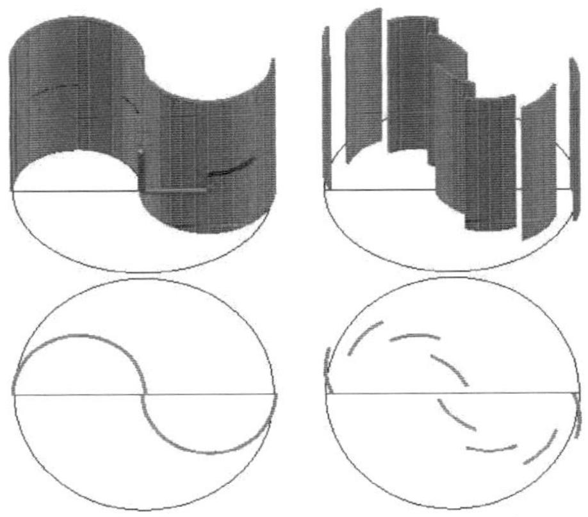

a) Closed strips b)Open strips
Continuous surface of blade Blade with gaps between strips

Figure 1. The concept of blade divided into strips.

The blade of the turbine is divided into several vertical strips, each of it, being able to rotate against its vertical axis. At repose, and low wind speed they are placed as to form a continuous surface, as a classic blade is. At a certain turbine speed, the strips begin to rotate against their own axis to form gaps between them. These gaps will let the wind escape through it, so a part of it dissipates, and no longer contributes to speeding up the turbine.

Figure 1 presents schematically the concept.

Rotation of strips is controlled by centrifugal force. Several ways are available to use centrifugal force to drive the strips rotation. One of them is based on the uneven distribution of the mass of strip against the rotation axis. That is obtained by moving the rotation axis of the strip from its symmetry axis, as can be seen in figure 2. This design has the disadvantage that does not benefit from any flexibility. The dissymmetry of the strip must be exactly determined theoretically to produce the wanted output. Even if the computations are accurate, once the strips are built and assembled on the turbine, no further adjustment is possible.

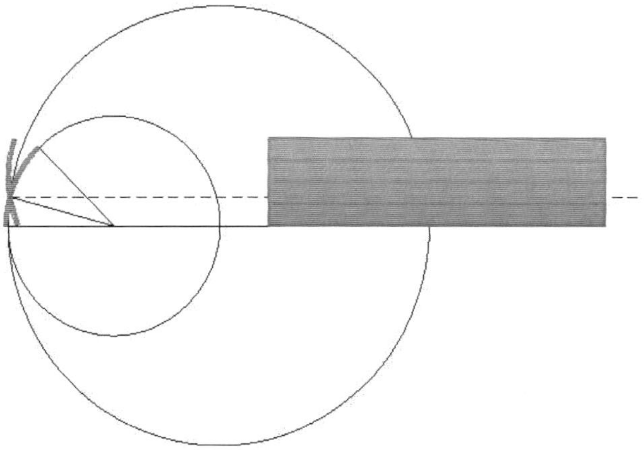

Figure 2. Rotation axis of the blade differs from the symmetry one.

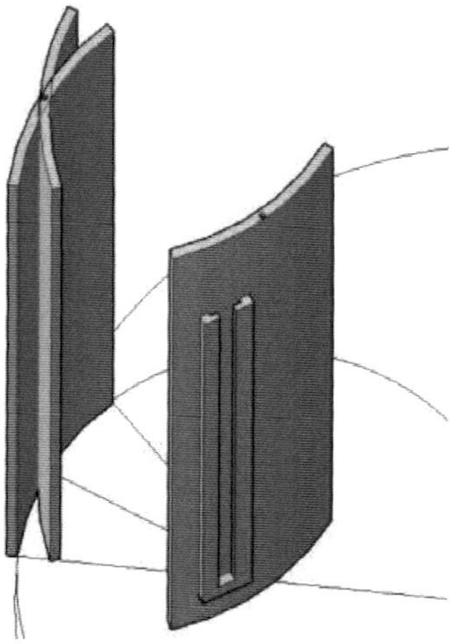

Figure 3. The pocket on the backside of the strip.

Figure 3 displays another way to unbalance the mass of the strip related to the rotation axis. This is placed even in the symmetry axis, but on the backside of the strip a pocket is placed. In this pocket several parts can be stored. This solution is better than the first presented one, because it allows a better flexibility: the mass imbalance can be controlled through the number and mass of the parts stored in the pocket. After the optimum is determined experimentally, its value can be applied to a single part strip. On the other hand, if the solution with pocket is kept as a final one, it allows controlling the turbine speed which determines the opening of strips by adding or removing parts from the pocket.

Another way to drive the rotation of strips is presented in figure 4. A rod that slides along a radial stem under the action of centrifugal force is connected by flexible cables to each strip. The flexible cables are guided by a reel.

Other systems used to convert sliding of a centrifugal mass into rotation of strips can use gears or levers.

Whichever the way to rotate the strips is chosen, they have to be led to their home position when the speed of wind slows down. This action is performed by torsion springs mounted on the axle of each strip. The extent to which the springs are strained can play a role in setting up the turbine speed that triggers the opening of strips.

The concept presented here can be adapted for other types of turbines, as well.

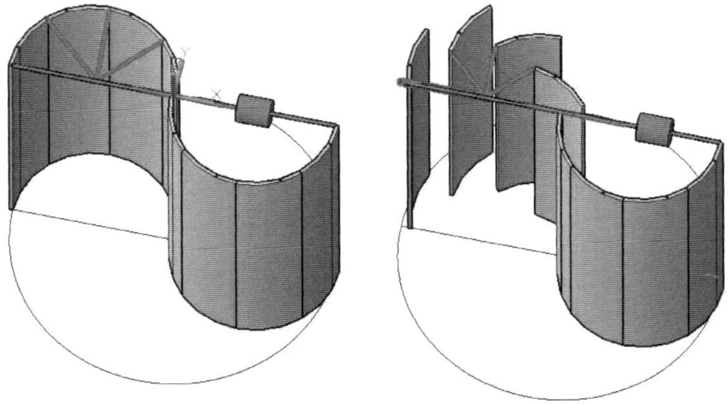

Figure 4. Central drive of the strips.

4. Concluding remarks

The turbine speed self-limitation system is a new, innovative one, which was not yet mentioned in the literature. Being based on centrifugal force that depends on turbine speed, one can conclude that it is a natural one. It prevents over-speeding the turbine, rather than counteracting high dangerous speed. The turbine speed self-limitation system offers some important advantages in opposition to using brakes that stop turbine from working if wind speed becomes too high:

- keeps turbine far from spindle overloading;
- the negative influence of high speed wind is avoided;
- losses in electrical power production are diminished, since the turbine does not need to stop when wind speed increases dangerously; problems that usually occur when restarting the work of turbine, are as well, avoided;
- the value of speed of turbine that triggers the protection system can be set up by means of springs.

The new concept of turbine blade can be applied to other types of wind turbines. For Savonius typed turbines, the new concept fits best to small turbines (up to 0.5 kw), suitable to be mounted and to work in insulated areas

What authors claim to be the novelty of the work is the concept itself, which splits the blade of the turbine into parallel strips, and uses the centrifugal force to rotate these strips against their own vertical axis. Wasting a share of wind stream through the gaps between the strips limits the turbine speed automatically.

The prototype can be improved experimentally, adjusting the mass that unbalances the mass distribution of the strips, as the limitation of turbine speed to be very effective. Scaling the prototype opens new research directions for different sizes of turbines. Research can be extended to identify some other systems to control the rotation of strips, more effective, reliable, and designed for manufacturability and assembling.

Acknowledgement

Building the first prototype of Savonius Turbine with self-limitation system of speed was supported financially by Transilvania University of Brasov.

References

[1] Chan C M, Bai H L and He D Q 2018 Blade Shape Optimization of the Savonius Wind Turbine Using a Genetic Algorithm *Apl. Energ* **213** 148
[2] Wenehenubuna F, Saputraa A and Sutanto H 2015 An experimental study on the performance

of Savonius wind turbines related with the number of blades *Energ Proc.* **68** 297

[3] Ogawa T, Yoshida H and Yokota Y 1989 Development of Rotational Speed Control Systems for a Savonius-Type Wind Turbine *J. Fluids Eng* **111(1)** 53

[4] Kumar P M, Purimitla S R, Shubhra S and Srikanth N 2017 3rd International Conference on Power Generation Systems and Renewable Energy Technologies (PGSRET), Univ Teknologi Malaysia, Johor Bahru, MALAYSIA 107

[5] Belabes B, Youcefi A and M Paraschivoiu 2016 Numerical investigation of Savonius wind turbine farms *J Renew Sustain Ene J Renew Sustain Ener* **8**

[6] Ducoin A, Shadloo M S and Roy S 2017 Blade shape optimization of the Savonius wind turbine using a genetic algorithm *Review Energ* **105** 374

[7] Ferrari G, Federici D, Schito P and Izoli F 2017 CFD study on Savonius wind turbine; 3D model validation and parametric analysis *Review Energ* **105** 722

[8] Liang X T, Ou Fu S C Bx, Wu C L, Chao C Y H and Pi K H 2017 *Review Energ* **113** 329

[9] Mereu R, Federici D, Ferrari G, Schito P and Inzoli F 2017 Parametrical numeric study of Savonius wind turbine interaction in a linear array *Review Energ* **113** 1320

[10] Roy S, Das R and Saha U K 2018 An inverse Method for optimization of geometric parameters of a Savomius-style wind turbine *Energ Convers Manage* **155** 116

[11] Pite H D 1986 *J Atmos Ocean Tech* **3 3** 487

IOP Conf. Series: Materials Science and Engineering **399** (2018) 012016 doi:10.1088/1757-899X/399/1/012016

Improving student's value in technical education

S S Duicu[1*]

[1] Manufacturing Engineering Department, Faculty of Technological Engineering and Industrial Management, Transilvania University, Colina Universitatii nr. 1, 500036 Brasov, Romania

[*] E-mail: simonaduicu@unitbv.ro

Abstract. This article is about value engineering, a method to improve the relationship between the function and the cost of a product. Industrial engineers can provide a better value of their products only if their own value becomes better and better, meaning not to reduce the cost of their education, but to increase their role in designing products and processes. The purpose of this analysis is to generate simple principles for understanding, modelling and optimizing this problem. The interaction between the components of the value provides the definition: the special relation between utility and cost catalysed by the wishes and resources of the employers at a specific moment. The article analyses the components that have effects on student's value and proposes a model harmonized with the growing ratio of the industrial area.

1. Introduction

The value engineering is a method to improve the relationship between the function and the cost of a product. Industrial engineers can provide a better value of their products only if their own value becomes better and better, meaning not to reduce the cost of their education, but to increase their role in designing products and processes.

Minimum cost is not the goal in technical education. If the term of „product" is assimilated to a student, then all the components of this notion will be analysed in consequence, including notions like „price" and „cost". The purpose of this analysis is to generate simple principles for understanding, modelling and optimizing this problem.

2. Value's components

It can be agreed that the components of a product value are: the market and the utility value given by the wishes and expectation of the employers, the rarity or the difficulty, the total cost to obtain the right product, and the possibility of the employer to choose, meaning the competition. Employers want to pay as little as possible for a good, useful, rare product not only against their competitors, but also in competition with their own resources. This is why, the interaction between these five components drives us to the simplest definition of the value: the special relation between utility and cost catalysed by the wishes and the resources of the employers at a specific moment.

This first classification appears to be the most realistic regarding the product „student". But talking about a continuing changing and alive product, different features of values must be taken into consideration. A "graduate" is considered to be the final form of the product "student" from the point of view of university education. But the educational process must never stop.

2.1. Utility and exchange values

One of the most formal and at the same time old classification was done in the antiquity by Aristotle. Understanding ancient Greeks' disapproval for trading and their love for individuality, it can be summarized that a useful engineer is more valuable than a well paid one. This exchange value linked with price can be measured only when the work market is flooded by specialists with the same education and/or qualification. Only in this moment a level of competence can be fixed and starting from this level a clear evaluation can begin.

Utility and exchange values substantiate the economic value. In modern time, new more interesting aspects are highlighted: the estimation value and the trading value. The estimated value can be isolated from the industrial context. All those components must interfere to create a living and functional model. Maynard says that [1] the most common mistake is to isolate the estimated value. The utility of a proper technical education and accessibility of a well prepared workforce contribute to a high reputation of the graduates.

2.2. Reputation value

In our days, this reputation value gives a competitive advantage to any product. It serves as a guide to the employers when the moment of choosing between different candidates comes. Finally, the utility value becomes firstly and mostly what counts in the CV. Employers will choose first from a CV database (which is more or less like a shape of the product named engineer), but the final decision will be made according the utility, so the shape will be subordinated to the function. A good example is the „swimmer illusion". Employers can make a confusion between selection factors and results and they become victims of an illusion. Students have a tendency to apply for very high reputation universities hoping they will get the best education possible and forgetting that top universities accept only the best candidates. It's understandable that after graduation the reputation of the university is borrowed by its graduates [2].

2.3. Market value

Employers have the possibility to choose from the human resources market the candidate they want. So, employers have the possibility to express their preferences, „buying" the work force. On a dynamic market, employers act like customers expressing their wishes by absorbing the graduates. Utility value and reputation value are linked with all the other physical features that make the product „engineer" what it has to be. But the market value and the exchange one reflect better the economic aspect of a product. Value engineering makes this connection between the economic characteristics and the real, physical ones such the number of graduates, gender and availability.

It is very clear that employers will set the criteria on economic evaluation and not on the educational system. There are two different points of view regarding student's value: one of their future employers and one of their educational system.

The educational system will examine this value quantitative and qualitative. Quantity means how many students are in the first year of study and how many of them graduate. Transilvania University is a state institution, so the ratio between investment and profit is not an issue because the profitability of the educational act is difficult to be approximated in money. Quality means the ratio of absorption on the workforce market and, sometimes, the degree of acceptance.

The employers will highlight the utility and the cost of engineers. The value engineering reflects the point of view of the employer, because all the aspects of the value starting with the profile and the cost, ending with the market value and the degree of workforce exploitation depend on his point of view.

Finally, the economic value of our students can be composed of:

- **Utility value** - how useful is the education of the graduates (not only their technical knowledge) to their future employers?

IOP Conf. Series: Materials Science and Engineering **399** (2018) 012016 doi:10.1088/1757-899X/399/1/012016

- **Exchange value** – beyond their technical education, how can graduates achieve better position or better jobs?
- **Market value** - how much can the employers pay for the graduates?
- **Estimation cost** – how can the market and exchange value be balanced when research, design and workmanship are very well established in the final cost of an industrial origin product?

3. Working model on value

The value's working model proposed by Carlos Fallon in Maynard's industrial engineering handbook combines five components that have effects on the graduate's value (figure 1).

The working model about the graduate's value needs: to ensure clear standards to evaluate the product „graduate", to provide all necessary information to the educational system and to ask questions to the employers.

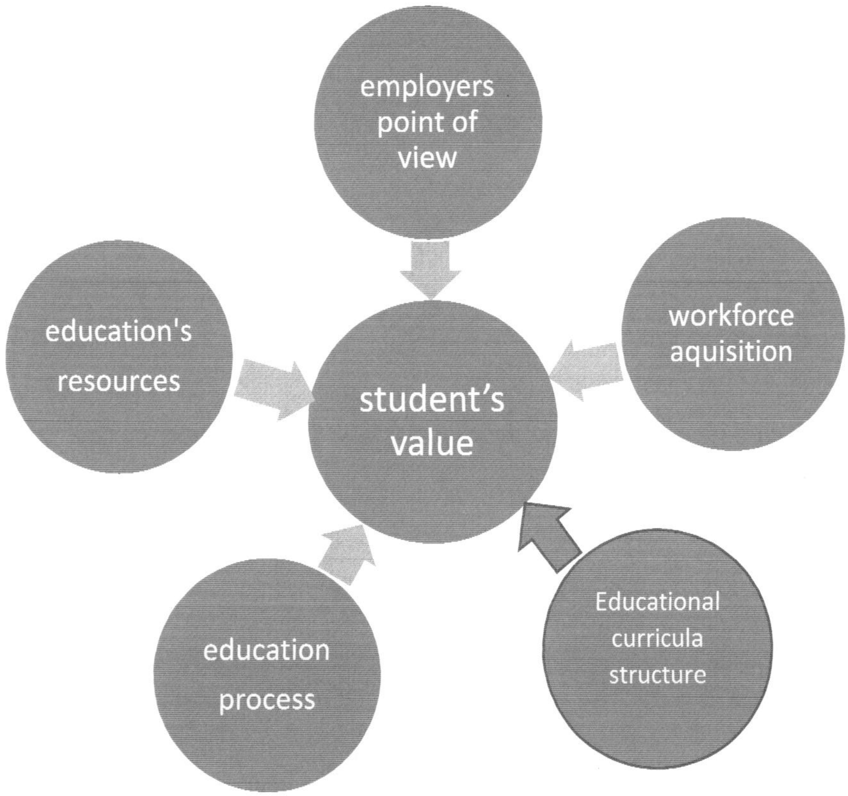

Figure 1. Value working model.

The value analysis team has to collect all information to be combined into a viable model. On the one hand we have the future graduates and on the other hand we have the future employers. A modality to compare different models is explained in table 1. The "criteria" are listed in the columns and the "choices" are listed in the rows.

Possibilities: What are the possibilities? How much does the improvement cost?

Possibilities to put into practice: What is the probability in % to put into practice the proposal according to the model? The measurement is made in % of accomplishment.

Information: Is there enough information to create the model? How much in % can be accessed to establish the best model?

Time: Is there enough time to put into practice the model? How much in % it is given to each optimization proposal?

Professional level: Is the professional level reached so as to put into practice the model? How much in % is it offered to each level of education?

Probability: Probability in % to put into practice the model.

The first three options are based on the point of view of the educational system, the last two, on the point of view of the employers.

Table 1. A modality to compare different models.

	Possibilities	Real possibilities	Information	Time	Professional level	Earnings
Improving curricula						
Combine theory with practice						
Increase requirements						
Training in the company						
Involvement in the educational system						

The best idea is to propose interdisciplinary models combining different choices. The advantages are: criteria must be adequate to both sides (educational system and employers), criteria must be written in the same language for both sides (in the language of "cost", "number of graduates", "payment of the trainee" etc), choices to make a dependency between sides.

For a better understanding of the actual situation, some data from Transilvania University report are connected with the technical workforce market.

A number of 6996 students from all technical faculties were enrolled at Transilvania University (which is a state university) in 2017, bachelor level (budget and fee), all forms of education: frequency, low frequency and distance learning [3]. The distribution of these students is: 6122 students with frequency, 528 low frequency, 346 distance learning. The total number of students in the last year of study in technical faculties, frequency form of study is 1167, in the following structure: 75 from Product Design (144 in the first year), 147 from Technical Engineering and Industrial Management (264 in the first year), 217 from Mechanical Engineering (305 in the first year), 43 from Wood Engineering (42 in the first year), 50 from Material Science Engineering (77 in the first year), 235 from Electrical Engineering and Computer Science (369 in the first year), 129 from Forestry and Forest Exploitation (194 in the first year), 109 from Construction Engineering (110 in the first year) and 162 from Food and Tourism (233 in the first year). If in the first year of study there are 1740 students, in the last one there are only 1167. Of course the comparison is not the best one because of the newly introduced programs of study in 2017, but 33% of those enrolling are lost. This study doesn't take into consideration the number of students enrolled in distance learning or low frequency because it is presumed that they are already working. Starting from the premises that 90% of these students will graduate, about 1000 engineers will enter the technical workforce market. Transilvania University is the most important regional university centre specialized in technical education, meaning, it prepares engineers coming from related areas. If only 25% of the graduates will decide to

IOP Conf. Series: Materials Science and Engineering **399** (2018) 012016 doi:10.1088/1757-899X/399/1/012016

stay in Brasov area, then, 250 engineers must meet the requirements of the technical workforce market.

Not only new graduates from Transilvania University technical faculties can be absorbed. An interesting variable in this problem can be the number of graduates from other universities, master degree graduates or already employed engineers moving from a company to another.

At the same time, the growing ratio of the industrial area is minimum 10% (for an organic grow) and the number of the job offer is of about150 engineers. Starting only from these two numbers it can be imagined the dynamics of the employment market. Facing new issues in education, a simple model of the student's value must take in consideration a lot of variables to harmonize demand and offer.

Finally, the educational process must ensure the progress of the existing situation. Strong competition in delivering good specialists becomes a catalyst in any value analysis team.

The effects of these components over the student's value are represented in figure 2.

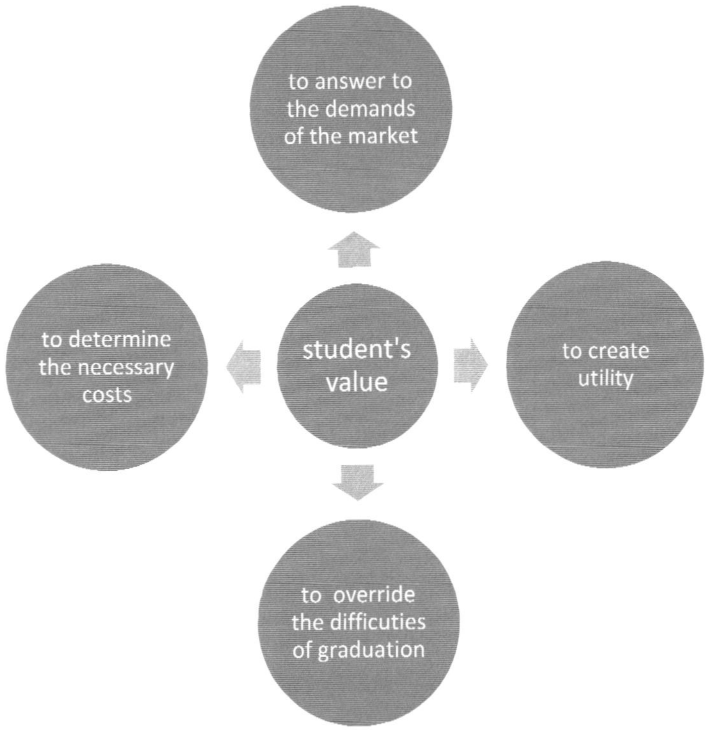

Figure 2. Effects on the student's value.

4. Working Plan

All models must have some goals mainly in the development of the educational process, not only in universities, but in the companies' training compartments. The key principles are [4]:

- to understand the end-use requirements;
- to select appropriate development approach;
- to establish for best methods;
- to measure the work;
- to apply allowances;
- to document standards and conditions;
- to validate standards.

Benefits can be adjusted for product "student":

- increase in productivity and elimination of waste – Increase the number of graduates, reducing the number of repeaters and abandons,
- basis for staffing, planning and scheduling production – Staffing, planning and scheduling educational process;
- basis for determining and controlling costs – Quality management system ARACIS;
- basis for managing performance – Performance quality system, number of innovations, brevets and research reports.

4.1. Proposal on the working plan

The steps of the working plan on the value model are the following.

Information step

- Identifying the choice or the combination of choices (table 1);
- Defining the lifetime of the product: 35 to 40 years;
- Defining the required qualities: technical, managerial, research, etc;
- Marketing conditions: how to promote and/or sell the workforce;
- Data about the cost: workforce cost linked with exchange, estimation and market value;
- New information: update information regarding employers, study programs in the University;
- Important aspects and features.

Analysis step must answer two questions

What is the function of the product "student"?

How is identified the beneficiary of this product?

- Defining the cost of the product "student": the cost of the education, the cost of the training in the companies, the cost of the change of the career path, social costs etc;
- Establishing how much should the product "student" cost: evaluation in comparison with other educational systems, the cost of imported workforce, cost per potential employee (given by the calculation of the price of an object produced by the future employee).

Creation step

- How can the student's value be improved? How can a proper atmosphere of study be installed, how can inventive personalities and inventive actions be encouraged, how can conditions for research and innovation be created?
- Identifying obstacles against innovation and remove them, improving the creation of mechanisms, creating the climate for change and adapting to change, looking for analogies, anchoring in the objective reality, finding the best solution;
- Optimization;
- Simplification.

Evaluation step

Because it is about a living product, none of the existing methods can be ad literam applied. A combination between the existing methods can be obtained.

- Choosing the educational system: establish anticipated benefits like graduation ratio, employment ratio or successful employment; advantages and disadvantages of the chosen educational system.
- The beneficiaries' criteria, superior and inferior limit for the student's value, correction for utility and efficiency, advantages and disadvantages for the employers.

Verification step

- What is the improvement brought by the new value to the employers? Do the employers accept the new value?
- How much is the cost of the improvement? Is there a syncope in the study program or delays?
- What are the risks? Could there exist problems in the performance of the educational system? Could there be risks in the acquisition of new educational means or other kinds of risks?

Recommendation step

- Making a full report with recommendations.

- Establishing levels of action.
- Authorization and approval.

Practice step

This last step must harmonize beneficiaries and the educational system. The factors that determine the acceptance must be keep in mind: what are the normal costs of the education and what unexpected costs could appear? Also, it is very important for both parts to make the right choice: what kind of education do the graduates need and is this training needed by the employers?

Another aspect that might influence putting in practice the model is represented by the conditions of acceptance. Both parts must be well informed about costs, risks and earnings, be honest with themselves about expectations and objectives and understand the final effect of the change on their structure.

The mechanism of putting the model into practice must be clear, simple, well documented, keeping in mind all the time the function of the product "student". For every case, some other features can be important and unexpected situations can appear.

The mechanism must establish:

- Tasks: Who is executing/ teaching/ responsible with practice or training; who is verifying and who is coordinating the educational program, who gives the funds and who controls the costs, who is centralizing the reports and establishes the priorities
- Affected areas: What are the modifications in the educational program and who will be affected?
- Deadlines: When does the action begin and how long will it be? Is it a continuous project and curricula will be affected? How long will it take until the first results will appear?
- Follow-up: Can follow-up procedures be established? Can some reports be done from time to time (establish the period)? Are there some landmarks or goals to be reached established by the future employers?
- Results of measurements: Can all earnings in time and money be measured? Can all costs in time and money be calculated? Can we make some comparisons of the total earnings with the total cost regarding the final effort?

4.2. COMBINEX Method

The well known COMBINEX method establishes a matrix that combines different advantages given by several input conditions. What must be taken into consideration:

- the analysis of the objective to establish demands and anticipated benefits;
- an evaluation method based on the existed resources;
- a scale with a specific measurement unit;
- measured factors whose contribution will affect the available solutions;
- final scoring table.

COMBINEX method may evidence the area in which an effort can bring a maximum advantage.

A scoring table choosing different conditions on the horizontal lines, and different choices on the columns can be proposed.

Hypothesis: Better absorption of the graduates on the technical workforce market

Conditions: strong technical education, flexibility and adaptation capacity, decrease the cost of employment.

Choices: improve technical education by modifying curricula, decreasing university dropout, increase the capacity of Transilvania University to face the competition with other universities.

The next step is to find out the combinations between conditions and choices, and this must be done through the idea of satisfaction and not to obtain maximum. This is the difference between well balanced effort and the maximum one. A good example in this direction: a strong technical education may be done by modifying the curricula, but the cost on short term and the effort could be maximum and ineffective.

It is very difficult to transform the combinations into numbers. This can be done only if a common property that can be mirrored in numbers can be found.

For example, for a future engineer, some technical skills can be measured in their grades: 5 for passing the exam, 9 very good, 10 for perfect. This scale can be approved of by the university and by the employers, too. Obtaining 10 can be significant or not regarding the effort and the costs. Also, maybe a 10 grade is not so important regarding the competition between different universities when high standards do not allow to get a 10 grade so easily.

Another measurement unit can be the cost, but not all the time a higher cost of the education will reflect a proper education. Sometime, a higher cost means too high for a state university and too expensive for the employers to buy (meaning to employ) a graduate.

Finally, a graduate from a technical faculty can value more for an employer than for another, or his utility can be bigger for one than for another. In the context of industrial growing, the economic value increases at the same time with the number of employers. There are a lot of theories regarding the mathematical approximation of the utility, but John von Neumann and Oscar Morgerstern [5] combine all the theories and create the modern analysis of the value. The main idea is that the product, in this case, the graduate must do something useful for the employer and this can be measured in the satisfaction degree produced to the employer.

A new hypothesis is translated into practice: technical universities must offer to the employers something more valuable than the price paid for its graduates.

In this case, innovation, modernization, simplification and a better use of the staff involved in education may offer to the future graduate employers something more valuable for a smaller price. So, the choices for the university are: to improve the efficiency of the staff using staff recalibration and qualification improvement.

Going back to the measurement scale, grade 5 can represent the lowest condition to be accepted by the employer, grade 7 can be equated with favourable, grade 9 with optimum and above grade 9 with additional benefits offer, and grade 10 with absolute maximum. Everything above 9 can be accepted by both parts only if the graduate's education does not involve excessive costs.

Table 2 proposes a COMBINEX matrix. "Conditions" are listed in the columns and "choices" are listed in the rows.

Table 2. COMBINEX matrix.

	Powerful technical knowledge (*w=0.5*)	Rv	Flexible thinking (*w=0.2*)	Rv	Cost employment (*w=0.3*)	Rv	*Relative results*
Modify curricula	9	4.5	7	1.4	7	2.1	8
Decrease dropout	7	3.5	5	1	9	2.7	7.2
Improve the efficiency of the staff	9	4.5	8	1.6	9	2.7	8.8

Conditions are assimilated with the benefits for the future employers of the technical faculty graduates. Grades from 5 to 9 were assigned as in the example given above. An essential aspect in the COMBINEX matrix is the importance of the factors or conditions. The total importance can be established to 1, so, if all conditions have the same importance, based on the presumption of summation, every condition will count as an equal percentage. If the relevance of the conditions is different the weights could be different but summing 1. In this case, each grade was multiplied by its weight "w" and the related value "Rv" was obtained specified in the next cell of the matrix. Finally, in the last column, there are the "relative results".

This is a very simple simulation for COMBINEX matrix. For each benefit obtained by the future employer, the university has to make a choice by improving something. Behind the grades, there is another submatrix or a chart that suggests or establishes (in the case of clear instruments like costs valuable in money) the relation between the variables. For example, when curriculum is modified, the costs of the university are very big at the beginning, and drops down in the following years (linear or asymptotically). This will have a big impact from the beginning on the technical knowledge, but the graduates will also be expensive. Decreasing dropout will affect mainly the employers because the number of the graduates will increase, and their cost will decline, and even the university will be more productive. As for the last choice, it is very clear that all variables are optimal.

For the studied matrix, in the last column there are the "relative results" obtained like sum, and it seems that the last choice could be the answer to the required problem.

The model is only a simple simulation to make a hierarchy of options. It is very important to also understand if the choices bring a plus value for the university, too (like increasing the prestige). It is important to split the question of funds: who is funding the employers and who is funding the university? Transilvania University is a state university, but important industrial companies for the related area fund in different forms the educational process.

5. Conclusions and personal contributions

In the value engineering the main question is: "What is the function of the product?" The idea of "function" will always be linked with the utility. If the educational system is a provider of human value, the principle of utility must be rewritten.

There are two important tasks: first, to establish the desire of the society in general and the desire of the employers in particular, and second, to design, to build and to apply a viable model based on which to relay the educational process. This complex and dynamic model has to fulfil the requirements of the workforce market in a rational proportion with the educational effort. Because no matter who will use the workforce, the product "student" or "graduate" must have a higher value for the society than the employer can pay. Otherwise, the employers will never be interested in making the exchange: money versus workforce.

My personal contribution to this paper is the adaptable model and working plan on the student's value. It is very difficult to put in a box this product. There are significant and sensible features that differentiate a material product from a living one: reputation, estimate, time delivery, personal features and personality may influence the whole educational project. The selling price of the workforce will remain the ultimate measurement, but how can the equitable value for utility be established? How can the function be defined? Can the abstracting degree of the model be increased and ask: why must the product "graduate" answer to a specific demand? Or, can the degree be decreased by asking: How must the product "graduate" do a specific job? Between these two questions there are a lot of possibilities to create models of evaluation.

This study was not concentrated on finding formulas but it puts important concepts like value, function and method into another context.

References

[1] Maynard H B1975 *Manual de inginerie industriala* (Bucuresti: Tehnica) p 250
[2] Duicu S S 2015 *About changing the paradigms in the technical education* J Ind. Design Eng. Graph. JIDEG **13**
[3] *Raportul anual privind starea universitatii Transilvania Brasov 2018* https://www.unitbv.ro/documente/despre-unitbv/regulamente-hotarari/hotarari/senat/HSnr.21
[4] https://www.hbmaynard.com/workforceperformance/wpm_engineeredstandards.asp
[5] Neumann J Von, Morgenstern 2007 *Theory of games and economic behaviour (*Princeton ed.)

IOP Conf. Series: Materials Science and Engineering **399** (2018) 012017 doi:10.1088/1757-899X/399/1/012017

Hybrid modelling for thermal deformation prediction of high speed motorized spindle

L T Fan[1,2], X R Jing[1], L X Zhang[1,2] and D Li[1,2*]

[1] School of Mechanical Engineering, Shenyang Jianzhu University, Shenyang 110168, China
[2] National-Local Joint Engineering Laboratory of NC Machining Equipment and Technology of High-Grade Stone, Shenyang Jianzhu University, Shenyang 110168, China

[*] E-mail: lidong@sjzu.edu.cn

Abstract. Thermal-induced deformation accounts for 40-70% of the total dimensional and shape errors that arise from various sources of machine tools. This paper suggests a hybrid model based on regression-based analysis and computation-based simulation to predict the thermal deformation of high speed motorized spindle. COMSOL Multiphysics was used to obtain the numerical solutions of the thermal deformation, and the PLS algorithm was used to modify the error between the simulation model and the experimental data collected from the spindle test system. The experimental results showed that the proposed hybrid model could predict thermal deformation effectively and accurately, and it can be used for real-time thermal deformation compensation to improve the machining accuracy of NC machine tool.

1. Introduction

Thermal-induced deformation accounts for 40 - 70% of the total dimensional and shape errors that arise from various sources of machine tools [1-3]. Thermal issues in machine tools have been studied extensively in the past decades, and detailed reviews can be found in literatures [4-11]. In order to improve the machining accuracy of finished parts, effective thermal error compensation measures are achieved via re-adjusting the axes positioning or via shifting the original coordinates based on the thermal deformation model.

In order to estimate thermal deformation, either computation-based or regression-based models are employed [12-24]. The computation-based models can be represented with thermal transfer functions. The finite element method (FEM) and finite difference method are widely used to model the thermal behaviour of motorized spindle under the influence of heat sources inside the structure and in its surroundings. Numerous computational models have been proposed as summarized in the literatures [25-31]. However, due to the low efficiency, the computation-based models were rarely employed in online thermal compensation, usually used only for simulation analysis and providing a design reference. As for regression-based models, the relationships between temperature rise and thermal deformation are studied based on the measured data. The regression-based models mainly include multivariate statistical models [19], neural network models [21], and gray theoretical models [22]. All the regression-based models play an important role in the thermal error compensation implementation. However, the accuracy of the data model depends on the quantity and quality of the modelling data.

IOP Conf. Series: Materials Science and Engineering **399** (2018) 012017 doi:10.1088/1757-899X/399/1/012017

Once the data exceeds the scope of modelling data, the model predictions could show a serious deviation, that is, weak generalization ability. Hybrid modelling can achieve the best of both computation-based and regression-based models. Combining computational analysis with experimental data, regression analysis is used to extract complex information that cannot be explained by the computation-based model, and the computation-based model could improve the generalization ability of the regression-based model.

This paper proposes a hybrid model based on PLS and COMSOL to predict the thermal deformation of the high-speed motorized spindle. Firstly, through the analysis of the temperature field of the motorized spindle, the geometric model and thermal boundary conditions for finite element modelling were obtained. The COMSOL Multiphysics software was used to simulate the motorized spindle to obtain the thermal deformation in three dimensions. Then, the PLS regression method was used to establish the relationship between the temperature rise and the errors of COMSOL model. Experimental data collected from the spindle thermal deformation test system were used for the model validation.

2. Thermal characteristics of motorized spindle

The structure of the motorized spindle is shown in figure 1, which includes the house, stator and rotor of the motor, shaft, coolant jacket and bearing set. In general, the motorized spindle is equipped with a built-in motor as an integrated component. The thermal deformation is induced by the electromagnetic loss between the stator and the rotor, and mechanical friction loss was generated in the bearing set. All the energy loss is converted into heat and then transferred to the shaft, leading to thermal deformation, especially in the axial direction, which heavily influences the machining accuracy.

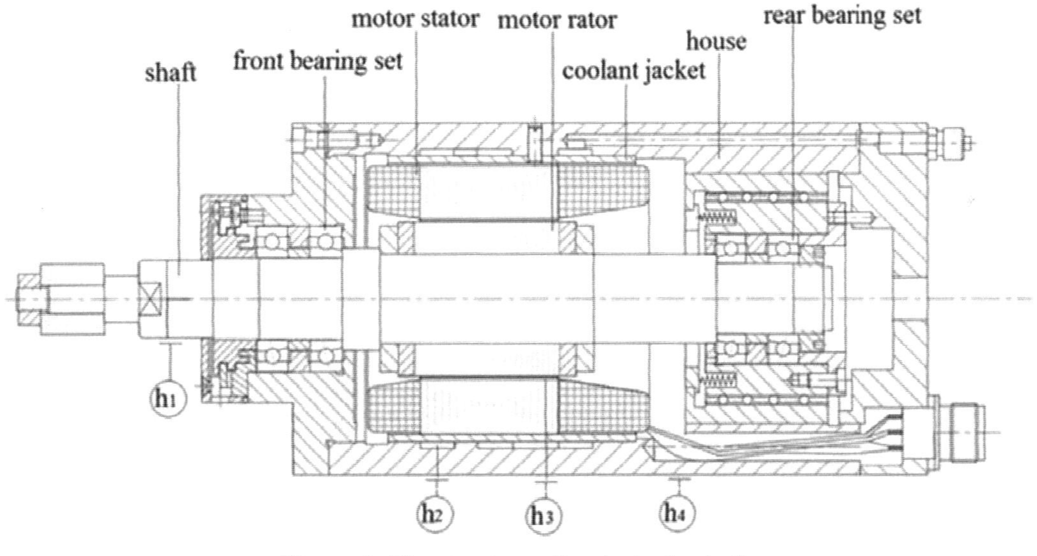

Figure 1. The structure of motorized spindle.

According to the energy conservation law:

$$\rho_1 C_{P1} \frac{\partial T}{\partial t} - \nabla \cdot (k \nabla T) = \Omega, \tag{1}$$

where $\Omega = \frac{Q}{V}$ is the heat generation rate (W·m^{-3}), Q is the heat source (W), V is the volume of the heat source (m^3), ρ_1 is the density (kg·m^{-3}), C_{P1} is the heat capacity (J·kg^{-1}·K^{-1}), T is the temperature (K), k is the thermal conductivity (W·m^{-1}·K^{-1}).

The thermal deformation equation of the spindle is:

$$\varepsilon_{th} = \alpha(T)(T - T_{ref}), \quad (2)$$

where ε_{th} is the thermal strain, α is the thermal deformation coefficient (1/K), T_{ref} is the ambient temperature (K).

The stator in the motorized spindle is cooled by water in the coolant jacket, and the rotors and bearings are cooled by compressed air in the oil mist lubrication device. As shown in figure 1, it is assumed that there are forced convection heat transfer in the air-gaps that between the stator and rotor, the rotor and the compressed air, the cooling water and the water jacket. There is natural convection heat transfer between the shell surface and environment air. For the convective heat transfer, according to Newton's Law, the relationship between the heat flow vector and the heat transfer coefficient is

$$q = h \cdot (T - T_{ref}), \quad (3)$$

where h is the heat transfers coefficient ($W \cdot m^{-2} \cdot K^{-1}$), q is the heat flux ($J \cdot m^{-2} \cdot s^{-1}$).

A finite element model can be established to predict the thermal deformation of the motorized spindle taking into account the structural parameters and operating condition based on the above equations.

3. Computation model based on COMSOL

The finite element model is obtained based on figure 1 by assuming that the bearing, the rotor and the stator are simplified to be assembled on the shaft, ignoring all the screws, the vent holes, the through holes, and integrating the fine structure, the bearing inner ring, outer ring and rolling elements. Because the motorized spindle is an axial symmetrical structure, 1/4 of the finite element model of motorized spindle is shown in figure 2. The total simulation time is 4800 seconds, and the step size is 300 seconds. A tetrahedral mesh structure of free partition is used and refined with 125,626 tetrahedral elements, 23,087 triangular elements, and 2,339 edge units. The material properties of the model are shown in table 1.

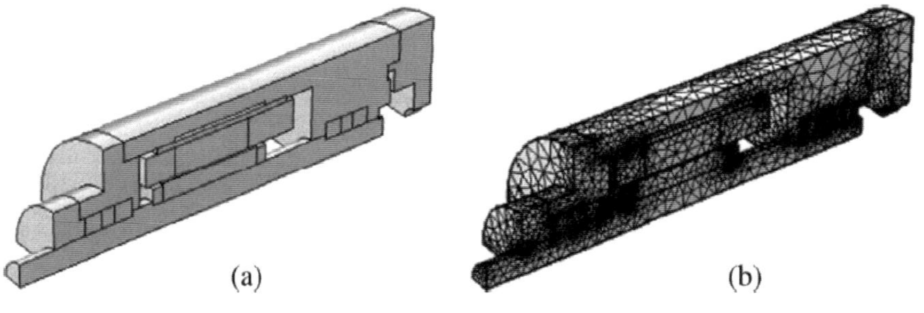

Figure 2. The finite element model of motorized spindle (a) geometrical model (b) meshing model.

The boundary conditions involve the stator heat, rotor heat and bearing heat obtained by the power loss test. In addition, there are four obvious different heat transfer processes of the 150MD24Z7.5 motorized spindle as shown in table 2, refer to figure 1. Therefore, the heat transfer coefficient between the shaft end and the environment air is h_1, the heat transfer coefficient between the air-gap and compressed air is h_2, the heat transfer coefficient between the cool water and the coolant jacket is h_3, and the heat transfer coefficient between the shell surface and environment air is h_4. The stator losses, rotor losses and bearing friction losses of the motorized spindle are obtained by the experimental method under the speed of 5,000 *rpm*, 8,000 *rpm*, 10,000 *rpm* and 12,000 *rpm*. The values of h_1~h_4 are listed in table 3.

The finite element method is used to obtain the numerical solutions to the temperature and thermal errors. The temperature field distribution of the motorized spindle system is shown in figure 3. The maximum temperature area distributes mainly over the rotor and the shaft core. The reasons are that the rotor accounts for one third of the motor heat dissipation, and there is no cool system to directly take out the heat of the rotor and the shaft core. The temperature field of the stator is lower than that of the rotor because the spindle sleeve connecting with the stator has a cool system which can greatly decrease the temperature of the stator and the spindle sleeve.

Table 1. Material properties for FEM.

	Material	Density (kg/m^3)	Linear expansion coefficient ($10^{-6}K^{-1}$)	Modulus of elasticity (Gpa)	Poisson ratio	Heat capacity J/(Kg·K)	Thermal conductivity W/(m·K)
Stator	Silicon steel	7.82	13	198	0.26	535	35
35Rotor	Cast aluminum	2.77	23.6	69	0.32	875	167
Winding	Copper	8.933	17	80	0.36	385	400
Bearing and other	Steel	7.872	10	208	0.3	434	60.5

Table 2. Heat transfer process of motorized spindle.

Numbering	position
h_1	Forced convective heat transfer between the shaft end and the environment air
h_2	Forced convection heat transfer between the air-gap and compressed air
h_3	Forced convection heat transfer between the cooling water and the coolant jacket
h_4	Natural convection heat transfer between the house surface and environment air

Table 3. Boundary conditions for FEM.

Speed(rpm)				Simulation conditions
5000	8000	10000	12000	
108.3	116	140.7	156.9	Stator heat
54.1	58	70.3	78.5	Rotor heat
2.3	17	28.7	41.3	Bearing heat
83.21	97.83	106.1	115	h_1
109.92	131.64	143.9	156.13	h_2
279	279	279	279	h_3
9.7	9.7	9.7	9.7	h_4

After the analysis of stable heat conduction has been completed, the element type is changed, the elastic module, poisons ratio and coefficient of thermal expansion are set, the analysis results of stable heat conduction serve as the temperature condition, and the DO constraints are attached to the bottom of the spindle. The results of thermal deformation analysis can be seen in figure 4.

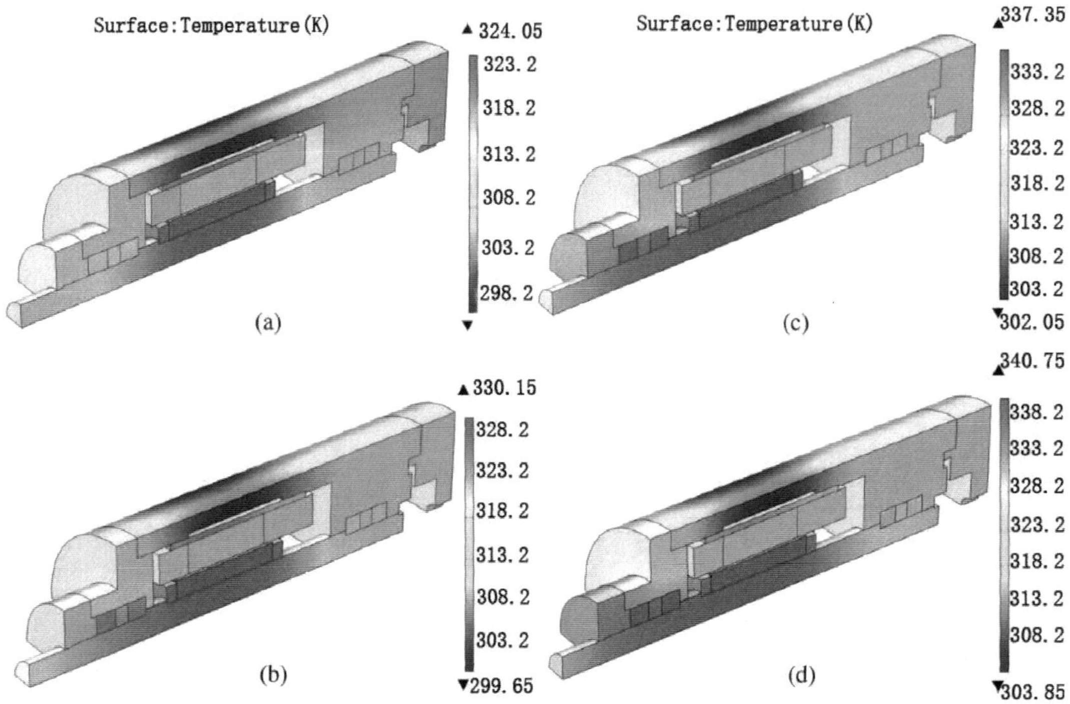

Figure 3. Temperature field distribution results (a) n=5000*rpm* (b) n=8000*rpm* (c) n=10000*rpm* (d)n=12000*rpm*.

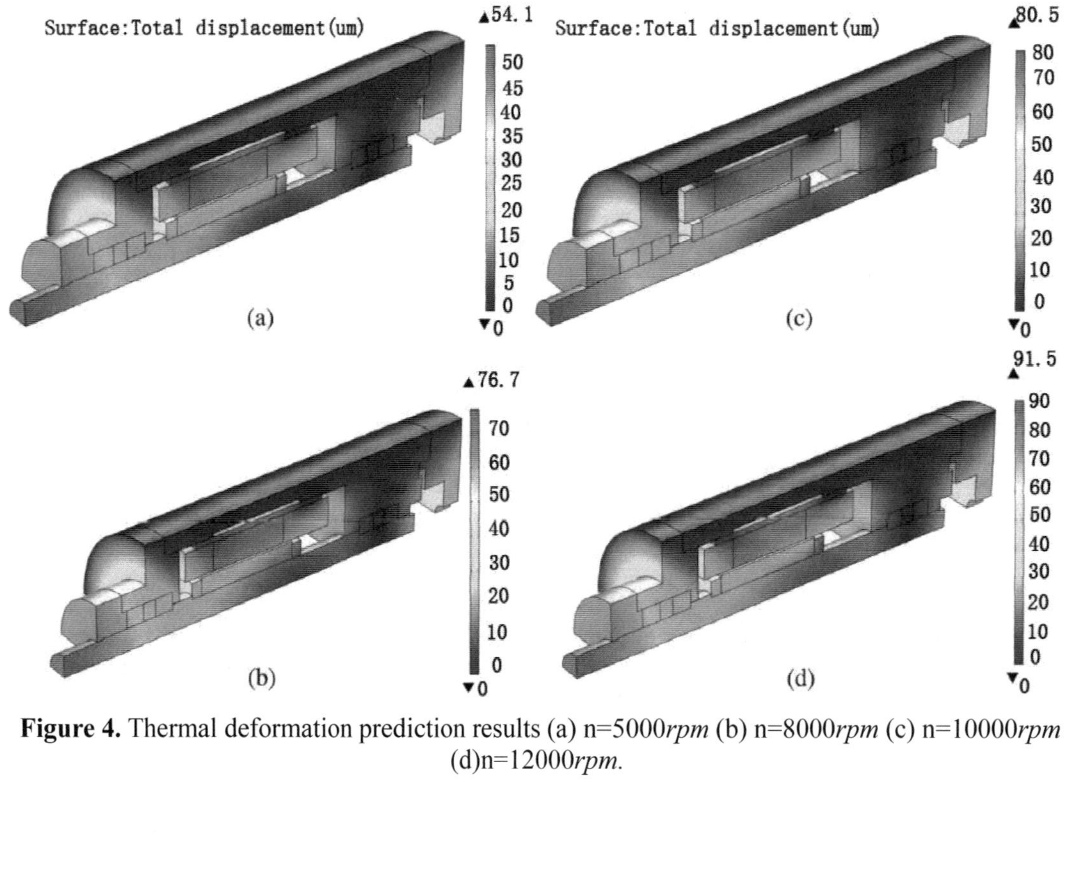

Figure 4. Thermal deformation prediction results (a) n=5000*rpm* (b) n=8000*rpm* (c) n=10000*rpm* (d)n=12000*rpm*.

The thermal deformations of the main components of the motorized spindle system are affected by the thermal characteristics and temperature field distribution, which causes the thermal error of the shaft and whole motorized spindle. In figure 4, in the axial direction, namely the Z direction, the maximum deformation is about 58*μm*, 70.4 *μm*, 72.7 *μm* and 89.6 *μm* when the speed is 5000 rpm, 8000 *rpm*, 10000 *rpm* and 12000 *rpm*, respectively. The axial deformation of the spindle is higher than the radial deformations.

The numerical solutions can approximate the analytical solutions very well as long as the spindle is correctly and finely meshed. The reliability of simulation results also depends on whether the thermal loads and boundary conditions are well defined. The accurate calculation of the thermal boundary conditions is a necessary condition for ensuring the accuracy of the finite element theory model. The inaccuracy of any of the material properties in the boundary conditions, the convection coefficient, the thermal contact resistance, and the heat generated by the heat source may cause inaccurate predictions. Moreover, the calculation process of the finite element model is relatively complex, generally requires loop iteration, and there is a case of non-convergence. Therefore, it is difficult to directly apply the thermal error compensation process. However, the finite element model describes the process of thermal deformation from the thermal characteristic mechanism. Therefore, the model has good generalization ability. Even if the operating conditions vary within a relatively wide range, the predictive result of thermal deformation can be obtained with limited accuracy. In order to improve the prediction accuracy of the model, the regression model can be used to supplement or modify the computation model. The regression model is a statistical model based on experimentally measured thermal deformation and the temperature measurement data of the motorized spindle. But the variation in the temperature measurement point of the motor spindle is similar, so there is a strong correlation problem in the data. Directly using the measurement data onto multiple regression modelling will inevitably produce multiple correlation problems. The existence of multiple correlations can seriously jeopardize the parameter estimation of the model, enlarge the model error, and destroy the stability of the model. However, the partial least-squares regression method can effectively overcome the adverse effects of multiple correlations in the model.

4. Regression model based on PLS

Partial least-squares regression is a new multivariate statistical data analysis method that integrates the basic functions of multiple linear regression analysis, canonical correlation analysis, and principal component analysis. It combines the modelling and forecasting data analysis method of the non-model data recognition analysis method and achieves the simplification of the data structure while modelling. The PLS method no longer directly considers the regression modelling of the dependent variable *Y* and the independent variable *X*. Instead, the information about the variable system is comprehensively screened again, and several new synthetic variables (principal components) that have the best explanatory power for the system are selected for regression modelling. It can effectively filter out the information overlap caused by multiple correlations and improve the accuracy of system modelling.

The PLS model consists of two parts: external relations and internal relations. There are l dependent variables $\{y_1, \cdots, y_l\}$ and m independent variables $\{x_1, \cdots, x_m\}$. Based on the n sets of observed data, the dependent variable matrix $Y_{n\times l}$ and the independent variable matrix $X_{n\times m}$ are constructed. According to the PLS algorithm, the principal components are extracted and X and Y are decomposed to get the external relation model:

$$X = TP^T + E = \sum_{i=1}^{T} t_t p_t^T + E , \tag{4}$$

$$Y = UQ^T + F = \sum_{i=1}^{r} u_i q_t^T + F , \tag{5}$$

where $P_{m\times r}$ and $Q_{l\times r}$ are the load matrices for X and Y, respectively, and $E_{n\times m}$ and $F_{n\times l}$ are the residual matrices for X and Y, respectively. The matrices $T = [t_1, \cdots, t_r]$ and $U = [u_1, \cdots, u_r]$ are the score matrices of the matrices X and Y, respectively, where the elements $t_1, \cdots, t_r$ and $u_1, \cdots, u_r$ are the extracted principal components.

The external relational model guarantees that T and U represent the data tables X and Y as closely as possible. The internal relational model ensures that the principal component T of the independent variable has the strongest explanatory power for the principal component U of the dependent variable. Therefore, the following regression model can be established for the principal components:

$$U = TB + R, \tag{6}$$

where $B = diag\{b_1 \cdots b_r\}$ is the regression factor matrix and R is the U residual matrix. The regression coefficients in the PLS model can be calculated by nonlinear iteration or using singular value decomposition.

As a regression model, the accuracy of the PLS model depends on the quantity and quality of the modelling data. Therefore, if the measurement data are inaccurate, or if the model prediction exceeds the range of modelling data, the model prediction results will be inaccurate. This means that the application scope of the PLS model will be limited by the capacity of the modelling data, and the generalization ability of the PLS model will be poor.

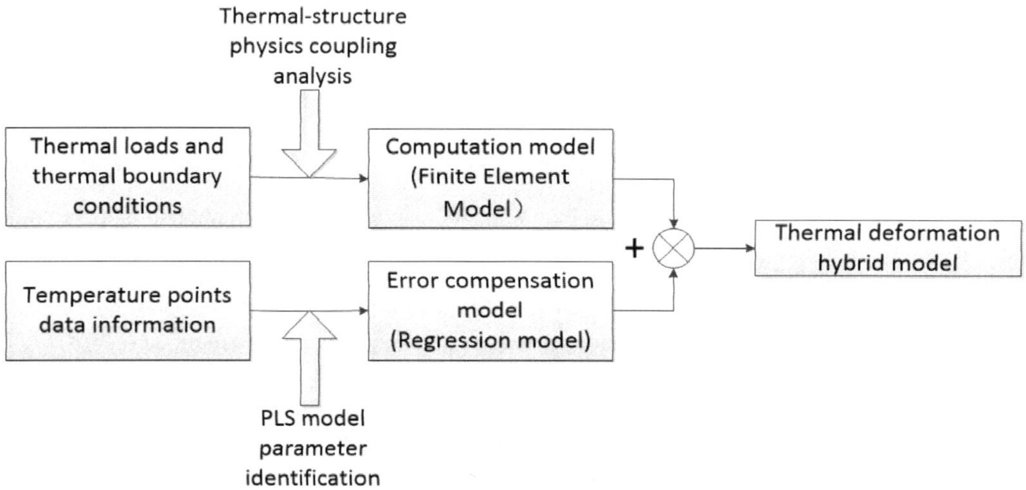

Figure 5. Structure of the thermal deformation hybrid model.

5. Thermal deformation prediction based on hybrid modelling method

The mixed use of computation models and regression models can achieve the effect of complementing each other. The PLS-based regression model can extract unexplained internal complex information from the computation model. And the computation model based on the finite element can provide the prior knowledge of thermal deformation and improve the generalization ability of the regression model. Based on the above design concept, the design framework of the thermal deformation hybrid model of the high-speed motorized spindle is shown in figure 5. The overall model consists of a finite element model and an error compensation model. The PLS method is used to predict the thermal deformation error between the output of the finite element model and the experimental data. Next, the output of the finite element model and the error compensation model are superimposed to obtain a hybrid model of thermal deformation. The hybrid model effectively combines the merits of the computation model and the regression model. Using the finite element model as the basis of the hybrid model, the extrapolation and generalization capabilities of the model have been improved, and the PLS model has been used to make up for the computational error, which has greatly improved the prediction results of the thermal deformation model.

The specific steps for the establishment of the thermal deformation hybrid model are as follows:

- Based on the heat loss experiment, the heat transfers coefficients and thermal loads calculated at different speeds of the motorized spindle are loaded as boundary conditions into the spindle model. According to the thermal stress analysis, the computation model of thermal deformation of the high-speed motorized spindle during thermal steady state is obtained.
- A regression model is established based on experimental data collected at different speeds. The input of the model is the temperature rise of each temperature measuring point, and the output of the model is the error between the experimental data and the computation results from the finite element model.
- After calculating the cross-validation, the principal components for regression modelling is selected, the principal component regression model is reduced to the original variable, and the thermal deformation error compensation model based on the PLS method is obtained.
- The performance of the thermal deformation hybrid model is evaluated based on the experimental data. If the predicted results are not good, then the PLS model is re-trained and the new principal components are selected until the result is good.

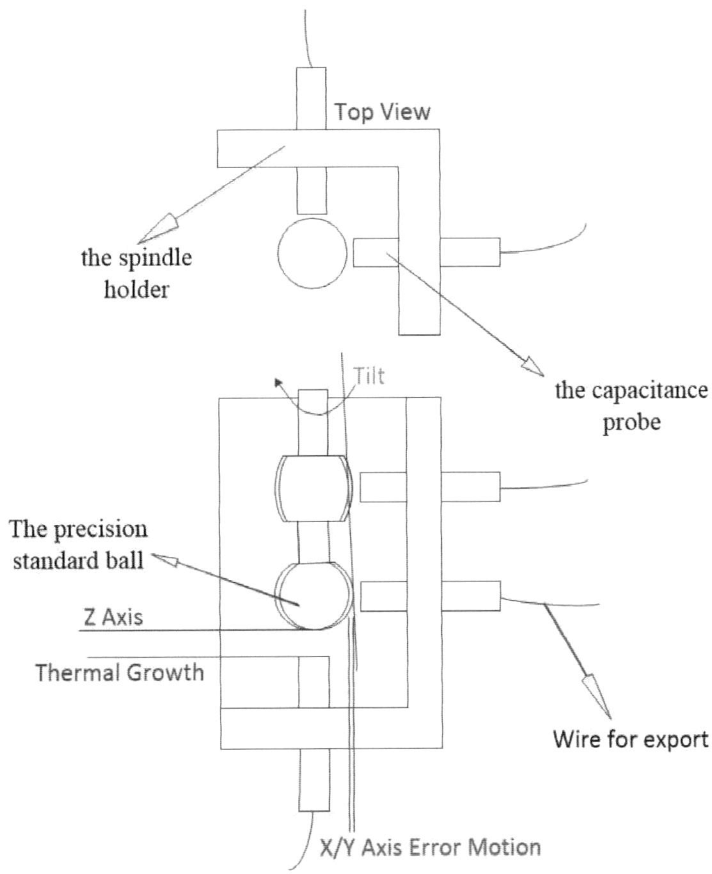

Figure 6. Displacement sensor position.

6. Thermal deformation experiment and result analysis

6.1. Experiment introduction

The experiment used the Lion Precision's spindle rotation error analyser for thermal deformation measurement. The entire measurement system includes high-resolution capacitive displacement sensors, precision standard balls, adjustable probe holders, and data acquisition components. The capacitive sensor driver has a resolution of up to 0.4 nanometers, which provides the most accurate spindle thermal deformation measurement. The precision standard ball is mounted on the spindle holder, the dynamic displacement of the standard ball is measured by the capacitance probe, and the data of each capacitive displacement sensor are combined and analysed to obtain the measurement result of thermal deformation. The displacement sensor installation position is shown in figure 6. A capacitive probe is installed on the bottom of the spindle to measure the axial Z-axis movement. A pair of right-angled capacitive probes is mounted on top of the precision standard balls to measure the radial X-axis and Y-axis movements.

A total of 7 temperature measurement points are arranged on the spindle to obtain the temperature field of the spindle. The Elite series TMP190 temperature sensor module is used. The position of the temperature measurement point is shown in table 4.

Table 4. Temperature sensor location.

No.	**Measuring position**
1	Spindle base
2	Spindle shaft head
3	Spindle housing middle
4	Spindle housing near front spindle bearing right side
5	Spindle housing near front spindle bearing left side
6	Spindle housing near rear spindle bearing right side
7	Spindle housing near rear spindle bearing left side

Considering that the temperature field and the thermal deformation are different at different rotational speeds, different spindle speed experiments are designed. The machine speed range is 2,000-12,000 *r/min*. In each set of fixed speed experiments, under the condition that the main shaft is fully lubricated and the cool system is turned on, the rotation speed is set until it reaches the thermal steady state

The laser sensor can achieve non-contact measurement of a measuring distance between 20mm.

Figure 7. TC series of multi-channel temperature tester.

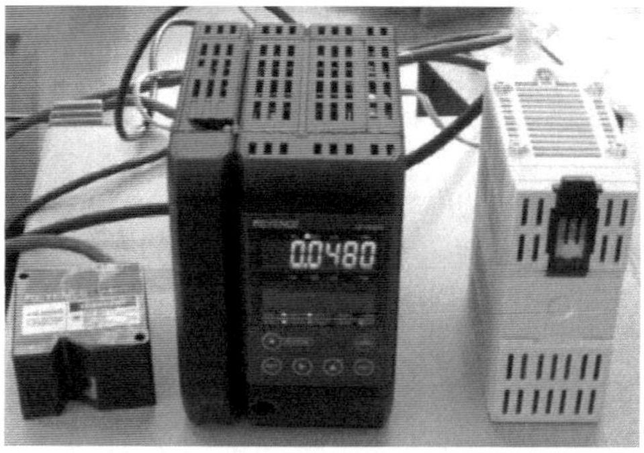

Figure 8. LK-G5000 CMOS laser sensor.

IOP Conf. Series: Materials Science and Engineering **399** (2018) 012017 doi:10.1088/1757-899X/399/1/012017

The principle of diffuse reflection is used for laser acquisition. The wavelength of the light source is 655 ηm and the diameter of the laser light is 50 μm. Select LK-G5000 high-accuracy CMOS laser sensor, TC series multi-channel temperature tester and seven temperature sensors to measure the thermal deformation and temperature rise of the spindle (see figures 7 and 8).

Experimental measurement conditions: room temperature 25 °C, using progressive speed. The temperature field data and spindle thermal distortion data at different rotation speeds are measured when the spindle is unloaded. The spindle speed starts from low speed and continuously increases the spindle speed. Keep running at a required speed until the relative stability of the heat balance and thermal displacement is reached at this speed. By comparing the thermal displacement of the motor spindle with the temperature rise and the simulation, the correctness of the simulation result is verified, and the experimental basis can be provided for further calculation of the thermal error.

6.2. Model simulation and verification

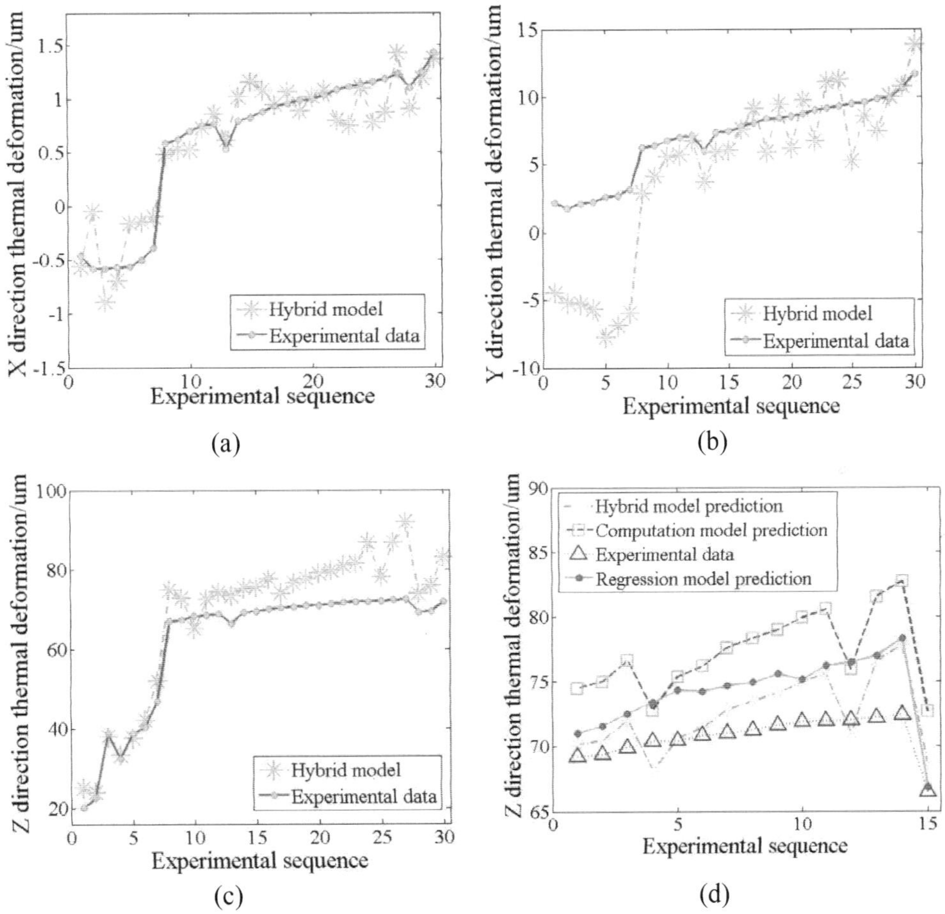

Figure 9. Hybrid modelling thermal deformation prediction results and comparison with other models in Z-direction. (a) Comparison between predicted X-direction thermal deformation and experimental data. (b) Comparison between predicted and experimental data of thermal deformation in Y direction. (c) Comparison between predicted and experimental data of thermal deformation in Z direction. (d) Comparison of the Z-direction thermal deformation hybrid model with other models.

This experiment measured the thermal deformation at different speeds from 2000-12,000 *r/min* and recorded the temperature rise at different speeds. A total of 50 sets of data were measured in the hybrid model results. 20 sets of data were used to establish a hybrid model, and the remaining 30 sets of data were used to test the accuracy of the hybrid model and the individual models. 15 sets of data were within the scope of model building data (internal data). The other 15 groups were outside the scope of the modelling data (external data).

Figure 9 (a), (b), (c) shows the overall prediction effect of the hybrid model in the three directions of X, Y, Z thermal deformation of the motorized spindle. It can be seen from the figure that the predicted thermal deformation of the hybrid model fluctuates within a certain range of experimental data, and the overall trend is similar to the experimental data. By calculation, the maximum prediction error in X direction is 0.6598 *um*, the minimum error is 0.1026 *um*, the maximum error in Y direction is 10.3142 *um*, the minimum error is 1.2752 *um*, the maximum error in Z direction thermal deformation prediction is 11.0347 *um*, and the minimum error is 0.8409 *um*.

The fitting curves of the predicted values and experimental data of each model of external data are shown in figure 9 (d). Since the thermal deformation in the Z direction is the largest, only the prediction of each model in the Z direction is shown, the other directions being similar to the Z direction. It can be seen from the figure that compared with the predicted value obtained by using the computation model and regression model alone, the predicted value of the hybrid prediction model corresponds better with the experimental data, and the error of the computation model is controlled between 6.433-11.725*um*. The prediction error of the regression model is controlled between 3.534-7.245*um*, and the prediction error of the hybrid modelling is within 2.312-6.356*um*. It can be seen from the figure that the hybrid model obtains higher prediction accuracy because the disadvantages of larger errors of the computation model are compensated by the PLS modelling method. The prediction results of the hybrid modelling can effectively follow the changing trend of the thermal deformation of the motorized spindle, basically around the experimental data distribution. The regression modelling and the computation modelling have large deviations from the experimental data and the error of the computation model is the largest. The hybrid modelling has higher prediction accuracy and improves the prediction accuracy. The data comparison results are shown in the table 5 below.

Table 5. Mean absolute error (MAE) of model predictions.

	Model	Regression modelling	Hybrid modelling	Computation modelling
MAE	Internal data (X/Y/Z)	0.5396um /0.9107um/ 1.1321um	0.1443um /1.3016um /1.4885um	0.1773um /1.5479um /3.7069um
	External data (X/Y/Z)	0.4528um /0.8443um /6.1225um	0.0938um /1.2437um /1.1549um	0.2256um /1.5532um /7.396um

By calculating the mean absolute error of the model, the internal and external data models are used to predict the effect of regression modelling, computation modelling and hybrid modelling. As can be seen from table 5, the hybrid model has not much effect on the internal and external data of the model. The mean value of the model prediction error is lower than that of the regression model. The overall prediction effect of the hybrid modelling is better. The PLS regression model has better prediction effect than the external data model in the prediction of the data, and there is a large difference in the Z axis. This is a limitation of the regression model modelling, and it has good prediction effect only within the scope of modelling data. By adding the theoretical model, the ductility of the model is enhanced, and the accuracy of thermal deformation prediction is improved.

7. Conclusions

In this paper, by means of the analysis of the advantages and disadvantages of the computation model and the regression model, a hybrid prediction model is proposed in order to improve the thermal deformation prediction of the motorized spindle and the following conclusions are obtained:

- the hybrid model can be used to analyse the thermal deformation of the motorized spindle within a large data range;
- the hybrid model combines the advantages of the computation model and the regression model, and hence achieves higher prediction accuracy;
- the hybrid model prediction is better than the computation model and the regression model; it is proved that the hybrid model can make up for the prediction error of other models, and make the real-time thermal deformation compensation be possible during operation.

Acknowledgments

This research is supported by General Project of Shenyang Jianzhu University（2017017）, Innovation Team Development Program of Ministry of Education (IRT_15R45), Science and Technology Project of Shenyang (F16-096-1-00).

References

[1] Wanli X, Xuebing Y, Lang L and Julong Y 2009 *J.Mech. Eng.* **45**1-18
[2] Zhu J 2008*Robust ThermalErrorModeling and Compensation for CNC MachineTools* (USA: Department of Mechanical Engineering/The University of Michigan) pp35-56
[3] Bryan J B 1990*Ann CIRP* **39** 645-56
[4] Weck M, Mckeown P, Bonse R and Herbst U 1995 *Ann. Manuf. Technol.* **44** 589-98
[5] Ramesh R, Mannan M A and Poo A N 2000 *Int. J. Mach. Tools Manuf***40** 1235-56
[6] Mayr J et al 2012 *CIRP Ann.Manuf. Technol.***61** 771–91
[7] Yang L, Wanhua Z, Shuhuai L, Jun N, Wenwu W and Bingheng L 2015 *Int. J. Mach. Tools Manuf* **95** 20-38
[8] Tseng P C andHo J L 2002 *Int. J. Adv. Manuf. Technol.* **19** 850-8
[9] Qianjian G, Hongmei W and Aijun L *J. Hebei Univ. Sci. Technol.* **36** 344-50
[10] Chunli L, Zhiyuan R and Emin L 2012 *J. Nanjing Univ. Sci. Technol.***36** 1021-5
[11] Chunli L, Zhiyuan R, Jun L and Ruicheng F 2011 *J. Xian Jiaotong Univ.* **45** 50-4
[12] Chibing H, Jing Y and Tingchuan L 2010 *Mach. Des. Manuf.* **12** 30-2
[13] Dongman Y, Yanhui H, Di W, Xiaojing L 2014 Appl. Mech. Mater. **556-562** 1170-3
[14] Shuangyi L, Chen C and Huijun Y 2011 *Manufa. Autom.* **34** 57-60
[15] Chi M, Jun Y,Liang Z, Xuesong M and Shi H2015 *Appl. Therm. Eng.* **86** 251-68
[16] Chunli L, Zhiyuan R, Jun Land Lina R 2012 *Adv. Mater. Res* **466-467** 961-5
[17] Chunli L, Zhiyuan R, Jun L and Jingfang F 2011 *Adv. Mater. Res* **291-294** 2991-4
[18] Weiqing L, Jianzhong J, Yazhou X and Zichen C 2008 *Comput. Integr. Manuf. Syst.* **14** 295-9
[19] Liangyu C, Dawei Z, Weiguo G, Xiangyang Q and Yu S 2011 *Adv. Mater. Res* **154-155** 1305-9
[20] Chunli L and Zhiyuan R2010*Adv. Mater. Res.* **129-131** 556-60
[21] Yufeng S, Wenxin Y, Deping L, Wufa L and Zhiyong D2013 *Modul Mach. Tool Autom. Manuf. Technol.* **1** 36-41
[22] Yi Z and Jianguo Y 2011 *J. Mech. Eng.* **47** 134-9
[23] Ke Z, Xiangnan M, Lixiu Z, Wenda Y and Yuhou W 2013 *Adv. Mater. Res.* **712-715** 1209-12
[24] Dongman Y, Yanhui H, Di W and Xiaojing Li 2014 *Appl. Mech. Mater.* **556-562** 1170-73
[25] Min Y, Di W, Peixin W and Xiaojing L 2012 *Appl. Mech. Mater.***155-156**273-7
[26] Chi M, Zhao L, Xuesong M and Jun Y Arch.2016 *Proc. Inst. Mech. Eng. Part C J. Mech. Eng. Sci.* (vols 203-210) pp 1989-96
[27] Yong L, Shengdong G and Zhaopeng H 2011 *Adv. Mater. Res.* **305** 340-3
[28] Shi H, Xuesong M, Jun Y and Tao T 2012 *J. Huazhong Univ. Sci. Technol.* **40** 26-30
[29] Guisong D, Feng Y, Yusheng Z, Jianguo Y 2014 *Mach. Tool Hydra.* **42** 9-13

[30] Xiaoling W, Yubing S, Yong L, Menghan L and Jianfeng W 2014 *J. Hebei Univ. Eng.* **31** 104-7
[31] Zhifeng L, Minghui P, Aiping Z 2015 *Int. J. Adv. Manuf. Technol.* **76** 1913-26

IOP Conf. Series: Materials Science and Engineering **399** (2018) 012018 doi:10.1088/1757-899X/399/1/012018

Integrating the Lean concept in sustainable manufacturing development

A Florescu[1*] and B Barabaş[1]

[1] Transilvania University of Brasov, Faculty of Technological Engineering and Industrial Management, Romania

[*] E-mail: fota.a@unitbv.ro

Abstract. In the current context of climate change in the world, it is increasingly necessary to implement strategies and ways of action to adapt industrial companies to sustainable production. The implementation of the Lean concept has a particular impact on creating added value, by reducing wastage, production waste and improving efficiency in order to obtain products/services of higher quality. Through a new approach, the authors propose the use of Lean techniques to develop sustainable manufacturing with an impact on the environment and their implementation in the production sectors. The main objective of the paper is to highlight the opportunities and challenges of Lean Manufacturing in the context of sustainable development. At the same time, the application of Green Manufacturing concept is the development of products/services/processes that save energy, natural resources, with beneficial effects on the environment. A starting point in the analysis will consist of the detailed study of the literature regarding the use of this two concepts, the limits that still occur in their implementation and the finding of a methodology for networking them. The results will demonstrate that integrating the two dimensions will lead to increased performance of the system and to the development of a sustainable company.

1. Introduction

The development of advanced production systems now involves profound changes that will affect both the basic technique and the management, organization and quality assurance techniques. The establishment of advanced production systems has been achieved in highly industrialized economies (such as Japan, USA, Germany, etc.), with the current development prefiguring profound mutations and transformations. The implementation of the concept of Industry 4.0 in the context of sustainability will, according to studies currently underway [1 - 3], have an important impact on the creation of added value in the industrial environment, together with the emergence of new opportunities related to the development of sustainable production, the characteristics of each company.

Competitive environment forces increasingly more organizations, industrial companies to review, restructure and redirect activities to achieve competitive advantage in terms of sustainable development. The concept of Lean manufacturing, as a new philosophy, successfully applied in the field of increasing production efficiency, can be developed and implemented together with the concept of Green manufacturing. Integration of the two dimensions can lead to increased competitiveness, the performance of the organization and, last but not least, the development of sustainable companies.

The Lean concept was developed by Taiichi Ohno in the 1950s and later developed by Shigeo Shingo and was first applied in the automotive industry, within the *Toyota Production System* (TPS), [4].

Lean philosophy was developed in the West by Womack and Jones (2003) [5], who considered this type of management as "an antidote against waste - any human activity that does not bring value". Currently there are two organizations that implement this management philosophy: Kaizen Institute Ltd. (Masaaki Imai, Switzerland, 1985) [6] and the Lean Enterprise Institute (J.P. Womack, USA, 1996) [7]. By efficiently applying Lean tools and techniques, the system has been rethought, the processes in the value chain have been controlled and improved to increase performance, the speed of processes and to minimize waste. To meet customer demand not only through Lean objectives, companies are rethinking modern manufacturing processes in terms of sustainable products / services ecologically sustainable, respecting government environmental regulations. At present, organizations have a proactive role in designing recyclable products, as well as developing cleaner production services and processes. Thus, the green paradigm has emerged as a philosophical and operational approach to reduce the negative environmental impact of an organization's products and services, as well as to improve the ecological efficiency of their operations, while at the same time fulfilling their financial goals [8].

1.1. The importance of research

EU waste management policies aim to reduce the impact of waste on the environment and health and to improve resource efficiency in Europe. The long-term goal is to turn Europe into a recycling society, avoiding waste and using inevitable resources as much as possible. According to Eurostat statistics [9] of 2014 and updated in 2017, total waste generated in the EU-28 by all economic and household activities amounted to 2 503 million tonnes; this is the largest quantity registered for the EU-28 in the period 2004-2014. It is noted that relatively large quantities of waste were generated in Romania, which requires the implementation of new methods of sustainable management and development.

The share of different economic activities and households in total waste generated according to the source [9] is shown in figure 1. At EU-28, in 2014, construction contributed 34.7% of the total, followed by extractive industry (28, 2%), manufacturing (10.2%), water and waste (9.1%) and households (8.3%); the remaining 9.5% was waste generated by other economic activities, especially services (3.9%) and energy (3.7%).

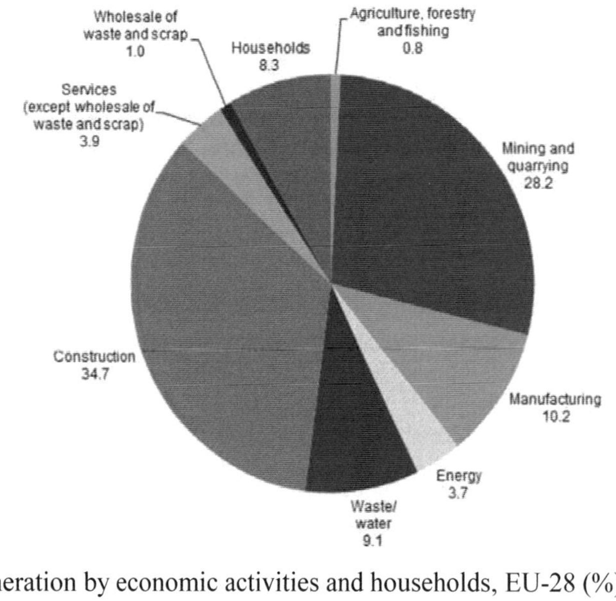

Figure 1. Waste generation by economic activities and households, EU-28 (%). Source Eurostat [9].

Also, 891 million tonnes of waste was generated in the EU-28, [9] except for the major mineral waste, equivalent to 36% of the total amount of waste generated (table 1). It is noticed that the production activity registered a relatively high level of waste, of 184.1 million tons, while the evolution over time recorded a decrease of 32.2%.

Table 1. Quantity of waste in different economic sectors in the EU (Eurostat source, [9]).

	2004	**2006**	**2008**	**2010**	**2012**	**2014**	**Rate of change% 2004-2014**
		(million tonnes)					
Total	**941.1**	**942.3**	**903.4**	**875.2**	**884.2**	**890.8**	**-5.3**
Agriculture, forestry and fishing	63.1	57.4	46.2	21.7	22.3	19.7	-68.7
Mining and quarrying	10.4	7.1	10.1	8.0	7.8	7.9	-24.1
Manufacturing	271.4	250.5	236.6	202.8	190.5	184.1	-32.2
Energy	92.2	100.0	88.8	81.9	93.6	90.7	-1.6
Waste / water	110.9	110.2	130.1	149.0	178.7	208.0	87.7
Construction	39.3	45.2	38.7	50.9	60.1	61.8	57.2
Other sectors	148.8	160.9	140.6	146.5	122.0	115.0	-22.7
Households	205.0	211.0	212.3	214.4	209.3	203.5	-0.7

Worldwide, as a result of studies in the USA, it is noted that manufacturing processes currently generate the largest amount of waste; the power consumption of the production system is 23% of total USA energy consumption [10]. As a result, one of the priority strategies for the development of sustainable production will have to be waste reduction, which will generate a favourable impact on the environment while reducing energy consumption and emissions.

Based on the literature [4, 5, 6, 7] and the remarkable results achieved by implementing the concept of Lean in various companies [11, 12], this research will identify opportunities to incorporate techniques and tools of Lean with current concept of Green manufacturing for the purpose of sustainable development. It will develop a methodology, an integrated model of sustainable practices, Lean and Green manufacturing, which will open future research directions.

2. Lean manufacturing vs. Green manufacturing

Lean manufacturing involves a team-based approach to identifying and eliminating losses, activities that do not add value through continuous improvement, ensuring a continuous flow of product / service, responding to customer demand, with the goal of perfection. Green Manufacturing focuses on eliminating waste and emissions. Developing organic production is often costly [10, 13, 14]. Applying new regulations and methodologies in the field of sustainable manufacturing involves the adoption of a new organizational culture by employees. Lean lines up perfectly with the organizational structure, culture and performance evaluation system to create long-term results. Although there are currently studies to develop the Lean concept integrated with the Green manufacturing concept [8, 10, 15], there is no methodology and no deployment practice in companies. Activities unsustainable that industrial companies should pay particular attention are: the waste of resources and high emissions from production, quality still inadequate products, the large number of defects / rejects and therefore increased number of reprocessing, the emergence waste correlated with the dissatisfaction of employees and customers. Application methods and tools to improve leading to eliminating waste in manufacturing simultaneously with regulations on development and protecting the environment can have a positive impact on reducing costs and emissions for long periods of time [10].

Strategies to implement Lean and Green Manufacturing concepts were analysed in several studies and recent research, observing that they can be implemented both sequentially and simultaneously [15, 16]. In the paper [17], the research was geared towards applying the Kaizen concept resulting in an improvement in environmental aspects with an environmental impact. This approach has resulted in a reduction in resource use by 20% and energy flow by 5%. A detailed research [18] analyses extensively (through 58 scientific papers) the possibility of integrating Lean and sustainable manufacturing.

2.1. Lean manufacturing. Economical Sustainable

Economic sustainability is linked to strategies to manage their resources in a way that results in financial benefits for a long time. Lean manufacturing practices are among the practices that are necessary for economic sustainability [10]. Lean concept is considered a business strategy with the main objective to increase customer satisfaction through better use of resources.

Research in the papers [19] states that Lean manufacturing is the most popular investment choice among companies that invest to remain competitive, flexible and profitable. By applying the Lean strategy in manufacturing systems, the effectiveness of the implemented system has been demonstrated, resulting in a number of benefits: reducing the direct and quality costs by 5-20%, reducing the waiting times by 80%, processing speeds, reducing production unfinished (WIP) and 50-80% finished product stocks, increase in operating profit by about 5%, waste reduction, productivity improvement, operational performance and quality, production flexibility, efficient use of labour, space and production capacity.

Lean Management methods can be summarized in three concepts: waste reduction / losses, maintaining quality of products / services and accelerating production. To implement the Lean concept, it is necessary to know the specific tools: Mapping Process, Value Stream Mapping (VSM), Six Sigma, 5S, Cellular Manufacturing, Kanban System, JIT Concept, Kaizen Continuous Improvement, Standardization, TQM (Total Quality Management), TPM (Total Preventive Maintenance), etc.

2.2. Green manufacturing. Sustainable manufacturing

Sustainability or sustainable development is generally defined as "development that meets the needs of today's generation, without compromising the ability of future generations to meet their own needs". Sustainability has three dimensions, namely economic prospects, ecological balance and social responsibility [20].

The paper [21] suggests that "a sustainable company is one that generates profits for its shareholders, while protecting the environment and improving the lives of those with whom it interacts". Also referred to as sustainable manufacturing, Green manufacturing has as major concern the consequences of production activities and the entire life cycle of the product on the environment. In another view of this concept there is a set of strategies, practices and behavior, to prevent, reduce and / or eliminate the negative impact on the environment [22, 23].

The synergy between the two Lean and Green manufacturing concepts [24] can be achieved from the waste patterns identified in the two systems (figure 2).

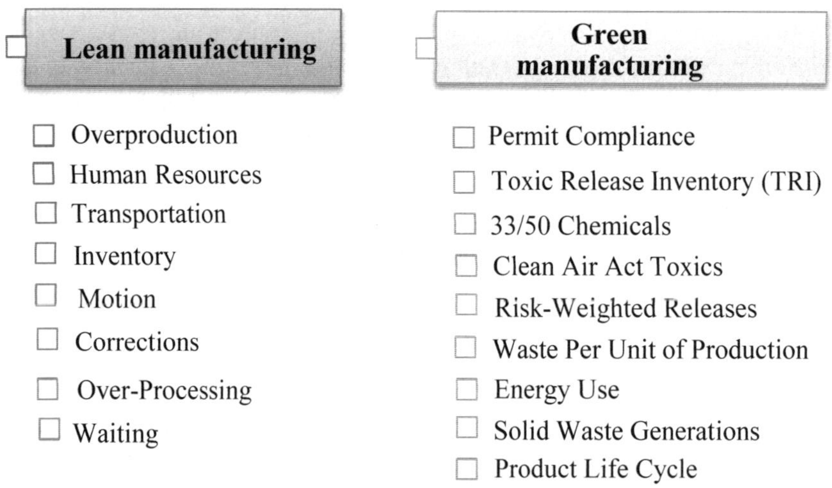

Figure 2. Forms of waste identified by Lean manufacturing & Green manufacturing.

The main operational activities [10] for waste reduction are: product redesign, replacement, reduction, recycling, remanufacturing, return, stock control, domestic consumption, reduction of transport, and reduction of defects.

Among the advantages of Green manufacturing can be listed: saving valuable resources, economic incentives, energy efficiency and environmental benefits by reducing waste and harmful emissions, competitive advantages, quality and reliability.

3. Implementation methodology

The identification and measurement of mass and energy flow, followed by the continuous improvement of the environmental value flow, is now a reflection, given the increasing environmental concern. These activities are similar to Lean Management principles to ensure information flows, material and energy continue in industrial processes. The integrated Lean & Green Manufacturing model will analyse the mass and energy flows of a plant as a final customer and will analyse the impact on sustainable development, [14]. The Lean Manufacturing and Green Manufacturing concept integration and implementation scheme is outlined in figure 3, where we can see the synergies between Lean thinking and the continuous improvement of the environment.

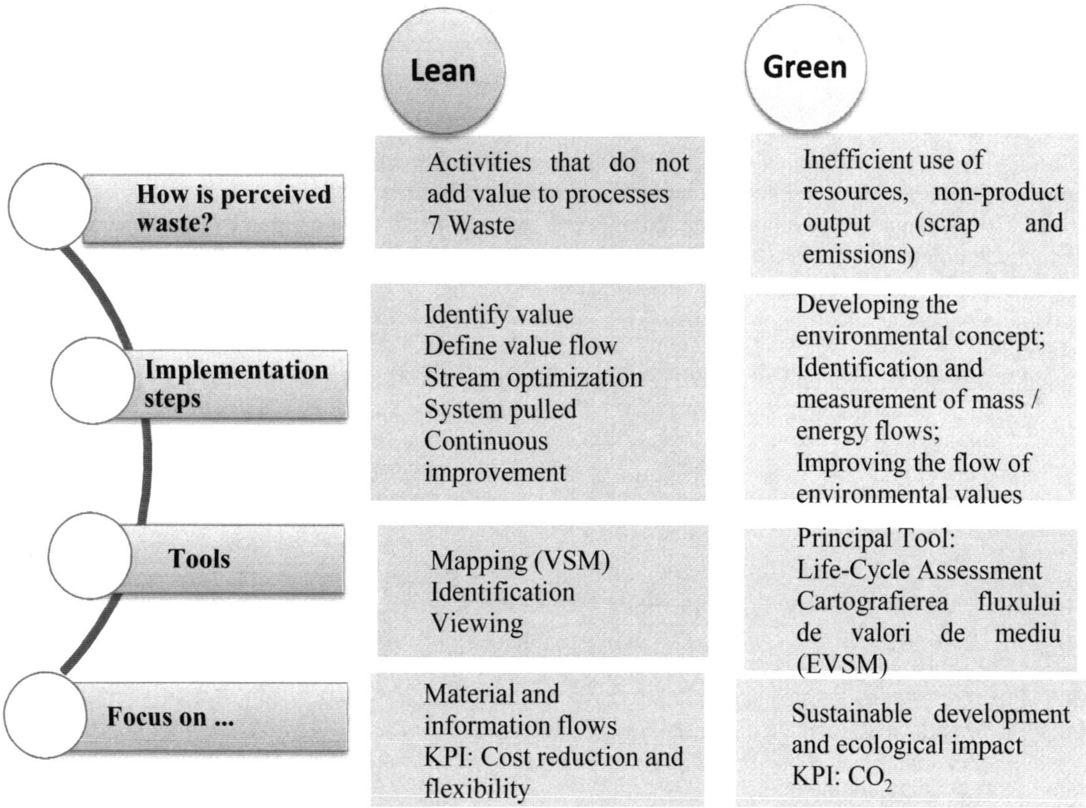

Figure 3. Lean & Green Manufacturing integrated implementation scheme.

The main stages of implementing a Lean & Green Manufacturing integrated model involves:

- identify the flow of values by identifying critical points that require improvement;
- identification of the environmental aspects and the impact of the value flow in manufacturing, according to the standards in force (ISO 14001);
- measuring the flow of environmental values, by identifying data about the environmental process;

- analysis of productivity in resource use and potential impact on the environment;
- lean management, organizing the team by involving the specialists, the managers;
- improving environmental value flows by identifying waste disposal opportunities;
- implementing plans for continuous improvement of the environment for strategic business projects; the improvement opportunities identified by the lean application will be integrated into the continuous improvement plan for the environment.

The implementation of Lean tools in production leads to the development of economic sustainability, through the efficient use of resources and the achievement of impressive benefits. Lean practices tend to lead to green results, being the first step in developing a green company.

4. Conclusions

Although it does not directly aim to reduce environmental impact, Lean methodology can provide environmental benefits by improving the environment while increasing productivity. The fundamental principle of Lean thinking, continuous improvement and employee empowerment creates opportunities for the development of sustainable green industries by reducing waste and emission elements that characterize Green manufacturing.

In the new integration concept for the development of sustainable manufacturing, Lean aims to link the core production targets, such as cost reduction and production times, to Green manufacturing waste disposal targets that affect the environment. It thus tends to interlink the objectives of the three pillars of sustainable development: economic sustainability, sustainable manufacturing and social sustainability. Although recent studies show a synergy of the two concepts, there is no common methodology yet, an agreement on their inter-implementation. In the development of a sustainable company, Lean and Green Manufacturing can mean an innovative. A future research activity is required to investigate real links, overlaps, advantages and disadvantages, positive impacts and even gaps, development and implementation of the Lean & Green Manufacturing concept, by developing an integrated model applicable to various organizations.

References

[1] Müller J M, Kiel D and Voigt K-I 2018 *Sustainability* **10** (1) 247
[2] Leyh C, Martin S and Schäffer T 2017 *Proc. Fed. Conf. on Comp. Sci. Inf. Sys.* 11 (ACSIS) 989-993
[3] Bartodziej C J 2017 *The Concept Industry 4.0: An Empirical Analysis of Technologies and Applications in Production Logistics* (Springer Gabler, Wiesbaden, Germany)
[4] Taiichi O 2009 *Toyota System Production* (Ikar Management Consult Press, Japan)
[5] Womack J and Jones D 2003 *Lean Thinking:* (Simon & Schuster, NewYork)
[6] Massaaki I 2012 *Gemba Kaizen* (Mcgraw-Hill Education)
[7] Rother M and Shook J 2009 *Learning to See* (The Lean Enterprise Institute, Inc.)
[8] Garza-Reyes J A et al. 2014 *Advanced in Prod. Manag. Syst.* (Springer) p 71-81
[9] Eurostat http://ec.europa.eu/eurostat/statistics-explained/index.php/Waste_statistics/ro
[10] Firozabadi S S F 2015 *An application of sustainable lean and green strategy with a six sigma approach on a manufacturing system.* Theses and Dissertations (The University of Toledo, Digital Repository)
[11] Liker J K and Rother M 2011 *Why Lean Programs Fail* (Lean Enterprise Institute Inc.)
[12] Jackson T and Jones K 1996 *Implementing a Lean Management System* (Productivity Press, New York)
[13] Dornfeld D A 2012 *Green manufacturing: fundamentals and applications* (Springer Science & Business Media)
[14] Murillo-Luna J L Garcés-Ayerbe C and Rivera-Torres P 2011 *J. Clean. Prod.* **19** (13) p 1417-1425
[15] Galeazzo A Furlan A and Vinelli 2013 A*J.Clean. Prod.* **85** p 191–200
[16] Yang M G Hong P and Modi S B 2011 *Int. J. Prod. Econ.* **129** (2) p 251–261

[17] Pampanelli A B Found P and Bernardes A M 2014 *J. Clean. Prod.* **85** p 19-30
[18] Hartini S Ciptomulyono U 2015 *Ind. Eng. and Serv. Science. Procedia Manufacturing* **4** 38 – 45
[19] Demeter K and Matyusz Z 2011 *Int. J. Prod. Econ.* **133** 154–163
[20] Lin Wu I et al. 2015 *Sustainability* **7** p 3841
[21] Wagner M Blom J *Bus. Ethics Eur. Rev.* **20** p 418–432
[22] Dilip Maruthi G Rashmi R 2015 Mat. Proceedings **2** (4-5) p 3350-3355
[23] Salvador R Piekarski C Mand Francisco A C 2017 *Int. J. Prod. Manag. Eng.* **5** (2) p 73-78
[24] Miller G Pawloski J and Standridge C 2010 *J. Ind. Eng. Manag.* **3** (1) p 11-32

Comparative analysis of thermal and acoustic performance of composites made from wood fibres, recycled rubber and ABS

S V Georgescu[1] **and C Cosereanu**[1*]

[1] Department of Wood Engineering and Wood Products Design, Faculty of Wood Engineering, Transilvania University of Brasov, B-dul Eroilor 29, Brasov, Romania

[*] E-mail: cboieriu@unitbv.ro

Abstract. The paper investigates the thermal and acoustic properties of composites made from wood fibres (WF) and recycled rubber (R) crumbs and acrylonitrile butadiene styrene (ABS) shavings resulted in the particleboards' edge banding. Panels with a target density of 300 kg/m^3 were manufactured for testing the thermal conductivity coefficient and sound absorption coefficient. Mixed panels WF:ABS, R:ABS with participation rates (in %) of 10:90, 20:80, 30:70; 40:60; 50:50 and WF:R:ABS with participation rates (in %) of 5:5:90, 10:10:80, 15:15:70; 20:20:60; 25:25:50 were investigated in this paper. The experiment simulated the indoor and outdoor temperature conditions for the winter and summer seasons, namely 20 °C for indoor and -10 °C to 35 °C for the outdoor temperatures. The results show that, with the increase of ABS share, the thermal performance of the panel increases. The presence of WF in the composition has a good influence on the thermal performance of the panels, whilst the presence of rubber brings a better acoustic performance of the composites. The analysis of the results show that the thermal conductivity coefficient (λ) experimentally determined depends on the outdoor temperature and recorded the best value for the composite WF-ABS (0.0434 $Wm^{-1}K^{-1}$) followed by WF-R-ABS (0.0460 $Wm^{-1}K^{-1}$) and R-ABS (0.0477 $Wm^{-1}K^{-1}$). Maximum values were recorded for R-ABS structure at a temperature of 35 °C (0.0573 $Wm^{-1}K^{-1}$), but this structure recorded the best sound absorption coefficient (0.87).

1. Introduction

Today's high consumption is one of the major causes of waste accumulation. Some materials, such as rubber and plastics decompose naturally over decades and centuries respectively, study estimates. A feasible way to reduce the inventory of such waste materials is to recycle them and design new added value products.

Acrylonitrile-Butadiene-Styrene copolymer (ABS) is a thermoplastic widely used in appliance housings, automotive interior parts, electrical devices, toys, medical equipment and furniture, because of its toughness, dimensional stability, heat and scratch resistance and excellent colour and aesthetic functions. Nowadays, the researchers investigate the possibility of recycling the used and aged ABS, by end-of-life treating and reprocessing it into new products [1, 2], or by using natural fillers such as agro-wastes [3, 4] or wood flour [5, 6] incorporated into ABS-matrix and obtaining composites with low density, improved strength and water resistance and good thermal performance. An unused resource of ABS results in the edge banding process of melamine faced particleboard in the furniture industry. Generally, the edge of melamine faced particleboard is preferably banded with ABS, which is 1.5 mm wider than the thickness of particleboard and it is finally removed by a levelling operation.

Even though ABS polymers are used largely for mechanical purposes, they also have good thermal insulation properties and the present research intends to use this waste resource to obtain composites with high thermal and acoustic performance.

Waste tires are significant raw materials for use in composites, because of rubber characteristics, such as good elasticity, sound insulating and light weight. The short life time of rubber and improvement of the living standards have as result a major demand for recycling the waste tires. Considering the thermal and acoustic properties as outputs, fine or coarse recycled tire rubber crumbs are used by researchers as fillers or matrix in manufacturing and testing composites with application in construction [7, 8, 9]. By adding wood flour, novel composites with tailored mechanical properties and optimized water resistance were obtained and tested by researchers in composites field [10, 11, 12]. Other recycled plastics such as high-density polyethylene (HDPE) or low-density polyethylene (LDPE) blended with rubber and wood dust resulted in strengthened and water resistant composites [13, 14].

In this study recycled ABS-rubber-wood fibres based composites were obtained, where ABS acts as a matrix, wood fibres and rubber as fillers. The obtained composites could be used as potential heat and sound insulating materials. The thermal and acoustic properties of the composites were characterized by performing tests for thermal conductivity and sound absorption.

2. Experimental

2.1. Materials

The materials used in designing and developing composites are presented in figure 1, and they are wood fibres (WF), recycled tire rubber crumbs (R) as fillers and acrylonitrile-butadiene-styrene waste particles (ABS) as matrix.

2.1.1. ABS waste. ABS used in the present research was obtained from the production facility of the Wood Engineering Faculty in Brasov, Romania, where ABS waste particles were collected in special bags attached to the edge banding machine. ABS waste consists of thin shavings of various colours (figure 1), having thicknesses between 0.2 mm and 0.5 mm, lengths in the range 2 mm - 16 mm and widths in the range 1 mm - 2.5 mm.

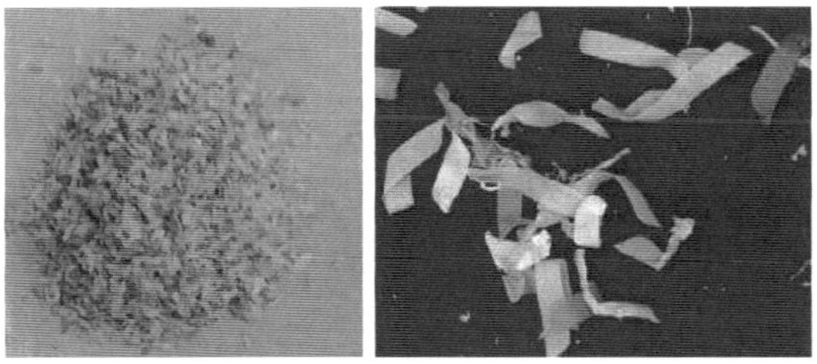

Figure 1. Sample of ABS shavings resulted in the furniture manufacturing process of edge banding melamine faced particleboard and used in the present research as matrix for new composites.

2.1.2. Recycled tire rubber crumbs. One size of crumb rubber was used to prepare the composites, namely coarse crumb rubber with particle size ranging from 2 mm to 4 mm. The rubber was provided by a local supplier and it was used in the present research as filler (figure 2).

Figure 2. Sample of crumb rubber with sizes between 2 mm and 4 mm used in the present research as filler for the new composites.

2.1.3. Wood fibres. The wood fibres were supplied by Kastamonu Romania, located in Reghin and they constitute the raw material used by the company to produce medium density fibreboard. The fibres are a mixture of 80% resinous wood and 20% beech wood (figure 3).

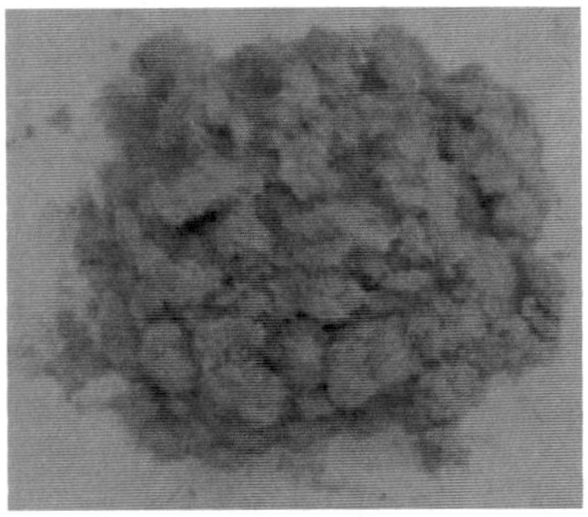

Figure 3. Wood fibres composed of 80% resinous wood and 20% beech wood, used in the present research as filler for the new composites.

2.2. Composites manufacturing

The composites were prepared using different ABS, R and WF shares, as shown in table 1. The target density of the panels was 300 kg/m^3. The components were mixed mechanically for 5 minutes and the mixture was then poured into suitable wood frames with the interior dimensions of 420 mm x 420 mm x16 mm. It should be mentioned that it was difficult to prepare homogenous composites with more than 50% rubber content. The mat thus formed had an initial thickness of 50 mm and resulted with the final thickness of 16 mm after hot pressing. A thermo-resistant foil was used to cover the faces and edges of the mat in order to prevent sticking of the sample with the hot press. The mixture thus formed was then introduced in the hydraulic hot press machine at a temperature of 160 °C and hot pressed between the plates for 20 minutes without applying a pressure from the plates. After conditioning the panels for 48 hours, they were sized to 410 mm x 410 mm x16 mm and then cut into specimens and subjected to different thermal and acoustic tests according to European standards. The number and sizes of specimens used for the tests are according to DIN EN 12667: 2001 standard and SR EN ISO 10534-1: 2002.

Table 1. Composites prepared for thermal and acoustic investigations and concentrations of ABS, R and WF.

	ABS (%)	R (%)	WF (%)
WF-ABS 1	90	0	10
WF-ABS 2	80	0	20
WF-ABS 3	70	0	30
WF-ABS 4	60	0	40
WF-ABS 5	50	0	50
WF-R-ABS 1	90	5	5
WF-R-ABS 2	80	10	10
WF-R-ABS 3	70	15	15
WF-R-ABS 4	60	20	20
WF-R-ABS 5	50	25	25
R-ABS 1	90	10	0
R-ABS 2	80	20	0
R-ABS 3	70	30	0
R-ABS 4	60	40	0
R-ABS 5	50	50	0

2.3. Test methods

2.3.1. Thermal conductivity. The tests were performed on HFM 436 Lambda equipment (Netzsch, Selb, Germany), according to ISO 8301: 1991 [16] and DIN EN 12667: 2001 [15]. This testing method is based on the determination of the quantity of heat that is passed from a hot plate to a cold plate through the sample. After recording the temperature difference between the cold plate and the hot plate, the thermal conductivity coefficient (λ) is automatically calculated based on Fourier's Law. The equipment is first calibrated depending on the temperature differences (ΔT) and mean temperatures (Tm) set for the experiment. Table 2 presents the values configured for temperature in order to simulate outdoor and indoor temperatures. In this respect, the indoor constant temperature was considered to be 20 °C. The dimensions of the specimens were of 300 mm x 300 mm x 16 mm. The average of three measurements was reported as result.

Table 2. Temperatures set on HFM 436 Lambda equipment for the determination of thermal conductivity coefficient (λ).

	Temperature T1 (°C)	Temperature T2 (°C)	ΔT = T2-T1 (°C)	Tm =(T1+T2)/2 (°C)
1	-10	20	30	5
2	-5	20	25	7.5
3	0	20	20	10
4	5	20	15	12.5
5	10	20	10	15
6	15	20	5	17.5
7	20	20	0	20
8	25	20	-5	22.5
9	30	20	-10	25
10	35	20	-15	27.5

2.3.2. Textural examination of composites. The examination of the composites was carried out using a HP Scanjet 7650 scanner. The samples with sizes of 50 mm x 20 mm x 20 mm were scanned with the highest resolution of 4800 dpi in order to achieve maximum magnification of textural and morphological characteristics of the composite edge. The scanned images were afterwards cropped and magnified so as to obtain a clear textural representation of the composite.

2.3.3. Sound absorption. The samples were tested using a SCS80FA impedance Kundt tube with the following characteristics: diameter of the tube of 100 mm and frequency range between 70 Hz and 1800 Hz. The standard Kundt's tube measurement method for the determination of the sound absorption coefficient under normal sound incidence was applied according to the SR EN ISO 10534-1: 2002 standard [17]. The circular specimens prepared for the experiment had a diameter of 100 mm and a thickness of 16 mm. The result reported in the research is the average value of three measurements.

3. Results and Discussion

3.1. Thermal conductivity measurements

The prepared composites with varying filler (WF and/or R) content ranging from 0 to 50% of the total mass were tested to determine their thermal conductivity coefficient (λ). The results, as seen in figure 4, show that the incorporation of fillers (R and/or WF) in the ABS matrix increases the value of the thermal conductivity coefficient. Thus, R-ABS composite has a higher λ as compared to both WF-R-ABS and WF-ABS composites, but all of them are good thermal insulators with thermal conductivity reaching values between 0.0434 $Wm^{-1}K^{-1}$ (for WF-ABS) and 0.0573 $Wm^{-1}K^{-1}$ for R-ABS (figure 5).

The same trend was noticed for WF-ABS sample, for which, with the increasing of the WF content, the thermal conductivity coefficient also increased. The higher values of the thermal conductivity coefficient both for R-ABS and WF-R-ABS samples are related to the big size of crumb rubber which occupies voids in the ABS matrix of composites and diminishes the air gaps of the structure and the porosity, affecting thus the thermal performances.

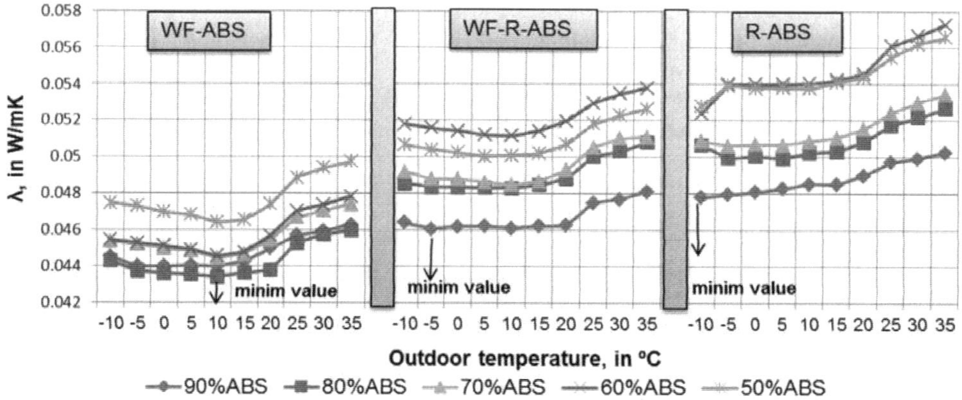

Figure 4. Thermal conductivity coefficient as function of temperature and structure.

The experimental results of all samples show that the thermal conductivity of the prepared composites depends on the temperature, as illustrated in figures 4 and 5. The thermal efficiency of the samples is different, depending on their components and structure. For example, the minimum thermal conductivity value obtained for the WF-ABS structure (0.0434 $Wm^{-1}K^{-1}$) was reached at an outdoor temperature of 10 °C and $\Delta T = 10$ °C respectively and an ABS content of 80% of the total mass. For the WF-R-ABS structure, the minimum value of λ was reached at an outdoor temperature of 5 °C (equivalent to $\Delta T = 15$ °C) and ABS content of 90% of the total mass, whilst for the R-ABS structure, the maximum efficiency was reached at a temperature of -10 °C (equivalent to $\Delta T = 30$ °C and a cold climate) and ABS content of 90% of the total mass. In general, the thermal conductivity of the prepared composites is promising and supports their utilization in constructive applications, their values being comparable with the thermal conductivity of commercial heat insulators such as EPS or XPS, for which the manufacturers provide technical sheets with values between 0.031 $Wm^{-1}K^{-1}$ and 0.045 $Wm^{-1}K^{-1}$. The maximum values of λ were recorded for an ABS content of 60% of the total mass in case of both R-ABS and WF-R-ABS structures, and 50% in case of the WF-ABS structure, for an outdoor temperature of 35 °C (equivalent to $\Delta T = -15$ °C and a hot climate).

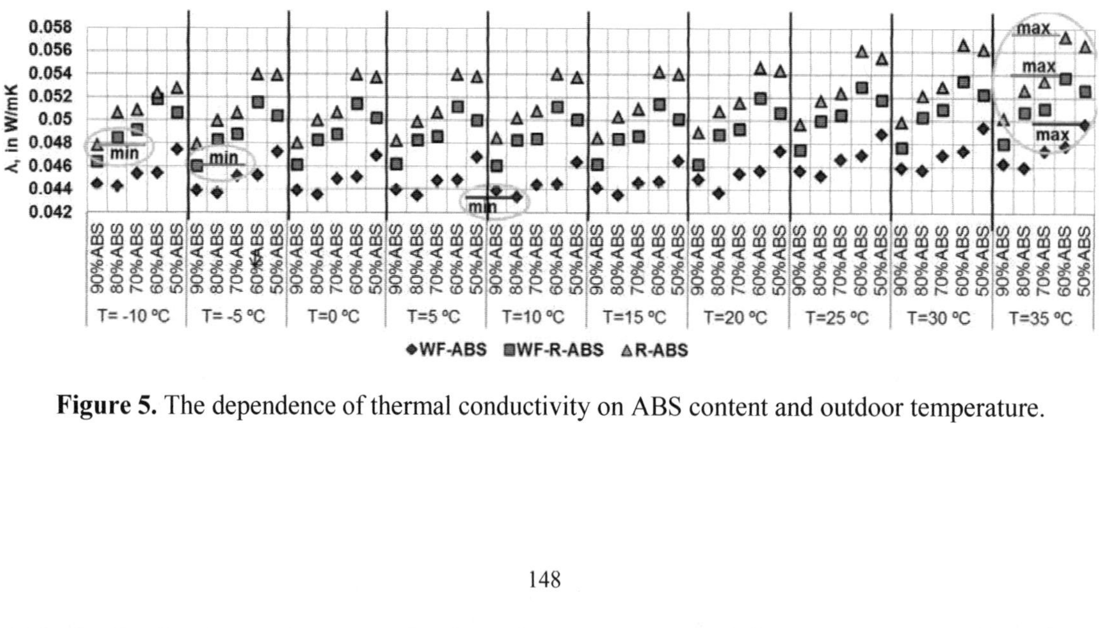

Figure 5. The dependence of thermal conductivity on ABS content and outdoor temperature.

3.2. Structure and morphology

Figure 6 presents the scanned structure and morphology of the prepared samples, showing differences between their cores and exterior faces. The structure is more compact on the faces, and the fillers are well incorporated in the ABS matrix, which practically melted at the contact with press hot plates and created a resistant surface of the composite. Instead, the core looks more porous, with a lot of voids created between the curled ABS shavings and filled from place to place by WF and R fillers.

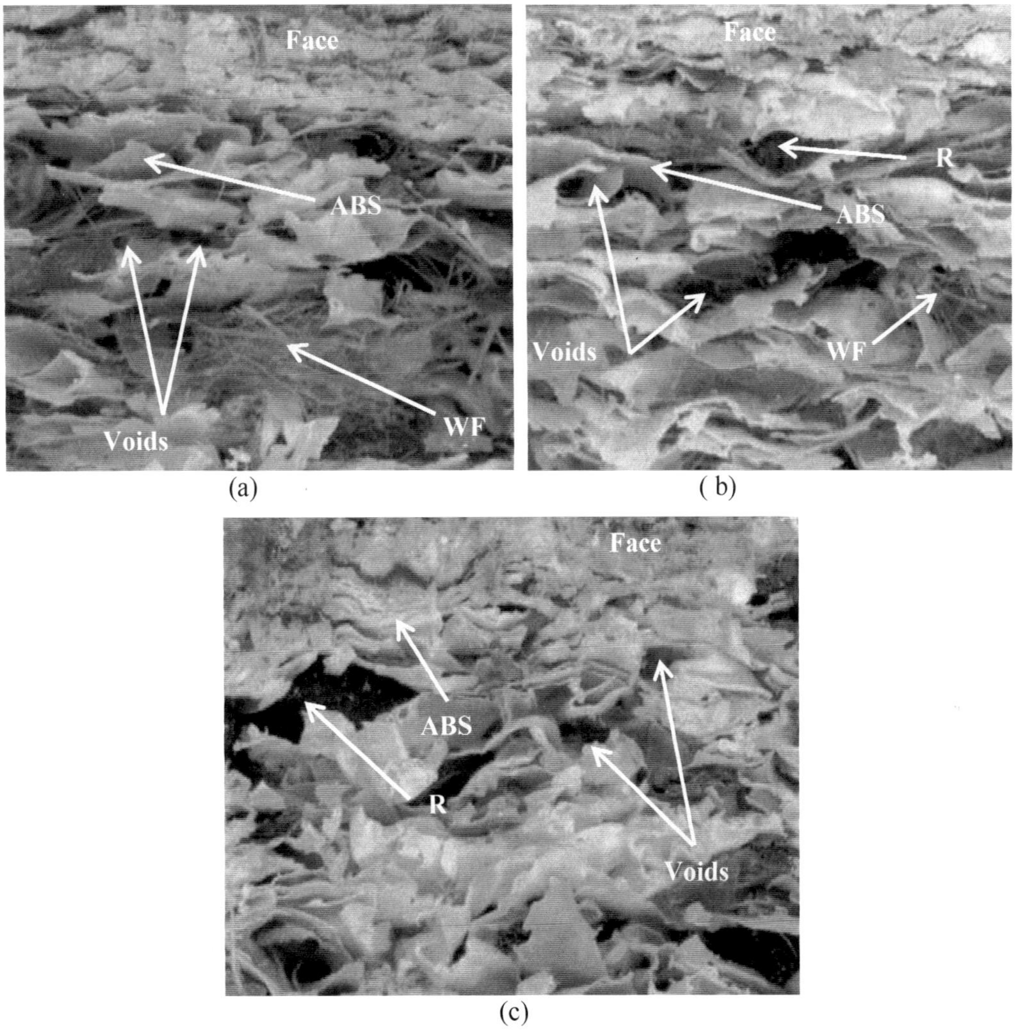

Figure 6. Scanned structures of composites with 50% ABS (a) WF-ABS, (b) WF-R-ABS, (c) R-ABS.

3.3. Sound absorption measurement

One sample of each composite panel was measured for sound absorption coefficient on the impedance tube. The results are shown in the diagrams in figures 7, 8 and 9. Contrary to the results of thermal conductivity, the rubber incorporation in the ABS matrix has a favourable influence on the acoustic performance of the composites. The best sound absorption coefficient was recorded for R-ABS sample with 50% R (0.87) followed by WF-R-ABS sample with 50% R (0.73) for frequencies around 1500 Hz.

IOP Conf. Series: Materials Science and Engineering **399** (2018) 012019 doi:10.1088/1757-899X/399/1/012019

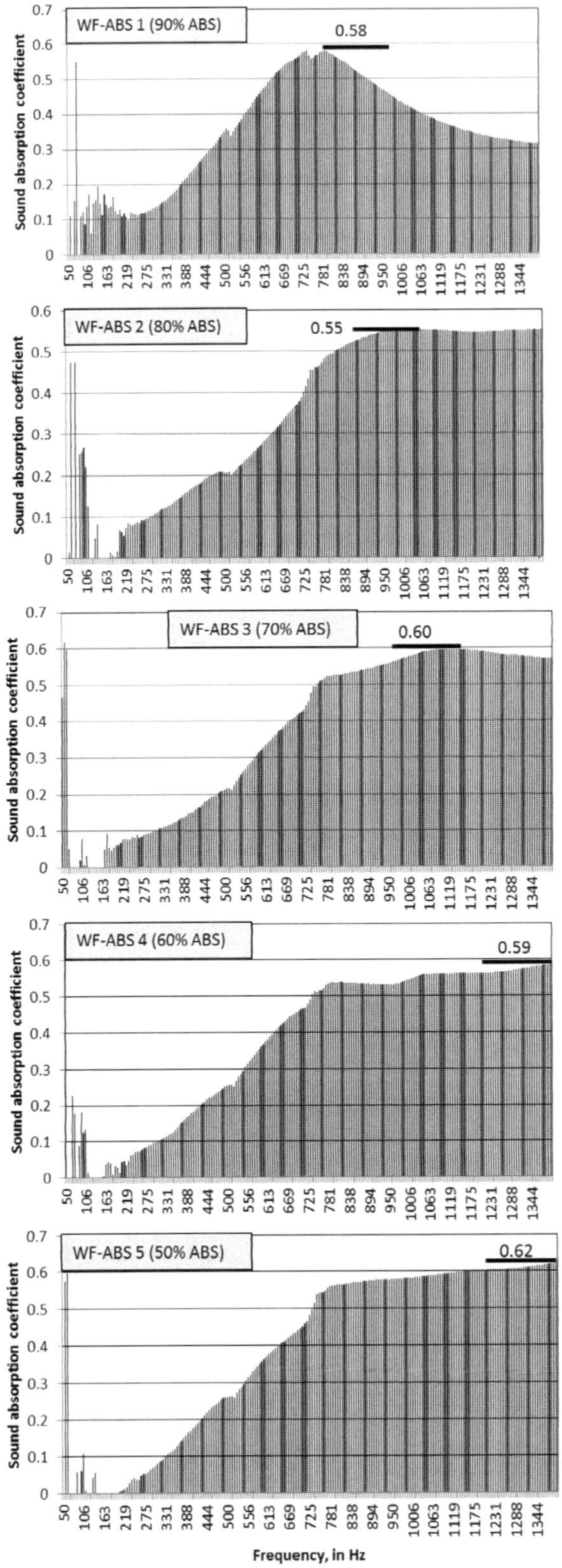

Figure 7. Sound absorption coefficient for WF-ABS structures.

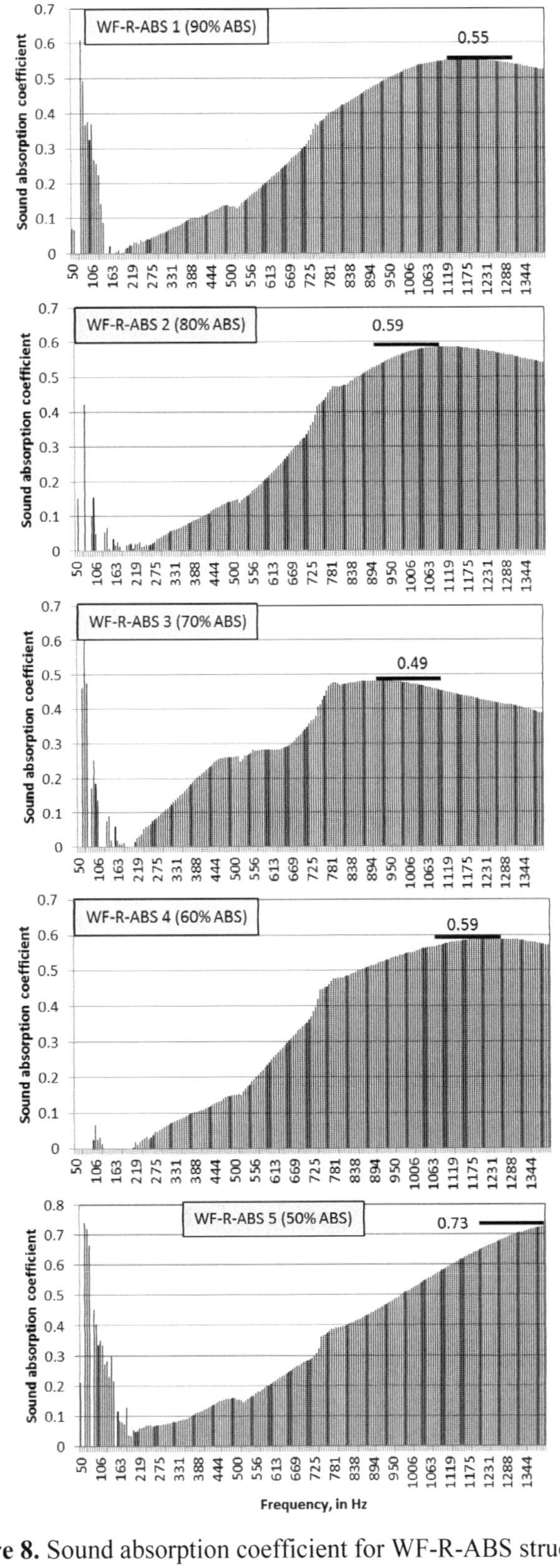

Figure 8. Sound absorption coefficient for WF-R-ABS structures.

IOP Conf. Series: Materials Science and Engineering **399** (2018) 012019 doi:10.1088/1757-899X/399/1/012019

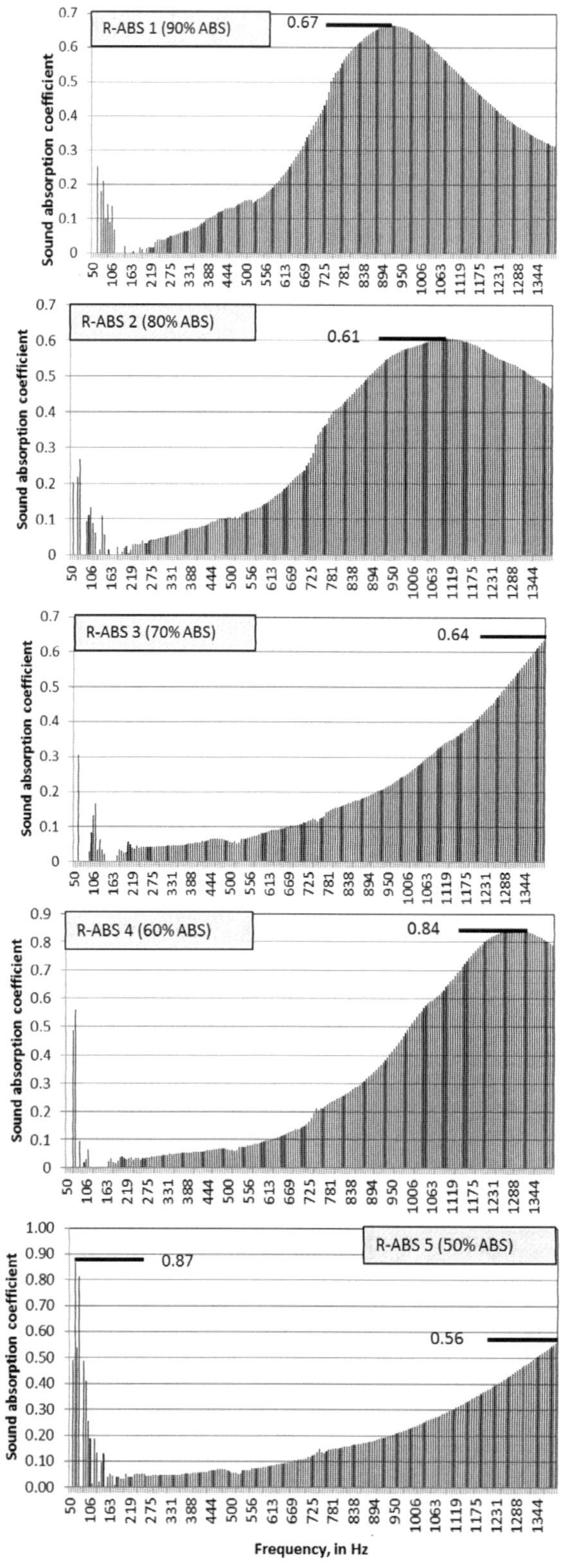

Figure 9. Sound absorption coefficient for R-ABS structures.

4. Conclusions

The aim of the present research was to create and test composites with good thermal and acoustic performances using wood fibres and recycled materials, such as rubber crumbs and ABS, where ABS acts as a matrix of the composite. The results show that the higher participation rate of recycled tire rubber crumbs and wood fibres as fillers reduce the thermal performance of the composites. Instead, incorporation of rubber in the ABS matrix increases the acoustic performance of the composite. The higher content of rubber, the better sound insulation coefficient was registered (0.87) for a rubber content of 50% of the total mass and for composite R-ABS. The best thermal performance was registered by the WF-ABS sample (0.0434 $Wm^{-1}K^{-1}$), followed by the WF-R-ABS sample (0.0460 $Wm^{-1}K^{-1}$) and the R-ABS sample (0.0477 $Wm^{-1}K^{-1}$). The analysis of the results shows that the thermal efficiency of the composites depends on the outdoor temperature. The potential of using recycled materials such as rubber and ABS to manufacture composites for thermal and acoustic insulation of buildings is promising. The testing of water permeability and water absorption of the composites presented in this paper is in progress.

5. Aknowledgement

The authors acknowledge the structural funds project PRO-DD (POS-CCE, O.2.2.1., ID 123, SMIS 2637, No. 11/2009) for providing the infrastructure used.

References

[1] Barthes M L, Mantaux O, Pedros M, Lacoste E and Dumon M 2012 Recycling of aged ABS from real WEEE through ABS/PC blends in the ABS-rich compositions *Advances in Polymer Technology* **31(4)** 343
[2] Mohammed M I, Das A, Gomez-Kervin E, Wilson D and Gibson I 2017 EcoPrinting: Investigating the use of 100% recycled Acrylonitrile Butadiene Styrene (ABS) for Additive Manufacturing *Proceedings of the 28th Annual International Solid Freeform Fabrication Symposium – An Additive Manufacturing Conference* pp 532-542
[3] Nikmatin S, Syafiuddin A, Naresworo N, Utama W and Wismogroho AS 2017 Thermo-Physical Properties of Kenaf-Filled Acrylonitrile Butadiene Styrene Composites *MATEC Web of Conferences* 95, 03001 (2017)
[4] Nikmatin S, Syafiuddin A, Naresworo N, Utama W and Irwanto DAY 2017 Properties of Oil Palm Empty Fruit Bunch-Filled Recycled Acrylonitrile Butadiene Styrene Composites: Effect of Shapes and Filler Loadings with Random Orientation *BioResources* **12(1)** 1090
[5] Threepopnatkul P, Teppinta W and Sombatsompop N 2011 Effect of Co-monomer Ratio in ABS and Wood Content on Processing and Properties in Wood/ABS Composites *Fibers and Polymers* **12(8)** 1007
[6] Rimdusit S, Atthakorn D, Damrongsakkul S and Saramas D 2012 Mechanical, thermal, and water uptake characteristics of wood flour-filled polyvinyl chloride/acrylonitrile butadiene styrene blends *J. Appl. Polym. Sci.* **124(2)** 943
[7] Adamopoulos S, Foti D, Voulgaridis E, and Passialis C 2015 Manufacturing and Properties of Gypsum-Based Products with Recovered Wood and Rubber Materials *BioResources* **10(3)** 5573
[8] Zhao J, Wang XM, Chang JM, Yao Y and Cui Q 2010 Sound insulation property of wood-waste tire rubber composite *Composites Science and Technology* **70(14)** 2033
[9] Abdu-Jdayil B, Mourad A-H and Hussain A 2016 Thermal and physical characteristics of polyester–scrap tire composites *Constr. Build. Mater.* **105** 472
[10] Kakroodi AR, Leduc S, Gonzalez-Nunez R and Rodrigue D 2012 Mechanical Properties of Recycled Polypropylene/SBR Rubber Crumbs Blends Reinforced by Birch Wood Flour *Polymers & Polymer Composites* **20(5)** 439
[11] Dongwei S, Min X, Liping C and Sheldon Q S 2016 Fabrication of Wood-Rubber Composites Using Rubber Compound as a Bonding Agent Instead of Adhesives *Materials* **9** 469

[12] Ayrilmis N, Buyuksari U and Avci E 2009 Utilization of Waste Tire Rubber in the Manufacturing of Particleboard *Materials and Manufacturing Process* **24(6)** 688
[13] Cosnita M, Cazan C and Duta A 2017 The influence of inorganic additive on the water stability and mechanical properties of recycled rubber, polyethylene terephthalate, high density polyethylene and wood composites *J. Clean. Prod.* **165** 630
[14] Chen L, Han JQ, Huang RZ, Xu XW and Wu Q 2017 Thermal decomposition properties of recycled tire rubber filled wood/high density polyethylene composites *Wood Research* **62(5)** 701
[15] DIN EN 12667: 2001 Thermal performance of building materials and products - Determination of thermal resistance by means of guarded hot plate and heat flow meter methods - Products of high and medium thermal resistance, *German Institute for Standardization* Berlin, Germany
[16] ISO 8301: 1991Thermal insulation - Determination of steady-state thermal resistance and related properties - Heat flow meter apparatus, *International Organization for Standardization* Geneva, Switzerland
[17] SR EN ISO 10534-1: 2002 Determination of sound absorption coefficient and acoustic impedance with the interferometer, Part 1: Stationary wave ratio method *Romanian Institute for Standardization* Bucharest, Romania

A study on the effects of the cement and mineral aggregates replacement with waste materials

C M Gradinaru[1], M Barbuta[1], V Ciocan[1] and A A Serbanoiu[1*]

[1] "Gheorghe Asachi" Technical University, Faculty of Civil Engineering and Building Services, 1 Prof. Dimitrie Mangeron Blvd., 700050, Iasi, RO

[*] E-mail: serbanoiu.adrian@tuiasi.ro

Abstract. Concrete is a large-scale building material with an important impact on environment mainly due to the production technology of cement and also because it consumes depleting mineral resources as aggregates. Therefore, it is necessary to find ways to replace some of its components in order to diminish its pollution effect. One of these is the partial replacement of cement with other cementitious materials and/or the replacement of mineral aggregates with renewable ones. The aim of this study was to investigate the effects on density, compressive strength, and split tensile strength of the concrete in the case of cement replacement by fly ash in a proportion of 10%, 20% and 30% of the volume, in the case of aggregates replacement by 20% of the volume with aggregates made of waste, and then these two cases combined. Even if the experimental results revealed the decrease of mechanical properties by replacing cement and the mineral aggregates, the smaller density of the developed concretes represents an important advantage.

1. Introduction

All over the world a general concern is to find solutions to decrease the environmental pollution and greenhouse gas emissions. In building materials industry this means the development of sustainable and ecological products. A sustainable and ecological building material is that in whose production are used non-depleting, low cost, and locally available raw materials, and its manufacture process implies reduced greenhouse gas emissions and a low level of energy consumption.

Concrete is a building material which has an important impact on the environment in all respects: its usual raw materials are of depleting category, the cement used in its composition is not very cheap or everywhere available, and its manufacturing process is a very high energy consumer, releasing very large quantities of CO_2. All these issues can be partially solved by finding alternative cementitious materials (instead of cement) and renewable raw materials for the partial or total mineral aggregates replacement. Among the cementitious materials, there are numerous alternatives, from industrial by-products such as fly ash (FA), blast furnace slag or silica fume, to plants ashes such as corn cob ash, wheat ash, rice husk ash, and others. For mineral aggregates replacement, there are even more possibilities, from new produced lightweight aggregates like expanded perlite, expanded clay or expanded polystyrene, to industrial by-products like blast furnace slag, or different types of waste (glass, plastics, lignocellulosic materials or agricultural waste). If the availability criterion is taken into account, FA can be used as cementitious material in Romania because it is a by-product derived from the thermal power plants, a still very large industry in this country. FA is an industrial waste generated by the coal combustion in thermal power plants. Its production efficiency/output coal: FA is 1:0.3-0.4 [1] which means that for every ton of burned coal, it results 0.3 - 0.4 tons of FA that is disposed as landfills, leading to important land occupation and environmental pollution [2].

FA physical and chemical properties vary according to the properties of the burned coal [1]. FA has some advantages over the Portland cement, such as lower heat of hydration, which can lead to a higher strength concrete (depending on the ash class), and it is widely available and cheaper than conventional cement. As disadvantage, it develops a slower rate of concrete strength, but by using additives, the hydration rates can be accelerated [3].

All over the world many investigations in very different directions have been made to find alternative solutions for mineral aggregates replacement. A relative new approach is the use of plants for this purpose, the main advantages this presents for concrete manufacturing being represented by their high renewability, availability, low cost and low density [4]. Among the plants used in building materials development are: hemp, kenaf, flax, jute, banana, sisal, pineapple, cotton, rice [4, 5].

Maize represents a very important crop in Romania, in 2015 this country being on the second position in EU-28 production [6]. Corn cob represents an agricultural waste, 15% of the total corn production. Its principal constituents are: 39.1% cellulose, 42.1% hemicellulose, 9.11% lignin, 1.7% protein and 1.2% ash [7]. Corn cobs density of 212 kg/m^3 and their high content of 77.52% oxygen promote them for being used in lightweight concrete production. They have a very high water absorption capacity of over 327% after 15 days of immersion, but also a quite good resistance of about 5 minutes to direct flame exposure, compared to a few seconds of the expanded polystyrene [8]. According to the above-mentioned reasons, corn cob may be considered an alternative material for mineral aggregates used in concrete composition if there are found solutions to minimize its main disadvantage, namely the high capacity of water absorption.

The aim of this study was to investigate the effects on density, compressive strength, and split tensile strength of the concrete in the case of cement replacement by fly ash in 10%, 20% and 30% by volume, in the case of aggregates replacement by 20% by volume with aggregates made of waste and then these two cases taken together. Although, there were many studies on the effects of FA on the concrete, the present research used this type of concrete as reference for the concrete recipes with 20% by volume of corn cob aggregates and FA. The novelty of this paper is represented by the analysis of the effects of FA on the vegetal concrete with 20% by volume of corn cob aggregates.

2. Materials and methods

The present research aimed to investigate the effects of fly ash and corn cobs on the concrete mechanical properties. For this purpose, starting from a reference concrete strength class 25/30, produced according to Romanian Standard NE012-1: 2007, were developed concrete recipes with 10%, 20% and 30% vol. cement replacement by FA, a concrete recipe with 20% vol. of mineral aggregates replacement by corn cob aggregates and then these two variants were combined in order to study the interconnection between fly ash and corn cobs.

The RC contained the following components:

- Portland cement CEM II/A-LL 42.5R, with granulated blast furnace slag and limestone, produced in Romania;
- natural sand (0-4 mm diameter) and gravel (sort 4-8 mm) extracted from riverbeds;
- water as to ensure a water/cement ratio of 0.43;
- additives: a policarboxilateter based superplasticizer (Sika Plast 140) and a rhodanid based accelerator (Sika BE5); the superplasticizer was used to obtain a smaller water/cement ratio and to improve the concrete workability; the accelerator was used to enhance the cement hydration process, necessary especially in the case of concrete with vegetal aggregates.

The study involved the development of eight concrete compositions, as follows:

- RC – a reference concrete of strength class 25/30;
- CFA10 – a concrete with 10% vol. of FA as cement partial replacement;
- CFA20 – a concrete with 20% vol. of FA as cement partial replacement;

- CFA30 – a concrete with 30% vol. of FA as cement partial replacement;
- CCC20 – a concrete with 20% vol. of corn cob granules as partial replacement of mineral aggregates;
- CCCFA 20-10 – a concrete with 20% vol. of corn cob granules as partial replacement of mineral ones and 10% vol. of FA as partial replacement of cement;
- CCCFA 20-20 – a concrete with 20% vol. of corn cob granules as partial replacement of mineral ones and 20% vol. of FA as partial replacement of cement;
- CCCFA 20-30 – a concrete with 20% vol. of corn cob granules as partial replacement of mineral ones and 30% vol. of FA as partial replacement of cement.

The source of the used FA was Holboca Thermal Power Plant, Iasi County, Romania.

The vegetal aggregates were made of corn cobs shredded in granules smaller than 5 mm and then treated with 40% sodium silicate solution in order to diminish their very high absorption capacity of 294% to 127% comparative to their dry weight. The apparent density of the corn cob aggregates was 398.43 kg/m^3.

The concrete was poured into cylinder moulds with 100 mm diameter and 200 mm length. There was measured the apparent density and then tests were performed to determine compressive and splitting tensile strength, according to SR EN 12350-6 and SR EN 12390 part 3 and SR EN 12390 part 7, respectively. Three samples were tested for each property mentioned above. Because the samples dimensions were different to those required by NE-012-1standard, and in order to obtain comparable values with those from the scientific literature, the values of compressive and splitting tensile strength were compensated.

3. Experimental results

3.1. The concrete density

The concrete density evolution from pouring up to the age of 28 days it is shown in figure 1. It can be observed that the slope degree of the density evolution is particular for each concrete recipe.

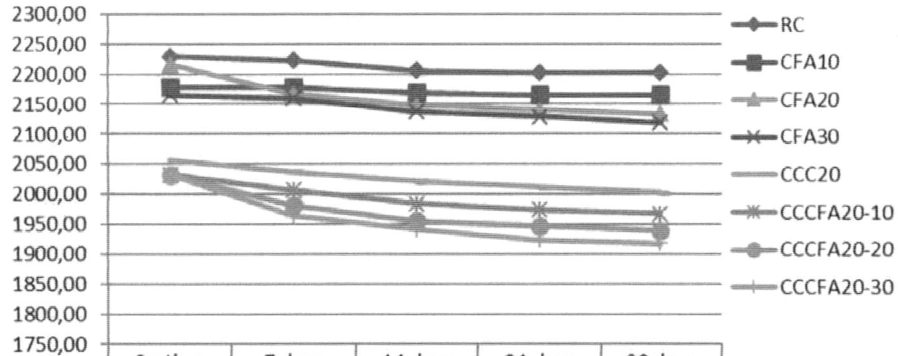

	Casting	7 days	14 days	21 days	28 days
RC	2229,30	2222,93	2205,73	2202,55	2202,55
CFA10	2178,34	2178,34	2168,79	2165,61	2165,61
CFA20	2216,56	2166,24	2147,13	2140,13	2133,76
CFA30	2165,61	2159,24	2136,94	2129,30	2118,47
CCC20	2056,28	2037,18	2021,27	2011,72	2002,17
CCCFA20-10	2031,85	2007,01	1984,08	1973,25	1966,88
CCCFA20-20	2031,85	1980,89	1955,41	1946,50	1939,49
CCCFA20-30	2031,85	1963,69	1941,40	1923,57	1917,83

Figure 1. The evolution of concrete density, from casting till 28 days of curing [kg/m^3].

RC registered a 1.2% density decreasing along the 28 days curing period, while the CCC20 a 2.63% smaller density. The smallest density decreasing of all was that of CFA10, by 0.58%, and the biggest one that of CCCFA 20-30, by 5.6%. Of the CFA category, the highest diminish was of CFA20, by 3.74%. Of the CCCFA group, the smallest diminish was of CCCFA20-10, by 3.20%.

According to figure 2, the increasing volume of the cement replacement by FA determined a decrease of the concrete density; the smallest density of the CFA group was registered by CFA30 with a value of 2118.47 kg/m^3, by 3.82% smaller than of the RC. The concrete recipe with only corn cobs registered a density diminish by 9.1% compared to RC. It seems that the association of FA with corn cobs aggregates had very positive effects, whereas the group of CCCFA registered the smallest density of all, under 2000 kg/m^3, with a decrease between 12.93% and 10.70% compared to RC.

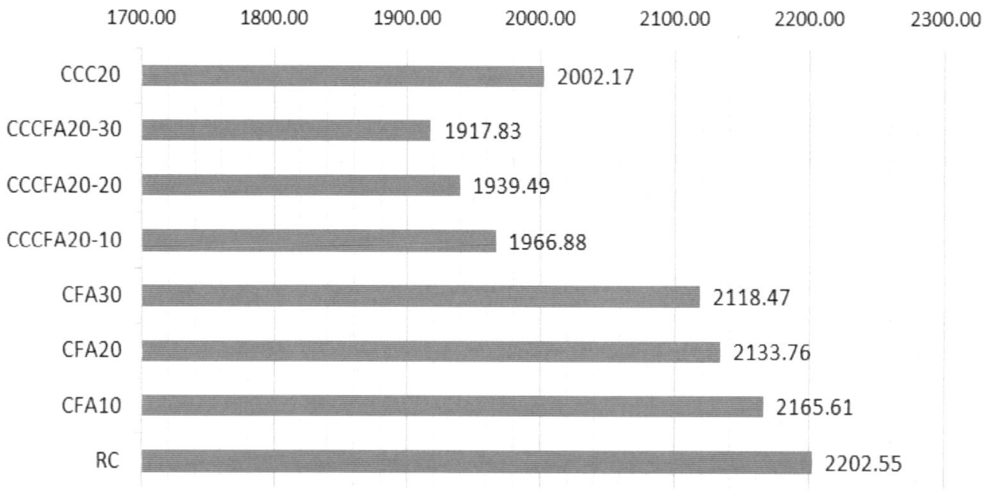

Figure 2. Concrete density at 28 days, [kg/m^3].

3.2. Concrete compressive strength

The values of compressive strength of the developed concrete recipes are presented in figure 3. One can notice that FA led to the decrease of this property by 7.98%, 17.47%, and 34.19% for 10%, 20%, and 30% vol. of cement replacement, respectively. The 20% vol. of mineral aggregates replaced by corn cob ones determined a strong compressive strength diminish, by 61.12%. FA association with corn cob aggregates revealed even smaller compressive strength than CCC20, between 2.55% and 24.72%, compared to CCC20.

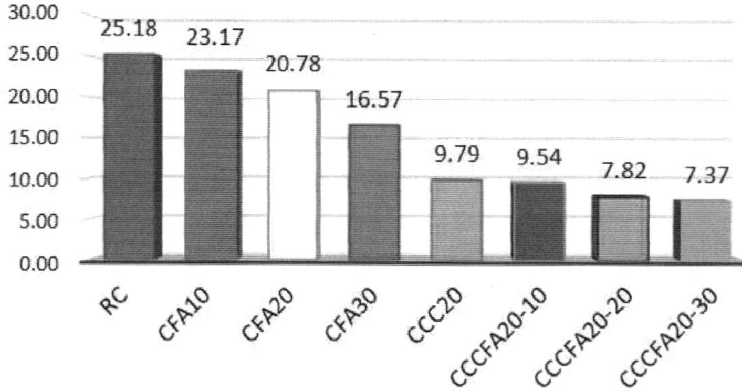

Figure 3. Compressive strength of the concrete [N/mm^2].

3.3. Concrete splitting tensile strength

The values of splitting tensile strength are presented in figure 4. Regarding this property, FA decreased the RC performance. The concrete recipe with 20% vol. corn cob aggregates registered a diminished splitting tensile strength by a little more than 50%. By using cement replacement with FA in the concrete with corn cobs, it was obtained an improvement of the above mentioned property by 4.9% and 23.53% in the case of using 20% vol. and 10% vol. of FA instead of cement, respectively.

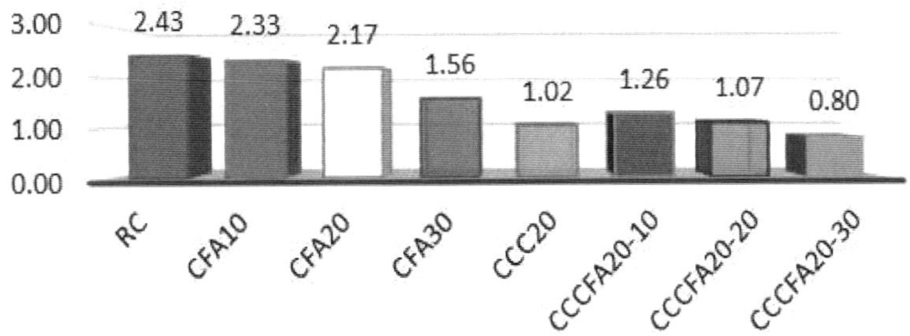

Figure 4. Splitting tensile strength of the concrete [N/mm^2].

4. Conclusions

The aim of this study was to investigate the effects on density, compressive strength, and split tensile strength of the concrete in the case of cement replacement by fly ash in 10%, 20% and 30% by volume, in the case of aggregates replacement by 20% by volume with aggregates made of waste and then these two cases combined. There were quantified their effects on density, compressive strength, and split tensile strength of the obtained concrete.

According to the obtained experimental results, the more fly ash was added for cement replacement, the more the density of the concrete decreased. The association of fly ash with corn cobs aggregates had very positive effects on reducing the concrete density after 28 days of curing. FA led to the decreasing of the concrete compressive strength but added a plus to the RC performance in terms of concrete splitting tensile strength, when using 10% vol. and 20% vol.; by using cement replacement with FA in the concrete with corn cobs, it was obtained an improvement of the splitting tensile strength.

Even if FA concretes registered smaller values for mechanical properties, but still appropriate for structural use, their advantage is the smaller density than RC, that lead to lighter structures, with positive effect on the seismic building behaviour. The concretes with fly ash and corn cobs were lightweight concretes, with possibilities of use in non-structural purposes like closures, wall finishes and concrete screeds.

Acknowledgement

Special thanks to Sika Romania representatives for their technical support and for providing the necessary additives in order to accomplish this research.

References

[1] Sun W, Handong Y and Binggen Z 2003 Analysis of mechanism on water-reducing effect of fine ground slag, high-calcium fly ash, and low-calcium fly ash *Cement and Concrete Research* **33** 1119-1125

[2] Aprianti E S 2017 A huge number of artificial waste material can be supplementary cementitious material (SCM) for concrete production - a review part II *Journal of Cleaner Production* **142** 4178-4194

[3] Meyer C 2009 The greening of the concrete industry *Cement and Concrete Composites* **31** 601-605

[4] Saxena M, Pappu A, Sharma A, Haque R and Wankhede S 2011 Composite Materials from Natural Resources: Recent Trends and Future Potentials *Advances in Composite Materials - Analysis of Natural and Man-Made Materials* online available: www.intechopen.com
[5] Valdés A, Mellinas A C, Ramos M, Garrigós M C and Jiménez A 2014 Natural additives and agricultural wastes in biopolymer formulations for food packaging *Front Chem* 2- 6
[6] Forti R and Henrad M (Eds) 2016 Agriculture, forestry and fishery statistics *Eurostat Luxembourg: Publications Office of the European Union* 230
[7] Ashour A, Amer M, Marzouk A, Shimizu K, Kondo R and El-Sharkawy S 2013 Corncobs as a Potential Source of Functional Chemicals *Molecules* **18** 13823-13830
[8] Pinto J, Cruz D, Paiva A, Pereira S, Tavares P, Fernandes L and Varum H 2012 Characterization of corn cob as a possible raw building material *Construction and Building Materials* **34** 28–33

IOP Conf. Series: Materials Science and Engineering **399** (2018) 012021 doi:10.1088/1757-899X/399/1/012021

The effects of sodium silicate on corn cob aggregates and on the concrete obtained with these agricultural waste

C M Gradinaru[1], M Barbuta[1], V Ciocan[1], E Antohie[1] and D Babor[1*]

[1] "Gheorghe Asachi" Technical University, Faculty of Civil Engineering and Building Services, 1 Prof. Dimitrie Mangeron Blvd., 700050, Iasi, RO

[*] E-mail: danbabor@yahoo.com

Abstract. In order to protect the environment and to reduce its pollution, the concrete greening is an important objective to accomplish this goal. Research across the world has expanded to replace some components of concrete with different kinds of alternative ecological materials, including agricultural materials that are readily renewable resources. They may partially replace the mineral aggregates, if they are used in a shredded form, and / or the cement from the concrete composition, if the ash resulting from their combustion is used. When used as an alternative of mineral aggregates, plant aggregates generally have a number of disadvantages, which requires finding a solution which reduces their water absorption, improves the interface between them and the cement matrix, and gives them durability in the resulting concrete. The present study investigated the effect of sodium silicate on plant aggregates made of shredded corn cobs, and on the concrete manufactured with this type of material as a replacement of mineral aggregates by 20%, 50% and 80% by volume. Experimental results revealed that sodium silicate reduced the water absorption of the grounded corn cobs, resulting a lightweight concrete with improved compressive and tensile strength compared to those of the concrete manufactured with untreated plant aggregates. The use of sodium silicate as a concrete additive resulted in an increase of the compressive strength of the concrete manufactured with 50% by volume corn cob aggregates, of the splitting strength in the case of concrete with 20% and 80% by volume vegetal aggregates, and of the density of the concrete manufactured with the highest percentages of plant aggregates which were investigated in this study.

1. Introduction

Research across the world has expanded to replace some components of concrete with different kinds of alternative ecological materials. So far, various natural, industrial or postconsumer waste has been used to improve the concrete properties and to reduce the environmental pollution risks by "saving" some of the raw materials [1 - 7].

The crop by-products were studied firstly as possible addition in concrete composition in order to improve its tensile strength, ductility and post-cracking behavior. But in the last years, an increasing interest to use this category of waste as a local available material emerged. Annual plants stems have a low density and high porosity that leads to the obtaining of a concrete with a lower density, lower thermal conductivity and higher permeability [8].

Agricultural materials may partially replace the mineral aggregates, being used in a shredded form, and / or cement from the concrete composition if the ash resulting from their combustion is used. When used as an alternative of mineral aggregates, plant aggregates generally have a number of disadvantages, which requires the adoption of a solution that reduces their water absorption, improves the interface between them and the cement matrix, and gives them durability in the resulting concrete. The limited compatibility between cementitious binder and polysaccharide carbohydrates of the plant structure inhibits the setting process of the binder [9], and the high water absorption capacity of vegetal aggregates leads to weak interfaces between vegetal aggregates and cementitious matrix and, implicitly, to lower compressive strength of the concrete [8 - 10].

To obtain a concrete with improved mechanical properties, interventions can be made on the plant aggregates and/or on the binder [9]. Some examples of improvement measures applied on plant aggregates are: treatment with cellulose ether that creates water retention and reduces the water transfer [9], distillation for extracting the sugars, waterproofing with linseed-oil [10] or other hydrophobic coatings, eliminating the deleterious components of the plant particles by alkaline solutions or water, pretreatment with the cementitious binder paste, surface treatment with silanes soaking in a silica fume slurry, or fibres homification (fibres are subjected to drying and rewetting cycles [11]. Another method is the addition in concrete of setting and hardening accelerators [11]. For this study, it was adopted the plant aggregates treatment with an alkaline solution, namely sodium silicate (SS) and an accelerator addition in the concrete composition. The plant aggregates treatment with sodium silicate solution was used in other researches, but not on the corn cob aggregates.

The present study investigated the effect of SS on plant aggregates made of shredded corn cobs and on the concrete with this type of material as a replacement of 20%, 50% and 80% by volume of mineral aggregates.

2. Materials and methods

In order to diminish the very high water absorption capacity of the corn cob granules, it was adopted the treatment of corn cobs, shredded in granules smaller than 5-6 mm in diameter, with SS solution, in two concentration variants, 20% and 40%. The treatment consisted in the immersion of the corn cob granules in the SS solution. The wet granules were spread on a surface warmed at 45-50°C, in a naturally ventilated room with air temperature of 23-25°C, and were let to dry until they reached constant weight. After drying, the water absorption capacity was measured. Afterwards there were made concrete compositions with untreated corn cob granules, with corn cob granules treated with 20% SS solution, and with corn cob granules treated with 40% SS solution. After testing and determining the compressive strength of these concrete types, the composition with the highest compressive strength obtained was established as the most efficient variant of corn cob treatment. The corn cob granules treated in this variant partially replaced mineral aggregates and three types of concrete compositions, with 20% vol., 50% vol. and 80% vol. replacement rate, respectively were made. In order to improve the corn cob concrete properties, the last three concrete compositions were cast again, but this time 100% SS solution as additive in fresh mixture was used.

As a summary, in this study were developed the following concrete compositions:

- RC – a reference concrete of 25/30 Strength (manufactured according to [12]);
- CUCC50 – a concrete with 50% vol. of untreated corn cob granules instead of mineral aggregates;
- CTCC50-20 – a concrete with 50% vol. of treated corn cob granules with sodium silicate sol 20%, instead of mineral aggregates;
- CTCC50-40 – a concrete with 50% vol. of treated corn cob granules with sodium silicate sol 40%, instead of mineral aggregates;
- CTCC20-40 – a concrete with 20% vol. of treated corn cob granules with sodium silicate sol 40%, instead of mineral aggregates;
- CTCC80-40 – a concrete with 80% vol. of treated corn cob granules with sodium silicate sol 40%, instead of mineral aggregates;
- CTCC20-40 SS – a concrete with 20% vol. of treated corn cob granules with sodium silicate sol 40%, instead of mineral aggregates, and sodium silicate solution 100% conc. as additive;

- CTCC50-40 SS – a concrete with 50% vol. of treated corn cob granules with sodium silicate sol 40%, instead of mineral aggregates, and sodium silicate solution 100% conc. as additive;
- CTCC80-40 SS – a concrete with 80% vol. of treated corn cob granules with sodium silicate sol 40%, instead of mineral aggregates, and sodium silicate solution 100% conc. as additive.

The RC composition contained the following ingredients:

- natural sand (diameter 0–4 mm);
- river gravel (diameter 4–8 mm);
- Portland cement CEM II/A-LL 42.5R, with granulated blast furnace slag and limestone, produced in Romania;
- water to ensure a water/cement ratio of 0.43;
- additives: a policarboxilateter-based super plasticizer (Sika Plast 140) in 2% of cement volume, and a rhodanid-based accelerator (Sika BE5) in 2% vol. of cement volume.

In this study, sodium silicate solution for vegetal aggregates treatment and as additive in the concrete composition was also used. As additive, it was used 100% SS solution in a quantity equal to 5% of the cement volume. Sodium silicate is the common name for a sodium metasilicate compound, Na_2SiO_3, also known as glass water or liquid glass. It is available in aqueous and solid form and it is used in concrete, passive protection against fire, refractoriness, textiles, woodworking, and automobiles. The use of sodium silicate for the impregnation of industrial wood has been described since 1920. For example, [13] discloses a method for treating wood and wood materials to obtain fire resistant wood. Glass water is also used to protect wood and wood products by improving resistance to termites, rot and degradation.

SS solutions are very used to prevent excessive evaporation of the mixing water from the fresh mix concrete if it is applied during the curing period [14]. SS-based nanosilica is a pore filling material for concrete, with excellent positive effects on chloride attack [15]. The powder of SS is an alkaline activator in alkali-activated cementless mortars [16]. SS addition in concrete fresh mixture is recommended as an accelerator for Portland cement setting [17].

The concrete was poured into cube molds with the side of 150 mm and cylinder molds with 100 mm diameter and 200 mm length. On the cube samples the apparent density was measured, according to SR EN 12350-6, and then tests were performed to determine the compressive strength, according to SR EN 12390 part 3; on the cylinder samples the splitting tensile strength was determined, according to SR EN 12390 part 7. Each test was performed on three samples. Due to the fact that the samples dimensions were different to those required by the NE-012-1 standard, and in order to obtain comparable values with those from the scientific literature, the values of compressive and splitting tensile strength were compensated.

3. Experimental results

3.1. Water absorption capacity of the corn cob granules

After the treatment application, the results revealed that the 20% SS solution decreased the water absorption capacity of the corn cob granules from 294% up to 181% compared to the untreated ones. The 40% SS solution usage obtained even better results, 127% water absorption capacity compared to the untreated granules.

3.2. Concrete density

The evolution of the concrete density along the 28 days curing period is presented in figure 1. It can be observed that RC density decrease by 3.23% has been exceeded by CTCC20-40 SS, which registered a value of 2.03% decrease and by CTCC20-40 with a value of 2.63%. This means that the water quantity from the fresh concrete was used in very high proportion for cement hydration and it was not sucked by the vegetal aggregates and then evaporated. Both minimums implied the smallest analyzed quantity of corn cob aggregates, 20% vol., and SS addition improved the cement hydration.

IOP Conf. Series: Materials Science and Engineering **399** (2018) 012021 doi:10.1088/1757-899X/399/1/012021

Next, the two recipes with 50% vol. corn cob aggregates were on a level below from cement hydration point of view, CTCC50-40 and CTCC50-40 SS, registering a density decreasing by 5.15% and 5.40%, respectively. In this case, SS didn't help more the cement hydration process.

The concrete compositions with the biggest quantity of corn cob aggregates, CTCC80-40 and CTCC80-40 SS, registered the biggest density decreasing by 13.46% and 11.85% respectively, this due to the fact that bigger quantity of vegetal led to a higher quantity of absorbed water, which was then evaporated during the curing period of 28 days. SS addition contributed to the cement hydration more than in the case of the concrete composition with 20% vol. corn cob aggregates.

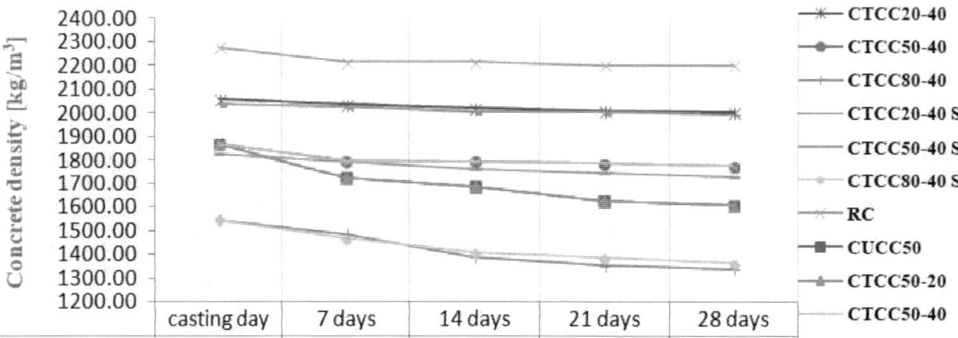

	casting day	7 days	14 days	21 days	28 days
CTCC20-40	2056.28	2037.18	2021.27	2011.72	2002.17
CTCC50-40	1871.43	1800.00	1796.43	1785.71	1775.00
CTCC80-40	1546.99	1487.14	1389.10	1354.73	1338.81
CTCC20-40 SS	2037.18	2024.45	2006.63	2005.35	1995.80
CTCC50-40 SS	1827.10	1792.08	1763.44	1744.34	1728.42
CTCC80-40 SS	1546.99	1467.41	1411.39	1385.28	1363.64
RC	2275.00	2214.29	2214.29	2201.62	2201.62
CUCC50	1867.86	1725.00	1685.71	1628.57	1607.14
CTCC50-20	1867.86	1725.00	1685.71	1628.57	1607.14
CTCC50-40	1871.43	1800.00	1796.43	1785.71	1775.00

Figure 1. The evolution of concrete density, from casting throughout the 28 days of curing [kg/m^3].

The final concrete density after 28 days of curing is presented in figure 2.

Figure 2. Concrete density at 28 days, [kg/m^3].

The application of 20% SS solution on the corn cob granules increased the concrete density by 9.77% compared to the one with untreated granules. The 40% SS solution treatment led to a supplemental 9.55% increase of the concrete density than the previous concentration. The SS addition led to a light decrease of the concrete density in the case of the recipes with 20% vol. and 50% vol. of corn cob aggregates, and a light increase in the case of the one with 80% vol.

3.3. Concrete compressive strength

The compressive strength values of the concrete compositions developed in this research are presented in figure 3.

From the graphic of concrete compressive strength values, it can be observed that the mineral aggregates replacement by corn cob aggregates led to concrete compressive strength decreasing.

The treatment of corn cob granules with 20% SS solution in a variant and with 40% SS solution in another variant improved the concrete compressive strength by 6.74 fold and 36.81 fold, respectively.

The 100% SS addition led to a decrease of compressive strength by 21.47%, 43.09%, and 9.97% for the concrete composition with 20% vol., 50% vol. and 80% vol., respectively, of corn cob aggregates.

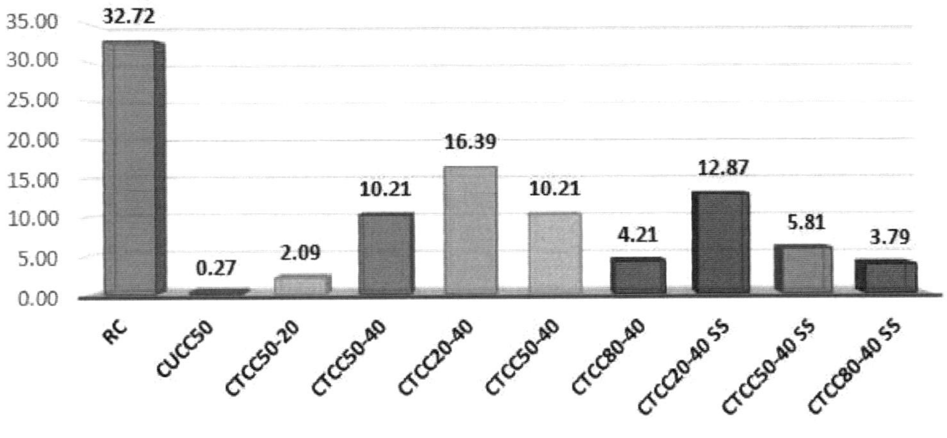

Figure 3. Compressive strength of the concrete [N/mm^2].

The compressive strength of the concrete made of 20% vol., 50% vol. and 80% vol. of corn cobs treated with the most efficient SS solution followed a decreasing trend corresponding to the increase of the replacement rate. The increase of the replacement rate from 20% vol. up to 50% vol. decreased the compressive strength by 37.70%. The next replacement rate increase by 30% led to a supplementary decrease by 58.76% compared to the previous level.

Figure 4 presents the interconnection between water absorption capacity of the vegetal aggregates and the compressive strength of the concrete obtained with these as 50% vol. replacement of mineral aggregates.

From the graphic an inverse proportionality between the water absorption capacity of the vegetal aggregates and the compressive strength of the concrete can be observed: the decreasing of the water absorption capacity of vegetal aggregates contributed to the increasing of compressive strength of the concrete.

A reduction of water absorption capacity by 113% led to an increase of the concrete compressive strength by 674.07%. The water absorption capacity decreasing up to 127% compared to the untreated granules means 54% supplementary reduction compared to the value obtained after the granules treatment with 20% SS solution. This led to an improvement of concrete compressive strength by 388.52%. Overall, a 167% decrease of the vegetal water absorption capacity contributed to an improvement of the concrete compressive strength by 3681.48%.

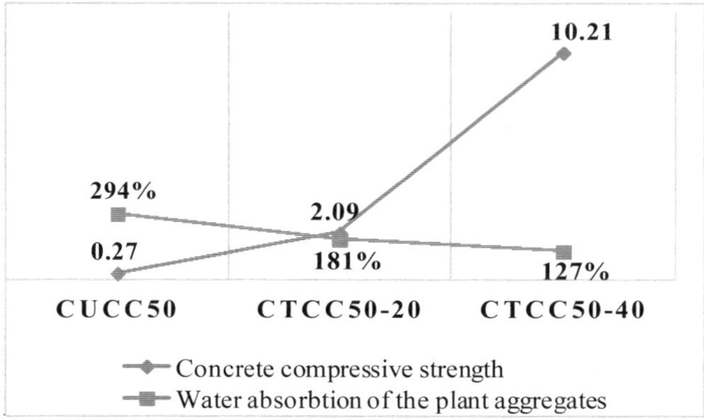

Figure 4. Evolution of compressive strength of concrete [N/mm^2] compared to water absorption of vegetal aggregates [%].

3.4. Concrete splitting tensile strength

In figure 5 are presented the values of the splitting tensile strength of the concrete compositions developed in this study.

As in the case of compressive strength, the treatment of corn cob granules with SS solution improved the concrete splitting test results. With the 20% SS solution treatment, it was obtained a splitting tensile strength of the concrete 20.33 times higher than the concrete composition with untreated corn cob granules. The 40% SS solution led to test results higher by 29 fold than the concrete with untreated vegetal aggregates and 40.62% higher than with the use of 20% SS solution. In conclusion, the most efficient treatment was the 40% SS solution.

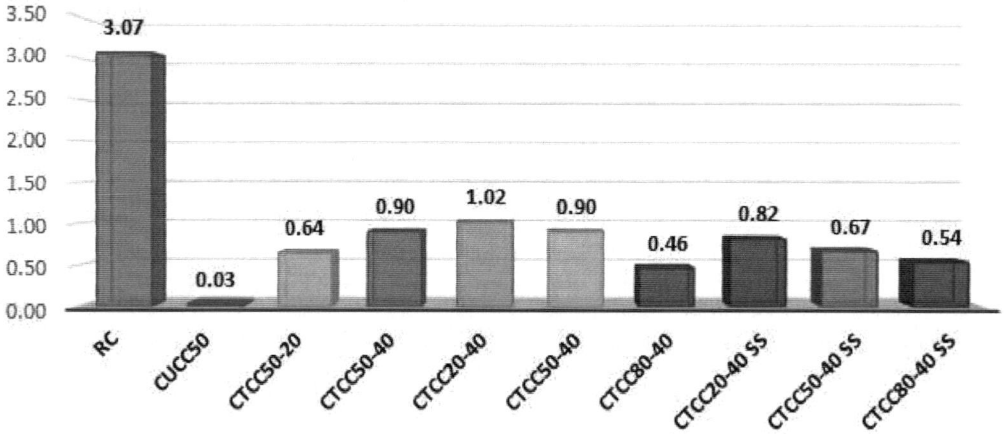

Figure 5. Splitting tensile strength of the concrete [N/mm^2].

Regarding the effect of the corn cob aggregates treated with the previously determined most efficient SS solution, a decrease of the splitting tensile strength along with the replacement rate increase was noticed. The increasing of the corn cob aggregate volume from 20% to 50% of the mineral aggregates quantity led to a reduction by 11.76% of the splitting tensile strength. The concrete composition with 80% vol. of corn cobs registered a 54.90% smaller value of the same property. The SS addition decreased the performances of the concrete compositions with 20% vol. and 50% vol. of corn cob aggregates by around 20% and 26%, respectively, but improved them in the case of 80% vol. of vegetal aggregates by 17.40%.

4. Conclusions

The chemical composition of vegetal aggregates has a high impact on the mechanical properties of the concrete, due to the deleterious interactions between bio-aggregates and mineral matrix: they affect the setting and hardening process of cementitious binders, can diminish the concrete mechanical properties and can influence its durability.

The effects of SS solution on corn cob aggregates and on the corn cob concrete are:

- the SS solution decreased the water absorption capacity of the corn cob granules from 294% up to 127%, compared to the untreated granules;
- in general, the SS addition improved the cement hydration of the concrete;
- the SS addition led to a light decrease of the concrete density in the case of the recipes with 20% vol. and 50% vol. of corn cob aggregates, and a light increase in the case of the one with 80% vol.;
- the treatment of corn cob granules with 20% and 40% SS solution improved the concrete compressive strength by 6.74 fold and 36.81 fold, respectively;
- the 100% SS addition led to a decrease of compressive strength of the corn cob concrete;
- the compressive strength of the concrete made with 20% vol., 50% vol. and 80% vol. of corn cobs followed a decreasing trend corresponding to the increase of the replacement rate of the mineral aggregates;
- decreasing the water absorption capacity of vegetal aggregates contributed to the increasing of compressive strength of the concrete;
- the treatment of corn cob granules with SS solution improved the concrete splitting test results; the 40% SS solution led to test results higher by 29 fold than the concrete with untreated vegetal aggregates and by 40,62% higher than the use of 20% SS solution;
- splitting tensile strength decreased with the mineral aggregates replacement rate increasing;
- the SS addition decreased the performances of the concrete compositions with 20% vol. and 50% vol. of corn cob aggregates by around 20% and 26%, respectively, but improved them in the case of 80% vol. of vegetal aggregates by 17.40%.

The concrete with corn cob aggregates can be used in non-structural purposes like closures, wall finishes and concrete screeds.

Acknowledgements

Special thanks to Sika Romania representatives for their technical support and for providing the necessary additives in order to accomplish this research.

References

[1] Barbuta M, Bucur R, Serbanoiu A A, Scutarasu S and Burlacu A 2017 Combined effect of fly ash and fibers on properties of cement concrete *Procedia Engineering* **181** 280-284

[2] Barbuta M, Serbanoiu A A, Teodorescu R, Rosca B, Mitroi R and Bejan G 2017 Characterization of polymer concrete with natural fibers *IOP Conference Series: Materials Science and Engineering* **246(1)** 012033

[3] Bucur R D, Barbuta M, Konvalina P, Serbanoiu A A and Bernas J 2017 Studies for understanding effects of additions on the strength of cement concrete *IOP Conference Series: Materials Science and Engineering* **246(1)** 012036

[4] Ciocan V, Serbanoiu A A, Dragoi E N, Curteanu S and Burlacu A 2017 Optimization of glass fibers used as disperse reinforcement of epoxy polymer concrete with fly ash *Environmental Engineering and Management Journal* **16 (5)** 1115-1121

[5] Gradinaru C M, Barbuta M, Ciocan V, Serbanoiu A A and Gradinaru A C 2017 Health and environmental effects of heavy metals resulted from fly ash and cement obtaining and trials to reduce their pollutant concentration by a process of combining-exclusion *SGEM2017 Conference Proceedings* **17(52)** 441-448

[6] Serbanoiu A A, Barbuta M, Burlacu A, Gradinaru C M 2017 Fly ash cement concrete with steel fibers – Comparative study *Environmental Engineering and Management Journal* **16 (5)** 1123-1128
[7] Timu A, Barbuta M, Serbanoiu A A, Babor D and Burlacu A 2017 Experimental researches on use of wastes in fly ash concrete *SGEM2017 Conference Proceedings* **17(62)** 163-168
[8] Nozahic V and Amziane S 2012 Influence of sunflower aggregates surface treatments on physical properties and adhesion with a mineral binder *Composites: Part A* **43** 1837–1849
[9] Chabannes M, Nozahic V and Amziane S 2015 Design and multi-physical properties of a new insulating concrete using sunflower stem aggregates and eco-friendly binders *Materials and Structures* **48** 1815–1829
[10] Sellami A, Merzoud M and Amziane S 2013 Improvement of mechanical properties of green concrete by treatment of the vegetals fibers *Construction and Building Materials* **47** 1117–1124
[11] Amziane S and Collet F (Eds.) 2017 Bio-aggregates based building materials State-of-the-Art *Report of the RILEM Technical Committee* **236** BBM 23
[12] NE012-1 2007 Practice Code for Execution of Concrete Works Reinforced Concrete and Prestressed Concrete Part 1 Concrete Production
[13] Hess H K 1920 Process of treating timber *U.S. Pat. No. 1344891* from June 29 1920
[14] Lenmerman P C and Remler R F 1933 Method of protecting concrete surfaces during curing *United States Patent No. US1899576* from Feb. 29 1933
[15] Kim K M, Kim H Y, Heo Y S, Jung S J 2014 Experimental evaluation of sodium silicate-based nanosilica against chloride effects in offshore concrete *Hindawi Publishing Corporation - The Scientific World Journal* 541035 11p
[16] Yang K H, Song J K, Ashour A F and Lee E T 2008 Properties of cementless mortars activated by sodium silicate *Construction and Building Materials* **22** 1981–1989
[17] Hall W C and Peterson J M 1983 Cementitious compositions with early high strength development and methods for controlling setting date *United States Patent No. US4402749* from Sep. 06 1983

A multi scale guided filter for illumination estimation in retinex algorithm

L Guo[1], D An[2,3], Y Xu[3], H Li[1] and M Shao[3*]

[1] School of Mechanical Engineering, Shenyang Jianzhu University, Shenyang 110168, China
[2] Research Centre for Analysis and Detection Technology, Shenyang Jianzhu University, Shenyang 110168, China
[3] Research Institute for Micronano Detection and Motion Control, Shenyang Jianzhu University, Shenyang 110168, China

[*] E-mail: mshao@sjzu.edu.cn

Abstract. The method of illumination estimation for images affects the effect of retinex algorithm and causes information loss and halo artefact. A multi scale guided filter is proposed to estimate the illumination. In this method, bidimensional empirical mode decomposition is combined with guided filter. Firstly, the image is decomposed into multiple bidimensional intrinsic mode functions and a trend function by bidimensional empirical mode decomposition. Secondly, we use guided filter to estimate illumination information of each bidimensional intrinsic mode functions. Then illuminance component is formed by combining the illuminance information and the trend function. Finally, guided filter is again used to achieve a more accurate illuminance component. The simulation results show that multi scale guided filter can estimate the illumination component deeply and ensure the integrity of the reflection component. In the effect image of retinex algorithm, the contrast is raised by 1, the luminance is raised by 1, the information entropy is raised by 0.18, the clarity is raised by 0.001, and the visual effect is improved as a whole.

1. Introduction

Retinex theory [1] understands the luminance of human perception of objects as the organic combination of the environment illumination and the reflection of the object surface, which is composed of the illumination component and the reflection component. The illumination component depends on the source of illumination and determines the dynamic range of the image pixels. The reflection component depends on the characteristics of the object and reacts the essential characteristics of the object. Retinex theory eliminates the influence of the illumination component from the original image, obtains the reflected component, processes the reflected component and achieves the better image which is not affected by the illumination.

In recent years, the methods of illumination estimation for images based on retinex theory have appeared in many kinds of variant forms. The estimated illumination methods used are different, such as Gauss filter [2, 3], guided filter [4], bilateral filter [5, 6], BEMD [7] and wavelet [8]. However, the adaptive ability of these methods is poor and there is information loss and halo artifact.

MSGF (multi scale guided filter) is proposed for shortcomings of the above methods. This method inherits the characteristics of BEMD (bidimensional empirical mode decomposition) and guided filter and takes into account the image information and adaptability. After BEMD, the guided filter, which can estimate the illumination component of the image in depth, is used to estimate each BIMF (bidimensional intrinsic mode functions). In order to accurately estimate the illumination component, the guided filter is again used to remove the residual details after reconstructing the illumination component. MSGF is applied to retinex algorithm to carry out simulation, which is compared with the enhancement results of Gauss filter, guided filter, bilateral filter and BEMD. MSGF effectively improves the brightness and contrast of the image, improves the information loss, avoids the halo artifact and has a strong visual sense.

2. Guided filter and bidimensional empirical mode decomposition

2.1. Guided filter

The guided filter [9, 10] is a novel filter with smoothing edge, which can effectively smooth the details of the image and keep the edge features of the image scene, and does not produce the gradient reversal effect. Its principle is to filter the input image T based on the guided image f. The output image Q does not only preserve the overall feature of the input image, but it can also fully obtain the change details of the guided image. Thus the guided filter can be represented as a local linear model:

$$q_m = a_k f_m + b_k, \forall m \in w_k \tag{1}$$

In the formula: w_k is the filter window, m is the center of the filter window w_k, q_m is the output of linear transformation at m, f_m is the value of m, a_k and b_k are the linear factors and the fixed values in the window m.

a_k and b_k are solved in the following ways:

$$a_k = \frac{\frac{1}{N}\sum_{m \in w_k} f_m t_m - \mu_k \overline{t_k}}{\sigma_k^2 + \varepsilon} \tag{2}$$

$$b_k = \overline{t_k} - a_k \mu_k \tag{3}$$

In the formula: N is the total number of pixels in the window w_k, t_m is the value of m, σ_k^2 and μ_k are the variance and the mean of w_k, $\overline{t_k}$ is the mean of the filter window w_k, ε is the smoothing factor.

2.2. Bidimensional empirical mode decomposition

BEMD [11], which is based on the nonlinear and non-stationary data analysis, which is a multi-scale structure and which it extends the empirical mode decomposition (EMD) from one dimension to two dimensional, is a method. BEMD relies on the properties of images to decompose the image into finite BIMF and a trend function in time domain. BIMF should satisfy two constraints: (1) The local mean of the original two-dimensional signal is symmetric and its mean is 0; (2) The maximum points of the BIMF are positive and the minimum points of the BIMF are negative.

After J layers of BEMD, the final decomposition process can be expressed in the following ways:

$$S(i,j) = \sum_{j=1}^{J} D_j(i,j) + K_J(i,j) \tag{4}$$

In the formula: $S(i,j)$ is the original image; $D_j(i,j)$ is the j th BIMF; $K_J(i,j)$ is the trend function after J layers of BEMD.

3. Multi scale guided filter for illumination estimation

3.1. Basic principles

The illumination components influencing the retinex algorithm can only be obtained from the original image. It is the core of the method of illumination estimation to get the illumination information. According to the core, MSGF is designed and applied to the retinex algorithm.

When BEMD is used to decompose images, the scale is automatically selected according to the characteristics of images and the details are mostly extracted. After decomposing the image by BEMD, BIMF and the trend function contain different frequency characteristics of the image. BIMF mainly contains the high frequency part of the images, which embodies some details of the images and the residual illumination information is retained. The trend function mainly contains the basic illumination information of the image. Therefore, it can result in the absence of illumination information and halo artifacts to use the trend function as the illumination component. For example, figure 2 (a) is a light image estimated by a trend function and figure 3 (a) is an image after retinex algorithm. In addition, it can effectively remove the influence of detail and illumination in the image to use guided filter estimating the illumination component. But the dynamic range of the image pixels is small, which leads to the poor visual effect of the object and cannot fully express the information of the original image. For example, figure 2 (b) is a light image estimated by guided filter and figure 3 (b) is an image after retinex algorithm.

MSGF combines with BEMD and uses guided filter to filter out the details of each BIMF and to extract illumination information. Then illuminance component is constituted by the illuminance information and the trend function. Most details of the illumination component are filtered out, but small details can be retained. Therefore, the guided filter is again used to remove the details of the illumination component. MSGF, which inherits the advantages of the guided filter, keep the edge of the images and smooth the details. And MSGF combined with BEMD to smooth the details and to remove the noise of different scales. It can effectively estimate the illumination component and reflect the illumination component. In retinex algorithm, MSGF will not cause the error of the illumination component and will not result in loss of information and halo artifact. For example, figure 2 (c) is a light image estimated by MSGF, and figure 3 (c) is an image after retinex algorithm.

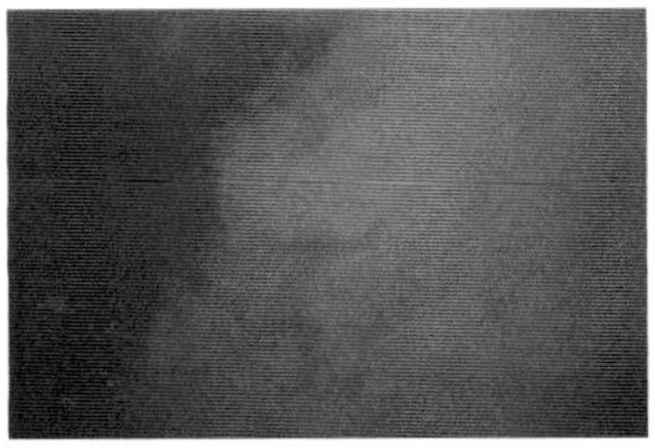

Figure 1. Original image.

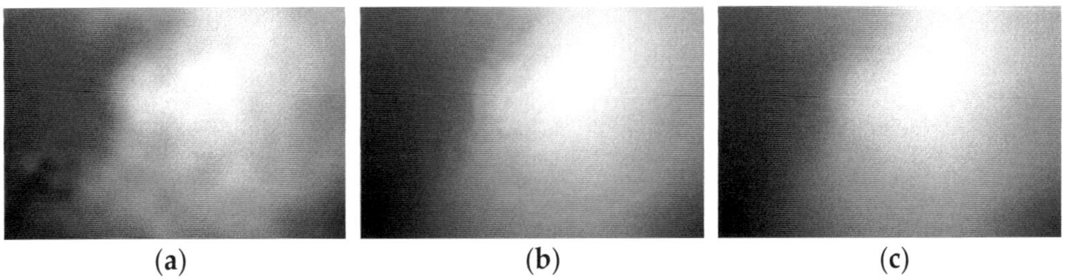

Figure 2. Illumination estimation images: (a) Trend function image; (b) Guided filter; (c) MSGF.

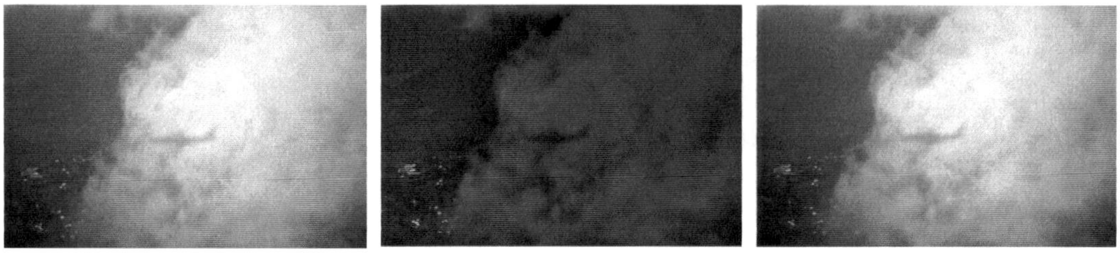

(a) (b) (c)

Figure 3. Retinex algorithm images: (a) Trend function image; (b) Guided filter; (c) MSGF.

3.2. Multi scale guided filter

According to the above principles, MSGF firstly deals images with BEMD, and gets the different components of the original images. The concrete steps are as follows:

(1) The trend image $K(i,j)$ is equal to the original image $S(i,j)$.

(2) If $K(i,j)$ is monotonous or reaches the number of decomposition layers, the algorithm stops. Otherwise, it will make $N(i,j)=K(i,j)$.

(3) Solve the extreme point of $N(i,j)$, find the regional maximum point set and the regional minimum point set and obtain the upper envelopes and the lower envelopes of the image according to the regional maximum point set and the regional minimum point set. Then calculate the envelope mean $M(i,j)$ of the image $N(i,j)$ according to the upper envelopes and lower envelopes.

(4) Order $N(i,j)=N(i,j)-M(i,j)$. If $N(i,j)$ satisfies the constraint condition, the algorithm goes on. Otherwise, go to step (3).

(5) Order BIMF $D(i,j)=N(i,j)$. Let $K(i,j)=K(i,j)-D(i,j)$ and go to step (2).

After the above steps, we can get the decomposition result of formula (4).

Secondly, each BIMF is filtered by guide filter and the illumination information is extracted. In MSGF, guided image is input image, then the formula in (2) is written as follows:

$$a_k=\frac{\sigma_k{}^2}{\sigma_k{}^2+\varepsilon} \tag{5}$$

$$b_k=\mu(1-\sigma_k) \tag{6}$$

Smooth BIMF $D_j(i,j)$ by formula (1) and extract illumination information $Q_{Dj}(i,j)$. It can be expressed as:

$$Q_{Dj}(i,j)=a_k D_j(i,j)+b_k \tag{7}$$

Then, the illuminance component is obtained by combining the illumination information $Q_{Dj}(i,j)$ and the trend function. It can be expressed as:

$$Z(i,j)=\sum_{j=1}^{J} Q_{Dj}(i,j)+K_J(i,j) \tag{8}$$

In the formula: $Z(i,j)$ is the illuminance component, $K_J(i,j)$ is the trend function filtered by guided filter.

Finally, guided filter is applied to the illuminance component, the details of the image are removed completely and the final illumination component $Q_Z(i,j)$ is obtained.

$$Q_Z(i,j)=a_k Z(i,j)+b_k \tag{9}$$

3.3. The application of MSGF in retinex algorithm

Through the images (f) of the following 5 sets, it is easy to find that MSGF inherits the advantages of guided filter and has a larger dynamic range of image pixels. MSGF improves the image contrast effectively, the brightness of the image and the dim image in the distance. MSGF highlights the details that are not easy to find in the image, has no halo artifact, has a strong visual sense and is more in line with the visual requirements of the human being.

4. Comparison of simulation experiments

In order to verify the effectiveness of MSGF, the simulation experiments of multiple images, which uses Gauss filter, bilateral filter, guide filter, BEMD and MSGF, are carried out on the computer, whose operating system is Windows10, whose main frequency of the processor is 2.6Hz, whose memory is 4GB, and whose software platform is Matlab2017. The result images are objectively evaluated by contrast, brightness, information entropy and clarity. The experimental images are 333×500, 390×597, 259×400, 234×350 and 234×350. The parameters of each method are set as follows: the standard deviation of Gauss filter is 250, the window of guided filter is 120×120 and BEMD decompose images 3 times which decompose images into 3 BIMF and a trend function. The results of the experiment are shown in figure 4~8.

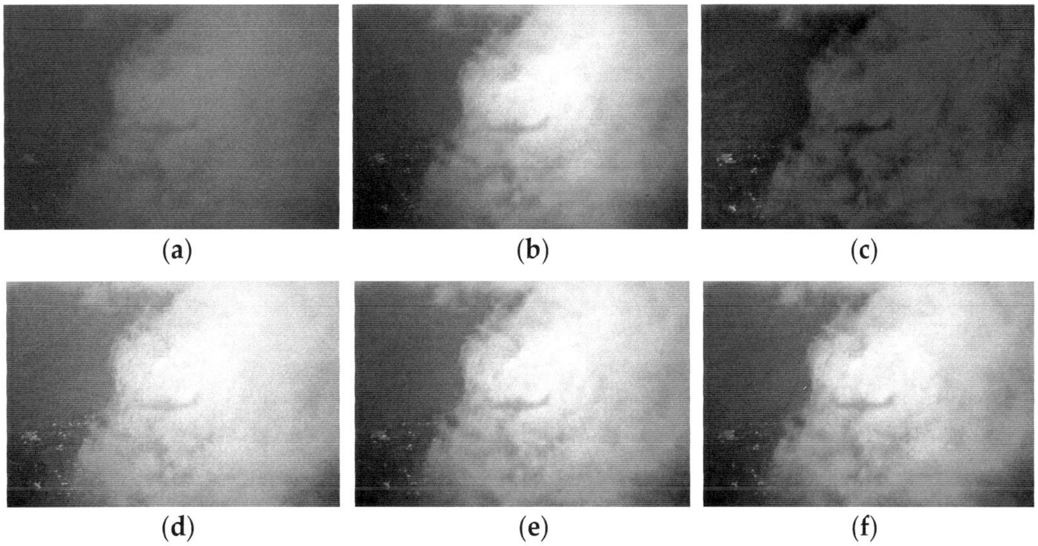

Figure 4. First group: (a) Original image; (b) Gauss filter; (c) Guided filter (d) Bilateral filter; (e) BEMD; (f) MSGF.

Contrast can well show the gradation level of images, and it can be expressed as:

$$c = \sum \delta(i,j)^2 P(i,j) \tag{10}$$

In the formula: $\delta(i,j)$ is the gray difference between two adjacent pixels, $P(i,j)$ is the probability of the gray difference $\delta(i,j)$.

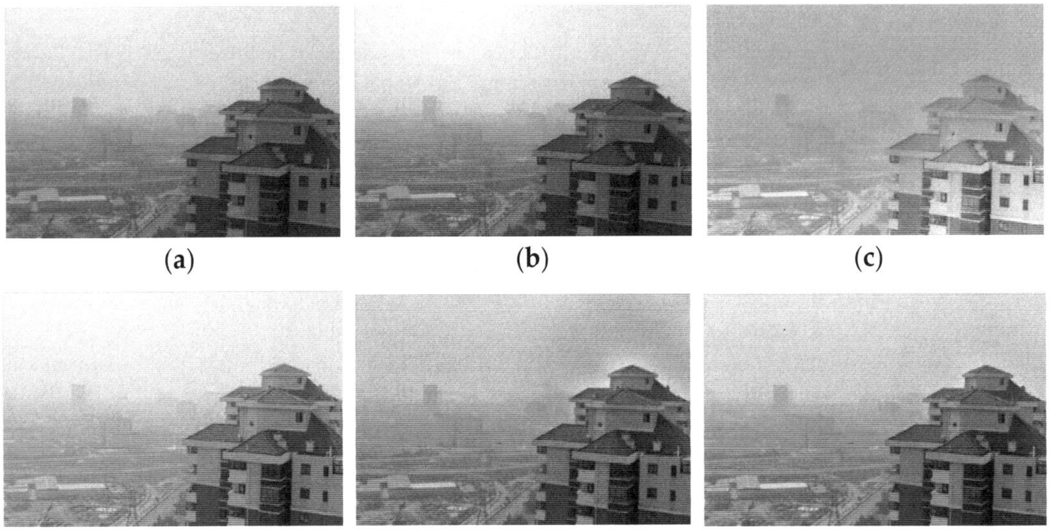

(a) (b) (c)

(d) (e) (f)

Figure 5. Second groups: (a) Original image; (b) Gauss filter; (c) Guided filter (d) Bilateral filter; (e) BEMD; (f) MSGF.

Brightness shows the degree of light and shade of images, and it can be expressed as:

$$a = \frac{\sum_{j=1}^{n}\sum_{i=1}^{m} g(i,j)}{mn} \tag{11}$$

In the formula: $g(i,j)$ is the pixel value at (i,j).

(a) (b) (c)

(d) (e) (f)

Figure 6. Third groups: (a) Original image; (b) Gauss filter; (c) Guided filter (d) Bilateral filter; (e) BEMD; (f) MSGF.

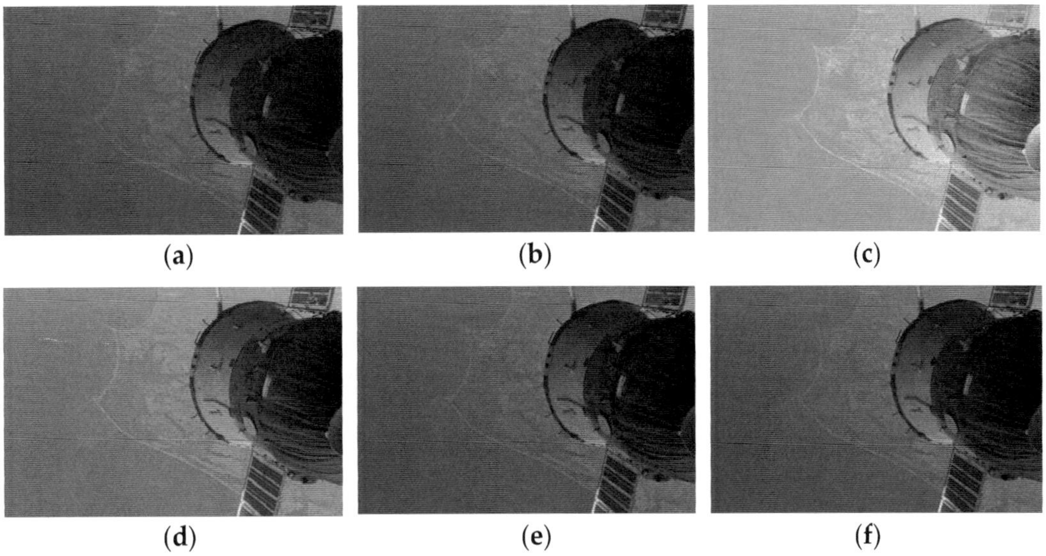

Figure 7. Fourth groups: (a) Original image; (b) Gauss filter; (c) Guided filter (d) Bilateral filter; (e) BEMD; (f) MSGF.

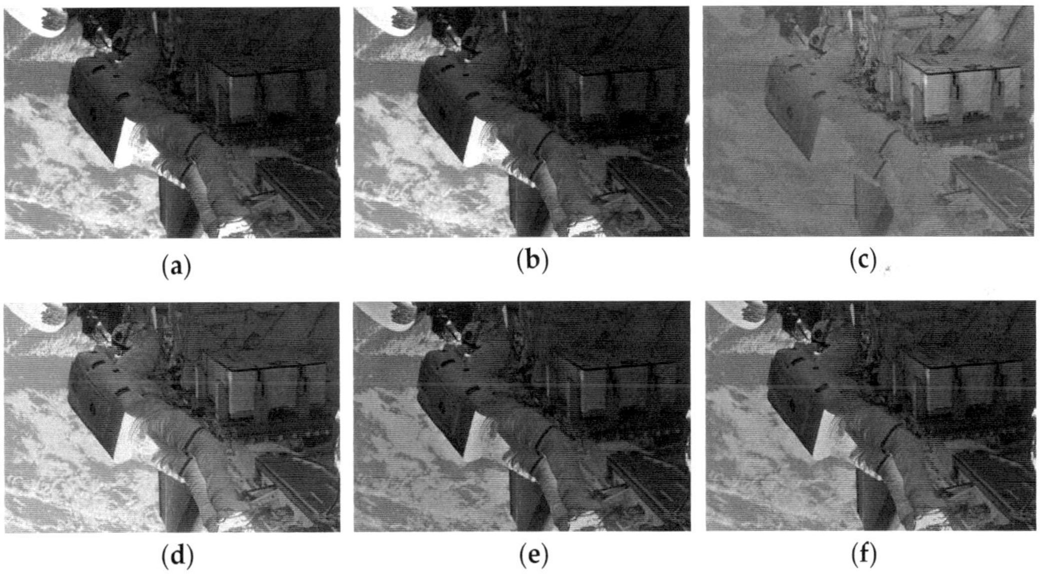

Figure 8. Fifth groups: (a) Original image; (b) Gauss filter; (c) Guided filter (d) Bilateral filter; (e) BEMD; (f) MSGF.

Information entropy represents the amount of information in an image, and it can be expressed as:

$$H(p) = -\sum_{i=0}^{255} p_i \log_2 p_i \tag{12}$$

In the formula: p_i is the probability of the i th gray level.

Clarity based on the *Tenegrad* gradient function shows the clarity of an image. It is defined as follows:

$$F_{Tenegrad} = \sum\sum|G(i,j)| \tag{13}$$

$$G(i,j) > Tenegrad \tag{14}$$

In the formula: $Tenegrad$ is 100, a given threshold of edge detection, $G(i,j)$ is the gradient of the image.

The results of various algorithms for figure 1~5 are shown in table 1~5.

Table 1. First group.

	original image	gauss filter	guided filter	bilateral filter	BEMD	MSGF
contrast	1.5481	7.2407	12.8425	21.3066	13.0440	12.0827
brightness	31.4558	44.5080	19.7783	56.1421	50.7910	49.3690
informatio n entropy	6.637	7.8541	5.9039	7.5596	7.7483	7.7552
clarity	0.0001	0.0007	0.0019	0.0024	0.0015	0.0014

Table 2. Second group.

	original image	gauss filter	guided filter	bilateral filter	BEMD	MSGF
contrast	39.1247	41.7655	62.8485	70.6183	62.7155	64.9710
brightness	51.5991	52.1697	58.4903	61.0883	51.2596	54.1183
informatio n entropy	7.4619	7.5958	6.0702	7.1444	7.2775	7.1907
clarity	0.0166	0.0177	0.0213	0.0267	0.0237	0.0248

Table 3. Third group.

	original image	gauss filter	guided filter	bilateral filter	BEMD	MSGF
contrast	54.8093	62.5503	75.4751	105.6953	75.2847	77.1326
brightness	18.7076	19.9591	36.5361	30.3010	21.8721	22.1615
informatio n entropy	6.9793	7.0996	6.7446	7.4586	7.1145	7.1374
clarity	0.0236	0.0252	0.0305	0.0400	0.0294	0.0295

Table 4. Fourth groups.

	original image	gauss filter	guided filter	bilateral filter	BEMD	MSGF
contrast	108.0476	118.2765	125.4374	156.5162	125.2910	112.3449
brightness	25.1383	25.4730	49.8711	36.7847	25.8264	25.7252
informatio n entropy	7.2050	7.2892	6.7322	7.4136	7.0578	7.1560
clarity	0.0164	0.0170	0.0299	0.0264	0.0188	0.0186

Table 5. Fifth group.

	original image	gauss filter	guided filter	bilateral filter	BEMD	MSGF
contrast	247.5912	243.2520	106.4342	275.2384	212.1725	216.2456
brightness	32.7923	31.1643	38.4612	40.7278	28.4515	30.7165
informatio n entropy	7.5167	7.4956	6.1890	7.6299	7.3181	7.4006
clarity	0.0544	0.0537	0.0304	0.0672	0.0489	0.0508

The details of the result images of guided filter are more prominent. Especially in the dark, a lot of details become clearly and contrast is improved. But the edge effect of the weaker gradient is poor. The brightness and contrast of the result images of bilateral filter are improved, but the noise in the original image is amplified and the brightness of the image is saturated. The partial residual illumination of the images is not removed in BEMD, resulting in halo artifact and color distortion. MSGF has clear details, good contrast, no halo artifact, obvious characteristics and a strong visual sense.

Compared with the other four methods from table 1~5, MSGF has a great advantage in contrast, brightness, information entropy and Clarity. The overall analysis shows that the objective evaluation of MSGF is higher than other methods. For contrast, although the value of the bilateral filter high, it has a poor edge effect of the weak gradient. Therefore, MSGF has advantage in contrast. For brightness, although the value of guided filter and bilateral filter are high, the mode of luminance enhancement of guided filter is not fixed and the brightness of the image may be reduced. While the brightness of bilateral filter is too brighter, which amplify the noise in the result images and cause the brightness of the result images is supersaturated. Aiming at information entropy, the guided filter is better, but its visual effect is poor. Aiming at clarity, the guided filter has high value, but its visual effect is also poor. For every test result, MSGF is stable and not the worst. Therefore, in contrast, brightness, information entropy and definition index, MSGF has a strong advantage on the whole and the effect of illumination estimation and evaluation index are better than other methods.

In the same parameter setting, MSGF not only inherits the advantages of guided filter, but also improves the disadvantage of the image in the distance and highlights the details that are not easy to find in the image. The whole image becomes clear and the details are gradually rich. The contrast and brightness of the effect image are relatively improved, and the detail and clarity of the image become prominent. It is more consistent with human vision requirements.

5. Conclusion

According to the characteristics of retinex theory, MSGF is proposed. BIMF of different scales is obtained by using BEMD. In the process of reconstruction, guided filter is used to filter each BIMF, most of the details in the image are smoothed and the integrity of the edge is maintained. In order to filter out all details, guided filter is again used to filter the reconstruction image. Finally, MSGF is applied to retinex algorithm and the quality of the result images is evaluated by contrast, brightness, information entropy and clarity. The simulation results show that MSGF improves the contrast of the result images of retinex algorithm by 1, brightness increases by 1, information entropy increases by 0.18 and the definition increases by 0.001. MSGF avoids halo artifact and information loss, highlights fine segments of the image and meets the visual vision of the human senses. Therefore, MSGF can estimate image illumination information effectively.

Acknowledgments: This study was supported by the Shenyang Jianzhu University (SJZU) Postdoctoral Innovation Fund Project (SJZUBSH201610); National-Local Joint Engineering Laboratory of NC Machining Equipment and Technology of High-Grade Stone (SJSC-2015-13; SJSC-

2016-3); Shenyang Jianzhu University Scientific Research Project (2017019); Colleges and Universities Basic Research Projects in Liaoning Province (LJZ2017035);National Science Foundation of China（51705341）; National Key R&D Plan (2017YFC0703903);Liaoning Province Key Research and Development Project（2017225016）

References:

[1] Kim B, Jo W and Park H M 2013 Integrated Multi-scale Retinex Using Fuzzy Connectivity Based on CIELAB Color Space for Preserving Color *International Conference on Neural Information Processing* pp 450-457

[2] Xiao J S, Shan S S, Duan P F, Chao-Ping T U and Ben-Shun Y I A 2014 Fast Image Enhancement Algorithm Based on Fusion of Different Color Spaces *Acta Automatica Sinica* **40** 697-705

[3] Xiao J, Peng H, Zhang Y, Tu C and Li Q 2016 Fast image enhancement based on color space fusion *Color Research & Application* **41** 22-31

[4] Xie S J, Lu Y, Yoon S, Yang J and Park D S 2015 Intensity Variation Normalization for Finger Vein Recognition Using Guided Filter Based Singe Scale Retinex. *SENSORS-BASEL* **15** 17089-17105

[5] Wei-Wei H U, Wang R G, Fang S and Qiong H U 2010 Retinex Algorithm for Image Enhancement Based on Bilateral Filtering *Journal of Engineering Graphics* **31** 104-109

[6] Yin J, Li H, Du J and He P 2014 Low illumination image Retinex enhancement algorithm based on guided filtering 639-644

[7] Cao W and Liu J 2012 A License Plate Image Enhancement Method in Low Illumination Using BEMD *Journal of Multimedia* **7, 6** 401-407

[8] Song H, Lin X and Liu Z 2014 A Face Recognition under Varying Illumination *Chinese Conference on Image and Graphics Technologies* pp 120-128

[9] He K, Sun J and Tang X 2013 Guided Image Filtering *IEEE Transactions on Pattern Analysis & Machine Intelligence* **35** 1397-1409

[10] Z L, J Z, Z Z, W Y and S W 2014 Weighted Guided Image Filtering *IEEE Trans Image Process* **24** 120-129

[11] He Z, Wang Q, Shen Y, Jin J and Wang Y 2013 Multivariate Gray Model-Based BEMD for Hyperspectral Image Classification *IEEE Transactions on Instrumentation & Measurement* **62** 889-904

IOP Conf. Series: Materials Science and Engineering **399** (2018) 012023 doi:10.1088/1757-899X/399/1/012023

Design and analysis of combined hinges using multi-tape spring

Z Han[1*], L Dong[1] and R Hao[1]

[1] School of Mechanical Engineering, Shenyang Jianzhu University, Shenyang, China

[*] E-mail: hanzeguang1@163.com

Abstract. In the space folding mechanism, the function of rotation can be provided by the traditional hinge, and additional driving and locking functions are completed by the unfolded and locking mechanism. Thus, such a mechanism is not only of high quality and volume, but also of complex structure. In this paper, a kind of folding hinge mechanism, which is composed of three springs for the expansion of solar canvas, has been designed and analysed. It can perform automatic expansion by the strain energy accumulated after bending and automatic locking by its inherent critical bending moment. The additional locking devices are not required. By using ABAQUS, the influence of tape spring's space layout and bending mode on the bending moment of the combined hinge with three springs in unfolding process has been researched. An optimal combined hinge which has better bending performance is then obtained. The problem of dead point which may exist in the process of expansion in this kind of hinge was successfully avoided.

1. Introduction

In the space folding mechanism, the function of rotation can be provided by the traditional hinge, and additional driving and locking functions are completed by the unfolded and locking mechanism. Thus, such a mechanism is not only of high quality and volume, but also of complex structure. Therefore, the development of light-weight, small-volume space folding mechanism has become a hot research topic.

The tape spring is a kind of single-layer open cylindrical shell structure similar to a steel tape gauge. It has special mechanical properties and can be automatically expanded by the strain energy accumulated after bending and automatically locked by its inherent critical bending moment. The additional locking devices are not required. Thus, the tape spring has been widely used in aerospace [1, 2]. Multi-tape spring combined hinges are used in large space deployable mechanisms, such as solar panels, spatial deployable antennas, and so on.

At present, the spatial deployable hinge based on the tape spring has several forms shown in figure 1. This kind of tape spring mainly composed of two tape springs has high stiffness, but there is an impact phenomenon at the end of expansion [3, 4]. Therefore, after finding out the optimal geometric parameters of a single belt spring, a new type of hinge consisting of three tape springs has been designed. The optimum spatial layout combination has been obtained after analyzing, and using a foundation for the design of the solar panel expansion mechanism for space.

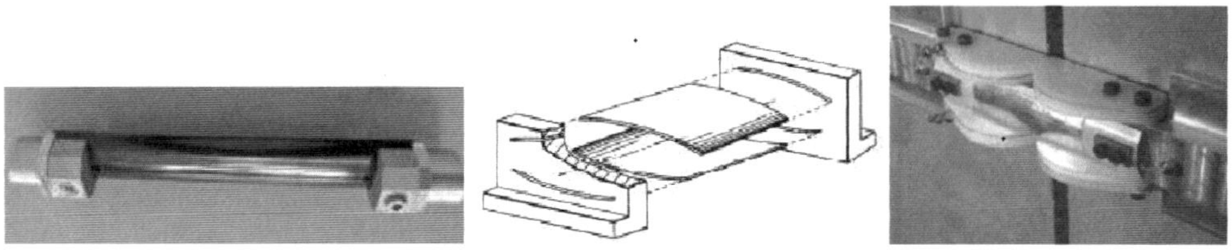

(a) Single spring hinge (b) Multi-tape spring combined hinge1 (c) Multi-tape spring combined hinge2

Figure 1. Several deployable hinge structures based on tape spring.

2. Theoretical introduction of single tape spring

Professor Pellegrino, University of Cambridge, has studied the bending process of a single spring and he found the moment-angle diagram of the bending process with spring which is shown in figure 2. The moment increases linearly to the maximum value $M_{+}^{\max}$ during reverse bending, and then jumps to the steady moment M_{+}^{*}. Similar to reverse bending, the moment increases linearly to the maximum value $\left|M_{-}^{\max}\right|$ in forward bending. And then it reduces linearly to the steady moment $\left|M_{-}^{*}\right|$. And it leaves the value unchanged [5].

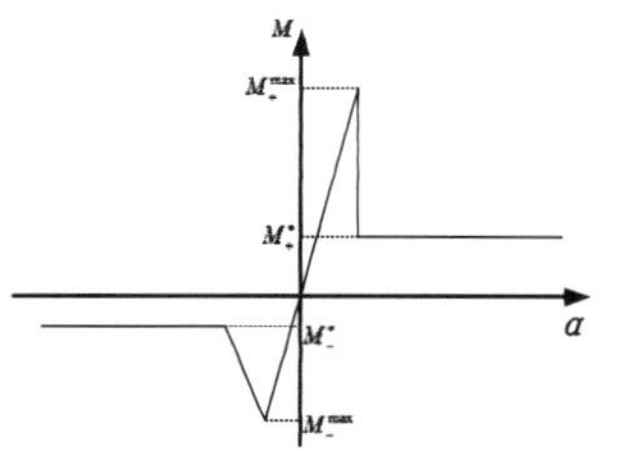

Figure 2. Relationship between bending moment and bending angle of single tape.

Assuming that the shape of the bending region after bending is close to a cylinder, the curvature of the cross section of the bending region changes from 1 / R to 0, and the curvature of the length direction from 0 to 1 /R[5-8]. The steady-state bending moment formula for the expansion process of single tape spring is:

$$\left.\begin{aligned} M_{+}^{*} &= \frac{Et^{3}(1+\mu)\theta}{12(1-\mu^{2})} \\ M_{-}^{*} &= \frac{Et^{3}(\mu-1)\theta}{12(1-\mu^{2})} \end{aligned}\right\} \qquad (1)$$

In the formula (1), M_{+}^{*} is the steady bending moment of the reverse bending, M_{-}^{*} is the steady bending moment of forward bending. E is the elastic modulus of the material of tape spring, ì is the Poisson's ratio, and t is the thickness of the tape spring.

3. Design and simulation of combined hinge

3.1. Design of combined hinges

Because the stiffness of single tape spring is poor, the driving ability and locking ability of unfolded spring are also weak. In general, the multi-tape spring combination hinge is used in engineering. On the one hand, it can improve the unrolling driving ability of the combined hinge, and on the other hand, it can greatly improve the ability of resisting the impact of the end after the hinge is unfolded [9].

In order to combine forward bending and reverse bending well, a combined hinge with three tape springs has been designed. The cross section arc direction of the two tape springs on both sides is the same, and the cross section arc direction of the middle band spring is opposite to that of both sides. There are two kinds of space layout for the three tape springs. There are two bending modes for each space layout. All spatial layouts and bending modes are shown in figure 3. The connection between the three tape springs and the solar panels will be described in detail later.

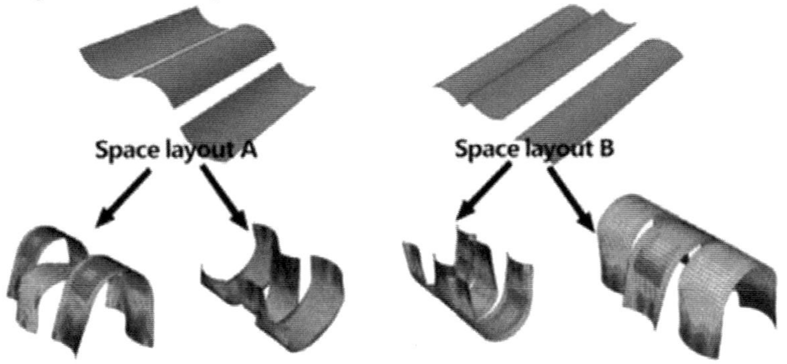

Bending mode 1 Bending mode 2 Bending mode 3 Bending mode 4

Figure 3. Spatial layout and bending mode of hinges.

For convenience of expression, the case of reverse bending of two springs on both sides and forward bending of one tape spring in the middle is defined as reverse bending. The forward bending of two tape springs on both sides and the reverse bending of one tape spring in the middle is defined as forward bending.

(a) Bending mode 1: Reverse bending on the basis of spatial layout A.
(b) Bending mode 2: Forward bending on the basis of spatial layout A.
(c) Bending mode 3: Reverse bending on the basis of spatial layout B.
(d) Bending mode 4: Forward bending on the basis of spatial layout B.

3.2. Finite element analysis of combined hinges

Taking bending mode 2 as an example, the whole structure of the combined hinge will be introduced and the finite element analysis of the working state of the combined hinge is also done.

The whole structure diagram of the combined hinge is shown in figure. 4. The combined hinge is composed of two parts: the spring and the chuck. The two ends of the side-by-side three tape springs are clamped by two chucks. In practice, the other end of the chuck will be connected to the solar canvas.

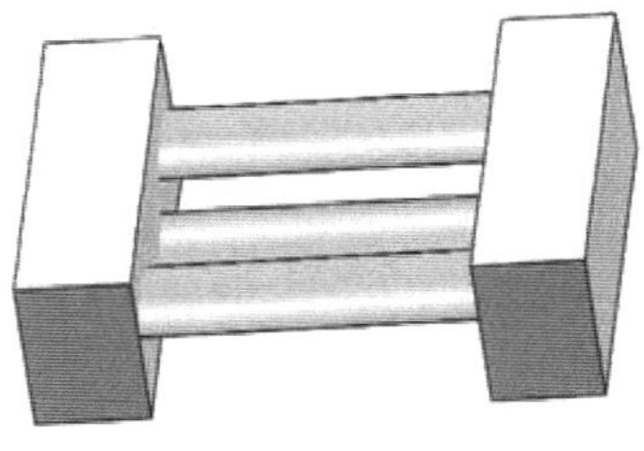

Figure 4. A schematic diagram of the whole structure of the combined hinge.

IOP Conf. Series: Materials Science and Engineering **399** (2018) 012023 doi:10.1088/1757-899X/399/1/012023

Due to the fact that only three tape springs are deformed during the operation of the hinge, the model can be simplified in ABAQUS, and the mechanical analysis of the three tape springs can be carried out after the equivalent model is established. The spatial layout of the three tape springs is shown in fig.5. The reference points RP1, RP2, RP3, RP4, RP5 and RP6 are respectively the center of the circular arcs across the two ends of the spring, and the six reference points are located on the same plane. Z to the view, RP2, RP4, RP6 are on the same line, and RP1, RP3, and RP5 are on the same line. The MPC node is established to couple the RP4 with three cross-sections at one end and the RP3 with three cross-sections at the other end, so that the synchronous constraint on the three tape springs can be carried out by imposing constraints on the RP3 and RP4, and the bending and unfolding of the hinge can be carried out. The 90 degree angular displacement of rotation around X was applied on RP3 and RP4 respectively to carry out the forward bending of hinges.

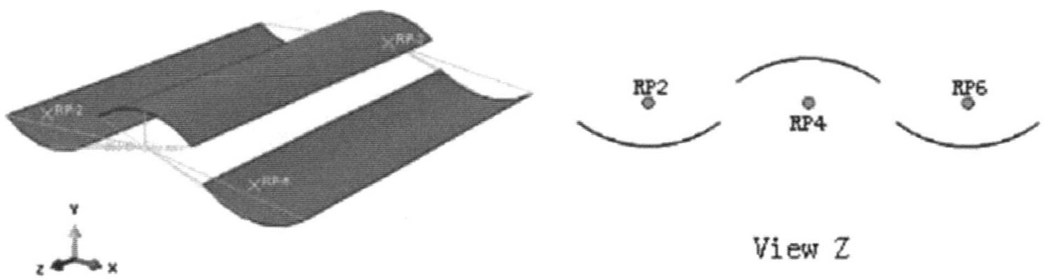

Figure 5. Equivalent model.

For the other three kinds of combined hinge, the simplified method and constraint conditions of the model are similar to that of bending mode 1, so there is no more repetition.

3.3. Effect of different spatial layout and bending mode on bending properties of combined hinges

The results of the finite element analysis of the expansion process of the combined hinges under four bending modes are shown in figure 6(a). Figure 6(b) is a magnification of the part near the X axis in fig. 6(a)

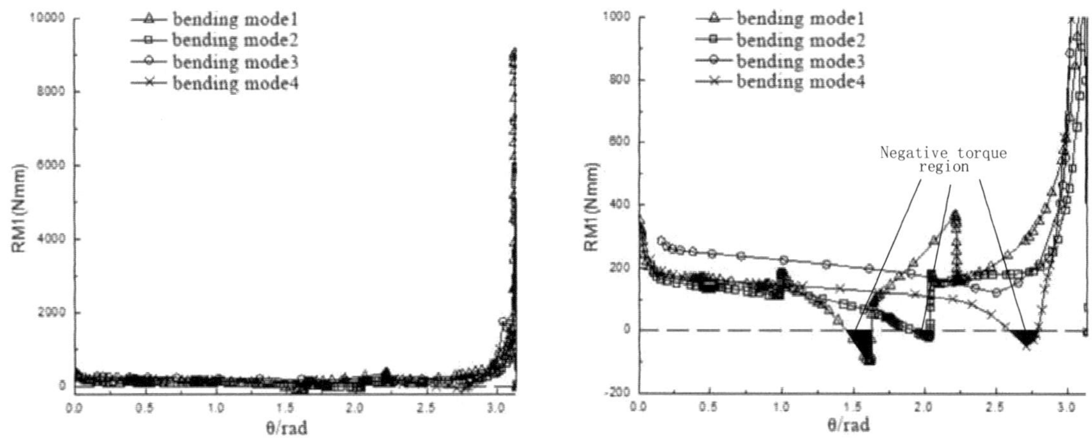

(a)Moment-rotation Diagram (b) Enlarged drawing

Figure 6. FEM analysis of combined hinge.

As shown in figure 6(b), the critical moment and the steady moment of the combined hinge are larger than those of the single belt spring, but combination hinges with some kinds of bending mode produce negative moments during expansion. That is, the driving moment of the hinge expansion process is negative, which is not conducive to the unfolding of the combined hinge. This shows that the dead center will occur on the hinge. The combination hinge will not be further expanded after it has expanded to the center point.

By comparing the torque variation of hinge expansion process with four bending modes, the following results can be obtained:

(1) With bending mode 1: The critical moment of the hinge is larger (9040N·mm), which indicates that the hinge has stronger self-locking ability. But, throughout the whole development process, the bending moment is not stable enough, and the moment is negative between 85.5°and 93°, and the minimum is -98.1146Nmm, which is very unfavorable to the expansion of the combined hinge.

(2) With bending mode 2: The critical moment of the hinge is 5995N·mm, and the moment is negative between 111.66°and 116.53°, and the minimum is -17.6423N·mm. Although its minimum negative torque is not small enough, the bending moment of the whole development process is not stable.

(3) With bending mode 3: The critical moment of the hinge is small (1764N·mm), and the moment value of the unfolding process is the most stable in the three bending modes, which is consistent with the theoretical analysis of the steady moment, and there is no negative moment in the unfolding process.

(4) With bending mode 4: The critical moment of the hinge is minimum (1119 N·mm), and the moment value of the unfolding process is more stable than that of the bending mode 1. However, there is also negative bending moment between 155 °and 160 °, with a minimum of -50 N·mm.

So, the hinge shown in bending mode 3 is an optimal solution.

4. Conclusion

The application of tape spring in aerospace is a major breakthrough in the research of space folding structure. Compared to the traditional deployable hinges, the proposed hinge is lighter in mass, better in unfolding performance and larger in unfolding torque, and self-locking can be achieved after unrolling. The finite element analysis method presented in this paper can also be verified by the theoretical calculation formula of single tape spring, and it can also be applied to the simulation analysis of other similar structures.

In this paper, the expansion process of multi-tape spring combined hinge has been mainly researched. The problem of dead point in the process of hinge expansion aiming is found and solved. The obtained conclusions are as follows:

(1) The bending and unfolding performances of multi-tape spring combined hinge are much better than that of single tape spring.

(2) The reverse bending performance of single tape spring is better than that of forward bending.

(3) When the number, material and shape of a single spring are fixed, the spatial distribution and bending mode of the spring are the main factors affecting the bending moment during the expansion process of the multi-tape spring combination hinge.

(4) The self-locking ability of multi-tape spring combination hinge in spatial layout mode is stronger. The unfolded process of the combined hinge based on spatial layout b is stable and when it bends reversely there is no dead point. It is a more optimized structure form and it has great engineering application value.

References

[1] Soykasap O 2009 Deployment analysis of a self-deployable composite boom *Composite Structures* **89** 374

[2] Hui Y H G 2016 Mechanical modeling and analysis for hyperelastic hinge in a triangular prism deployable mast *Journal of Astronautics* **37** 275

[3] Ayesha K 2017 Review on technological significance of photoactive, electroactive, ph-sensitive, water-active, and thermoresponsive polyurethane materials *Polymer-Plastics*

Technology and Engineering **56** 606

[4] Jiménez F L and Pellegrino S 2012 Folding of fiber composites with a hyperelastic matrix *Int.l J.f Solids and Struct* **49** 395

[5] Hoffaita.S, Brulsa O, Granville D et al 2010 Dynamic analysis of the self-locking phenomenon in tape-spring hinges *Acta Astronautica* **66** 1125

[6] Sofiyev A H 2009 The vibration and stability behavior of freely supported FGM conical shells subjected to external pressure *Composite Structures* **89** 356

[7] Schenk M, Viquerat A D, Seffen K and Guest S 2014 Review of inflatable booms for deployable space structures packing and rigidization *Journal of Spacecraft and Rockets* **51** 762

[8] Jun W, Fuling G and Zhigang Z 2007 Design and analysis of tape hinge for deployable space structure *Journal of Astronautics* **28** 720

[9] Yuqing W and Yanghong S2011Finite Element analysis on bending processes of the tape spring used in space hinges*Electro- Mechanical Engineering* **27** 29

IOP Conf. Series: Materials Science and Engineering **399** (2018) 012024 doi:10.1088/1757-899X/399/1/012024

An intelligent household greenhouse system design based on Internet of Things

Z Han[1,2], Z Wu[1*], S Lin[1] and F Luan[1]

[1] Faculty of Informationand Control Engineering, Shenyang Jianzhu University, Shen yang 110168, China
[2] Department of Digital Factory, Shenyang Institute of Automation, the Chinese Academy of Sciences (CAS), Shenyang 110016, China

[*] E-mail: shuaizhenbo@163.com

Abstract. In order to combine indoor greenery conservation with Internet of Things Technologies, this paper designs an intelligent household greenhouse project with the features of comprehensive sensing, reliable transmission and intelligent processing. Through the analysis of functional requirements of the intelligent household greenhouse system, an intelligent household greenhouse system is designed with the functions of greenhouse environmental data detection, greenhouse environmental control regulation, data remote transmission and human-computer interaction. Its sensor layer collects environmental data in real time based onthe ZigBee wireless sensor network. The network layer STM32 intelligent gateway coordinates with network server, so as to exchange data from sensor layer to application layer, and solve the problems of non-blocking of data sending and receiving as well as concurrent requests of multiple mobile terminals. The application layer is designed into two types. One is a desktop management system as a data storage and analysis center, and the other is a mobile terminal APP. At the same time, we design a communication protocol that is applicable to the interaction of the three-layer structure of the Internet of Things, with the characteristics of simplicity, stability, readability, and scalability. It can avoid the mutual influence of multi-level data exchange and ensure the correctness of data circulation. In the design, the system sensor layer ensures stable transmission of various data and instructions, and the network layer has a high degree of concurrency and real time. And various measurement and control data of the sensor layer can interact with the data of mobile-terminal equipment of the application layer. What's more, the desktop management system and mobile terminal APP can monitor greenhouse data in real time and control various actuators in the greenhouse.

1. Introduction

At present, intelligent greenhouses are mainly used in agricultural greenhouses. As for the proposal of household greenhouse as well asdesign and implementation of intelligent household greenhousesystem, we can drawon research results of agricultural intelligent greenhouse, combine with the spatial limitations of household greenhouse, and simultaneously meet the requirements of nurturing various plants, easy dismounting, strong practicality, and high cost performance.In this way, we can transform intelligent greenhouse from agricultural field to home life, designing an intelligent greenhouse with reasonable cost and complete functions.

Referring to the research on structure system of the IOT at home and abroad, the intelligent household system based on the ITO adopts a layered architecture, which mainly includes the sensor layer, network layer and application layer [1-3]. Currently, most sensor layers of the IOTemploy wireless transmission technology. Referring to Zhang Meng's Greenhouse Remote Monitoring System Design Scheme Based on Zigbee [4] in 2012, and Liang Ruihua's proposal to replace traditional wired networks with ZigBee technology in 2016, both of the margue that Zigbee technology has the features of low-energy consumption, reliable communication, low cost and data security, which is a feasible solution for sensor layer of intelligent household greenhouse [5].

At present, the gateway design and development for intelligent household based on theIOT focus on designingan embedded network server as the only device in the network layer. Referring to Chang Yingliang's design scheme for the IOT gateway in 2014 [6], and the idea that uses wireless to connect data-acquisition sensor with the Internet, proposed by H Messer and P Alpert [7],they come up with a gateway design scheme for embedded chips in the basis of ARM11 or ARM Cortex-A8 and other high-performance cores. These schemes can basically meet the IOT requirements of data circulation, analysis and processing in the IOT system. And the cost and development of such schemes are relatively high. At the same time, there are no specific solutions proposed when the system has multiple client requests and when the data block. Considering the gateway server based on STM32 designed by Zhang Wenjing in 2015 [8], and the embedded Web server under the uC/OS-II operating system designed by Huang Xiaofen and Liu Damao in 2006 [9], we find that such design solutions are low-cost and easy-development after a comprehensive analysis. However, its processor performance cannot fully meet the functional requirements in the case of big data volume analysis and multiple client requests. By analyzing the above-mentioned research on the IOT's functional design at the network level, we propose a network layer solution for the coordination and cooperation between the intelligent gateway and the server, aiming at the problem that the current network layer design cannot balance performance and cost. The network layer gateway implements the multi-task concurrency function under the real-time operating system. In the server design, we combine with multithreading, socket communication and other technical means, realizing the data flowing across the sensor layer, network layer, and application layer.

At present, the intelligent household app control system has been widely used in China. The control terminal in intelligent household system employs Android platform equipment. The Android system provides a strong technical support and good user experience, and it is easy to transplant. Consequently, the mobile phones, iPads and other mobile devices can become the control terminal of the intelligent household system, which reduces the costs and improves the usability. In 2016, Zhao Wenbing's intelligent greenhouse remote monitoring system design based on Internet of Things proposes a remote monitoring strategy for mobile terminals with Android as the operating system [10]. It has the functions of user login, data query, and greenhouse control. In 2012, Yan Wei and others brought up the application of multi-threading technology in the development of Android phones, which can avoid the non-response of applications, and ensures that the program transforms smoothly and responds quickly [11].

According to the above pieces of research of intelligent greenhouses based on the Internet of Things, the intelligent household greenhouse has a typical IOT hierarchy. In the system, the sensor layer mainly completes the acquisition and processing of household environment and equipment status. The principal role of the network layer is to achieve data interoperability from the underlying sensor to the user's mobile client, ensuring the reliability and real-time nature of data exchange. The application layer chiefly includes types of mobile terminals, realizing real-time data acquisition and control of household devices and related devices in the intelligent greenhouse. The system design scheme covers the following: ZigBee wireless communication network is used as a short-range wireless communication method for various sensors, executors and controllers in the sensor layer; the core of the network layer is constructed through the intelligent gateway and server, and the mobile terminal APP is used in the application layer which designs the communication protocols among layers.

2. Requirement analysis of the overall system function

The intelligent household greenhouse system is designed for the users who want to grow flowers and plants at home. In the household environment, the indoor parameters are small in range and variation with respect to the outdoors, but they are not necessarily suitable for the growing environment of flowers and plants.

Firstly, in the process of intelligent household greenhouse design, we should first consider the problem of greenhouse placing due to the limitations of the indoor space. Most of the buildings face southeast, making it cool in the summer and better daylighting in the winter. As a result, the greenhouse is usually placed on a south-facing balcony or room. Therefore, the intelligent household greenhouse should be flexible in size, transparent, and leak-proof.

Secondly, different plants need different growth environments. In the same greenhouse, we need to nurture plants with diverse habits. It is necessary to make full use of the limited space in the greenhouse, to properly place various plants and to customize personalized planting scheme.

Thirdly, the environmental regulation ability of greenhouses is an important performance index of greenhouses. Above all, each sensor and actuator must be rationally designed and deployed at the sensor layer, so as to collect the environmental parameters in the greenhouse in real time and make corresponding adjustments according to control instructions.

Fourthly, the network layer plays an important role in data connectivity between sensor layer and application layer, and has high real-time performance for sending and receiving data. In the light of the situation that network layer needs to handle numerous network connection requests, the designed server should have concurrent processing capabilities. In the entire system, the server, on the one hand, establishes a network connection with the application layer mobile client to provide services, and on the other hand, it needs to set up a connection with the gateway so as to realize bidirectional communication of data from the application layer to the gateway, then to the sensor layer.

Fifthly, if users want to monitor and adjust the greenhouse environment, it is necessary to design a desktop management system with data storage and analysis functions and a mobile terminal APP that can be easily carried and operated. The user can manually and automatically control the greenhouse through the current greenhouse environmental data and the operating status of the actuator displayed on the desktop management system and the mobile terminal APP.

3. Overview of system structure

The intelligent household greenhouse is designed as a self-contained greenhouse. As an absolute construction with flexible and variable size, it is attached to the building's orientation and it is typically placed on a south facing balcony or room to maximize the lighting. The intelligent household greenhouse is an upper and lower two-story structure. The main body adopts a transparent acrylic removable panel, which has good light permeability, rich colors, long service life, outstanding rigidity and strength, and excellent chemical resistance.

The intelligent household greenhouse system is divided into sensor layer, network layer and application layer. The system transmits the greenhouse environmental parameters to servers and mobile terminals through ZigBee, STM32 intelligent gateways and routers. At the same time, the control instructions of the mobile terminals should be sent to the devices in the greenhouse through the server.

3.1. System sensor layer

The system sensor layer is composed of ZigBee wireless sensor control network, which realizes the collection and processing of greenhouse environmental data.

3.2. System network layer

The network layer in intelligent greenhouse system consists of STM32 intelligent gateway and server. The intelligent gateway can achieve real-time and quick interaction among data in each layer, ensuring the data circulation between the sensor layer and the application layer. The intelligent gateways have the ability to analyze and process data. They can analyze alarm data in abnormal environment detected by the sensor layer, control the actuators, handle environmental anomalies, and eliminate alarms.

3.3. System application layer

The application layer refers to the desktop management system and mobile terminal APP. The user can obtain the greenhouse environment data and actuator's working status through the desktop management system and the mobile terminal APP, and controls the actuator.

The mobile phone APP acts as a mobile control terminal. Through the APP, the user regulates greenhouse environment by adjusting the setting value of greenhouse environment (automatic mode) and actuator's gear (manual mode).

As the greenhouse environmental monitoring center and data processing center, the desktop management system can store communication-related information, such as user's information, IP addresses and port numbers, environmental data, actuators' working status, data instructions, and control instructions. The communication data between sensor layer and application layer of mobile phone APP is stored in the database of desktop management system. After being processed by the desktop management system, the data is forwarded to the gateway and the mobile phone APP.

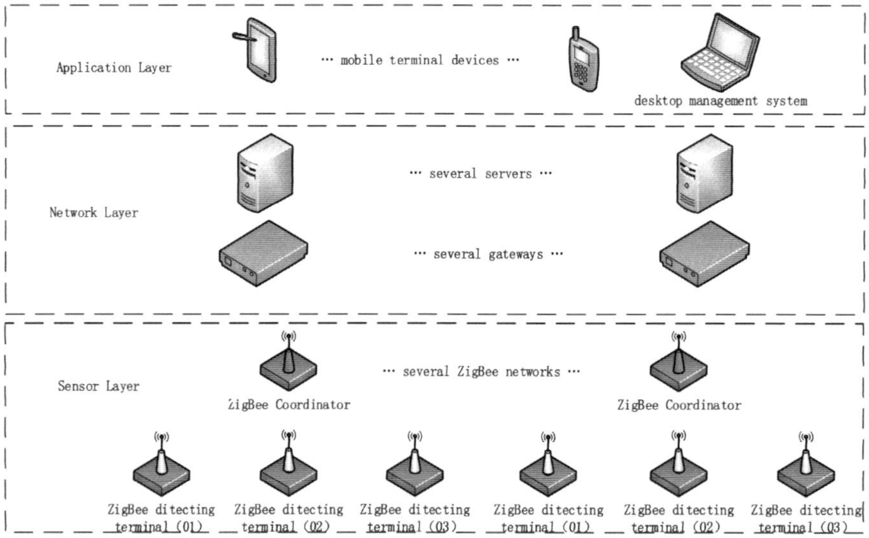

Figure 1. Intelligent Household Greenhouse System Structure.

4. Overview of system structure

4.1. ZigBee network design

The sensor layer uses ZigBee wireless technology to achieve short-range communication. The ZigBee features low-energy consumption, low cost, short delay, large network capacity, and reliable security. The ZigBee coordinator is directly connected to the ZigBee terminal nodes that are equipped with diverse sensors and actuators to establish the entire ZigBee network. In this ZigBee wireless sensor control system, the ZigBee network is designed to consist of nine ZigBee modules. One of the ZigBee modules serves as the coordinator of the ZigBee network, and the remaining eight pieces act as the terminals of the ZigBee network, six of which are detecting terminals equipped with sensors and two of which are executive terminals equipped with executors. The ZigBee communication module design includes ZigBee coordinator module, ZigBee detecting terminal module, and ZigBee executive terminal module. The core adopts CC2530 chip, and carries 8051CPU and 2.4GHZ RF transceiver. The CC2530 has different operating modes and meets the requirements for low-energy consumption.

4.1.1. ZigBee Coordinator Design. The software design of the ZigBee wireless sensor control system is developed by using Z-Stack protocol stack. The ZigBee coordinator is responsible for establishing the ZigBee network and interacting with the STM32 intelligent gateway in network layer. When the ZigBee coordinator receives a query sent from the serial port of the STM32 intelligent gateway in network layer, it will package the queried sensor data into a communication protocol and return it to the STM32 intelligent gateway in network layer through the serial port; when the ZigBee coordinator receives control instructions from the serial port of STM32 intelligent gateway in network layer, it will send the control instructions to the ZigBee executive terminal, and receive the executors' working status returned from the control terminal. What's more, the executor will be packaged into a communication protocol and return to STM32 intelligent gateway in network layer through the serial port.

The ZigBee coordinator module includes RS232 serial communication module, power module, reset module, Debug module, I/O and RF antennas. The CC2530 chip integrates CPU, memory, and RF transceiver, which simplifies the circuit design. And the CC2530 chip contains decoupling circuit, in order to remove the noise on the chip's power pin.

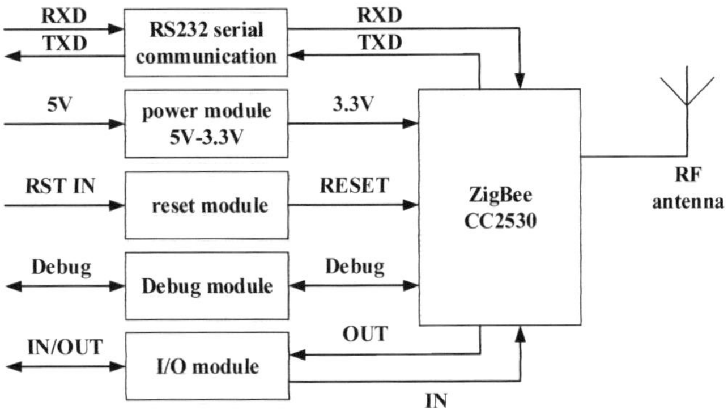

Figure 2. Module Structure of ZigBee Coordinator.

4.1.2. ZigBee Detecting Terminal Design. The environmental factors that affect plant growth and can be regulated in the greenhouse are temperature, humidity, light, gas environment and soil environment. In the household greenhouse environment, the temperature ranges from -10°C to 40°C, the humidity ranging from 10% to 100%, and the light intensity ranging from 0lux to 30000lux. In order to obtain more accurate environmental parameters, we must choose sensors with right size, high accuracy and suitable range. The ZigBee detecting terminal module includes sensor input module, power module, reset module, Debug module, I/O and RF antennas.

A temperature and humidity sensor DHT11 is used to detect the temperature and humidity of the greenhouse. The DHT11 sensor is a temperature and humidity conforming sensor that contains a calibrated digital signal output, the advantages of fast response, strong anti-interference, low-energy consumption, and high cost performance. The power input VCC of the DHT11 temperature and humidity digital sensor is powered by 3.3V DC, and the reset circuit is reset by a low level. Using an external power supply, we add a 4.7kΩ pull-up resistor to the data bus of the DTH11, for the purpose of increasing circuit stability by preventing interference.

The environmental light intensity sensor of BH1750FV digital output is used to detect light intensity of the greenhouse. The BH1750FV makes communication based on the I2C bus communication protocol. During data transfer, SCL is high level, while the data on the SDA line must remain stable. Only when SCL is low level are data on the SDA line allowed to change.

The greenhouse soil moisture is detected by FC-28 soil moisture sensor, which employs LM393 as a comparator to export soil moisture analogue quantity signal output and soil moisture threshold digital quantity output. The soil moisture threshold is adjusted by a potentiometer, being able to define the current soil moisture as wet and dry state. When lower than the setting value, the soil moisture is in dry output high level. When higher than the setting value, the soil moisture is in wet output low level. The STM32 directly detects the digital quantity signal of soil moisture, setting low level of soil moisture as the alarm signal. In this ZigBee network, six detecting terminals are set, whose functions are the same, all equipped with DHT11 temperature and humidity sensor, BH1750FVI light intensity sensor, and FC-28 soil moisture sensor. These six ZigBee detecting terminals have different number which are 01~06, corresponding to six different detecting positions. And the number corresponds to the address in the communication protocol. The final data of temperature, humidity, light intensity, and soil moisture are obtained by averaging the sensor data from six different locations. After the ZigBee coordinator establishes the network, the ZigBee detecting terminal will access the network and drive the installed sensors to work. The ZigBee detecting terminal collects the environmental data in the greenhouse in real time, integrating and packaging the data and location number, and sending them to the ZigBee Coordinator.

4.1.3. ZigBee Executive Terminal Design. The environmental factors that affect plant growth and that can be regulated in the greenhouse are temperature, humidity, light, gas environment and soil environment. In the household greenhouse environment, the temperature ranges from -10°C to 40°C, the humidity ranging from 10% to 100%, and the light intensity ranging from 0lux to 30000lux. In order to obtain more accurate environmental parameters, we must choose sensors with right size, high accuracy and suitable range. The ZigBee detecting terminal module includes sensor input module, power module, reset module, Debug module, I/O and RF antennas.

The irrigation equipment uses 12V vertical submersible pumps with 4.5W rated power. And there are 6 outlets circularly placed on the surface of the soil layer. Each floor of the greenhouse is a drawer sink, providing water for submersible pumps and humidifiers. At the same time, the excess water after irrigation will flow back to the sink through the holes at the bottom of the greenhouse, which not only solves the problem of water leakage, but also reduces the waste of water resources. The temperature of household greenhouse will not fluctuate. For one thing, the sun can provide about 25% heating. For another, according to different plant requirements for different temperatures, we will use 12V DC ceramic heating plate to warm the greenhouse, located in the greenhouse's front and rear bottom sides. Ventilation can promote gas exchange in the greenhouse, lower the temperature and supplement CO2. The heat dissipation and ventilation equipment employs a 12V, 3.9W DC fan with 8cm diameter.The heat dissipation fan is divided into an inlet end and an exhaust end. Because the hot air generally goes upwards and the cold air goes down, the fan is installed in a low-inlet and high-exhaust way. Unsuitable humidity will make plants get diseases. As a result, we use 12V, 2.5W DC fan with 7cm diameter for ventilation and dehumidification, employing 24V atomization humidifier for humidification. In northern China, the duration of sunshine is short. When the light intensity is insufficient, plants will grow weak and even die. Therefore, supplementing light is one of the indispensable tasks in the greenhouse. The 5V blue-violet LED featuring with a small amount of heat, is installed on top of the greenhouse, and the executive equipment is controlled by electric relay.

Number 07 and number 08, these two executive terminals are set in this ZigBee network. Each executive terminal is equipped with drip irrigation pump, heat dissipation fan, dehumidification fan, supplementing light LED, heating lamp, and humidifier, which can expand and increase executors. After the ZigBee coordinator creates the network, ZigBee executive terminal accesses to the ZigBee network, waiting for the broadcast from the ZigBee Coordinator. When the ZigBee coordinator receives a control command from the serial port, the ZigBee coordinator will broadcast the control instruction to the ZigBee network. After the executive terminal receives the control instruction, it will first determine whether the instruction is valid, and it will control the executor under the control instruction if the determination is correct. Meanwhile, the current working state of the executor is packaged into a communication protocol and sent to the ZigBee coordinator. Then, the ZigBee coordinator sends the communication protocol to the STM32 gateway through the serial port.

4.2. STM32 intelligent gateway design

The STM32 intelligent gateway is the core device of the intelligent household greenhouse network layer. In this intelligent household greenhouse system, the intelligent gateway is designed to have a network client that connects the ZigBee network coordinator in sensor layer and network layer server. That is, the main work of gateway is interacting data through the serial port and ZigBee network coordinator in sensor layer, which is under author's design "instruction protocol and data protocol of sensor layer, as well as data protocol of network layer sensor and device status", and analyzing different types of sensor layer data.

In the design process, full consideration should be given to the real-time requirements of the intelligent gateway in processing the data uploaded by the sensor layer and the instructions issued by the application layer. The intelligent gateway of the network layer serves as the core part of connecting sensor layer and network layer, which requires high real-time performance for sending and receiving data. The abnormal data alarm of the sensor layer is first delivered to the intelligent gateway. The identification and processing of the alarm require the characteristics of non-blocking and high real time. At the same time, there should be no data interference with the process of sending and receiving data by the server, and there should be no process blocking [12].

In the process of designing the gateway, we must follow the idea of modular programming, respectively establishing two independent tasks of data receiving and sending of the network communication module, so as to avoid interfering with the normal work caused by blocking in the process of sending and receiving data, and ensure the real-time data circulation. We also establish an independent task to deal with data alarm in sensor layer, which is not affected by network connections and concurrently processed with other processing tasks. In this way, the alarm condition can be handled in real time. In addition, we employ interrupted mode execution to communicate with the ZigBee serial in sensor layer.

The μC/OS-II embedded operating system is transplanted in the gateway. Its function of supporting multi-task concurrency is mainly implemented by the system-supplied OSTaskCreate() API function. After the gateway completes hardware resource initialization and system initialization, a new task can be created by calling OSTaskCreate().

After the initialization of hardware, protocol stack and operating system, we will establish multi-process for concurrent processing. The ZigBee coordinator and the STM32 intelligent gateway communicate through the serial port whose data sending and receiving are performed separately. The serial port receiving data thread is an evaluation operation to the data global variable. The STM32 intelligent gateway receives the environmental parameters sent by the ZigBee coordinator and the working state of the executor, and then it assigns the received data to the global variable. The serial port sending data thread reads the global variable of the control instruction, and relays the control instruction sent from the server to the ZigBee coordinator. First, we have to judge whether the flag bit is 1. If the flag bit is 1, we should read the global variable of the control instruction. Then, we should package the control instruction into a communication protocol that communicates with the ZigBee Coordinator, send it to the ZigBee Coordinator to control the executor.

The STM32 intelligent gateway communicates with the server through the network whose data sending and receiving is performed separately. The network data receiving thread is an evaluation operation to the global variable of the control instruction, it receives the control instruction sent by the server to the STM32intelligent gateway, assigns the control instruction to the global variable of the control instruction after analysis, and then sets the flag bit to 1. The network data sending thread is a reading operation to data global variable, it reads data global variable, and packages the data into a communication protocol that communicates with the server, sending it to the server at regular intervals.

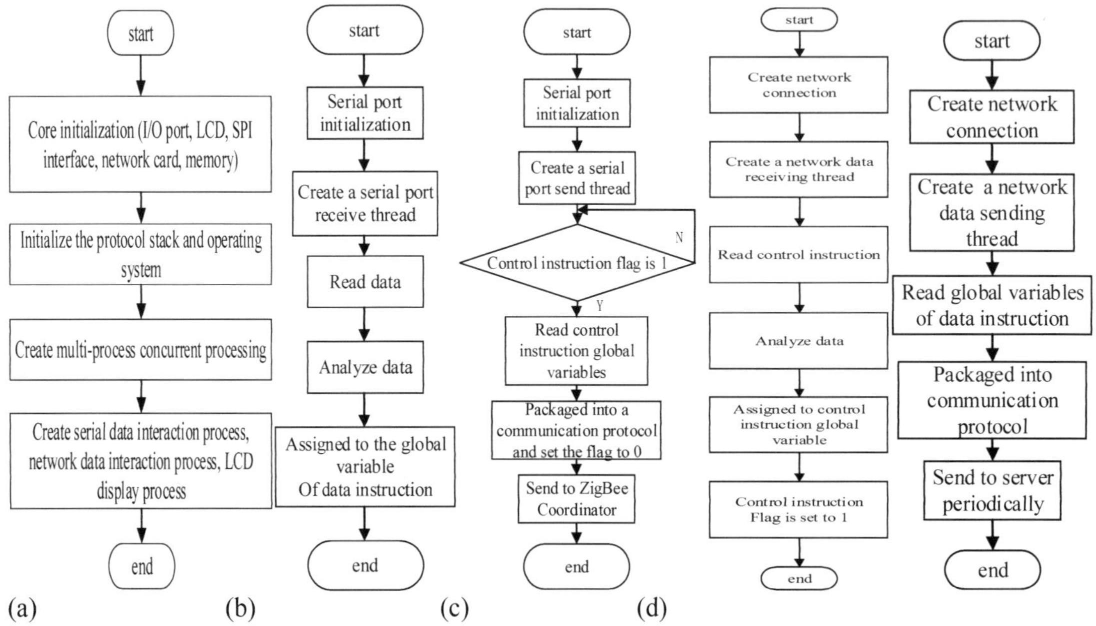

Figure 3. STM32 Intelligent Gateway Software Process.

4.3. Software design of desktop management system

The application layer is divided into desktop management system and mobile terminal APP. The desktop management system software of the intelligent household greenhouse designed in this paper includes user login management, greenhouse environmental monitoring, and historical data statistical analysis and communication services.

The data of temperature, humidity, light intensity and soil moisture in the greenhouse obtained by sensor layer, as well as active status of actuator are periodically sent to the desktop management system for storage and updating to the interface. The manual-control instructions and automatic-control programs sent by mobile terminal APP are all stored in the database of desktop management system. Through the analysis and comparison of data, the actuators are manually and automatically controlled. The large amount of historical data in the desktop management system and the curve diagram formed from these data can facilitate the users to analyze the current control scheme and form a better control strategy.

The front-end interface design of the desktop management system uses the WPF user interface framework, which contains both program code and XAML. Firstly, we employ XAML to define the initial interface of the application, and then write the corresponding implementation code of the function. After that, we will embed the logic code directly into each XAML file [13].The database development adopts the SQL Server 2008 database which allows to use the data in custom applications developed by Microsoft .NET and Visual Studio [14]. The desktop management system database of the intelligent household greenhouse has 13 tables used to record communication-related information, such as user's information, IP addresses and port numbers, environmental data, actuators' working status, data instructions, and control instructions.

4.3.1. Design of desktop management system's function. The software design of desktop management system uses the WPF user interface framework, containing both program code and XAML. Firstly, we employ XAML to define the initial interface of the application, and then write the corresponding function implementation code. After that, we will embed the logic code directly into each XAML file.

The user login management is applied to authenticate and log in user's account and password. When the user enters the account and password, they will be compared with the information stored in the database, so as to achieve secure login.

The environmental monitoring is employed to display the real-time environmental parameters in the greenhouse and the working state of the actuators, and it regulates the actuators. This module has the user management, mobile phone management, gateway management, and network settings.

The communication service program establishes a network connection with intelligent gateway client in network layer, and also it needs network connection with mobile client in application layer to achieve data interaction.

The historical data analysis draws the curve of historical data by reading the data of temperature, humidity, and lighting in the database. Historical data is a true and comprehensive record of the past in the greenhouse, as well as the data source for analyzing the correlation between parameters, which is of great importance for users to maintain and manage the plants [15]. Correlation between parameters is analyzed on historical data. From the analysis results, we know that when one parameter changes, the other parameters will be influenced. By analyzing plenty of historical data, we conclude the experience and counter measures for plant management. And combined with growth pattern of the plant, a well-focused knowledge base is established to provide the basis for user decisions and device control.

As to the automatic-control mode of the intelligent household greenhouse system, the fuzzy neural network is used for intelligent control. In the greenhouse environment, the variables that have a greater impact on plants are temperature and humidity. Therefore, in the design of fuzzy neural network controller in this paper, the inputs are temperature error, temperature error variation, humidity error, and humidity error variation, while the outputs are temperature controlled variable and humidity controlled variable. For the temperature controlled variable, the positive value is the heating control and the negative value is the cooling control. For the humidity controlled variable, the positive value is the humidification control and the negative value is the dehumidification control.

For temperature error e_T, its actual range is set to [-5, 5] in this paper, and its fuzzy set domain is set to {-3, -2, -1, 0, 1, 2, 3}. And the quantitative factor of temperature error is $k_{et}=3/5$. Based on the quantitative factor, the temperature error can be mapped from the basic domain to the fuzzy set domain. At the same time, the fuzzy quantity of temperature error is designed to 7 levels, and the corresponding fuzzy set is Et = {NB, NM, NS, ZO, PS, PM, PB}.

For the temperature error variation Δe_T, its actual range is defined as [-2, 2] in this paper, and its fuzzy set domain is set to {-2，-1，0，1，2. And the quantitative factor of temperature error variations $k_{ect}=2/2$. Based on the quantitative factor, temperature error variation can be mapped from the basic domain to the fuzzy set domain. Meanwhile, the fuzzy quantity of temperature error variation is designed to 5 levels, and the corresponding fuzzy set is ECt={NM,NS,ZO,PS,PM}.

For the humidity error e_H, its actual range is defined as the basic domain of the humidity error, setting to [-20%, 20%] in this paper. Its fuzzy set domain is set to {-3，-2，-1，0，1，2，3} and its quantitative factor is$k_{et}=3/20\%$. Based on the quantitative factor, the humidity error can be mapped from the basic domain to the fuzzy set domain. At the same time, the fuzzy quantity of the humidity error is designed to 7 levels, and the corresponding fuzzy set is Eh={NB, NM, NS, ZO, PS, PM, PB}.

For the humidity error variation Δe_H, its actual range is defined as [-5%，5%] in this paper, and its fuzzy set domain is set to {-2，-1，0，1，2}. And the quantitative factor of humidity error variation is $k_{ech}=2/5\%$. Based on the quantitative factor, humidity error variation can be mapped from the basic domain to the fuzzy set domain. At the same time, the fuzzy quantity of the humidity error variation is designed to 5 levels, and the corresponding fuzzy set is ECh= {NM，NS，ZO，PS，PM}.

In the fuzzy neural network controller of the intelligent household greenhouse system, the outputs are temperature-control adjustment and humidity-control adjustment. The temperature-control adjustments are divided into heating adjustment and cooling adjustment, and humidity-control adjustments are divided into humidification adjustment and dehumidification adjustment. The heating adjustment is a heating plate with 5-gear mode, and the cooling adjustment is a cooling fan with 5-gear mode. In addition, the humidification adjustment is a humidifier with 1-gear mode, and the dehumidification adjustment is a dehumidifying fan with 5-gear mode.

For the temperature-control adjustment, its actual range is defined as [-5，5] in this paper, and its fuzzy set domain is set to ｛-3，-2，-1，0，1，2，3｝. And the quantitative factor of temperature-control adjustment is $k_{tcn}=3/5$. Based on the quantitative factor, temperature-control adjustment can be mapped from the basic domain to the fuzzy set domain. At the same time, the fuzzy quantity of temperature-control adjustment is designed to 7 levels, and the corresponding fuzzy set is Tcon = ｛NB，NM，NS，ZO，PS，PM，PB｝.

For the humidity-control adjustment, its actual range is defined as [-5，1] in this paper, and its fuzzy set domain is set to ｛-3，-2，-1，0，1｝. And the quantitative factor of humidity-control adjustment is $k_{hcn}=3/5$. Based on the quantitative factor, humidity-control adjustment can be mapped from the basic domain to the fuzzy set domain. Meanwhile, the fuzzy quantity of humidity-control adjustment is designed to 5 levels, and the corresponding fuzzy set is Hcon= ｛NB，NM，NS，ZO，PS｝.

The control strategy for temperature and humidity in the greenhouse is analyzed and summarized. And we design 35 control rules of the temperature adjustment and humidity adjustment respectively, as shown in table 1 and table 2.

Table 1. Fuzzy rules of temperature adjustment.

ECt	*Et*						
	NB	NM	NS	ZO	PS	PM	PB
NM	NB	NB	NM	NS	PS	PS	PM
NS	NB	NM	NS	NS	PS	PM	PM
ZO	NM	NS	NS	ZO	PS	PM	PM
PS	NS	NS	NS	PS	PM	PM	PB
PM	NS	NS	PS	PS	PM	PB	PB

Table 2. Fuzzy rules of humidity adjustment.

ECh	*Eh*						
	NB	NM	NS	ZO	PS	PM	PB
NM	NB	NB	NM	NS	PS	PS	PS
NS	NB	NM	NS	NS	PS	PS	PS
ZO	NM	NS	NS	ZO	PS	PS	PS
PS	NS	NS	NS	ZO	PS	PS	PS
PM	NS	NS	ZO	PS	PS	PS	PS

4.3.2. Communication service program on desktop management system. When the communication service program sends and receives data with gateway client and mobile client, it is required to implement real-time communication, and to avoid the mutual influence and interference between gateway client data and mobile client data. In order to meet this demand, the author puts forward using socket communication and multi-thread technology to design the server side with two sockets. One socket is used to establish a connection with the gateway client, handling the connection request of the gateway client. The procedure of desktop management system and gateway communication is shown in figures 4(a) and (b). The other is applied to process the connection request of the mobile client. These two sockets bind the same IP address with different port number, responsible for data analysis and forwarding between each other. Meanwhile, the data, actuator's working state, control instructions, and communication data detected by the sensor will be stored. The procedure of desktop management system and mobile phone APP communication is shown in figures 4(c) and (d). In the experimental environment, the port number connected to the gateway is 8080, and the port number connected to the mobile client is 8088. When the server communicates with these two clients, the data-receiving thread and data-sending thread are established. These two threads perform independently, so as to prevent blockage when the network-communication data are interacting.

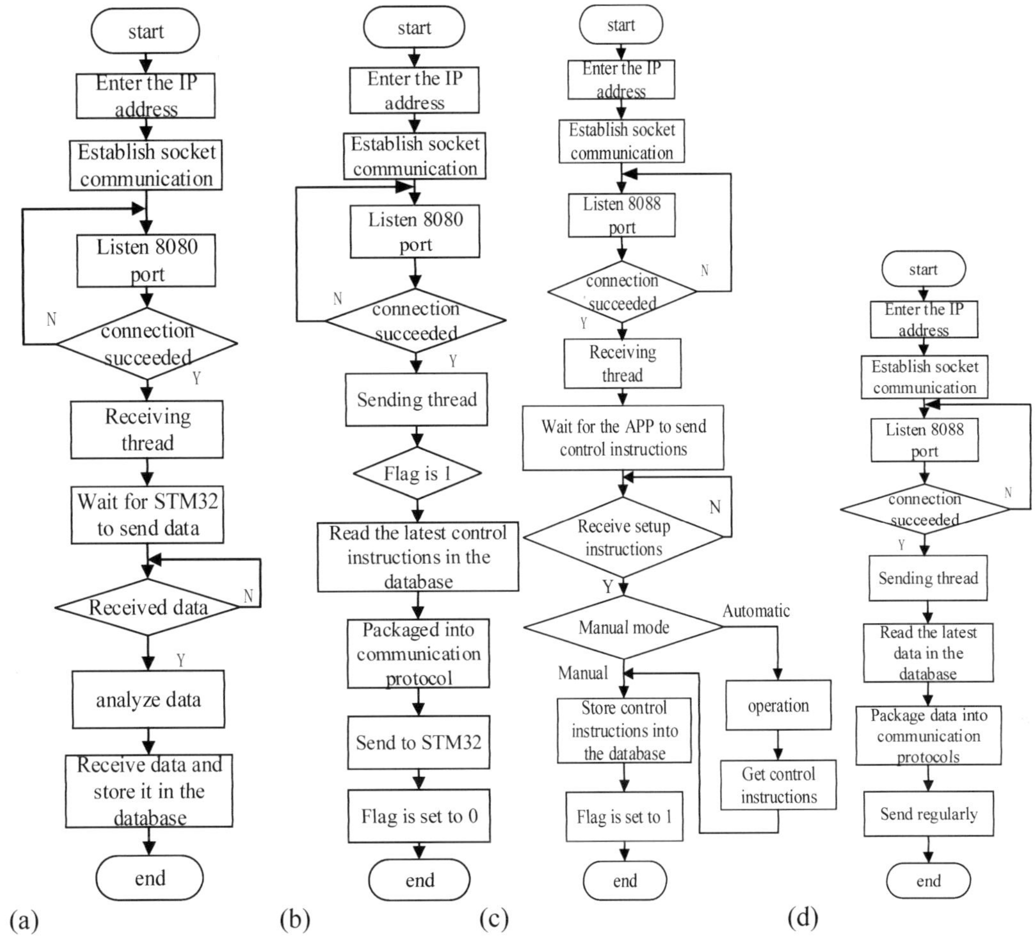

Figure 4. Flow chart of communication service software program of desktop management system.

4.4. APP software design

The App software design of system application layer is based on the Android Studio development platform. It is mainly designed by various component API interfaces provided by Android. The APP interface of intelligent household greenhouse system includes three parts: network connection interface, monitoring interface and adjustment interface.

The APP communicates with the server through the TCP-based Socket. The server first initializes the Socket, then binds to the port and monitors the port, waiting for client's connection by calling Accept(). The client creates a Socket object based on the server's IP address and port number, connecting to the server. The APP obtains the Socket output stream through the getOutputStream method provided by the Socket class, and the Socket input stream through getInputStream method, so as to interact data with the server. At the same time, in order to avoid the ANR (Application Not Responding) problem, we apply multi-threaded programming, define the handler in the main thread, and implement the handle Message method for this handler. Afterwards, the main thread's handler is called in the child thread, and sent message through its sendMessagemethod [16].

When the APP is running for the first time, the user needs to add an IP address and a port number to create a socket communication to connect to the server. The information added by the user can be saved to the mobile client for later operation. After the user clicks the monitor in the Tab bar at the bottom of the App interface, then going to the monitoring interface, the APP will first judge whether the network connection is successful. If the network connection works, the APP will display the packaged communication protocol, which is sent from server at regular time, in the APP monitoring interface. The interface's adjustment function is designed for automatic and manual mode for the user to select. Under the automatic mode, according to the setting value input by the user, the APP follows an automatic control mode instruction to the server. Then, the server automatically controls the actuator to perform an action through data comparison. In the manual mode, the user manually controls the executor by sliding the progress bar. Following the manual control instruction protocol from application layer, the APP sends the manual mode control instruction to the server, so as to control executor's action.

4.5. Communication protocol design

The author designs a complete set of communication protocol for the following reason: to help data circulate from sensor layer to application layer; to avoid the mutual influence of data exchange among multiple layers; to prevent wrong data in the complex environment from interfering the system; to ensure the accuracy of data. The role of each domain in the communication protocol should be clear. The length and location of the data domain are fixed and clear. The length of the protocol is appropriate, with complete information. And the type of protocol package should be reduced, so as to improve the readability. In addition, it also includes a check mechanism so that the receiver can determine the validity and integrity of the protocol package. When adding functionality or implementing equipment in the future, the protocol will only increase in quantity without changing the structure, and it will still be able to perform its job. The communication protocol is designed by using six fields: frame header, lengths, addresses, types, instruction (data), and end bit.

The frame header is located in the lead byte of the protocol, used to determine the legitimacy of the data. First of all, the frame header is judged. If not comply with the protocol, both interactive sides will directly determine it as illegal data, not making process and jumping out to prepare for the next data interaction. In this communication protocol, the frameheaders are all defined as 0xA5.

The length indicates the total number of bytes in data interaction process. According to the protocol, the interactive side should determine whether the received data length is consistent with that of protocol. If the data does not meet the protocol specified length, it will be judged as illegal data and it will not be processed.

The address represents the corresponding numbers of the detecting terminal and the executive terminal in the ZigBee network. All executors are on the same ZigBee terminal. If you want to increase or decrease executors, you only need to add or subtract the addresses.

The types respectively indicate which node of the sensor layer and the corresponding operation of the sensor layer. Different executors correspond to different properties such as temperature, humidity and lighting. Different types of data represent different properties.

The instruction (or data) indicates the commands issued to the nodes of the sensor layer. As for the executor, the instruction is the gear, and the data are the working status. For the sensor, the instruction is the query data, and the data are the detected real-time data.

The end bit includes high and low bits, namely 0x0d and 0x0a respectively, which is used to mark the end of the read data.

This part of protocol is followed by data interaction between the ZigBee node in sensor layer and the intelligent gateway in network layer through serial connection. The instruction protocol (part) of sensor layer is shown in table 3, and data protocol (part) of sensor layer is shown in table 4.

Table 3. Instruction protocol (part) of sensor layer.

Instruction	Frame Header	Length	Address	Type	Instruction	End bit
supplemental lighting LED on1[st]gear	A5	06	07	30	01	0D
supplemental lighting LED on 2[nd]gear	A5	06	07	30	02	0D
CLED	A5	06	07	30	00	0D
Heating plate on 1st gear	A5	06	07	31	01	0D
Heating plate on 2nd gear	A5	06	07	31	02	0D
Heating off	A5	06	07	31	00	0D
Collect temperature on location1	A5	06	01	20	10	0D
Collect temperature on location 2	A5	06	02	20	10	0D

Table 4. Data protocol (part) of sensor layer.

Data	Frame Header	Length	Address	Type	Data	End bit
supplemental lighting LED on 1st gear	A5	06	07	50	01	0D
supplemental lighting LED on 2nd gear	A5	06	07	50	02	0D
supplemental lighting LED off	A5	06	07	50	00	0D
Heating plate on 1st gear	A5	06	07	51	01	0D
Heating plate on 2nd gear	A5	06	07	51	02	0D
Heating plate off	A5	06	07	51	00	0D
temperature on location 1	A5	06	01	01	Data	0D
temperature on location 2	A5	06	02	01	Data	0D

Table 5 and table 6 are the data protocols of network layer followed by the STM32 when it sends data to the server. The "locations 1 to 6" represent the data detected by sensors at different locations. The "gear" indicates the operating capacity of the current actuator. The switch 0 means work suspension and switch 1 represents on work. table 7 is instruction protocol of network layer followed by the server when it sends automatic and manual mode instructions to the STM32 gateway.

Table 5. Data protocol of network layer sensor.

Frame Header	Length	Temperature		Humidity		Light intensity		Soil moisture		End
		Type	Location1~6	Type	Location1~6	Type	Location 1~6	Type	Location 1~6	
A5	2E	60	data1~6	61	data1~6	62	data1~6	63	data1~6	0D

Table 6. Data protocol of executor status in network layer.

Frame Header	Length	LED		Heating plate		Cooling fan		Dehumidify fan		Drip irrigation system		Humidifier		End
		Type	Gear 0~3	Type	Gear 0~5	Type	Gear 0~5	Type	Gear 0~5	Type	Gear 0~1	Type	Switch 0~1	
A5	2E	50	00~03	51	00~05	52	00~05	53	00~05	54	00~01	55	00~01	0D

Table 7. Instruction protocol of network layer.

Frame Header	Length	LED		Heating plate		Cooling fan		Dehumidify fan		Drip irrigation system		Humidifier		End
		Type	Gear 0~3	Type	Gear 0~5	Type	Gear 0~5	Type	Gear 0~5	Type	Switch 0~1	Type	Switch 0~1	
A5	0E	30	00~03	31	00~05	32	00~05	33	00~05	34	00~01	35	00~01	0D

Table 8 is the data protocol of application layer followed by the server when it sends data to the APP, where "result" represents the average of the following temperature, humidity, lighting and soil moisture. Table 9 and table 10 is the instruction protocol of application layer followed by the mobile phone APP when it sends the automatic and manual mode control instruction protocol to the server.

Table 8. Data protocol of application layer.

Frame Header	Length	LED		Heating plate		Cooling fan		Dehumidify fan		Drip irrigation system		Humidifier		Temperature		humidity		Light intensity		Soil moisture		End
		Type	Gear 0~3	Type	Gear 0~5	Type	Gear 0~5	Type	Gear 0~5	Type	Switch 01	Type	Switch 0~1	Type	res	Type	res	Type	res	Type	res	
A5	1A	50	00~03	51	00~05	52	00~05	53	00~05	54	00~01	55	00~01	60	data	61	data	62	data	63	data	0D

Table 9. Automatic control instruction protocol of application layer.

Frame Header	Length	Temperature		Humidity		Light intensity		End
		Type	Setting value	Type	Setting value	Type	Setting value	
A5	09	40	data	41	data	42	data	0D

Table 10. Manual control instruction protocol of application layer.

Frame Header	Length	LED		Heating plate		Cooling fan		Dehumidify fan		Drip irrigation system		Humidifier		End
		Type	Gear 0-3	Type	Gear 0-5	Type	Gear 0-5	Type	Gear 0-5	Type	Switch 0~1	Type	Switch 0~1	
A5	0E	30	00~03	31	00~05	32	00~05	33	00~05	34	00~01	35	00~01	0D

5. Conclusion

The article analyzes the users' needs of the intelligent household greenhouse system, proposing a design scheme with greenhouse environmental data detection and transmission, human-computer interaction system, greenhouse environmental control and regulation functions. In the meanwhile, we also consider the advantages of low cost, easy installation and aesthetics. The intelligent household greenhouse system can realize the real-time and accuracy of data collection through the design of the ZigBee network. The network layer where the intelligent gateway and server coordinate and cooperate has a high degree of concurrency and real time, satisfying the functions of multi-client requests and data storage. The system can monitor greenhouse through desktop management system and mobile terminal APP, with manual and automatic control. At the same time, the author standardizes the communication protocol to avoid the impact of multi-layer data interaction, which can effectively handle the data interaction among multiple layers. Through the experiment, the system performance is reliable and stable. With the continuous development of the Internet of Things and intelligent household, the intelligent household greenhouse has a broad space for development.

Acknowledgments
This work was supported by Project of Liaoning Province Education Department（LJZ2017015）and Shenyang Science and Technology Planning Project (Z18-5-015).

IOP Conf. Series: Materials Science and Engineering **399** (2018) 012024 doi:10.1088/1757-899X/399/1/012024

References

[1] Zhang W Q 2017 *Electronic Tech. Soft. Eng.* **2017** 38
[2] Datta S K2014 *Internet of Things IEEE* **2014** 514
[3] Zhang P J, Guo G Z and Yu Z Z 2015 *App. Mech. & Mat.* 738 247
[4] Zhang M and Fang J L 2012 *Automation Techn. and App.* **31** 32 42
[5] Liang R H 2016 *J. Henan Agricultural Univ.* **50** 346
[6] ChangY L 2014 Beijing Jiaotong University
[7] David N, Sendik O, Messer H and Alpert P2015 *Bulletin of the American Meteorological Society* **96** 141218100836009
[8] Zhang W J,Li G H and Wu Z 2015 *J. of China Agric. Mechanization* **36** 96
[9] Huang X F, Liu D M *Modern Electronic Techn.* **29** 78
[10] Zhao W B, Mao H P and Ma W Z 2016 *J. of China Agricultural Mechanization* **37** 230
[11] Yan W and Ye J X 2012 *Information and Communications* **2012** 46
[12] Wang F, LouWZ, Guo M R and Lu YF 2015 *Key Eng. Mat.* **3928** 869
[13] Wang J G, Ji Q X and Jiang Y 2012 *Adv. Mat. Research* **1549** 2296
[14] Wang D Y 2010 *App. Mech. Mat.* **893** 748
[15] Cañadas J, Sánchez-Molina J A,Rodríguez F and Águila I M D 2017 *Information Processing in Agriculture* **4** 50
[16] Liu X P and Zhou Y B 2014 *App. Mech. Mat.* **2802** 576
[17] Zhou J, Hu L, Wang F, Lu H and Zhao K 2013 *Tsinghua Science and Technology* **18** 369

IOP Conf. Series: Materials Science and Engineering **399** (2018) 012025 doi:10.1088/1757-899X/399/1/012025

Calculation and analysis of critical speed of high speed motor spindle rotor system

S Huaitao [1,3*], Z Jizong [1], Z Yu[1] and H Gang [2]

[1] School of Mechanical Engineering, Shenyang Jianzhu University, National and Local Joint Engineering Laboratory of High-grade Stone Numerical Control Processing Equipment and Technology, Shenyang 110168, China, Hunnan East Road on the 9th,Hunnan New District,S henyang, China.Postcard:110168

[2] Shenyang Yadong Technology Co. Ltd., 110000, China

[*] E-mail: 651554653@qq.com

Abstract. The traditional transfer matrix method used to calculate the critical speed of high-speed spindle motor rotor system will produce large errors. Taking the 170SD30 electrospindle as the analysis object, a high-speed electric spindle dynamics theoretical model was established based on the Riccati method, using Matlab to program and calculate the kinetic parameters of the first three critical speeds and natural frequencies of the high-speed spindle. Finally, the simulation data was compared with experimental data. The results show that the Riccati transfer matrix method is 2.3% more accurate than the traditional transfer matrix method, and it verifies the accuracy and feasibility of the Riccati transfer matrix method, and is of great significance to effectively guide the analysis and design of high-performance and high-speed motor spindle.

1. Introduction

The high-speed motor spindle is the core component of CNC machining and is the carrier of advanced manufacturing technology. Its performance will directly affect the development of CNC machine tools and even the entire manufacturing industry. With the development of high-speed electric main shaft high-speed, high-precision, and high-efficiency direction, higher requirements have been put on the dynamic characteristics of the electric spindle itself and on the electric spindle[1-4].When the spindle is running at high speed, it will produce different degrees of centrifugal force, inertial force and various unbalanced responses, and when the frequency of excitation force is close to or equal to the natural frequency of the rotor system, the rotor will resonate. This speed is the critical speed of the rotor, which shows that, when the rotor system is running at critical speed, the rotor will experience severe vibration, and if the speed is greater or less than a certain speed value or speed range, the rotor will run smoothly. The rotor runs at critical speeds, which can make the rotor's vibration worse, seriously affecting the processing precision and quality of the processed parts, and causing serious accidents [5, 6].For high-speed motorized spindles, it is necessary to analyse and study the critical speed of the rotor system due to its large load, high speed, and other characteristics.

At present, there are mainly three methods for calculating the critical rotational speed of high-speed electric spindle: formula method, transfer matrix method and finite element method [7-9].The formula method is applicable to the driveshafts of equal section, and it is more accurate to calculate the critical speed [10];the traditional transfer matrix method is widely used in the calculation of the critical speed of a chain structure, such as a rotor with the advantages of simple program, less time and memory required. However, with the trial frequency increasing, the accuracy of calculation will decrease [11]; the finite element method can quickly and accurately determine the critical rotational speed of the rotor with the development of computer technology, but the finite element method variable occupies a large amount of memory, it requires a high degree of computer hardware, and a long calculation time [12, 13].The above methods for calculating the critical rotational speed all have different degrees of inadequacy and cannot accurately and quickly calculate the critical rotational speed of the rotor system.

In this paper, the Riccati transfer matrix method is used to propose a new electrospindle dynamics model to calculate the critical rotational speed of high-speed electric spindle, and the influence of critical rotational speed on the dynamic characteristics of high-speed electric spindle is analysed. In order to carry out the above work, firstly, the principle and solving method of the Riccati transfer matrix method is introduced, then, Matlab was used to compile the Riccati transfer matrix method and the traditional matrix method program and calculate the critical speed of the electric spindle. Finally, the simulation data and experimental data were compared and analysed.

2. High-speed electric spindle dynamics modelling

The transfer matrix method first simplifies the structure to a lumped parameter mechanical model, in the process of simplification, in accordance with the principle of invariable center of mass, so as to maintain all the dynamic characteristics of the original structure as much as possible, the rotor system of the high-speed electric spindle system is simplified to a model consisting of lumped masses and several shaft sections [14].

2.1 Model simplification

The electric spindle is a high-speed precision spindle for machining centers in the new generation of integrated machining, electro-hydraulic and hydraulic machining, which is mainly composed of housings, spindles, bearings, stators and rotors [15]. Figure 1 shows the structure of the 170SD30 electric spindle, the front and rear bearings are all angular contact ball bearings, the front end of the spindle bearing is model 7015C, and the rear model is 7012C.

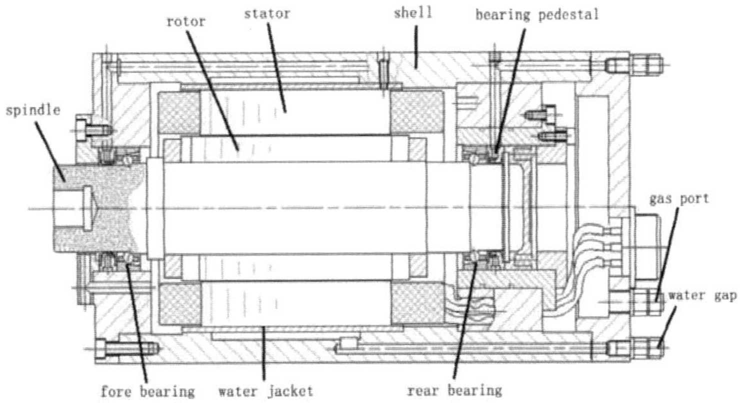

Figure 1. Spindle structure.

The support of the main shaft is generally an angular contact ball bearing. Due to the initial pre-tightening force, the working load capacity and the rigidity of the spindle can be increased. Therefore, the rigidity of the bearing needs to be taken into account in simplifying the high-speed spindle model. Bearing stiffness can be divided into radial stiffness, axial stiffness and angular stiffness, under actual conditions, with the increase of the rotational speed, both the axial stiffness and the angular stiffness will reach a relatively stable value, and the influence on the dynamic characteristics of the electric spindle will be small [16].When simplifying high-speed motor spindle models, bearings can be reduced to four radially-distributed springs, only considering their radial stiffness, simplified as shown in figure 2. After radial preloading through the bearing, the radial stiffness is calculated as [17]:

$$K_r = \delta \times 17.7236\sqrt[3]{Z^2 D_b} \frac{\cos^2 \alpha}{\sqrt[3]{\sin \alpha}} \sqrt[3]{F_{\alpha 0}} \quad (1)$$

δ is the angular contact ceramic ball bearing correction factor, δ=1.8 at light preload, δ=1.9 at preload, and δ=2.0 at heavy preload; D_b is the rolling element diameter (8mm); Z is the number of bearing rolling elements (18); α is contact angle (15^0); $F_{a0}$is an axial preload, the electric spindle studied in this paper is δ=1.8 for light load and 400N for front and rear preloads. Substituting the above parameters into (1) K_r=2245.42N·mm^{-1}.

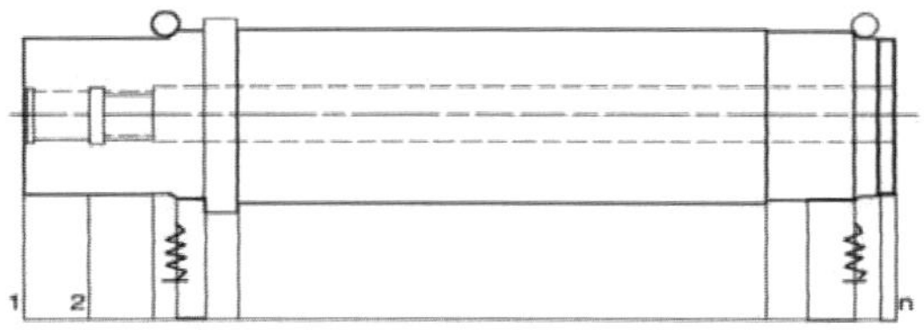

Figure 2. Simplified model of the spindle

2.2 *Node division*

When the transfer matrix method is used to simplify the model of the high-speed motor spindle, the motor spindle needs to be divided into nodes. The number of nodes will seriously affect the calculation result of the critical speed of the electric spindle. The number of nodes is too small and the result of calculation is inaccurate and too many nodes will lead to complicated calculation, long time-consuming, and increased error. According to the calculation result of the equal section beam, if the calculated critical speed error is less than 1%, then the total number of nodes N should satisfy the following relationship [18]:

$$N \geq 1+5.34r \quad (2)$$

r is the highest order of the required natural frequency (or critical speed),according to the rotor model, the motor spindle is divided into 25 nodes, then r=4. The electrospindle dynamics model is shown in figure 3.

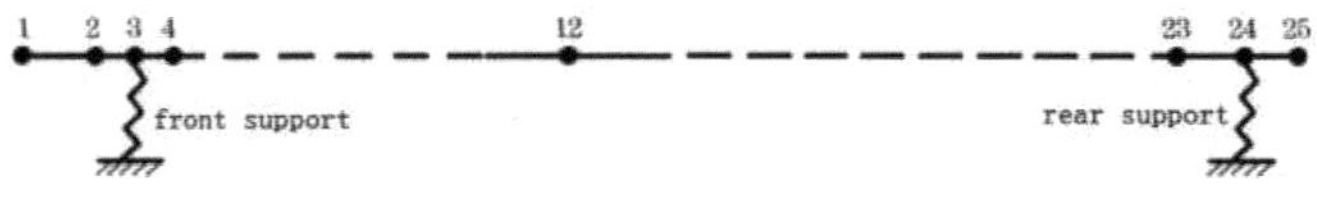

Figure 3. Simplified mechanical model of electric spindle.

3. Riccati transfer matrix method

The Riccati transfer matrix method is used to establish a new high-speed spindle model for calculating the critical speed of an electric spindle. When using matlab programming, considering the gyro torque, shear effects and cross-section coefficients and other influencing factors, the calculation results are more accurate and the calculation speed is faster [19].

The calculation principle of the Riccati transfer matrix method is: first dividing the shaft into several equal shaft sections, then according to the boundary conditions at both ends of the shaft, recursively from one end of the axis to the other to determine the state vector of the starting end and to seek the state vector of each section in reverse; in the process of increasing the speed from small to large, the point of maximum response is the critical speed [20].

The cell transfer matrix has four valid parameters for each node *Z*: radial displacement of the cross-section of the section: cross-section radial displacement *X*, angle *A*, bending moment *M*, and shear force *Q*.*Marked as* $Z=[X,A,M,Q]^T$. The relationship between the two nodes of the ith axis segment is Zi+1=TiZi, and Ti is the transfer matrix.

$$T_i=\begin{bmatrix} 1+\frac{l^3}{6EJ}(1-v)(m\omega^2-K_{sj}) & 1+\frac{l^2}{2EJ}(I_P-I_d)\omega^2 & \frac{l^2}{2EJ} & \frac{l^3}{6EJ}(1-v) \\ \frac{l^2}{2EJ}(m\omega^2-K_{sj}) & 1+\frac{l}{EJ}(I_P-I_d)\omega^2 & \frac{l}{EJ} & \frac{l^2}{2EJ} \\ l(m\omega^2-K_{sj}) & (I_P-I_d)\omega^2 & 1 & 1 \\ m\omega^2-K_{sj} & 0 & 0 & 1 \end{bmatrix}$$

E is the elastic modulus; *J* is the axial section moment ; l is the length between two units; $v=\frac{6EI}{k_t GAl^2}$ is the shearing coefficient between the nodes ,*J* is the moment of inertia of the cross section to the neutral axis, k_t is the section coefficient, ,k_t is the section modulus ,*G* is the shear modulus of the material, *A* is the cross-sectional area, and *l* is the length of the shaft section; *m* is the quality focused on the node;ω is the angular velocity of the rotor; *K* is the bearing stiffness at the support; I_p and I_d are the rotor's polar moment of inertia and the moment of inertia of the rotor.

The Riccati transfer matrix method divides *r* elements of the state vector of each node *Z* into two groups of *f* and *e*.

$$Z_i=\begin{Bmatrix} f \\ e \end{Bmatrix}_i \tag{3}$$

In formula (3), *f* consists of *r*/2 elements corresponding to zero values in the starting interface state vector Z_1; *e* consists of the remaining *r*/2 complementary elements. Due to the fact that the left end of the spindle is free, there are boundary conditions $M_1=0$, $Q_1=0$.So $f_i=[M,Q]_i^T$, $e_i=[y,\theta]_i^T$. The relationship between the state vectors of two adjacent sections can be rewritten as:

$$\begin{Bmatrix} f \\ e \end{Bmatrix}_{i+1}=\begin{bmatrix} u_{11}u_{12} \\ u_{21}u_{22} \end{bmatrix}_i\begin{Bmatrix} f \\ e \end{Bmatrix}_i \tag{4}$$

In formula (4), $u_{11i}=\begin{bmatrix} 1 & l \\ 0 & 1 \end{bmatrix}_i$; $u_{12i}=\begin{bmatrix} l(m\omega^2-K_{sj}) & (I_P-I_d)\omega^2 \\ m\omega^2-K_{sj} & 0 \end{bmatrix}_i$

$$u_{21i}=\begin{bmatrix} \frac{l^2}{2EJ} & \frac{l^3}{6EJ}(1-v) \\ \frac{l}{EJ} & \frac{l^2}{2EJ} \end{bmatrix}_i \quad u_{22i}=\begin{bmatrix} 1+\frac{l^3}{6EJ}(1-v)(m\omega^2-K_{sj}) & L+\frac{l^2}{2EJ}(I_P-I_d)\omega^2 \\ \frac{l^2}{2EJ}(m\omega^2-K_{sj}) & 1+\frac{1}{EJ}(I_P-I_d)\omega^2 \end{bmatrix}_i$$

Expand equation (4), introducing the following Riccati transformation:

$$f_i=S_i e_i \tag{5}$$

S_i in formula (5) is called Riccati matrix, which is a square matrix of $r/2\times r/2$, substituting (4) into (3)

$$e_i=[u_{21}+u_{22}]_i^{-1}e_{i+1} \tag{6}$$

$$f_{i+1} = [u_{11}S + u_{12}]_i [u_{21}S + u_{22}]_i^{-1} e_{i+1} \quad (7)$$

Comparisons (5) and (7):

$$S_{i+1} = [u_{11}S + u_{12}]_i [u_{21}S + u_{22}]_i^{-1} \quad (8)$$

This is the recursion formula of the Riccati transfer matrix.

From the starting boundary conditions, the condition for the nonzero solution of equation 3 is:

$$|S_{N+1}| = \begin{vmatrix} S_{11} & S_{12} \\ S_{21} & S_{22} \end{vmatrix} = 0 \quad (9)$$

This is the frequency equation of the system. The root ω that satisfies equation (9) is the angular velocity of the critical speed required.

4. Experiment and result analysis

4.1 Experimental analysis

In order to verify the correctness of natural frequency and critical speed of high-speed motorized spindle obtained by the Riccati transfer matrix method, a dynamic performance testing system for high-speed spindle was built. In this experiment, the 170SD30 high-speed electric spindle was struck with hammering and the input and output signals were excited and collected. Experimental equipment includes hammers, charge amplifiers, acceleration sensors, force sensors, and signal collectors. The experimental platform is shown in figure 4 below.

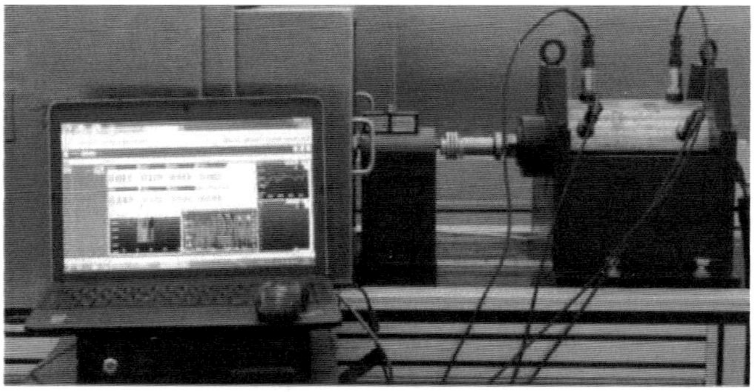

Figure 4. Spindle dynamic performance test system.

In order to reduce the experimental error and obtain more accurate experimental data, this experiment mainly selects three hammering points, namely the front section of the main shaft, the middle part of the main shaft and the rear end of the main shaft. The experimental method is single-point response and multi-point vibration. The specific steps are as follows: first, start the spindle and start running at a certain low speed; then, tapping each point with the hammer 6 times, the impact of the hammer on the front end of the spindle is a force of 126 N. Secondly, the speed is increased step by step in certain stages. At the same time, the vibration response value of the motor spindle is measured and recorded for each rotation speed. Finally, the experimental frequency and mode shape of the motor spindle are obtained, as shown in table 1 below.

Table 1. Natural frequency and critical speed of electric spindle.

order	frequency/HZ	critical speed/(r/min)
1	1767.6	106056
2	2214.0	132840
3	3186.6	191196

4.2 Simulation calculation and result analysis
Using Matlab to compile the traditional transfer matrix method and the Riccati method to calculate the critical rotational speed of the high-speed spindle, the rotor's natural frequencies and critical speeds can be obtained. The initial input spindle configuration parameters are as follows:outer diameter of stator D_1=130mm, stator inner diameter D_2=80mm,Shaft length l=11mm,Support stiffness K=2245.42N·mm,unit quality m=6.0×10^{-9}t/mm^3 .Shaft material parameters down: elastic modulus E=20600N·mm^{-2}, poisson's ratio μ=0.3,density ρ=7.85g·cm^{-3}. The first three primary natural frequencies and critical speeds are obtained by solving the equations, as shown in tables 2 and 3 below.

Table 2. Comparing the natural frequency obtained by the riccati method with the experimental results.

order	Riccati method calculation result		Experimental results		error/%
	frequency/HZ	critical speed/(r/min)	frequency/HZ	critical speed/(r/min)	
1	1667.9	100074	1767.6	106056	5.6
2	2058.1	123486	2214.0	132840	7.0
3	2956.5	177390	3186.6	191196	7.2

Table 3. Comparing the natural frequency obtained by the prohl method with the experimental results.

order	Prohl calculation results		Experimental results		error/%
	frequency/HZ	critical speed/(r/min)	frequency/HZ	critical speed/(r/min)	
1	1661.5	99690	1767.6	106056	6.0
2	2032.5	121950	2214.0	132840	8.2
3	2890.2	173412	3186.6	191196	9.5

Comparing the data from table 2 and table 3, it can be clearly seen that the critical speed that can be calculated by the Riccati transfer matrix method is 2.3% higher than the critical speed accuracy calculated by the traditional transfer matrix method, which reduces the error between the actual operating condition and the critical speed of the high-speed motor spindle using the transfer matrix method, which verifies the accuracy and feasibility of the Riccati transfer matrix method. It is of great significance to guide the development of high-performance high-speed spindles.

5. Conclusion

(1) The Riccati transfer matrix method for calculating the critical speed is by 2.3% more accurate than the traditional transfer matrix method for calculating the critical speed accuracy, which overcomes the problem of the reduced accuracy of traditional transfer matrix methods.

(2) The Riccati transfer matrix method preserves all the advantages of the traditional transfer matrix method and improves the computational speed and computational stability of the transfer matrix method.

(3) The Riccati transfer matrix method can accurately and rapidly solve the critical rotation speed and natural frequency of the electric spindle and analyze the dynamic characteristics of the electric spindle.

Acknowledgments
The article is supported by National Key R&D Program of China: 2017YFC0703903, National Science Foundation of China 51705341, Liaoning Natural Science Foundation 2016010623 and Shenyang Science and Technology Plan Project 17-231-1-28.

References
[1] Wu Y H and Zhang L X 2013 *High Speed CNC Machine Tool Spindle Control Technology* (Beijing: Science Press)
[2] Jędrzejewski, Kowal Z, Kwaśny W and Modrzycki W 2005 High-speed precise machine tools spindle units improving *J Mater Process Tech.* **2 (162)**615
[3] Abele E, Altintas Y and Brecher C 2010 Machine tool spindle units *Cirp Ann-Manuf Techn.* **59 (2)** 1
[4] Cao H, Holkup T and Altintas YA 2011 Comparative study on the dynamics of high speed spindles with respect to different preload mechanisms *Int J Adv Manuf Tech.* **57 (9)** 871
[5] Wen B, Gu J , Xia S et al. 2000 *Advanced rotor dynamics* (Beijing: Machinery Industry Press)
[6] Shan W, Chen X, Wang H et al 2017 Research on milling stability of high speed motorized spindle *Shock Vib.* **36 (19)**242
[7] Li C, Liu Y and Ai L 2010 Calculation of rotor critical speed based on hybrid model *Shock Vib.* **29 (11)** 246
[8] Wu Y, Zhang J, Zhang L et al 2014 Dynamic Analysis of rotor of ceramic electric spindle based on transfer matrix and finite element method *J Shenyang Jianzhu Univ.* **30 (3)** 510
[9] Huang W, Gan C, Yang S et al 2017 Analysis of the stiffness of high speed motorized spindle angular contact ball bearings and its influence on the critical speed of electric spindle *Shock Vib.* **36 (10)** 19
[10] Zhong Y H Y and Wang Z 1987 *Rotor dynamics.*(Beijing: Qinghua University Press)
[11] Wan H,Li Y and Zheng L 2013 Improved structural vibration transfer matrix method *Shock Vib.* **32 (9)**173
[12] Ou Y and Li Y 1994 Comparison and analysis of common methods used in calculation of rotor dynamic characteristics *J Aeros Power* **9 (2)**142
[13] Wang J and Sun K 2012 Calculation of rotor critical speed based on finite element method *J Mech Design.* **29 (12)** 10
[14] Du C, Ding Y, Zhao W et al 2010 Analysis of dynamic characteristics of electric spindle based on transfer matrix method *Manuf Tech Mach Tools* **(10)** 135
[15] Li S and Wu Y 2013 High-speed non-inner-type ceramic electric spindle design and development of experimental research *J Dalian Univ Tech.* **53 (2)** 214
[16] Chen X, Zhang P, He Y et al 2013 Study on axial vibration of high-speed spindle *Shock Vib.* **33 (20)** 70
[17] Dai S 1993 *Rolling Bearing Application Manual* (Beijing: Machinery Industry Press) p152
[18] Li Z and Cui W 2014 Spindle Dynamics analysis of wind turbine based on transfer matrix method *Mech Engin Autom* **(5)** 44
[19] Cao S and Liang Z 2008 Riccati transfer matrix method for calculation of imbalance response of rotor system *Mech Engin Autom* **148 (3)** 54
[20] Han Q, Wang D et al 2010 *Non-linear vibration analysis and diagnosis method of faulty rotor system (*Beijing: Science Press)

The impact of green building principles in the sustainable development of the built environment

D N Isopescu[1*]

[1] „Gheorghe Asachi" Technical University of Iasi, Romania

[*] E-mail: dorina_isopescu@yahoo.co.uk

Abstract. Climate change and the exhaustion of natural resources is a topical issue and represents a "to be solved immediately" goal in the development of built environment. The increasing development of buildings affects the natural environment more than any other industrial process. In Europe, the construction sector is responsible for consuming approximately 50% of the natural resources extracted per year, 40% of the energy produced, 16% of the treated water, and also for producing around 36% of the CO_2 emissions and 40-50% of the solid waste. Nowadays, specialists involved in building design have to think, more and more, about the strategies toward high performance buildings, and apply the principles of sustainability to find ways to coexist with the natural conditions, conserving natural resources, preventing land pollution, protecting the environment, and reducing energy consumption. This paper attempts to investigate the results of SWOT analyses on two aspects of the implementation of the principles of eco-innovative design of buildings in Romania: technical regulations for design and execution technologies. Then, analysing cases of green buildings design in the world, and their impact on the environment, it is concluded with the measures to be taken into account for the creation of healthy built areas, by applying the eco-friendly principles in all forms of their manifestation, and also with the protection of the environment surrounding.

1. Introduction

It is scientifically proven that 250 million years ago the earth underwent a dramatic global warming event, and the global temperature increased in that period by 9 degrees Celsius over a period of about 20,000 years. Nature managed to restore balance in a period of 200,000 years, and due to the fairly long period of global warming, almost all species have managed to adapt to the new tropical environment [1].

Nowadays, the global temperature increase rate is over 20 times faster. The average global temperature has already increased by more than one degree Celsius between 1880 and 2000, and by the end of 2100 there is a risk of an increase of even more than 4°C [1]. So the question arises whether ecosystems can adapt to this climate change.

One cause of global warming is the greenhouse effect produced by the pollutant emissions of industries that increase the amount of gases in the atmosphere. The issue of global warming can be prevented by minimizing greenhouse gas (GHG) emissions in the natural environment.

Supported by the United Nations, the COP21 Summit, Paris in 2015 brought together over 195 nations around the world, aiming at concluding an international treaty of anti-pollution rules, measures and laws to limit global warming by 2 degrees Celsius at the end of the year 2100. Above this limit of

2 degrees Celsius, the world of scientists warned that we could face an extremely severe climate change as a result of too much pollution. According to the report „Climate Change 2014 - Mitigation of Climate Change" of WG3 issued by the Intergovernmental Panel on Climate Change (IPCC), the total anthropogenic GHG emissions reached 49 (±4.5) $GtCO_2eq$/year in 2010, where the CO_2 accounts for 76% (38±3.8 $GtCO_2eq$/year) [2].

In Europe, the commercial and residential building sector accounts for almost 36% of the carbon dioxide (CO_2) emissions per year. The construction sector is responsible for consuming approximately 50% of the natural resources extracted per year, 40% of the energy produced, 16% of the treated water, and also for producing around 36% of the CO_2 emissions and 40-50% of solid waste. Most of the CO_2 emissions come from the combustion of fossil fuels to provide heating, cooling and lighting, and to power appliances and electrical equipment. This includes also the CO_2 generated through the entire building process: extraction, manufacturing (including the energy to manufacture capital equipment, heating and lighting of factories and so on), transportation, construction, maintenance, and disposal.

Nowadays, specialists involved in building design have to think, more and more, about the strategies toward high performance buildings, and apply the principles of sustainability to find ways to coexist with the natural conditions, conserving natural resources, preventing land pollution, protecting the environment, and reducing energy consumption.

One of the current basic strategies is the implementation of green buildings which aims to reduce the environmental impact of building. In line with the definition given by the U.S. Environmental Protection Agency, a green building is the result of a complex work in which a design team creates structures where using processes and products that are environmentally responsible and resource-efficient throughout a building's life-cycle. A green building is conceived, from the early stage of its creation, paying special attention to all required factors for design, construction, operation, maintenance, renovation and deconstruction. The classical building design is expanded and complemented by these new principles related to economy, utility, durability, and comfort, and, therefore, a green building or an eco-friendly building is also known as a sustainable or high performance building [3]. The green building has developed in response to the knowledge that buildings often have a negative impact on the environment and natural resources.

The green building brings together a vast array of practices, techniques and skills to reduce and, ultimately, eliminate the impacts of buildings on the environment and human health [4, 5]. The multi-criteria system developed by Ding and Langston [6] is now considered a useful tool for setting sustainability performance to new constructions and renovation, and is a key factor in the implementation of green buildings. The multi-criteria system is based on four fundamental criteria:

- maximizing economic efficiency with the goal of maximizing profits, which includes aspects of maintenance and sustainability;
- maximizing utility, which can be linked to external benefits such as social inclusion;
- resources minimization analysed over the lifetime of the construction, in terms of energy used in execution and consumed for operation;
- impact minimization on environment.

Today, the green buildings, recycling and eco-labelling of building materials have captured the attention of professional builders around the world. In the construction industry, a sustainable building can make a difference in terms of global environmental sustainability by reducing the consumption of natural resources and energy-intensive materials. According to Plessis [7], ignorance and lack of information about sustainable construction are a major obstacle which the construction industry has to overcome.

Eco-design and energy efficiency are concepts that need to find new building materials and technologies that are environmentally friendly and lead to a decrease in material and energy consumption. In other words, reducing the impact of buildings on the environment is possible by rethinking and redesigning materials and technologies used in the construction sector.

This paper attempts to investigate the results of the SWOT analyses on two aspects related to the green building implementation in Romania: the technical regulations for design and execution technologies. The results were obtained during the activities of the implementation of the project "Eco-Innovative Products and Technologies for Energy Efficiency in Construction" in "Gheorghe Asachi" Technical University of Iasi. The project has been developed in the program co-financed by the European Regional Development Fund: Operational Program Competitiveness 2014-2020, POC / 71/1/4 / Knowledge Transfer Partnerships, Priority Axis 1 - Research, Technological Development and Innovation in Support of Economic Competitiveness and Business Development, Action 1.2.3: Knowledge Transfer Partnerships [8].

2. The implementation of green building principles in Romania. EU and national legislation

Governments around the world rely on building codes to protect their citizens from harm due to structural collapse and other building related risks. The codes should meet these needs by specifying at least the minimum requirements for both construction materials and construction practices. Events such as earthquakes highlight the importance of implementing building codes which should continually evolve from the point of view of their consistent enforcement.

Generally, the strategies offered for green building design are voluntary and presented as design options or ideas to be taken into consideration. The strategies do not necessarily support one another, and the use of one strategy may preclude the use of others. Designers, owners, developers and other people involved in the processes of green building maintenance will need to weight the benefits of using any particular set of strategies. There may be phases and particular analyses needed to achieve the diverse design goals of any particular project.

The EU's legislative framework for energy efficient buildings was strengthened in 2002 with the adoption of the Energy Performance of Buildings Directive (EPBD – Directive 2002/91/EC), which introduced Energy Performance Certificates (EPCs) and minimum energy performance requirements. The original EPBD 2002/91/EC included the following main aspects: establishment of a calculation methodology, minimum energy performance requirements, and the requirement of energy performance certificates.

Directive 2010/31/EU on the energy performance of buildings (EPBD recast) widened the scope of the EPBD and raised the standards. It introduced the concept of Nearly Zero Energy Buildings (NZEB) and a target of 2018/2020 for their introduction. It also introduced the idea of "cost-optimal levels" of energy performance in buildings, meaning the "energy performance which leads to the lowest cost during the estimated economic life cycle". Furthermore, the EU Directive encourages the creation of national plans to lay in practice the definition of nearly zero energy buildings, the definition of intermediate targets for improving the energy performance of new buildings by 2015. In November 2016, the European Commission submitted a package of legislative proposals under the title "Clean Energy for All Europeans", with the goal of providing a stable regulatory framework, needed to facilitate the clean energy transition, and, thus, a significant step was taken towards the creation of the Energy Union. The "Clean Energy for All Europeans" proposals are intended to help the EU energy sector to become more stable, more competitive, and more sustainable, suited to the 21st century as well as enabling the EU to deliver on its Paris Agreement commitments.

Related to the products used in the construction industry, mention must be made of the Directive 2009/125/EC of the European Parliament and of the Council issued on 21 October 2009 which establishes a framework for setting the eco-design requirements of energy-related products [Eco-design Directive 2009].

In Romania, the annual specific energy consumption registered in buildings is extremely high, mainly due to the heat losses through the construction elements, as well as to the poor exploitation of the heating systems. In this case, the thermal rehabilitation of the buildings and the use of renewable energy have became key objectives at national level, in order to ensure energy savings and environmental protection. In order to facilitate the environmental protection, a series of policies started to be developed in Romania.

In order to reach the targets set by the EU in Directive 2009/28/EC of the European Parliament and of the Council issued on 23 April 2009 on the promotion of the use of energy from renewable sources, which amended and repealed Directives 2001/77/EC and 2003/30/EC, the Member States have to increase the share of renewable energy sources in the energy mix, [Renewable Energy Directive 2009].

After transposing Directive 2010/31/EU into the Romanian legislative regulation, the Law no. 372/2005 was amended and completed in 2016, and the Law no.121/2014 was implemented. The energy performance of the building, as defined in the Romanian Law 372/2005 updated and republished in 2013, represents the actual energy, consumed or estimated, to meet the needs of the normal use of the building which mainly includes: heating, hot water, cooling, ventilation and lighting.

The Romanian Government Decision no. 55/2011 on establishing the eco-design requirements for energy-related products specifies the requirements that these products must meet for the increase of the energy efficiency and the environmental protection level, in order to contribute to a sustainable development of built environment. This decision is represented by Directive 2009/125/EC.

Directive 2009/28/EC is implemented by the Romanian Law no. 220/2008, with subsequent amendments, establishing the system to promote energy production from renewable energy sources.

The green building design strategy challenges the specialists involved in design and execution, to consider both the inside and the outside of the house with the aim to achieve: proper location selection, efficiency in water consumption, increased day lighting, use of eco-power, and use of environmentally friendly materials. The first segment in the green building design strategies is, therefore, subjected to "eco-innovation". Eco-Innovation refers to all forms of innovation – technological and non-technological – that create opportunities for the development of economic activities and that have a positive effect on the environment, by preventing or reducing the impact of these activities, or by optimizing the use of resources.

The measurement of innovativeness of EU countries, through various parameters, aims to promote a global view on economic, environmental and social performance. Two surveys, which complement other parameters, are relevant: the Eco-Innovation Scoreboard (Eco-IS) and the Eco-Innovation Index. They illustrate eco-innovation performance across the EU Member States and aim at capturing the different aspects of eco-innovation by applying 16 indicators grouped under five categories: eco-innovation inputs, eco-innovation activities, eco-innovation outputs, resource efficiency, and socio-economic outcomes. The Eco-Innovation Index shows how well individual Member States perform under different classes of eco-innovation compared to the EU average, and indicates their strengths and weaknesses [9, 10].

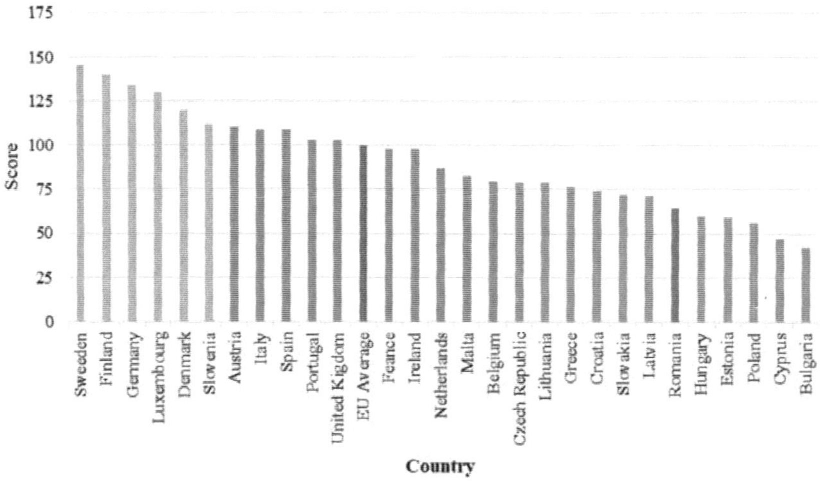

Figure 1. The chart of European Eco-Innovation Index by countries, 2017, [10].

In figure 1, the chart for Eco-Innovation Index in EU [10], highlights Romania's low position, which is mainly due to the lack of efficient management of the available resources. The Eco-Innovation Scoreboard (Eco-IS) tool also shows that, in Romania, the indicator for "the value of green early stage investments (per capita) is almost nearly zero [10].

3. SWOT analyses

3.1. SWOT analysis of the eco-innovation design of buildings

Figure 2 presents the strategic planning technique with the inputs for the Strengths, Weaknesses, Opportunities, and Threats in a SWOT analysis related to the eco-innovation design of buildings. It aims to identify the internal and external factors that are favourable and unfavourable to achieving the specified objective in the goals.

Strengths: S1 - positive environmental balance S2 - rehabilitation of structures by functional reconversion of industrial spaces S3 - the existence of technologies in the market in this field S4 - use of local materials S5 - site rehabilitation	Weaknesses: W1 - insufficient popularization of strategies W2 - lack / insufficiency of existing norms W3 - lack of collaboration in the fields involved (architecture, construction, sociology, installations) W4 - insufficient training of specialists W5 - lack of national strategy
Opportunities: O1 - updating the legislation O2 - orientation towards eco-sustainable materials O3 - tax deduction for designers adopting eco-innovation O4 - modern technologies development for useful traditional materials O5 - high demand of investors / beneficiaries O6 - stimulating designers to these solutions	Threats: T1 - high costs T2 - depreciation of long-term investments T3 - reluctance of designers to leave the comfort zone professionally T4 - lack of information for beneficiaries (public)

Figure 2. Chart with internal and external factors for the SWOT analysis of the eco-innovation design of buildings, [11].

The results, after a statistical processing of responses received from stakeholders are presented in figure 3.

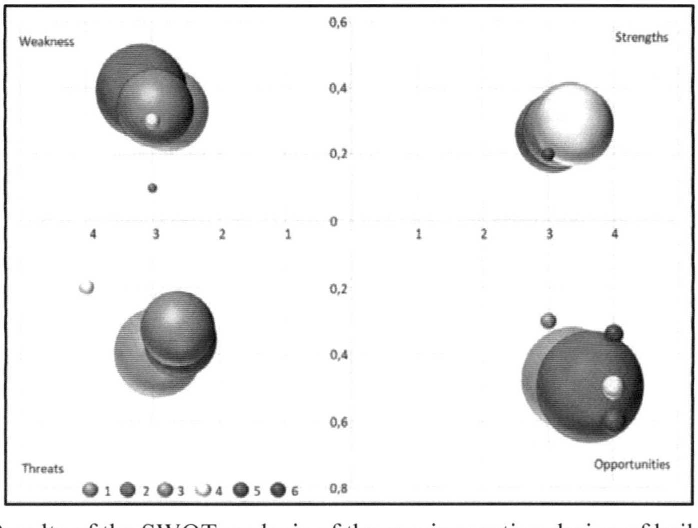

Figure 3. Results of the SWOT analysis of the eco-innovation design of buildings, [11].

For the eco-innovation design of buildings, following the SWOT analysis, the results highlighted that it is necessary to prioritize the setting of objectives for strategic planning, as follows:

- emphasis should be placed on positive environmental balance (S1) and use of local materials (S4);
- the problems identified in relation to the insufficient training of specialists (W4) and lack of a national strategy in domain (W5) should be minimized;
- the legislation should be updated (O1) and the owners and developers should be stimulated to accept these solutions (O6);
- measures should be taken to create financial compensations for the high costs (T1) at the construction stage;
- data bases shoudl be created to overcome the lack of information for beneficiaries (T4) and to improve the knowledge in domain.

3.2. SWOT analysis of the eco-sustainable technologies for the manufacturing of construction materials

Strengths: S1 - re-incorporation of materials / waste into the production cycle S2 - reducing the amount of energy embedded S3 - high insulation capacity (thermal, hydro and acoustic to fire action) S4 - reduction / elimination of waste emissions in the manufacturing process S5 - implementation of circular economy principles	Weaknesses: W1 - lack / insufficiency of necessary arrangements; W2 - lack / insufficiency of collection points for construction waste W3 - lack of information W4 - lack of recycling, W5 - high costs, W6 - lack of government strategy, W7 - lack of capital of private investors
Opportunities: O1 - increased recycling and bio-degradation of used materials O2 - reducing CO2 emissions and pollutants during the manufacturing process O3 - improving production yield O4 - rational / responsible management of natural resources O5 - involving local people in production using local materials	Threats: T1 - lack of specialists in the field T2 - lack of marketing actions T3 - depreciation of long-term investments

Figure 4. Chart with internal and external factors for the SWOT analysis of the eco-sustainable technologies for the manufacturing of construction materials, [11].

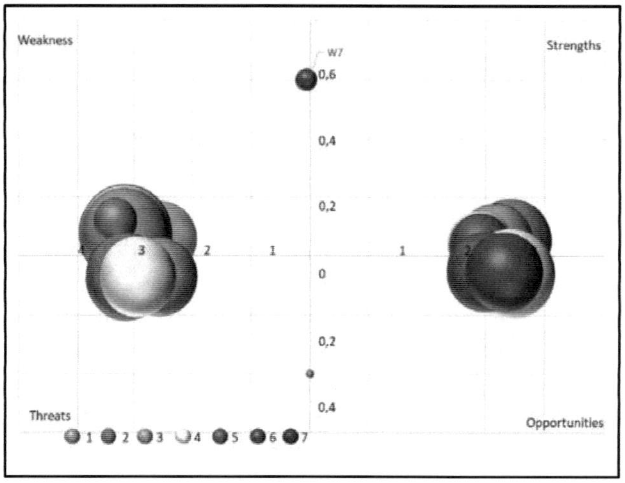

Figure 5. Results of the SWOT analysis of the eco-sustainable technologies for the manufacturing of construction materials, [11].

The strategic planning technique with the inputs for the Strengths, Weaknesses, Opportunities, and Threats related to the eco-sustainable technologies for the manufacturing of construction materials is presented in figure 4. Figure 5 presents the statistically processed answers received from stakeholders According to the SWOT analysis, the results highlighted that in the cases of eco-sustainable technologies for the manufacturing of construction materials, it is necessary to prioritize the setting of strategic planning objectives. Therefore, the following prioritizations have to be implemented:

- emphasis should be placed on the re-incorporation of materials / waste into the production cycle (S1), on the decrease of the amount of energy embedded (S2) and on the implementation of circular economy principles (S5);
- the problems identified related to the lack of government strategy (W6) and the lack of capital of private investors (W7) should be minimized;
- the use of recycled and bio-degradation materials (O1) should be increased;
- policies for the rational / responsible management of natural resources (O4) should be implemented;
- the lack of specialists in the field (T1) and the lack of marketing actions (T2) in the domain should be solved .

3.3. Discussions

The SWOT analyses were developed in the project "Eco-Innovative Products and Technologies for Energy Efficiency in Construction" implemented in "Gheorghe Asachi" Technical University of Iasi [11]. The period of project implementation is between 2016 and 2021, and the estimated project results are:

- creating eco-innovative products and improving the energy performance of existing products used in civil, industrial and agricultural buildings;
- developing and implementing an integrated system of design and execution of passive buildings (energy efficient);
- optimizing the energy performance of buildings through the use of software modelling energy consumption;
- making an interactive energy management tool in construction products;
- implementing renewable energy sources of high efficiency in manufacturing products used in buildings, and also, throughout the phases of construction and use of buildings;
- setting up the European technical assessment body for construction products;
- setting up the platform of information and transfer of knowledge about eco-innovative products and technologies for energy efficient constructions.

Some project actions were dedicated to identifying the actual situation of the factors which influence the Romanian's targets in green buildings implementation. The SWOT analyses developed in the project have involved identifying and mapping the internal and external factors that support or hinder the achievement of Romanian's goals in this domain. The SWOT analyses provided a good framework for reviewing the current strategies and directions of implementing the green building goals in Romania. These analyses lead to a better understanding of what the existing strategies can offer at present, and the key weaknesses that need to be solved in the future in order to succeed.

The charts presented in figures 3 and 5, through the distribution of the analysed parameters, highlight the fact that in Romania the problems related to the design and the execution technologies of the green building are poorly perceived by the actors involved in the implementation of the principles that are defining for this type of buildings. From the analyses of the charts, a set of questions arises: is it a problem related to poor information in the field or is it cumbersome to overcome certain stereotypes formed over time by the traditional design of the buildings? An affirmative answer to these questions leads to the urgent need for national support policies which, by providing tax benefits to the owners, which will lead to increasing the use of the environmentally friendly buildings.

4. Eco-innovative principles applied in the design of green buildings

The design principles of green buildings are applied in many countries. In the technical literature one can identify multiple examples of green buildings in which one or more eco-innovative solutions for products and/or execution technologies have been implemented successfully. In the author's opinion, the most important strategies which must be observed in the design of green buildings are:

- Passive solar design, which is a cumulative method of organizing the parts of a building in such a manner as to exploit and control the solar radiation; this is the key to take the best advantage of the local climate. The goal is to maximize the positive inputs of solar radiation in the cold season while reducing the effects of overheating during the warm season. Solar design is part of a wider approach, integrated design, which involves several aspects that need to be considered from the conception phase and the collaboration of the specialists involved.
- During the design stage, care must be taken to balance the architectural aesthetics with the shape factor. The shape factor is defined as the ratio between the wall and roof surface and the inner volume (SA: V), determining the compactness of a building, and influencing the energy efficiency. This ratio is inversely proportional to the size of the building - the larger the construction, the smaller the ratio. Small buildings have a SA: V ratio higher than large-scale constructions of identical shape. Compactness is closely related to energy efficiency, and therefore this parameter is important and must be considered when designing green buildings.
- "Life cycle" estimates of a building are increasingly included in the design phase as fundamental sustainability assessment methods and include two major components: Life Cycle Costing (LCC), and Life Cycle Assessment (LCA). The "LCC" procedure allows cost-optimization design, taking the form of an investment estimate in an overall approach. The "LCA" methodology evaluates inputs and outputs from a system, in particular a construction, over a life cycle, in terms of the system interaction with the natural environment. Systemic modelling takes into account all inputs, including extraction of raw materials and fuel use, water consumption, and so on. Also, this modelling takes into account all outputs, including mainly air emissions, water pollution and soil pollution, as an environmental impact assessment over the building lifetime, from construction to demolition.
- The study of green buildings through the evolution of digital technologies is an important concern at global level. In parallel with the advancement of engineering sciences, most of the concepts underpinning traditional architecture are reconsidered by computer, which becomes a central factor. Advanced technologies for green buildings include: innovative materials in order to ensure low carbon dioxide emissions and better thermal insulation qualities, complex structural systems to reduce the use of materials through engineering conceptual solutions, and social evaluations to predict the requirements of the building owners.

The implementation of these design strategies has led to certain benefits during life of the buildings. Therefore, the case studies analysed confirm the need to implement specific solutions to each region but with certain results in achieving the main purpose of green buildings, namely energy efficiency and low impact on environmental pollution.

5. Conclusion

This synthetic analysis of the implementation of the principles of design and execution of green buildings, as well as their influence on the sustainable development of the built environment, highlights the key directions for the implementation of the existing strategies in the creation of healthy built areas, by applying the eco-friendly development principles in all forms of their manifestation.

The construction sector has a significant environmental impact, being one of the most important consumers of natural resources and, at the same time, having a significant carbon footprint. Therefore, civil engineering specialists should develop and implement specific solutions with a reduced environmental negative influence such as the following: to stop the unsustainable urbanization of the cities, to reduce the natural resources consumption rates applying the principles of the sustainable management of raw materials, to use the renewable energy sources, and to develop new eco-friendly materials, technologies, and design principles.

Two factors are decisive: the national governments that will have to create adequate legislation and provide financial support for the research on this topic, and the local authorities that should promote a sustainable urbanization through local tax policies.

In Romania the need to develop technical, sustainable and legislative supported strategies for the green buildings implementation is obvious. A comparison with the stages undergone by other states in Europe or other continents, shows that in Romania there is a delay in the implementation of the green buildings principles in the built environment, with a gap of approximately 20 years.

The steps taken so far indicate that there is interest at government level, but the education of the population on this matter is delayed, thus generating negative effects in understanding the consequences of global warming phenomenon and in taking actions. The Romanian specialists, through all means of mass information, are permanently involved in overcoming these negative aspects and are constantly trying to comply with the requirements implemented at international level.

The efficient use of resources and the development of circular economy will be challenges for Romania's construction sector on its way to becoming more sustainable and more "green". The implementation will stimulate investment and bring benefits both in the short and long term for the built environment, environment and citizens alike.

Acknowledgements
This paper was realized in the framework of the project "Eco-Innovative Products and Technologies for Energy Efficiency in Construction", POC/71/1/4 - Knowledge Transfer Partnership, Cod MySMIS: 105524, ID: P_40_295, Project co-financed by the European Regional Development Fund.

References
[1] Mărculescu O 2017 Global warming is faster than estimated *Science and Technology Review*
[2] IPCC-Working Group III 2014 Contribution to the IPCC's Fourth Assessment Report (AR4) *Climate Change 2014: Mitigation of Climate Change* Cambridge University Press
[3] Rinkesh K 2016 What is a green building? *Green & Sustainable Business,* http://www.businessfeed.sunpower.com
[4] U.S. Environmental Protection Agency 2009 Buildings and their Impact on the Environment: A Statistical Summary http://www.epa.gov/greenbuilding/pubs/components.htm
[5] WBDG (accessed 2018) Whole Building Design Guide *Public Library of The National Institute of Building Sciences* Washington http://www.wbdg.org/building-types/libraries/public-library
[6] Langston C and Ding G 2002 A methodology for assessing the sustainability of construction projects and facilities *Proceedings of the 3rd World Congress on Cost Engineering, Project Management and Quantity Surveying* Australian Institute of Quantity Surveyors Deakin, A.C.T.
[7] Du Plessis C 2001 *Agenda 21 for Sustainable Construction in Developing Countries* CSIR Building and Construction Technology, Pretoria
[8] Project (accessed 2018) *Eco-Innovative Products and Technologies for Energy Efficiency in Construction,* "Gheorghe Asachi" Technical University of Iasi http://www.efecon.tuiasi.ro/
[9] EU Commission 2011 Innovation for a Sustainable Future - The Eco-innovation Action Plan (Eco-AP) Document 52011DC0899, Brussels
[10] Eco-Innovation Observatory (accessed 2018) The Eco-Innovation Scoreboard and the Eco-Innovation Index https://ec.europa.eu/environment/ecoap/indicators/index_en
[11] Workshop (accessed 2018) Innovative eco-technologies to ensure energy efficiency in construction http://www.efecon.tuiasi.ro/rezultate/workshopuri/workshop2/index.html

IOP Conf. Series: Materials Science and Engineering **399** (2018) 012027 doi:10.1088/1757-899X/399/1/012027

The characteristics of studying high speed motorized spindle dynamic magnetic coupled

Z Ke[1], W Zinan[1*] and W Qingyuan[1]

[1] School of Mechanical Engineering, Shenyang Jianzhu University, National and Local Joint Engineering Laboratory of High-grade Stone Numerical Control Processing Equipment and Technology, Hunnan East Road on the 9th, Hunnan District, Shenyang, China. Postcard: 110168

[*] E-mail: wzn591672678@126.com

Abstract: The vibration of high speed motorized spindle directly influences the processing quality; through the finite element method, the dynamic magnetic coupling model of high speed motorized spindle rotor system is established. The coupling relationship between the field parameters of electromagnetic and the characteristics of dynamic are considered. At the same time, the influence of electromagnetic coupled on the dynamic vibration characteristics of the rotor system is analysed and the experimental verification is carried out. The results indicate the stator and rotor magnetic density, the magnetic field energy distribution about motorized spindle are also periodic. The skin effect of magnetic field and air gap eccentricity have great influence on radial vibration. With the increase of speed, the centrifugal force of the bearing is increasing; the softening effect of bearing stiffness is obvious; the natural frequency is decreasing; the amplitude of vibration is increasing. The experimental results are in good agreement with the simulation results. The accuracy of the dynamic magnetic coupling model is illustrated.

1. Introduction

The motorized spindle basic characteristics of high speed, high rotation accuracy and small vibration are the important factors to ensure the reliability of machine tools. Scholars at home and abroad have done a lot of motorized spindle research on the electric field, magnetic field and dynamics. Masoud Razban and Mohammad R. Movahhedy [1] considered the gyroscopic effect, an automatic variable pressure preload system was put forward. Atsushi Matsubara [2], through non-contact excitation, evaluated the method of dynamic stiffness on the motorized spindle. O. Özs¸ Ahina [3] predicted the dynamic performance of motorized spindle and bearing through the FRF response function. CaoHongrui [4] put forward the six concepts of motorized spindle balance, motorized spindle health and so on. Chen Xiaoan [5] analysed the vibration shape of rotor rigid body and proposed the theory of active inhibition. Zhang Ke [6, 7] designed a vector controller based on slip frequency control. The coupling relationship between the thermal characteristics and the vibration characteristics of the motorized spindle is obtained. Fourier transform, polynomial curve fitting and Hooke's law are combined to design and develop an algorithm for measuring the magnitude of dynamic imbalance force, vector position and direction of motorized spindle. Zhang Lixiu [8, 9] established that the a multi-field coupling model is to predict the dynamic performance of motorized spindle, and the control performance of motorized spindle is improved by using ANN and CBR. Wang Bo [10] completed the finite element modelling of motorized spindle system based on the modelling idea of the parallel rotor. All the above studies aim at single dynamic analysis. The influence of electromagnetic force and electromagnetic torque on motorized spindle dynamics are neglected. Especially, the electromagnetic model is not integrated into the dynamic model, which is not a comprehensive consideration of the dynamic characteristics.

In view of the above shortcomings, a new dynamic magnetic coupled model is established in this paper. The coupling relationship between motorized spindle system dynamics and electromagnetism is analyzed. The influence of different electromagnetic force and torque on dynamics is discussed. At the same time, the unbalanced magnetic pull and the centrifugal force of the bearing are used to establish the dynamic model, which is closer to the actual working condition. Finally, the test verifies the accuracy of the model.

2. The establishment of dynamic magnetic coupled model

2.1. The structure diagram of dynamic magnetic coupled model
Due to the change of air gap between the rotor and stator, electromagnetic parameters, such as current, magnetic field, magnetic flux and so on are motivating magnetic force and torque influencing the dynamics of motorized spindle. Figure 1 is the structure diagram of high speed motorized spindle dynamic magnetic coupled.

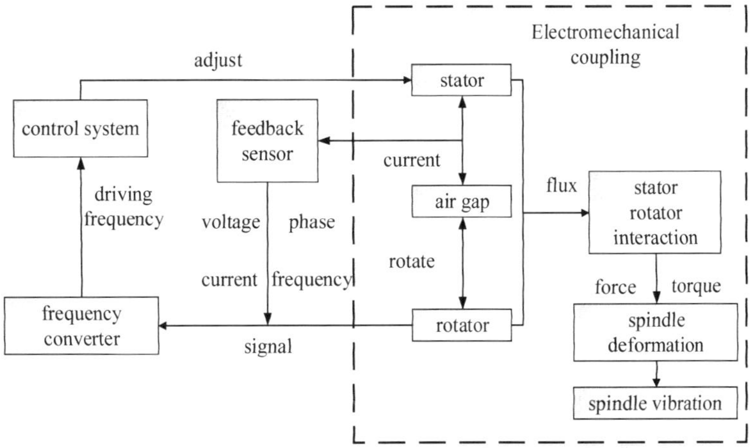

Figure 1. The structure diagram of high speed motorized spindle dynamic magnetic coupled.

The air gap of motorized spindle is the gap of between stator fixed and rotor rotated pole. On the one hand, the air gap affects the dynamic performance of motorized spindle, the unbalanced magnetic pull produced by eccentric operation as a non-negligible external load is the important factor of motorized spindle vibration. On the other hand, the magnetic flux leakage and loss generated by air gap determine the electromagnetic parameters. Electromagnetic force, torque and other electromagnetic factors are the basic conditions of the velocity and other parameters in dynamics.

2.2. The establishment of dynamic magnetic coupled model for motorized spindle

2.2.1. The establishment of electromagnetic model. The main components of the high-speed motorized spindle, the shell, stator and rotor are simplified as cylinders, and the magnetic field model of the motorized spindle is established.

The Bessel function is applied to make the motorized spindle rotor equivalent to the magnetic field of the cylinder [11]. Using the m order imaginary Bessel equation, we can obtain:

$$x^2 \frac{d^2R}{dx^2} + x\frac{dR}{dx} + (x^2 - m^2)R = 0 \tag{1}$$

$$x = \sqrt{\mu\rho} \tag{2}$$

Because the motorized spindle is a symmetrical circular annular current, the vector potential A of the air gap is solved.

$$A = \frac{\mu}{4\pi}\int_V \frac{J'(r')dV'}{R} \tag{3}$$

$$J(r') = e_\varphi J_\varphi(\rho', z') \tag{4}$$

There are different magnetization effects between the air gap of the motorized spindle. The medium is magnetized, generating the intensity of the magnetic field, which is:

$$B = \mu_0(1+\chi_m)H = \mu_0\mu_r H \tag{5}$$

μ_0—Vacuum magnetic permeability, χ_m—medium magnetic susceptibility, μ_r—relative magnetic permeability, H—medium magnetic field intensity vector

The motorized spindle rotate speed of rotor:

$$v = 2p\tau\frac{n_1}{60} = 2\tau f_1 \tag{6}$$

l—the effective length of stator and rotor, v—the velocity of the rotating magnetic field in the air gap, τ—the center point distance of adjacent magnetic poles in rotating magnetic field

According to the dynamic equation, motorized spindle induction strength and electromotive force, the radial force of XY axis about rotor system is obtained.

$$\begin{bmatrix} F_{1x} \\ F_{1y} \end{bmatrix}_{i+1, j+1} = \begin{bmatrix} f_{1x} \\ f_{1y} \end{bmatrix}_{i, j} \begin{bmatrix} D_{ii} & D_{ij} \\ D_{ji} & D_{jj} \end{bmatrix} \begin{bmatrix} S_i \\ S_j \end{bmatrix} \tag{7}$$

Unbalanced magnetic pull and radial force in rotor system

$$\begin{cases} f_e = k_{xx}x + k_{xy}y + c_{xx}\dot{x} + c_{xy}\dot{y} \\ f_r = k_{yx}x + k_{yy}y + c_{yx}\dot{x} + c_{yy}\dot{y} \end{cases} \tag{8}$$

ε_{ij}—The eccentricity of air gap, S—contact surface area of stator and rotor

2.2.2. The establishment of loss model. According to the input parameters of frequency converter, the motorized spindle input power is known.

$$P_1 = 3U_0 I_0 \cos\varphi_1 \tag{9}$$

$\cos\varphi_1$—the power factor of motorized spindle, U_0、I_0—the effective value of phase current and voltage, φ_1—the angle between phase voltage and phase current

The stator and rotor are made of copper wire winding, copper loss is

$$P_{cu} = 3(I_1^2 + I_2^2)R_2 \tag{10}$$

Mechanical power is:

$$P_m = P_1 - P_{cu} \tag{11}$$

Electromagnetic torque is:

$$T_{em} = \frac{P_m}{\Omega} = \frac{30pP_m}{\omega\pi n_1} \tag{12}$$

p—pole of the rotor Ω—the angular velocity of synchronous mechanical

2.2.3. Dynamics equation. Calculation of dynamic characteristics of motorized spindle based on the Timoshenko beam element theory [12]. The dynamic balance equation of the motorized spindle structure is established.

$$M\ddot{x}(t) + (C+G)\dot{x}(t) + Kx(t) = T_{em} + F_e + F_r \tag{13}$$

M—Mass matrix of motorized spindle, C —Damping matrix of motorized spindle
G—Gyroscopic effect matrix of bearing, K—Support stiffness matrix of motorized spindle
T_{em}—Electromagnetic torque of motorized spindle, F_e—Unbalanced magnetic pull
F_r—Electromagnetic radial force

Natural frequency of motorized spindle

$$\omega = \sqrt{\frac{K}{M}} \tag{14}$$

Rayleigh damping

$$C = a_0 K + a_1 M \tag{15}$$

$$\begin{pmatrix} a_0 \\ a_1 \end{pmatrix} = 2\frac{\omega_m \omega_n}{\omega_n^2 - \omega_m^2}\begin{pmatrix} \omega_n & -\omega_m \\ -1/\omega_n & 1/\omega_m \end{pmatrix}\begin{pmatrix} \xi_m \\ \xi_n \end{pmatrix} \tag{16}$$

ξ_m and ξ_n are the different frequencies of the motorized spindle material ω_m, ω_n, Respectively the corresponding damping ratio are 0.05and 0.08. ω, which is the natural frequency of a motorized spindle.

3. The analysis of electromagnetic field

The magnetic field model of the motorized spindle is established through the finite element ANSYS software electromagnetic module, and the magnetic density of the stator and rotor, electromagnetic energy distribution diagram and torque diagram are calculated. The establishment of a fixed rotor model is assumed as follows, ignoring the external magnetic field interference and axial variation of the magnetic field. Only considering the variation of the radial magnetic field, a two dimensional model is selected.

3.1. The establishment of rotor model for motorized spindle

Parameter setting of electric motorized spindle: pole of the rotor is 2, number of stator slots is Z_1=24, number of rotor slots is Z_2=28, external diameter of stator is 130mm, internal diameter of stator is 80mm, air gap is 0.3mm, external diameter of rotor is 79.4 mm, internal diameter of rotor is 51mm.
According to the electromagnetic model of the motorized spindle, the rotor is uniformly rotated in the stator, so the time stepping method is used to calculate the two time's stator and rotor magnetic dense cloud diagram, as is shown in figure 2. It can be seen that figure 2 shows a circumferential symmetry distribution of magnetic flux density. The interior of the stator and the outside of the rotor are densely distributed. The magnetic density is mainly concentrated in the magnetic pole and the surrounding area. The interior of the rotor and the outside of the stator are less affected by air gap. The magnetic density distribution is uniform and more and more scarce. Because of the high speed frequency conversion of the motorized spindle, the eddy current of the magnetic field generated by cutting the magnetic line, which results in the skin effect leading the maximum magnetic density of the contact magnetic pole on rotor and stator. Firstly, it is in the state of saturation, the magnetic flux is limited by air gap magnetic density. The magnetic lines of force are bending. The magnetic flux enters the inner

edge region of the rotor to form a magnetic circuit closure. The internal magnetic density of the rotor is the least. The magnetic flux is permeable to the adjacent magnetic pole. The magnetic flux density of the adjacent poles increases and the reduced phenomenon of magnetic appears in the radial direction of the stator and rotor.

As compared to figure 2(a) and (b), it is known that the transfer rate becomes larger. The distribution of the magnetic pole density of the fixed rotor is gradually reduced. The magnetic flux density line diffuses along the outer side of the stator and the inner part of the rotor. The magnetic flux eddy current phenomenon is more obvious. The close magnetic density periphery of the magnetic stator and rotor pole is increased. The density of the adjacent magnetic poles of the adjacent magnetic poles is reduced.

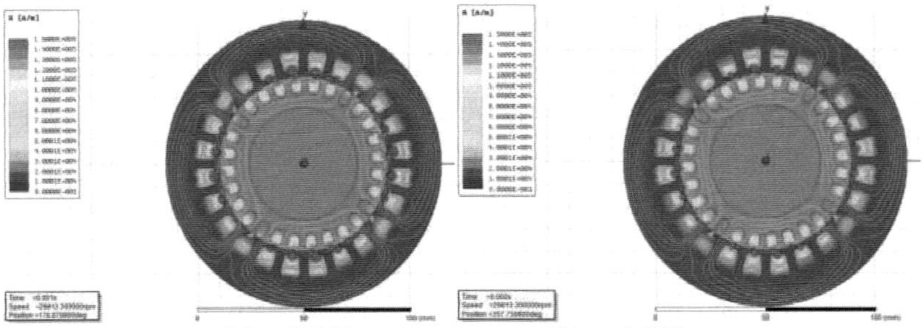

(a) t=0.001s (b) t=0.002s

Figure 2. The cloud diagram of stator and rotor magnetic flux density.

3.2. Calculation and analysis of electromagnetic field performance of motorized spindle

According to the electromagnetic model of motorized spindle, the energy diagram of motorized spindle which is obtained, as is shown in figure 3. At the beginning of 0.001-0.003s, the peak value of the electric field and the magnetic field energy appears. The peak value of electromagnetic energy coincides with the maximum value of electromagnetic energy in figure 3. The accuracy of the electromagnetic model of the electric motorized spindle is reflected. The electric field energy decreases rapidly in time. After running smoothly, the electric field tends to be stable and no longer attenuates. The regularity fluctuate is periodicity. The energy of the magnetic field starts to increase rapidly. Rapid magnetic circuit saturation occurs between the magnetic poles of stator and rotor. The magnetic flux diffusion is limited. After the smooth operation, the magnetic circuit slows down.

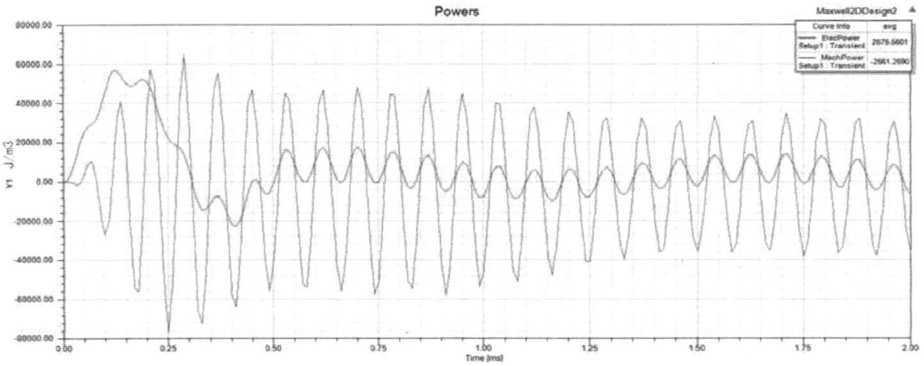

Figure 3. The energy curve of motorized spindle.

The main reason that affects the magnetic density is the difference between the poles of the stator poles and the uneven magnetic poles of the air gap. The uneven air gap poles the main reason for the radial force. Combined with figure 2, due to the distribution of the symmetrical magnetic density, analysing a periodic region of magnetic density distribution is needed.

IOP Conf. Series: Materials Science and Engineering **399** (2018) 012027 doi:10.1088/1757-899X/399/1/012027

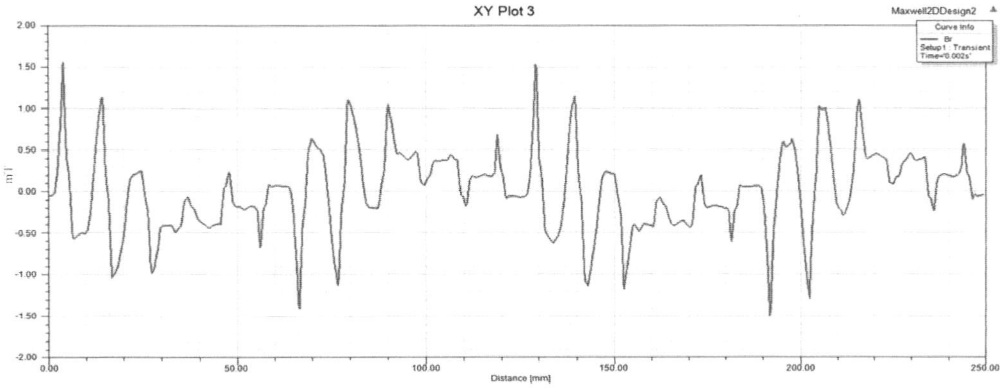

Figure 4. The radial magnetic density diagram of motorized spindle.

4. Dynamic performance analysis of motorized spindle

The exciting force of motorized spindle is mainly composed of random excitation force and steady state exciting force. The steady state exciting force mainly consists of electromagnetic torque, radial force, bearing centrifugal force and preload force, etc. The steady state exciting force of motorized spindle is loaded into the dynamic model of motorized spindle. The influence of multiple factors on motorized spindle vibration and bearing stiffness and frequency is calculated by means of the Kriging algorithm.

$$\bar{y}(x) = \sum_{i=1}^{k} \beta_i f_i(x) + Z(x) = f^T(x)\beta + Z(x) \tag{17}$$

$f^T(x)$ is base function vector of radial force regression motorized spindle model, β is the centrifugal force, gyroscopic moment and other parameters of the regression motorized spindle model. The approximate equivalent of electromagnetic torque is the normal distribution function. $N(0,\sigma^2)$ is a random function that obeys the normal distribution, the function covariance is:

$$\mathrm{cov}(Z(x_1), Z(x_2)) = \sigma^2 R(x_1, x_2) \tag{18}$$

$R(x_1, x_2)$ is the correlation equation between x_1 and x_2 two electromagnetic data points. Gauss function is applied to find the dependencies between variogram and covariance function to guarantee the smoothness of the semi variant function.

The Monte Carlo method is used to generate motorized spindle dynamic data samples

$$y(x) = \beta + r^T(x)R^{-1}(Y - f\beta) \tag{19}$$

$y(x)$ is the vibration fitting curve of the electric motorized spindle.

The B7009C/P4 bearing of the front bearing of the ceramic motorized spindle is taken as an example. The parameters of the front bearing are shown in table 1.The front bearing is preloaded with constant pressure. The bearing preload force is 600N. The process of the experiment is that the motorized spindle speed is 0-18000rpm.

Table 1. The parameters of front bearing.

Bearing type	Ball material	d_m (mm)	D_b(mm)	N	A(°)	υ
B7009C/P4	Si_3N_4	60	7.5	19	15	0.3

The air gap eccentricity of the motorized spindle is affected by the centrifugal force, the gyroscopic moment of the bearing and assembly. The air gap magnetic field energy and electromagnetic torque of the motorized spindle are changed. The simulation results show that the normal and eccentric vibration spectrum under no-load operation condition of the motorized spindle is shown in figure 5.The trend of vibration track for normal and eccentric operation is basically the same. But the vibration amplitude of eccentric operation is much larger than that of normal operation. The results show that the air gap eccentricity causes the torque of the fixed rotor, the magnetic field distribution is not uniform, the torque and radial force amplitude is much larger than the normal running condition. The unbalanced magnetic pull causes the radial vibration of motorized spindle shell to increase.

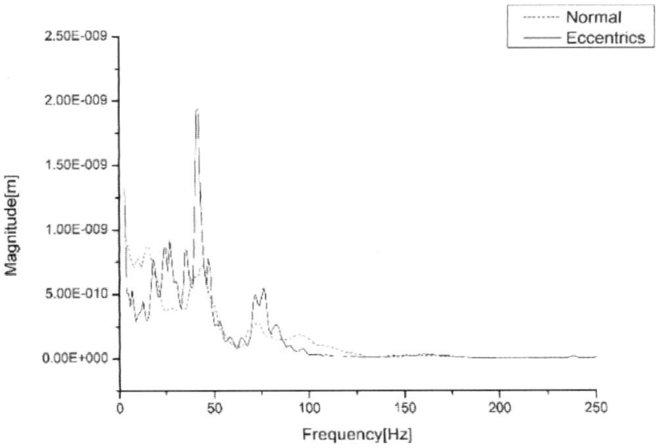

Figure 5. Normal and eccentric vibration spectrum of motorized spindle.

In the motorized spindle high-speed test, the natural frequency of the motorized spindle is mainly affected by the centrifugal force of the bearing. The stiffness of the bearing is changing as the speed changes. With the increase of the rotating shaft, the centrifugal force of the bearing increases the contact deformation of the angular contact ball bearing outer ring and outer ring groove. The contact between inner ring and inner race decreases, and the stiffness of radial support is softened. The stiffness softening causes the reduction of the natural frequency of the motorized spindle. At the speed of 6000rpm, the experimental data were measured by hammering. From figure 6, it can be seen that the experiment is in general in agreement with the simulation data. The first natural frequency of simulation is 1125HZ, and the data obtained from the experiment is 1110HZ, the difference between them is 15HZ. It can be seen that the bearing stiffness is softened and, together with the gyroscopic effect, it is the important factor that causes the natural frequency of the test to be lower than the natural frequency of simulation. It can be seen from the diagram that the simulation results agree well with the experimental values, indicating the accuracy of the motorized spindle bearing system model.

Three experiments have been taken to verify the reliability of the electromagnetic coupling simulation system of the motorized spindle. The rotational speed is designed for 6000rpm, 12000rpm, and 18000rpm to collect the displacement sensor signals to form the vibration displacement diagram of the motorized spindle. In figure 7 (a), it can be seen that the motorized spindle speed is the direct factor determining the amplitude of vibration. Due to the fact that the centrifugal force acts on the bearing inner ring, the inner ring displays expansion and deformation. The inner ring expansion is greater than the rotor expansion, and the interference amount of the bearing inner ring and the rotor is reduced. The radial stiffness of the bearing decreases with the increase of the speed, and the radial vibration of the bearing becomes larger. The contrast in figures 7 (a) and (b) show that, despite the simulation and experiment in motorized spindle 18000rpm, the maximum amplitude value difference is a 10^{-8}m error. There are many electromagnetic interfering factors, such as external magnetic field, magnetic leakage and so on. The trend of simulation and experimental model can be consistent. The reliability of dynamic magnetic coupling model of high-speed motorized spindle is verified.

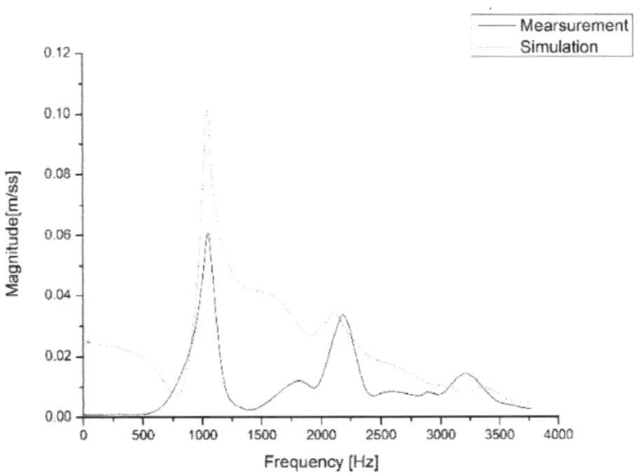

Figure 6. The diagram of natural frequency response function.

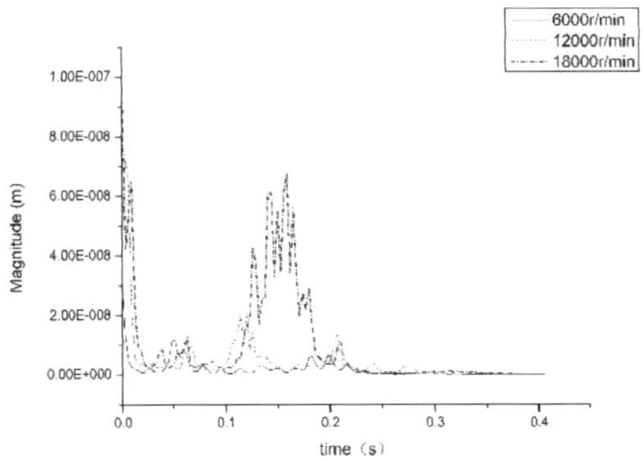

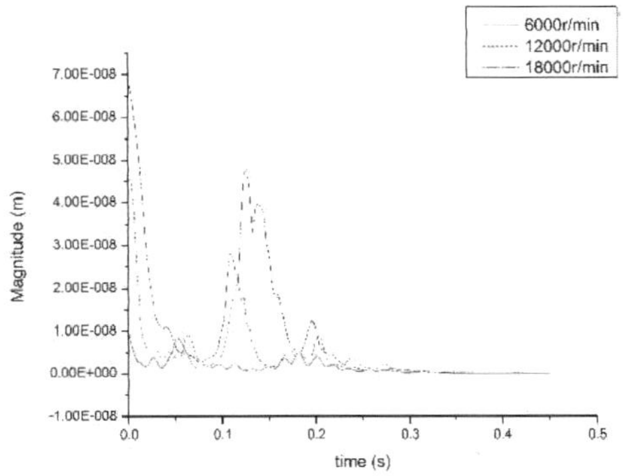

(a) The diagram of test vibration displacement (b) The diagram of simulate vibration displacement

Figure 7. The vibration displacement diagram of motorized spindle.

5. Conclusion

(1) The magnetic flux density and electromagnetic energy of high-speed motorized spindle stator and rotor are periodic distribution with time. The influence of the skin effect of the magnetic field and air gap eccentricity on the distribution of magnetic field is obvious. Unbalanced magnetic tension formed by gas gap eccentricity has a direct connection with the radial vibration of the motorized spindle.

(2) During the high-speed operation of the rotor and the inner ring of the motorized spindle, the centrifugal force makes the contact deformation of the angular contact ball bearing outer ring and outer ring groove increase. The contact between inner ring and inner race decreases, the inner ring of the bearing expansion is deformed, and the expansion of the inner ring is larger than that of the rotor. The interference between bearing inner ring and rotor decreases, the radial stiffness of bearing decreases with the increase of the rotational speed, and the radial vibration becomes larger.

(3) The electromagnetic field and other parameters are coupled into the dynamic model. The coupling relationship between the dynamic and the magnetic model has been analysed. The influence of multi factors coupling on dynamic vibration characteristics has been studied. The error between simulation results and experimental results is small, which reflects the accuracy of the dynamic magnetic coupling model.

Acknowledgments
The article is supported by Shenyang science and technology program project F16-096-1-00, National Natural Science Fund 51675353 and The innovation team project about the Ministry of Education IRT-15R45.

References

[1] Razban M and Movahhedy M R 2015 A speed-dependent variable preload system for high speed spindle *Prec. Eng.* **40** 182
[2] Matsubara A, Tsujimoto S and Kono D 2015 Evaluation of dynamic stiffness of machine tool spindle by non-contact excitation tests *CIRP Annals-Manuf. Tech.* **64** 365
[3] Özsahina O, Budakb E and Özgüven H N 2015 Identification of bearing dynamics under operational conditions for chatter stability prediction in high speed machining operations *Prec. Eng.* **42** 53
[4] Cao H, Zhang X and Chenba X 2017 The concept and progress of intelligent spindles: A review *Int. J. of Machine Tools & Manuf.* **112** 21
[5] Xiaoan C, Peng Z, Ye H and Jun-feng L 2014 Axial vibration of high-speed motorized spindles *J. of vibration and shock* **33 (20)** 70
[6] Ke Z, Tong L, Huabo D and Yuhou W 2016 A Motorized Spindle Online Dynamic Balance Testing Algorithm *J. of Shenyang Jianzhu Univ. (Natural Science)* **32 (1)** 148
[7] Ke Z, Zhixue L, Lixiu Z, Ying X and Yuhou W 2012 Modeling of High- Speed Motorized Spindle and Design and Simulation Research of Vector Controller Based on Slip Frequency Control *J. of Shenyang Jianzhu Univ. (Natural Science)* **28 (6)** 1114
[8] Lixiu Z, Yuhou W and Jinxiang P 2013 The Method about Stator Resistance Hybrid Intelligent Identification of Motorized Spindle *J. of Shenyang Jianzhu Univ. (Natural Science)* **29 (6)** 1099
[9] Lixiu Z, Ming Y, Yuhou W and Feng L 2016 Multi-field coupled model and dynamic performance prediction for 150MD24Z7.5 motorized spindle *J. of vibr. and shock* **35 (1)** 59
[10] Bo W, Wei S and Bangchun W 2012 The Finite Element Modeling of High-speed Spindle System Dynamics with Spindle-holder-tool Joints *J. of Mech.Eng.* **48 (15)** 83
[11] Zhongkui Z and Lixia C 2012 The Application of Bessel Function in Solving Variable Coefficient Differential Equation *J. of Jiamusi Univ. (Natural Science Edition)* **30 (6)** 926
[12] Xiaofeng W and Qing-shan Y 2008 Coupled bend and torsion analysis of the spatial thin-walled beam using timoshenko theory *Eng. Mech.* **25 (5)** 12

IOP Conf. Series: Materials Science and Engineering **399** (2018) 012028 doi:10.1088/1757-899X/399/1/012028

Solutions for using renewable energy sources to produce electricity and to provide water for domestic supply and irrigation

B Lepadatescu[1] **and A E Dumitrascu**[1*]

[1] Department of Manufacturing Engineering, Transilvania University of Brasov, M. Viteazul Street, No. 5, Romania

[*] E-mail: dumitrascu_a@unitbv.ro

Abstract. Environmental pollution, greenhouse effect, intensification due the wide-ranging of organic fuel, as well as care of energy supply to the future generations caused the accelerated development of technologies oriented to the use of renewable energy sources. Some of these technologies are shown in the paper where some examples of the methods to produce electricity and to provide water for irrigations in the places where there is no access to a large electricity grid like small isolated communities, remote consumers, rural villages, inquiring centers, light houses, inaccessible farms, shelters, telecommunication stations are presented. In order to produce electricity a solution of a wind turbine with vertical axis helical rotor shape with good energetic performances is presented. Also, in the paper some considerations regarding the use of renewable energy sources are presented in order to pump water that can be used in different ways including supply water for domestic needs or for irrigation in small-scale area of lands. In this direction two solutions of using renewable energy sources with a wind turbine to pump water for houses and to pump water that is used to irrigate land of small farmers are shown.

1. Introduction

All around the world there is a number of small isolated communities, like islands and rural villages without access to a large electricity grid. The absence of an electrical network in their major area or the prohibitively high connection cost, force these remote consumers to cover their urgent electrification needs to use small diesel electric generators with all the problems of pollution of the environment and noise.

The development of renewable energy sources is a central aim of European Union energy policy reflecting the clear benefits that clean, sustainable and secure energy supplies will bring to current and future generations of Europeans, as well as other people worldwide. Clean-because they reduce the pollution that scars our cities and countryside. Sustainable –because they are renewable and do not contribute to the buildup of greenhouse gases that cause climate change. Secure-because they are sourced within Europe, not imported, and thus they reduce our dependency on events elsewhere in the world that it cannot control [1].

Because the desire was to build a wind turbine which was meant to pump water, it is necessary to know first the percentage of renewable energy sources in total consumption and the energetic potential of renewable energy sources.

Wind turbines are classified as vertical or horizontal axis machines depending on the axis of rotation of the rotor. The energy available in the wind is proportional to the cube of the wind speed: if the wind speed is doubled, the available power is eight times higher [2].

Wind turbine has many applications as:

- stock water – sheep, cattle, horses;
- domestic applications – household water for drinking, sanitation, and washing;
- irrigation – for house and vegetable gardens;
- filling dams, reservoirs, fish ponds, lakes;
- dewatering wet lands;
- water for small industry;
- waste water removal;
- sewage;
- remote locations;

The selecting of a wind turbine should be based on the next criteria: wind conditions, source of water supply, water requirements, total pumping head, and governing.

As regards the wind conditions, the right combination of wind turbine and pump is one which will work easily in light winds (about 10 km/h). This will also depend on the total pumping head [3].

The use of wind energy is an important issue for Romania to meet the target of the Renewable Energy Fact Sheet, with 24% renewable energy consumption by 2020 (figure 1), [4].

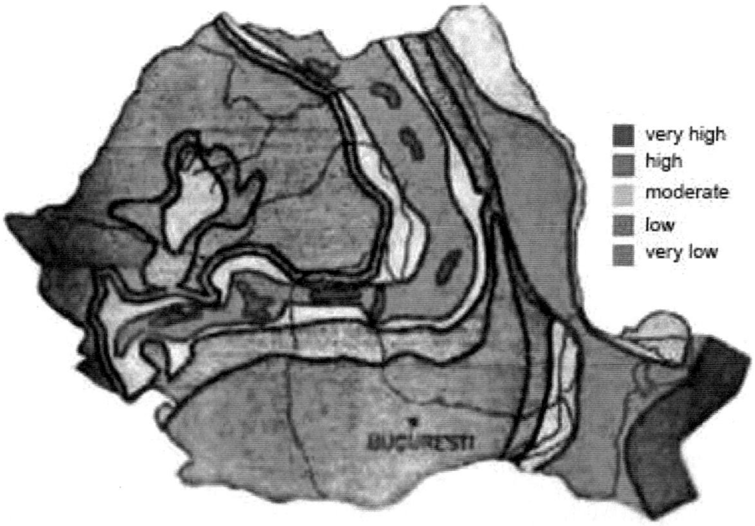

Figure 1. Chart with energetic potential wind of Romania [kWh/m^2/year].

2. Wind turbine with vertical axis helical type

For the remote locations, and when it is desired to be independent in terms of the electric energy source, a wind turbine with vertical axis helical type was designed. This type of wind turbine is easy to mount on the roof of a house, having the main advantage of not necessitating to be pointed into the wind direction with a system as other types of wind turbines. In the same time it works without any noise and the shape of the rotor make these turbines inoffensive for the birds.

The rotor of this wind turbine was made by the FINEX Company, patented in Romania, and it is with three blades with fiber glass material. The main advantages of this type of wind turbine with helical type rotor whose main rotor shaft runs to the flow streamlines, from other type of wind turbines with horizontal axis are [3, 5]:

- simplicity in construction and good strength;
- high reliability;

- smaller cost by 20 % compared to similar turbines;
- specific power bigger on the active surface;
- big torque moment at starting;
- at the wind speed bigger than 20 m/sec it is self-braking without mechanical components, due to its original shape of rotor;
- doesn't need orientation after the wind direction;
- it can work at high wind speed, such as 50 m/s;
- it is the only one wind turbine that is accepted by environmental agencies, because it doesn't kill birds;
- it doesn't make noise during its functioning;

The rotor was mathematically designed and tested in experimental conditions on the car adapted as a mobile laboratory, where the next parameters were measured:

- wind speed [m/s];
- rotational movement of rotor shaft [rpm];
- generator power [W];
- air temperature [^{0}C].

This rotor was made after long-lasting tests for the wind turbine at small power of 0.5-2 kW. For the determination of the maximum theoretical power that can be extracted from a laminar air flow, a wind turbine rotor with a horizontal axis subjected to an air flow shown in figure 2 is taken into account [5].

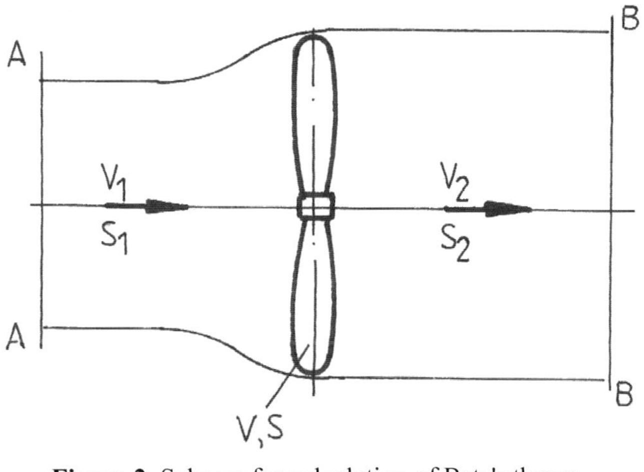

Figure 2. Scheme for calculation of Betz's theory.

where: v_1- air flow velocity in A-A section;
S_1 – size of A-A section;
S_2 – size of B-B section;
V_1 – air flow velocity through A-A section;
V_2 – air flow velocity through B-B section

Producing energy in rotor makes that kinetic energy of the air mass which goes out from rotor to be smaller than that that enters the rotor, thus $v_2 < v_1$. it can be considered that air is not compressible and the results are:

$$S_1V_1 = S \cdot V = S_2V_2 = Q \tag{1}$$

If the force which interacts between rotor and air flow is evaluated, it can be written:

$$F = \rho \cdot Q(V_1 - V_2) = \rho \cdot S \cdot V(V_1 - V_2) \tag{2}$$

The power absorbed by the rotor is:

$$P = F \cdot V = \rho \cdot S \cdot V^2(V_1 - V_2) \tag{3}$$

$$P = \rho \cdot S \cdot V^2(V_1 - V_2) = \frac{1}{2}\rho \cdot S \cdot V \cdot (V_1^2 - V_2^2) \tag{4}$$

From where will result:

$$V = \frac{V_1 + V_2}{2} \tag{5}$$

It is possible to write F and $P = f(V_1, V_2)$:

$$F = \rho \cdot S\frac{(V_1+V_2)^2}{4}(V_1 - V_2) = \frac{1}{2}\rho \cdot S(V_1^2 - V_2^2) \tag{6}$$

$$P = \rho \cdot S\frac{(V_1+V_2)^2}{4}(V_1 - V_2) = \frac{1}{4}\rho \cdot S(V_1^2 - V_2^2)(V_1 + V_2) \tag{7}$$

To determine the maximum of the function related with V_2:

$$\frac{dP}{dV_2} = 0 \Leftrightarrow V_1^2 - 2V_1V_2 - 3V_2^2 = 0 \tag{8}$$

With the solutions:

$$V_2 = -V_1 \tag{9}$$

$$V_2 = \frac{V_1}{3} \tag{10}$$

From where the value is taken $V_2 = \frac{V_1}{3}$

So,

$$P_{max} = \frac{1}{4}\rho S\left(V_1^2 - \frac{V_1^2}{9}\right)\left(V_1 + \frac{V_1}{3}\right) = \frac{8}{27}\rho S V_1^3 \tag{11}$$

For air density $\rho = 1.25$ kg/m^3 results: $P_{max} = 0.37SV_1^3$.

The power factor of the turbine is determined by:

$$C_p = \frac{P}{\frac{1}{2}\rho S V_1^3} \tag{12}$$

The maximum power coefficient will be given by:

$$C_{pmax} = \frac{P_{max}}{\frac{1}{2}\rho S V_1^3} = \frac{\frac{8}{27}\rho S V_1^3}{\frac{1}{2}\rho S V_1^3} = \frac{16}{27} = 0{,}593 \tag{13}$$

It can be concluded that the maximum power to be obtained by means of a wind rotor in relation to the power of a free air flow passing through the same surface is approx. 0.6. This figure is called Betz's coefficient and it represents a maximum theoretical limit.

In practice, this coefficient decreases and values of good values of $C_p = 0.4$ are considered. An important parameter defining a wind turbine is the specific speed coefficient that varies according to the number of blades: 1 blade, $\lambda = 9$; 2 blades, $\lambda = 6$; 3 blades, $\lambda = 5$; 10 blades, $\lambda = 3$).

In the graph shown in figure 3, the variation of the power coefficient, C_p, is given according to the parameter λ, called the specific speed coefficient, where R is the radius of the rotor, ω is its angular velocity, and v is the wind speed. From this figure it can be seen that the turbine with blades has the parameter C_p, close to that of the Savonius turbine. For λ closely related values, it is noted that the turbine with blades has the higher C_p power factor, and besides this advantage, the cost price is lower compared to the Savonius type rotor due to lower quantities of materials that are part of the component.

The wind turbine, made in the experimental variant, is placed after testing at a good power coefficient C_p= 0.4 and the parameter λ has the value of 3.5 making the parameters particularly good for this new type of rotor [6].

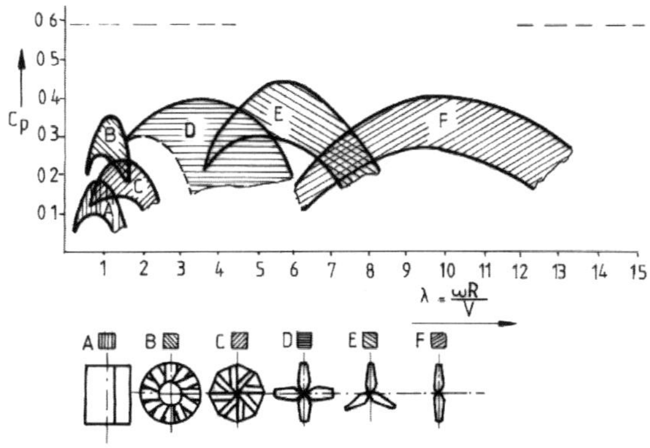

Figure 3. The variation of the power coefficient C_p according with the parameter λ.

The turbine is acting a current generator with permanent magnets (PMG) that has the characteristics shown in figure 4.

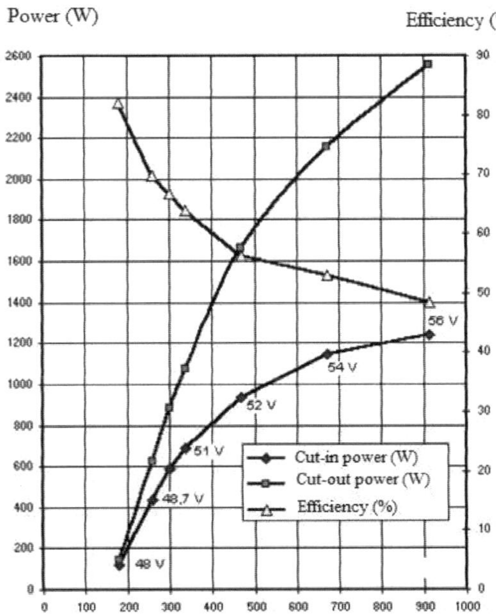

rot/min	P-IES (W)	P-IN (W)	EFIC %
181	118	143	82
260	438	628	70
301	590	885	67
336	689	1077	64
467	936	1663	56
673	1145	2157	53
913	1238	2555	48

Figure 4. The characteristics of the current generator type PMG 600.

The rotor was mathematically modeled and tested in experimental conditions mounted on a car adapted as a mobile laboratory, where the next parameters were measured (figure 5), [6]:

- wind speed [m/s];
- spindle revolutions [rpm];
- power of generator [W];
- air temperature [^{0}C].

IOP Conf. Series: Materials Science and Engineering **399** (2018) 012028 doi:10.1088/1757-899X/399/1/012028

Because the rotation of this type of rotor is maximum 120 m/min, it is necessary to use a multiplicator of rotation, which ensures an optim regime in function, for the current generator with permanent magnets. The current generated is threephasic, and it needs to use a rectifier to accurately measure the current tension at different rotations of spindle.

For the values obtained, wind speed [m/s], spindle speed [rpm] and current tension of the generator, it is possible to calculate the main parameters of the turbine. These main parameters are shown in the graph of power variation [W] according to the wind speed [m/s] (figure 6). In order to do experimental data processing the Excel Program was used.

Figure 5. Wind turbine with vertical axis helical type mounted on a car during experimental tests.

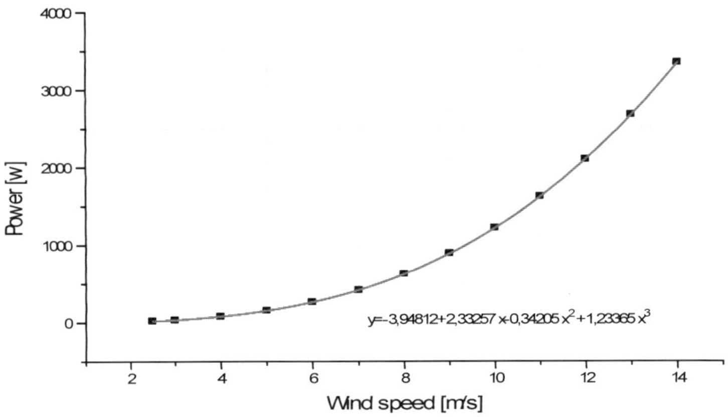

Figure 6. Graphic representation of power variation according to the wind speed.

This graph is compared with the values obtained by other companies which produce wind turbine with vertical axis and similar power.

After the experimental tests the formula of power given by this wind turbine was obtained:

$$P = 0.5 \cdot \rho \cdot C_p \cdot A \cdot \omega^3 \ [\text{W}] \tag{14}$$

where:

P – power obtained [W];

ρ – air density (1.225 kg/m^3);

C_p – power coefficient;

A – rotor area ($4m^{2)}$;

ω – wind speed [m/s].

The power coefficient obtained was $C_p = 0.33$.

Experimental tests of producing electricity using energy of wind with turbine vertical axis with two rotors helical type were also made. In these tests the values of electricity produced increased nearly two times comparing with the case when the turbine had one rotor (figure 7) [6].

Figure 7. The experimental tests of two rotors helical type on vertical axis wind turbine.

After many experimental tests this wind turbine was mounted in different places where the wind has a good energy during the days and it was observed how it works and the level of electric current that was produced. One of these places is shown in figure 8, where the wind turbine with vertical axis was placed in a garden.

Figure 8. Wind turbine with vertical axis helical shape mounted in a garden.

3. Renewable energy sources to provide water for domestic supply

The wind is a natural resource that is available everywhere. As a general rule, the wind blows for between 8 to 10 hours per day, at a speed that is useful for the wind turbine. Wind turbine will be useful for pumping water for stock and domestic supplies in isolated rural areas. It can be an economical alternative where:

- wind conditions are reliable;
- unattended pumping is required for long periods;
- there is no other viable power source;
- the user requires environmentally clean power.

The principal advantages of using a wind turbine are outlined below:

- run unattended for long periods;
- low maintenance;

- suits isolated locations;
- no energy costs;

Some of their disadvantages are:

- high capital costs;
- intermittent pumping in very light winds;
- requires auxiliary storage;

The idea was to build a wind turbine which is used to pump water for domestic supply necessities in a holiday location where there is a number of 10 houses. The holiday place is near a river and they intend to use the water from the river for their daily needs.

In Figure 9 the principle of the idea of taking water from a river and to send it to the tank from where it can be used anytime is shown.

The main elements of the wind turbine used to extract water from the river shown in Figure 9 are [7]: 1-the water tank; 2-wind turbine body; 3-diaphragm pump; 4- pipe system; 5-the river.

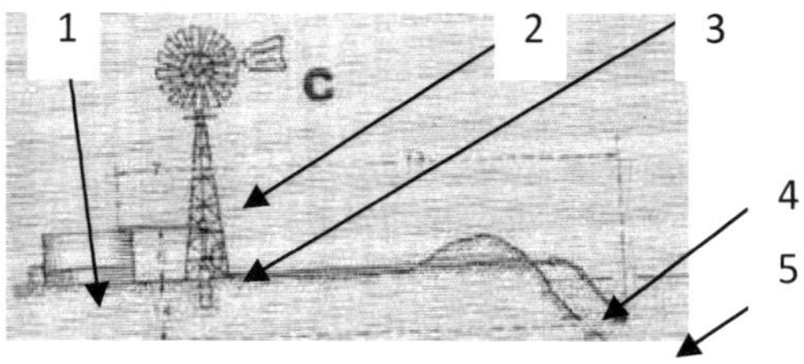

Figure 9. Principle of wind turbine to pump water.

Table 1. Determination of the necessary of water volume.

Water use	Estimated daily use (Litres)
Domestic–drinking (per person)	2
Domestic–hygiene, bathing (per person)	10
Domestic–toilet (per person)	10
Domestic-laundry (per person)	15
Domestic-household cleaning	10
Swimming pool	750
Machinery wash-down	100

To achieve the needs regarding the domestic water supply a wind turbine with horizontal axis which is presented in figure 10 was designed and realized, [7]. The main parts of the wind turbine are: 1- the rotor with blades; 2 - the tower or mast made by tubular steel with two sections of pipe coupled together; 3 -the guy wires attached to the tower in three points and with an underground concrete anchor for each set; 4 - the diaphragm pump; 5 - the concrete base.

The rotor is made by 12 blades and it has a diameter of 1.5 m. The height of the entire turbine is 6 m. The wind turbine is located at 5 m distance from the river whose water was supposed to be used for domestic needs.

IOP Conf. Series: Materials Science and Engineering **399** (2018) 012028 doi:10.1088/1757-899X/399/1/012028

Figure 10. A wind turbine with horizontal axis to pump water.

To pump the water a diaphragm pump shown in figure 11 was chosen [7]. The pump is acted by a cable (4) that has a reciprocating motion which is received from the rotor mechanism. The rotor mechanism transforms the revolution motion into reciprocating motion through a cam. In a cam operated wind turbine, the lift of the water occurs during more than half the cycle. Using a cam that allows for three-quarters of the cycle lifting and one-quarter return, starting torque is reduced by 60 percent. The wind turbine will start in light winds, and if combined with counterbalancing, starting torque is reduced by 72 percent.

A diaphragm pump (5) is a pump that uses a combination of the reciprocating action of a membrane made by rubber and non-return check valves to pump water. When the volume of a chamber is increased – the diaphragm moving up- the pressure decreases, and fluid is drawn into the chamber. When the chamber pressure increases from decreased volume – the diaphragm moving down- the fluid previously drawn in is forced out. After that, the diaphragm is moving up again, it draws fluid into the chamber, completing the cycle.

The entire perimeter of the diaphragm is tightly held and sealed in place between the side of the displacement chamber and a flange, which is tightly bolted to the chamber. Each displacement chamber has two valves, usually spring-loaded ball valves made from the same material as the diaphragm. One of the valves admits the water being pumped to the diaphragm chamber, and the other allows the material to exit the diaphragm chamber. It is a driving mechanism that flexes the diaphragm in and out of the chamber to suck in and force out the water. The water is taken from the river through the device 7 and is sent to the water tank by the flexible pipe 3. The body of the pump 7 is connected to the tower 1 by the plate 2 by a bolt and bush-bearing which helps when the mast is erected in position.

Figure 11. Diaphragm pump used to pump water.

4. Water wheel powered by hydro resource used for irrigation

In figure 12 a constructive solution of a water mill used to take the water from a river and to send further to a reservoir is shown [7]. The main elements of this water mill are: the primary reservoir (1), the pipe (2) from where the water is conducted to the main reservoir; the rotor axis (3); the screws (4) that permit setting of the water mill in a vertical plan; the counterweight (5); the concrete pillar (6) that supports all the construction; 24 collecting tubes (7); the rotor ring (8) where are mounted the collecting tubes.

Water mill has a diameter of 4000 mm which allows lifting water from the river at a height of 3500 mm, where it is stored in a primary reservoir 1. From the primary tank (1), water is passed through the pipe 2 to the main tank located at a distance of 2 m from the concrete pillar (6). Rotor ring (8) is made of two rings of steel sheet of thickness 0.5 mm and is connected to the main shaft (3) by means of profiled spokes. On the two rings of the rotor 24 tube collectors, made by polypropylene, with a length of 500 mm and diameter of 75 mm are fixed. The impeller is submerged in the river water at a depth of 200 mm. This allows complete filling the tube collector with water from the river. Turbine construction allows adjustment of horizontal and vertical plan to a range of 150 mm which allows being adapted to specific conditions of the installation in the site near the river.

In the period between harvesting crops and starting a new season in agriculture, the turbine is decommissioned by tilting it 20 degrees. In this mode the life of the plant is extended and it reduces maintenance costs. Keeping the turbine in this inclined position is realized by a lock system [4, 5].

Turbine works without interruption throughout the period between the start of the season in farming and till harvesting period ends. Throughout this time water is stored in the main tank, and then is distributed through the pipes to areas to be irrigated. Irrigation system is different from one culture to another depending on the terrain where the turbine is installed.

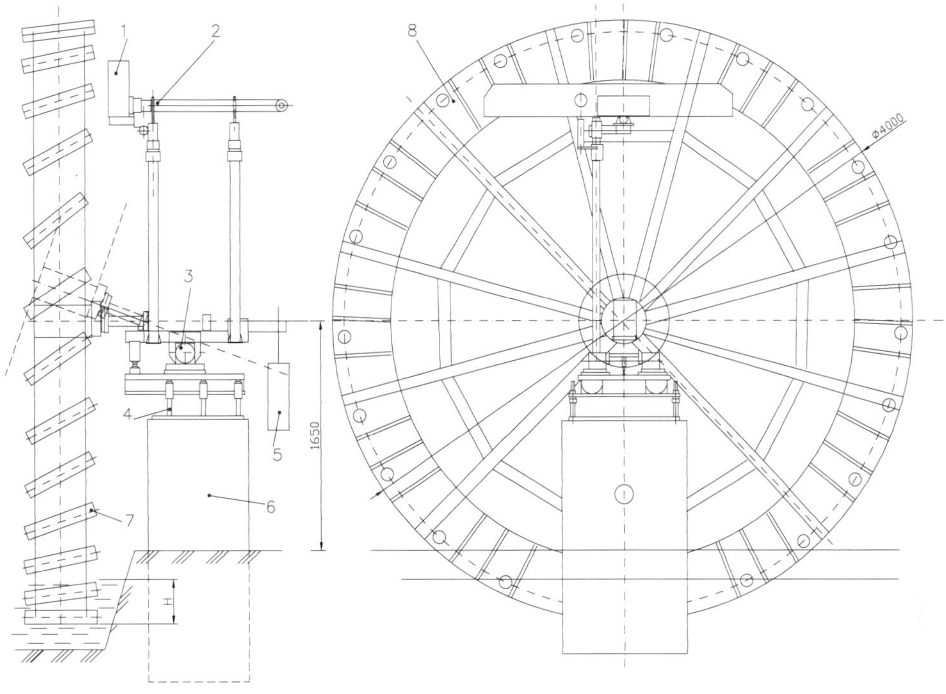

Figure 12. Schematic illustration of a water mill used for irrigation.

To calculate the turbine parameters are used figure 13. This figure shows schematically the position of the collector tube fixed on the rotor in the time when its upper end comes out from the water [7].

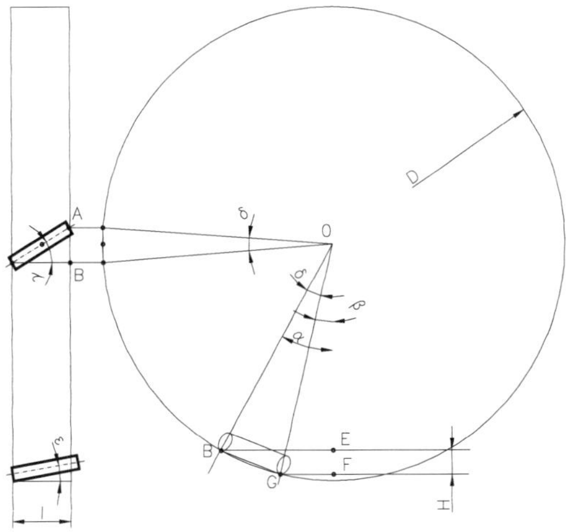

Figure 13. Sketch used to calculate the water mill parameters.

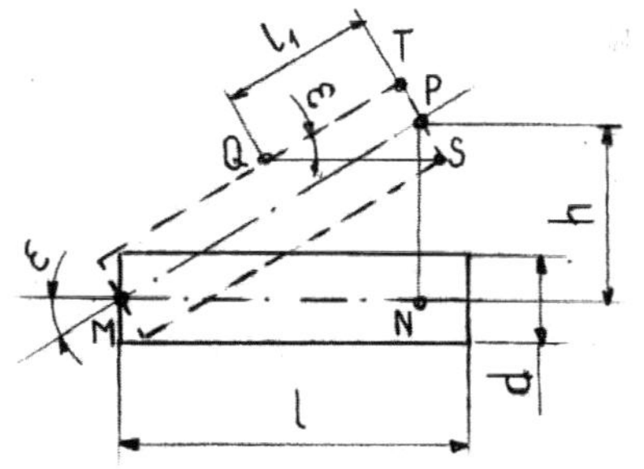

Figure 14. Calculation of the amount of water carried by the tube collector.

At this point, the collector tube is almost full with water and continues the lifting movement once the rotation of the rotor. Water collection tube is mounted on the rotor's circumference at an angle γ.

Due to tube collector mounting angle, water can never fill its entire cavity. To calculate the amount of water that is transported by one collector tube figure 14 is used [7].

After a series of calculations equation (15) is obtained which shows the amount of water that is taken from the river by one tube collector. Since the rotor has 12 tube collectors, this value must be multiplied by 12 to determine the amount of water that comes to one rotation of the rotor.

$$V_2 = \frac{\pi d^2}{4}\left(l - \frac{d\sqrt{4l^2-\{D[cos(\alpha-\delta)-1]+2H\}^2}}{2\{D[cos(\alpha-\delta)-1]+2H\}}\right) \tag{15}$$

Equation (15) gives the amount of water that is taken from the river by one tube collector. Since the rotor has 12 tube collectors, this value must be multiplied by 12 to determine the amount of water that comes from one rotation of the rotor.

Below is the calculation of water volume that is taken from the river for the next elements of the rotor:

l = 500 mm; D = 3900 mm; H = 200 mm; $\gamma = 45^0$; d = 75 mm.

Using formula (15) results:

$$V_2 = \frac{\pi \cdot 75^2}{4}\left(l - \frac{d\sqrt{4\cdot 500^2-\{3900[cos(26^010'-10^020')-1]+2\cdot 200\}^2}}{2\{D[cos(26^010'-10^020')-1]+2\cdot 200\}}\right) \approx 1{,}57\ l \tag{16}$$

So, the amount of water that will be taken from the river and transported in uppermost position during one rotation of the rotor will be: 1.57x12=18,84*l*.

An important factor for plant efficiency is the speed of the water of the river. As the speed of the water is higher, the amount of water extracted by the rotor will be greater and therefore, higher plant efficiency.

Number of revolutions of the rotor in a minute n_{min} is given by:

$$n_{min} = \frac{v_W}{\pi \cdot D} \tag{17}$$

Where v_w is the velocity of water's river, and D is the diameter of the tube collector disposed on the circumference of the rotor.

Volume of water extracted from the river in one minute V_{min} is given by:

$$V_{min} = n_{min} \cdot V_2, \tag{18}$$

where V_2 is the volume of water extracted from the river by the rotor in one rotation given by equation (15).

Relation (17) can be written:

$$V_{min} = C \cdot v_W, \tag{19}$$

where C is a coefficient given by: $C = \frac{V_2}{\pi \cdot D}$

Coefficient C was calculated for the following angles γ of mounting tube collectors on the rotor circumference, and volume of water V_2 extracted from the river water at one rotation (Table 2).

Table 2. Coefficient C depending on the angle γ and the water volume V_2.

γ	15^0	20^0	25^0	30^0	45^0	60^0
V_2	14.64	24.20	27.12	36.24	37.68	40.08
C	1.196	1.979	2.215	2.959	3.077	3.273

Using C coefficient values (table 1), the volume of water extracted from the river in one minute V_{min}, can be calculated depending on the speed of river v_w for different mounting angles γ of the collector tubes on the rotor circumference.

Having all these equations it is possible to calculate the volume of water extracted from the river one day (V_d), depending on the angle of installation of the collecting tubes on the rotor circumference (angle γ) and the speed of the river water (V_w) (figure 15).

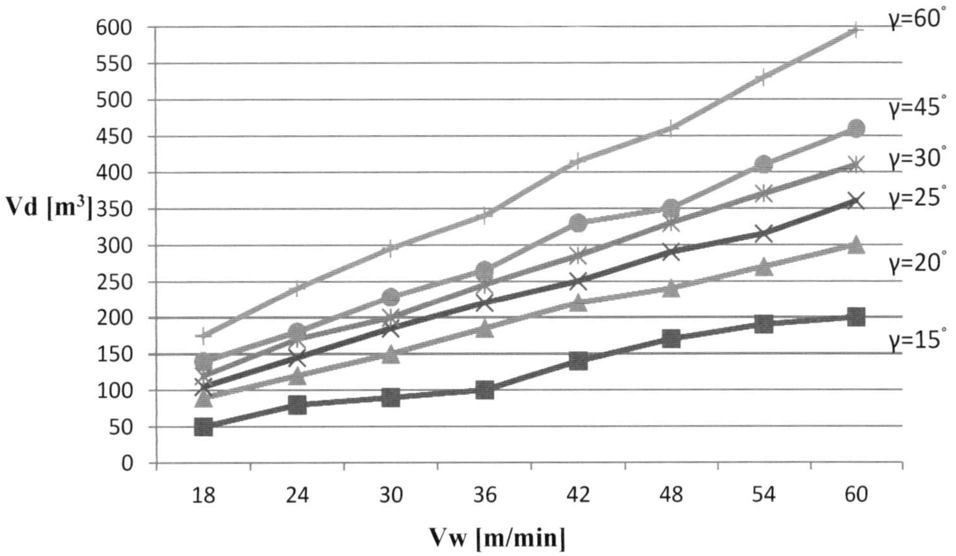

Figure 15. Volume of water extracted from the river in one day (V_d), depending on the angle of collecting tubes on the rotor circumference (angle γ) and the speed of the water' river (V_w).

In figure 16 the undershot water wheel that was made to irrigate a land of farm from a small river with the dimensions: width 2 m, depth 0.5 m and the speed of the water is 35 mm/min is presented. The rotor has a diameter of 4 m, [7].

Figure 16. Undershot water wheel used for irrigation of dry lands.

5. Conclusions

All the experiments presented in this paper are personal contribution of the authors to the field of renewable energy sources.

The main contributions consist of:

- design and testing of the wind turbine with vertical axis helical type;
- determining the power variation of wind turbine with vertical axis helical type according to the wind speed;
- the experimental tests of two rotors helical type on vertical axis wind turbine;
- the experimental tests of wind turbine with vertical axis helical shape mounted in a garden;
- the experimental tests of wind turbine to provide water for domestic supply;
- design and testing of water wheel powered by hydro resource used for irrigation by determining the volume of water extracted from the river in one day depending on the angle of collecting tubes on the rotor circumference and the speed of the water's river.

The renewable energy sources have clear benefits that distinguish them from fossil fuels. Renewable energy sources can be exploited in more ways and in more places than fossil fuels. They are environment – friendly origin, being known as "green electricity". That is way the wind energy was used to pump water in a location that is remote from electricity grid and that is unattended for long periods.

One of the advantages of using this type of irrigation system is to function as a green technology concept having no negative effect on the environment. It has a big impact on the rural economy, increasing the productivity of the cultivated fields. The development of energy systems based on water energy resources is vital to enhance the livelihood of people and it has a great potential in terms of economic impact due to its low cost, sustainability and energy renewability.

In country agriculture, water use is of great importance to obtain yields proportionate to the population. Drought is a natural very harmful factor that affects a large agricultural area from the total agricultural area of a country. Drought effect is observed mainly in the plains and hills with a moisture deficit and uneven water distribution cultivated area.

Considering the technical information from the literature in the field of renewable energy sources, the main purpose of irrigation is to supply water during the growing agricultural areas located in dry areas or in areas with unfavorable distribution of rainfall. If there are larger surface that need irrigation multiple installations mounted on the river can be used. Some other advantages of using these types of installations for the irrigation of dry areas are:

- it improves the general health of the villages;
- zero pollution;
- it operates 24 hours a day, 7 days a week without supervision;
- no fuel or electricity cost;
- low maintenance and repair cost;
- repairs are done locally;
- installation is up to 80% cheaper than other water system models;
- local manufacturing and training generates employment.

Acknowledgement

This paper is based on the international project SEE - EEA Grants – no. 3BIL/29.08.2016.

References

[1] Lopez A 2014 Water Related to Biorefinery and Agriculture *Conference on the Indo-European Research and Innovation Partnership*, Brussels

[2] Manwell JF 2002 *Wind energy* (New York: Wiley).

[3] Thierheimer W Wet al. 2010 Reducing Environmental Pollution from Mobile Sources *Environmental Engineering and Management Journal* Iasi

[4] Energy is life. Renewables Consulting, Available at: http://renewables-invest.com/index.php?p=1_19_Energie-eoliana

[5] Hofkes EH and Visscher J T1986 *Renewable Energy Sources for Rural Water Supply in Developing Countries,* International Reference Centre for Community Water Supply and Sanitation. The Hague, The Nederland

[6] Lepadatescu B, Mugea N, Dumitrascu A-E and Enescu I 2011 Studies Regarding Designing and Testing of a Multicolor Helical Wind Turbine with Vertical Axis for a Power of 1000 W. *Recent Advances in Fluid Mechanics and Heat & Mass Transfer*, p. 43-46, Florence, Italy, August 23-25, 2011

[7] Lepadatescu B and Ghimbaseanu I 2016 Study regarding the use of wind energy in the area of water supply, *Recent Advances in Energy and Environmental and Biological Sciences*, available at: http://www.wseas.us/e-library/conferences/2016/barcelona/EAB/EAB-26.pdf

[8] Prakha M 2014 Sustainable *energy solutions for irrigation and harvesting in developing countries*, PhD thesis, California Institute of Technology, Pasadena, California

[9] Watt S B 1995 *Manual on the Hydraulic for Pumping Water* (Intermediate technology publication, London)

Research on multi-agent control system for concrete distribution

D Li[1,2], P Zhou[1,2], L T Fan[1,2*], D F Zou[1,2], W D Yu[1,2], J J Guo[1,2] and Q Y Wang[1,2]

[1] School of Mechanical Engineering, Shenyang Jianzhu University, SY 110168,China
[2] National-Local Joint Engineering Laboratory of NC Machining Equipment and Technology of High-Grade Stone, Shenyang Jianzhu University, SY 110168,China

[*] E-mail: fanliting@sjzu.edu.cn

Abstract. A concrete distributor is the key equipment in the production of precast concrete, abbreviated PC, components. Due to the lack of the automatic and intelligent distribution control system for the most domestic concrete distributor, the quality index is poor and the efficiency is low in the concrete distributing process. In order to solve the above problem, the method of constructing a control system is proposed in this paper. This method is based on the theory of multi-agent and the production technology of PC components. The model structure of complex control systems of multi-agent is studied, and the partition mechanism and coordination mechanism of multi-agent for concrete distribution are analysed. Then, the new model of a multi-agent control system for concrete distribution is designed, and the simulation platform of the system is built. The model of the multi-agent control system for concrete distribution designed is beneficial to theory research in the process control of concrete distribution and its application in the field bus control system. Therefore, the foundation for realizing the automation and intelligence of production in concrete distributing is established.

1. Introduction

In the concrete distributing production link, precast concrete (abbreviated PC) components are mainly produced with specialized and industrialized production methods by using the concrete distributor. Abroad, the control system of the concrete distributor is usually designed with the field bus control system [1]. For the control system, it is not enough to realize the digitization in the form, but also to pursue the deeper automation and intelligence in the structure of the control system, integration strategies and control methods. However, in our country, manual control of the concrete distributor is widely used and the distributing process is too dependent on artificial auxiliary distribution. These artificial factors lead to poor quality index and low efficiency in the concrete distributing production. Therefore, a control system with a digital, automatic and intelligent function for the concrete distributor is urgently needed to meet the quality stability and production demand in the production of PC components.

Multi-agent theory is one of the frontier directions in the field of artificial intelligence. With this theory, a complex system, such as a software system, a hardware system, or a hybrid system of the above two systems, can be divided into agents of small, interconnected, coordinated and easy-to-manage [2, 3]. These agents constitute a multi-agent system, which is well compatible with the control system, and can significantly improve the intelligence level of the latter. So, multi-agent theory is applied to various fields gradually, for example the rolling process automation, robotics, distribution networks, and so on [4-8]. These application cases provide a solid theoretical support and practical basis for the optimization design and advanced control of complex systems. Therefore, the direction for the intelligent development of the complex industrial process of concrete distribution production is pointed out by the multi-agent theory.

In the production of PC components, in order to realize the automatic and intelligent distribution of concrete distributor, and to ensure the accuracy of PC components in size and weight, it is not only necessary to adopt a variety of setting functions, such as pass schedule planning, discharge quantity prediction, landing point model, screw speed setting and self-learning model, etc., but also to employ multiple control functions at the same time, for example the speed control of the small cart of concrete distributor, the pre-calibration position of the concrete distributor, weight control, thickness control and so on. From the point of view of concrete distribution control, all the above setting functions and the control functions can be regarded as agents of the multi-agent system. With the help of multi-agent technology, the mathematical model, expert system, neural network and so on can be integrated into the corresponding setting functions the control functions. Then, the platform of the multi-agent control system for concrete distribution is formed and the automatic and intelligent control for the whole process of concrete distributing is achieved.

In this paper, the control system model for the production of concrete distribution is established with the multi-agent theory. The model structure, partition and coordination mechanism of the multi-agent system are studied, and the intelligent control system for concrete distribution production based on multi-agent is designed. After that, a multi-agent simulation platform of concrete distribution is built. Then, the partition of the functions, tasks, knowledge and data sources of the control system based on the multi-agent theory is completed. A theoretical and practical foundation for the automatic and intelligent development of concrete distributor is laid.

2. Model structure of control system based on multi-agent

For a complex control system, it is easier to handle, debug, and maintain by breaking it down into multiple subsystems which are relative and independent. After that, the fault tolerance of the control system can be also improved [9, 10]. The multi-agent system is composed of several single agents coordinating with each other, and it is easy to form a corresponding relationship with a complex control system. Therefore, the system design method in the multi-agent system can provide ideas for the decomposition and design of complex control systems [11, 12]. Based on the hierarchical design attributes of the multi-agent, the model structure of a complex control system based on the multi-agent is constructed, as shown in figure 1.

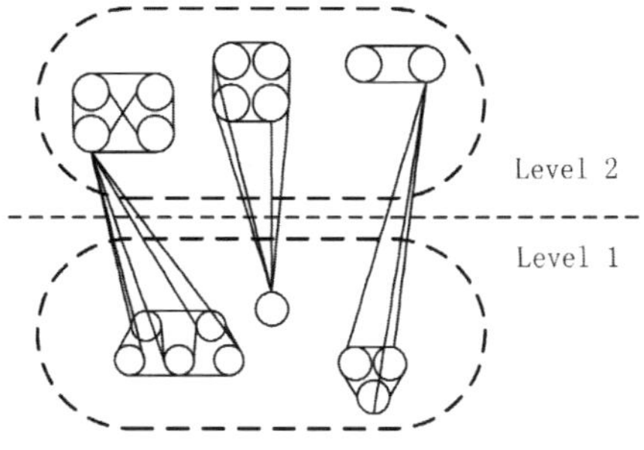

Figure 1. Model structure of a complex control system based on multi-agent

As we can see from figure 1, the multi-agent control system can be composed not only of multiple agents with low capability or single capability, but also of several complex agents, as well as of the two kinds of agents mentioned above. In the two adjacent layers of the multi-agent control system, a unit consisted of multiple agents in Level 2 can be considered as a single entity in Level 1 and a single entity in Level 1 can be regarded as a unit consisting of multiple agents in Level 2.

Based on the model structure shown in figure 1, the multi-agent control system of the concrete distributor is planned according to the control function of the concrete distribution. Level 2 is the control function of the distribution process, including the positioning function, the speed setting function, the weight control function and so on. On the basis of Level 2, partial control functions are refined to get sub-functional agents represented by Level 1. For example, the weight control function can be divided into the sub-functions about calculation, control and setting, including the prediction calculation of the discharge quantity, the calculation of the number of opening doors, the real-time weight control, and the screw speed setting and so on. The positioning function can be divided into the sub-functions about the landing point calculation, the position pre-calibration, etc. In addition, it is necessary to consider the relatively independent control function when the model structure of the multi-agent control system for concrete distributor is established, such as thickness control, model self-learning, pass schedule planning, coordinated control, etc.

3. Multi-agent partition mechanism of concrete distribution

The cooperation of each agent is required in order to realize the control of complex systems in multi-agent control systems. When a multi-agent system is partitioned, each agent can be matched to an actual physical unit. For instance, in accordance with the physical structure of a concrete distributor, the small cart, the big cart, or each screw can be defined as an agent separately. Alternatively, each agent can also be matched to an abstract unit. For example, according to the function needed in automatic concrete distributing, the speed setting, weight control and thickness control can be designed as an agent respectively. The divided agents can respond reason and make decisions about the environment, solve the given problem and achieve the goal by cooperating with each other.

When a complex system is divided into multi-agent systems, the divided agents can neither be too coarse nor too fine. If the agents are divided too coarsely, the complexity of a single agent increases; If the agents are divided too finely, the single agent will be simplified, but the number of agents will be increased, the system will be complex, and the workload of collaboration will be increased [13]. So it is necessary to consider the complexity of single agent and multi-agent systems in a comprehensive way when a multi-agent control system is divided.

In the control system of a concrete distributor, each function for weight control can be used as an agent and the agents, such as the pass schedule planning agent and self-learning agent, should also be considered. Then, the multi-agent system of weight control is built for automatically and intelligently controlling the distributing weight. The model of the weight control system of multi-agent is shown in figure 2.

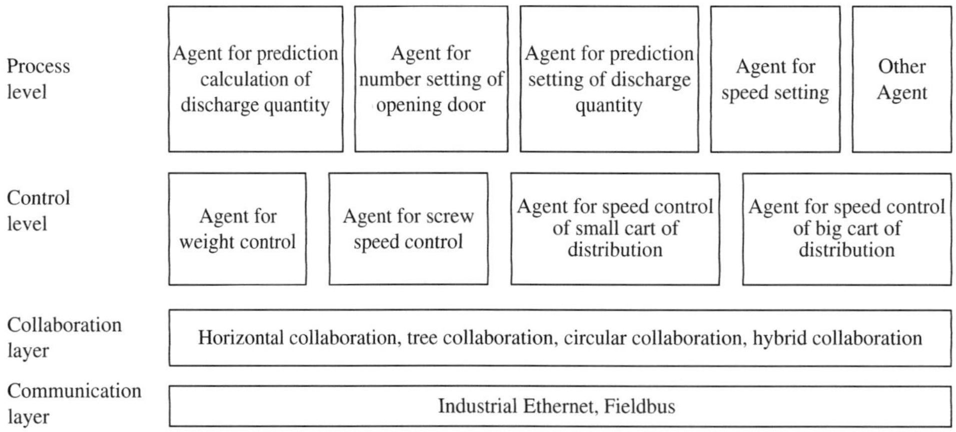

Figure 2. Model of weight control system of concrete distributor based on multi-agent.

4. Multi-agent control coordination of concrete distribution

The process of concrete distribution is characterized by a fast, time variation and high reliability demand. The multi-agent control system for concrete distribution needs to use finite time and conditions to solve the problems about the coordination and cooperation, such as task assignment and action coordination, under this production process.

Based on the current production characteristics of PC components, according to the degree of mutual influence and restriction of each mechanical device, the system should have a cooperation and coordination mechanism of self-adaption, self-learning and fault-tolerant ability in the design of multi-agent control system for concrete distribution. The self-adaptive capabilities are mainly reflected in that the system can coordinate and control each mechanical device independently according to the latest state of concrete distributing. The self-learning ability is mainly reflected in: the system's ability to improve the accuracy of the model according to the real-time feedback information of the production link, such as screw speed, distribution weight, etc. Fault tolerance is mainly reflected in: the remaining agents' ability to work normally to avoid "paralysis" of the equipment when agents of relatively low importance are in trouble.

5. Model of the multi-agent control system for concrete distribution

In the multi-agent control system for concrete distribution production, each agent has its own task. The model of multi-agent control system, which can deal with the problem of distributing process, is constructed reasonably through the division of work and cooperation, mutual exchange and sharing of resource information, as shown in figure 3.

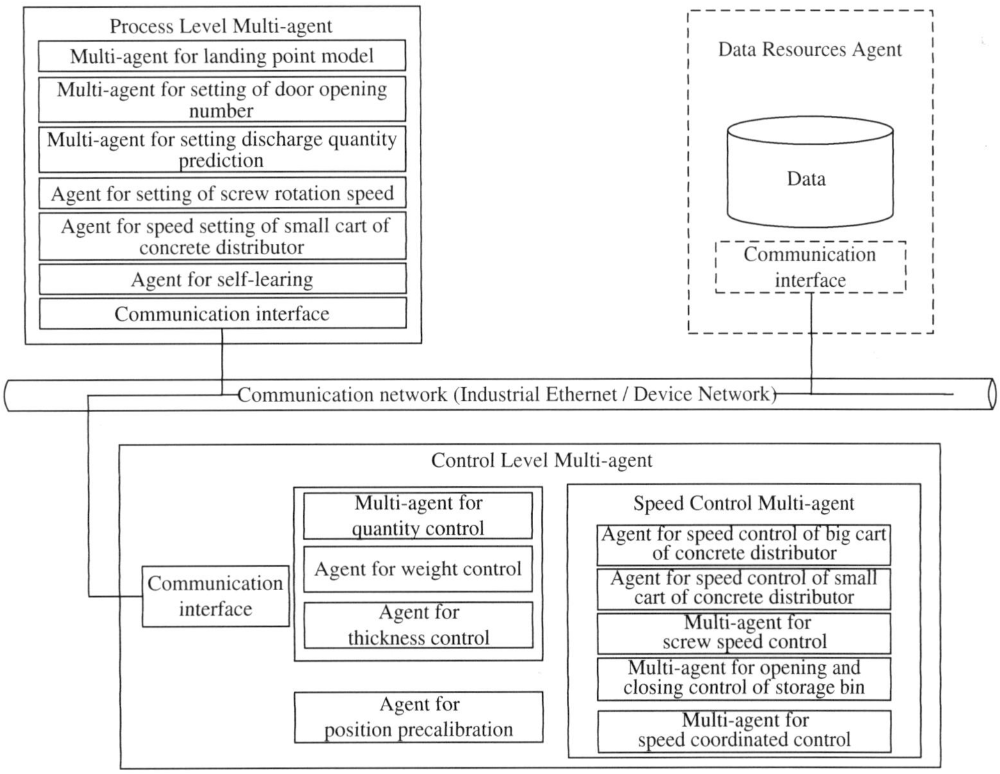

Figure 3. Model of the multi-agent control system for concrete distribution.

The theoretical and practical basis for the design of the control system model shown in figure 3 is provided by both the process engineering of concrete distributing production and the guidance method for designing the multi-agent control system in literature [14-16]. With the help of this model, agents at different network levels can easily communicate and cooperate with each other and successfully complete the corresponding tasks in the distributing process. At the same time, the model also lays a foundation for the realization of the multi-agent control system of concrete distribution.

6. Establishment of the simulation platform for the multi-agent control system

Based on the model of the multi-agent control system shown in figure 3, the simulation platform of the multi-agent control system for concrete distribution is designed, as shown in figure 4.

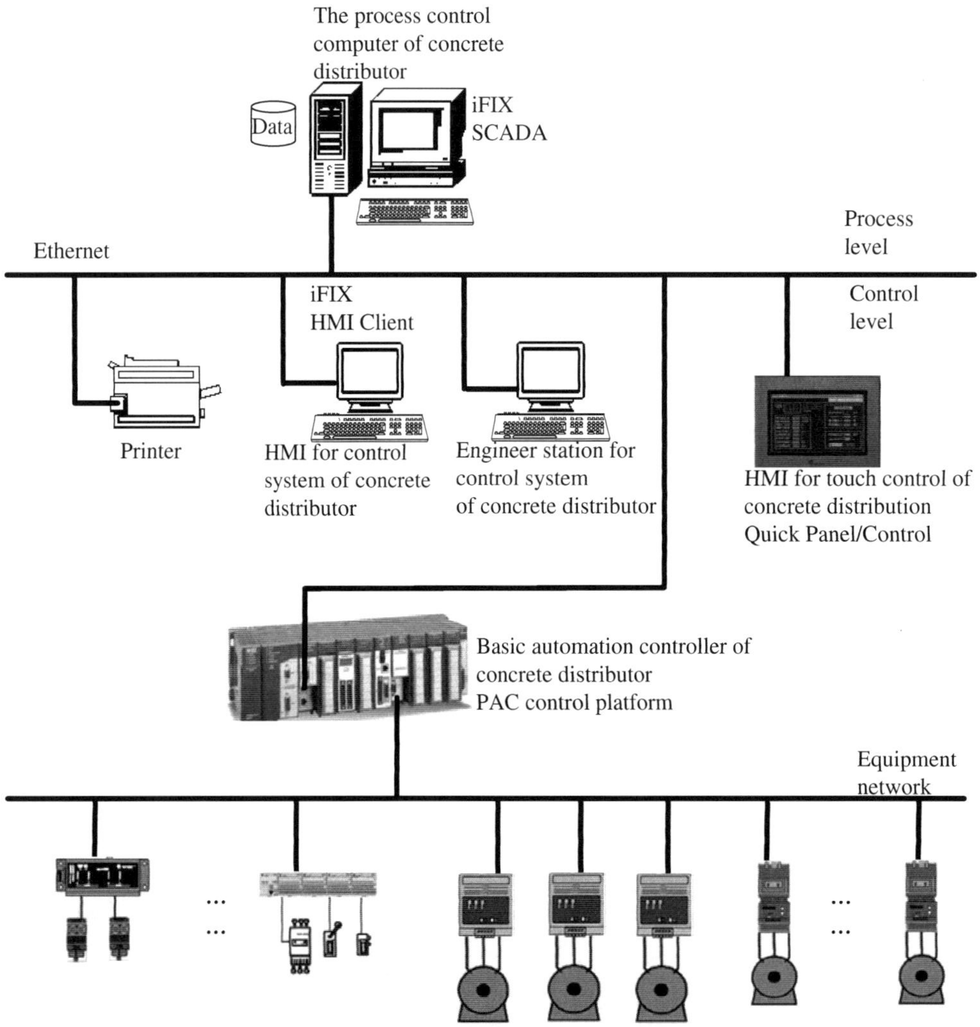

Figure 4. Simulation platform of the multi-agent control system for concrete distribution.

Figure 4 shows that the simulation platform is mainly composed of the process control computer, the human-machine interface, the programmable-automation-controller (abbreviated as PAC) the control platform and the communication network, and it is easily realized with the field bus control system. The process control computer is the realization foundation of the agents with the function of calculation, setting and self-learning. The PAC control platform is the running platform of the agents with autonomous control functions and adaptive functions. The human-machine interface is mainly used to provide the interface between human, machine and platform for monitoring and controlling for the whole system. The Ethernet network is used to establish communication between the process level and the control level. The equipment network is mainly used to connect the controller and the actuator for realizing the control of the controller to the actuator. The simulation platform of the multi-agent control system for concrete distribution is built to lay the foundation not only for further theoretical research and practical application of the process control of the concrete distribution, but also for the intelligent development in the future.

7. Conclusions

With the improvement of customers' requirements for the quality and output of PC components, the automation and intelligence level of the control system for the domestic concrete distributor is difficult to meet the production demand of components. Therefore, the method of constructing control system based on the multi-agent theory is presented, the new model for the multi-agent control system of the concrete distributor is designed, and the simulation platform of the system is built in this paper. The multi-agent control system for the concrete distributor designed in this paper is clear in structure and reasonable in function division. It is beneficial to the theoretical research and application integration of the control function of the concrete distributor, and also to the realization of the automation and intelligence in the process control of concrete distribution production.

Acknowledgments
Project fund: National Key R & D Program of China (2017YFC0704003); General Planning Project of Liaoning Education Department (LJZ2017031); General project of Shenyang Jianzhu University (2017016).

References

[1] Sun H, Sun J and Wu Y H 2015 Brief introduction of large scale intelligent automatic production line of PC components *China Concrete and Cement Prod.* **(3)** 35
[2] Monostori L, Váncza J and Kumara S R T 2015. Agent-Based systems for manufacturing *CIRP Annals-Manuf. Techn.* 55(2) 697
[3] You X, Hua C and Guan X. 2017 Distributed adaptive event triggered control for leader following consensus of multi-agent systems *Asian J. of Control* 19(6) 2155
[4] Zhao J, Zhan H R, Wang X L She G L and Wang W 2008 Integrated scheduling model for continuous casting-hot rolling based on multi-agent *Chinese J. of Sci.Instr.* **29(7)** 1540
[5] Chen Y, Lu J, Yu X et al 2013 Multi-agent systems with dynamical topologies: consensus and applications *IEEE Circuits & Systems Magazine* **13(3)** 21
[6] Wen G, Duan Z, Su H et al 2012 A connectivity-preserving flocking algorithm for multi-agent dynamical systems with bounded potential function *Iet Control Theory & Applications* **6(6)** 813
[7] Sampaio R F, Melo L S, Leão R P S, et al. 2017 Automatic restoration system for power distribution networks based on multi-agent systems *Iet Generation Transmission & Distribution* **11(2)** 475
[8] Kim S, Son Y J, Tian Y et al 2017 Cognition-based hierarchical en route planning for multi-agent traffic simulation *Expert Systems with Applications* **85(1)**335
[9] Dong T T, Li L J, Xiong L and Xv O G 2015 Distributed Predictive Control based on Associated Subsystems *Control Eng. of China* **22(6)** 1201
[10] Zeng Q, Yuan H W, Zhang L, Zhao G H and Zhang X L 2012 Modeling and application of multi-agents orented to test systems *Chinese J. of Sci. Instr.* **33(6)** 1427
[11] Wang X K, Li X and Zheng Z Q 2013 Survey of developments on multi-agent formation control related problems *Control and Decision* **(11)** 1601
[12] Min H B, Liu Y, Wang S C and Sun F C. 2012 An overview on coordination control problem of multi-agent system *Acta Automatica Sinica* **38(10)**1557
[13] Tan S B and Liu J C 2008 Research on multi-agent based control system for continuous rolling process *CAAI Trans.on Intelligent Systems* **3**(2) 150-154
[14] Cai J, Kim D, Putta V K et al 2015 Multi-agent control for centralized air conditioning systems serving multi-zone buildings *American Control Conference* p 986
[15] Liu J K and Wang S Q 1999 Design and realization of multiagent control system *Control Theory And Applications* **16(4)** 580
[16] Huang J G 2006 Design of Multi-Agent Control system for thermal power plant based on profibus *Computer & Information Tech.***(6)** 43

Problems and countermeasures of green building development in cold areas

X Li[1*] and Y Zhang[1]

[1] School of Management, Shenyang Jianzhu University, Shenyang, Liaoning CH032

[*] E-mail: 323165171@qq.com

Abstract: In the process of rapid development of the society, there is inevitably a waste of resources especially for construction industry. Thus, the demand for green buildings increases swiftly. In the cold areas, the winter lasts for a long time, and several months' heating results in the increase of carbon emissions, which has a great impact on our environment, so it is necessary to develop green buildings. Starting from the concept of green building, this paper expounds the development status of green buildings in the cold area, illustrates the importance of developing green buildings in cold areas, and analyzes the problems faced by the development of green buildings in cold areas, and puts forward some suggestions.

1. Green building concepts

In recent years, China's social economy has developed rapidly accompanied by some serious environmental pollution and ecological damage. The premise of sustainable development is to protect the ecological environment and reduce pollution, and green building is the key to achieve the goal of sustainable development. Green building is the environmentally friendly trend for the development of the construction industry [1].

The green building evaluation standard defines the concept of green building as: "green building means to maximize the conservation of resources, environmental protection and pollution reduction, and to provide health for people throughout the life cycle of the building. Applicable and efficient applicable space, harmonious and symbiotic architecture with nature. The term "green" in "green buildings" does not refer to the general sense of three-dimensional greening, roof gardens, but represents a concept or symbol that buildings are environmentally friendly and can make full use of environmental natural resources. Under the condition of not destroying the basic ecological balance of the environment, building can be called sustainable development building, ecological building, returning to nature building, energy saving and environmentally friendly building, etc.

2. The development of green buildings in cold areas

2.1. The development status of green buildings abroad

In the early 1960s, the American architect Paul first put forward the new concept of ecological building, and then it was gradually accepted by various countries, the development of green building becoming a consensus in all the countries. In the mid-1970s, some countries began to implement the building energy saving standards, and gradually improved the energy saving standards, green buildings began to be a government behavior. The convening of the United Nations Conference on

Environment and Development in 1992 resulted in a worldwide consensus on the important idea of "sustainable development". Green Building has gradually become a system and has been popularized in many countries, which has become an important tendency of architecture development in the world. To make the concept of green building practical and operable, developed countries have also developed a green building evaluation system adapted to the characteristics of different countries one after another. Through the quantitative description of energy saving effect, environmental impact and economic performance of green building, a basis for policymakers and designers was provided. The American green building association established and implemented the LEED system in 2003, and now it is one of the most comprehensive criterion for evaluating green buildings. LEED (Leadership in Energy and Environmental Design) is a tool for evaluating green buildings, and some states and countries in the United States have listed it as mandatory standards. The scores required for LEED certification of the corresponding grading scale are shown in table 1.

Table 1. LEED rating scale.

LEED Grade	Score
Platinum Certification	52-69
Gold Medal Certification	39-51
Silver award certification	33-38
Conformity certification	26-32
Unqualified	Below 25

2.2. The development status of green buildings in China

In the last decade, green buildings in our China have developed from less to more, from local to national, from single building to urban area. Since the release of "green building action plan" in 2013, governments at all levels have constantly issued incentive policies for green building development [2]. The number of green signs in the country has been growing in a blowout trend, and several green ecological demonstration urban areas have emerged, which means green buildings in China have entered a new stage of development. At the end of December 2015, a total of 3979 green building evaluation and identification projects have been evaluated in China. The total area is 460 million square meters and the total LEED certificated building area is 34 million 620 thousand square meters. The cumulative LEED certificated and the registered building area is of 163 million 720 thousand square meters, and a total of 3144 projects in China have participated in the LEED rating. Among them, the number of logo design projects is 3775, a construction area of 430 million square meters, accounting to 94.9% of the total. Shenyang has 8 LEED certified buildings currently, including the Shenyang brilliance BMW factory main office building obtaining LEED-NC registration, Shenyang Huang Cheng Heng Long LEED-CS gold, Shenyang Olympic sports Wanda plaza hotel one-star design, etc. On January 29th, 2018 the "work Plan for promoting the Sustainable and healthy Development of the Construction Industry in Shenyang" was officially announced. According to the plan, the proportion of green buildings in cities and towns in Shenyang will reach 40% in 2018, and more than 50% by 2020. The project area of annually green building is shown in figure 1, and the development of green building evaluation logo in 2008-2015 is shown in figure 2.

There are many local accreditation agencies in Jiangsu, Shenzhen, Shandong, Shaanxi and Hebei, but the number of green buildings in the cold areas is very low, as shown in figure 3.

LEED assessment certificate in China started in 2003, and the project began to register in 2004. In 2005, the Century Wealth Center of Beijing won the first LEED certificate in 2005, and in 2006, Shanghai Intel Exhibition Hall has become the first LEED certificate project in China (LEED gold grade). As of March 2015, China had received 654 applications for LEED certificate. Since 2003 LEED assessment standard has made great progress, but the LEED certificate is not completely in line with the national conditions in China. It still has its limits, and we should draw lessons from the essence of the LEED evaluation system, to develop and adapt to the situation of our country's green building evaluation standard.

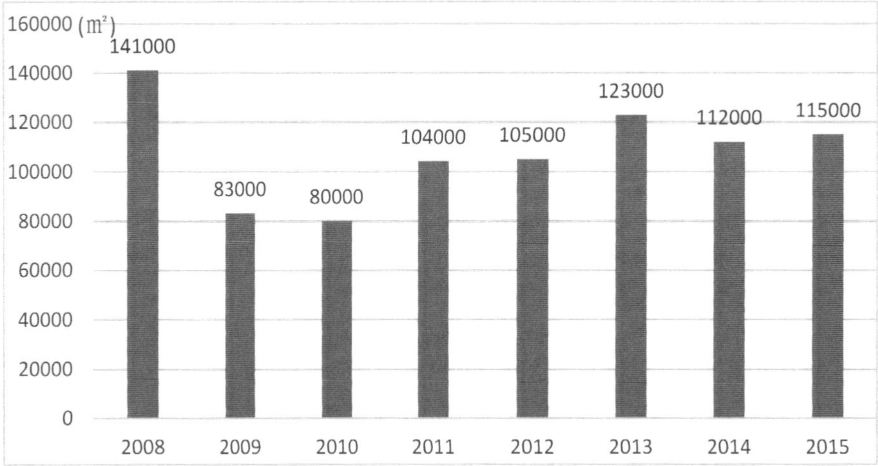

Figure 1. Area of declared green building project in 2008-2015.

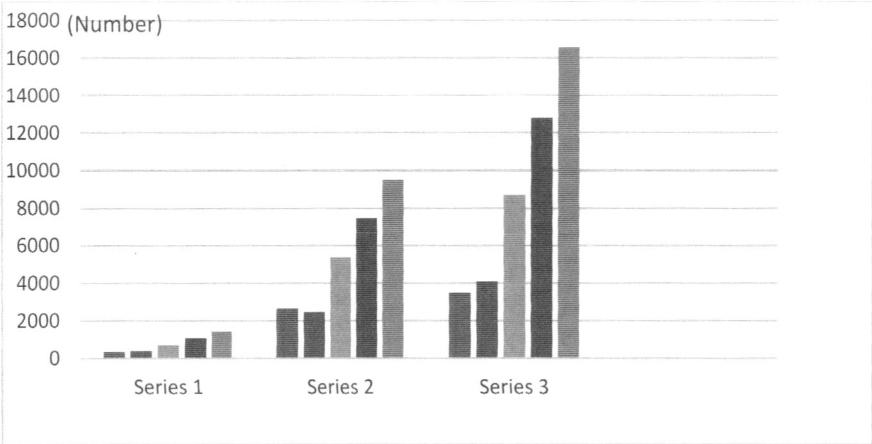

Figure 2. Development of green building evaluation and marking project in 2008-2015.

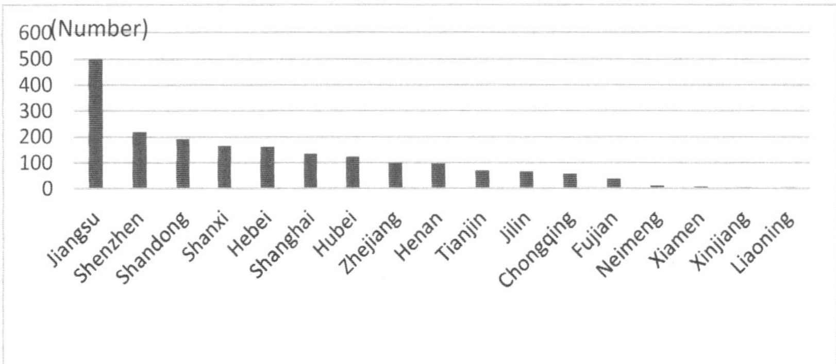

Figure 3. The national green building marks the number of evaluation institutions in 2008-2015.

Due to the unique climatic and environmental characteristics, residential buildings in cold areas in China ought to achieve maximum economic and social benefits in accordance with the local climate, environment characteristics and technical and economic levels, with limited resources and minimum energy consumption costs to meet the growing demand. Meanwhile, we should try our best to reduce or eliminate the inherent energy consumption of buildings, and more importantly, the joint efforts of the whole society.

3. The importance of green building in cold areas

3.1. Promoting sustainable development

Promoting green building design can effectively lessen the waste of resources and waste of energy, and can promote the development of science and technology of green building, and the application of new building methods, in the construction process, effectively reduce the pollution to the environment, so as to make the green building design to get effectively promoted, eventually realize the sustainable development advocated by the Chinese government [3].

3.2. Effective ways to improve environmental quality

In the cold areas, due to the low temperature, it is very different from the southern part of China. With the global warming and the influence of urban heat island, the air conditioners popularity in the cold areas has increased dramatically, and the energy consumption of air conditioners has increased year by year. Therefore, the energy efficiency of residential buildings in cold areas should be based on the actual needs of building energy efficiency, taking into account the diversity and instability of climate factors, energy-saving design methods and technical measures integrated into the overall construction design as much as possible should be considered seriously, and such problems as the shape coefficient, the palisade structure heating preservation, thermal-resistance, window wall area ratio and other issues supposed to be in mind as well, so that energy saving ideas for buildings will spread into a large number of buildings [4]. Green building is of great significance in energy conservation and emission reduction, improving people's living environment and promoting the construction of resource-conserving and environment-friendly society. The application of renewable energy buildings and green building materials are beneficial to the sustainable development of the society.

3.3. Important measures to transform urban and rural construction patterns

At present, the construction of urban and rural areas in China is still in an extensive and inefficient form, and there are some problems in the development of urbanization. The first is the areas development level difference, the second is the quality development difference. The housing departments in charge of urban and rural construction should strictly control quality certificate standards management, strengthen the quality of green building assessment certificate criteria, and check the certificated project during the process and afterwards to enforce the implementation of the project and promote to build up the green building assessment certificating market in which all the companies are self-disciplined, the industrial level supervised, the government inspected with other management means of combining the green building evaluation quality supervision system. Take the green building concept to guide urban and rural construction industry.

4. Developing green building problems in cold areas

4.1. Lack of awareness of green building concept and professionals

Although some scholars in China are very enthusiastic about green building, quite a lot of designers are not familiar with green building design, meanwhile the green building design will incur the construction cost and make the design work more complicated, all these will reduce the practice of green building enthusiasm by the designer. For the real estate developers, it may be possible to raise prices in the name of green buildings. Consumers also have misunderstandings about green buildings.

They believe that green buildings are just high cost, high green coverage and high technology with the buildings. This is pretty a one-sided view. The incremental cost of green building is indeed higher, but from the perspective of the whole life span, its cost will not necessarily be higher than the ordinary building and even be much lower [5]. Considering the comprehensive ecological benefits and comfortable living conditions, the green buildings have higher performance price ratio.

4.2. Cost problem of green building

From the perspective of the development of green building in the cold areas in China, the initial investment is higher, the economic benefit is relatively lower, and the cost recovery period is longer. For the risk of investment, investors tend to chase short-term profits and quick return of their capital, they will not really apply green buildings to actual construction. The contructors lack of the theoretical knowledge of green building technology reduces the possibility of technology implementation and increases the investment cost of green building. In all the stages from the investment, design to construction, the green buildings require more strict standards than ordinary buildings. More attention should be paid to resources at all the time all the way, this will lead to the risk of the green building problems in a more pregnant way.

4.3. Green building technology is not mature yet

There are rarely many researches on green building in cold areas, and nor are the energy saving technology, energy-saving building materials and the products. Plus, the high using cost, uncompleted variety of products and the complicated supporting construction technology will arise some engineering problems make developers lack interest to try the green buildings. Not to mention the new technology, new material of supporting the green building. The more foreign advanced equipment and technologies adopted, the higher cost of green buildings will incur, this will go against the promotion of the green buildings.

5. Countermeasures

5.1. Strengthen employee training and the green building concept disseminated

In the cold areas and even in the whole country, the concept of green building is not fully popularized. Therefore, it is necessary for the whole society to have a clear understanding of the green building, to strengthen the propaganda and to disseminate the green building concept. We should also enhance the independent research and development and technological innovation ability to let people realize the importance of green building, which can improve the status of China in the international community [6]. It can also carry out the construction of green building related disciplines through colleges and universities, increase the related courses of green buildings, and cultivate various professional and technical talents who can meet the needs of green building development. By regularly organizing green building related training, we can improve the green building application ability of employees at different levels in the whole industry, and in the assessment of building qualification or industry qualification, we should appropriately increase the assessment of green building knowledge and improve the green building quality of employees.

5.2. Establish incentive policy system by government

Since the operation and technical equipment cost of green building is high, the developers are not likely to participate. The local governments should make the green building as indicators and standards into the overall planning, controlled detail planning, construction detailed planning and planning that should be implemented into the specific projects. Although our country has already issued the building energy saving code, the enforcement of building energy saving laws and regulations is still not widespread, and all aspects need to be further improved. So, we should set up the law, the enforcement and supervision agencies or departments to create and encourage a good environment for the development of green buildings. The government also should have some incentive policies and financial support to the green buildings.

5.3. Learn advanced technical experience
From the point of view of overall green building design in cold areas, many of the building designs have not reached the standards of the green building. Because China lacks green building construction technology, we need to learn the advanced technology and technical experience from developed countries like European countries, America, Japan and so on, who have already attached great importance to energy saving building, such as the solar energy utilization technology, natural and mechanical ventilation, a new type of heating and air-conditioning equipment, new thermal insulation materials and so on. We should cooperate with foreign advanced green building areas and use new building energy saving materials to improve Chinese green building technology as soon as possible.

5.4. Avoid risks and increase profits
To developers, project orientation and design concept should be clearly defined before signing the contract with the design institute. The designer should carry on the detailed investigation on the construction site to establish a perfect green building design database, strengthen employee training, consult and collect all kinds of information, exercise brainstorm and innovation. To supervising engineers, standardized supervising criteria and the capability should be enhanced to check the contractor's green construction scheme, ensure constructors to follow the green construction plan in an orderly manner, and guarantee the quality and quantity to be completed on time. At the same time, take the advantage of modern technics like BIM to practice contract management between the developer and the designer, to predict and avoid the future risks and variations [7]. For all parties in the construction process, strengthen management to reduce construction mistakes increase profits is a win-win goal forever.

6. Conclusion
Above all, green building is the inevitable developing trend in cold areas, it has a great significance to establish a green building system in the cold areas, using new energy saving technology to make environmentally-friendly buildings is consistent with the strategy of sustainable development. Green building development in cold areas is not easy, with the help of the state and local government, the effort of developer, designer, engineer and contractor, the buildings in cold areas will be more energy saving and environmentally friendly.

References
[1] Xian S 2018 *The Development Meaning of Green Building and Its Design Concept* (Jiangxi Building Materials)
[2] Zhang M 2017 *On the Sustainable Development of Green Building* (Urban and Rural Planning Center)
[3] Chengting M 2017 *The development of green building in China* (Northwestern university)
[4] Meng K 2014 *A brief analysis of the application of green buildings in the development trend* (Northwest research and design institute)
[5] Bielek B 2016 Green Building Towards Sustainable Architecture *Applied Mechanics and Materials*
[6] Anil Kumar P P 2015 *Overview of Natural Stones as an Energy Efficient and Climate Responsive Material Choice for Green Buildings* (Department of Architecture)
[7] Ron C, Hsi C Y and Hsiao C C 2012 The Application of Green Architecture to Residential Building Development *Applied Mechanics and Materials*

Analysis for hysteresis of piezoelectric actuator based on microscopic mechanism

H Li[1], Y Xu[3], M Shao[3], L Guo[1] and D An[2,3*]

[1] School of Mechanical Engineering, Shenyang Jianzhu University, Shenyang 110168, China
[2] Research Center for Analysis and Detection Technology, Shenyang Jianzhu University, Shenyang 110168, China
[3] Research Institute for Micronano Detection and Motion Control, Shenyang Jianzhu University, Shenyang 110168, China

[*] E-mail: andong@sjzu.edu.cn

Abstract. The hysteresis widely existing in piezoelectric ceramic actuators seriously affects the positioning accuracy in practical applications. Based on the microscopic displacement mechanism of piezoelectric actuators, this paper studied the displacement mechanism of the electrostrictive effect and the inverse piezoelectric effect from the microscopic polarization mechanism and, at the same time, the displacement mechanism of electric domain reversal in ferroelectric effect. This paper clarified that the inverse piezoelectric effect and the ferroelectric effect are the main causes for the displacement of the piezoelectric actuator. The contribution of the electrostrictive effect microscopically is extremely weak and can be ignored. This paper shows that the relationship between the voltage and displacement of the piezoelectric actuator in the inverse piezoelectric effect is linear, and the hysteresis characteristic mainly exists in the ferroelectric effect. In this paper, we pointed out that there is energy loss during the piezoelectric domain transition of piezoelectric ceramics, and some irreversible non-180° domain steering is the root cause of the hysteresis of the piezoelectric actuator. This paper provides a scientific basis for further hysteresis curves modelling or correcting hysteresis modelling errors of piezoelectric ceramic and for improving the control accuracy of piezoelectric ceramic actuators in practical applications.

1. Introduction

With the rapid development of nano and micro-level electronic technology, people's requirements for electronic devices are gradually turning to the microscopic field. The research and application of micro-displacement technology has received more and more attention, such as the case of piezoelectric ceramics as a nanometer-level driving element, due to its high displacement resolution, high electromechanical coupling efficiency, large output force, small size, fast response, etc. In the field of micron and nanometer, drive has an excellent performance, thus, it is widely used in precision positioning, precision machining, biomedical, robotics, aerospace and other micro-displacement technologies [1, 2].

However, in the practical application of piezoelectric ceramics, the widely-occurring hysteresis characteristics in piezoelectric materials seriously affect the control accuracy of piezoelectric ceramic actuators, thus greatly limiting its application in micro-displacement technology [3-5]. The hysteresis characteristic of piezoelectric ceramics means that the corresponding displacement curves of piezoelectric ceramics in the voltage lift and return intervals do not overlap, and there is a displacement difference between this two curves. The hysteresis characteristic of piezoelectric ceramics is the non-local storage type inherent characteristic of piezoelectric ceramics [6]. Figure 1 shows the relationship between the displacement and the voltage across the piezoelectric actuator. It can be seen from the figure that when the voltage is applied to the piezoelectric ceramic, the voltage rising displacement curve and the voltage falling displacement curve are inconsistent, and the displacement does not return to zero after the applied voltage is reduced. This property is called the hysteresis characteristic of the piezoelectric ceramic. This indicates that the output displacement of the piezoelectric actuator is determined not only by the current input or applied voltage, but also by the historical input voltage [7].

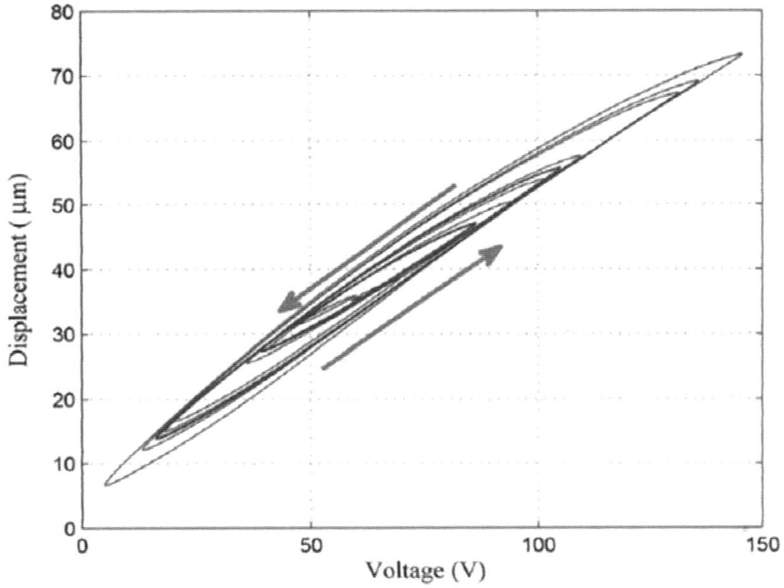

Figure 1. Hysteresis characteristic of a piezoelectric ceramic.

As more and more researchers focus on piezoelectric ceramic actuators, many workers have tried to compensate for hysteresis by using modelling methods. They can be roughly divided into two parts: physics-based models and phenomenological-based models [8]. The physics-based model refers to the scientific concept abstracted from a large number of experiments for the purpose of facilitating research. It is based on the physical meaning of hysteresis characteristics and can be strictly verified. One of the advantages of physics-based models is that their physical meaning is clear. However, due to the complexity of the form, physics-based models are not often used for the control of piezoelectric ceramic actuators.

The commonly used physics-based hysteresis modelling methods for piezoelectric ceramics include the Jiles-Atherton (J-A) model [9] and the Maxwell model [10].

Robert Malczyk et al proposed an extension of the Jiles–Atherton (J–A) magnetic hysteresis model to describe the hysteresis curve narrowing phenomenon in ferrite ZnMn material. Their new model permits the inclusion of a wide variety of additional effects observed for ferromagnetic materials without invalidating the well-known and broadly used J–A model parameters. The experiment proves the feasibility of this method [11].

Yanfang Liu et al. presented a Maxwell model to describe the hysteresis in a piezoelectric actuator. They studied the effect of the number of elements and presented both the forward and inverse algorithms. Moreover, they used the inverse Maxwell model obtained an almost linear performance of hysteresis compensate. The result of their experimental validates the effectiveness of the proposed algorithm and show that hysteresis nonlinearity reduces from13.8 to 0.4% [12]. The Phenomenon-based models refers to the models that the researchers obtain by generalizing and summarizing the input and output data as well as the actual experimental phenomena. These models use mathematical methods which directly establish mathematical models to satisfy the experimental rules, regardless of physical meanings, such as Preisach model [13], Prandtle-Ishlinskii (PI) model [14], Duhem model [15] and Bouc-Wen model [16].

Song et al. proposed a novel modified Preisach model to identify and simulate the hysteresis phenomenon observed in a piezoelectric stack actuator. Their approach can handle a varying-frequency dependence by employing a time-derivative correction technique. Parameter estimation and model verification demonstrate high accuracy of the derived model, keeping the deviation in a low percentage range (about 2–3%) [8].

Wang et all. proposed a novel modified Bouc–Wen (MBW) model to describe the asymmetric hysteresis of a piezoelectric actuator. They used a polynomial-based non-lag component to realize the asymmetric hysteresis property. The results demonstrate that their model is superior to its competitors' models in describing the asymmetric hysteresis of a piezoelectric actuator [17]. However, the lack of physical meaning makes the above models difficult to understand. At the same time, neither of these two types of models reveals the cause of hysteresis characteristics from a microscopic perspective, so the modelling errors in these modelling methods are inevitable.

Conducting a research about the hysteresis characteristics of piezoelectric ceramics is the basis for hysteresis modelling of piezoelectric ceramics and for improving the control accuracy of piezoelectric ceramic actuators.

Based on the microscopic displacement mechanism of piezoelectric actuators, this paper analysed the electrostrictive effect and the inverse piezoelectric effect displacement mechanism from the microscopic polarization mechanism and analysed the ferroelectric's effect domain reversal displacement mechanism.

We explained that the inverse piezoelectric effect and ferroelectric effect are the main causes of the displacement of the piezoelectric actuator. The energy loss during the micro domain reversal of piezoelectric ceramics was analysed. The results show that some irreversible non-180° domain steering is the root cause of the hysteresis in the piezoelectric actuator.

This paper provides a scientific basis for further hysteresis curves modelling or correcting hysteresis modelling errors of piezoelectric ceramic and improving the control precision of piezoelectric ceramic actuators in the practical application process.

2. Micro-mechanism for displacement of piezoelectric actuator

Piezoelectric ceramic is a kind of ferroelectric material. In the absence of an external electric field, its internal structure is the same as that of a ferroelectric material, and it is in a highly polarized state, therefore, the piezoelectric ceramic is consistent with ferroelectric material in performance. When an external electric field is applied across the piezoelectric ceramic, the internal structure of the piezoelectric ceramic will follow the change of the applied electric field to generate the electrostrictive effect, the inverse piezoelectric effect and the ferroelectric effect [18].

Piezoelectric ceramics are obtained after ferroelectric ceramic materials pre-polarization, generally piezoelectric ceramic working area used for precision positioning as shown in figure 2.

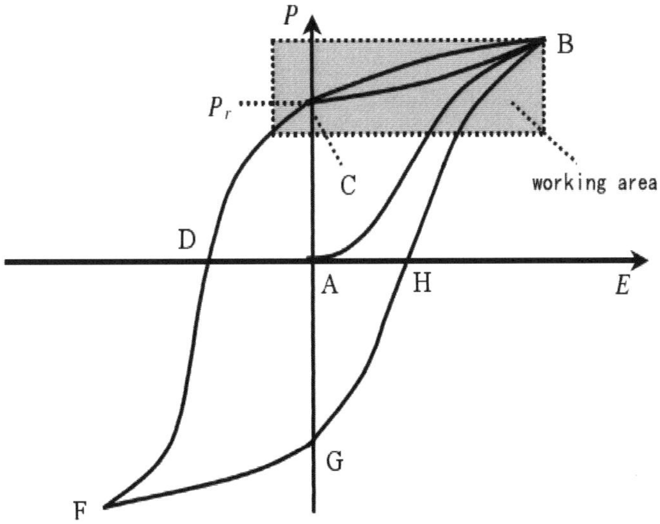

Figure 2. Curves for the electric polarization with electric field.

For piezoelectric ceramics, after pre-polarization, the residual polarization P_r is fixed. The mechanism of displacement deformation of piezoelectric ceramics can be understood as occurring on the basis of residual polarization strain. Without applying an external electric field, the total polarization strain of the piezoelectric ceramic is equal to its residual polarization strain, i.e. $X = X_r$. The total polarization of piezoelectric ceramics is mainly composed of remnant polarization P_r and irreversible polarization P_{ir}. According to the ferroelectric phenomenological theory of ferroelectric crystals, the Gibbs free energy power series expansion is:

$$G_1 = \frac{T - T_0}{2C} P^2 + \frac{1}{4}\beta P^4 + \frac{1}{6}\gamma P^6 - \frac{1}{2}sY^2 - GP^2Y \tag{1}$$

In the formula: T is temperature, T_0 is Curie-Weiss temperature, s is elastic modulus, C is Curie constant, G is electrostriction coefficient, Y is external stress of piezoelectric ceramics, β、 γ is irrelevant coefficient.

The strain of piezoelectric ceramic is:

$$X = -\frac{\vartheta G_1}{\vartheta X} = sX + GP^2 \tag{2}$$

When there is no external stress apply to the piezoelectric ceramic, its strain is:

$$X = GP^2 = G(P_r + P_{ir}) = GP_r^{\ 2} + gP_{ir} + GP_{ir}^{\ 2} \tag{3}$$

In the formula: g is piezoelectric coefficient, $g = 2GP_r$. The formula (3) shows that the strain of the piezoelectric ceramic is determined by the strength of the polarization. When the ferroelectric material is pre-polarized, its internal structure has changed and will exist in the form of a piezoelectric ceramic. The change of the strain displacement is determined by the following three:

Residual polarization strain:

$$X_r = GP_r^{\ 2} \tag{4}$$

When the external electric field is zero, the remanent polarization P_r is considered to be reversible, so the relationship between the residual polarization strain X_r and the applied electric field is also reversible, and there is no hysteresis. In this paper we defined that the strain of the piezoelectric ceramic produces a certain amount of elongation and contraction based on the residual polarization strain.

Electrostrictive effectstrain:

$$X_k = GP_{ir} \tag{5}$$

For a piezoelectric ceramic, when the inverse piezoelectric effect and the electrostrictive effect exist simultaneously, because the magnitude of the electrostriction coefficient is much smaller than that of the piezoelectric coefficient, the contribution of electrostrictive effect to the macroscopic displacement of the piezoelectric ceramic is very weak and often ignored.

Inverse piezoelectric effectstrain:

$$X_p = gP_{ir} = dE \tag{6}$$

In the formula, d is the piezoelectric strain coefficient.

The inverse piezoelectric effect was discovered when the Curie brothers studied quartz crystals in 1880 [19]. In this paper, we give the piezoelectric effect and inverse piezoelectric effect strain mechanism: When the mechanical force is applied to the outer surface of the piezoelectric ceramic, the surface of the crystal will generate a polarization charge. This phenomenon of the piezoelectricity of piezoelectric ceramics caused by mechanical forces is called the direct piezoelectric effect, as shown in figure 3a. In contrast, if an external voltage is applied across the piezoelectric crystal, the electric dipole inside the crystal is polarized, causing the piezoelectric ceramic to deform macroscopically. This phenomenon of piezoelectric ceramic deformation caused by an applied electric field is called inverse piezoelectric effect, as shown in figure 3b. From the derivation of the above equation, it can be seen that the relation of output displacement in the inverse piezoelectric effect with the strength of the applied electric field is linear, and there is no hysteresis characteristic.

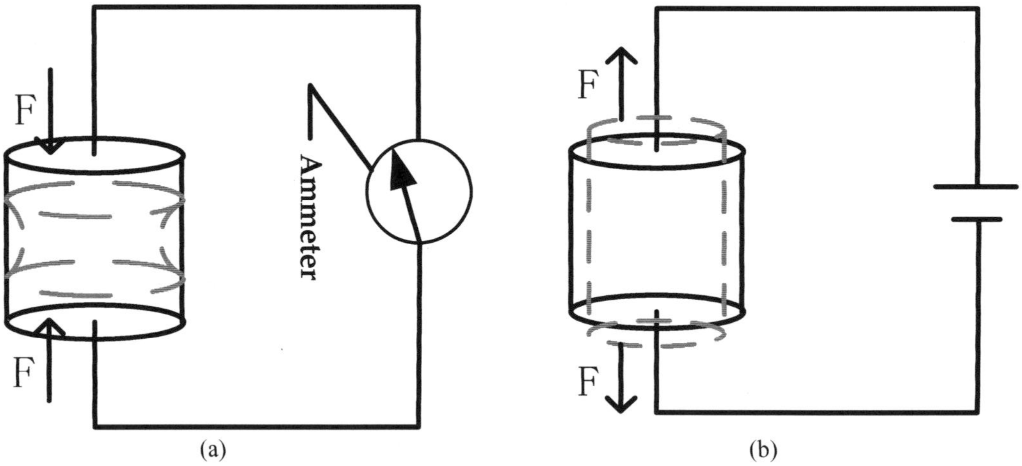

Figure 3. Piezoelectric effect diagram (Red dashed lines indicate after deformation): (a) Direct piezoelectric effect diagram; (b) Inverse piezoelectric effect diagram. The black rectangle represents the original shape of the piezoelectric ceramic block, and the red dashed rectangle represents the deformed shape.

Piezoelectric ceramic is a kind of ferroelectric material. In the case of no external electric field inside the piezoelectric ceramic, the dipole distance between the internal crystal molecules is fixed and aligned in a certain direction. At this time, the piezoelectric ceramic crystal is in the high polarization state. In a highly-polarized ferroelectric material, it is always spontaneously split into a series of small regions with different polarization directions, so that the electric field intensity generated by the spontaneous polarization inside the crystal and the electric field applied by the external space cancel each other out. Therefore, there is no electricity in the piezoelectric ceramic. These spontaneously polarized microdomains with uniform orientation are called domains. The internal domains of piezoelectric ceramics usually have four directions: 71° domains, 90° domains (as shown in figure 4), 109° domains, and 180° domains. It should be noted that for the piezoceramic crystal strain, only the non-180° domain steering contributes to the displacement of the piezoceramic actuator, while the steering of other electric domains has no effect on the displacement. The spontaneous polarization direction of the domains will be redirected under the influence of an external electric field. In the presence of an external electric field, the phenomenon that the spontaneous domain in the piezoelectric ceramic reorientation is called ferroelectric effect.

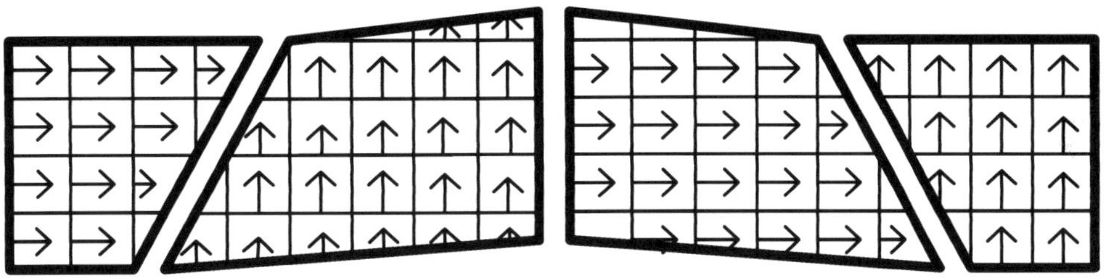

Figure 4. Piezoelectric crystal domain diagram.

From the above analysis, we summarized as follows: The displacement of piezoelectric ceramic is caused by the electrostrictive effect, the inverse piezoelectric effect and the ferroelectric effect. The strain mechanisms of the electrostrictive effect and the inverse piezoelectric effect are the electric dipole polarization in the piezoelectric ceramic based on the residual polarization strain, but the contribution of the electrostrictive effect to the macroscopic displacement of the piezoelectric ceramic is very weak, and often negligible. The relation between the output displacement of the inverse piezoelectric effect and the applied electric field is linearly, there is no hysteresis. The ferroelectric effect displacement mechanism is due to the internal domain reversal of the piezoelectric ceramic. When a fixed electric field is applied across the piezoelectric ceramic, the electric domains inside the piezoelectric ceramic will generate a certain degree of turn and elongation along the direction of the electric field, and the boundaries of the domains will also generate elongational deformation. Therefore, the piezoelectric ceramic will generate elongational deformation along the direction of the electric field (as shown in figure 5).

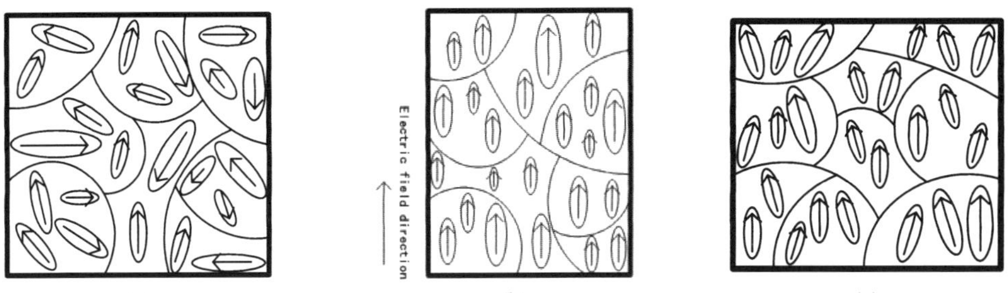

(a) (b) (c)

Figure 5. Schematic diagram of the spontaneous polarization alignment: (a) before; (b) during; (c) after presence of an electric field.

3. Analysis for hysteresis cause

In the above discussion, we know that the displacement of the piezoelectric ceramic is caused by the electrostrictive effect, the inverse piezoelectric effect, and the ferroelectric effect. Theoretically, the three effects of piezoelectric ceramics exist more or less at the same time in practical applications, but the microscopic performance of the electrostrictive effect has a very weak influence on the displacement of the piezoelectric actuator, so it is negligible. We can know from equation (6): in the inverse piezoelectric effect, the relation between the piezoelectric ceramic output displacement and the electric fieldis linear, and there is no hysteresis, so the ferroelectric effect is the root cause of piezoelectric ceramic hysteresis.

In this paper, we analyse the causes of piezoelectric ceramic hysteresis based on the energy loss model during microscopic domain reversal. In 2007, Labey et al. proposed that when the domains are rotated under the influence of an electric field, the entire domain is not oriented like a dipole. On the contrary, there will be the following four stages: the nucleation of new domains, the vertical growth of new domains, the lateral expansion of new domains and the consolidation of new domains [20]. Piezoelectric deformation is due to the internal domain transition. Previous experiments have confirmed that the physical mechanism of domain reversal is a nucleation process, and the nucleation rate of domains is a function of the applied electric field. Therefore, through the change of the nucleation rate, the volume change rate of the domain reversal can be obtained, and the deformation rate of the piezoelectric ceramic can be obtained. It is generally believed that in the low electric field range, the nucleation rate has the following relationship with the electric field:

$$n_1 = k \exp \frac{-\delta}{E} \tag{7}$$

Where the n is the number of nucleation inside the piezoelectric ceramic per unit time, δ is the activation field strength, k is a constant. In the high electric field, the nucleation rate has the following relationship with the electric field:

$$n_2 = hE^{1.4} \tag{8}$$

Where h is a constant. Here we assume that the saturation field strength of the ferroelectric ceramic is, the electric field changes uniformly, the total number of domains contained in the piezoelectric ceramic crystal is:

$$N = \int_0^{E_c} k \exp \frac{-\delta}{E} dE + \int_{E_c}^{2E_c} hE^{1.4} dE \tag{9}$$

Theoretically, the total number of domains before and after the deformation of the piezoelectric ceramic should be the same, i.e. $N_1 = N_2$ (N_1 is the number of non-180° domain steering in the piezoelectric ceramic when the applied field strength across the piezoelectric ceramic increases, N_2 is the number of non-180° domain steering in the piezoelectric ceramic when applied field strength across the piezoelectric ceramic decreases). When the applied electric field strength exceeds a certain critical field strength (when the direction of the electric field begins to change), the piezoelectric ceramic strain besides the contribution of the inverse piezoelectric effect. Non-180° domain steering began to dominate. During the turning of the domain, there are some obstacles inside the crystal that prevent the domains from turning, which causes energy loss in the domain turning process. According to the method proposed by Jiles [21], it is assumed that the surface area of the domains is A, the movement distance of the domains is x, the energy consumed by the domain crossing the obstacle is:

$$u = -\int_0^r \frac{n < u_\pi >}{2} (1 - \cos\theta) A dx \tag{10}$$

Here, n is the average number of obstacles, u_{π} is the average energy required for the non-180° domain to cross the obstacle.

Since this part of the energy loss is not recoverable, when the field strength decreases, some non-180° domains cannot recover to the same level as when the field strength is increased and resulting in the hysteresis of the piezoelectric ceramic. From the above analysis, it can be concluded that the hysteresis of the piezoelectric ceramic is due to the irreversible energy loss in the domain turning process, resulting in the number of the non-180° domain rotation during the process of the applied electric field increases and decreases is different, i.e. $N_1 > N_2$, then, resulting in the hysteresis in the process of piezoelectric ceramic deformation. In addition, the greater the field strength, the greater the irreversibility of the non-180° domain steering and the larger the hysteresis displacement of the piezoelectric ceramic actuators.

4. Conclusion

In this paper, we analysed the micro displacement mechanism of piezoelectric actuators systematically. The electrostrictive effect displacement mechanism and inverse piezoelectric effect displacement mechanism of piezoelectric ceramics is analysed in detail based on the microscopic polarization mechanism. The displacement mechanism of the ferroelectric effect is analysed from a microscopic view based on the domain reversal theory. The energy loss model of piezoelectric ceramics during domain reversal is given, the energy loss in the domain reversal process of piezoelectric ceramics is explained and then we propose that the partially irreversible non-180° domain steering is the root cause of piezoelectric ceramic hysteresis. This paper provides a scientific basis for further hysteresis curves modelling or correcting hysteresis modelling errors of piezoelectric ceramic and improving the control accuracy of piezoelectric ceramic actuators in practical applications.

Acknowledgments: This study was supported by the Shenyang Jianzhu University (SJZU) Postdoctoral Innovation Fund Project (SJZUBSH201610); National-Local Joint Engineering Laboratory of NC Machining Equipment and Technology of High-Grade Stone (SJSC-2015-13; SJSC-2016-3); Shenyang Jianzhu University Scientific Research Project (2017019); Colleges and Universities Basic Research Projects in Liaoning Province (LJZ2017035);National Science Foundation of China（51705341）; National Key R&D Plan (2017YFC0703903);Liaoning Province Key Research and Development Project（2017225016）

References

[1] Fan W, Lin Y Y and Zhong-Shen L I 2016 Hysteresis characteristics of piezoelectric ceramic actuators *Optics & Precision Engineering* **24** 1112-1117

[2] Bahrami A, Tafaoli-Masoule M and Bahrami M N 2013 Active Vibration Control of Piezoelectric Stewart Platform Based on Fuzzy Control. *International Journal of Material & Mechanical Engineering* **2** 17-22

[3] Qin Y, Tian Y, Zhang D, Shirinzadeh B and Fatikow S 2013 A Novel Direct Inverse Modelling Approach for Hysteresis Compensation of Piezoelectric Actuator in Feedforward Applications *IEEE/ASME Transactions on Mechatronics* **18** 981-989

[4] Rakotondrabe M 2011 Bouc–Wen Modelling and Inverse Multiplicative Structure to Compensate Hysteresis Nonlinearity in Piezoelectric Actuators *IEEE Transactions on Automation Science & Engineering* **8** 428-431

[5] Chang-Hai R U, Wang ZH, Chen L G and Sun L N 2010 A Hysteresis Control Model of Piezoelectric Actuator Based on Microscopic Polarization Mechanisms *Control Engineering of China* **17** 1107-1114

[6] Gu G Y, Zhu L M, Su C Y, Ding H and Fatikow S 2016 Modeling and Control of Piezo-Actuated Nanopositioning Stages: A Survey *IEEE Transactions on Automation Science & Engineering* **13** 313-332

[7] Al Janaideh M F 2009 Generalized Prandtl-Ishlinskii hysteresis model and its analytical inverse for compensation of hysteresis in smart actuators. *Mechanical & Industrial Engineering* **09** 307-312

[8] Song X, Duggen L, Lassen B and Mangeot C 2017 Modeling and Identification of Hysteresis with Modified Preisach Model in Piezoelectric Actuator *IEEE International Conference on Advanced Intelligent Mechatronics* **2017** pp 1538-1543

[9] Gao X, Liu Y and Pei Z 2016 Minor loop dynamic Jiles-Atherton model in giant magnetostrictive actuator *Journal of Beijing University of Aeronautics & Astronautics* **42** 2648-2653

[10] Wang B S, An WG and Zhang D 2008 Electrical breakdown mechanics and reliability analysis for PZT piezoelectric ceramics *Journal of Harbin Engineering University* **29** 129-337

[11] Malczyk R and Izydorczyk J 2015 The frequency-dependent Jiles–Atherton hysteresis model *Physica B Condensed Matter* **463** 68-75

[12] Liu Y, Liu H, Wu H and Zou D 2015 Modelling and compensation of hysteresis in piezoelectric actuators based on Maxwell approach *ELECTRON LETT* **52** 188-190

[13] Liu L, Tan K K, Chen S L, Huang S and Lee TH 2012 SVD-based Preisach hysteresis identification and composite control of piezo actuators *ISA T* **51** 430-438

[14] Hassani V, Tjahjowidodo T and Do T N 2014 A survey on hysteresis modeling, identification and control *MECH SYST SIGNAL PR* **49** 209-233

[15] Chen H, Tan Y, Zhou X, Dong R and Zhang Y 2011 Identification of Dynamic Hysteresis Based on Duhem Model *Fourth International Conference on Intelligent Computation Technology and Automation* **2011** 810-814

[16] Zhu W and Rui X T 2015 Online parameter identification of Bouc-Wen model for piezoelectric actuators *Optics & Precision Engineering* **38** 921-927

[17] Wang G, Chen G and Bai F 2015 Modeling and identification of asymmetric Bouc–Wen hysteresis for piezoelectric actuator via a novel differential evolution algorithm *Sensors and Actuators A: Physical* **235** 105-118

[18] Huang X, Zeng J, Ruan X, Zheng L and Li G 2017 Structure, electrical and thermal expansion properties of PZnTe-PZT ternary system piezoelectric ceramics *J AM CERAM SOC* **101** 274-282

[19] Weihua X U, Hai B, Yang Y and Wei X 2010 Voltage Signal Transmission Principle Based on the Inverse Piezoelectric Effect of Piezoelectric Ceramic *Automation of Electric Power Systems* **34** 80-83

[20] Rabe K M, Ahn CH and Triscone J M 2007 *Physics of Ferroelectrics* Springer Berlin Heidelberg p 203-234

[21] Jiles DC and Thoelke J B 1989 Theory of ferromagnetic hysteresis: determination of model parameters from experimental hysteresis loops *Magnetics IEEE Transactions on* **25** 3928-3930

IOP Conf. Series: Materials Science and Engineering **399** (2018) 012032 doi:10.1088/1757-899X/399/1/012032

Low-carbon building design and practice in severe cold areas

S Li[1*], T Shi[2], B Han[2], T Li[2], T Hao[2] and Y Dong[2]

[1] Ecological planning and Green Building Research Institute of Shenyang Jianzhu University
[2] Shenyang Jianzhu University

[*] E-mail: lisui0517@126.com

Abstract. With a vast territory and seven climate zones, China's urbanization is constantly improving while building energy conservation and emission reduction are facing severe challenges, especially in severe cold areas. In China's Liaoning Province, urban public buildings and rural residential buildings consume huge amounts of energy. According to the characteristics of Liaoning, "passive technology advanced, active technology second" is the best choice for energy saving and emission reduction. Taking Sino-German Energy Conservation Demonstration Center in Shenyang Jianzhu University as an example in this paper, through analysis of the operation monitoring data of energy consumption, the objective is to provide a good reference method and a design strategy for the near zero energy buildings in the severe cold areas and to verify its feasibility. At the same time, the parametric design technique is applied to feature-based design system, such as multi objective optimization and genetic algorithm combined simulation technology, Rhino Grasshopper Energy plus energy saving technology and cost-benefit analysis model to obtain the optimal solution. It provides innovative technology demonstration for building energy conservation in the severe cold areas.

1. Background

China has a vast territory and contains seven climate zones. In the past few decades, China's urbanization has been advancing, buildings energy-saving and emission-reduction are facing tough challenges, especially the ones in the severe cold areas. The energy consumption of urban public buildings is so high and the rural residences' emission is high as well [1]. According to 2016's statistics data, urban and rural residential buildings' electricity consumption was 21.25 billion kilowatt hours [2], coal consumption was 3.15 million tons and the annual emission of CO2 was about 10.17 million tons [3].

The public buildings have many problems such as too large in size, weak insulation, low energy utilization ratio and lack of intelligent control. The large CO2 emissions have caused a series of environmental problems. Rural residential buildings have problems too, such as lack of energy planning, too much coal-fired, environment pollution etc. According to the climate characteristics of severe cold areas, Passive House Technology was first laid out by Professor Wolfgang Feist, it is fit for the buildings in these severe cold areas. The principles of passive technology include: to strengthen external heat preservations, to reduce thermal bridge improve air tightness, to recover waste heat of air, a little heat supplement.

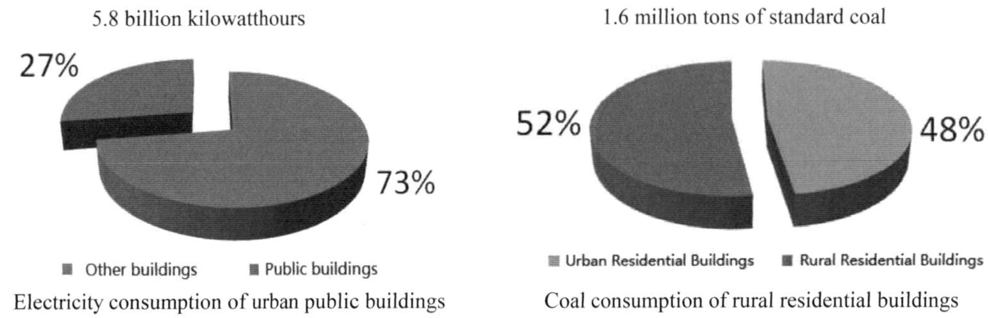

Figure 1. Urban public buildings and rural residential buildings energy consumption in 2016.

Near zero energy building is a kind of building which does not consume conventional energy, it fully relies on solar energy or other renewable energy to achieve building energy consumption and its indoor environment is comfortable [4-6].

According to the definition of energy consumption and carbon emission, the definition of near zero energy building is a little different from it in the international academic circle. According to the definition of Torcellini [7], the near zero energy building is a building whose produces and consumes are in a balance and it can be connected with the public power grid. Its measurement time is generally based on the year.

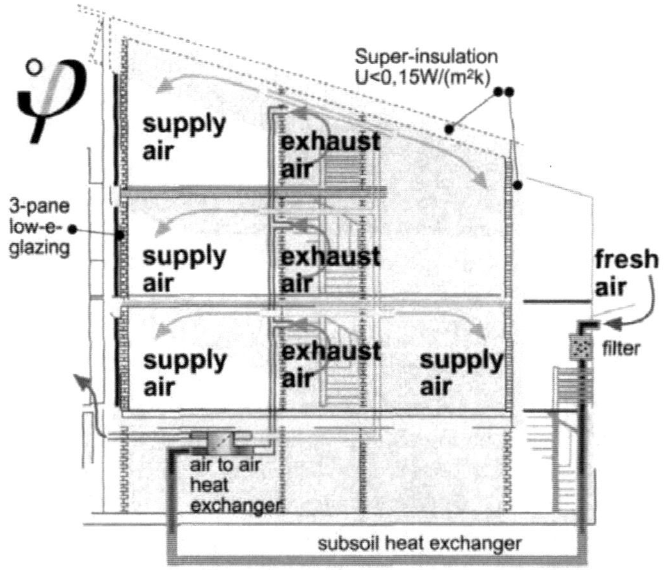

Figure 2. The technical route of the passive house building.

2. Our reasearch work in Green Building

Sino-Germany Energy Conservation Demonstration Center is designed by Shenyang Jianzhu University cooperated with three German universities. This center's technical system is in accordance with the standard of 'Chinese green building + German passive house'. In 2015, it received the 3-star green building certification of China, the first 3-star public green building in Liaoning Province. In 2016, it received the 'Ultra-low Energy Passive Building' certification by Chinese Passivehouse Committee.

Figure 3. Sino-Germany Energy Conservation Demonstration Center.

By building and assessing of this building, we want to research the energy-saving way in the severe cold areas in China, including building form, insulation, double-resource heat pump, primary air system of underground tunnel, intelligent control and sustainable energy. By recording the data of energy use and analysing the data of indoor environment, the results have verified the feasibility of near zero energy building in severe cold areas. And we hope this case will provide a valuable experience for other areas in high latitudes.

Figure 4. Roofing and Skylight

The specific practice of the roofing and skylight part is roof insulation, 300mm (millimeter) extruded sheet, divided by three layers. The Veluk Skylight is used, the window frame material is aluminium packaged wood, the window is made of triple glass, double vacuum. The external shading system at the windows on the facades has an automatic reaction of light, can cut off bright light and prevent glare [8].The insulation of external wall is composed of two layers of 280mm graphite polystyrene board insulation on the east, north and west direction; In the south, the 300mm rock wool board is used in an inner layer and a vertical air interface layer and photovoltaic panel wall are in an outer layer. The outside window consists of three layers of low-E glasses. For common exterior windows on the east, west and north facades, the heat transfer coefficient is U=0.95W/(m^{2}k). The south glass wall heat transfer coefficient is U=1.0W/(m^{2}k). The double-resource heat pump and phase transition of water tank heating system use renewable energy, they can provide cold and heat sources, water tank accumulates and stores energy during the daytime and heat the indoor air at night [9].

Winter Summer Equipment pipelines

A. Double-resource heat pump

Heat pump

Tank heating

B. Phase change tank heating system

Figure 5. Air source - Double-resource heat pump and Phase change tank heating system.

According to the electricity consumption monitoring of the center, the air-conditioning cooling energy consumption was 10.8 KWh/m^2·a, the heating energy consumption was 26.6 KWh/m^2·a, in the summer of 2017, energy efficiency of this building was 83.4 and CO2 reduction was 120 t/a.

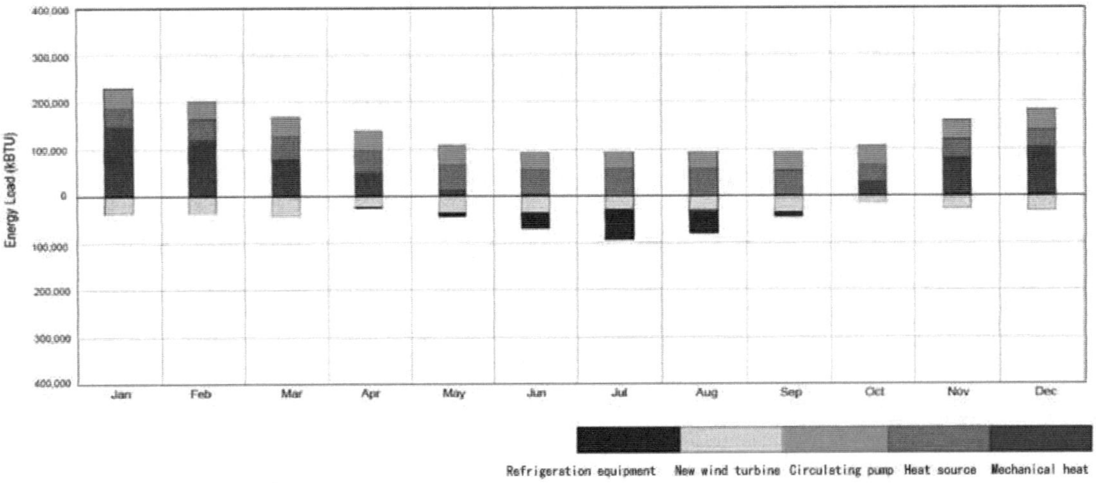

Figure 6. Annual energy load of Sino-German Energy Conservation Demonstration Center in 2017.

3. Multi objective optimization

In China, the social and economic basis is climate related. Multi objective optimization of energy saving and cost control is very important for severe cold areas. In the process of accounting for the whole life cycle of the building, the CO2 emissions generated by the use stage are up to 60-80% and the CO2 amount of heating emissions accounts for the total emissions are 50-80% [10]. Multi objective optimization is used to optimize and compare with HVAC system, hot water system, lighting system, renewable energy and intelligent control system. The basic idea is to select and determine the energy saving technology of key buildings through the incremental cost benefit analysis of the whole life cycle of energy saving technology and to solve the multi-objective optimization model and finally generate the best energy saving technology system [11]. The optimization of energy saving and cost control in cold ground is studied by the genetic algorithm and the optimal solution is obtained by the parametric design technique, using Rhino Grasshopper Energyplus to analyse the energy saving incremental cost benefit of the energy saving technology. Taking the east wing building of Shenyang Jianzhu University Sino-Germany Energy Conservation Demonstration Center as an example, the assembly steel structure passive housing technology system was used in the design process. The cost of building is low and the energy saving effect is good. The building uses the hot water heating system, which is less than 15kWh/ (m²a), close to the heating energy standard of German passive house.

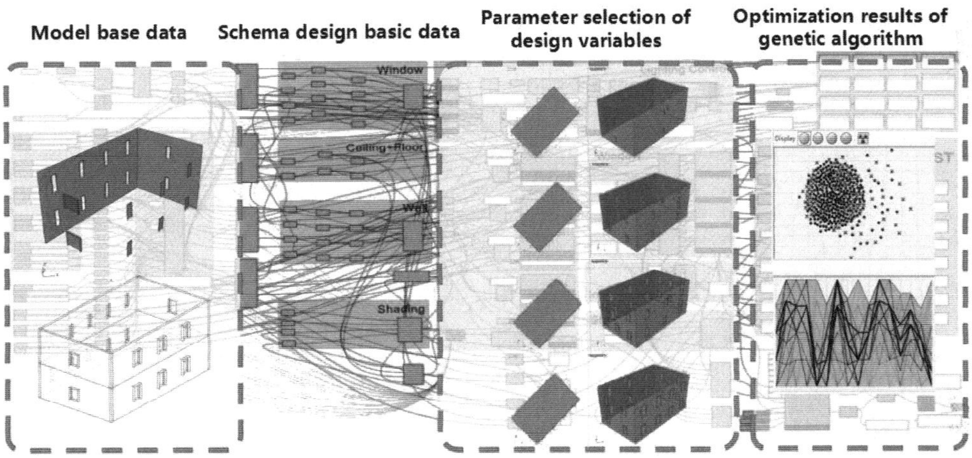

location	The main structure parameters and cost optimization	Cost per unit area (¥)
	Heat transfer coefficient(W/(m²K))	
roof	0.09	500
Exterior wall	0.105	450
The ground	0.11	400
Door and Windows	1.5	1000

Figure 7. The main structure parameters and unite area cost optimization.

In January 2017, the temperature of the heating and heat pump in the east wing of the center was 40°C, the average indoor temperature was 28 - 29°C. When the water temperature of the pump was adjusted to 31°C, the average indoor temperature was 23 - 24°C. Before it was put to use, we conducted an experiment of stopping heating for 3 days and the indoor temperature was 19-20°C. The tests we made showed that, after the heating had been stopped for 3 days, the indoor temperature was reduced by only 4°C. This building technology system provides an innovative technology demonstration for building energy efficient utilization in severe cold areas such as high cost-effective, high assembly rate and high energy saving rate, etc. This technology has six advantages (industry standard statistical data):

1) low construction costs: lower cost than reinforced concrete building by 15% - 25%;
2) high degree of prefabricated: prefabricated assembly rate up to 60% - 75%;
3) short construction period: shorter construction period than reinforced concrete buildings by 30% - 50%;
4) strong seismic capacity: giving full play to the advantages of flexible structural systems;
5) high energy efficiency: insulation effect can reach the standard of German passive house;
6) quality living environment: only with a little heat, the indoor temperature can be achieved over 20°C without burning coal in winter.

4. Prospect

At present, all countries vigorously promote green building projects. No matter the system certified by the International Energy Agency (IEA), or the passive house in Germany, or the green building in China, the purpose is the same: "energy conservation".

Finally, we hope we can work together with other countries, make contributions to develop the energy-saving building and low-carbon city.

References

[1] Yu Z, Ding W and Cheng Y 2009 A brief analysis of the layout of energy development and construction in Liaoning *J. Contemporary economy* **07** 96-97
[2] World Business Council for Sustainable Development. Transforming the Market: Energy Efficiency in Buildings Switzerland: Atar Roto Presse SA, 2009: 1
[3] Li X 2013 Simulation study on energy efficiency optimization of an office building in Chongqing *J. Acta building* **10** 164-168
[4] Tsinghua University Building Energy Efficiency Research Center. Annual report on China's building energy efficiency Beijing: China Construction Industry Press, 2014
[5] Research on sustainable design strategy and energy saving technology of Sino German Energy Saving Demonstration Center Shi Tiemao Shan Rudai Xiaxiao Dong, Architectural Journal 2016 S1
[6] Xiaxiao D, Shan R and Shi T 2015 Design and practice of ultra-low energy consumption green building in severe cold area - Shenyang Construction University Sino German Energy Saving Demonstration Center project, Construction Science and Technology
[7] Shan R, Liuguannan, Xiaxiao D and Shi T 2018 Optimization design of envelope based on genetic algorithm, Architecture Technology
[8] World Business Council for Sustainable Development. Transforming the Market: Energy Efficiency in Buildings Switzerland: Atar Roto Presse SA, 2009: 1
[9] Li X 2013 Simulation study on energy efficiency optimization of an office building in Chongqing *J. Acta building* **10** 164-168
[10] Tsinghua University Building Energy Efficiency Research Center. Annual report on China's building energy efficiency Beijing: China Construction Industry Press, 2014
[11] Konis K and Kensek K 2016 Passive performance and building form: an optimization framework for early-stage design support *J. Solar Energy* **02**

IOP Conf. Series: Materials Science and Engineering **399** (2018) 012033 doi:10.1088/1757-899X/399/1/012033

The dynamic stability analysis of telescopic booms of the crane based on the energy method

S Liu[1*], J Liu[1], K Zhang[1] and L Meng[1]

[1] School of Mechanical Engineering, Shenyang Jianzhu University, Shenyang 110168, China

[*] E-mail: liushiming_1983@163.com

Abstract. According to the energy method, the dynamic stability of a crane's telescopic boom under the periodic load is studied, which is known as parametric resonance, based on the design code for a crane (GB/T3811-2008), the stability analysis model of the box telescopic boom can be regarded as variable-section stepped columns. The parametric vibration equation of n-stepped columns which expressed as Mathieu Equation is deduced by the Hamilton Principle, then the Critical frequency equation of dynamic instability regions of the telescopic booms are derived, finally, the effects of damping to the dynamic stability of telescopic boom are discussed. The results show that, the dynamic instability regions are reduced when the damping coefficient is increased and the effect on the second dynamic instability region is more obvious, that means damping improves the structure's dynamic stability.

1. Introduction

The telescopic boom is the main load-carrying part of the truck crane and is the most important working part. When the lifting quality rises or falls off the ground during the braking process or suddenly unloads, the inertia of the hoisting crane and the flexibility of the structure can cause vibration of the telescopic boom, especially when the telescopic boom is in a fully extended state and at minimum amplitude, the telescopic boom structure may suffer from dynamic instability. Based on the crane design specification (GB/T3811-2008), the calculation model of telescopic boom static instability is a variable cross-section stepped column, and its instability critical force can be solved by the energy method, the exact finite element method, etc [1-3] . Therefore, in the study of the dynamic stability of the crane telescopic boom, the telescopic boom is also equivalent to a variable cross-section stepped column model, which is in line with the actual engineering requirements. To solve the dynamic stability under cyclic loading, the method of the former Soviet Union scholars, Bolotin is often used to determine the dynamic stability and dynamic instability regions of the structure [4]. Sun Qiang [5] used the Mathieu-Hill equation to study the dynamic stability of isometric straight bars under cyclic loading. Li Xiaodong [6] determined the dynamic instability region of the buckling restrained supporting bar under cyclic loading according to the Bolotin method. Ratko Pavlovic [7] studied the dynamic stability of an isometric straight rod under the combined effect of axial and bending moments by using the differential equation method; MA De Rosa [8] studied the dynamic stability of the continuously variable beam with the DQM method. However, the above studies on the dynamic stability under axial cyclic loading mainly focus on single components with continuous section changes. The dynamic stability of the stepped columns with discontinuous changes in the cross-section and the multiple structures composed of multiple individual components are investigated. In [9-11], the finite element method was used to study the dynamic stability of the crane's telescopic boom under cyclic loading. Although this method can effectively solve the dynamic stability problem of the telescopic boom under cyclic loading, it has the disadvantage of a complicated solution which is not conducive to the practical application of engineering. Therefore, in this paper, the energy method combined with the Hamilton principle is used to establish the parametric vibration equation of the telescopic boom which is expressed in the form of the Mathieu equation. The critical frequency equation of the boundary of the dynamic instability region of the telescopic arm is deduced, and the influence of the damping on the unstable region of the dynamics is discussed. The proposed method provides a basis for design calculations and can guide the practical application of engineering.

2. Establishment of parametric vibration equation of telescopic boom

The dynamic stability analysis model of the telescopic boom of a truck crane can be equivalent to a multi sectioned column, the mechanical models are shown in figure. 1. Assuming l_i as the total length of i sections telescopic boom. L is the total length of the telescopic boom namely when $i=n$, $L=l_n$. m_i is the mass per unit length of section i, E is the elastic modulus of the material ,J_i is the inertia moment of section i, the axial resonant force $P_0+P_t\cos\theta t$ is applied at the top of the column, where P_0, P_t are the amplitudes of the resonant force respectively, θ is the resonant circular frequency.

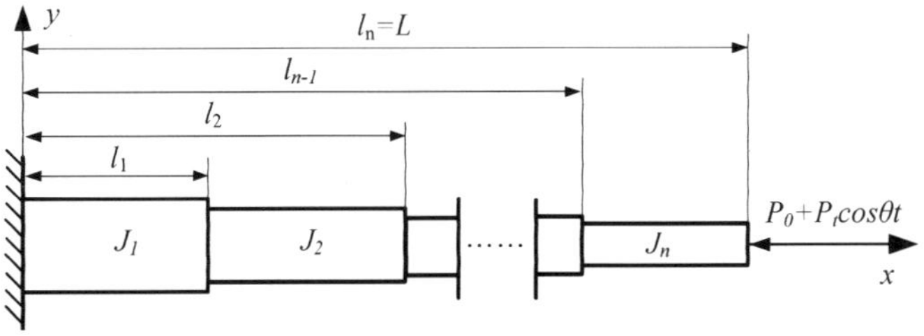

Figure 1. The mechanical model of dynamic stability for multi sectioned telescopic boom.

Assuming the lateral vibration displacement curve of the telescopic arm is $V(x,t)=f(t)(1-\cos\frac{\pi x}{2L})$ which conforms to the boundary conditions of figure.1. Then the kinetic energy of the telescopic boom $T(t)$ is obtained by using the energy principle

$$T(t)=\frac{1}{2}\int_0^{l_1} m_1(\frac{\partial V}{\partial t})^2 dx+\frac{1}{2}\int_{l_1}^{l_2} m_2(\frac{\partial V}{\partial t})^2 dx+\cdots+\frac{1}{2}\int_{l_{n-1}}^{l_n} m_n(\frac{\partial V}{\partial t})^2 dx$$
$$=\dot{f}^2(t)\sum_{i=1}^{n}\frac{m_i L}{4}(\frac{l_i-l_{i-1}}{L}-\frac{\sin(\frac{2\pi l_i}{L})-\sin(\frac{2\pi l_{i-1}}{L})}{2\pi}) \quad (1)$$

The potential energy of the telescopic boom is expressed as $U(t)=U_1(t)+U_2(t)$, where $U_1(t)$ is the bending strain energy ,$U_2(t)$ is the external potential energy.

$$U_1(t)=\frac{1}{2}\int_0^{l_1} EJ_1(\frac{\partial^2 V}{\partial x^2})^2 dx+\frac{1}{2}\int_{l_1}^{l_2} EJ_2(\frac{\partial^2 V}{\partial x^2})^2 dx+\cdots+\frac{1}{2}\int_{l_{n-1}}^{l_n} EJ_n(\frac{\partial^2 V}{\partial x^2})^2 dx$$
$$=\frac{\pi^4 f^2(t)}{4L^3}\sum_{i=1}^{n}EJ_i(\frac{l_i-l_{i-1}}{L}-\frac{\sin(\frac{2\pi l_i}{L})-\sin(\frac{2\pi l_{i-1}}{L})}{2\pi}) \quad (2)$$

$$U_2(t)=-\frac{1}{2}\int_0^{L}(P_0+P_t\cos\theta t)(\frac{\partial V}{\partial x})^2 dx=-\frac{\pi^2(P_0+P_t\cos\theta t)}{4L}f^2(t) \quad (3)$$

The damping of the telescopic boom in vibration is equivalent to linear viscous damping, then, the energy consumed by the damping $D(t)$ is gained.

$$D(t)=\frac{1}{2}\int_0^{l_1}C_1(\frac{\partial V}{\partial t})^2dx+\frac{1}{2}\int_{l_1}^{l_2}C_2(\frac{\partial V}{\partial t})^2dx+\cdots+\frac{1}{2}\int_{l_{n-1}}^{l_n}C_n(\frac{\partial V}{\partial t})^2dx$$
$$=\dot{f}^2(t)\sum_{i=1}^{n}\frac{C_iL}{4}(\frac{l_i-l_{i-1}}{L}-\frac{\sin(\frac{2\pi l_i}{L})-\sin(\frac{2\pi l_{i-1}}{L})}{2\pi}) \tag{4}$$

where: C_i is the equivalent damping coefficient of section i; EJ_i is the flexural rigidity of section i. According to the Hamilton Principle,

$$\begin{cases}L=T(t)-U(t)\\ \frac{d}{dt}(\frac{\partial L}{\partial \dot{f}})-\frac{\partial L}{\partial f}+\frac{\partial D}{\partial \dot{f}}=0\end{cases} \tag{5}$$

Substitute equation (1) - (4) into equation (5)

$$f''(t)+2\beta f'(t)+\Omega^2(1-2\mu\cos\theta t)f(t)=0 \tag{6}$$

Equation (6) is the famous Mathieu equation, assuming the dimensionless coefficient $\alpha_i=l_i/L(i=0,1,2,\cdots n)$, $\lambda_i=J_i/J_1(i=1,2,\cdots n)$, $\alpha_0=0$, then the symbols in equation(6) can be further expressed as follows

$$\Omega=\omega\sqrt{1-\frac{P_0}{\rho}} \tag{7}$$

$$\mu=\frac{P_t}{2(\rho-P_0)} \tag{8}$$

$$\omega^2=\frac{\frac{\pi^4EJ_1}{(2L)^4}\sum_{i=1}^{n}\lambda_i((\alpha_i+\frac{\sin(\pi\alpha_i)}{\pi})-(\alpha_{i-1}+\frac{\sin(\pi\alpha_{i-1})}{\pi}))}{\sum_{i=1}^{n}m_i((3\alpha_i-\frac{8\sin(\pi\alpha_i/2)}{\pi}+\frac{\sin(\pi\alpha_i)}{\pi})-(3\alpha_{i-1}-\frac{8\sin(\pi\alpha_{i-1}/2)}{\pi}+\frac{\sin(\pi\alpha_{i-1})}{\pi}))} \tag{9}$$

$$\rho=\frac{\pi^2EJ_1}{(2L)^2}\sum_{i=1}^{n}\lambda_i((\alpha_i+\frac{\sin(\pi\alpha_i)}{\pi})-(\alpha_{i-1}+\frac{\sin(\pi\alpha_{i-1})}{\pi})) \tag{10}$$

$$\beta=\frac{\sum_{i=1}^{n}\frac{C_iL}{4}(\alpha_i-\alpha_{i-1}-\frac{\sin(2\pi\alpha_i)-\sin(2\pi\alpha_{i-1})}{2\pi})}{2\sum_{i=1}^{n}\frac{m_iL}{4}(\alpha_i-\alpha_{i-1}-\frac{\sin(2\pi\alpha_i)-\sin(2\pi\alpha_{i-1})}{2\pi})} \tag{11}$$

where: Ω is the natural frequency of the lateral free vibration of the telescopic arm under the axial resonance force amplitude P_0, and the unit is Hz, μ is the excitation coefficient, ω is the natural frequency of lateral free vibration of the stepped column without any load, and the unit is Hz, ρ is the static destabilizing critical load of the stepped column, the unit is N, β is the dimensionless damping coefficient.

Equations (6) - (11) are the parametric vibration equations of n sections crane telescopic boom.

3. Determination of dynamic stability and instability regions of telescopic boom

The equation (6) is a Mathieu equation, we judge it must have periodic solutions of T and $2T$ based on the properties of the mentioned equation. In order to get the critical frequency equation of the dynamic stability region, the Fourier series expansion is used to get solutions of period $2T$ and T, and they can be expressed respectively as follows:

$$f(t)=\sum_{n=1,3,5\cdots}^{\infty}a_n\sin\frac{n\theta t}{2}+b_n\cos\frac{n\theta t}{2} \tag{12}$$

$$f(t)=b_0+\sum_{n=2,4,6\cdots}^{\infty} a_n\sin\frac{n\theta t}{2}+b_n\cos\frac{n\theta t}{2} \tag{13}$$

Substituting equation (12) and equation (13) into equation (6), in order to make the establishment of equations, each coefficient of sin ($n\theta t/2$) and cos ($n\theta t/2$) must be equal to zero. The linear equations in terms of $\{a_1,b_1,a_3,b_3,\cdots a_{2k-1},b_{2k-1}\}^T$ and $\{b_0,a_2,b_2,a_4,b_4,\cdots a_{2k},b_{2k}\}^T$ can be derived, the necessary and sufficient condition for a system of linear equations to have a nonzero solution is that the determinant of its coefficient matrix must be zero. Therefore, let the mentioned determinant of coefficient be equal to zero, then this infinite determinant is convergent and it can be expressed as follows.

The corresponding critical frequency equation for the $2T$ cycle is as follows:

$$\begin{vmatrix}
1+\mu-\frac{\theta^2}{4\Omega^2} & -\frac{\theta\beta}{\Omega^2} & -\mu & 0 & 0 & 0 & \vdots & \vdots \\
\frac{\theta\beta}{\Omega^2} & 1-\mu-\frac{\theta^2}{4\Omega^2} & 0 & -\mu & 0 & 0 & \vdots & \vdots \\
-\mu & 0 & 1-\frac{9\theta^2}{4\Omega^2} & -\frac{3\theta\beta}{\Omega^2} & -\mu & 0 & \vdots & \vdots \\
0 & -\mu & \frac{3\theta\beta}{\Omega^2} & 1-\frac{9\theta^2}{4\Omega^2} & 0 & -\mu & \vdots & \vdots \\
0 & 0 & -\mu & 0 & 1-\frac{25\theta^2}{4\Omega^2} & -\frac{5\theta\beta}{\Omega^2} & \vdots & \vdots \\
0 & 0 & 0 & -\mu & \frac{5\theta\beta}{\Omega^2} & 1-\frac{25\theta^2}{4\Omega^2} & \vdots & \vdots \\
\cdots & \cdots & \cdots & \cdots & \cdots & \cdots & \cdots & \cdots \\
\cdots & \cdots & \cdots & \cdots & \cdots & \cdots & \cdots & \cdots
\end{vmatrix}=0 \tag{14}$$

where: n=1, 3, 5, 7,$\cdots$2k-1.

The corresponding critical frequency equation for the T cycle is as follows:

$$\begin{vmatrix}
1 & 0 & -\mu & & & & & \vdots \\
0 & 1-\frac{\theta^2}{\Omega^2} & -\frac{2\theta\beta}{\Omega^2} & -\mu & 0 & & & \vdots \\
-2\mu & \frac{2\theta\beta}{\Omega^2} & 1-\frac{\theta^2}{\Omega^2} & 0 & -\mu & 0 & & \vdots \\
0 & -\mu & 0 & 1-\frac{4\theta^2}{\Omega^2} & -\frac{4\theta\beta}{\Omega^2} & -\mu & 0 & \vdots \\
0 & 0 & -\mu & \frac{4\theta\beta}{\Omega^2} & 1-\frac{4\theta^2}{\Omega^2} & 0 & -\mu & \vdots \\
0 & 0 & 0 & -\mu & 0 & 1-\frac{9\theta^2}{\Omega^2} & -\frac{9\theta\beta}{\Omega^2} & \vdots \\
0 & 0 & 0 & 0 & -\mu & \frac{9\theta\beta}{\Omega^2} & 1-\frac{9\theta^2}{\Omega^2} & \vdots \\
\cdots & \cdots & \cdots & \cdots & \cdots & \cdots & \cdots & \cdots
\end{vmatrix}=0 \tag{15}$$

where: n = 2, 4, 6, 8,$\cdots$2k.

Equation (14) and equation (15) are the critical frequency equations of the boundary of the dynamic unstable region of the structure. The critical frequency refers to the external load frequency corresponding to the boundary of the dynamic unstable region. However, the first and second areas of dynamic stability are some of the most dangerous in practical engineering.

The critical frequency equation of the first dynamic unstable region is obtained by substitute n=1 into equation (14), that also means keeping the first order solution of the determinant of $2T$ cycle

$$\begin{vmatrix} 1+\mu-\dfrac{\theta^2}{4\Omega^2} & -\dfrac{\theta\beta}{\Omega^2} \\ \dfrac{\theta\beta}{\Omega^2} & 1-\mu-\dfrac{\theta^2}{4\Omega^2} \end{vmatrix}=0 \tag{16}$$

By solving equation (16), we can find:

$$\theta_{1,2}=2\sqrt{\Omega^2-2\beta^2\pm\sqrt{4\beta^2\left(\beta^2-\Omega^2\right)+\mu^2\Omega^4}} \tag{17}$$

Using the same method, the critical frequency equation of the second dynamic unstable region is obtained by substitute n=2 into equation(15), that also means keeping the second order solution of the determinant of T cycle, then we can get:

$$\theta_{3,4}=\sqrt{\Omega^2\left(1-\mu^2\right)-2\beta^2\pm\sqrt{4\beta^4-2\beta^2\Omega^2\left(1-\mu^2\right)+\mu^4\Omega^4}} \tag{18}$$

Under the action of a given axial resonance force $P(t)=P_0+P_t\cos\theta t$, the boundary region composed of critical frequencies θ can be determined, the region of dynamic stability is limited by the critical frequency of different cycles, and the region of dynamic instability is determined by two critical frequencies of the same cycles. That is, the dynamic unstable region is separated from the stable region by the solution of the cycles T and $2T$, and the critical frequency between the two cycles is the dynamic stable region. By substitute equation (7), equation (8) and equation (11) into equation (17) and equation (18), then the dynamic stable region and unstable region of crane's telescopic boom under periodic load can be obtained.

4. Examples

Taking the telescopic boom of a typical five-section crane as an example, the dynamic stability calculation model is a cantilever stepped beam which is shown in figure 2, the total length of the stepped column L=30m, Young's modulus E=200GPa, Sectional moment of inertia $J_1=1.1786\times10^2\text{m}^4$, density 7800kg/m^3, $l_1=0.24L$, $l_2=0.43L$, $l_3=0.62L$, $l_4=0.81L$, $l_5=L$, $\lambda_1=1.0$, $\lambda_2=J_2/J_1=0.77$, $\lambda_3=J_3/J_1=0.59$, $\lambda_4=J_4/J_1=0.46$, $\lambda_5=J_5/J_1=0.35$, analyze the dynamic stability of the telescopic crane when the cyclic axial loads are $P(t)=P_0+P_t\cos\theta t$.

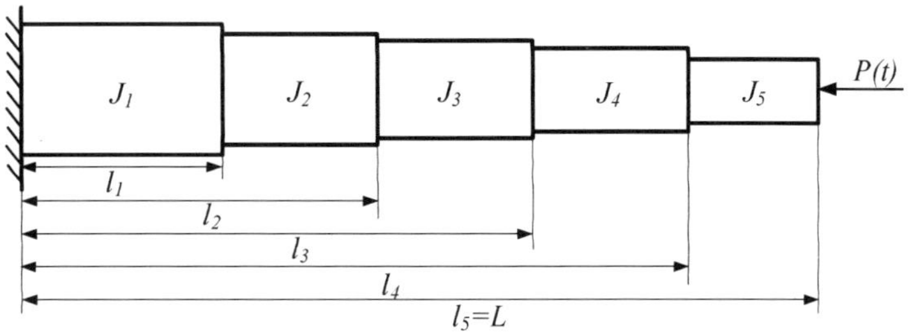

Figure 2. The calculation model of 5-section telescopic boom.

In order to analyze the influence of different values P_0 on the dynamic instability of the telescopic boom, the value of P_0 is chosen by 0 N, 0.5×10^6 N, 1.0×10^6 N, 1.5×10^6 N, 2.0×10^6 N respectively, assuming the damping coefficient β=0, the amplitude of the vibration force of the periodic load P_t is taken as the horizontal coordinate, and the critical vibration frequency $\theta/2\Omega$ is taken as the vertical coordinate. Drawing the first and second dynamic stability curves of a five-section telescopic arm under different axial resonant force amplitudes P_0 and P_t , which is shown in figure 3. The figures show that with the increase of P_0, the area of dynamic instability increases, when P_0 is equal to the static destabilizing critical load ρ, no irrespective of the value P_t is, the structure will still lose bearing capacity.

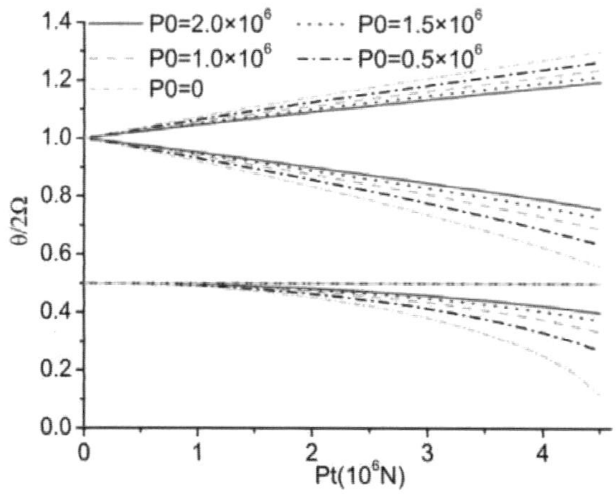

Figure 3. Dynamic curves of clamped 5-sections stepped column under different axial resonant forces.

In order to study the influence of damping on the dynamic unstable region, taking the 5-sections crane telescopic boom shown in figure 2 as an example, assume $P(t)=P_0+P_t\cos\theta t$, $P_0=1.0\times10^6$ N. Making P_t as the horizontal coordinate, and the critical vibration frequency $\theta/2\Omega$ as the vertical coordinate, according to the energy method, the first and second dynamic stability region curves of a five-section telescopic arm under different damping coefficient β=0, 0.2, 0.4, 0.6 are drawn, which is shown in figure 4.

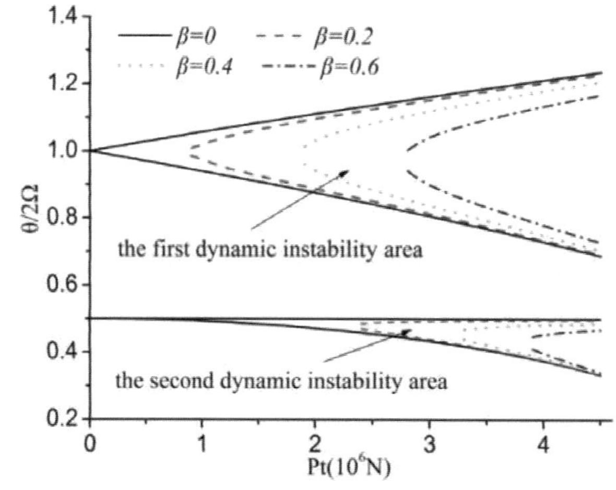

Figure 4. Parametric vibration curves of telescopic boom with different damping coefficients.

According to figure 4, it can be seen that as the damping coefficient gradually increases, the instability region of the telescopic boom decreases, the existence of damping makes the instability region smaller, and the influence on the second instability region is more obvious. At the same time, the existence of damping "cuts offs" the unstable region connected with the ordinate, which makes it impossible for the structure of the crane's telescopic boom to be unstable when the load amplitude P_t is small. The results show that the existence of damping reduces the possibility of dynamic instability of the crane telescopic boom, that is, the presence of damping is conducive to structural dynamic stability.

5. Conclusions

In this paper, based on the energy method, the Hamiltonian principle is used to establish the parametric vibration equation of the crane telescopic arm expressed in the form of Mathieu equation. The critical frequency equation of the dynamic instability boundary of the telescopic boom is deduced, and the dynamic stability region and the dynamic instability region of the telescopic boom are determined, even more, the impact of damping on dynamic stability is discussed. The results show that the dynamic instability region increases with the increase of the amplitude of the axial resonance force P_0. When $P_0=\rho$, the structure loses the bearing capacity regardless of the value of P_t; With the gradual increase of the damping coefficient, the dynamic instability region decreased, and the impact on the second dynamic instability area is more obvious.

Acknowledgments

This work was financially supported by the National Natural Science Foundation of China (51505304), the Natural Science Foundation of Liaoning Province (201602639), the Science Research Project of Liaoning Provincial Education Department (L2015444) and the Key Science and Technology Research and Development Program of Shenyang (17-49-2-00). And the reviewers are thanked for thought-provoking suggestions that helped to improve the paper.

References

[1] Du L, Lu N and Lan P 2014 Accurate analysis of lateral displacement and stability of elastically supported step columns *J. of Harbin Eng. Univ.* **35(8)** 993
[2] Lu N and Du L 2015 Accurate analysis of lateral stiffness and axial force of multi-stage stepped column and its practical calculation formula *Eng. Mech.* **32(8)** 217
[3] Lu N, Lan P and Bai H 2000 The Accurate Theoretical Solution to the Stability Analysis of Crane Box Type Telescopic Arm *J. of Harbin Univ. of Civil Eng.and Arch.* **33(2)** 89
[4] Bolotin V V 1964 *The dynamic stability of elastic system*(San Francisco: Holden-Day)
[5] Sun Q 1996 Analysis dynamical stability of rod *J. of Anhui Inst. of Arch. and Ind.* **4(1)** 38
[6] Li X and Wang X 2008 Dynamic Stability Analysis of Buckling Restrained Brace *J. of Huazhong Univ of Sci. and Tech.***25(4)** 223
[7] Pavlovic R, Kozic P 2007 Dynamic Stability of a Thin-walled Beam Subjected to Axial Loadsand End Moments *J. of Sound and Vibr.* **301** 690
[8] De Rosa M A, Auciello N M 2008 Dynamic Stability Analysis and DQM for Beams with Variable Cross-section *Mech. Res. Com.***35** 187
[9] Liu S, Lu N and Wang Y 2013 Nonlinear three-junction point beam element in dynamic stability analysis of boom *J. of Harbin Eng. Univ.* **24(3)** 358
[10] Sakar G and Sabuncu M 2003 Dynamic stability of a rotating asymmetric cross-section blade subjected to an axial periodic force *Int. J. Mech. Sci.* **45(9)** 1467
[11] Saravia C M and Machado S P 2011 Free vibration and dynamic stability of rotating thin-walled composite beams *Europ. J. Mech.-A/Solids* **30(3)** 432

IOP Conf. Series: Materials Science and Engineering **399** (2018) 012034 doi:10.1088/1757-899X/399/1/012034

The lateral displacements analysis of space lattice beam with variable section based on the flexibility coefficients method

M Lixia[1*], Y Aoqun[1] and L Shiming[1]

[1] School of Mechanical Engineering, Shenyang Jianzhu University, Shenyang 110168, China

[*] E-mail: menglixia1983@163.com

Abstract. The space lattice beam is a complex statically indeterminate rod structure which, composed with several beams, is an important process for calculating the internal force and lateral stiffness in the design analysis. First, taking the single-chip truss structure with variable section as research object, each member's force, lateral displacements and flexibility coefficients of the tapered truss with different arrangement in the form of webs are given. Finally, based on the flexibility coefficient of monolithic truss structure, considering it comprehensively with the effects of lateral stiffness caused by the chords and webs, the lateral displacement expressions are deduced for rectangular space truss structure. Result of analysis of examples show that the application of the method mentioned in this paper to calculate the lateral displacement of space lattice beam with variable section is correct and effective which can be used for practical engineering.

1. Introduction

Due to the reasonable stress, material saving and light weight, the lattice structure has become one of the main structural forms of crane metal structures and steel structures. Tower cranes' body and booms, outrigger of gantry crane, power transmission towers, and other steel structures are all complex statically indeterminate structures composed of numerous members, its internal force and stiffness calculation are important for designing and analysis in engineering. The small flexural rigidity of the lattice members will result in large lateral displacement when the structure is working, and its overall stability will also become weaker. Therefore, the lateral deformation is clearly required to be controlled in the design specification of steel structure [1]. The space lattice structure with variable cross-section is a typical statically indeterminate rigid frame structure, and the connecting form of the members is a rigid connection, which makes the internal force of each member become difficult to be obtained. However, through theoretical studies, scholars found that, when the ratio of the length of the rod to its span is large enough, the bending stress caused by the joint stiffness is less than 5% of the axial stress, and the effect of the stiffness of the joint can be ignored [2], the analysis of spatial lattice structure can be solved as spatial truss structure.

There are two basic methods to solve the internal force of the lattice member: One is to decompose the space truss into plane truss to calculate it, and then the internal force of the same rod should be superimposed to find the final answer [3]. Another method is to establish the finite element model, which is used for solution analysis [4, 5]. Although it is convenient to use finite element software to solve the internal forces of the structure, it has to establish different finite element models for various structures and various sizes; the modelling is complex and complicated and therefore cannot be applied to the subsequent theoretical research.

The lateral displacement of the lattice structure which is composed with several members is an important parameter of stiffness control in structural design. How to calculate the lateral displacement of such lattice structures quickly and accurately is an important issue which is worthy of attention. The lateral displacement of the truss is related to the arrangement form of the webs, so different arrangements will occur due to different answers. In the previous algorithms, due to the incomplete consideration of the connection between the chords, webs, and the relationship of internodes and internodes, a considerable error was always caused [6-8], and, moreover, the mentioned problem of space lattice beam with variable cross-section is researched rarely.

Based on the above reasons, the internal forces and lateral deformation of space lattice beam with variable cross-section are studied in this paper, the single-chip truss structure with variable section is used as the research object, according to the balance relationship between the forces of the internodes, the expressions of internal force and corresponding structural flexibility coefficients of plane truss are deduced. Then, the expressions of the lateral displacement of the space variable cross-section lattice member are obtained, which provides a new, simple and effective calculation method for the design calculation of the space lattice beam with variable cross-section.

2. Internal force analysis of lattice beam with variable cross-section

2.1. Web arrangements of planar lattice structure

The lattice structure with variable cross-section is usually composed by four monolithic planar structures, according to the arrangement of webs, the most used types of planar truss are shown in figure 1.

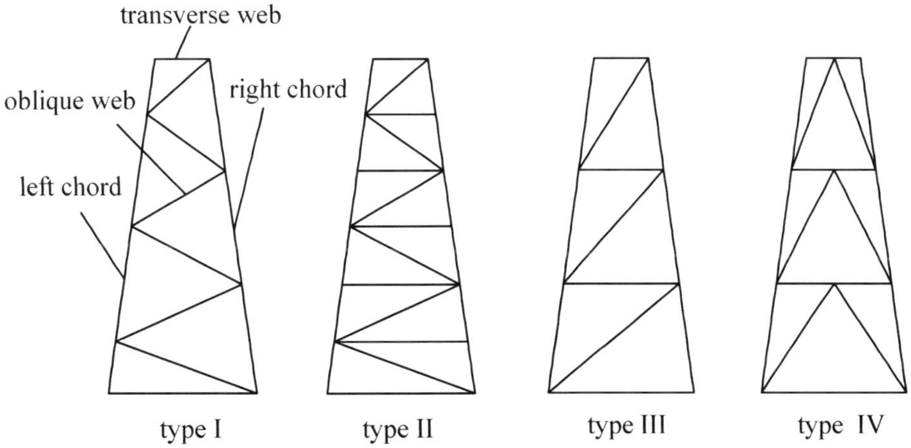

Figure 1. Sample of truss with different web arrangements.

The single-chip truss can form different types of space truss structures with different combinations; the expressions of the internal force of each type are different, but they have the same analytical method. Therefore, taking type I as an example, the derivation of the internal force expression of the chords and webs under the horizontal concentrated force are given. By similar use, the internal force expressions of other types are gained.

2.2. Internal force calculation of planar lattice structure

In order to save space, the truss structure of type I in figure 1 is rotated for 90° counterclockwise. The mechanical model is shown in figure 2.

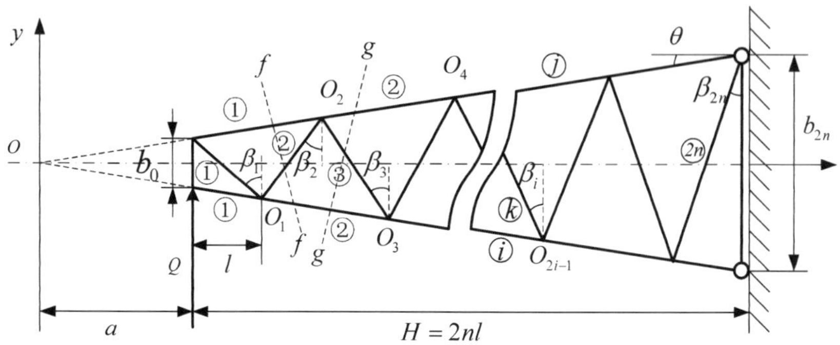

Where, n—number of internodes; θ—oblique deflection angle of chord, namely the angle between the chord and the horizontal direction; l—spacing of web, distance between each internodes is equal; β_k—the angle between oblique web and vertical direction ; b_0,b_{2n}—small end width and big end width of truss; b_k—width of truss cross section corresponding to the k_{th} oblique web; i, j, k—the number of left chord, right chord and oblique web, it is numbered from left to right; N_i, N_j. N_k—the internal forces of the i_{th}, j_{th} and k_{th} member under lateral concentrated force Q.

Figure 2. Mechanical model of monolithic tapered truss structure.

Assuming the pulled force of the member as positive, and the pressured force as negative, the section method is used to solve the internal force of each member.

(1) Right chord force

Take the section f—f for analysis, taking moments on O_1, O_3, O_{2j-1} points respectively, then, the internal force of the right chord of the tapered truss can be expressed as:

$$N_j = -\frac{Q \cdot l \cdot (2j-1)}{b_{2j-1} \cos\theta} \quad (j = 1,2,3...n) \tag{1}$$

(2) Left chord force

Take the section g—g for analysis, taking moments on O_2, O_4, O_{2i-2} points respectively, then, the internal force of the right chord of the tapered truss is deduced.

$$N_i = \frac{Q \cdot l \cdot (2i-2)}{b_{2i-2} \cos\theta} \quad (i = 1,2,3...n+1) \tag{2}$$

(3) Oblique web member force

a) As shown in figure 3(a), when k=1， 3， 5…2n-1， in the corresponding cross section i=j=(k+1)/2, according to the equilibrium condition of force

$$N_i \sin\theta + N_k \cos\beta_k = Q + N_j \sin\theta \tag{3}$$

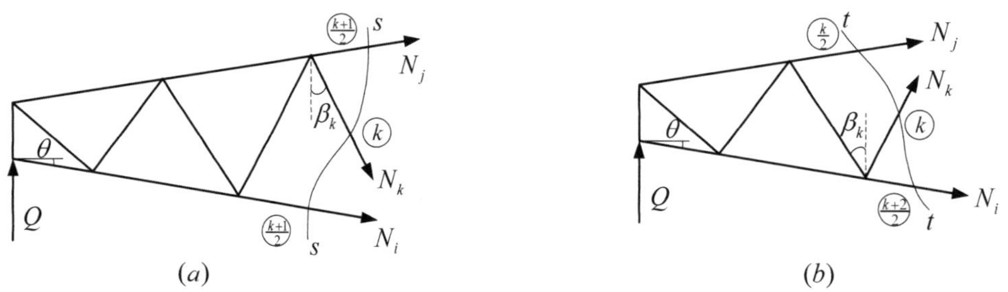

Figure 3. Component's cross-sectional view of oblique webs numbered odd-even.

Substitute the equation (1) and (2) into equation (3), we have:

$$N_k = \frac{Q}{\cos\beta_k}\left[1 - \left(\frac{k}{b_k} + \frac{k-1}{b_{k-1}}\right) l \tan\theta\right] \quad (k = 1,3,5\ldots 2n\text{-}1) \tag{4}$$

b) As shown in figure 3(b), when k=2, 4, 6…2n, in the corresponding cross section $i=(k+2)/2$, $j=k/2$, based on the equilibrium condition of force, we find:

$$N_i \sin\theta - N_k \cos\beta_k = Q + N_j \sin\theta \tag{5}$$

Taking the formula (1) and (2) into formula (5), and we can get:

$$N_k = \frac{-Q}{\cos\beta_k}\left[1-\left(\frac{k}{b_k}+\frac{k-1}{b_{k-1}}\right)l\tan\theta\right] \quad (k=2,4,6\ldots 2n) \tag{6}$$

The synthesis of formula (4) and (6) are:

$$N_k = \frac{(-1)^{k+1}\cdot Q}{\cos\beta_k}\left[1-\left(\frac{k}{b_k}+\frac{k-1}{b_{k-1}}\right)l\tan\theta\right] \quad (k=1,2,3\ldots 2n) \tag{7}$$

Also known from figure 2,

$$\tan\theta = \frac{b_k - b_0}{2kl} = \frac{b_{k-1}-b_0}{2(k-1)l} \tag{8}$$

$$\tan\beta_k = l/(b_k - l\tan\theta) \tag{9}$$

Substituting equations (8) and (9) into equation (7), the internal force of the oblique webs of the variable cross section truss can be expressed as:

$$N_k = (-1)^{k+1}\cdot\frac{b_0}{l}\cdot\frac{Q\sin\beta_k}{1-\sec^2\theta\sin^2\beta_k} \quad (k=1,2,3\ldots 2n) \tag{10}$$

In the same way, the tapered truss structures of II, III and IV types in figure 1 are studied, the internal forces of the members under the horizontal force Q are calculated, and the expressions are listed in table 1.

Table 1. Truss's force under different arrangements of webs.

Type	I	II	III	IV
Left chord N_i	$\frac{Ql(2i-2)}{b_{2i-2}\cos\theta}$ $(i=1,2\ldots n+1)$	$\begin{cases}\frac{Ql(i-1)}{b_{i-1}\cos\theta} \\ (i=1,3,\ldots 2n-1) \\ \frac{Ql\cdot i}{b_i\cos\theta} \\ (i=2,4,\ldots 2n)\end{cases}$	$\frac{Ql(i-1)}{b_{i-1}\cos\theta}$ $(i=1,2,\ldots n)$	$\frac{Ql(i-1)}{b_{i-1}\cos\theta}$ $(i=1,2,\ldots n)$
Right chord N_j	$-\frac{Ql(2j-1)}{b_{2j-1}\cos\theta}$ $(j=1,2,\ldots n)$	$\begin{cases}\frac{-Ql\cdot j}{b_j\cos\theta} \\ (j=1,3,\ldots 2n-1) \\ \frac{-Ql(j-1)}{b_{j-1}\cos\theta} \\ (j=2,4,\ldots 2n)\end{cases}$	$\frac{-Ql\cdot j}{b_j\cos\theta}$ $(j=1,2,\ldots n)$	$-\frac{Ql(j-1)}{b_{j-1}\cos\theta}$ $(j=1,2,\ldots n)$
Left oblique web N_k	$\frac{b_0}{l}\frac{Q\sin\beta_k}{1-\sec^2\theta\sin^2\beta_k}$ $(k=1,3,\ldots 2n-1)$	$\frac{b_0}{l}\frac{Q\sin\beta_k}{1-\sec^2\theta\sin^2\beta_k}$ $(k=1,3,\ldots 2n-1)$	$\frac{b_0}{l}\frac{Q\sin\beta_k}{1-\sec^2\theta\sin^2\beta_k}$ $(k=1,2,\ldots n)$	$\frac{Qb_0\tan\beta_k}{4l(\cos\beta_k-\tan\theta\sin\beta_k)}$ $(k=1,2,\ldots n)$

Right oblique web N_k	$-\frac{b_0}{l}\frac{Q\sin\beta_k}{1-\sec^2\theta\sin^2\beta_k}$ $(k=2,4,\ldots 2n)$	$-\frac{b_0}{l}\frac{Q\sin\beta_k}{1-\sec^2\theta\sin^2\beta_k}$ $(k=2,4,\ldots 2n)$	——	$\frac{-Qb_0\tan\beta_k}{4l(\cos\beta_k-\tan\theta\sin\beta_k)}$ $(k=1,2,\ldots n)$
Left transverse web N_h	$-Q$	$\begin{cases}-Q \\ (h=0) \\ 0 \\ (h=1,2,\ldots n)\end{cases}$	$\begin{cases}-Q \\ (h=0) \\ -Q\cdot\frac{b_0}{b_h} \\ (h=1,2,\ldots n-1)\end{cases}$	$\begin{cases}-Q \\ (h=0) \\ -Q\cdot\frac{b_0}{2b_h} \\ (h=1,2,\ldots n-1)\end{cases}$
Right transverse web N_h	——	——	——	$Q\cdot\frac{b_0}{2b_h}(h=1,2,\ldots n-1)$

Note：①Left（Right）oblique web：That is the oblique member which lower end point is on the left (right) chord. ②Left (right) transverse web：That is, the transverse member connected with the left (right) chord. ③The internal force of rest unlabeled members default to zero.

3. Analysis of lateral displacement of lattice structure with variable cross-section

3.1. Lateral displacement of plane structure

Taking the truss structure of the type I web arrangement shown in Fig. 1 as an example, a specific solution calculation is performed. Unit force in the *j* direction $F_j=1$ produces a displacement δ_{ij} along the *i* direction, then δ_{ij} is called the flexibility coefficient. Therefore, according to the definition of the flexibility coefficient, the unit load method can be used to obtain:

$$\delta_{ij}=\sum_{k=1}^{num}\frac{\overline{N}_{ki}\cdot\overline{N}_{kj}\cdot l_k}{EA_k} \tag{11}$$

where *num* is the total number of members; $\overline{N}_{ki}$ is the internal force of the k_{th} member caused by the unit force acts in the *i* direction (N); $\overline{N}_{kj}$ is the internal force of the k_{th} member caused by the unit force acts in the *j* direction (N); l_k is the length of the k_{th} member (m); A_k is the section area of the k_{th} member (m^2).

Then, the lateral displacement Δ_1 under the corresponding force F can be expressed as:

$$\Delta_1=F\delta_{11} \tag{12}$$

That is, the lateral displacement of the top of the truss depends on the flexibility coefficient of the structure.

In sum, combining the equation (11) with the equation (12), the top lateral displacement of planar tapered truss is gained:

$$\Delta=\sum_{i=1}^{n_x}\frac{\overline{N}_iN_il_i}{EA_x}+\sum_{k=1}^{n_f}\frac{\overline{N}_kN_kl_k}{EA_f}+\sum_{h=0}^{n_h}\frac{\overline{N}_hN_hl_h}{EA_h} \tag{13}$$

where A_x, A_f, A_h is the cross section area of truss chord, oblique webs and transverse webs respectively (m^2); n_x, n_f, n_h is the total number of truss chord, oblique webs and transverse webs respectively; $\overline{N}_i$, $\overline{N}_k$, $\overline{N}_h$ is the internal force of the i_{th} chord, the k_{th} oblique web and the h_{th} transverse web under the influence of the unit horizontal force respectively(N); l_i, l_k, l_h is the length of the i_{th} chord, the k_{th} oblique web and the h_{th} transverse web (m).

Equations (1), (2) and (10) are substituted into equation (13), then the lateral displacement of the monolithic truss structure in the webs arrangement of type I is obtained:

$$\Delta = \frac{2Ql^3}{EA_x \cos^3\theta} \cdot \sum_{i=1}^{n} \left[\left(\frac{2i-1}{b_{2i-1}} \right)^2 + \left(\frac{2i-2}{b_{2i-2}} \right)^2 \right] + \frac{4Ql^3 n^2}{EA_x b_{2n}^2 \cos^3\theta}$$
$$+ \frac{Qb_0^{\ 2}}{EA_f l} \cdot \sum_{k=1}^{2n} \left[\frac{\sin\beta_k}{\left(1-\sec^2\theta \sin^2\beta_k\right)^2} \right] + \frac{Qb_0}{EA_h} \tag{14}$$

where $b_k = b_0 + 2kl\tan\theta$; $\beta_k = \arctan\left(\dfrac{l}{b_k - l\tan\theta} \right)$.

The same method is used to obtain the tip lateral displacements of the tapered monolithic truss of various web arrangements as shown in figure 1, the corresponding expressions are shown in table 2.

Table 2. Tip lateral displacements of tapered monolithic truss of different webs arrangements.

	Tip lateral displacements
I	$\frac{2Ql^3}{EA_x \cos^3\theta} \cdot \sum_{i=1}^{n} \left[\left(\frac{2i-1}{b_{2i-1}} \right)^2 + \left(\frac{2i-2}{b_{2i-2}} \right)^2 \right] + \frac{4Ql^3 n^2}{EA_x b_{2n}^2 \cos^3\theta} + \frac{Qb_0^{\ 2}}{EA_f l} \cdot \sum_{k=1}^{2n} \left[\frac{\sin\beta_k}{\left(1-\sec^2\theta\sin^2\beta_k\right)^2} \right] + \frac{Qb_0}{EA_h}$
II	$\frac{2Ql^3}{EA_x \cos^3\theta} \cdot \sum_{i=1}^{n} \left[\left(\frac{2i-1}{b_{2i-1}} \right)^2 + \left(\frac{2i-2}{b_{2i-2}} \right)^2 \right] + \frac{4Ql^3 n^2}{EA_x b_{2n}^2 \cos^3\theta} + \frac{Qb_0^{\ 2}}{EA_f l} \cdot \sum_{k=1}^{2n} \left[\frac{\sin\beta_k}{\left(1-\sec^2\theta\sin^2\beta_k\right)^2} \right] + \frac{Qb_0}{EA_h}$
III	$\frac{Ql^3}{EA_x \cos^3\theta} \cdot \sum_{i=1}^{n} \left[\left(\frac{i-1}{b_{i-1}} \right)^2 + \left(\frac{i}{b_i} \right)^2 \right] + \frac{Qb_0^{\ 2}}{EA_f l} \cdot \sum_{k=1}^{n} \left[\frac{\sin\beta_k}{\left(1-\sec^2\theta\sin^2\beta_k\right)^2} \right] + \frac{Qb_0^{\ 2}}{EA_h} \sum_{h=0}^{n-1} \frac{1}{b_h}$
IV	$\frac{2Ql^3}{EA_x \cos^3\theta} \cdot \sum_{i=1}^{n} \left(\frac{i-1}{b_{i-1}} \right)^2 + \frac{Qb_0^{\ 2}}{8EA_f l} \cdot \sum_{k=1}^{n} \left[\frac{\sin\beta_k}{\cos^4\beta_k \left(1-\tan\theta\tan\beta_k\right)^2} \right] + \frac{Qb_0^{\ 2}}{4EA_h} \sum_{h=1}^{n-1} \frac{1}{b_h} + \frac{Qb_0}{2EA_h}$

3.2. Lateral displacement of space structure

According to the different arrangement of web members, the unfolded schematic diagram of the commonly used tapered spatial lattice structure is given in figure 4.

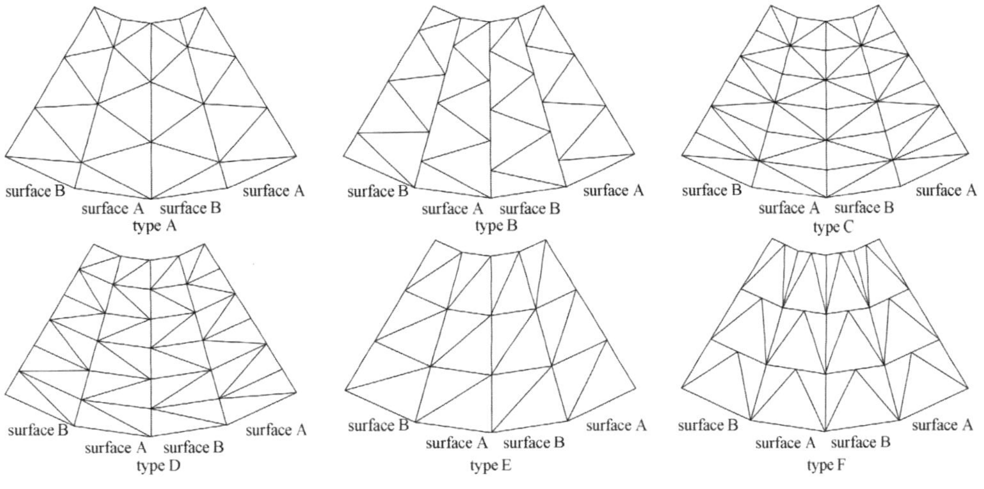

Figure 4. Unfolded schematic diagram of tapered spatial truss structure.

The mechanical model of space truss structure is shown in figure 5, when the spatial truss bear horizontal concentrated load Q at the top, the force of the entire structure can be decomposed into two plane trusses on the A surface, each bearing a horizontal load of $Q/2$, that is, the lateral displacement of the entire structure is equivalent to the lateral displacement caused by the horizontal force of the monolithic truss of A which subjected to $Q/2$.

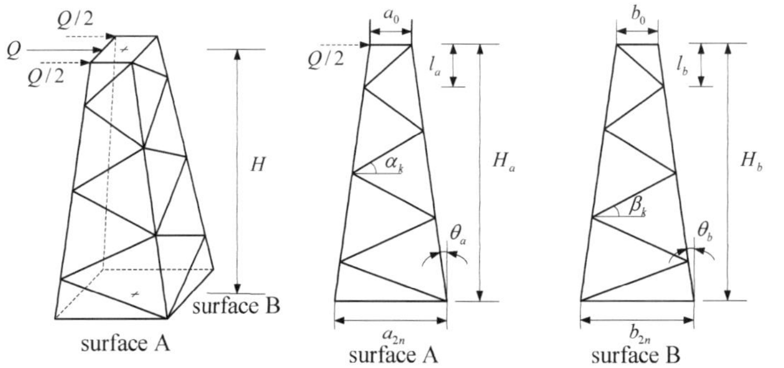

Figure 5. Mathematical model of spatial truss.

Taking the form of type A shown in figure 4 as an example, the mentioned method is used to calculate the displacement of the variable-section space truss structure, and, as is shown in figure 5, the webs arrangement type of A-plane belongs to the monolithic I type structure, therefore, based on the lateral displacement equation (14) of the monolithic truss, the lateral displacement of the spatial truss can be obtained as follows:

$$\Delta' = \frac{Ql_a^3}{EA_x \cos^3\theta_a} \cdot \sum_{i=1}^{n}\left[\left(\frac{2i-1}{a_{2i-1}}\right)^2 + \left(\frac{2i-2}{a_{2i-2}}\right)^2\right] + \frac{2Ql_a^3 n^2}{EA_x a_{2n}^2 \cos^3\theta_a} + \frac{Qa_0^2}{2EA_f l_a} \cdot \sum_{k=1}^{2n}\left[\frac{\sin\alpha_k}{\left(1-\sec^2\theta_a \sin^2\alpha_k\right)^2}\right] + \frac{Qa_0}{2EA_h} \tag{15}$$

where l_a, l_b is the span between each internodes of plan-A and plane-B respectively (m), $l_a = \frac{H_a}{2n}$, $l_b = \frac{H_b}{2n}$; H_a, H_b is the total height of plan-A and plane-B respectively (m), $H_a = \sqrt{H^2 + [(b_{2n} - b_0)/2]^2}$, $H_b = \sqrt{H^2 + [(a_{2n} - a_0)/2]^2}$; a_k, b_k is the cross-sectional width of the truss corresponding to the k_{th} oblique web of plan-A and plane-B respectively (m), $a_k = a_0 + 2kl_a \tan\theta_a$, $b_k = b_0 + 2kl_b \tan\theta_b$; α_k, β_k is the angle between the k_{th} oblique web of plan-A and plane-B with the horizontal direction respectively (rad), $\alpha_k = \arctan\left(\frac{l_a}{a_k - l_a \tan\theta_a}\right)$, $\beta_k = \arctan\left(\frac{l_b}{b_k - l_b \tan\theta_b}\right)$.

4. Examples

The verification model is the A type of web arrangement shown in figure 4, the main model parameters are as follows: the section of chords are $\phi 108 \times 10\text{mm}$, the section of webs are $\phi 70 \times 3.5\text{mm}$, the top spans of each plane are $a_0 = 1.0\text{m}$, $b_0 = 1.0\text{m}$ respectively, the bottom spans of each plane are $a_{2n} = 2.5\text{m}$, $b_{2n} = 2.5\text{m}$ respectively, and the elastic modulus $E = 200\text{GPa}$. Assuming the spacing of each internodes is constant, $l = 1.0\text{m}$, the total height H of the structure is changing, while other parameters keep unchanged. Taking transverse load as $Q = 1.0 \times 10^4 \text{N}$, the finite element results of truss-frame and rigid-frame are compared with the results of the proposed formula, which are listed in table 3.

Table 3. Comparison results of spatial truss with height changing.

Number of Internodes n	Total Height H (m)	Proposed formula value(1) (mm)	Finite Element Analysis value (mm)		Error	
			Truss-frame(2)	Rigid-frame(3)	$\frac{\lvert(2)-(1)\rvert}{(2)}$	$\frac{\lvert(3)-(1)\rvert}{(3)}$
4	8.0	1.011	1.011	1.003	0	0.7976%
8	16.0	5.938	5.938	5.921	0	0.2871%
12	24.0	18.771	18.771	18.734	0	0.1975%
16	32.0	43.464	43.464	43.390	0	0.1705%
20	40.0	83.969	83.969	83.834	0	0.1610%

As is shown in table 3, the results of the proposed method are exactly the same with the values of the truss-frame structure, which is analysed by finite element software. The results show that the proposed equations are correct, and, the same time, the proposed method which divides the spatial structure into 2 plane truss for force calculation, then composing the plane truss to form space truss for calculating structural displacement is feasible. It can also be seen that, with the increasing of height, the error between the lateral displacement of space truss-frame structure and space rigid-frame structure is gradually reduced, the errors are all within 1%, which means that it is satisfactory for practical engineering when making the spatial lattice structure a special truss for lateral displacement calculation.

5. Conclusions

In this paper, the solving of lateral displacement of the complex spatial lattice structure is transformed into a simple planar truss force analysis problem, the influence of the chord and web on the stiffness of the space truss is fully considered, and a high-precision expression of the lateral displacements of the variable-section space truss is obtained. The results show that the mentioned method can be used for lateral displacement analysis of tapered spatial lattice structure, which has the advantages of being a clear concept, of simple calculation and high precision, and of being easily applied in engineering.

Acknowledgments

This work was financially supported by the National Natural Science Foundation of China (51505304) and the Science and Technology Research Project of Liaoning Provincial Education Department (LJZ2016022). We are also thankful to the reviewers for thought-provoking suggestions that helped to improve the paper.

References

[1] European Standard (Eurocode 3) 2005 *Design of steel structure-Part 1-1: General rules and rules for building. EN1993-1-1:2005 CEM* (Brussels: CEM)
[2] Chuqing W, Huchang W and liang L 2013 Multi-scale finite element analysis of steel tube tower for large crossing transmission lines *Eng. Mech.* **30(7)** 147-152 166.
[3] Wang J, Lu N and Luo B 2007 Equivalent inertia moment method to stability analyze of lattice type beam: *Proc. of the 3rd Int.Conf. on Steel and Composite Structures*, ICSCS07 - Steel and Composite Structures pp 311-315

[4] Baocheng Z, Qiang G, Lin S, Ruoquan H and Tianbing H 2010 Finite element analysis load-bearing capacity of concrete-filled tubular steel column-staggered truss structure *J. Harbin Institute of Technology* **42(1)** 244
[5] Xia L 2012 The finite element analysis of truss beam structure based on ANSYS *Adv. Mat. Res., Adv. Ind. Civil Eng.* 594-597 2939-2944
[6] Yuan X, Jun L, Nianli L and Daqi W 2013 The analysis of stiffness and stability of luffing jib cranes based on theory of second order: *App. Mech. Mat., Mech. and Mechatronics* **419** 197
[7] Lixia M, Nianli L and Shiming L 2011 Exact expression of element stiffness matrix for a tapered beam and its application in stability analysis *Adv. Mat. Res., Adv. Ind. Civil Eng.* 255-260 1968-1973
[8] Wattanamankong N, Petchsasithon A and Dhirasedh S 2017 Analysis of Lateral Buckling of Bar with Axial Force Accumulation in Truss: *IOP Conference Series: Materials Science and Engineering* **216(1)**

Molecular dynamics simulation and experimental study on Si_3N_4-GCr15 of thin film lubrication

Z Lixiu[1,2], L Bing[2*], W Junhai[1,2], T Tian[2] and S Qinghua[3]

[1] Testing and Analysis Center, Shenyang Jianzhu University, Shenyang, China
[2] Mechanical College, Shenyang Jianzhu University, Shenyang, China
[3] Party and government office, Shenyang Jianzhu University, Shenyang, China

[*] E-mail:70492819@qq.com

Abstract. The molecular dynamics analysis and experimental investigation of thin film lubrication of silicon nitride (Si_3N_4) and bearing steel(GCr15) under the n-hexadecane-based lubricant were studied. A molecular dynamics method was used to establish a molecular model with Materials Studio7.0 software. Forcite module for Geometry Optimization, Dynamics and Confined Shear in three steps. The effects of pressure and velocity on lubrication properties such as van der Waals energy and shear stress of film lubrication were explored. The Rtec friction and wear tester was used to test the Si_3N_4-GCr15 friction pair with the change of speed and load.

1. Introduction

Si_3N_4-GCr15 friction pair is hybrid ceramic ball bearings, which is the most widely used in high-speed electric spindle bearings, that is the rolling body uses hot isostatic pressing Si_3N_4 ceramic, the inner and outer rings still are the steel ring. Silicon nitride has high strength, hardness and fracture toughness in engineering ceramic systems. Compared with the traditional steel bearing, the hybrid ceramic ball bearing with hot isostatic pressing Si_3N_4 ceramic has the advantages of high speed performance,high dynamic stiffness, low temperature rise and good thermal stability. Liquid lubricant is now commonly used as a common high-speed bearing lubricant, the lubricant forms micrometer lubricant film on the surface of parts, in-depth study of the characteristics of lubricating films, the molecular dynamics method were introduced. Recent years, molecular dynamics has made great progress in the lubrication, it can essentially reveal the principle of friction lubrication. Shang Shunshi et al [1] used self-design friction tester to test the film lubrication. It was demonstrated that the coefficient of friction was remarkably influenced by rotate speed and contact area of friction under light load, but the influence of the contact area could be neglected under heavy load. Hu Yuanzhong et al [2] used the molecular dynamics simulation to explore the properties of lubricants under thin film constraints which was carried out with spherical molecular fluid. Li yiya et al [3] used molecular dynamics simulation to study the lubrication performance of mineral oil base oil, and found that the role of the base oil molecules to stabilize the lubricating oil film mainly came from van der waals interaction. Bai Minli et al [4] simulated the film lubrication with n-heptane lubricant and investigated the effect of temperature and pressure on the composition, slippage and friction characteristics of the lubricant film. Lv Jizu et al [5] applied molecular dynamics method to study the lubricating effect of nanofluids between two parallel flat walls. Through the analysis of the movement and morphology of the nanoparticles, it found that the nanoparticles do rotational movement and translational movement under the influence of the wall. The agglomeration of the nanoparticles is very obvious when the pressure is relatively large. The nanoparticles in the fluid effectively played a supporting role in order to separate the friction pair. Sun Yi at el [6] who applied a molecular dynamics simulation established an interaction modle of n-hexadecane long-chain molecules used as lubrication to study the friction of the special physical phenomena. The results show that with the change of the shear velocity, the temperatures and shear stresses of the lubricant vary continuously. The phenomena of the delamination and the slippage between the layers of the lubricant appeared at each shear velocity. Zeng Fanlin et al [7] applied molecular dynamics simulations to study the nano-scale thin film lubrication behaviour of two lubricants, pure tridecane and n-tridecane with polyhedron siloxane material T_8H_8 added (the mole fraction of T_8H_8 is 10%),at different shear velocities. Rhykerd et al [8] Hoheisel et al [9]used the molecular dynamics method to calculate the film lubrication with the LJ potential energy function. The results show that the viscosity of the lubricating oil film increases with the increase of molecular dispersion. J.Gao et al [10] used classical molecular dynamics simulation and free energy calculation method to calculate the ultra-thin oil film lubrication, and found that the oil film diffusion and rheological properties are due to the relatively thick layer of 5 or 6 layers in the oil film. I.M.Sivebaek et al [11] studied the kinetic friction calculations of closed hydrocarbon "polymer" solid molecules with molecular lengths between 20 and 1400 carbon atoms to analysed the relationship between frictional shear stress and the shear rate on the different molecular length with two assumptions are made that the polymer slides on the hard substrate and the polymer slides on the polymer. U. Tartaglino et al [12] used the molecular dynamics method to study the lubrication effect of crude surfaces with n-butane and isobutane as lubricants. Tanaka et al [13] studied the dynamic characteristics of the lubricating film between two wall surfaces with the non-equilibrium molecular dynamics simulation (NEMD). JN Ding et al [14] studied the effect of surface roughness on the nanorheological properties of ultra-thin films sandwiched between two solid walls with molecular dynamics simulations. Pinzhi Liu et al [15] used the non-equilibrium molecular dynamics method with branched alkanes as model, studied the relationship between shear viscosity of lubricating oil with temperature and pressure, and received the pressure/viscosity coefficient of base oil molecules. In this paper, n-hexadecane was used as a lubricant to study the influence of pressure and velocity on the lubrication characteristics of Si_3N_4-GCr15 friction film with the molecular dynamics method, and the correctness of the simulation was verified by experiments.

2. Molecular dynamics model and simulation calculation process

2.1. Molecular dynamics model and potential energy function

The molecular model of Si_3N_4, GCr15 and lubricating oil was established by using the Materials Visualizer module. N-dozen alkanes (CnH_{2n+2}) are often used as lubricants for their research on lubricants because their molecular models are relatively close to those commonly used in engineering. The n-hexadecane was established in this paper. Figure 1 is the structure of the model, the two contact surfaces of GCr15 and Si_3N_4 with n-hexadecane are the upper wall surface and the lower wall surface, the positions of the two walls are 29 angstroms and 5 angstroms respectively. In this paper, the shear movement is required to add a vacuum layer of 20 angstroms above the GCr15.

Figure 2 is a molecular dynamics model, the size of the model is a = 10.62 Å, b = 10.62 Å, c = 59.98 Å.

The key to accuracy or not of the result of molecular dynamics simulation lies in the selection of the potential function of the interaction between atoms in the system, the potential function is the force field. Many notable discoveries in molecular dynamics simulation of material tribology are based on the L J potential energy function. The L-J potential function is a classical counter-potential function, which can reflect the trajectory of particles. This paper uses the function LJ / 126, the expression of the potential energy function being:

$$V_{LJ}=4\varepsilon[(\frac{\sigma}{r})^{12}-(\frac{\sigma}{r})^{6}] \quad (1)$$

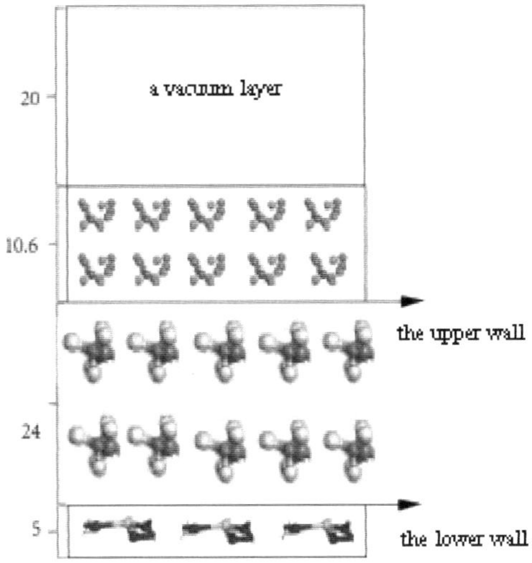

Figure1. Model structure.

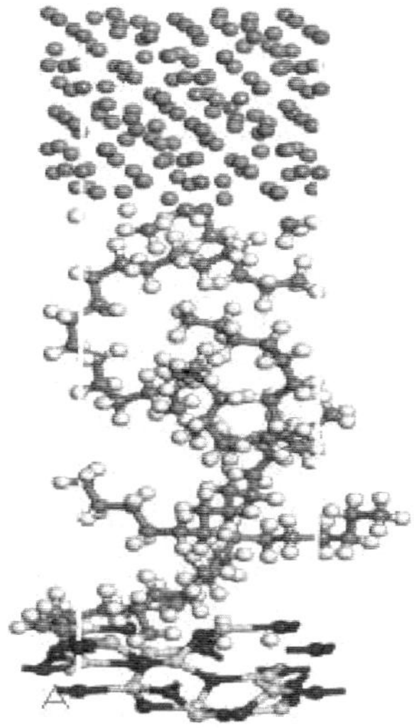

Figure 2. Molecular dynamics model.

In the formula: V_{LJ}-system potential energy, which is van der waals energy; ε-the energy parameter($r=2^{1/6}\times\sigma$ the minimum energy of the equilibrium position);σ -the length parameter (zero potential energy distance); r - the distance between atoms.

2.2. Simulation and calculation process

The calculation is based on the Forcite module in Materials Studio7.0, the force field is Universal. Non-bonded truncation of electrostatic interactions and van der Waals interactions using Atom based methods, with a truncation distance of 12.5Å. The pressure and temperature control functions are the Berendsen and NHL (Nosé-Hoover-Langevin) methods, respectively.

The simulation process is divided into three parts. Firstly, the constructed model is optimized by Geometry Optimization. The algorithm uses the accurate calculation method (Smart) to find the lowest energy conformation of the model, which is used as the simulation and calculation model. Secondly, the optimized model is subjected to kinetic relaxation, which is the equilibrium molecular dynamics simulation process, and the molecular thermal motion is considered. The potential energy model chooses Universal, and the constant temperature and pressure (NPT) is the ensemble. The initial temperature of the simulation is 298K, the pressure is respectively 0.025, 0.051, 0.076,0.102, 0.127, 0.153, 0.178, 0.201GPa. The time step is 0.5fs, the total simulation time is 600ps. The number of simulation steps is 1.2 million steps and one frame is output every 5000 steps. The process of structural optimization and kinetic relaxation is to balance the model so that the system is in a quasi-stable state for the next calculation. Finally, the non-equilibrium molecular dynamics simulation is performed, that is the shear process. Applying the shear function of Confined shear in Forcite module, the shear rate is applied on the contact surface between n-hexadecane molecules with GCr15 and Si_3N_4 molecules in the model. the speed of the contact surface is the same and the direction is opposite. Shear velocities are 1.88, 3.77, 5.65, 7.53, 9.42, 11.3, 13.19, 15.07, 16.96, 18.84Å / ps, respectively. The upper Cr steel and the lower silicon nitride move in opposite directions at a relative speed, which is similar to a frictional process. The potential energy model chooses Universal. The time step selection 0.5fs, the total simulation time is 6000ps. The number of simulation steps is 12 million steps and one frame is output every 5000 steps.

Through the combination of different pressure and velocity, various lubrication characteristics of the Si3N4-GCr15 friction pair film lubrication are obtained. After the simulation is completed, the Analysis function in Forcite is used to analyse the dynamic results and obtain the required data.

2.3.Simulation results and analysis

2.3.1.The lubrication of n-hexadecane. Lubrication action of lubricating oil usually comes from the lubrication film formed on the surface of the parts. According to the model of molecular dynamics, it can be considered that the adsorption of n-hexadecane molecules on the surface of GCr15 and Si_3N_4 molecules. This adsorption is the intermolecular force, which is van der waals forces. Because the polarity of hydrocarbon molecules is weak, van der Waals forces play a main role, while static electricity can be neglected.

In this paper, the shear movement of n-hexadecane on the surface of GCr15 and Si_4N_3 is simulated by the confined shear. Figure 3 is a simulation result model under the conditions of 0.102 GPa and 1.88 Å / ps. The distribution of n-hexadecane molecules on the surface of GCr15 and Si_3N_4 can be observed. Figure 4 shows the distribution of n-hexadecane relative concentration on the surface of chrome steel and silicon nitride with the speed of 9.42 Å / ps and the pressure of 0.102 GPa. The abscissa indicates the distance and the ordinate indicates the relative concentration of n-hexadecane molecules. It can be seen from the figure that the shearing motion has no significant effect on the molecular concentration distribution of n-hexadecane molecules on the surface of GCr15 and Si_3N_4 either before or after shearing. The two peaks in the graph are approximately at 4.266 and 22.069 angstrom (The change in the wall is due to the pressure exerted by the kinetic relaxation that compresses the entire structure), exactly the surface where n-hexadecane molecules are in contact with the GCr15 and Si_3N_4 molecules. The two surfaces have the highest concentration of molecules, the concentration between the two is relatively low. Because n-hexadecane produced an adsorption phenomenon on the contact surface, the phenomenon of quasi-solidification appeared, while the middle part still retained the characteristics of the liquid. It is also proved that the alkane molecule can adsorb on the contact surface under the action of van der Waals force and have an important lubricating effect on the part.

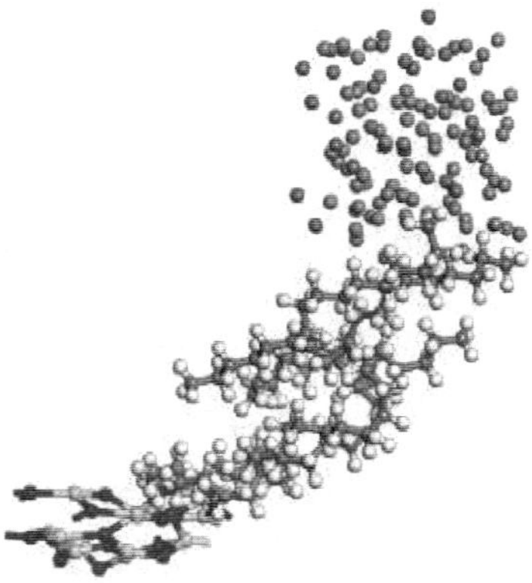

Figure 3. Simulation results.

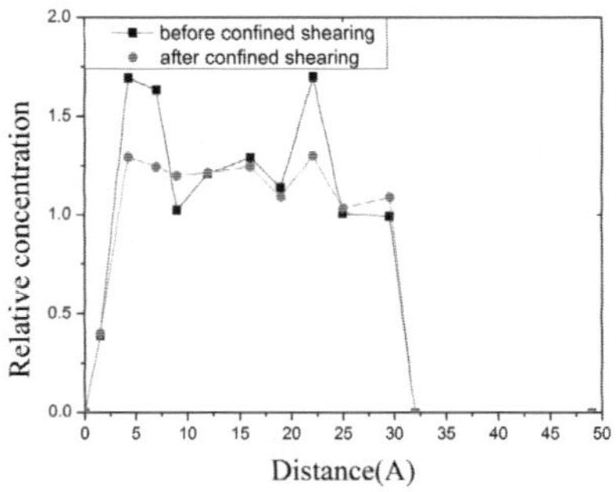

Figure 4. Relative concentration distribution of n-hexadecane before and after shearing at GCr15 and Si_3N_4 surfaces.

2.3.2. The effect of pressure on thin film lubrication. The effect of pressure on the lubrication performance of the friction pair was set at 9.42 Å/ps and the pressures were 0.025, 0.051, 0.076,0.102, 0.127, 0.153, 0.178, 0.201, respectively.

(1) The effect of pressure on Van der Waals

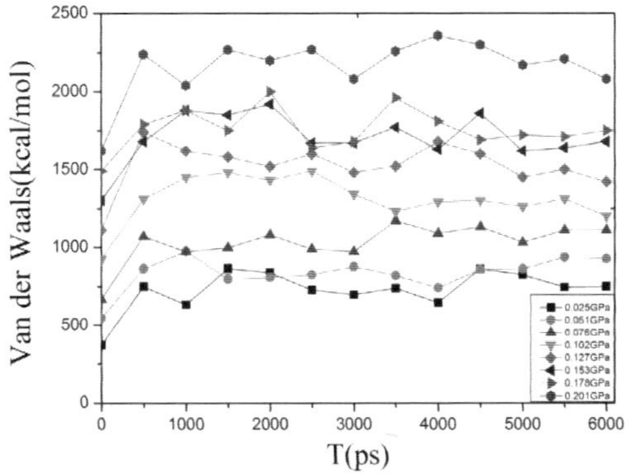

Figure 5. The curve of van der Waals change with time under different pressures.

In this paper, Van der Waals energy after the final molecular simulation is studied by combining different pressures and speeds. Changing the pressure with the shear rate of 9.42Å / ps. Figure 5 shows the changes of van der Waals energy with time under different pressures. The horizontal axis shows the time and the vertical axis shows the van der Waals energy. It can be seen from the figure that Van der Waals can have some fluctuations with the growth of time, but the growth is not much. And with the increase of pressure, Van der Waals can show a rising trend. Because the adsorption of n-hexadecane on the two contact surfaces is gradually enhanced with the increase of pressure, and the lubricating effect of the lubricating oil film on the contact surface is getting bigger and bigger.

(2) The effect of pressure on the shear stress

The n-hexadecane has a lubricating effect, but the friction will not be eliminated. Under the shearing motion, shear stress will be generated on the upper wall surface and the lower wall surface in the same direction in the x-direction and in opposite directions.

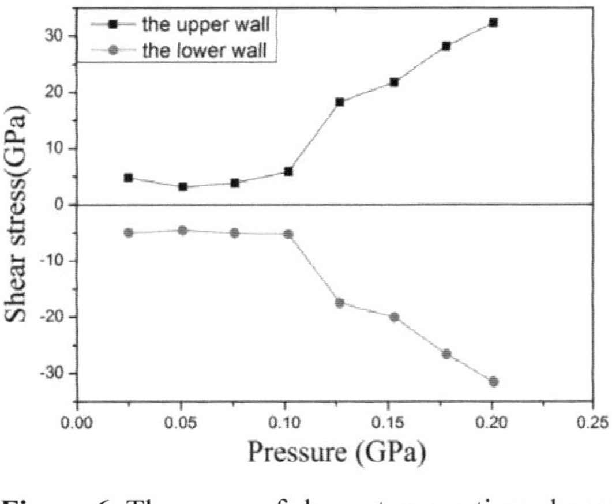

Figure 6. The curve of shear stress as time changes.

Figure 6 illustrates the curve of the shear stress with pressure change. It can be seen from the figure that the shear stress on the upper and lower walls is symmetrical. When the pressure is 0.025GPa to 0.102 GPa, the shear stress changes very little. The tendency of shear stress changes almost linearly and rises rapidly when the pressure is 0.102GPa to 0.201 GPa. The frictional resistance mainly comes from the internal friction of the lubricating film, that is, the shear force, so the shear flow characteristics of the lubricating film determines the tribological characteristics of the lubrication system. The greater the shear stress, the greater the flow resistance, so the greater the friction.

(3) The effect of pressure on the type of solid film thickness

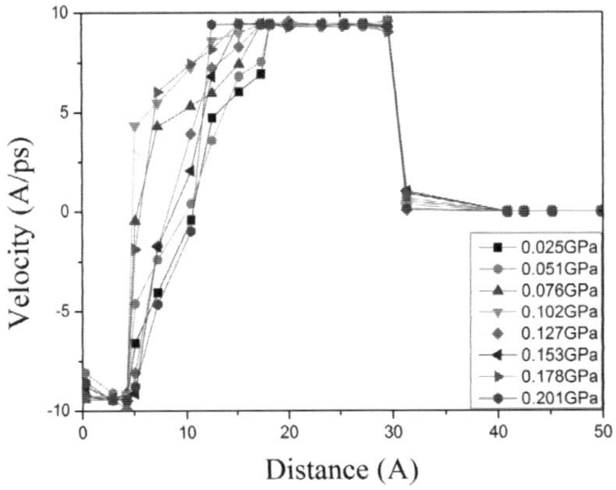

Figure 7. Velocity distribution along the thickness of the film at different pressures.

Figure 7 shows the velocity profile along the thickness of the film at different pressures. The abscissa indicates the distance in the film thickness direction and the ordinate indicates the speed. It can be seen from the figure that the curves of speed distribution under different pressures are very similar. At the lower wall 4.266 angstroms and the upper wall 22.069 angstroms, the speed is maintained at -9.42Å / ps and 9.42 Å / ps respectively. This is because there is a strong adsorption on the two walls, and the velocity is almost the same as the given shear rate. Between 4.266 and 22.069 angstroms, the velocity increases linearly, and the velocity tends to be stable at a certain time, which is consistent with the speed of the upper wall. As the pressure increases, the distance becomes less stable. Because the pressure increases, the liquid film thickness decreases, the thickness of the type of solid film increases.

2.3.3. The influence of speed on film lubrication. The effect of speed on the lubrication performance of the friction pair was set at 0.102 GPa, and the speed was 1.88, 3.77, 5.65, 7.53, 9.42, 11.3, 13.19, 15.07, 16.96, and 18.84 Å/ps, respectively.

(1) The influence of speed on van der Waals energy

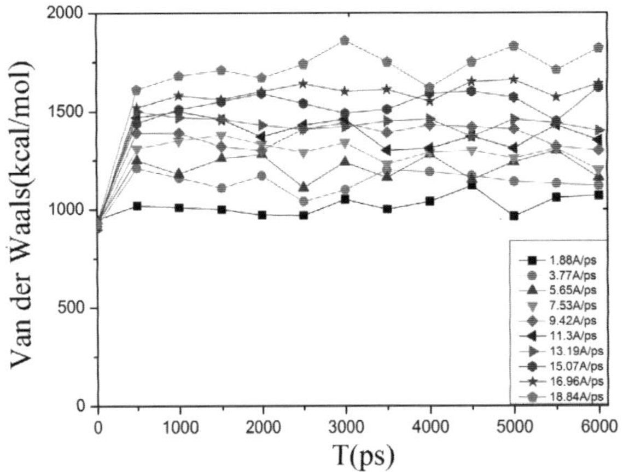

Figure 8. Curves of van der Waals at different speeds over time.

Figure 8 is the curve of van der Waals at different speeds over time, that is different speeds. The abscissa indicates the time, and the ordinate indicates the van der Waals energy. The condition of Figure 8 is a fixed pressure of 0.102 GPa, varying at different speeds. It can be seen from the figure that there will be some fluctuations in the curve with the increase of time, but van der Waals can be seen as unchanged. As the speed increases, van der Waals can show an upward trend. And when speeds respectively at 1.88Å/ps to 7.53Å/ps,9.42Å/ps to 13.19Å/ps,15.07Å/ps to 16.96Å/ps, van der waals can be very close. The greater the role of Van der Waals, the greater the adsorption effect.

(2) The influence of speed on shear stress

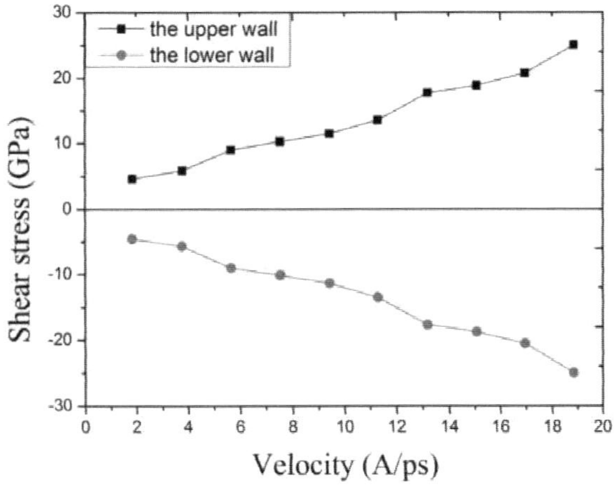

Figure 9. The curve of shear stress with the change of velocity.

This paper deals with the shear motion, which produces the same magnitude of shear stress in the opposite direction. Figure 9 shows the shear stress changes with the velocity changes, the abscissa is the velocity, the vertical axis of the shear stress. With the increase of velocity, the shear stress increases. The greater the shear stress, the greater the flow resistance, so the greater the friction.

(3) The influence of velocity on the type of solid film thickness

Figure 10 shows the velocity distribution along the thickness of the film at different velocities, the abscissa shows the distance along the film thickness, and the ordinate shows the velocity. It can be seen from the figure that, as the speed increases, the distance from the point of equilibrium tends to be further away from the upper wall surface, and the thickness of the solid film gradually decreases. At 1.88, 3.77, and 5.65 Å/ps, 7.53, 9.42, and11.3Å/ps, 13.19, 15.07, 16.96, and 18.84 Å/ps, the points that tend to balance are similar, and the thickness of the solid-like film is the same.

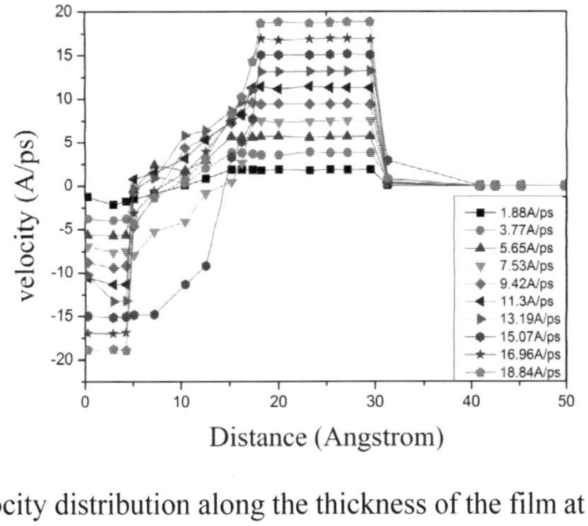

Figure 10. Velocity distribution along the thickness of the film at different velocities.

3. Experimental investigation

3.1. Test materials, instruments and processes

Figure 11. Test materials and oil.

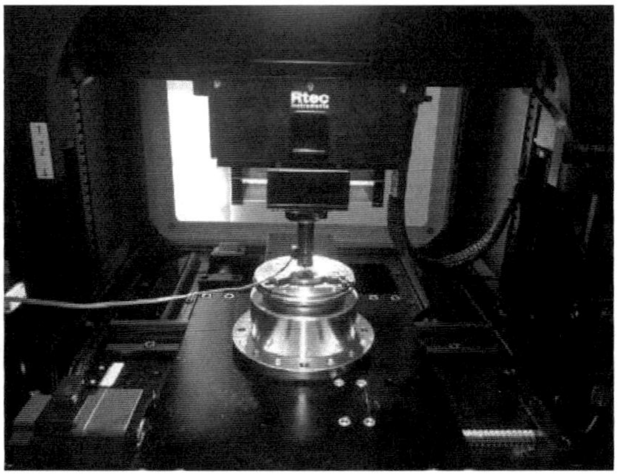

Figure12. Rtec friction and wear testing machine.

On the left in figure 11 is a GCr15 bearing steel disc with a diameter of 50 mm and a thickness of 6.5 mm. In the middle of the hot isostatic pressing Si_3N_4 ceramic balls, the diameter is 4.763mm, which is the standard bearing rolling ball size. On the right is the oil used in the test is #32 turbine oil, which has good lubricating properties and proper viscosity. Figure 12 shows the Rtec friction and wear testing machine, this paper applies the rotating module of the testing machine. The top of the graph is the force sensor where the load is applied. The bottom is the rotating module, GCr15 bearing steel disc is fixed on the rotating module. In the middle is the clamping rod, which holds the ceramic ball. In the experiment,Si_3N_4 ceramic ball is not moving, the force sensor to impose a certain load on the ceramic ball, making the ceramic ball and bearing steel disc contact. The rotating module takes the bearing steel disc as the radius of the 5mm rotation, and the two forms friction. Test run time is 50 min, the ambient temperature is room temperature, 25 ^{0}C or so. The load of test load is from 20N to 160N, which is recorded every 20N. When studying the effect of speed, the load is fixed at 80N. When studying the effect of load, the speed is fixed at 500 r / min. Fixed load and speed are taken in the middle, to avoid the impact of the limit value. Test results can be obtained friction force and friction coefficient by the force sensor.

Table 1. Test plan.

Project	Load(N)	Speed(r/min)
first	80	100, 200, 300, 400, 500, 600, 700, 800, 900, 1000
second	20 ,40, 60, 80, 100, 120, 140, 160	500

3.2. Test results and analysis

Figure 13 shows the wear scars of different test pieces. The experimental conditions are 500r/min and the loads are 40N, 60N and 100N, respectively. Wear scars of all three test pieces are about 1mm. Although the hardness of Si_3N_4 is significantly greater than that of GCr15, the GCr15 disc is worn by the Si_3N_4 balls, but the running time is not long, and the contact area does not change greatly from the wear scar, so there is no influence on the test results.

IOP Conf. Series: Materials Science and Engineering **399** (2018) 012035 doi:10.1088/1757-899X/399/1/012035

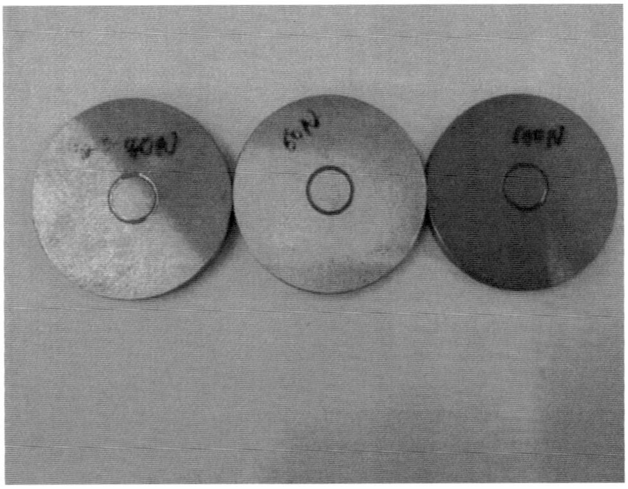

Figure 13. Wear scars of different test pieces.

(1) the friction result of the load Change

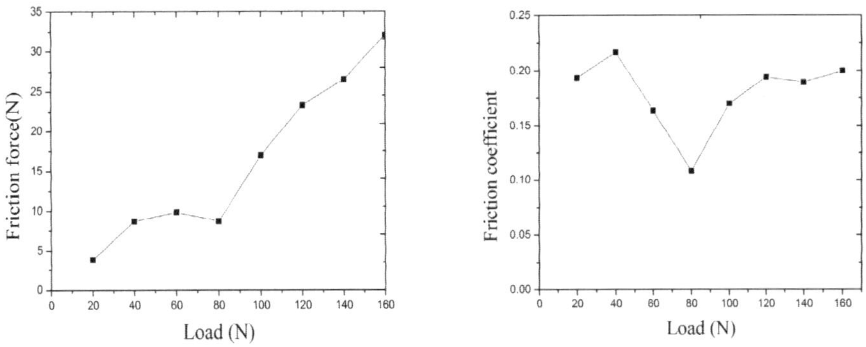

(a) the curve of friction as the load change.

(b)the curve of friction coefficient as the load change.

Figure 14. The curve of friction results with load change.

Figure 14 (a) represents the curve of friction as the load changes. The abscissa shows the load, and the ordinate shows the frictional force. (a) shows that the friction increases gradually as the load increases. The curve of friction increases before 80N is gentle, and very steeply after 80N, almost linearly. This is in agreement with the simulation, which initially showed a slow increase in friction and later a rapid growth.

The friction coefficient changes in Figure 14 (b), due to the change of friction. Figure (b) shows the change of the friction coefficient under different loads when the speed is fixed at 500r / min. The abscissa represents the load and the ordinate the friction coefficient. It can be seen that the friction coefficient shows the trend of decreasing firstly and then increasing with the increase of load, and the friction coefficient is the smallest at 80N. Because when the load is small, the lubrication film between the friction pairs of silicon nitride and bearing steel is good, and the friction is reduced, so the friction coefficient is declining; When it drops to a certain value, the increase of load affects the roughness of the contact surface of friction pair, and the friction force increases, so the friction coefficient increases with the increase of load.

IOP Conf. Series: Materials Science and Engineering **399** (2018) 012035 doi:10.1088/1757-899X/399/1/012035

(2) the friction result of the speed Change

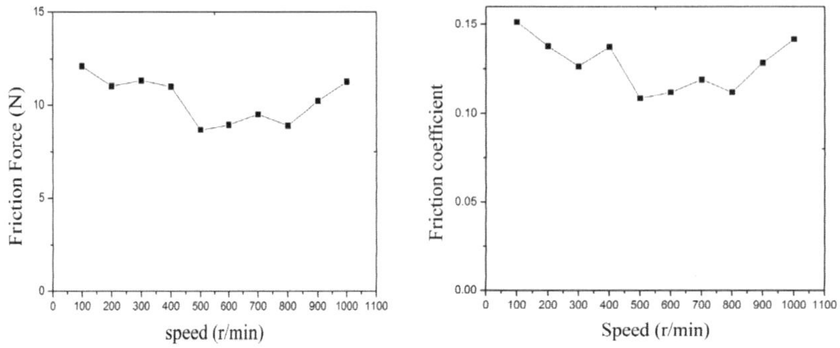

(a) the curve of friction as the rotational speed change (b)the curve of friction coefficient with the change of rotational speed

Figure 15. The curve of friction result with rotational speed.

Figure 15 (a) shows the curve of the friction with the rotation speed, the abscissa is the rotation speed, and the ordinate is the friction force. It can be seen that the friction decreases first and then increases with the increase of rotational speed, and the friction is the least at 500r / min. And the friction force of 1000r / min is no bigger than 100r / min, this change rule is consistent with the simulation result.

Figure (b) is the curve of friction coefficient with the change of rotational speed and the load is fixed at 80N. Abscissa for the speed, the vertical axis for the friction coefficient. It can be seen that the friction coefficient shows the trend of decreasing first and then increasing with the increase of the rotational speed, and the friction coefficient is the least when the rotational speed is 500r / min. Because of silicon nitride and bearing steel friction pair is dry friction when the speed is very slow. Van der Waals can be smaller, N- hexadecane in the chromium steel and silicon nitride surface does not have a strong adsorption, the film has not been fully formed, the two friction pairs in direct contact, the friction coefficient is bigger. It gradually becomes the boundary friction and fluid friction as the speed increases. Van der Waals can be gradually increased, and the role of intermolecular adsorption increased, forming a stable oil film. Therefore, the friction coefficient of these two kinds of friction is small. The speed continues to increase, the dynamic pressure oil film formation. At this point the friction coefficient is lowest. The speed then continues to increase to reach the mixed friction state, and the friction coefficient increases. However, due to the fact that the final van der Waals can be larger than at the beginning, the friction coefficient is smaller than the dry friction coefficient.

4. Conclusion

Based on the method of molecular dynamics simulation and experiment, this paper studies the lubrication characteristics of the lubrication of Si_3N_4-GCr15 friction film by using pressure and velocity.

(1) The test results show that with the change of pressure, friction force and friction coefficient are the minimum at 80N. With the change of speed, friction and friction coefficient are the minimum at 500r/min. Therefore, the lubrication effect is best when the condition is a load of 80 N and the rotation speed 500 r/min.

(2) The simulation results show that when the pressure is 0.025GPa~0.102GPa, the van der Waals energy and the solid film thickness gradually increase, the friction force remains almost unchanged, and the lubrication effect gradually becomes better; When the pressure is 0.102GPa~0.201GPa, the solid film thickness is very close, although van der Waals can increase, while the friction force increases, the lubrication effect gradually deteriorates. When the speed is 1.88Å/ps~9.42Å/ps, Although the friction force gradually increases, Van der Waals can gradually increase, and the solid

film thickness is larger, so the lubricating effect becomes better; When the speed is 9.42Å/ps~18.84 Å/ps, Van der Waals can increase, However, the thickness of the solid film decreases, the friction force increases, and the lubrication effect gradually deteriorates.

References

[1] Shunshi S, Jialin G, Feiyu G and Donglin J 2004 Primary influence factors on coefficient of friction in thin film lubrication *Petroleum Processing and Petrochemicals*

[2] Yuzhong H, Hui W, Yan G, Kun Z and Linqing G 1997 Molecular dynamics simulation of nano-scale liquid lubrication: simulation results of the liquid with spherical molecule *Chinese Journal of Materials Research* **11 (2)** 131

[3] Yiya L, Jun L, Qinghua D, Zhenyu D, Yi Z and Suo S 2017 Molecular dynamics simulation on the lubricating property of mineral base oil *Acta Petrolei Sinica*

[4] Minli B, Mei L, Yu W, Jizu L, Zhicheng H and Wang Peng 2017 Molecular dynamics simulation of structure and friction properties of thin filn lubrication *Lubrication Eng.* **42 (4)**

[5] Jizu L, Minli B, et al 2011 The molecular dynamics simulation on impact and friction characters of nanofluids with many nanoparticles system *Nanoscale Research Letters* **6 (1)** 200

[6] Zeng F and Sun Y 2006 Molecular dynamics simulation of the nano-scale thin film lubrication *Journal of Harbin institute of technology* **38 (9)**

[7] Zeng F and Sun Y 2006 Molecular dynamics simulation of nano-scale thin film lubrication and its modification *Chinese Journal of Mechanical Engineering* **42 (7)** 138

[8] Rhykerd C L Jr, Schoen M, Distler D J and Cushman J H 1987 Epitaxy in simple classical fluids in micropores and near-solid surfures *Nature* **6147** 461

[9] Hoheisel C, Vogelsang R and Schoen M 1987 Bulk viscosity of the Lennard-Jones fluid for a wide range of states computed by equilibrium molecular dynamics *J. of Chem. Phys.* **87 (12)** 7195

[10] Gao J, Luedtke W D and Landman U 1997 Layering Transitions and Dynamics of Confined Liquid Films *Phys. Review Letters* **79 (4)** 705

[11] Sivebeak I M, Samolov V N and Persson B N J Velocity dependence of friction of confined hydrocarbons *Langmuir* **26 (11)** 8721

[12] Tartaglino U, Sivebeak I M, Persson B N J and Tosatti E 2006 Impact of molecular structure on the lubricants queeze-out between curved surfaces with long range elasticity *J. of Chem. Phys.* **125 (1)** 469

[13] Tanaka K, Kato T and Matsumoto Y 2003 Molecular Dynamics Simulation of Vibrational Friction Force Due to Molecular Deformation in Confined Lubricant Film *J. of Trib.* **125(3)** 587

[14] Ding J.N, Chen J and Yang J C 2006 The effect of surface roughness on nanotribology of confined two-dimensional films *Wear* **260 (1-2)** 205

[15] Liu P, Yu H, Ren N, Lockwood F E and Wang J Q 2015 Pressure-viscosity coefficient of hydrocarbon base oil Through molecular dynamics simulations *Trib. Letters* **60 (3)** 1

IOP Conf. Series: Materials Science and Engineering **399** (2018) 012036 doi:10.1088/1757-899X/399/1/012036

Research on preparation of HFCVD diamond coated milling cutter and stone cutting performance

F Lu[1*], L Zha[1] and X Wang[1]

[1] Shenyang Jianzhu University, College of Mechanical Engineering, Shenyang, 110000, China

* E-mail: lufeng72@163.com

Abstract. The study is on the properties of diamond coating prepared on milling cutter and its effect on the cutting performance of stone. The method of hot filament chemical vapour deposition HFCVD was used to deposit diamond film coatings on the surface of the cemented carbide milling cutter. The surface morphology, composition and grain orientation of the diamond films on the cemented carbide milling cutter were analysed by using scanning electron microscopy, Raman spectroscopy and X-ray diffraction; the prepared diamond coated milling cutter was used to process the stone experiment, and the cutting performance of the diamond layer cutter with different carbon sources was analysed. The prepared diamond-coated cutter has a small diamond grain size on the surface, typically <111> crystal face disappears, and the grain orientation presents nano-crystalline diamond. Compared to the uncoated cutter, the surface roughness decreases. After coating, the diamond coating on the surface of the cutter did not show obvious shedding, and the amount of wear on the back face of the uncoated cutter was consistent.

1. Introduction

Chemical vapor deposition diamond films have excellent properties such as hardness, high modulus of elasticity, low coefficient of friction, high thermal conductivity, good self-lubrication, and chemical stability that are very close to that of natural diamonds. The field of molds and wear-resistant devices has broad application prospects. With the continuous improvement of people's material and cultural level, people's demand for stone art products is also growing. Stone is a typical hard-brittle material. The commonly used diamond sintered cutters are not suitable for high-precision machining of the stone because they cannot ensure the machining accuracy. The electroplating tools cannot be used for a long time due to excessive wear. The CVD diamond-coated tool is a new type of cutting tool. On the basis of the cemented carbide forming tool, a layer of diamond coating is deposited on the surface of the tool by chemical vapor deposition to improve the wear resistance and surface hardness of the tool. After the coating of the tool, the machining accuracy can be significantly improved, and the service life of the tool is extended to a certain extent, thereby reducing the processing cost and improving the processing efficiency.

1.1. Substrate surface pre-treatment method

The test adopts hot wire chemical vapour deposition method with methane as carbon source to deposit diamond coating on YG6 cemented carbide ball end mills. Prior to deposition, pre-treatment of the cutter surface was performed. The specific process of surface pre-treatment is as follows: First, ultrasonic cleaning in acetone solution, deionized water and ethanol solution for 5 minutes to remove impurities such as oil stains on the substrate surface, followed by alkali solution (solution mass ratio KOH:K3(Fe(CN)6):H2O=1:1:20). Etching for 10 minutes in order to remove tungsten carbide grains from the substrate and expose Co on the surface of the substrate. Then, use an acid solution (the volume ratio of the solution is H2SO4:H2O2=1: 10) Etching for 1 minute to remove Co from the surface of the substrate and ultrasonically sonicating in acetone solution, deionized water, and ethanol solution for 5 minutes to remove residual acid and alkali solution from the surface of the substrate and put it into a diamond suspension solution. Crystallization will take place for 30 minutes. A diamond-coated tool was then prepared on the HFCVD apparatus using the deposition process parameters in table 1.

Table 1. Deposition parameters of diamond films.

Temperature (℃)	Pressure (Mpa)	H2 (sccm)	CH4 (sccm)	growth time (h)
850-900	5	800	8	10

1.2. Preparation and characterization of diamond-coated tools

Scanning electron microscopy (SEM) was used to observe the surface morphology of the diamond-coated tool. The Raman spectrometer was used to analyze the surface composition of the diamond-coated tool. The X-ray diffractometer was used to study the grain orientation of the diamond-coated tool.

2. Test results and analysis

2.1. Diamond coated tool surface morphology

Figure 1. Diamond coated cutting tools surface morphologies of the as-deposited films.

Figure 1 shows the surface morphology of the diamond film of the cutter. When the methane concentration is 1%, the grain size of the film is in the order of micrometer, and the structure between the diamond grains is dense. Although the grain size is different, the exposed crystal planes are typical. 111> crystal surface, exposed crystal edges are clearly visible. Most of the grains are buried in thin films and exhibit typical octahedral features.

2.2. Raman analysis of diamond films

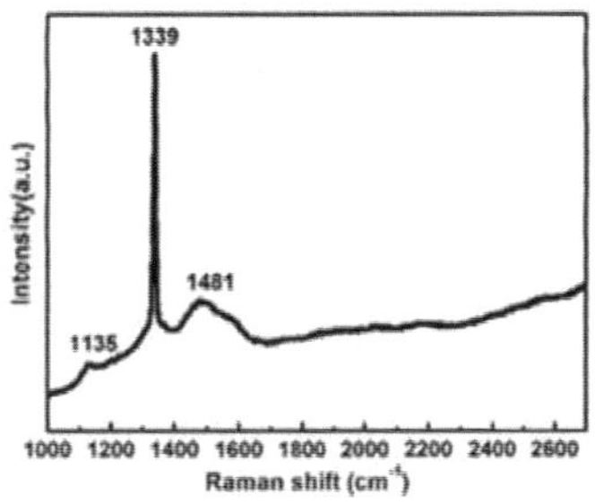

Figure 2. Raman spectra of different carbon source concentrations.

Raman spectroscopy was used to detect and analyse the diamond film. The quality of the diamond crystal and the purity of diamond can be obtained from the intensity of the Raman peak. In the Raman spectrum, the Raman characteristic peak of diamond is at 1332.5 cm^{-1}. The scattering peak of the crystalline graphite structure component generally appears at 1580 cm^{-1}, and the scattering of amorphous carbon component occurs. The peak is located at 1350-1600 cm^{-1}. The specific location is determined by the relative content of non-diamond impurities in the diamond film. The Raman spectra of diamond films are shown in the figure 2. Raman spectra of diamond films with 1% carbon source concentration have obvious diamond Raman features at 1339 cm^1, indicating that the lattice of the film is mainly with the combination of sp^3key, the Raman characteristic peak of the diamond film with 1% carbon source shifts to high wave number relative to the characteristic peak of natural diamond 1332.5cm^{-1}, indicating that there is residual compressive stress inside the film.

2.3. XRD phase analysis

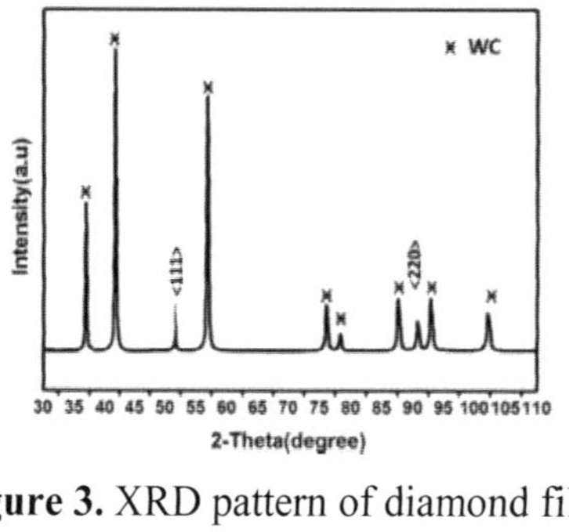

Figure 3. XRD pattern of diamond film.

The lattice constant of hot filament chemical vapour deposited diamond film is close to that of natural diamond, indicating that the film is a high quality diamond, but, due to the presence of stress in the film, lattice distortion is caused, resulting in a certain amount of deviation of the lattice constant. Shift. Figure 3 shows the diamond film XRD pattern. It can be seen from the figure that the <111> and <220> crystal planes of the diamond correspond to the diffraction peaks with 2θ of 43.9° and 75.2°, respectively. When the methane concentration is 1%, the intensity of the <111> and <220> crystal plane diffraction peaks is significant. The other obvious WC peaks in the figure are the characteristic peaks of the carbide cutting tool substrate. This is mainly due to the fact that the thin diamond film is easily penetrated by X-rays.

By analysing the half-width of the diffraction peak of the <111> plane of the film, it can be found that the <111> surface diffraction peak of the 1% methane concentration film has the smallest FWHM, the diamond crystallinity is high, and the <111> surface of the 1% methane concentration film. The ratio of the characteristic peak to other characteristic peak intensities is much higher than that in natural diamond, indicating that a 1% methane concentration film is a <111> oriented diamond film.

2.4. Bonding strength of membrane

Figure 4. Surface scanning electron microscopy diagram of diamond-coated tools.

Using a diamond-coated cutter, conducting a milling test study and cutting a white marble stone ($CaCO_3$、$MgCO_3$、SiO_2). Figure 4 shows the wear of diamond-coated ball end mills' cutting edges for one hour in the milling process. Due to the continuous impact and friction of the white marble stone on the tool flank, the diamond-coated tools all exhibit typical flank wear. Grain wear?. Figure 5a shows a 1% carbon source diamond coated tool. The flank wear of the tool is normal wear with no obvious peeling of the coating. The diamond coating reduces impact and friction during cutting. Figure 5b shows an uncoated tool. The flank wear of the tool is severe and the cemented carbide substrate is significantly damaged.

From Raman analysis under 1% methane concentration, the purity of diamond under 1% methane concentration is very high, and only a small amount is sp^2 bond carbon. Therefore, the binding force at the interface is the strongest, and there is no obvious shedding phenomenon after one hour of cutting.

3. Conclusions

By preparing a diamond-coated tool and performing a milling test, the following conclusions can be drawn:

(1) 1% carbon source concentration diamond coating, diamond grain size, sp2 bond less, <111> crystal surface characteristic peak intensity, grain orientation is microcrystalline diamond, the film base binding force is high.

(2) Diamond-coated cutters, after wear, analysis of flank wear shows that: from the perspective of diamond coating wear after cutting, the coating of 1% carbon source concentration diamond-coated cutter is combined with the matrix. When high, the wear of the back face of the milling cutter is small, there is no obvious peeling of the diamond coating, and the diamond coating has a certain effect of reducing impact and friction. The flank wear of the uncoated tool is severe and the matrix is significantly damaged.

References

[1] You H 2013 Research on Special Shapes Stone Viodeling and Machining Technology Eased on the Reverie Engineering *Shenyang Jianzhu University*

[2] Wu Y H, Yan G Y, Zhao D H et al 2014 Wear Characteristics of CVD Composite Coating Tools in Natural Marble Cutting *J. of Shenyang Jianzhu Univ. (Natural Science)* **4** 712 (in Chinese)

[3] Lv F X, Tang W Z, Song J H et al 2004 Research Progress of Diamond Film Coated Carbide Tools *Metal processing (thermal processing)* **5** 62. (in Chinese)

[4] Lv F X 2014 *Preparation and Application of Diamond Films* (Beijing: Science Press) p 3-7

[5] Liu L S, Chen X D, Tian X Q et al 2016 Research Progress and Prospect of Diamond Coating Tools *Vacuum* **53(2)** 18 (in Chinese)

[6] Ma Y P2007 *Preparation and Application of Diamond Coating Tools*](Shanghai: Shanghai Jiaotong University) (in Chinese)

[7] Wang H W. 2014 *Preparation and Application of Diamond - coated Cutting Tools for Zirconia Ceramic Processing* (Shanghai: Shanghai Jiaotong University) (in Chinese)

[8] Xue H G 2014 *Preparation and Cutting Experimental Research on High Performance and Complex Shape Diamond Film Coated Tool* (Shanghai: Shanghai Jiaotong University) (in Chinese)

[9] Polini.R 2004 Cutting force and wear evaluation in peripheral milling by CVD diamond dental tools. *China Surface Engineering* **469(50)** 161

[10] Hanyu H 2002 .Development of high performance diamond- coated drills for cutting high silicon aluminium alloy. *China Surface Engineering* **413(1)** 139

[11] Gao D 2018 Diamond and Carbon Nitride Coated Tools for Processing High Silicon aluminium alloy *China Plant Engineering* **6**

[12] Zhou Y 2017 Carbide Tool diamond-like diamond coating and its performance evaluation *Cemented Carbide* **6**

IOP Conf. Series: Materials Science and Engineering **399** (2018) 012037 doi:10.1088/1757-899X/399/1/012037

Hybrid effects on effective mechanical properties of CF/FF and BF/FF epoxy-based composites

D Luca Motoc[1*]

[1] Department of Automotive and Transport Engineering, Transilvania University of Brasov, 500024 Str. Politehnicii no. 1, Brasov, Romania

* E-mail: danaluca@unitbv.ro

Abstract. The paper explores the synergistic effects due to hybridization of different synthetic reinforcements with flax fibres on the effective mechanical properties of resulted compounds. A commercial epoxy polymer DGEBF is depolyed as embedding material. Accounting for different stacking sequences, from symmetrical to unsymmetrical architectures, the samples were subjected to 3-point bending and their effective mechanical properties retrieved. Positive hybrid effects in the range of 9,5% and 35% for BF/FF and CF/FF architectures, were encountered in the hybrid composite specimens, respectively.

1. Introduction

Natural fibre-reinforced composites captured researchers' attention through-going over the last decade. Driven either by economic and social reasons or inherent yearning to state and bring an argument on their sustainability, these materials were intensively investigated both for materials' properties and behaviour with a relatively large area of engineering applications.

Literature survey provides evidence on various combinations between natural fibres and polymer matrices. Particular to the mechanical performance of these material types, the outstanding reviews of Shah (2013), Gurunathan (2015), Jauhari (2015) and more recently of Pickering (2016) and their co-authors can be used for further insights and issue development [1-4]. Readers get accustomed not only to a large spectrum of reinforcement and matrix materials' combinations and factors limiting their performance in action, but experimentally gathered data on mechanical properties from different testing configurations.

Moreover, arguments on natural materials' potential benefits, deployment of surface conditioning measures to improve the fibre/matrix adhesion, addition of chemical modifiers as cross-linking promoters and other procedures to limit the penalties of the resulting composite material performances were intensively tested and extensively debated [5-8]. Despite these, the issues of commercially available and practical use remain an intricate and yet to be implemented at production level desire.

In addition, since through hybridization improvements on the combination's effective properties were mostly achieved by individual material selection, both fibres and matrix, by smart reinforcement layering or intimately connecting, predictability about the preferences on the composite architectures adopted by different researchers' groups and lately by various industry players worldwide can be easily identified [9-12].

The green polymer-based composites developed hitherto were deploying natural fibres acquired from cellulose/ lignocellulose sources embedded mainly within affordable polymer resins. Scaling to hybrid architectures augmented various combinations with synthetic reinforcements to address minimum requirements as structural elements or functional materials. The influence of constitutive, layering sequences, debates on fibre/ matrix interface issue upon the developed architectures' effective properties (e.g. mechanical, dynamical, thermo-physical, electrical, degradation, etc.) can be traced from already published reports [13-20].

The herein author has already approached some of the above issues on several hybrid composites out of glass and carbon fibres reinforcements and unsaturated polymer resin while investigating the influence of layering sequencing and fibre orientation upon their mechanical, dynamic and thermal effective material properties [21-22].

The present paper explores the feasibility of tailoring different hybrid architectures from natural- and synthetic-fiber reinforced polymer based prepregs, other than selected materials and tested combinations reported or delivered to the public at large. The synergetic effect revealed through hybridisation will be emphasised individually for each stacking sequence and material combination.

2. Experimental procedure

Basalt (n. BF), untreated flax (n. FF) and plain 1/1 weave carbon-fibre (n. CF) fabrics were used as reinforcements for the hybrid composite architectures. All reinforcements can be ranked as having a balanced distribution along the fabric's warp and weft directions. A commercial DGEBF epoxy resin (i.e. Epikote™ 04434) was cured with its delivered hardener under a ratio of 100 to 45 parts by weight of each constituent. The resin was chosen due to its wide availability and high thermoforming stability during laminate manufacturing.The hybrid composite laminates were produced by differently stacking nine pre-impregnated sheets out of natural and synthetic reinforcements.

Concerning the stacking sequence, materials exhibiting the highest strength values (i.e. CF, BF) were layered as external and external/ middle layers, giving rise to symmetrical and unsymmetrical hybrid architectures. Flax fibres were embedded in-between due to intrinsic poor material properties comparatively with the synthetic reinforcements. Table 1 lists the stacking layering codes used to further address the hybrid composite architectures and summarise their individual and total volume fraction within the final laminate.

The hybrid composite specimens were subjected to the flexural testing, according to DIN EN ISO 14125:1998 standard, using a mechanical testing machine INSTRON model 3369 running in a 3-point bending mode with a length span of 48 mm and 10 kN load cell. A bending rate of 1.5 mm/min was used to apply the load and an average out of five representative samples is provided. All mechanical tests were conducted at room temperature (approximately 21 ± 2 °C). Flexural strength and modulus were retrieved as claimed by the ISO 178:2010 standard.

The morphologies of composite architectures were examined deploying a scanning electron microscopy technique (SEM) on an EVO MA 25 (Zeiss, D) device, at room temperature, running in the 2.00 K x magnification mode (see figure 1 and figure 2 on representative specimens).

Table 1. Details on hybrid composites stacking sequences, assigned codes and volume fractions.

Stacking sequence	Laminate codes	Reinforcements volume fraction (vol %)		Total fibre loading (vol %)
		nf	*sf*	
□□□□□□□□□	9FF	40	-	40
■□□□□□□□■	BF/7FF/BF	21	14	35
■□□□■□□□■	BF/3FF/BF/3FF/BF	19	21	40
■□□□□□□□■	CF/7FF/CF	24	11	35
■□□□■□□□■	CF/3FF/CF/3FF/CF	23	12	35

IOP Conf. Series: Materials Science and Engineering **399** (2018) 012037 doi:10.1088/1757-899X/399/1/012037

Figure 1. SEM micrograph on FF reinforced DGEBF laminate.

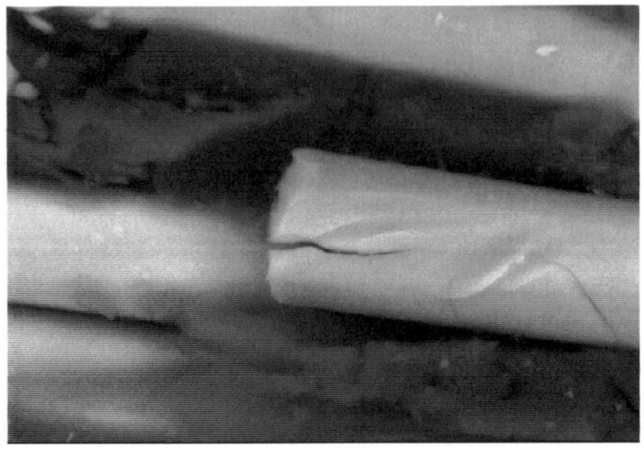

Figure 2. SEM micrograph on unsymmetrical BF/FF reinforced hybrid laminate.

3. Rule of mixtures (RoM) and hybrid rule of mixtures (RoHM)

Rules of mixtures (RoM) and hybrid mixtures (RoHM) were used to predict the effective stiffness of individual substrates and different stacked sequences of FF and CF and BF combinations, respectively. Expressions are functions of the composites' constitutive volume faction and corresponding to individual material properties:

- RoM for natural-fibre reinforced composites

$$E_{nfc} = E_{nf}V_{nf} + E_m\left(1 - V_{nf}\right) \tag{1}$$

- RoM for synthetic-fibre reinforced composites

$$E_{sfc} = E_{sf}V_{sf} + E_m\left(1 - V_{sf}\right) \tag{2}$$

- RoHM for natural/synthetic hybrid composites

$$E_c = E_{nfc}V_{cnf} + E_{csf}V_{sfc} \tag{3}$$

In the above expressions, the indices stand for: *nf* – natural fibre, *sf* – synthetic fibre, *nfc* – natural fibres reinforced composite, *sfc* – synthetic fibres reinforced composite and *c* – hybrid composite, respectively.

Departure from the RoHM predictions of the experimental values from the symmetrical/unsymmetrical combinations contributes to the hybrid effect quantification, which in turn can be ranked as positive or negative [19]. This hybrid effect highlights the synergy reflected on the mechanical properties by natural and synthetic fibres combinations.

4. Results and discussions

Polymer composite behaviour in flexure can be regarded as depending both on the reinforcements and matrix material performance as well as on the fibre/matrix interface [14, 20]. The more optimum the fibre content, the more enhanced the stress transfer from polymer matrix to the fibres, and thus, the overall mechanical property aimed.

Prior insights in the mechanical behaviour of herein composite specimens, recalling the SEM images from figures 1 and 2, respectively, the enhanced adhesion of epoxy polymer to the BF against the fragility of FF fibres after the manufacturing process can be acknowledged.

As can be seen in figure 3, hybrid composite specimens, irrespective of CF or BF layers and stacking sequence, perform better in flexure compared with the FF reinforced composite samples. The load vs. elongation curves reveal increased slope with the increase of the synthetic reinforcement content and thus an increase of the effective elastic modulus and flexural strength.

Data collected in table 2 or represented in figure 4 reflect the above statements. Thus, the outperforming attributes of CF/FF hybrid architectures has to be assigned to the presence of CF layers. An increased content of CF layers results into both enhanced modulus and flexural strength.

On the other hand, an increasing tendency was expected on the percentage elongation at break values with the increase of synthetic reinforcement's content. The latter seems to reveal a negative effect and it has to be regarded to the FF and not to the CF or BF contents.

Table 2. Flexural properties of FF and hybrid composites.

Laminate codes	Flexural strength (MPa)	Strain at break (%)	Max. load (N)
9FF	169.55 ± 14.77	2.79 ± 0.27	226 ± 17.03
BF/7FF/BF	251.65 ± 7.61	4.07 ± 0.36	411 ± 19.98
BF/3FF/BF/3FF/BF	317.59 ± 23.63	3.74 ± 0.56	552 ± 36.46
CF/7FF/CF	350.95 ± 12.92	3.16 ± 0.65	467 ± 21.51
CF/3FF/CF/3FF/CF	358.15 ± 34.79	2.18 ± 0.10	488 ± 56.62

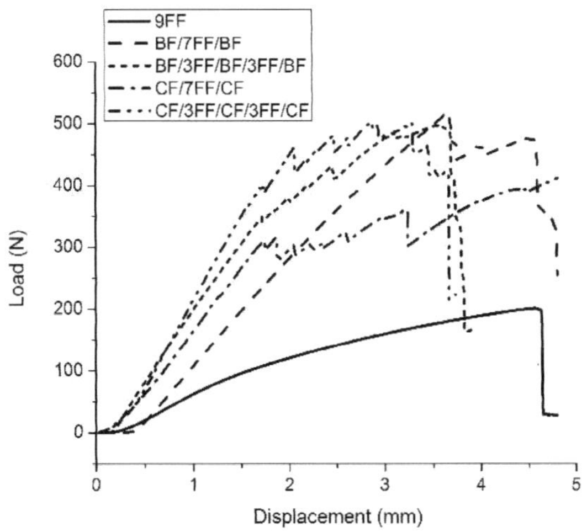

Figure 3. Load vs. displacement curves of composite specimens.

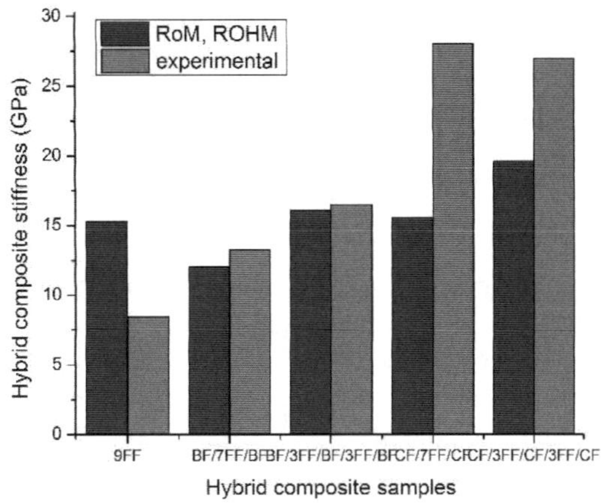

Figure 4. Theoretical vs. experimental stiffness values of samples.

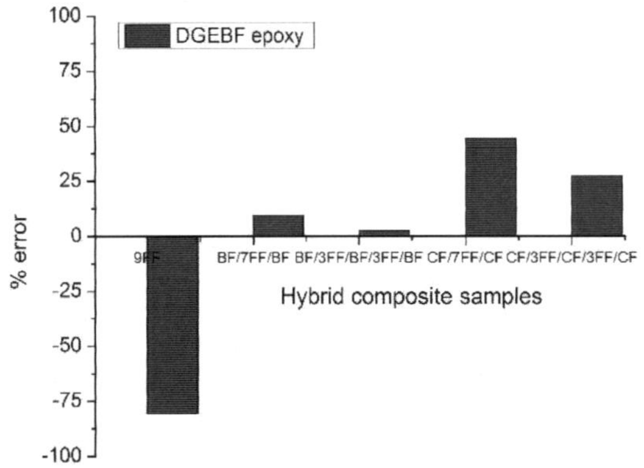

Figure 5. Percent error to aid hybrid effect characterisation.

Fibre/matrix interface has to be accounted further for these discrepancies in the mechanical properties. Thus, improper wetting and poor adhesion could be responsible for the above in addition to bending and shear failure mechanisms present with the outer and inner individual layer surfaces during the 3-point bending testing [17]. In this study, those above can be traced from accompanying SEM images with clear evidence on the inadequacy of FF fibres' outer surface.

Positive departures are obtained on BF and CF reinforced hybrid composites, either symmetrically or unsymmetrical stacked, in connection with flexural strength and elastic modulus, followed as performance indicators in the debate. Thus, the overall elastic moduli of BF hybrid architectures exhibited higher values compared with those from the FF laminate referential, with approximately 45% and 65%, respectively.

On the other hand, with CF/FF hybrid architectures, the synergetic effects are more pronounced in comparison with the BF/FF combinations, highlighting the efficiency of CF as reinforcement material. Little discrepancies were found between the symmetrical and unsymmetrical architectures, particularly the different individual and total volume fractions of reinforcements. Thus, property improvement increments by approximately 107% and 105%, for symmetrically and unsymmetrical stacked CF reinforced DGEBF composites over the FF laminate was identified, as it can be seen from data plotted in figure 4.

Further on the previous and concerning the hybrid effect, it can be accounted for from percent error values between experimentally retrieved and theoretically predicted values represented in figure 5. As it can be sized, all hybrid composite architectures reveal a positive hybrid effect on their performance indicator, with the same decreasing trend as the increase of the synthetic fibre content (e.g. from 9.3 % to 2.4 % for the BF/FF specimens). Once more, CF/FF hybrid architectures disclose their mechanical performance over the FF and BF/FF composites, owing to the CF reinforcement.

5. Conclusions

The paper aimed at debating on the effective mechanical properties and hybrid effects of several stacked sequences of synthetic (i.e. CF or BF) and natural (i.e. FF) fibre reinforced laminates based on a commercial DGEBF resin.

The synergetic effects, due to individual synthetic or natural reinforcements and various stacking sequences were tackled based on their flexural mechanical properties.The following conclusions have been drawn based on the predicted and retrieved data:

- improvements in the mechanical performances can be underlined for all hybrid composite architectures herein, irrespective of the constitutive stacking sequence and reinforcement material deployed;
- as expected, CF reinforced hybrids performed better in comparison with BF reinforced combinations for the same stacking sequence, whereas the latter appears to be an insurmountable competitor when cost issues come into focus;
- the results on overall material properties showed that symmetrically stacked CF reinforced architectures reveal similar values with unsymmetrical BF hybrid composites.

This research is expected to contribute effectively to CF and BF reinforced polymer composites market in search for novel routes for obtaining outstanding performance/cost ratios from their applications that can be sought in the area of constructions, automotive industry, transports, etc.

Acknowledgments
The corresponding author gratefully acknowledges the financial assistance of German Academic Exchange Service – DAAD that enabled and supported the internship with Fraunhofer Research Institution for Polymeric Materials and Composites - PYCO, Germany.

References
[1] Shah D 2013 *J. of Material Science* **48** 6083
[2] Gurunathan T, Mohanty, S and Nayak S 2015 *Composites: Part A* **7** 1
[3] Pickering K, Efendy M and Aruan L T 2016 *Composites Part A: Appl. Sci. Manuf.* **83** 98
[4] Jauhari N, Mishra R and Thakur H 2015 *Materials Today: Proceedings* **2** 2868
[5] Jawaid M and Khalil A H 2011 *Carbohydrate Polymers* **86(1)** 1
[6] Swolfs Y, Gorbatikh L and Verpoest I 2014 *Composites Part A: Appl. Sci. Manuf.* **67** 181
[7] Jawaid M, Khalil A H and, Alattas O 2012 *Composites Part A: Appl. Sci. Manuf.* **43(2)** 288
[8] Dicker M P M, Duckworth P F, Baker A B, Francois G, Hazzard M K and Weaver P M 2014 *Composites Part A: App. Sci. Manuf.* **56(0)** 280
[9] Pil L, Bensadoun F, Pariset J and Verpoest I 2016 **83** 193
[10] LeGault M 2016 *Composites World* 68
[11] Cheung H, Ho M, Lau K, Cardona F and Hui D 2009 *Composites Part B: Engineering* **40(7)** 655
[12] Dittenber D and Ganga Rao H 2012 *Composites Part A: App. Sci. Manuf.* **43(8)** 1419
[13] Praveen R, Jacob S, Murthy C, Balachandran P and Rao Y 2011 *Cryogenics* **51(2)** 95
[14] Muralidhar B 2013 *Mat. & Design* **52** 835
[15] Zhu J, Njuguna J, Abhyankar H, Zhu H, Perreux D and Thiebaud F 2013 *Ind. Crops and Prod.* **50** 68
[16] Lazko J, Landercy N, Laoutid F, Dangreau L, Huguet M and Talon O 2013 *Polymer Degradation and Stability* **98(5)** 1043
[17] Assarar M, Zouari W, Sabhi H, Ayad R and Berthelot J 2015 *Composite Structures* **132** 148
[18] Bertomeu D, García-Sanoguera D, Fenollar O, Boronat T and Balart R 2012 *Polymer Composites* **33(5)** 683
[19] Marom G, Fischer S, Tuler F and Wagner H 1978 *J. of Mat. Sci.* **13(7)** 1419
[20] Ozturk S 2005 *J. of Mat. Sci.* **40(17)** 4585
[21] Motoc Luca D, Ferrandiz Bou S and Balart R 2015 *J. of Comp. Mat.* **49(10)** 1211
[22] Motoc Luca D, Ferrandiz Bou S and Balart R 2018 Thermal properties comparison of hybrid CF/FF and BF/FF cyanate ester-based composites *J. of Thermal Analysis and Calorimetry* 10.1007/s10973-018-7222-y

Tools to quantify environmental impact and their application to teaching: projects City-zen and HEREVEA

M Marrero[1,4,*], C Martin[2], R Muntean[3], P González-Vallejo[1] and M D Rodríguez-Alba[1]

[1] University of Seville, Spain
[2] Delft Netherlands University of Technology, Netherlands
[3] Transilvania University of Brasov, Romania
[4] Department of Building Construction II. School of Building Engineering. Av. Reina Mercedes 4A, 41012, Seville, Spain

[*] E-mail: madelyn@us.es

Abstract. This article presents several strategies to teach university students and professionals in the sector how to reduce the environmental impact of our cities. Firstly, the European City-zen project is summarized and its application to the city of Seville, more specifically to Tirode Linea, a working class neighbourhood, is described; the viability and functionality of the neighbourhood are analyzed and improvements are proposed so that it can become an area that attains zero emissions. Secondly, the HEREVEA project is presented, which developed software for the feasibility analysis and proposals to improve neighbourhoods. Its usefulness is discussed in a case study carried out in the same neighbourhood. Finally, the experience of how to transmit all this knowledge through university teaching is presented.

1. Introduction

A new conception of learning has been imposed in Higher Education, which is more focused on the learner than on teaching [1-3]. One of the elements involved is the process of promoting autonomous, critical, and reflective learning, which is one of the main objectives of the European Higher Education System. The subjects and the curricula should be designed from a perspective that facilitates the training process and the future professional practice of the students and, in this sense, the selection of the most suitable didactic methods to achieve this goal depends on the scientific knowledge of its strategic profile [4]. College students must be proficient in learning to learn. Learning in the university and in subsequent professional activities requires self-regulated learners, capable of managing themselves in an environment that demands making appropriate decisions regarding what and how to learn according to the context and learning objectives. This implies awareness, management, and control of one's abilities [5].

Another important aspect, with the advent of information and communication technology, is that people and, in particular, young generations, use technological tools intensely, which requires constant attention. The disposition of students towards learning has changed; many of these student changes can be attributed to technological developments, which teachers have to face and adapt to [6]. The growing digital technologies have given rise to new learning scenarios that offer people multiple opportunities and resources to learn. This diversity of learning possibilities allows each person to build their own learning trajectory: set goals and objectives, develop their own plans, and choose among the different ways of learning offered [7].

The various types of university students also determine the best teaching methodology that must be taken into account. Currently, students are usually between the ages of 18 and 25. They are influenced by the internet, which is used to obtain the necessary information quickly. So they are very impatient students, who get bored easily, and have little capacity for concentration, as well as few communication and verbal writing skills [6]. The use of new technologies is fundamental, but it must be taught in well guided classes by teachers. There are many studies on learning strategies which show that using new technologies alone, such as computer resources, the internet or complex programs, will not guarantee good results for students [6-7]. There are studies carried out so that students have their own online internet platform with which to work on subjects, which show that the best results are related to students who have more frequent classroom classes a week, and make more use of the platform [7]. Therefore, it must be borne in mind that any new technology to be used as a new learning method must be accompanied by the correct motivation on the part of the teacher, since by itself it is not a guarantee for success. The way that groups of higher education students capitalize on technological resources is fundamental; by using them, the teacher helps students build their own personal learning environment. The key is not only to offer technological resources, but to design activities that invite, guide, and support their construction [7].

In the particular case of teaching sustainability in the construction sector, professionals and methodologies can bring the architecturally qualitative and scientifically quantitative aspects together to reveal the latent potential of our cities and people. These experts must have the necessary communication skills, personalities, and backgrounds to firmly place city stakeholders at the heart of this local and global challenge [2]. This strategy was developed within the framework of the City-zen Project 'Roadshow' (a European Union FP7 funded initiative to develop and demonstrate Zero Energy Cities) and the HEREVEA project (funded by Andalusia´s Regional Government), which has developed a software tool for the economic and environmental assessment of rehabilitation projects. Based on the experience applied with university students through the projects City-zen and HEREVEA, a way to reduce the environmental impacts in cities was proposed starting from the transformation, improvement, and recovery of neighbourhoods, and from the way those objectives were transferred to university students and professionals in the sector.

Last fall in Seville, the European project City-zen celebrated two activities, namely the Sustainable Workshops in Architecture & Technology (SWAT) Studio and the City-zen Roadshow. SWAT Studio took place at the School of Engineering of Building Engineering (ETSIE) of the University of Seville, hosting 20 international students who, along with others from the University of Seville who were studying in the Architecture Doctoral Program and the Integrated Building Management Master's Program within the School of Building Engineering, developed proposals aimed at achieving a more sustainable Seville. In the second stage, the City-zen Roadshow took place between 20th-24th November, and international sustainability experts who have already worked in other cities such as Belfast, Izmir, Dubrovnik, and Menorca, addressed sustainability issues in Tiro de Linea, Seville.

The work developed within SWAT was presented by the students at the event "Seville Sustainable City". The solutions were discussed and developed through the participation of global experts of the City-zen project, and students and local actors drew a plan to obtain a zero energy city, with high environmental quality and energy security. The Roadshow project invited people who knew the city, university staff, master / doctoral students, and professionals (architects, engineers, designers, artists, urban planners, and energy experts) to participate. Different groups dedicated between one hour and 5 days to "Seville Sustainable City" and the participants brought their experience, passion, and energy. The results were finally delivered to the city.

The HEREVEA project was funded by Andalusia´s Regional Government. As part of the activities within the project, a software tool was developed to evaluate the cost and environmental impact of the rehabilitation projects [8]. The software is easy to use and is proposed as a teaching tool that can complement the SWAT and Roadshow activities. HEREVEA uses the Ecological Footprint (EF) indicator [9] to assess the amount of land required to provide resources (grain, feed, firewood, fish, and urban land) and absorb the emissions (CO_2) of humanity. The EF and the Carbon Footprint have become two of the most widespread indicators thanks to the simplicity of their concept and their ability to place sustainability on the world agenda. The EF has been employed in the development of various assessment systems for projects in the construction sector [10-18].

A building in the Tirode Linea neighbourhood of Seville was selected, and the HEREVEA tool applied. The existing pathologies were considered and a study was carried out to determine if it was more viable to repair or demolish it and build a new one. All this was posed in class with undergrads, master's, and PhD students. The teaching strategies and tools are summarized in the following sections of the paper, and the results from these experiences presented.

2. Materials and methods

The teaching strategies to be considered should also take into account to whom they are addressed, more specifically, the university level they will be applied to and the course level, namely 1st, 2nd, or last years of the bachelor's, master's, or doctoral degree students. There are studies that raise these learning differences based on the students' grade [4]. Knowledge of these learning strategies followed by university students belonging to different courses and the effects of the course on the use of the learning strategies are the objectives of these studies. For the approach of a specific methodology, it is essential to know the students to whom it is addressed in order to adapt it to each level.

There are also other factors to be taken into account, as shown in other studies based on surveys about learning applied to university students by other research groups [5]. These studies show that students who have acquired knowledge or who have prepared for learning strategies have better results than other students who have not. In addition, the professors worked in their subjects learning to learn processes, with consensual programming in each degree, and in case it is well organized, other initiatives would not be necessary.

Other researchers [19] have carried out learning experiences based on organizing work groups and practical activities related to technical studies in engineering, construction, and materials. For this purpose, the students were divided into different groups and different universities were also involved. The results obtained and the answers to the surveys applied to the students are very satisfactory. The learning on the part of the students has been very positive. This way, the following section explains how the City-zen and HEREVEA projects can introduce such learning strategies.

2.1. City-zen project

2.1.1. SWAT. The 'Pre-Roadshow Analysis', the SWAT Studio (Sustainable Workshop Architecture +Technology), began 2 months prior to Roadshow. Both SWAT and Roadshow were developed to be intensive by optimizing time, simplifying communication and explanation, and maximising participation (lectures, site excursions, design workshops, and mini-master classes). SWAT is a Master's level student workshop where Building Technology students from the Faculty of Architecture at Delft University of Technology develop and propose innovative, sustainable, contextually sensitive urban design interventions, together with local students of the visiting city. A key ambition of the workshop is to always demonstrate that, through building interventions at all scales ranging from façade, building, street, neighbourhood and district, sustainable lifestyles are possible within existing cities. SWAT is a precursory educational event to the later specialist Roadshow. At the University of Yasar in Bornova (Izmir), SWAT forged pre-Roadshow relationships with key city stakeholders, allowing project sites to be evaluated and selected [2].

2.1.2. Roadshow. City-zen Roadshow travels with a team of internationally recognized experts in the field of energy planning and design to help develop a sustainable agenda for cities and their neighbourhoods. It will visit a total of 10 cities over a 4-year period, cities which are seeking expert guidance on how to become sustainable and wish to move towards energy neutrality. The overall aim of the Roadshow team, who are known as the 'Roadies', is to work closely with people from host cities, whether they are city leaders, neighbourhood associations, energy planners, architects, academics, students, and citizens themselves. The event stays in each host city for 5 days, pre-Roadshow preparations taking at least 2 to 3 months. Local stakeholders are welcomed and encouraged to join and take ownership of the final outcomes. The outcomes will allow the cities' resources, their people, the knowledge and renewable energy potential to be directed effectively by first highlighting the neighbourhoods' lifestyle and energy challenges [2].

2.2. HEREVEA project

The EF indicator adapted to buildings is used for the environmental evaluation of the projects. This indicator has already been used in other publications [20-22], in which the methodology of Solis-Guzman [23] was applied to assess the environmental impact of residential buildings. In the methodology, impact sources are identified in accordance with the bill of quantities of each project, including food expenditure by operators and their subsequent generation of municipal solid waste. Building materials consume fuel or electricity through production processes.

The consumptions produced on site (electricity and water consumptions) are also evaluated [24]. In summary, the EF calculation model determines the total footprint, which in turn consists of different partial impacts: energy (fossil fuel), pastures, fisheries, cropland, forests, and built-up land. These come from the impacts generated by the resources in terms of energy and the generation of CO2 emissions and waste.

Based on this methodology developed by the collaborating team in Seville, ARDITEC, the computer tool "HEREVEA" was implemented to calculate the ecological footprint (EF) and the cost of the rehabilitation projects. It was also used as a teaching tool (see figure 1).

The HEREVEA tool works as a complement to QGIS, an open-source Geographical Information System (GIS) which provides direct access to the data from the Cadastral Electronic Database [25], in order to automatically obtain the initial information (building typology, year of construction, number of floors, etc.) (figure 1).

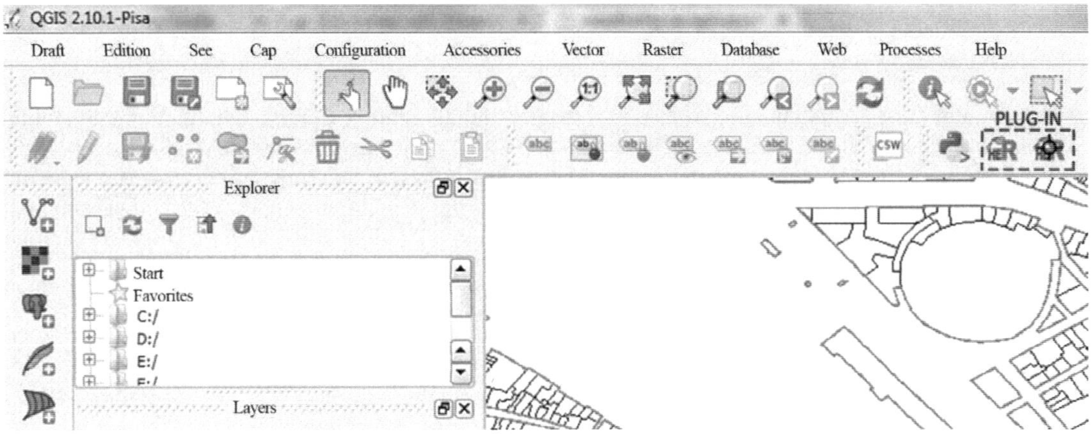

Figure 1. The HEREVEA tool in QGIS.

The user enters other constructive information that is not in the database, such as the type of foundation, structure, height, etc., as well as the interventions to be carried out and the degree of damage (figure 2).

The tool finally calculates the economic and environmental impact of the rehabilitation works. The application for the installation of the plug-in can be downloaded via the following link: http://www.aopandalucia.es/innovacion/principal.asp?alias=con3_s.

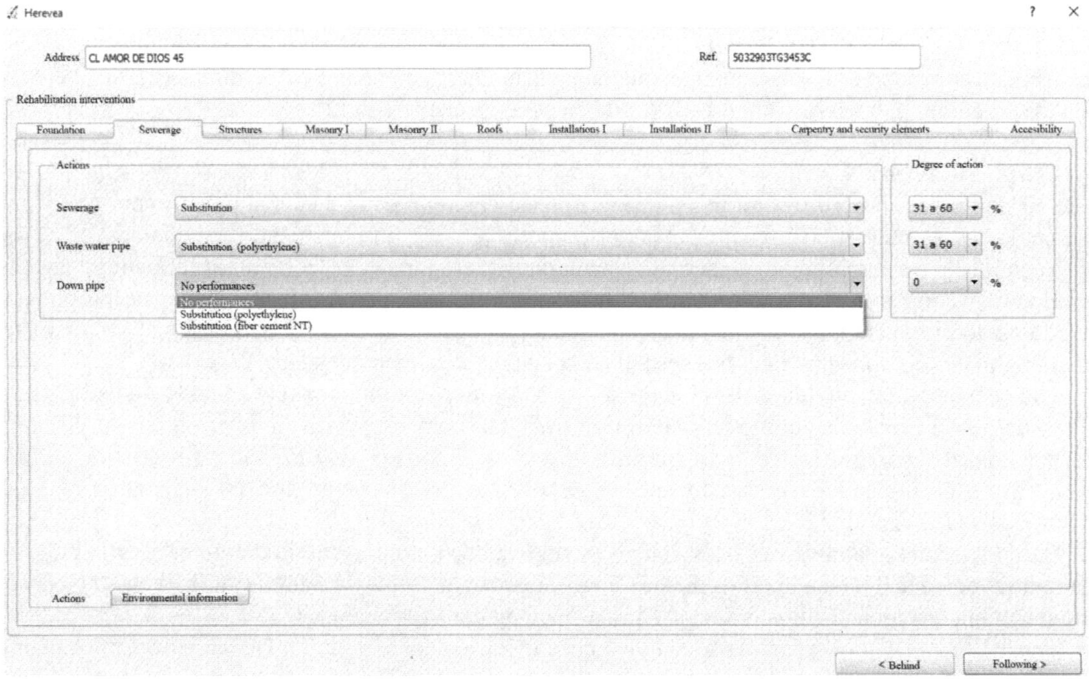

Figure 2. Constructive information and degree of damage per chapter (HEREVEA tool).

3. Methodology

3.1. Teaching process: SWAT

SWAT Studio took place at the School of Engineering of Building Engineering (ETSIE) of the University of Seville, hosting 20 international students who, along with others from the University of Seville, studying in the Architecture Doctoral Program and the Integrated Building Management Master's Program within the School of Building Engineering, developed proposals aimed at achieving a more sustainable Seville (see figure 3).

The activities can be summarized as follows:

1) Acquiring knowledge on the neighbourhood, the architectonic and social aspects.
2) First drafts on the improved neighbourhood, plans, design, installations, etc.
3) Correcting and discussing the proposals with the city town hall representatives.
4) Final oral presentations.

Figure 3. Working groups during SWAT in Seville. Clockwise: professors and students of the Building Engineering School, University of Seville, Urbanism Office of Seville and Delft Technological University, Netherlands; the neighbourhood map divided into working areas; students designing the projects; and the final presentations of their proposals.

3.2. Teaching process: Roadshow

In a 5-Day period, student focussed workshops facilitated an extended and detailed discussion with city stakeholders. 'FIVE DAYS' describes the methodology on a day-to-day 'themed' basis that guides the evolution of the vision in which expert input would be delivered at key points throughout the five days. 'PARALLEL' visually communicates what activities took place. It must be noted that a Roadshow is not intended to be a one-way stream of information and ideas. On the contrary, the method aims to activate, convince, and openly invite and encourage 'the City' to be part of the process at any level that they feel comfortable with. The method includes going out of the studio and into the community to engage with various initiatives and to meet and talk with their members.

The final outcomes of Roadshow are graphically presented in 'VISION'. The timetable is summarized in figure 4. The 5-Day programme was devised in such a way to encourage participants to 'drop-in' and 'drop-out' so that the Roadshow workshops and mini-master classes could fit into their professional and family schedules, a strategy that would increase city involvement and bolster involvement later in the week.

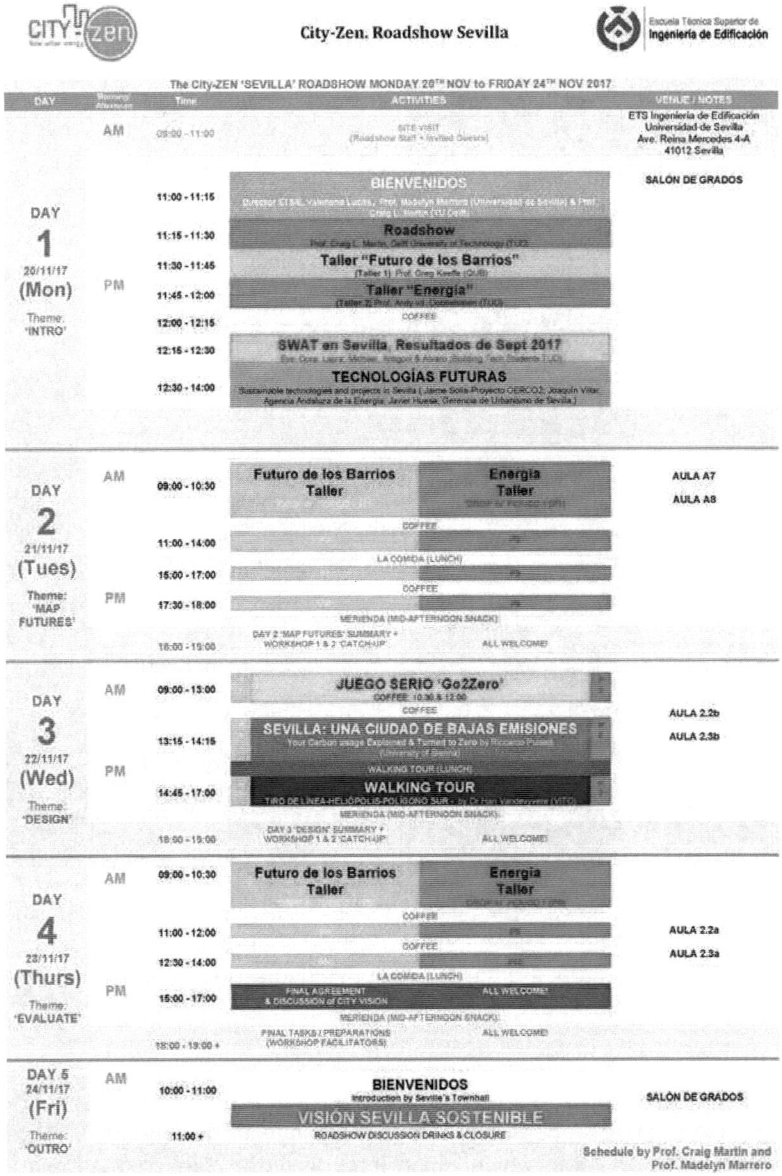

Figure 4. Timetable of the activities taking place during Roadshow.

Typical co-creative and intensive scenarios from a workshop session, as depicted in figure 5, are incredibly effective in translating what is to come. Coloured markers, rolls of tracing paper, laptops, and notebooks are the tools of choice for the Roadshow participants. Activities have the same aim, namely energy neutrality. However, each component is enjoyably diverse and offers new perspectives and skills on how to attain it. While the two parallel workshops run continually over the week, participants sign up to play a 'Serious Game'. This allows them to have playful fun and experience the cause and effect of energy strategy decisions made at the regional, neighbourhood, and family household scale.

The 'City Vision' takes the form of three overlapping presentations on Day 5. The first briefly outlines the objectives of the City-zen project. The second and third presentations form the major body of the presentation and represent the integrated outcomes of both workshops. The presentation of Workshop 1 would be qualitative in nature and include urban planning intervention proposals at the neighbourhood scale, together with spatial, social, and building regulation strategies. The third and final presentation would be quantitative, focused on energy strategies, scenarios, and carbon offsetting measures.

Figure 5. Roadshow activities in Seville.

3.3. *Teaching process: HEREVEA*

The students were graduates with architectural or engineering technical careers, which is important for the previous knowledge necessary to use the HEREVEA tool. Teaching was divided into two parts: theory and practice. The theory develops concepts about sustainability and environmental indicators. The students come from technical careers, so they must have sufficient knowledge about the rehabilitation and budgeting of construction projects. However, concepts on rehabilitation and repair of buildings are introduced and the systematic classification and criteria of construction cost databases are explained.

In the practical class, an exercise is proposed in which students evaluate the costs and environmental impact with the necessary software and detailed rehabilitation project. The practical session needs, in the first place, a laptop and internet connection in order to be able to perform the practice satisfactorily.

The materials needed for the exercise are:

- the free access program QGIS in its exact version, Pisa2.10;
- the HEREVEA Plugin;
- the user's manual with download and installation instructions;
- the project report, generated automatically by the software, whereby the student can consult the previous concepts necessary for the correct use of the tool.

The practical exercise is developed, which consists of choosing a residential type building, describing its current status and pathologies, and establishing the appropriate repairs. Once the report is obtained, an alternative solution is proposed to the one which suggests that improvements be made from an economic and environmental point of view, i.e., using other more sustainable materials, according to the choices in the software.

A building of the Tirode Linea neighbourhood of the Roadshow was selected and the HEREVEA tool was applied. The existing pathologies were considered and a study was carried out to determine if it was more viable to repair or demolish it and build a new one, considering the inclusion of energy improvements. This was done in class with undergrads, master's, and PhD students.

The HEREVEA tool was applied to a social housing located in the aforementioned neighbourhood; the rehabilitation project is summarized in table 1, where details can be found of the interventions for each of the projects, their pathologies, and the level of damage, ranging from 0 to 100%. The house presents a very serious pathological condition. The damage to the foundations and structural elements presents a high degree of deterioration. Given the integral nature of the action in this case, the interventions included the energy-efficiency improvement of the building envelope and systems, and the accessibility conditions of the common areas of the building.

Table 1. Summary of the pathologies in terms of chapter classification and level of damage.

Chapter/Sub-chapter/Sections	Works	Case of study
03. FOUNDATION / 03R. REHABILITATION		
m^3. Reinforced concrete pads	Foundation underpinning	61 to100
04. SEWER / 04R.REHABILITATION		
u. Sewer	Repair of waterproofing	61 to100
m. Waste water pipe	Substitution	61 to100
m. Down pipe	Substitution	31 to60
05. STRUCTURES / 05R.REHABILITATION		
m^2. Reinforced concrete slab	Repair of compression level	61 to100
06. MASONRY / 06R.REHABILITATION		
m^2. Envelope of brick walls from the outside	Substitution. Energy efficiency improvement	61 to100
	Repair of dampness	61 to100
m^2. Internal partitions	Fissure repair	31 to60
m^2. Envelope of brick walls from the inside	Crack repair. Staple, mesh and plaster	61 to100
	Repair due to fungi and moisture	1 to 30
07. ROOFS /07R.REHABILITATION		
m^2. Horizontal roofs		
Complete reparation. Energy efficiency improvement		61 to100
08. INSTALLATIONS / 08R.REHABILITATION		
Air conditioning		
Air-conditioning equipment	New installation	100
Electricity		
m. Circuits		61 to100
m. Lines	Replacement	61 to100
u. Light switches/Plugs		61 to100
m. Electric line to ground	New installation	100
Plumbing		
m. Hot water pipelines	Replacement	31 to60
m. Cold water pipelines		31 to60
u. Water heater	Thermal solar heater	100
11. CARPENTRY/ 11R.RECUPERATIONS		
m^2. Doors and windows	Replacement (double glazing)	31 to60
ACCESSIBILITY		
Stairs	Substitution	100
Ramp	New installation	100
u. Elevator		100

The quantification of the basic resources (%weight and emission of materials and working hours) is summarized in table 2. The materials selected are the most representative in the rehabilitation activities carried out. The total working hours are distributed between workers from the different work units (direct costs), auxiliary labour from the indirect costs of the budget, and the operators handling heavy machinery.

Table 2. Quantification of construction materials and working hours.

MATERIAL	Weight (%)	Emission (%)
Sand	40.02	0.16
Cement	25.04	38.87
Concrete	13.52	4.33
Ceramic	12.96	19.46
Steel	4.43	26.55
Paint	0.32	2.66
Aggregates	2.08	0.27
Polyethylene	0.32	1.35
PVC	0.33	1.56
Aluminium	0.15	3.08
Polystyrene	0.21	1.67
Others	0.61	0.03
WORKING HOURS(H)		
Direct cost		68.92
Indirect cost		22.95
Heavy machinery		8.13

The results after applying the HEREVEA tool to the selected project are shown in figure 6, where they are compared economically and environmentally in terms of budget chapters. The EF of the foundations chapter is highlighted; this is due to the major consumption of materials during the process of underpinning, caused by the demands for cement slurry by the poor carrying-capacity of the soil. However, comparatively, the economic result is less important in this chapter.

The damage in the foundation is associated with the need to intervene in other elements that have suffered collateral damage, such as the repair of cracks and fissures in enclosures and interior partition walls, as can be observed in the EF results of the masonry chapter, which stands out with a high environmental impact (figure 6). This is due to the high energy consumption necessary for the transformation and implementation of the materials used in this chapter. From an economic point of view, the masonry chapter has the highest budgetary cost too.

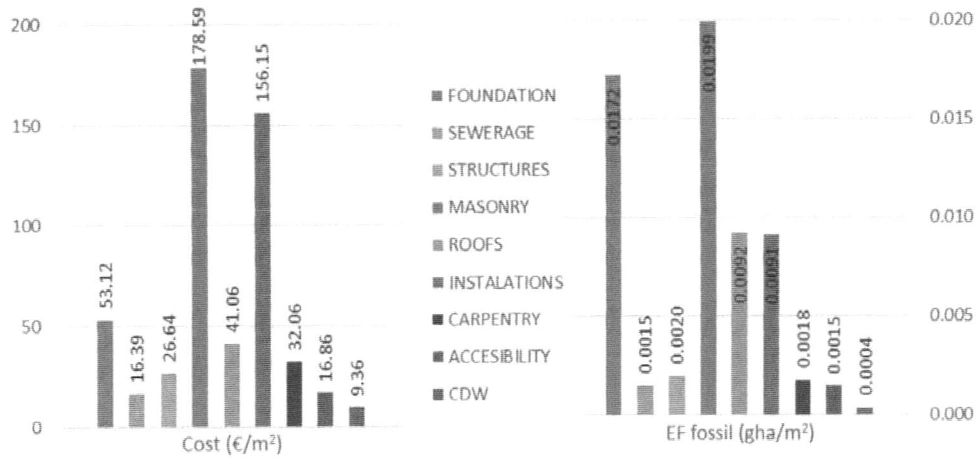

Figure 6. Comparison of the economic and environmental impacts per chapters.

3.3.1. Energy-efficiency improvement comparison. In figure 7, the energy simulation by the CE3 software can be observed [26]. The improvements made to the case study generated a significant reduction in energy consumption during a 25-year period. The initial energy rating E from 0.150 gha/m2 improved to a B rating of 0.018 gha/m2 in the rehabilitated state. This reduction is due to the fact that energy-efficiency improvement actions were carried out in all the elements that influence energy consumption, such as enclosures, roofs, carpentry, and installations.

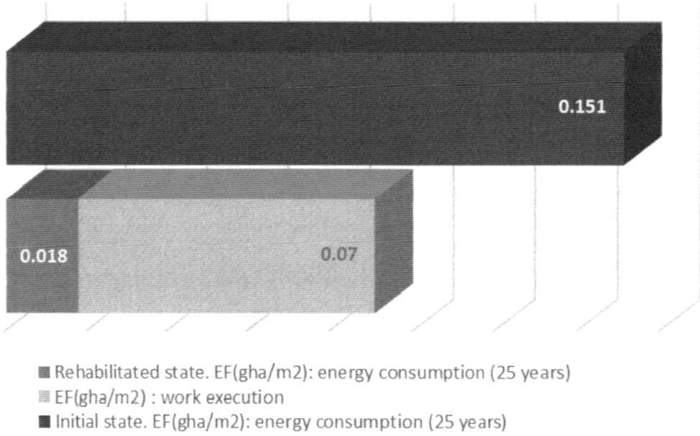

Figure 7. Environmental-impact comparison: total EF of energy consumption and rehabilitation work.

4. Results and discussion

The results of these experiences and in other studies [7] are coincident in that teaching environments which use new technologies and computer tools are strictly necessary to carry out the proposed activities. From an educational point of view, the key is not to offer university students technological media that allow them to define their method of learning, but to design activities that guide and support them in the construction of their own learning pass, and that help to establish synergies between different educational contexts, formal and non-formal, and professional, where they learn and work. The challenge is not technological, but pedagogical [7].

The ability to analyse the influence of the academic activities on the strategic profile of university students is limited. Even due to the importance in the development of learning processes in university education and, especially, of future teachers, research on the strategies used by students, in our case, is scarce. The findings suggest that in order to promote or intervene in the improvement of the strategic profile of university students, emphasis must be placed, above all, on meta-cognitive strategies and on those related to the knowledge and regulation of the affective components involved in the learning process [4].

The society of the future will be highly technological and new generations are becoming deeply familiar with electronic tools. These changes have affected the university. Digital technologies democratize learning by making knowledge accessible to many and by weakening barriers of time and space. The classroom as a place of learning is being challenged and, in some cases, replaced by informal modes of learning [6]. Learning at the university level should involve a variety of teaching methodologies so that teaching will respond to the multiple forms of communication that produce not only text, but also images, movement, and sound [27].

5. Conclusions

Through the City-zen, SWAT, and Roadshow experiences, the students have successfully collaborated with city stakeholders across Europe. Their internationally wide reputation as innovative and impactful event models is growing. They recently became an example of EC 'Best Practice', building upon the visits to 10 cities in total until 2019. In addition, interpreters and student 'facilitators' will join future events.

In the HEREVEA experience, students who have more prior knowledge and who, following the teacher's guidelines, do their homework prior to the practical class, obtain very positive results and have a real advantage from the classes. The teacher must be prepared to correct errors in the installation of the program and the plugin, as well as possible errors that can be made while using the tool. It is important that the teacher insist on the importance of the previous concepts, thus achieving a better understanding and use of the HEREVEA tool.

The methodology proposed in this paper, which is directed at bachelor, master, and doctorate students, aims to teach how to make cities more sustainable, transforming neighbourhoods with zero emissions. For this purpose and taking into account how students need new forms of learning strategies, a different approach is proposed in terms of teaching these concepts. Different work methodologies are proposed, which can obtain better results from the students:

- meetings with professors, city council technicians, in relation to sustainability issues and environmental indicators applied to cities and achieving zero emissions (City-zen project) while also reducing the impacts on building rehabilitation (HEREVEA tool);
- guided tours of the city areas to be studied;
- work groups are organized and guided by teachers;
- the areas to be studied are distributed;
- the tools and computer programs related to the reduction of environmental impact in the rehabilitation of neighbourhoods are provided, and applied to the buildings of the area studied, implementing energy improvements which will contribute to the city having zero emissions;
- support material is provided, such as official reports, general data related to the city, etc.;
- the works are presented in public at the university and in front of city council staff, who can integrate the new trips into real city projects.

In future editions of Roadshow, this previous knowledge acquired will help improve the learning process of students and transmit the innovative ideas to the cities visited, to society and its representatives.

Acknowledgements

This research is developed as part of the project Ecological footprint of the recovery of buildings: economic and environmental feasibility (HEREVEA), financed by the Public Works Agency of the Junta de Andalucía (Project Type: Contract 68/83 Ref: 2434/0604). Complete Project available on: www.aopandalucia.es/innovacion/principal.asp?alias=con3_s.

City-zen project is funded by the European Commission to help cities that we live in, work in and love, to become more sustainable, enjoyable, and healthier (www.cityzen-smartcity.eu).

References

[1] Biggs J 2005 Calidad del aprendizaje universitario, Narcea, Madrid

[2] Martin C L, van den Dobbelsteen A, Keeffe G 2017 The Societal Impact Methodology – Connecting Citizens, Sustainability Awareness, Technological Interventions & Co-creative City Visions, in: PLEA 2017, Edimburg. UK

[3] Samuelowicz J D, K & Bain 2001 Revisiting academics' beliefs about teaching and learning, High. Educ. **41** 299–325

[4] Aizpurua A 2017 Learning strategies: Differences according to year of study in undergraduate psychology students and Master's students, Cult. Y Educ. **29** 31–61

[5] Gargallo B, Campos C and Almerich G 2016 Learning to learn at university. The effects of an

instrumental subject on learning strategies and academic achievement / <i>Aprender aaprenderen la universidad. Efectos de una materia instrumental sobre las estrategias de aprendizaje yel rendimiento a *Cult. Y Educ.* **28** 771–810

[6] Stromquist N P 2009 The impact of information and communication technologies on university students: A tentative assessment *Cult. Y Educ.* **21** 215–226

[7] Coll C, Engel A, Saz A and Bustos A 2014 Los entornos personales de aprendiza jeen l educación superior: deldiseñoaluso *Cult.YEduc.* **26** 775–801

[8] Marrero M 2015 Huella Ecológica de la Recuperación de Edificios: Viabilidad Económicay Ambiental (Ecological Footprint of the Recovery of Buildings: Economic and Environmental Viability). Final report. HEREVEA, http://www.aopandalucia.es /innovacion/principal.asp?alias=con3_s.

[9] Wackernagel M and Rees W E 1996 Our Ecological footprint, reducing Human Impact on the Earth *Ind. Ecol.*

[10] Samadpour P and Faryadi S 2008 Determination of ecological footprints of dense and high-risedistricts, case study of Elahie neighbourhood, Tehran *J. Environ. Stud.* **34** 63–72

[11] Zhao X and Mao X 2013 Comparison environmental impact of the peasant house hold in han, zangandhui nationality region: Case of zhangye, Gannanand Linxiain Gansu Province, Shengtai Xuebao *Acta Ecol. Sin.* **33** 5397–5406

[12] Li B and Cheng D-J 2010 Hotel ecological footprint model: Its construction and application *Chinese J. Ecol.* **29** 1463–1468

[13] Bin G and Parker P 2012 Measuring buildings for sustainability: Comparing the initial and retrofit ecological footprint of a century home – The REEP House *Appl. Energy.* **93** 24–32

[14] Olgyay V 2008 Greenfoot: A tool for estimating the carbon and ecological footprint of buildings. American Solar Energy Society-SOLAR 2008, in: *Incl. Proc.37th ASES Annu. Conf.,33rd Natl. Passiv. Sol. Conf., 3rd Renew. Energy Policy Mark*

[15] Li D Z, Hui E C M, Leung B Y P, Li Q M and Xu X A 2010 methodology for eco-efficiency evaluation of residential development at city level, *Build. Environ.* **45** 566–573

[16] Teng J and Wu X 2014 Eco-footprint-basedlife-cycleeco-efficiency assessment of building projects, *Ecol. Indic.* **39** 160–168

[17] Bastianoni S, Galli A, Pulselli R M and Niccolucci V 2007 Environmental and economic evaluation of natural capital appropriate on through building construction: practical case study in the Italian context, *Ambio.* **36** 559–565

[18] Solís-Guzmán J 2010 Evaluación de la Huella Ecológica del sector de la edificación (usoresidencial) en la comunidad andaluza, http://fondosdigitales.us.es/tesis/tesis/ 2051/evaluación-de-la-huella-ecologica-del-sector-edificacion-uso-residencial-en-lacomunidad-andaluza/

[19] Cho C S, Mazze C E, Dika S L and Gehrig G B 2015 Enhancing Construction Education: Implementing Habitat for Humanity Projects as Service-Learning for Construction Materials, *Int. J. Constr. Educ. Res.* **11** 4–20

[20] Marrero M, Puerto M, Rivero-Camacho C, Freire-Guerrero A and Solís-Guzman J 2017 Assessing the economic impact and ecological footprint of construction and demolition waste during the urbanization of rural land *Resources, Conservation & Recycling. Resources, Conservation and Recycling* **117**, pages 160-174

[21] González-Vallejo P, Marrero M and Solís-Guzmán J 2015 The ecological footprint of dwelling construction in Spain *Ecol. Indic.* **52** 75–84

[22] Alba Rodríguez M, Marrero Meléndez M and SolísGuzmán J Economic 2013 and Environmental Viability of Building Recovery in Seville (Spain) Phase1: Database in Arcgis, *Rodowisko Mieszk.* **11** 297–302

[23] Solís-Guzmán J, Marrero M and Ramírez-de-Arellano A 2013 Methodology for determining the ecological footprint of the construction of residential buildings in Andalusia (Spain) *Ecol. Indic.* **25** 239–249

[24] Solís-Guzmán J and Marrero M 2015 Ecological Footprint Assessment of Building Construction *Bentham Science Publishers*

[25] M. de H. y A. Públicas, Sede Catastral de la Dirección General del Catastro del (Cadastral Headquarters of the General Directorate of the Cadastre of the Ministry of Finance and Public Administrations) [online]. Query: 2013. Available in: (n.d.). http://www.sedecatastro.gob.es/

[26] Spain Government, Ministerio der energía, turismo y agenda digital, Certificación eficiencia energética de edificios (Ministry of energy, tourism and digital agenda. Energy efficiency certification of buildings), 2018 http://www.minetad.gob.es/energia/desarrollo/eficienciaenergetica/certificacionenergetica/documentosreconocidos/paginas/documentosreconocidos.aspx

[27] Carey J 2008 Multimodality and Literacy in School Classrooms *Rev. Res. Educ.* **32** 241–267

IOP Conf. Series: Materials Science and Engineering **399** (2018) 012039 doi:10.1088/1757-899X/399/1/012039

Standard economic bioclimatic solar individual houses – four hypostases

T Mihăilescu[1*]

[1] Transilvania University of Braşov, Romania

[*] E-mail: t.mihailescu@unitbv.ro

Abstract. In the present Romanian architectural practice for individual houses and/or residential areas, designing the architectural objects in order to function together with the nature seems to be neglected. This happens despite a great variety of bioclimatic solutions materialized in the traditional houses of all the Romanian geographical regions, in an old and rich history of traditional architecture. The approach starts with a historical overview, analyzing several examples of traditional houses in all the regions of Romania so as to identify the traditional bioclimatic solutions used to better adapt the architecture to the environment. This constitutes the source of inspiration for the four modern standard economic bioclimatic (solar) houses designed for the geographical and climatic parameters of Braşov, Romania, and destined for the middle class families. The concept is simple, based on modesty, involving minimum costs for building and using the houses, as well as for a comfortable life. The projects also offer all the premises for an efficient use of complementary systems for producing and storing energy from the environment (if opted for). The aim is to optimize the relation between the natural and the built environment in order to maximize the comfort, efficiency and economy in all the aspects of a family living in a bioclimatic house (form, function, building materials, equipments, living space, quality of life, etc.). In this respect, the architecture of the four standard economic bioclimatic (solar) individual houses presented represents a neutral and simple so-called "extruded box", as an interface between humans and environment, as a chance to better explore ourselves.

1. Premises, historical and contemporary context, problems

In the Romanian architectural practice for individual houses and/or new residential areas, designing the architectural object in order to function together with the nature seems to be neglected in the majority of the situations. This may be related to five main factors: 1) a certain reticence of the people to live in a bioclimatic (solar) house (the absence of information, trust, interest, cultural and/or mentality reasons), 2) a design process ignoring by default the conditions/principles involved by the bioclimatic (solar) architecture;, preferred are instead : an architecture of representation of the clients' social status and dreams rather than an architecture of the economy, and/or an architecture as a manifest of the architect's dreams, accepted by the clients sometimes despite their real needs, 3) inadequate sites, 4) local/regional real estate realities, and 5) the absence of a real financial help from the state for complementary systems creating energy from natural and clean sources. This happens in spite of a great variety of bioclimatic solutions materialized in the traditional houses of all the Romanian geographical regions in an old and rich history of architecture adapted to the environment.

The historical overview, which analyses almost all the examples of the traditional houses in all the regions of Romania, despite their local/regional particularities, shows that all the houses seem to share the same passive/solar/bioclimatic characteristics to better adapt the architecture to the environment: 1) simple structures, smart-simple but efficient details and local building materials (mainly wood, but also mud, stone, and/or bricks), 2) north-south orientation: maximal openings/main living areas towards south, minimal openings/secondary functions towards north, 3) a very simple and efficient function and a smart way to use all the interior available space, 4) cheap and effective solutions for insulation (against water, cold, heat and wind), 5) exterior areas, placed close to the house, for storage of provisions (firewood, rain water, food, food for animals) and shelter for domestic animals, and 6) the north-south section of the volume designed to use at its best the sun angles, with the "invention" of *prispa / the porch*, having a) an "open" facade towards the south, using the roof to have complete shadow in the summer, and as much sun as possible in the winter in order to maximize the sun heat, and b) a "blind" facade towards the north, to better isolate the house and minimize the heat loss.

Above all, the local differences (also a result of other kinds of adaptations), the (Romanian) traditional architecture seems to demonstrate convergent adaptation: the same response to the same problems.

Ten examples of traditional houses, chosen from significant historical/geographical regions of Romania (conserved within *Dimitrie Gusti* Village Museum in Bucharest, Romania; photo by Teofil Mihăilescu, 2015), are emblematic for the result of the historical approach leading to the identification of the traditional bioclimatic solutions used to better adapt to the environment (figures 1 to 7).

Figure 1. Traditional houses in Dobrogea (SE Romania, at the Black Sea, Danube and Danube Delta): (left) in Jurilovca (Tulcea), and (right) in Ostrov (Constanța); 19th century.

Figure 2. Traditional houses (hills regions) in Romania: (left) Rușețu (Buzău), 20th century, in Muntenia (S Romania), and (right) Nereju Mic (Vrancea) in Moldavia (W Romania), 19th century.

Figure 3. Traditional houses (hills regions) in: (left) Stănești (Argeș) in Muntenia (S Romania), and (right) in Curtișoara (Gorj), in Oltenia (SW Romania); 19th century.

Figure 4. Traditional houses in Transylvania (central-northern Romania, mountain/hills region): (left) Clopotiva (Hunedoara), 20th century, and (right) Şanţ (Bistriţa-Năsăud), 19th century.

Figure 5. Traditional houses: (left) Berbeşti (Maramureş) in Maramureş (N Romania), and (right) Straja (Suceava) in Bucovina (NE Romania, mountain region); 18th century.

Figure 6. *Prispa* / (aprox.) *the porch* - The southern exposure in the history of the traditional Romanian architecture, from the Black Sea/plains/Danube (left) to the Carpathian mountains (right).

Figure 7. The northern exposure in the history of the traditional Romanian architecture, from the Black Sea/plains/Danube Delta (left) to the Carpathian Mountains (right).

2. Inspiration, goals, keywords

The inspiration for the concept of the Standard economic bioclimatic (solar) individual house (SEBIH) came from: 1) the history of the (Romanian) traditional architecture, 2) the bioclimatic/passive modern architecture, 3) a personal/professional change of paradigm concerning the philosophy of life and a sustainable way of life, 4) the need of independence in a world of dependence (having the "plan B"), 5) living as a statement: a neutral, simple exterior, for a complex interior personal/intellectual life [1].

The Eco-Goals tend to be close to the Passivhaus German Standard: 1) for heating: less than 15 kWh/m^2/year, 2) for thermal insulation: U<=0.15W/ (m^{2}k), 3) for windows and doors: U<=0.80 W/ (m^{2}k), 4) for final energy: less than 50 kWh/m^2/year, within a concept governed by seven keywords: economy, independence, efficiency, modesty, sustainability, functionality, statement.

3. Proposal, context, architectural concept

The approach starts from the premise that cheap, efficient individual houses could be realized, which would be independent or multiplied within real estate investments, with local resources (materials, workers), for middle class families. They would be designed with simple solutions and within the frame of the bioclimatic architecture's principles, also using (at choice) complementary systems for producing and storing energy from the environment (sun, wind, water and/or earth). The approach here developed is based on a previously published research of the author, *Cheap type solar bioclimatic individual houses for residential areas* [2].

3.1. Context: geo-climatic parameters, sun angles, site

The houses are designed for the geographical and climatic parameters of Brașov (Transylvania), Romania: latitude: +45.66 (45°39'36"N), longitude: +25.61 (25°36'36"E), time zone: UTC+2h, altitude ~540 m, temperature variation summer/winter +/- 23°C. The sun angles are: a) 19°-24° (considered 22°) at 13.00 h on December, 21, at the Winter Solstice, with the sunrise at E and sunset at W, b) 42,5°-45° (considered 44°) at 12.00 h on March, 21, at the Spring Equinox and on September, 23, at the Autumn Equinox, with the sunrise at E and sunset at W, and c) 66°-70° (considered 66°) at 13.00 h on June, 21, at the Summer Solstice, with the sunrise at N-E and sunset at N-W [3] (Figure 8).

The SEBIH - Standard economic bioclimatic (solar) individual house projects require a small dedicated north-south oriented site (about 500 m^2, even less), preferably flat or having a light slope towards south, or a bigger site (about 1000 m^2 or more), offering the freedom to orientate the houses on the north-south direction, in a proper context and (predictable) neighbourhood, allowing the sun to reach the house all day long, from sunrise to sunset.

3.2. Concept: architectural invariants, building materials, philosophy

The SEBIH - Standard economic bioclimatic (solar) individual house concept is simple, based on economy and modesty. The main invariants in their architecture and design are: 1) a compact, modular, linear, "extruded" volume for a minimum physical and visual impact within the environment, with an architectural expression given by its linear south façade, 2) a north-south orientation: the biggest façade, maximal openings and main living areas towards south and the smallest façade, minimal openings and secondary functions towards north, 3) a very simple wooden structure, realized as much as possible with local building materials and local human resources (not necessarily a very qualified human resource), 4) a very simple and efficient function, with a smart way to use the interior available space, 5) an improved, efficient and smart insulation (against water, cold, heat and wind), 6) a section of the volume designed to use at its best the sun reference angles at solstices/equinoxes, 7) adaptive options for living with the nature as well as isolating the house against elements when relevant, 8) multiple role active and passive heating sources to ensure a complete independence during the local cold months (November-February), 9) multiple role exterior areas for resources storage close to the house (firewood; rain water or used but "clean" water used within the house, for recycling it to the toilets and/or for watering the garden), 10) vegetation within the site for the benefit of the house by having a "barer" of perennial trees (resinous), permanently offering protection from the winter winds of the north, and a seasonal vegetation (trees with seasonal foliage, but also vine), disposed at the south to provide shadow from the sun during the summer and offering the sun warmth for additional heating of the house during the winter, and 11) freedom of aesthetic options concerning the façades.

In order to use at its best the sun energy, the houses should be placed at the north side of the site in order: 1) to receive the biggest amount of sunlight during the winter months from sunrise to sunset, and 2) to prevent being in the shadow of other architectural objects built nearby at S, E, W.

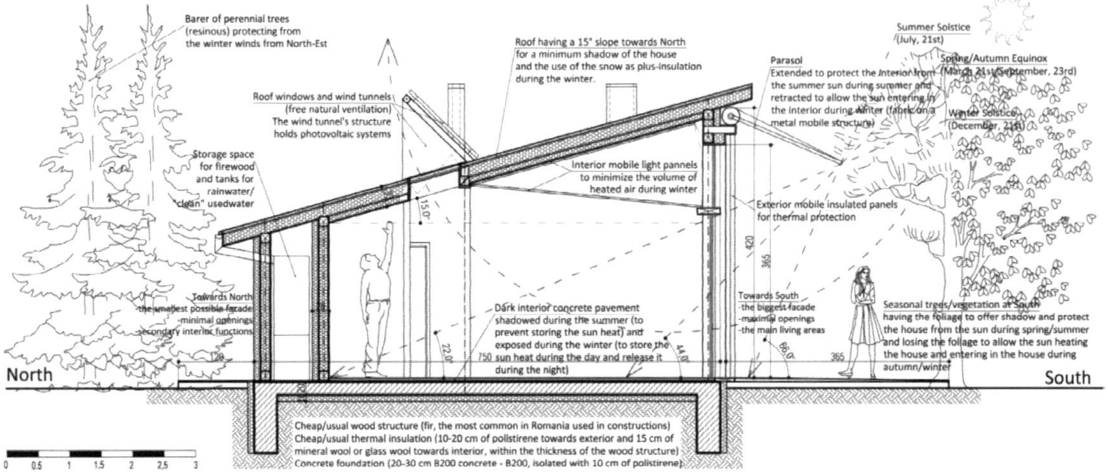

Figure 8. General concept-section for the Standard economic bioclimatic (solar) individual house, covering all the seasonal situations within a year and showing the adaptive response of the house.

The houses presented also have a symbolic value within their dimensions: lots of the interior spaces are dimensioned using the golden ratio, the height of the south windows and the width of the front terrace have 365 cm (the number of the days in a year), the rooms for the children are positioned to east (sunrise, "present in the future") and the matrimonial bedroom is positioned to west (present as "future past", sunset) as each day starts with its symbolic "birth" at sunrise and ends with its symbolic "death" at sunset (figure 8).

3.3. Heating/ventilation, water, electric power

The heating/ventilation for the bioclimatic (solar) individual house should be realized both passively and actively, on condition that the entire volume is almost perfectly isolated.

The passive heating is realized using the large windows, the dark interior pavement of the concrete foundation to store the heat from the sun and release it during the night (as it is exposed to the sun up to 80% of its surface), and using the additional heat generated by the electricity and kitchen equipments. In the good sunny winter days, up to 50-80% of the total heat requirements of the house can be supplied this way [4].

During the cold periods, active heating is used as follows: the traditional tiled/glazed stoves positioned to heat two spaces (D), a traditional kitchen stove (F) and a chimney (E) with firewood, and/or regular, common efficient appliances based on natural gas and/or electric power. In the bathrooms, the heating and the hot water could be provided by the little traditional iron/bronze stoves with a water boiler (C) using an insignificant quantity of any kind of firewood with a maximum result. For an extra-insulation of the volume of the house and to minimize the loss of the interior heat during the winter nights, the windows have an exterior plus-protection offered by a series of mobile insulated panels (G), which could "seal" the house.

The series of the roof windows (H) of the house are both sources of natural light and wind tunnels [5] (for a free natural ventilation), and have a parasol structure which could hold photovoltaic systems. As alternative, a ventilation system with heat recuperating systems could be used during the winter.

IOP Conf. Series: Materials Science and Engineering **399** (2018) 012039 doi:10.1088/1757-899X/399/1/012039

The house tends to use drinking water only for the human subsistence, and the rain water, collected gravitationally from the roof in the water tanks (A) nearby the bathrooms and kitchen, for other needs. Heated water is produced as much as possible by the solar systems [6].

The electric power could be partially produced by photovoltaic systems [7] and/or by small turbines using the wind power [8]. The house is equipped with electric equipments of energy class A or superior. The led technology provides all the artificial light.

3.4. The four Standard economic bioclimatic (solar) individual houses A, B, C, and D

3.4.1. Standard economic bioclimatic (solar) individual house A. It is a house designed for a family with two adults and two children, having a total built area of 135.00 m^2, of which 94.00 m^2 for living (figures 9, 10). Of a total of 74.70 m^2 of the south facade, 36.50 m^2 are windows (48.86%). To minimize the volume of heated air, as well as the loss of heat through the big windows of the main facade during the winter, of the total of 317.84 m^3 of the interior air volume, only 275.72 m^3 is heated (86.75%), the rest of 42.12 m^3 (13.25%) being isolated at will by the light interior mobile panels. The house has: an access wind fang (1), functional and efficient circulations (2, 5, 8), a living room and a dining area (3), a kitchen (4), a matrimonial bedroom (7), two bathrooms (6, 9), two children rooms (10, 11), a pantry (12). The storage spaces for firewood is designed at the north (B) of 6.00 m^2/15.20 m^3, and for water tanks (A) of 4.35 m^2/9.50 m^3, collecting it from the roof.

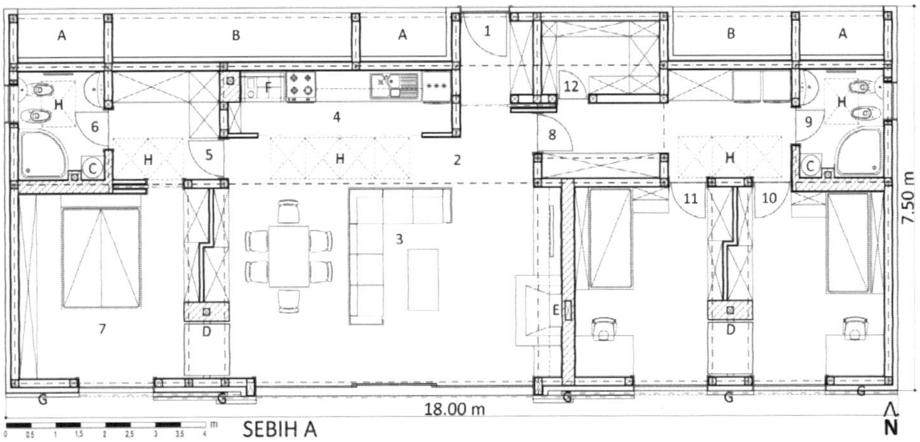

Figure 9. Plan of the Standard economic bioclimatic (solar) individual house A.

Figure 10. Concept perspectives of the Standard economic bioclimatic (solar) individual house A.

3.4.2. Standard economic bioclimatic (solar) individual house B. It is a house designed for a family made up of two adults and one child, having a total built area of 105.00 m^2, of which 79.00 m^2 for living (figures 11, 12). Out of a total of 58.10 m^2 of the south facade, 29.05 m^2 are windows (50.00%). To minimize the volume of heated air as well as the loss of heat through the big windows of the main facade during the winter, out of the total of 194.24 m^3 of the interior volume, only 168.50 m^3 are heated (86.75%), the rest of 25.74 m^3 (13.25%) being isolated at will by the light interior mobile panels. The house has: an access wind fang (1), functional and efficient circulation spaces (2, 5, 8), a living room and a dining (3), a kitchen (4), a matrimonial bedroom (7), two bathrooms (6, 9), a room (10), a pantry (11). To the north, the house has a storage spaces for firewood (B) of 4.05 m^2/10.12 m^3, and for water tanks (A) of 4.35 m^2/9.50 m^3, collecting the water from the roof.

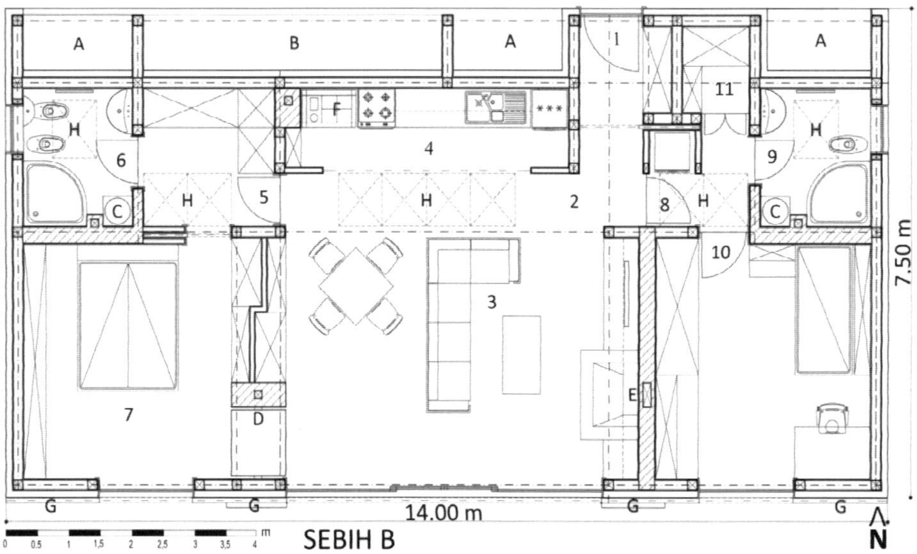

Figure 11. Plan of the Standard economic bioclimatic (solar) individual house B.

Figure 12. Concept perspectives of the Standard economic bioclimatic (solar) individual house B.

3.4.3. Standard economic bioclimatic (solar) individual house C. It is a house designed for a family with two members, having a total built area of 82.50 m², of which 62.00 m² for living (figures 13, 14). Out of a total of 46.65 m² of the south facade, 21.68 m² are windows (46.47%). To minimize the volume of heated air, as well as the loss of heat through the big windows of the main facade during the winter, out of the total of 247.22 m³ of the interior volume, only 214.46 m³ is heated (86.75%), the rest of 32.76 m³ (13.25%) being isolated at will by the light interior mobile panels. The house has: an access wind fang (1), functional and efficient circulation spaces (2, 5), a living room and a dining (3), a kitchen (4), a matrimonial bedroom (7), a bathroom (6). To the north, the house has the storage spaces for fire wood (B) of 4.05 m²/10.12 m³, and for the water tanks (A) of 2.90 m²/6.38 m³.

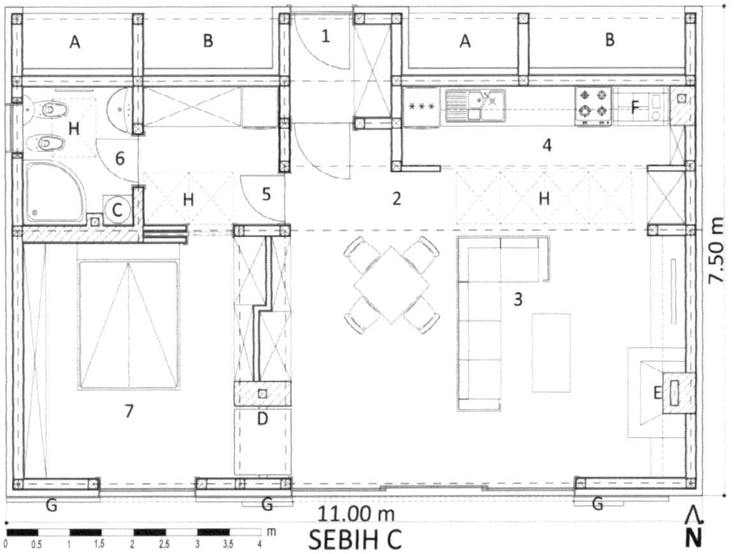

Figure 13. Plan of the Standard economic bioclimatic (solar) individual house C.

Figure 14. Concept perspectives of the Standard economic bioclimatic (solar) individual house C.

3.4.4. Standard economic bioclimatic (solar) individual house D. It is a house designed as a little country holiday cottage for a couple, having a total built area of 52.50 m^2, of which 38.00 m^2 for living (figures 15, 16). Out of a total of 29.05 m^2 of the south facade, 16.24 m^2 are windows (56.52%). To minimize the volume of heated air, as well as the loss of heat through the big windows of the main facade during the winter, of the total of 123.61 m^3 of the interior volume, only 107.23 m^3 is heated (86.75%), the rest of 16.38 m^3 (13.25%) being isolated at will by the light interior mobile panels. The house has: an access wind fang (1), a functional and efficient circulation (2), a living room and a dining (3), a kitchen (4), a bathroom (5). To the north, the house has the storage spaces for firewood (B) of 2.51 m^2/5.62 m^3, and for the water tanks (A) of 1.45 m^2/2.24 m^3.

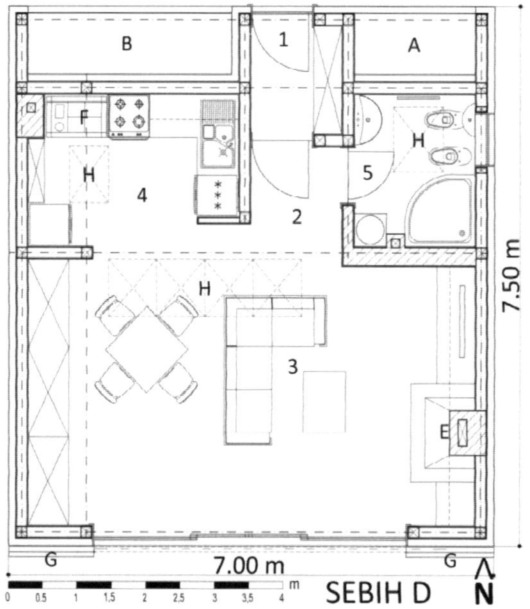

Figure 15. Plan of the Standard economic bioclimatic (solar) individual house D.

Figure 16. Concept perspectives of the Standard economic bioclimatic (solar) individual house D.

4. Conclusions

The estimated average price of such a standard economic bioclimatic (solar) individual house in Romania, realized with local building and human resources, could be about 250 Euro/m^2 without V.A.T. This is considerably less than the average estimated building price of a similar house of the same built area, in a similar site, and realized with different building materials (concrete, bricks) and different technologies (estimated at about 400 Euro/m^2 without V.A.T. in present) [9], or even realized with similar materials and technology but not paying attention to specific (solar) bioclimatic design and philosophy generating a significant economy. Above the affordable price, the bioclimatic architecture involves also a way of life. Therefore, an entire way of thinking about living should be changed, involving a change of paradigm in mentalities and the help of the professionals, media and governments/local administrations for the urban actors to be motivated in choosing the bioclimatic architecture as a step forward towards sustainable development [10].

The main premises in both designing these SEBIH - Standard economic bioclimatic (solar) individual houses and living in them are: 1) to learn from the valuable lessons and experiences of the history of the (Romanian) traditional architecture, 2) to accept a way of life involving a set of different life and living principles than the ones we are used to within the post-modern society/the contemporary matrix, 3) to optimize the relation between the natural and built environment in order to maximize the efficiency and economy in all the aspects of (personal) life, 4) to adopt a sustainable way of thinking and of life, offering yourself the chance of a relative or almost complete independence in a world of dependence: having the famous "plan B".

Living in a Standard economic bioclimatic (solar) individual house is a statement and means choosing a way of life. Traditional adaptive solutions are integrated in such a house, both using the qualities of the elements and offering protection from elements, as well as modern solutions translated in these simple, economic, efficient, modest "green" buildings, having a "life" almost together with the life of their inhabitants.

In this respect, the architecture of the Standard economic bioclimatic (solar) individual houses means a neutral, decent, modest, a simple so-called "extruded box", apparently lost in the nature, open to the southern sky, as a (bright) interface between humans and environment, and a chance to develop a very complex personal/family/intellectual interior life and better explore ourselves.

References

[1] Mihăilescu T 2012 *Projecting with the nature: pleadingfor the bioclimatic architecture Artitudine* (Braşov: Editura Universităţii Transilvania) pp 47-55

[2] Mihăilescu T 2016 Cheap type solar bioclimatic individual houses for residential areas *Renew. Energy Environ. Stustain.* **1** 7

[3] http://www.gaisma.com/en/dir/ro-country

[4] Guedes P 1979 *The Macmillan Encyclopedia of Architecture and Technological Change* (London: The Macmillan Press Ltd.) pp 224-225

[5] Guedes P 1979 *The Macmillan Encyclopedia of Architecture and Technological Change* (London: The Macmillan Press Ltd.) pp 12-13

[6] Dabija A M, Gheorghe S, Mortu A and Buzatu L 2010 *Photovoltaic systems in architecture* (Bucharest: Editura Universitară Ion Mincu) pp 43-44

[7] Antony F, Durschner C and Remmers K-H *Photovoltaics for Professionals: Solar Electric Systems Marketing, Design and Installation* (Germany: Solarpraxis AG)

[8] http://saferenvironment.wordpress.com/2008/11/03/wind-energy-renewable-energy-harnesses-natural-wind-power-%E2%80%93-effective-answer-for-emission-problem-towards-cleaner-safer-and-greener-environment

[9] The Order of Romanian Architects 2006 *Exercising the profession of architect* (Bucharest: Simetria) pp 45-55

[10] http://architecture.about.com/od/greenconcepts/g/sustainable.htm

Lost shuttering for columns as possible applications of dispersely reinforced concrete with polypropylene fibers

R Muntean[1*], C Cazacu[1], G C Chitonu[1] and T Galatanu[1]

[1] Faculty of Civil Engineering, Transilvania University of Brasov, Turnului no 5street, Brasov, Romania

[*] E-mail: radu.m@unitbv.ro

Abstract. There are various practical applications of dispersely reinforced concrete with polypropylene fibers; one of them could be achieving formworks for concrete columns. This paper presents some practical tests showing the combined effect of a lost shuttering made by dispersely reinforced concrete with fibers and a steel reinforced concrete core, for round and square shaped columns. The test results could be considered a contribution to the development of knowledge in the dispersely reinforcement area using one type of polypropylene fibers for which information on their application field are little or missing.

1. Introduction

The rapid, dynamic and diversified evolution of society is compelling the construction sector to follow this trend, thus responding to the strength, safety and comfort requirements that ensure the smooth running of different human activities.

The constructions development of both vertical (office buildings, dwellings, hotels) and horizontal (industrial halls, shops with surface areas of thousands of square meters) requires precise techniques and technologies that can capitalize a range of new materials. In this context, the use of simple concrete or reinforced or pre-stressed concrete is somewhat restricted by specific phenomena such as cracking, fire resistance, shrinkage, shock resistance, wear resistance, durability, and so forth.

For this reason, various in-depth studies and research have been carried out showing that an improvement in concrete performance can be achieved by adding in their masses dispersed reinforcement in the form of fibers made from different materials [1 – 9].

Consequently, the realization of dispersely reinforced concrete is one way to maintain the competitiveness of concrete as one of the main building materials [10, 11].

However, fiber-reinforced concrete cannot replace conventional reinforced one. There are areas of application in which fiber reinforced concrete can be used alternatively or in addition to classic reinforced one (regular), providing constructive and economic advantages.

The use of fibers of any kind will improve the properties of regular concrete [12].

The opportunity to use fibers as reinforcement occurs in the case of using a small percentage of steel reinforcement or in the case of constructive steel reinforcement of regular reinforced concrete. The possibilities of use increase due to the improved cracking behaviour, the reduction of the deformations due to drying contractions or the increase in the shear resistance.

The fields of use for fiber-reinforced concrete have an extensive range, including: concrete pipes, supporting walls, thin facade elements, prefabricated steps, airport runways, car foundations, lost formwork, concrete guniting consolidation works for tunnels, etc. [13, 14].

Most dispersely fiber reinforced concrete applications are based on the idea of improving the strength properties. However, the role of fiber reinforcement is not so much in the improvement of the resistance, but in controlling the fracturing process and thereby in improving the ductility, the properties of energy absorption in case of an impact, shock and temperature variations.

From this point of view, C.D. Johnstone's observations are relevant [15]: "only if engineers stop thinking about resistance, it will not be difficult to find many other areas of application of these new materials, and improvements will also be made".

2. Experimental method

Practical applications of dispersely reinforced concrete with polypropylene fibers are varied; one of them could be achieving formworks (shuttering) for concrete columns.

This study is part of a larger research which focuses on finding efficient building elements made of dispersed reinforced concrete with polypropylene fibers, such as: concrete tiles for pavements, concrete pillars and boards for fences, lost formwork for lintels, beams, etc. [10, 11, 12, 16].

Figure 1. Different building elements made of dispersely reinforced concrete with polypropylene fibers.

The efficiency of some reinforced concrete elements with polypropylene fibers was pursued not only from the point of view of the final mechanical characteristics (increased strengths at different stresses), frost resistance - thaw, permeability, etc., but especially in relation to the price to be paid. It was considered that if the improvements to conventional concrete (simple or reinforced with steel bars) through fiber-dispersed reinforcement are more than the additional cost of fibers, without affecting other characteristics such as workability, ease of operation, visual appearance, etc., these elements could be considered effective.

2.1. Polypropylene fibers used

Polypropylene fibers are varied and have different uses. The chosen type of fibers was Fibrofor® Multi, produced by *BruggContec AG*, CH-8590 Romanshorn, Switzerland, some of the cheapest on the market and with a limited area of application.

Figure 2. Fibrofor® Multi fibers.

The choice of fiber type was based not only on the lack of information regarding its limited application domain, but also on other criteria, such as: price per kilogram, fibers quantity recommended by manufacturer to be added to mixture, their physical and mechanical characteristics.

This study aims to develop the applicability domain of these cheap polypropylene fibers.

Table 1. Characteristics of polypropylene fibers (according to manufacturer) [13].

Dosage:	minimum	0.6
[kg/m^3]	maximum	0.9
	fire-resistant	2.0
concrete		
Form factor		multifilament
Diameter		34 μm
Length (+/- 5%)		6.3mm (type 63)
		12.7mm (type 127)
Density		0.91 g/cm^3
Resistance to acids / alkali		inert
Tensile strength		300 – 400 MPa
Modulus of elasticity		4900 MPa
Melting point		150°C
Colour		white
Packing		bags 0.9 kg

2.2. Composition of concrete used

Before starting the experimental program, it was necessary to establish the preliminary composition of the concrete to be used. To establish the composition of the concrete, the following information must be known:

- the class of concrete: C20/25 class was chosen;
- the characteristics of the elements which will be executed, fact which implies a specific granulometric curve of aggregates and appropriate workability;
- the conditions for transportation – transportation was not necessary as the mixing was made where the placing of concrete took place, in a mixing machine with a 50 liters load;
- the conditions of concrete placing – manually by compacting rod;
- the conditions of hardening – normal, the store and trial of the elements are made in the lab.

Considering all factors and after preliminary trials, the following concrete compositions were obtained:

Table 2. R1 – cement mortar formula.

river aggregate 1720 kg/m3: type 0 … 4	1720 kg/m^3
cement CEM IV/B 42.5N	390 kg/m^3
water	250 litres/m^3
A/C ratio	0.64
plasticiser	no
polypropylene fibres type Fibrofor®Multi	0.9 kg/m^3

Table 3. R2 – concrete formula.

river aggregate 1720 kg/m3:	type 0 … 4	771 kg/m^3
	type 4 … 8	422 kg/m^3
	type 8 … 16	422 kg/m^3
cement AV I42.5R		456 kg/m^3
water		235 litres/m^3
A/C ratio		0.51
plasticiser		Viscocrete 20HE
polypropylene fibres type Fibrofor® Multi		0.9 kg/m^3

2.3. Experimental elements

The components for the fibrous dispersed-reinforced concrete thin walls (tubes) were produced in order to be used as circular or rectangular columns shuttering. Two types of lost shuttering were produced: some of circular-shaped tubes and others which were rectangular-shaped, figure 3.

The circular ones have an external width of 200 mm, the walls' thickness of 25 mm and the height of 500 mm. The cross-section of the rectangular ones is square-shaped, with the side of 150 mm, the walls' width of 25 mm and the height of 700 mm.

Figure 3. Circular, respectively square-shaped lost shuttering.

All shuttering components, both circular and rectangular, were produced from micro-concrete based on R1 formula by means of a single type aggregate, with the grain between 0 – 4 mm.

These components were produced in order to determine the input regarding the safe bearing capacity growth of the regular concrete column cast inside the ring (of the lost shuttering).

The circular internal column was produced from reinforced concrete with 6 longitudinal PC52 steel bars with Ø10mm and OB37 steel hoop enclosure with Ø4mm, figure 4 [17].

The rectangular internal pole was produced from reinforced concrete with 4 longitudinal PC52 steel bars with Ø10mm and OB37 steel marginal hoop enclosure, figure 4 [17].

The concrete used to produce the internal columns was prepared according to formula R2.

Parallel to these poles, cast inside the dispersed-reinforced concrete shuttering, other (full) circular-section, respectively (full) square-section concrete poles of the same type were cast and enclosed similarly, having the external size equal to that of the ring (either circular or rectangular).

Therefore, two models were produced, namely a classic one, with a homogenous cross-section (named SC1, SC2, SP1), cast in a regular wood shuttering and the second one, with a mixed cross-section (named SCC1, SCC2, SPC1), produced from two concrete classes (an external layer of dispersed-reinforced concrete, practically representing the equivalent of the concrete coating layer of the first component and a core of regular reinforced concrete).

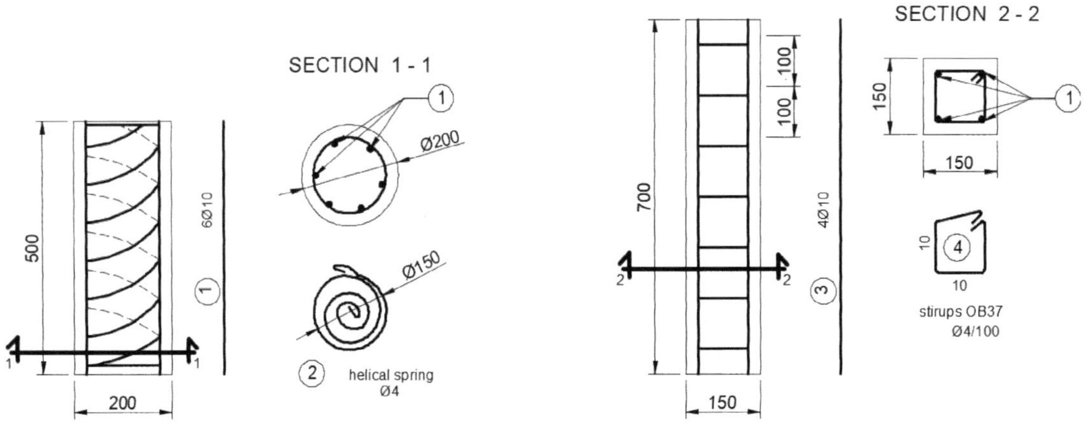

a b

Figure 4. Plan of shuttering and reinforcement for circular (a) and rectangular columns (b).

These elements were centric compression tested. So, since the casting, metallic plates were provided centrically and in a slightly higher position than their lip (approx. 4 mm). As a result, the components' core only received energy.

For other two elements (circular shaped) the force applying was done across the whole cross section.

3. Results

For the elements tested using metallic plates to center the applied force, the following were observed during trials:

- the circular column SC_1, with regular reinforced concrete cross-section, collapsed when charged with 40.0 kN; the collapse was due to cracks along the compressive stress (classical collapse) [17], figure 5;
- the circular column SCC_1, with mixed cross-section, made of a regular reinforced concrete core and a fibrous dispersed-reinforced concrete ring, collapsed when charged with 36.7 kN; the collapse was due to longitudinal cracks based on ring bending stress, figure 5;
- the rectangular column SP_1, with regular reinforced concrete cross-section collapsed when charged with 31.5 kN; the collapse was due to cracks along the compressive stress, in the centre of the square sides (classical collapse) [17], figure 6;
- the rectangular column SPC_1, with mixed cross-section, made of a regular reinforced concrete core and a fibrous dispersed-reinforced concrete ring, collapsed when charged with 27.3 kN; the collapse was due to longitudinal cracks located on the edges of the square sides (due to the bending stress specific to the sides of the square representing the lost shuttering), figure 6;

Figure 5. Columns $SC_1$and SCC_1 during trials.

Figure 6. Columns SP_1 and SPC_1 during trials.

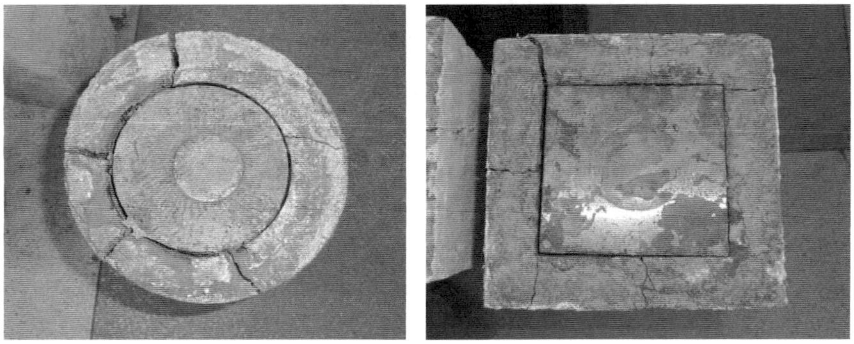

Figure 7. Collapse of fibrous dispersed-reinforced concrete columns.

- the confined columns had a similar reaction to the benchmark ones (margin of 87-91% of the bearing capacity), their efficiency being economically assessed as follows: the costs relative to the materials and the specialized labor force and the placement time were reduced by replacing the wooden or metallic classical shuttering with these dispersed-reinforced concrete lost shuttering;
- for a bearing capacity equal or even greater than the classically-executed poles, lost shuttering can be produced having a wider diameter so that the internally cast component has the same cross-section as a classically executed one; in this case, the efficiency is ensured by the greater (concrete coating) protection layer leading to longer durability of the component and the reduction or even elimination of costs relative to the materials and labor corresponding to the finishing / overcoating of the element, because the fibrous reinforced concrete surface does not have damages, faults, coring etc.

For the elements tested by applying the force across the whole cross section, the following were observed during trials:

- the circular-shaped SC_2 column with the regular reinforced concrete cross-section loaded on the entire transverse surface collapsed at a 46.75 kN load; the failure occurred due to the occurrence of longitudinal cracks (classic failure), figure 8;
- the circular shaped SCC_2 column with the mixed cross-section made of conventional reinforced concrete in the core and the dispersely reinforced concrete as an outline ring, loaded over the entire transverse surface, collapsed at a load value of 45.0 kN; the failure also occurred because of the appearance of longitudinal cracks due to the tensile stresses in the ring, figure 8.

Figure 8. Failure mode for columns when the force was applied over the entire cross-section

- the first cracks occurred around the force of 20.0 kN for both columns, with the distinction that in the case of the regular reinforced concrete one they developed faster, having larger openings at the same values of the load than those from the confined column;
- as mentioned above, by the almost equal bearing capacity between the two columns (the confined pillar reaching 96.7% of the standard one value), it would appear that the solution for the execution of the columns in lost formwork made of reinforced concrete with

polypropylene fibers can be efficient by reducing material and labor costs and shortening execution time; it removes the need to purchase or make wooden or metal shuttering and reduces the workforce for their execution, installation and dismantling;

- the outer surface of the columns made using lost formwork of dispersed reinforced concrete did not show defects, segregation, asperity, so it can be an economical solution (no need for finishing, plastering) for the apparent elements.

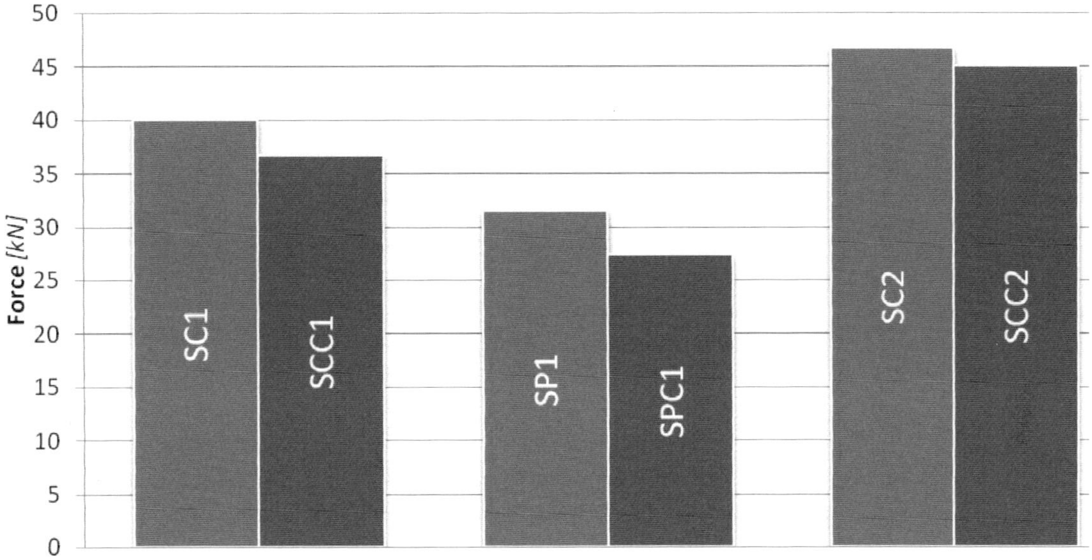

Figure 9. The bearing capacity of the confined columns compared to the benchmark ones.

4. Conclusions

A final conclusion of the experimental program is that it has led to convergent results which, on the one hand, have confirmed the research so far, and on the other hand, created the premises for their continuation and opened new roads and directions for seeeking solutions to improve the materials and building elements currently being used.

Dispersely reinforcement with polypropylene fibers is beneficial between certain limits such as the used concrete class, the type and granulation of the sand and grit, the nature and size of the loadings to which the element is subjected, the physical dimensions and the possibility of suitable compacting of the concrete elements made with this material.

References

[1] Bemac Laboratories PTY LTD Carlingford 2118, Australia, Investigation of the Use of Polypropylene Fibre in Concrete. Report No. MR 344, 1984

[2] Barbuta M et al 2017 Combined effect of fly ash and fibers on properties of cement concrete, *Procedia Engineering* **181** 280-284

[3] Barbuta M, Harja M and Baran I 2010 Comparison of mechanical properties for polymer concrete with different types of filler *Journal of Material in Civil Engineering* 696-701

[4] Boland J E 2008 Fracture of fiber-reinforced cement composites: effect of fiber dispersion, *Springer*

[5] Mustea A, Manea D, Muntean R and Miron O 2016 Evaluating the adherence to the substrate of fiber reinforced plaster mortars *Bulletin of Transilvania University of Brasov* **9** (58) 101-108

[6] Mustea A and Manea D L 2017 Influence of polypropylene fibers upon the mechanical characteristics of reinforced composite mortars, *Procedia Engineering* **181** 338-345

[7] Orban Y A, Manea D L, Aciu C and Mustea A 2018 Virtual manufacturing and mechanical properties of synthetic fiber-reinforced mortars *Procedia Manufacturing* **22** 262-267

[8] Ramakrishan V, Golapudi S P and Zellers R C 1987 Performance Characteristics and Fatigue Strength of Polyproypelene Fiber Reinforced Concrete, South Dakota School of Mines and Technology

[9] Serbanoiu A A, Barbuta M, Burlacu A and Gradinaru C M 2017 Fly ash cement concrete with steel fibers-comparative study *Environmental Engineering and Management Journal* **16** 1123-1128

[10] Muntean R and Muntean G 2012 Practical applications of dispersely reinforced concrete with polypropylene fibers: shuttering *Bulletin of the Transilvania University of Braşov* **5** 219-222

[11] Muntean R, Cazacu C, Mizgan P, Galatanu T and Tamas F L 2017 Practical applications of dispersely reinforced concrete with polypropylene fibers: beams *Procedia Engineering* **181** *285-292*

[12] Muntean R and Muntean G 2008 Proprietatilebetoanelorarmate cu fibre din polipropilena (The properties of polypropylene fiber reinforced concrete), Session of scientific communications with international character - CIB 2008, Braşov

[13] Documentary materials from fibers' producer, BruggContec AG, Switzerland (http://www.bruggcontec.com)

[14] Rudnov V, Belyakov V and Moskovsky S 2016 Properties and Design Characteristics of the Fiber Concrete *Procedia Engineering* **150** 1536-1540

[15] Johnston C.D. 1992 Fibre Reinforced Cement and Concrete - Advances in Concrete Technology, Canada, pp.629-697

[16] Muntean R and Muntean G 2015 The concordance between theoretical and experimental model for formworks made of dispersely reinforced concrete with polypropylene fibers; *Bulletin of the Transilvania University of Braşov* **8** (57) 127-132

[17] Kiss Z and Onet T 2008 Proiectarea structurilor de beton după SR EN 1992-1 (Design of concrete structures by SR EN 1992-1), Ed. Abel

A customization-oriented framework for product development. Case study- energy-efficient windows

A L Parv[1*], B A Deaky[1], R Stasiak-Betlejewska[2] and J Stawicka[2]

[1] Department of Manufacturing Engineering, Faculty of Technological Engineering and Industrial Management, Transilvania University of Brasov, Romania
[2] Institute of Production Engineering, Faculty of Management, Czestochowa University of Technology, Poland

[*] E-mail: luminita.parv@unitbv.ro

Abstract. This paper presents the authors' approach to producing customised products. The authors address the support of information sharing and activity synchronization within and between teams. The paper discusses a methodology for modelling data and original software for product data management that includes customers' preferences and specifications. The developed system has the ability to select an activity, as well as machining processes and parameters based on a set of design and production parameters, and to estimate product cost throughout the entire product development cycle. The case study refers to a product type called "energy-efficient windows". Window panes must be delivered to customers in a variety of shapes, dimensions, and materials according to certain specifications. The Access database provides the geometrical data for the next module, the CAD module that is achieved using Solid Works.

1. Introduction

In order to create higher value for a variety of customer needs and wants [1, 2], manufacturers often customise their offerings [3] and integrate services more and more with their products [4]. Nowadays, product development is impossible without taking into account customers' requirements. When managers understand the jobs and outcomes their customers are trying to achieve and the constraints they are trying to overcome, they dramatically improve their chances for growth, as they know exactly where to create value – leaving less to chance.

It is important for managers to know just what types of information they obtain when capturing the voice-of-the-customer.

Development and marketing managers are responsible for identifying opportunities for growth, segmenting markets, conducting competitive analysis, generating and evaluating ideas, communicating value to customers, and measuring customer satisfaction. Interestingly, they are dependent on obtaining information from customers in order to successfully perform these activities, making the process of gathering "customer requirements" one of the most critical business processes [5, 6].

Product development might ever more take into account improved customers' tastes and requests in a shorter time-to-market [7].

This paper presents a methodology for modelling data that includes customers' preferences and specifications.

The case study refers to a product type called "thermo-insulating windowpanes". Windowpanes must be delivered to customers in a variety of shapes, dimensions, and materials according to certain specifications.

The system consists of four modules:

- the Customer requirement module;
- the CAD module;
- the CAM module;
- the ABC module.

The customer requirement module provides the geometrical data for the next module, the CAD module that was achieved using Solid Works. The CAD module provides the data for the CAM module. The CAD and CAM modules provide the data for the ABC module (Activity Based Costing).

This paper describes the module regarding customers' requirements management. The implementation of the above presented model was carried out using Microsoft Access.

2. Considerations regarding the systemic designing approach

Systemic methods allow visualizing and understanding data organization. These methods are abstractions that present the real world as a collection of entities and associations established between those entities [5, 8]. Figure 1 shows the system architecture.

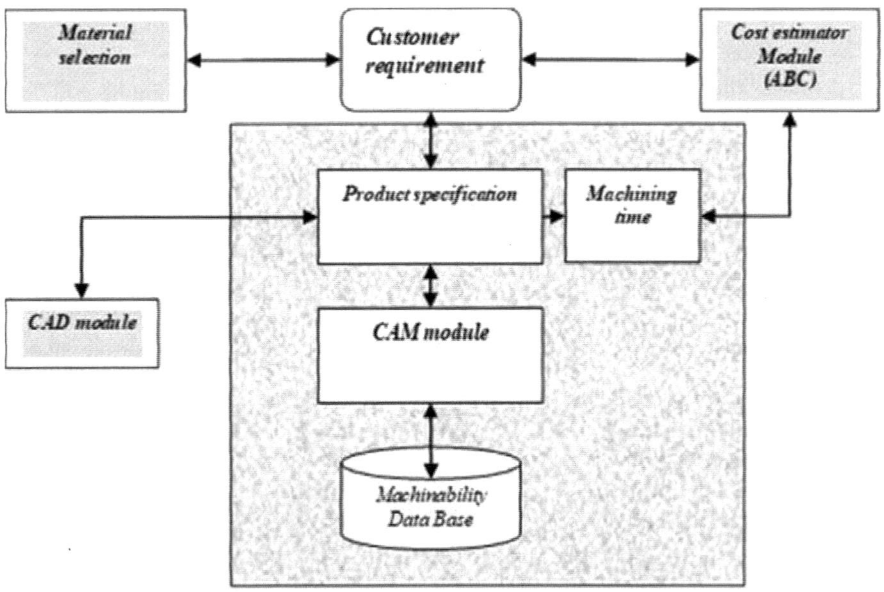

Figure 1. The system architecture.

Most systemic methods allow defining restrictions that describe static aspects, dynamic aspects or even temporal characteristics of entities.

They are legible formalisms of required specifications. Two methods are usually references for semantic representation: the individual method that will be integrated Merise and the entity-association method. Combining basic concepts of entity-association model (entity, attribute, association, cardinality, integrity restrictions) with Codd's normalizing rules was defined as a dictionary of data, entities, associations, cardinalities, structural, and semantic integrity restrictions.

The matrix of functional dependencies between attributes, as an additional element for validating the models of the studied system from a conceptual, relational (logical), and physical point of view was also defined.

These Codd´s rules provide the theoretical (although sometimes not practical) underpinnings for modern database design [8]. The rules may be summarized as follows:

- all database management must take place using the relational innate functionality of the database;
- all information in the database must be stored as values in a table;
- all database information must be accessible through the combination of a table name, a primary key, and a column name;
- the database must use null values to indicate missing or unknown information;
- the database schema must be described using the relational database syntax;
- the system must be able to update all updatable views;
- the database must provide single-operation insert, update, and delete functionality;
- changes of the physical structure of the database must be transparent for applications and users.

These are the rules used for the proposed database model.

3. Defining the thermo-insulating windowpane product

Specialized studies show that heat in a building is lost in many ways: 41% through the windows, 9% through the floor, 24% through the roof, 26% through the walls [9].

The thermo-insulating windowpane is an assembly of two window sheets that seal between them an air cushion with zero humidity or an inert-gas, usually argon for perfect thermo-insulating. If phonic insulation is wanted the argon is replaced by sulphur hexafluoride.

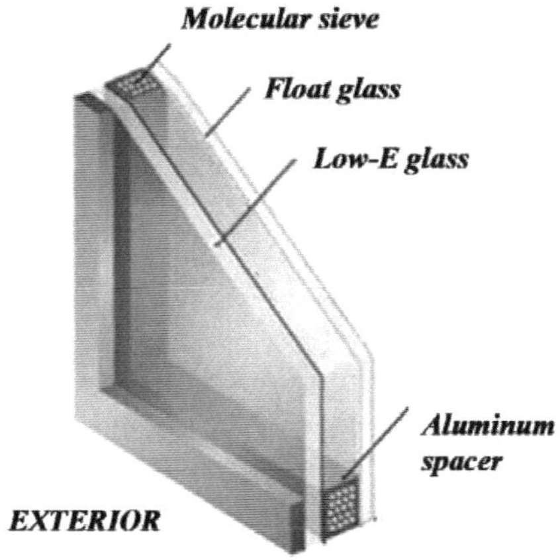

Figure 2. Studied product – Windowpane.

The degree of thermo insulation is determined according to the type of glass. For better results, treated glass is recommended, for example, LowE glass (low emission glass).

Emissivity is the property of a material to release the absorbed heat [9]. Normal glass surfaces allow significant heat loss caused by conduction and convection phenomena. Low emission glass allows improving the insulation property of double sheet windows by reducing heat convection.

This kind of window works like a mirror, reflecting back the heat in the environment where it was produced (figure 2).

4. Modelling and implementing technical and non-technical data

Data Dictionary

Database modelling is an important part of the database design process. It provides a structured, systematic approach that supports the development of well-structured and high-performance databases.

If databases are well designed, potential problems associated with data redundancy and poor performance due to ill-structured links between tables can be avoided or minimized.

A data dictionary will be developed according to the modelling methodology presented before and based on the enumerated data administration rules.

The matrix of the functional dependencies is used for the determination of the relationships between the attributes.

Conceptual model

The conceptual model defines the entities and relationships in the system being modelled.

Logical model

Having as starting-point the data dictionary and the conceptual data model, the logical or relational model is created.

For this purpose, the well-known rules regarding the normal forms of Codd (FN1, FN2, FN3) and the rules for passing from the conceptual data model to the relational (logical) data model were used.

Software implementation

Microsoft Access is widely used by small businesses and hobby programmers to create ad hoc customized systems for handling small tasks. Its ease of use and powerful design tools give non-professional programmers a lot of power for little effort.

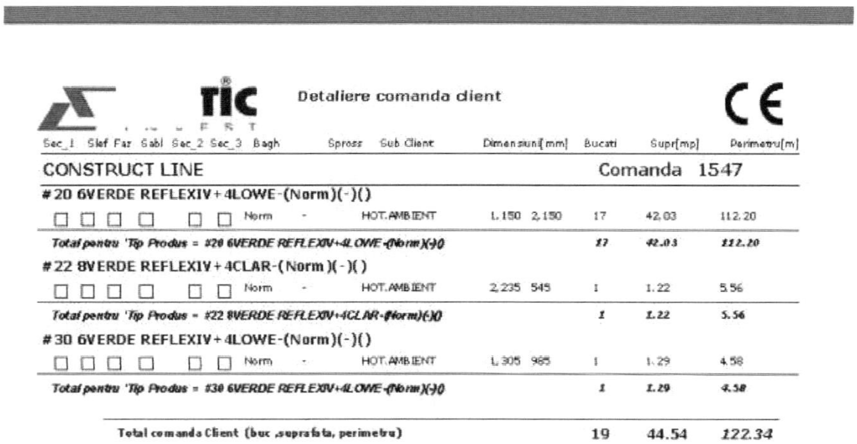

Figure 3. Customer requirements.

However, this ease of use can be misleading. This sort of power user is often an office worker with little or no training in application or data design.

Because of Access, it is possible even for such developers to create usable systems and many are misled into thinking that the tool itself is limited to such applications.

Creating a module for data administration in Access (figure 3) allows data transfer in Excel tables, which facilitates the integration of the two modules (this module and the CAD module which contain the parametric design of the windowpane).

The main menu contains five sub-menus. These sub-menus manage:

- customers with principal data;
- requirements;
- types of glass;
- types of baguette;
- types of window bar.

5. The CAD module

As each CAD system has its own method of describing geometry, both mathematically and structurally, there is always some loss of information when translating data from CAD data format to CAM.

The intermediate file formats are also limited in what they can describe, and they can be interpreted differently by both the sending and the receiving system [10].

It is therefore important when transferring data between systems to identify what needs to be translated.

A model of the component is constructed in Solid Works by the designer via the CAD system. The component dimensions and its volume are retrieved from the database of CAD.

The designer must specify all the features of the component and their attributes.

The system then prompts the user to select the material of the product.

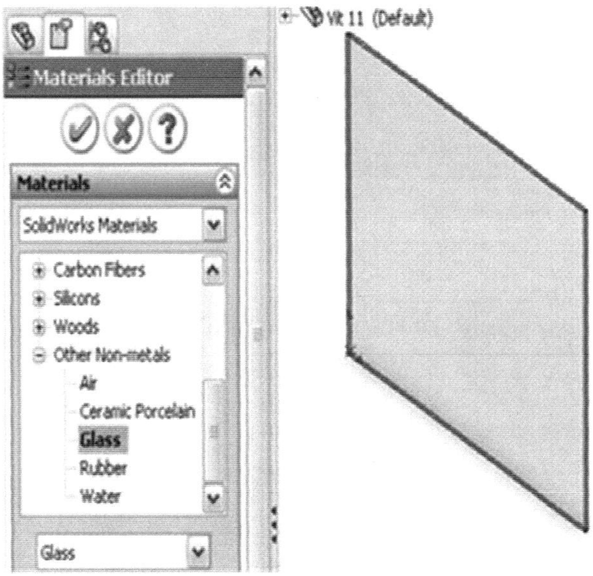

Figure 4. Material selection.

There are many constraints for material selection (figure 4), such as product functionality, material cost, and the type of manufacturing process.

6. The CAM module

The compatibility of the modularly designed system components is assured while facilitating a continuous production process – from storage to cutting solutions, to automatic breaking and sorting systems.

The extensive product variety ranges from manual plants up to fully automatic complete solutions [11, 12].

Figure 5 shows the CAM software (X-cut) would output code for the Bystronic CNC machine.

Bystronic glass cutting systems (figure 6) guarantee efficiency at the highest level.

IOP Conf. Series: Materials Science and Engineering **399** (2018) 012041 doi:10.1088/1757-899X/399/1/012041

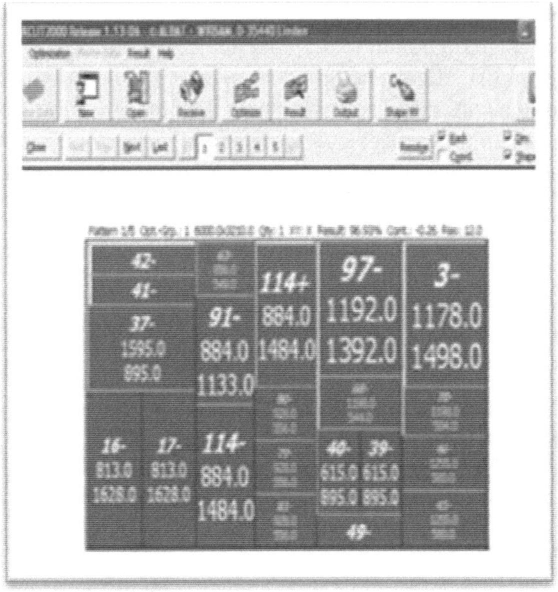

Figure 5. Optimization window.

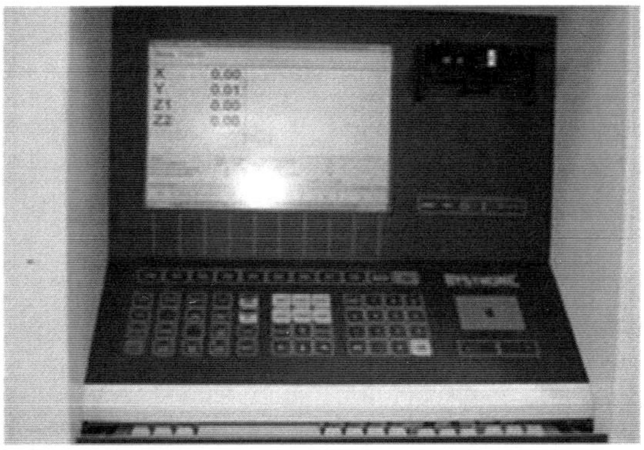

Figure 6. Bystronic CNC window.

7. The cost estimation module

The cost estimation according to the ABC method assumes covering the main following stages [13]:

- sharing the costs in direct and indirect costs;
- indirect costs separation on activities connected to survey centers;
- the identification of cost determinants for each type of activity;
- activities realignment based on cost determinants in realignment centers and the unitary cost estimation/determinant;
- the total cost estimation of the product through totalizing the direct cost with that of the cost determinants deriving from the realignment centers necessary for the achievement of the product.

In conformity with the carried-out survey upon the costs situation on activities, authors made the cost model shown in the relation (1).

$$C = F + \sum_i b_i * x_i + \sum_{ij} D_i * B_{ij} \quad (1)$$

where:

C - product cost;
b_i - direct costs of the product unit (including human resource costs, direct costs with materials, etc);
x_i - number of type i of cost drivers;
D_i - cost driver rate of i activity;
B_{ij} - number of cost drivers on i activity used by j resource.
The relation (1) will become:

$$C = F + VD + AD \tag{2}$$

where:
VD- total direct variable cost;
AD - total indirect activity cost.

The previous model can be shown as a matrix in relation (2), so:

$$VD = \begin{bmatrix} b_1x_1 & b_2x_2 & b_3x_3 \dots & \dots & b_mx_m \end{bmatrix}^{\mathrm{T}} \tag{3}$$

where:
b_i - direct costs of the product unit (including human resource costs, direct costs with materials, etc.) ;
x_i – number of type i of cost drivers.

The total indirect cost of activities will be presented in relation (4):

$$AD = \left[\frac{A_{ij}}{\sum_{i=1}^{m} A_{ij}}\right]_{m*n} * \left[\frac{R_{jk}}{\sum_{j=1}^{n} R_{jk}}\right]_{n*s} * p_k \tag{4}$$

where:
A_{ij} -number of activity j consumed of product;
R_{jk}- resources consumed by j activity;
p_k - price of type k .

Adding the parameter time t to this formula leads to the determination of the cost drivers.

8. Conclusion

Over the past few years, two issues have become mantras for corporations all around the world: ″mass customization″ and ″meeting or exceeding the customer's needs″.

The original software presents a way of implementing a customer module with a CAD module, a CAM module, and a Cost management module (ABC).

The integration of the module referring to customer requirements is absolutely necessary in the context of the presented product typology.

The module may be generalized and its use extended to other products.

Customer requirement management has been well recognized as one of the principal factors in product development for achieving success in the marketplace.

The developed system has the ability to select a material, as well as machining processes and parameters based on a set of design and production parameters, and to estimate the product cost throughout the entire product development cycle, including the assembly cost.

References

[1] Sakao T, Song W and Matschewsky J 2017 Creating service modules for customising product/service systems by extending DSM*CIRP Annals - Manufacturing Technology* **66** 21

[2] Hu S J, Ko J, Weyand L, ElMaraghy H A, Lien T K, Koren Y, Bley H, Chryssolouris G, Nasr N and Shpitalni M 2011 Assembly System Design and Operations for Product Variety. *CIRP Annals—Manufacturing Technology* **60** 715

[3] ElMaraghy H, Schuh G, ElMaraghy W, Piller F, Schönsleben P, Tseng M and Bernard A 2013

Product Variety Management. *CIRP Annals—Manufacturing Technology* **62** 629

[4] Lau G T and Tan W K 2000 A case study of buyer-seller relationships during the industrial product development process *Proceedings of IEEE Conference on Management of Innovation and Technology* **2** 832

[5] Meier H, Roy R and Seliger G 2010 Industrial Product-Service Systems—IPS2 *CIRP Annals—Manufacturing Technology* **59** 607

[6] Xuehong D, Jianxin J and Tseng M M 2000 Architecture of Product Family For Mass Customization *Proceedings of IEEE Conference on Management of Innovation and Technology*

[7] Song W, Sakao T 2017 A Customization-oriented Framework for Design of Sustainable Product/service System *JCLEP* **140** 1672

[8] Fong A C M (2000) Managing resources in the software development process *Proceedings of IEEE Conference on Management of Innovation and Technology* **2**5 30

[9] Hoffer J A, Venkataraman R and Topi H 2015 *Modern Database Management* (12th Edition) Addison Wesley, USA

[10] Binder K 2002 Numerical Investigations of the Glass Transition and Glassy State of Structural Glasses and Spin Glasses *Proceedings of Neumann Institute for Computing Symposium*, NIC Series **9** 247

[11] Schönsleben P 2012 Methods and tools that support a fast and efficient design-to-order process for parameterized product families *CIRP Annals - Manufacturing Technology* **61** 179

[12] Schuh G, Lenders M and Hieber S 2008 Lean Innovation: Introducing Value Systems to Product Development Portland *Int. Conf. on Management of Engineering and Technology*, PICMET '08, Piscataway, USA

[13] Parv L and Urdea M 2009 Building a software cost estimation based on Activity-based methodology *Annals of DAAAM for 2009 & Proceedings of 20th DAAAM International Symposium*

Assessment of energy savings achieved in tourism industry through innovative management of preparation techniques

G Pasca Pascariu[1,*], V Pădureanu[2], B Arsene[1] and G Calefariu[1]

[1] Department of Engineering and Industrial Management, Transilvania University of Brasov, Brasov, 500174, Romania
[2] Departament Engineering and Management Food and Tourism, Transilvania University of Brasov, Brasov, 500174, Romania

[*] E-mail: pascariu.gabriela@unitbv.ro

Abstract. With the increase in the income of the population, with the exception of periods of economic crisis, tourism industry continues to grow year after year. An increasing number of people benefit from the products offered by this industry. This observation is valid for Romania as well as for leading states like China. Without innovations in the field, energy consumption in tourism industry will naturally also grow. This paper proposes a management innovation which whose effect will be both the increase of tourism product quality and significant energy savings. It is an innovative way to prepare and serve breakfast. The savings made at the scale of a large five-star hotel, but also in smaller hotels, as well as in tourist boarding houses, will be evaluated. These results will be scaled to the whole of Romania.

1. Introduction

Countries have to cut gas emissions and raise energy-efficiency. It is a requirement that electricity production should be provided more and more from non-fossil renewable energy (such as wind, solar, geothermal, fuel cells, biomass) [1].

High-efficiency systems are associated with low emissions. The installation of combined heat and power schemes replacing less efficient systems leads to economic and environmental benefits, as well as significant optimisation of primary energy consumption [2].

The effects of retrofit strategies, such as cogeneration systems, often integrated with absorption chillers on primary energy saving and pollution reduction have been investigated in several applications characterized by relevant energy demands.

Combined heat and power, also known as cogeneration, offers considerable reduction in greenhouse gas emissions by making use of heat that is otherwise wasted during electricity production. These include improving the efficiency with which energy is produced and reducing emissions from energy. Research and development stimulate climate-friendly technologies. Currently, the desire to alleviate the problem of global warming has resulted in environmental concerns over the reduction of greenhouse emissions from industrial sources. The greenhouse gases include methane (CH_4), carbon dioxide (CO_2), nitrous oxide (N_2O), halogens such as chlorofluorocarbons and hydro fluorocarbons. From among them, CO_2 is the primary contributor to the problem due to its abundance, thus becoming a major target for reduction. Concern over global climate change has led to a need for reducing CO_2 emissions from chemical plants [3].

IOP Conf. Series: Materials Science and Engineering **399** (2018) 012042 doi:10.1088/1757-899X/399/1/012042

The tourism sector is changing along with the changes in demand. People have greater interest in experiences and creativity and therefore, the product that the tourism industry is offering is developing from mere services to experiences [4].

It is essential for service providers in hospitality industry to comprehend what customers of various market segments expect from a service firm in order to deliver high quality services. This study aims at measuring and comparing differences in the desired service quality expectations of hotel customers, as well as identifying the service dimensions that matter most to hotel guests.

Food and Energy are the inseparable connection that links these two commodities to the hospitality industry. This project focuses on uncovering points in culinary product processing where tangible steps can be taken to reduce energy consumption. Baseline information allowed to determine alternative actions that would lead to energy reductions. Therefore, this research addresses both the reduction of resources and the economic factors related to it. This research will lead to tangible impacts on the high energy consumption of the food system [5].

This paper presents an innovative method through which 5 star restaurants may modify their breakfast service through the bettering of the finger buffet serving system, by establishing a thermal treatment module in the serving hall, thus limiting the culinary products made in the kitchen.

In tourism industry breakfast service is performed in a planned and systematic manner, ensuring its conformity to the hotel's class category. Breakfast service is a component of the accommodation service but it can be purchased independently by the hotel non-resident customers.

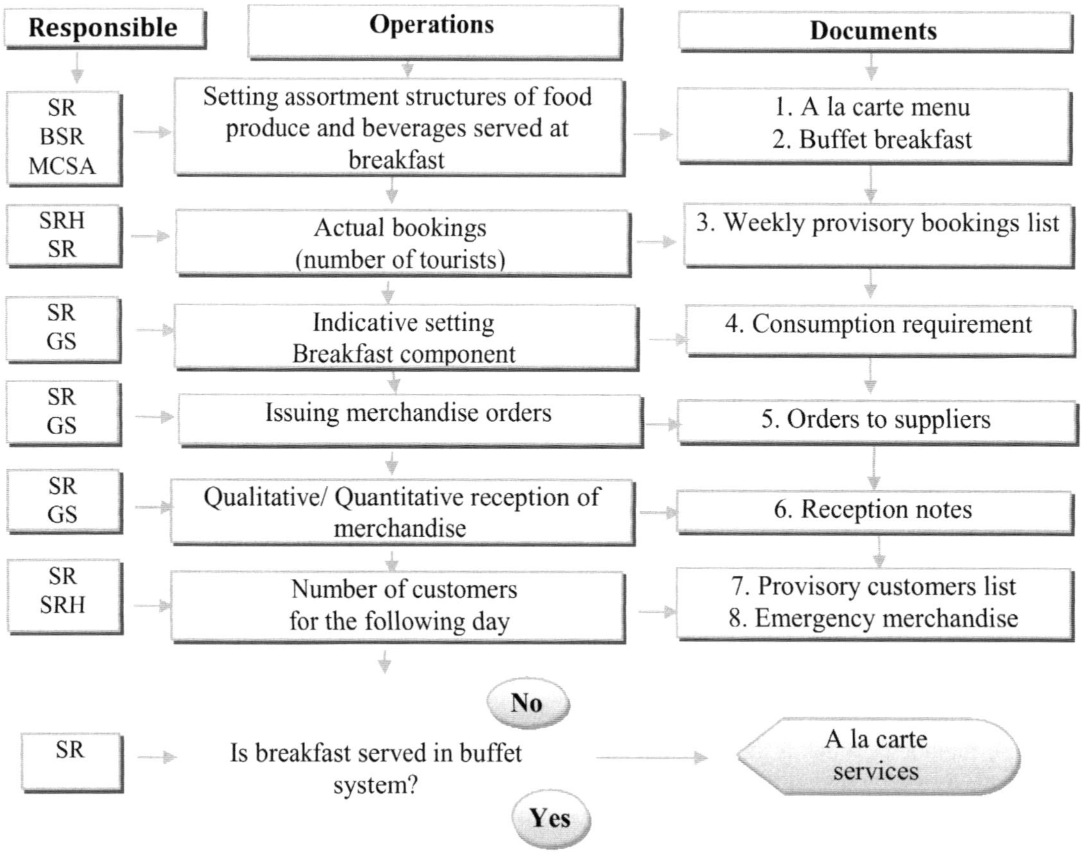

Figure 1. Scheme regarding actions flow at serving buffet breakfast [6].

Abbreviations used: SR = restaurant head of staff, BSR = restaurant chef, MCSA = quality management responsible, SRH = hotel head receptionist, GS = department administrator, SSR = head waiter.

Regarding servicing, breakfast can be room-serviced, buffet system or at tables; by component, there are two breakfast types: continental (full) and English/ American [7-10].

For continental breakfast only non-alcoholic warm drinks are indicated, the rest of the dishes served being fixed: butter, jam, croissants, muffins, rolls, toast. The British breakfast is more consistent than the continental one, including, besides the dishes above, one or more choices of buffet, kitchen, confectionery, pastry or bar types of food [11, 12].

The organization of the traditional breakfast service needs the simultaneous approach of different phases: the providing of necessary equipment (furniture, hardware, serving and work inventories), the preparing of the rooms and sections, the effective accomplishment of the services abiding by the work techniques at the same time. The equipment of the rooms with furniture, hardware, service and work inventory is all right, since it is the same as the one needed for lunch and dinner in all the rooms of the restaurant, this being influenced by the way in which breakfast serving is made [13].

Breakfast service resides in fulfilling planned activities in controlled and documented conditions by: setting breakfast assortment structure, ensuring raw material stock, preparing breakfast service, operations preliminary to breakfast serving, breakfast serving, breakfast closing (figure 1).

1.1. The organizing of breakfasts service in finger buffet system

At the moment our finger buffet stand has only equipment for storing food, bakery products, pastry, confectionery products, warm and cold drinks, all of which are made in the departments of the restaurant: kitchen, buffet, bar, cafeteria or confectionery lab and transported to the serving hall of the restaurant.

The equipment used for the finger buffet is: warm display cases (work temperature $+30^0$C to $+90^0$C), cold display cases (work temperature: -1^0C to 0^0C), multifunctional cart – display and serving (equipped with adjustable thermostat from 0^0C to $+15^0$C, hinged board, transparent protective cover, compartments or shelves at the bottom for the transport of the necessary inventory), instant drink dispenser machine, coffee machine, juice, cereal dispenser, toaster, chafing dish, milk thermos, frozen display case (work temperature from -18^0C to $+5^0$C), neutral display cases for fruit, bakery and pastry products, jam, comfiture, etc., cup warmer, electric boiler for water/tea, coffee filter, display platters with chilling device.

At present, part of the dishes served are made the day before, respectively the slicing of cold meats and cheeses, the preparation of salads, baking of pastry, confectionery and bakery products. These are placed on platters or in salad bowls, covered with clear sheets and labeled and stored in refrigerators. The next day they are put on the display stands at 6: 45 a.m.

Ready-to-eat produce requires limited processing prior to consumption and may harbor human pathogens that can present a significant food safety risk [14].

1.2. Setting assortment structure

The assortment structure of produce served at breakfast has to satisfy the customers' expectations in respect of the accommodation class category. According to the number of customers, breakfast service is organized in buffet or 'a la carte' system.

The Breakfast Menu will offer customers the possibility of choosing dishes by preference as well as some pre-set menu options at a fixed price within the budget limits allocated to breakfast from the accommodation costs.

In the case of the buffet breakfast system, a various assortment of dishes will be prepared, culinary, confectionery-pastry and juices to satisfy the customers' expectations within the costs frame limit determined by the multiplication of the number of customers to the costs allocated to breakfast form the accommodation total sum [15].

The quantities of food products, culinary produce and beverages will be calculated to the number of customers, after the closing of the previous accommodation day in the IT management system, through the Daily breakfast consumption sheet. Possible customer preferences will hereby be registered. Breakfast assortment structure will be designed thematically from one day to another, in order to avoid monotony or a negative perception from the customers on a longer stay.

The extended breakfast structure is set in the Buffet breakfast assortment, which contains non-alcoholic hot beverages, butter, marmalade, jam, honey, meat products, dairy products, egg-produce, fresh vegetables, bakery produce, refreshments, mineral water and which can be served as breakfast, grouped on categories [16].

2. The evolution of trade margin

The evolution of the average mark-up quota for breakfast serving was analysed in a public food service unit the 5*by comparing the months of January 2018 and December 2017.

In the restaurant operational registration, in the month of January, the consumption cost rose up to 58 103.17 lei (with bar produce rising to a total value of 10 015.66 lei and culinary produce of 48 087.57 lei), at the price index (purchase price without VAT). The ratio of bar produce in total consumption was 17.2 % and that of culinary produce was 82.8%. A number of 4 769 people were served, out of which: 4 648 people at 45 lei/ person, 66 people at 44 lei/ person, 4 children at 28 lei/ person and 51 people free of charge (guides, drivers, tourism agencies representatives).

By comparing these consumption costs to those of December 2017 we notice a downward trend of the mark-up quota. Compared to the previous month the number of served customers declined to 6.3% and the average mark-up quota fell 6 percent points from 245% to 239% in January 2018. The mark-up quota decline is due to the increase in individual consumption value of bar produce from 1.92 lei/ customer to 2.10 lei/ customer and of culinary produce from 9.95 lei/ customer to 10.08 lei/ customer (table 1).

Table 1. Average mark-up quota evolution between December 2017 – January 2018.

Breakfast Consumption 2017	December 2017	January 2018
Bar Consumption	9 779.03	10 015.66
Kitchen Consumption	50 670.75	48 087.51
Total	60 449.78	58 103.17
No. of people at breakfast	5 092.00	4 769.00
Bar consumption/person	1.92	2.10
Kitchen consumption/ person	9.95	10.08
Consumption/person (lei)	11.87	12.18
Average mark-up quota	245%	239%

Total costs per person were 12.18 lei, exceeding the standard cost. The average revenue per customer was 41.02 lei, with an insignificant variation as compared to December 2017 when the indicator value was 40.96 lei. If the positive aspect of the average revenue increase per customer served at breakfast is the result of the hotel reception service concern, which adjusts the exchange rate to the fluctuations of the exchange market intending to reduce the number of customers free of charge and those who pay in lei, the increase of average consumption per customer served is a negative aspect related to the restaurant operational activity, meaning that the quantity and structure of culinary and bar produce on breakfast stands is not rigorously measured as against the allocated budget.

The number of customers served at breakfast was by 58% larger than the same month in 2016 but the number of customers dynamic was exceeded by the total consumption index (161%). In comparison to the same month, the average consumption per customer served was by 0.17 lei more expensive (+1%), the fluctuation rising in bar produce (+0.27 lei; +15%) and falling in culinary produce (-0.1 lei; -1%), making the total influence over the average mark-up quota insignificant (-1%) (see table 2).

Table 2. Average mark-up quota in January 2018 as compared to January 2016.

Breakfast Consumption 2018	January 2017	January 2018	Absolute fluctuation	Relative fluctuation
Bar consumption	5 496.23	10 015.66	4 519.43	182%
Kitchen consumption	30 663.36	48 087.51	17 424.15	157%
Total	36 159.59	58 103.17	21 943.58	161%
No. of people at breakfast	3 011.00	4 769.00	1 758.00	158%
Bar consumption /pers.	1.83	2.10	0.27	115%
Kitchen consumption /pers.	10.18	10.08	-0.10	99%
Consumption/person (lei)	12.01	12.18	0.17	101%
Average mark-up quota attained	240%	239%	-1%p	100%

Daily evolution of the number of customers served in January 2018 shows important fluctuation at a maximum of 300 customers on Jan 16th 2018 and Jan 30th 2018 reflected in consistent fluctuation of mark-up quota, from 310% on Jan 1st 2018 (291 customers) to 141% on Jan 30th 2018 (66 customers). Breakfast was served in buffet system every day (figure 2).

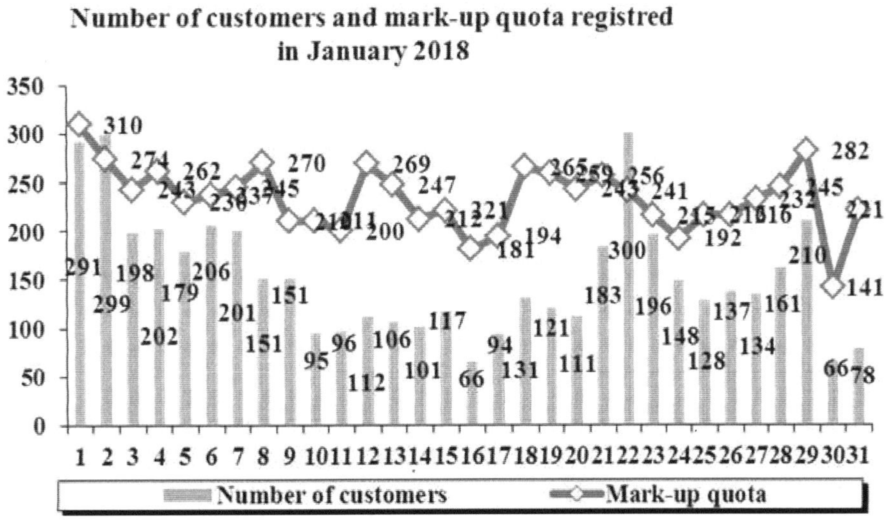

Figure 2. Number of customers and mark-up quota registered in January 2018.

From the prior analysis, one may observe the preservation of the decrease tendency of the trade margin obtained due to the increase of the value of individual consumption for bar products and kitchen products. The increase of the consumption is influenced also by the losses recorded by thermally treated culinary products which are prepared and placed on the presentation stand before the customers arrive.

Taking into account the large volume of culinary products within the structure of the buffet, they are prepared an hour before the assembly and arrangement of the presentation stand.

In this situation the following nonconformities may appear: culinary products made of eggs lose their freshness (pancakes, omelets, poached eggs dry and form a crust at the surface) or some culinary products may not be eaten, other items being preferred by customers and those need supplementation.

The products which depreciate, or which are not consumed within 2 hours maximum of their production, are removed from the stand and replaced with the culinary products which have been solicited by the customers. Thus, losses are registered because these products can no longer be used, the raw materials consumption used for their preparation increases, along with energetic consumptions.

3. Methodology

Regarding the drawbacks presented above, this paper aims to elaborate an innovative method, through the completion of the buffet stand, which is placed in the serving hall together with an area of thermal preparation.

This system's purpose is the ability to cook dishes in front of the customers according to their personal preferences. One may be able to cook: minute eggs (omelets with different side ingredients – ham, Swiss cheese, mushrooms, vegetables, herbs etc.) pancakes with different fillings (cottage cheese and cream, jam, Swiss cheese, chocolate, etc.). The customer may also select the ingredients used to prepare these dishes (for example omelet only from egg whites, chocolate filled pancakes without toppings etc.) or the thermic preparation (soft eggs etc.).

The suggested method leads to the decrease of electrical energy consumption through the reduction of the activity in the kitchen section and the selection of high-performance equipment, specific to the breakfast service.

The ability to cook such dishes in the dining hall considerably improves the quality of the products/dishes being served in relation to their freshness, taste and aspect. The customer may verify organoleptically the raw materials used. The transport of the products from the kitchen section through the office to the dining hall is eliminated, and the risk of physical, chemical and microbiological contamination is significantly reduced.

For the selection of the thermal treatment equipment an analysis was made on five different professional equipments for the HoReCa segment (hotel, restaurant, catering), noted with E, respectively: E1 – multi function tilting bratt pan 800, E2 – electric bench cooking stove with 4 hobs, E3 – multi function tilting bratt pan 600, E4 – electric cooking stove with 6 round hobs and E5 – convection oven (self cooking center).

By applying the multi-criteria analysis, the following criteria have been chosen [17]:

1. the costs of equipment acquisition (C);
2. the safety of using the equipment (S);
3. the effect of cooking food in the hall (E);
4. the accessibility of the price for clients (A);
5. the ease with which equipment is used (U);
6. the quality of the products (O);
7. energy consumption (N).

In [18] can be noticed that through multi-criteria analysis (MCA) one obtains rankings of solutions, subjectivism is removed and there is a higher degree of objectivity because the order of criteria is established by comparing each two criteria among themselves.

Step 1. The establishing of the criteria is done as follows: one criterion is more important than the other (=1), just as important (=1/2), less important (=0).

Step 2. Determining the weight parameters (p: the sum of points obtained in line; Δp: the difference between the points of the considered element and the points of the element of the last level; m: the number of the criterion surpassed by the considered element; Ncrt: number of criteria considered; Δp′: the difference between the points of the considered element and the points of the first element).

$$\mathrm{Yi} = \frac{\mathrm{p} + \Delta\mathrm{p} + \mathrm{m} + 0{,}5}{-\Delta\mathrm{p}' + \frac{\mathrm{Ncrt}}{2}} \qquad (1)$$

Step 3. Identifying the implementation variants.

Step 4. Giving a mark (to each variant according to each criterion).

Step 5. Calculating the product of the marks and weight parameters.

Step 6. The matrix of the consequences – determining the sums of the products.

Step 7. Final ranking.

Table 3 and Table 4 show the results of the quality comparison of the importance of criteria in the analysis of the variants/options for improving breakfast serving and the results of the calculation of the Yi weight for analysis criteria.

Table 3. The qualitative comparison of the importance of the criteria.

Criterion	C	S	E	A	U	O	N	Points	Level	Yi
C	0,5	1	1	1	1	1	0,5	6	1,5	4.714
S	0	0,5	1	1	1	1	0	4,5	3	2.500
E	0	0	0,5	1	1	0	0	2,5	5	0.930
A	0	0	0	0,5	0,5	0	0	1	6,5	0.176
U	0	0	0	0,5	0,5	0	0	1	6,5	0.176
O	0	0	1	1	1	0,5	0	3,5	4	1.590
N	0,5	1	1	1	1	1	0,5	6	1,5	4.714

Table 4. The results of the importance Yi weight factor.

Criterion	Points	Level	Δp	m	Ncrt	Δp’	Yi
C	6	1,5	5	5	7	0	4.714
S	4,5	3	3,5	4	7	-1,5	2.500
E	2,5	5	1,5	2	7	-3,5	0.930
A	1	6,5	0	0	7	-5	0.176
U	1	6,5	0	0	7	-5	0.176
O	3,5	4	2,5	3	7	-2,5	1.590
N	6	1,5	5	5	7	0	4.714

In table 5 marks are given to the seven variants in compliance with each criterion.

Table 5. The mark awarded based on the criterion.

Alternative/ Equipment	The mark awarded based on the criterion						
	C	S	E	A	U	O	N
E1	10	10	5	2	5	9	9
E2	8	9	7	6	7	8	7
E3	10	10	8	4	8	10	9
E4	7	7	8	4	7	10	7
E5	8	7	6	5	6	7	8

In table 6 the matrix of the consequences is presented.

Table 6. Matrix of consequences.

Alternative	Partial value of the coefficient Fij							Value
	I=1(C)	I=2 (S)	I=3(E)	I=4(A)	I=5(U)	I=6(O)	I=7(N)	
	YC=4.714	YS=2.500	YE=0.930	YA=0.176	YU=0.176	YO=1.590	YN=4.714	
	FJC= NjcYc	FJS= NjsYs	FJE= NjeYe	FJA= NjaYa	FJU= NjuYu	FJO= NjoYo	FJN= NjnYn	
E1	47.140	25.000	4.650	0.352	0.880	14.310	42.426	134.758
E2	37.712	22.500	6.510	1.056	1.232	12.720	32.998	114.728
E3	47.140	25.000	7.440	0.704	1.408	15.900	42.426	140.018
E4	32.998	17.500	7.440	0.704	1.232	15.900	32.998	108.772
E5	37.712	17.500	5.580	0.880	1.056	11.130	37.712	111.570

Within the analysis regarding the selection of the equipment we took into account the energetic consumption of each item of equipment, respectively their power (kW/h), usage duration (2 hours/day), watts (1kW=1000w) and the consumption per hour, day, month and year. We calculated the consumption and price of electric power, taking into account the average price or 0.6 lei/kWh (table 7).

Table 7. Analysis regarding the selection of the items of equipment according to the energetic consumption.

(E)	Power (kW/h)	Watt	Energetic consumption (kW/h)				Price (lei)			
			hour	day	month	year	hour	day	month	year
E1	10.5	10 500	10.5	21	638.7	7 665	6.3	12.6	383.2	4 599
E2	10.4	10 400	10.4	20.8	632.6	7 592	6.2	12.4	379.6	4 555.2
E3	7	7 000	7.0	14	425.8	5 110	4.2	8.4	255.5	3 066
E4	15.6	15 600	15.6	31.2	949	11 388	9.3	18.7	569.4	6 832.8
E5	18.6	18 600	18.6	37.2	1 131	13 578	11.1	22.3	678.9	8 146.8

The equipment with the lowest energetic consumption is the multi function tilting bratt pan 600 (E3), with an energetic consumption of 7 kW/h, the yearly cost of electrical power being 3 066 lei. At the opposite end we find the convection oven (E5) with an energetic consumption of 18.6 kW/h, the yearly cost of electrical power being 8 146.8 lei. On the second and third places, at a very small distance, we find the electric bench cooking stove with 4 hobs (E2) and the multi function tilting bratt pan 800 (E1), the price of electric power per year being 4 500 lei, for a yearly consumption of 7 600 kW. On the fourth place we find the electric cooking stove with 6 round hobs (E4), with a consumption of 11 388 kW per year, the annual price of the electrical power being 6 832.8 lei.

4. The effects of the application of the solution

From the analysis of the score obtained by each item of equipment we conclude that the most indicated variant is number 3 (figure 3).

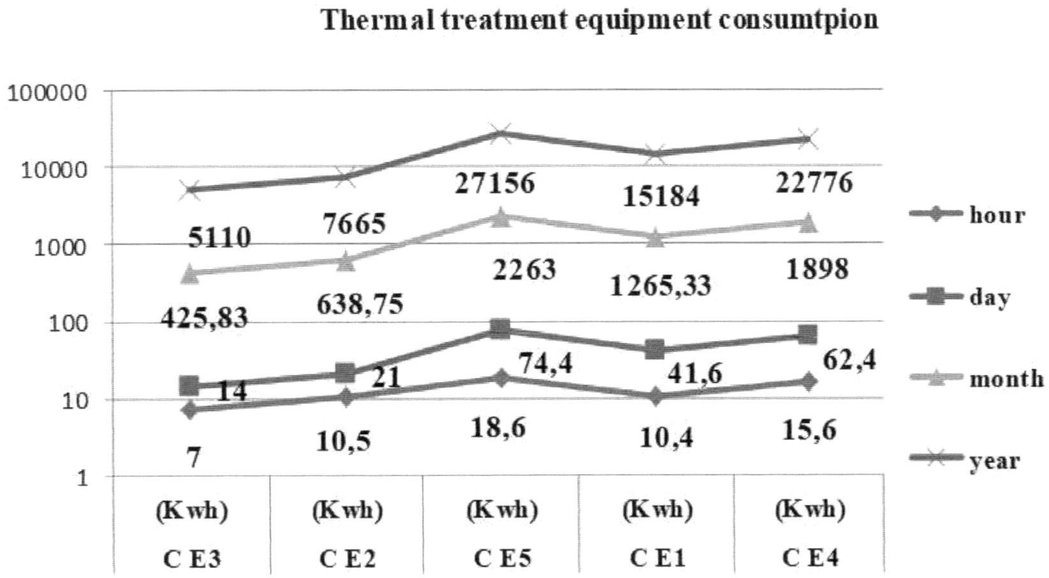

Figure 3. Thermal treatment equipment consumption.

This system's purpose is to prepare an assortment of culinary products in front of the customer according to their options while they can see the ingredients used and the preparation process.

The equipment which is used for preparing a wide range of culinary products in front of the customer, according to their options, is the multi function tilting bratt pan for frying, grilling, boiling and steaming.

The cooking surface is very smooth, for perfect cooking results, with a low thermal radiation and easy to clean even when it is functioning. The special surface for frying heats quickly and the heat distribution is even. The tilting bratt pan is assembled without welding, through casting, with rays that allow efficient cleaning. The depth of the pan does not allow spattering around. The temperature sensors are incorporated in the plate for optimal temperature control and safe manipulation. It also has a large drain plug for the effortless discharge of cooked products in gastronomic recipients placed at its base. The base is similar to a cabinet open at the front part and closed at the other three parts. The tilting bratt pan has the following dimensions: 600 x 800 x 750/900h mm (L x l x H cm); it also has two cooking areas with individual heating, a total power of 7 kW, electric supply: 400V-3N/50Hz, the size of the cooking surface: 530 x 569 x 75 mm, 21 liters capacity, frying/grill surface 530 x 470 mm, temperature: 95-250^{0}C.

For the time being the production process for breakfast products expands on a 4-hour duration between 6:00 and 10:00 a.m. The items of equipment used for the thermal treatment of the culinary products are: electric cooking machine with two hobs ($E^{a}1$) for cooking short orders (8kW power, 4 hour functioning time), convection oven (self cooking center) - $E^{a}4$ – for cooking breadstuff and confectionery products (18.6 kW, 1 hour functioning time), bain-marie with two basins ($E^{a}3$), for keeping food warm (2 kW power, 4 hour functioning time) and rapid salamander ($E^{a}2$), for the gratin (4.8 kW power, 2 hour functioning time).

In table 8 we point out the energetic consumptions and the price of the electric power of the utilized items of equipment during the breakfast service. Thus the total daily energetic consumption is 68.2 kW, 2 074.41 kW per month, 24 893 kW per year and the yearly price of the electric power is 14 936.3 lei.

Table 8. Energetic consumption for the present equipment.

(E^a)	Power (kW/h)	Watt	Energetic consumtion (kW/h) hour	day	month	year	Price (lei) hour	day	month	year
E^a1	8	8 000	8	32	973.33	11 680	4.80	19.20	584	7008
E^a2	4.8	4 800	4.8	9.6	292	3 504	2.88	5.76	175.2	2 102.4
E^a3	2	2 000	2	8	243.33	2 920	1.20	4.80	146	1 752
E^a4	18.6	18 600	18.6	18.6	565.75	6 789	11.6	11.6	339.45	4 073.9
Total E^a			33.4	68.2	2 074.41	24 893	20.48	41.36	1 244.65	14 936.3

By applying the innovative method proposed in this paper the equipment used for thermal treatment is the convection oven and the multi-functional tilting bratt pan. The total energetic consumption per day is 32.6 kW, per month 991.55 kW, per year 11 899 kW, and the yearly price of the electric power used is 7 139.9 lei. The savings thus realized would be 35.6 kW a day, 1 082.86 kW a month and 12 994 kW a year. (table 9)

Table 9. Energetic consumption of proposed equipment.

Equipment	Energetic consumtion (kW/h) hour	day	month	year	Price (lei) hour	day	month	year
Present equipment (E^a)	33.4	68.2	2 074.41	24 893	20.48	41.36	1 244.65	14 936.3
Proposed equipment (E^p)	25.6	32.6	991.55	11 899	15.8	20	594.95	7 139.9
Saving up	7.8	35.6	1 082.86	12 994	4.68	21.36	649.7	7 796.4

In tourism industry the problems of costs, trade margin, sale price and obtained profit represent the basic information for the analysis of activity-based results. The purchase price of raw materials, the addition of the trade margin, energy consumption, the total value or the profitability criteria are elements of managerial decision and therefore must be known.

Sales prices used in catering enterprises must cover all costs and achieve a profit justifying the efficient use of the invested capital [16].

In Romania there are 1 551 functioning hotels, out if which 35 are classified according to the Ministry of Tourism's Order 65/2013 to 5 stars. By applying the proposed method the results mentioned in figure 4 are obtained.

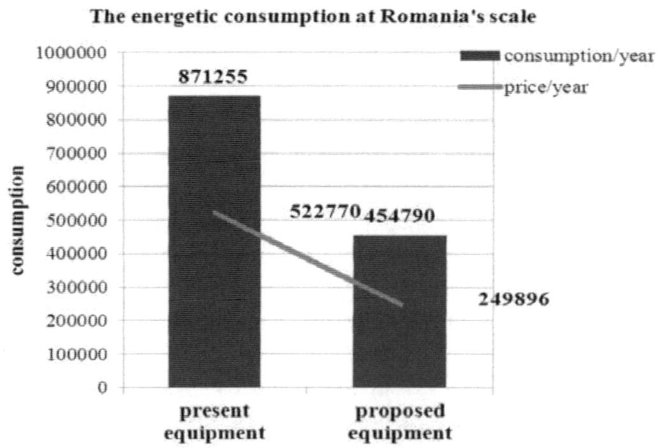

Figure 4. The energetic consumption at Romania's scale.

In tourism industry the production of culinary products is a complex activity, due to the variety of the operations which transform edible raw materials and culinary products which are made according to the specifications of the production recipes and menu plans.

Just like in other areas, the efficiency of the processes must be a major concern for the managers; the profit must be an incentive in the execution of the production program, as well as the improvement of the quality and the diversification of the brand. The ways of increasing profit are: reducing production costs without affecting the quality, having a correct correspondence between sale price level and profit, the growth of the volume of the sold products and services as well as the bettering of the quality of the products and services.

By analyzing the energetic consumption at Romania's scale, one can notice a decrease of 48%. In the present system of cooking and serving breakfast, the annual energetic consumption registered by the thermal treatment equipment used in the kitchen section is 871 255 kW, and through the proposed method the consumption is reduced to the value of 454 790 kW, thus obtaining a saving of 416 465 kW.

5. Conclusions

This paper presents an innovative method through which 5 star restaurants may modify their breakfast service through the bettering of the finger buffet serving system, by establishing a thermal treatment module in the serving hall, thus limiting the culinary products made in the kitchen.

The goal of the effective activities in tourism industry is to provide balanced, nutritional food with aesthetic and hygienic characteristics, at affordable prices and in sufficient quantities, under reducing costs concerning the raw materials, utilities, technological losses through preliminary processing and heat treatment and of the labor cost.

This system's purpose is to prepare an assortment of culinary products in front of the customer according to their options while they can see the ingredients used and the preparation process.

The preparation of these products in the serving hall improves the quality of the food by making it fresher, with a better aspect and better taste. The customer may check visually and olfactory the raw materials used during the cooking process. The transportation of the product from the kitchen, through the office, to the restaurant hall is eliminated. During the transportation the product may cool and may also become physically, chemically and microbiologically contaminated.

The method's aim is to reduce the value of the consumed energy through adequate management, using high-performance thermal treatment equipment and reducing the consumption of electric power from 4 hours to 2 hours maximum, by improving the energy efficiency. In order to complete the finger buffet stand professional equipment items for the HoReCa sector have been identified and selected.

The proposed method leads to the decrease of the electric power consumption through the reducing of the activity of the kitchen sector and cooking ready to eat products, eliminating in this way equipment used for maintaining food warm, gratin and re-heating, as well as the selection of high performance thermal treatment equipment specific to the breakfast service.

Through the application of this innovative method at a 5 star restaurant the following advantages ensue: there are important economies in the electric consumption, the losses of non-conformable products are eliminated, because they have not been solicited by the customers or they have not been consumed during the 2 hours after their preparation. The key criteria used to reduce raw materials consumption and electric power were: energetic efficiency, environmental impact, customers' health and the quality of the products.

References

[1] Davis M L and Cornwell D A 1991 *Introduction to environmental engineering: 2nd edn*, (McGraw-Hill Inc.)

[2] Chinese D, Meneghetti A and Nardin G 2005 Waste-to-energy based greenhouse heating: exploring viability condition through optimisation models *Renewable Energy* **1573**

[3] Kovac Krali A 2018 Energy-Saving by Adding Vinegar or all Acidity Components in Food, at

the End of Cooking *Innov Ener Res* **7(1)** 183
[4] San Salvador R and Calvo-Soraluze J 2013 Tourism Policy Makers and Managers as Generators of Meaningful Leisure Experiences Knowledge, Skills and Values Needed by the New Professional *J Tourism Res Hospitality* **126**
[5] Wonglorsaichon P and Wiriyakitjar R 2013 Hotel Customer Expectations of Service: A Provincial Analysis of Family Business in Chiangmai, Thailand *J Tourism Res Hospitality* **123**
[6] Pasca Pascariu G and Mărăscu-Klein V 2016 Specific Consumption Efficiency in Tourism Industry *Recent- Transilvania University of Brasov* **17** 577
[7] Jamison C A and Jamison B 2002 *Real Amern Breakfast* (Harper Collins)
[8] Vollstedt M 2012 *The big Book of Breakfast*: Serious Comfort Food for Anytime of the day (Chronicle Books)
[9] Duda C 2008 *Completely Breakfast* (Cedar Fort)
[10] Moos D 2011 *The Art of Breakfast* (Down East Books)
[11] Lupu N 2005 *The Hotel – economics and management* (All Beck)
[12] Berechet G 2004 *The manager of the restaurant* (THR-CG)
[13] Pasca Pascariu G and Mărăscu-Klein V 2017 The optimization of consumption and costs in breakfast services *Review of Management and Economic Engineering* **64** 333
[14] Dewangan AK, Patel AD and Bhadania AG 2015 Stainless Steel for Dairy and Food Industry: A Review *J. of Mat. Sci. & Eng.* **191**
[15] Nicolescu R 1981 *Technology of work in the restaurant and bar* (Bucuresti:Ed Sport-Turism)
[16] Stavrositu S 2008 *The art of services in restaurants and bars, culinary technology, hotel services* (Bucuresti: Ed Fundaţia Arta serviciilor în turism)
[17] Bobancu Ş 2015 Analiza multicriteriala avansata *Creativity and Innovation (C&I)- Transilvania University of Brasov* **10** 61
[18] Pasca Pascariu G 2016 Optimization of breakfast service in finger-buffet. System by multi-criteria analysis *Creativity and Innovation (C&I)- Transilvania University of Brasov* **8** 35

Homogeneous nested bearings

C Pisarciuc[1*] and R Cioara[1]

[1] Engineering and Industrial Management Department, Faculty of Technological Engineering and Industrial Management, Transilvania University of Brasov, Colina Universitatii nr. 1, corp A, 500036 Brasov, Romania

[*] E-mail: pisarciuc.c@unitbv.ro

Abstract. The solutions presented in the paper are based on the possibility to obtain stacked bearings by mounting two bearings if the outer diameter of one corresponds to the inner diameter of the second. Another variant is to specifically design an inner ring with two raceways, one for the lower bearing and the other one for the upper bearing. The paper presents solutions of this type by using bearings of the same type (homogeneous). Regardless of the chosen variant, the final solution presents two layers of rolling bodies or, in other words, a single intermediate ring. The construction of nested bearings must primarily take into consideration the balancing of the static and dynamic basic load rating ensured by the two layers of rolling bodies. This objective can be achieved by adequately establishing the radius and number of balls in the two layers. The advantage of such a construction is that it allows obtaining high rotational speeds. From a theoretical point of view, one can conclude that the maximum speed results as the sum of the values for the two bearings.

1. Introduction

Although the use of bearings is varied, in the machine tools domain, medium and large dimensions are used. In order to obtain the high speeds required by modern cutting processes, specially designed bearings are used or a replacement of this type of motion transmission with another variant which, in most cases, is more expensive.

To increase the performance of bearings, rolling elements are made from non-metallic materials such as ceramics (Silicon Nitride-Si_3N_4 or Zirconium Oxide-ZrO_2). Ceramic bearings are developed for industrial applications under special operating conditions or in corrosion environments [1]. The silicon nitride material is associated with the longest life of materials. However, the dynamic capacity of a full silicon nitride bearing will be only 5% to 12% compared to that of an all-steel bearing, of similar geometry [2].

Because balls made of ceramic materials are lighter (up to 40% less than steel ones) [3], speeds of about 10~40% greater are possible. The low density of ceramics makes them attractive as ball or roller materials for very high speed bearings. This benefit is due to the fact that the fatigue life of very high speed ball bearings can be reduced as a result of excessive centrifugal force on the balls and subsequent increased stress at the outer race [4]. Lower mass balls can diminish this fatigue life reduction. Concluding, ceramic balls are lighter and the ball could easily achieve a higher speed and, in operation, less energy is needed to maintain the rotation. The drawback is that ceramic rolling elements are usually harder than the steel raceway. Under the circumstances, in time, a groove will form in the raceways. In the same way, because ceramic is frequently harder than steel rings, it generates an increase of stresses, and an overall decrease of load capacity. Another variant used is a full ceramic bearing. Because of the above-mentioned ceramic characteristics, such bearings are noisy at high speeds and prone to cracks at high loads.

If any bearing catalogue is analysed, it is possible to outline the following idea. The maximum speed [min^{-1}] at which a bearing is operating is a decreasing function in relation to the size increase and width decrease of the bearing. In the case of steel ball bearings, higher speeds, for example over 10,000 min^{-1}, are only possible for small size bearings. According to [5], starting from the size of 30 mm of inner diameter for a radial ball bearing, a 15,000 min^{-1} speed limitation is reached. Growing only one-step, for the inner diameter of 32 mm, the maximum speed limit is 15,000 min^{-1}.

2. Concept origin

Many inventors have made recourse to a solution meant to increase the maximum bearing speed limit, namely that of making single-bearings using common bearings or producing equivalent special bearings designed as independent entities.

The concept of "stacked bearings" has been proved since 1924, being the subject of a French patent [6]. The patent focuses on a bearing composed of two bearings. The constructive solution shown in figure 1 proposes a bearing composed of two radial ball bearings 11 and 12, distinctly dissimilar in size and implicit having different static and dynamic loading capacities. To minimize the radial size of the assembly, a bush 14, with two steps, is used. The patent does not explicitly state that the two bearings can be assembled into one another. The assertion that this compound carries a maximum limit speed equal to the double limit speed of the component bearings is not true. The maximum limit speed is obtained as the sum of the maximum limit revolutions of the bearings comprised by the bearing as further presented.

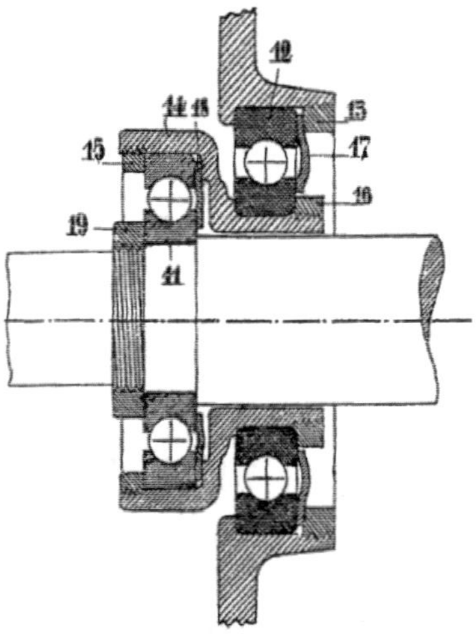

Figure 1. Bearing assembly with two-stage intermediate bushing [6].

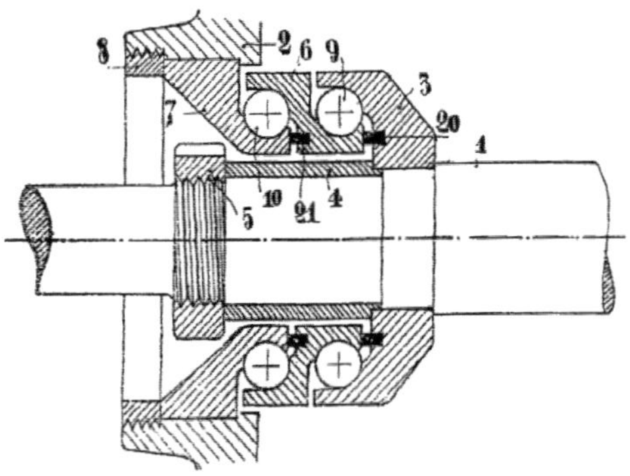

Figure 2. Stacked radial axial bearing [6].

The solution presented in figure 1 may have been a starting point for the solution presented in figure 2. In this case, the bearing does not have two distinct bearings, but represents a radial-axial bearing, even if the author does not emphasize this in the description and in his claims [6]. The bush 6 is an intermediate ring of the bearing, cumulating the functions of the bush 14 of the assembly from figure 1, the outer ring of the bearing 11 and the inner ring of the bearing 12.

An obvious example of "stacked bearing" given in [6] is shown in figure 3. This is in fact a combination of a two-row radial ball bearing on the inner floor of the construction and a radial ball bearing on the outer floor. The outer ring of the two-row ball bearing from the inner floor is modified; on the outside, there is the rolling path for the ball bearings in the outer layer. The inner ring of the

IOP Conf. Series: Materials Science and Engineering **399** (2018) 012043 doi:10.1088/1757-899X/399/1/012043

bearing in the outer layer is not presented. The differentiation of the ball diameters from the two floors of the bearing highlights the intention of the author to ensure equalization of the static and dynamic loading capacities of the two floors.

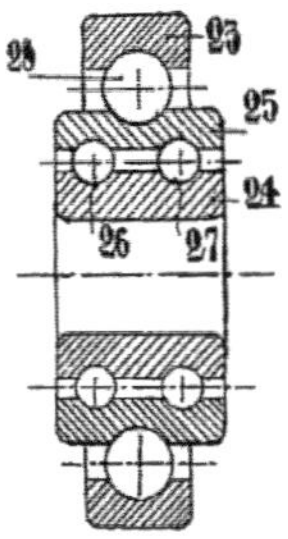

Figure 3. Stacked radial bearing [6].

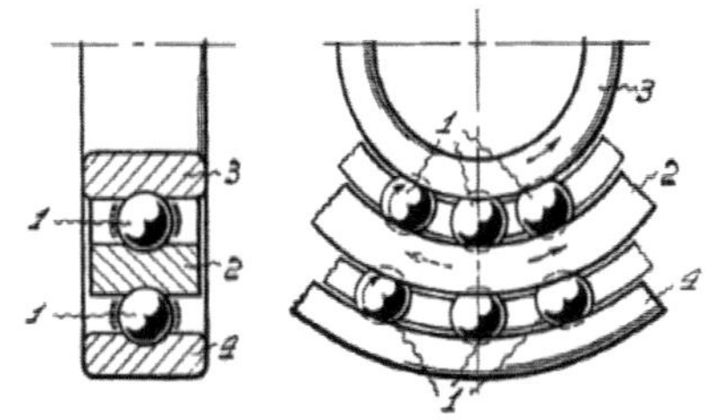

Figure 4. Stacked radial, single row ball bearing [7].

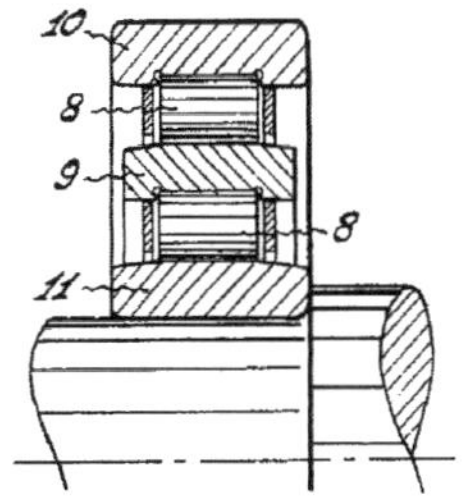

Figure 5. Stacked bearing with rollers on one row [7].

The concept of "stacked bearings" is used in [7] to designate bearings similar to those in figure 3. This patent discloses a claim of a very general nature "bearing, characterised in that at least two rings are separated from one another by a free (free-rotation) ring disposed between an outer ring and an inner one". The description and drawings accompanying the invention explicitly convey the concept by several constructive solutions, all of which have a single intermediate ring.

The stacked radial, single row ball bearing shown in figure 4 is a bearing equivalent to the assembly shown in figure 3. Balls 1, from figure 4, are the same in size, suggesting particular concern for the construction principle and not for the equalization of loading capacities in the two layers.

Figure 5 shows the construction of a "stacked bearing" obtained by combining two radial roller bearings. The intermediate ring 9 and the rollers in the inner bearing layer can axially translate relatively to the inner ring 11 and to the rollers in the outer bearing layer. This detail should be taken into consideration because, if unsolved, it inevitably leads to the failure of the bearing. This is not a universal solution because this type or bearing cannot be obtained by combining NU or N type bearing. The only possibility is to combine one NU bearing with a NUP bearing, the last one being placed on the inside.

3. Stacked bearings obtained by assembling

The solutions presented in the article address the issue of manufacturing bearings capable of carrying high speed, but available at reduced cost and without significant technological modifications in relation to currently used bearings. The main objective that led to the conceiving of these types of bearings, regardless of their constructive type, was the increase of the maximum limit speed that these bearings are capable of carrying. The developed constructive solutions are achievable without or with minimum modifications of the current manufacturing technology of bearings. As the significant increase of the maximum limit speed was the main objective pursued by the authors, subsequent research addressed the development of specific constructive solutions, establishing of optimization modalities of these solutions and refining of the mathematical model, as a prior step to experimental research.

A realistic solution satisfying largely the imposed requirements is given by stacked bearings. This original concept is based on the combination of two bearings, one mounted within the other, thus yielding a bearing with two intermediary layers of roller bodies. The exterior race of the interior bearing and the interior race of the exterior bearing form a single body, the intermediary race, a machinery part characteristic for planetary bearings. Based on the particularity that the inner ring is in fact a combination of the two rings, and if we take into account the resulting motion trajectory for the elements of the ensemble, this new type of bearing can also be called a planetary bearing similarly to a planetary gear. Planetary bearings can be obtained by combining different types of existing bearings.

IOP Conf. Series: Materials Science and Engineering **399** (2018) 012043 doi:10.1088/1757-899X/399/1/012043

A homogeneous planetary bearing is obtained by using two similar type bearings. By using two different types of bearings a mixed planetary bearing is obtained.

Because bearings could be identified with a nominal exterior diameter, the same as the interior diameter of other bearings of the same or different types, the idea emerged to assemble two bearings one within the other, as shown in figure 6. The next step was to identify a model that should describe the operation of such an assembly and the performance study. The developed model highlighted that the maximum limit speed of the assembly is theoretically equal to the sum of the maximum limit speeds of the component bearings.

Figure 6. Assembly of homogenous stacked radial ball bearing.

In a stacked bearing, either the interior or the exterior ring is fixed, depending on the mounting diagram, figure 7. The intermediary ring rotates in the same sense as the mobile ring. The rolling bodies rotate both around their axes and around the axis of the bearing in the same sense as the mobile rings.

The name "planetary bearing" assigned by the authors to this new class of bearings may be motivated by the aforementioned movements of the components of such a bearing, even though the name as such may not be entirely accurate.

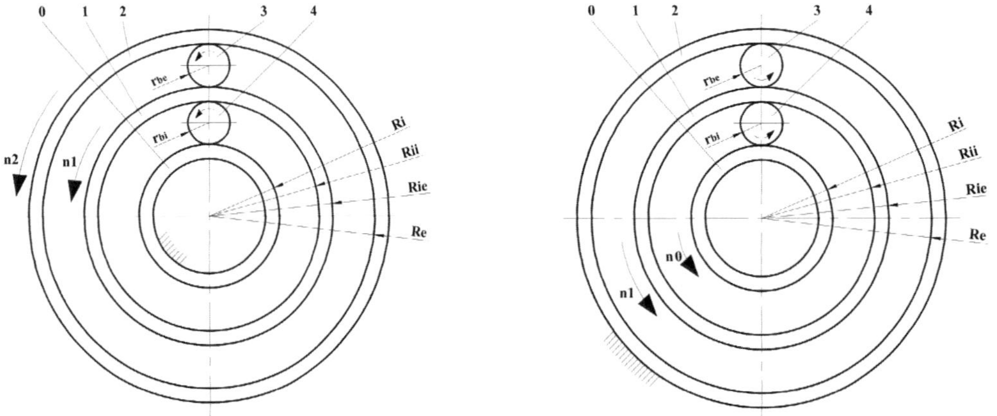

a. fixed inner ring b. fixed outer ring

Figure 7. Motions of the component elements of a stacked bearing.

In planetary bearings, under real conditions, the value of the intermediary race speed - as well as that of the rolling bodies - is not constant and determined, being dependent on several aspects: friction coefficients, clearances, the magnitude of the pre-stressing, lubrication, temperature, etc.

From the point of view of commercial bearings, one can identify practical solutions that correspond to a pair that balances either identical widths or similar loading capacities. The data presented in table 1 correspond to a perfect match regarding the width of bearing.

There are many typo-dimensions of radial or radial-axial bearings available where the (nominal) exterior diameter of one is equal to the interior diameter of bearings of the same or different type. When assembling one bearing within another one of the same type, a homogenous bearing is achieved. Figure 8 a, b, c, d shows some examples of homogenous bearings obtained by assembling. Considering the mode of obtaining of a planetary bearing, their symbolisation can be of "exterior bearing symbol"//"interior bearing symbol" type.

Table 1. Example of matching of two ball bearings [8].

Bearing type		Dimensions [mm]			Basic load ratings [kN]		Limiting speed $[min^{-1}]$
		d	D	B	C_r	C_{0r}	
interior	6200	10	30	9	5.10	2.40	24000
exterior	6906	30	47	9	7.25	5.00	14000

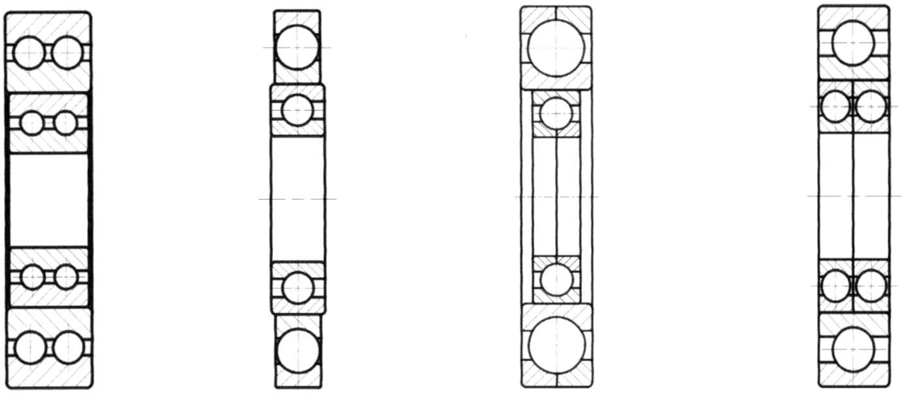

a. 4//4 b. 6//6 c. Q//QJ d. 6//2×6 (6011//2×16006)

Figure 8. Examples of homogenous planetary (ball) bearings obtained by assembling.

The manufacturers' catalogues indicate a certain performance for each typo-dimension of bearing. From the viewpoint of stacked bearings, the values of the basic loads and of the capable maximum limit speeds are of importance. The ratio of the basic loads for exterior and interior bearings indicates that, in the great majority of combinations, the basic load - both static and dynamic - of the exterior bearing is significantly higher than that of the interior bearing. As the basic load of the planetary bearing is equal to the smallest of the basic loads of the component bearings, equalising measures of the basic loads of the two layers of the planetary bearing is recommended. A solution could be the utilisation of two bearings in the interior layer as presented in figure 8.d.

According to [9], a well-known bearings manufacturer, there is no precise method to indicate at what maximum speed a bearing may operate. To complicate even more the speed assessment, in the case of duplex mounting, as frequent in high-speed machine tool spindles, bearing preload and contact angle affect (decrease) the permissible speeds. The estimated values of the maximum limit speeds of the stacked bearings, obtained by assembling, are determined based on the maximum limit speed values given in the catalogue for the component bearings of the respective bearings.

4. Nested bearings

In stacked bearings, the interior ring of the exterior bearing and the exterior ring of the interior bearing form a single body, the "intermediary ring" and, typically, the same width is selected for both rings. The intermediary ring does not need to have the same thickness as the one resulted from the assembly of the two bearings. In order to overcome the rise in radial dimensions, it is possible to design a common intermediary ring used simultaneously by the inside and outside bearing. In the absence of

IOP Conf. Series: Materials Science and Engineering **399** (2018) 012043 doi:10.1088/1757-899X/399/1/012043

another criterion, on a nested bearing the values of the diameters of the balls in the two layers are determined such that the static and dynamic loading capacities are equal.

Considering what has been stated above, the authors of this article have developed a concept based on three rings bearings, presented in figure 15. The invention relates to a radial-axial planetary ball bearing, figure 15.a. This bearing has an inner ring 1 made of two symmetrical elements (a, b), an intermediate ring 2 and an outer ring 3, as well made of two symmetrical components (c, d). The rolling bodies are of the same diameter and are arranged in concentric layers.

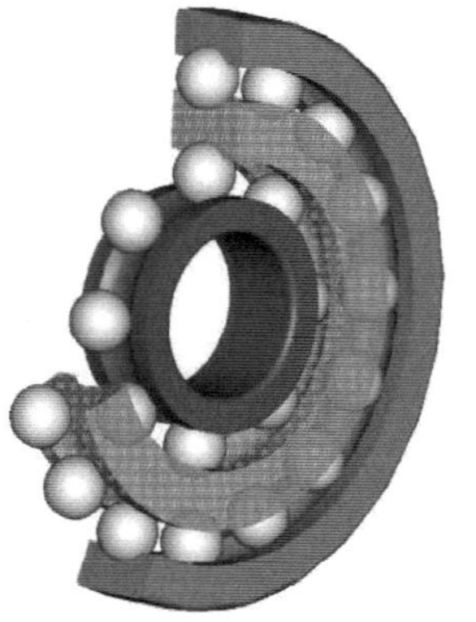

Figure 9. Nested radial ball bearing, homogenous, type 6//6, with equally sized balls on the two layers.

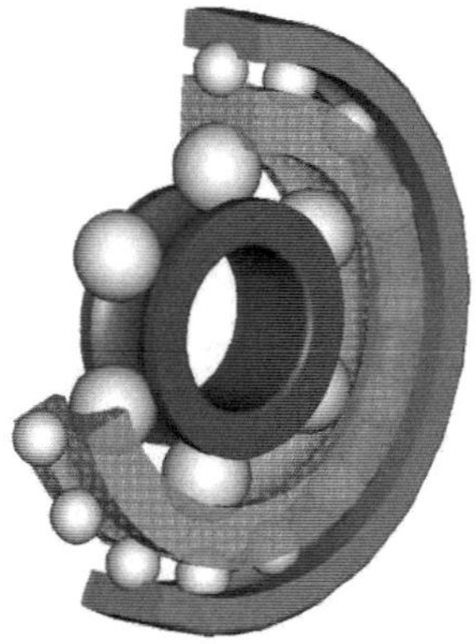

Figure 10. Nested radial ball bearing, homogenous, type 6//6, with the balls' radius on the inner layer bigger than the one on the external layer.

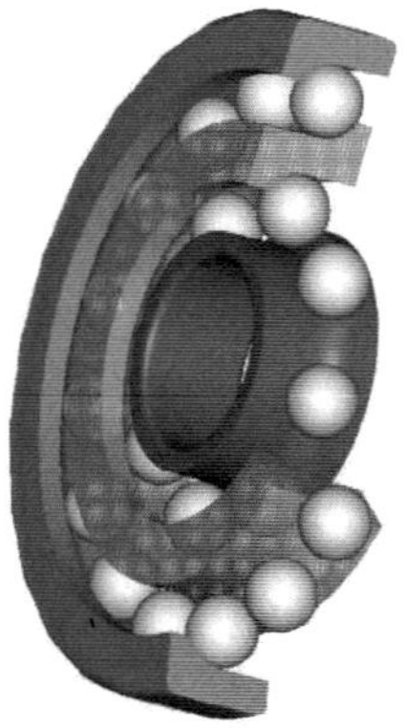

Figure 11. Nested angular ball bearing, homogenous, type 7//7, with equally sized balls on the two layers.

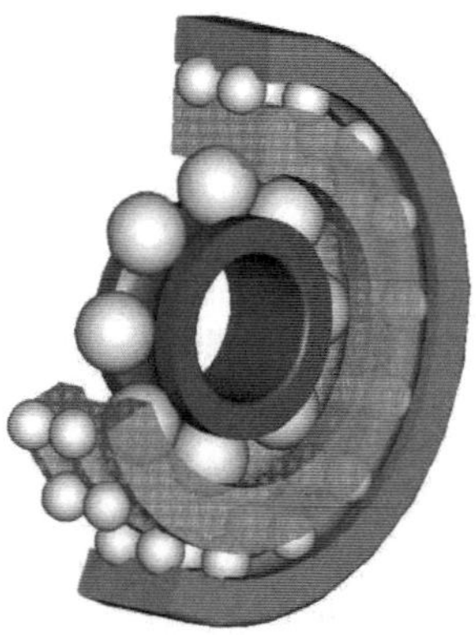

Figure 12. Nested ball bearing, mixed, type 4//6, with the balls' radius on the inner layer bigger than the one on the external layer.

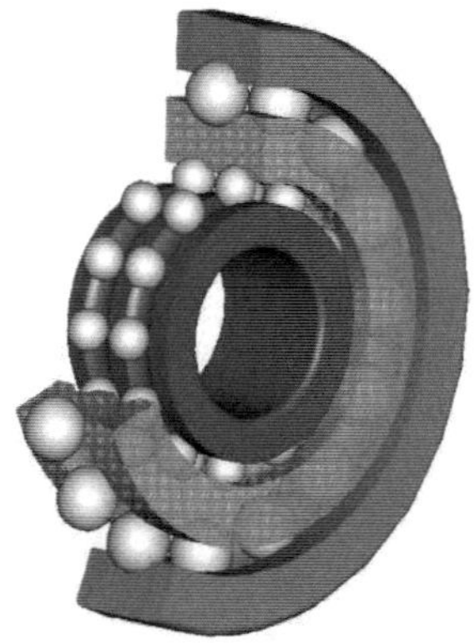

Figure 13. Nested ball bearing, mixed, type 6//4, with the balls' radius on the inner layer smaller than the one on the external layer.

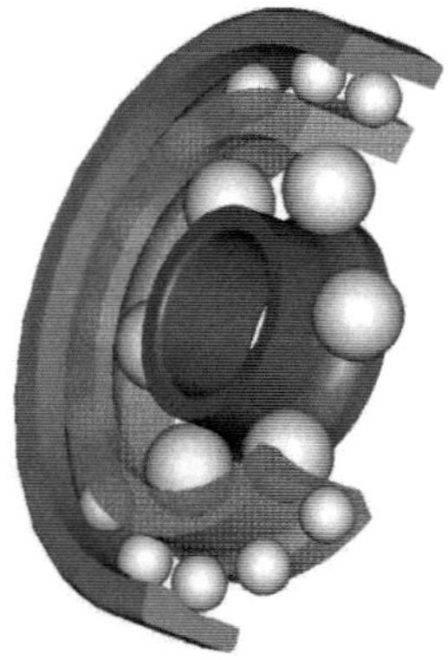

Figure 14. Nested angular ball bearing, homogenous, type 7//7, with the balls' radius on the inner layer smaller than the one on the external layer.

In another variant, figure 15.b, the bearing has an inner ring 6, made of symmetrical elements c and d. The outer ring consists of two lateral parts (g, h) and an intermediate element i. The major difference is related to the number of ball rows and to their size. In this case, the balls are placed in two rows at the exterior.

The variants from figure 15.c and d are practically variations of those presented in the first place, in figure 15.d. with the balls having the same size. This invention is endorsed by Romanian patent RO 123080 [10].

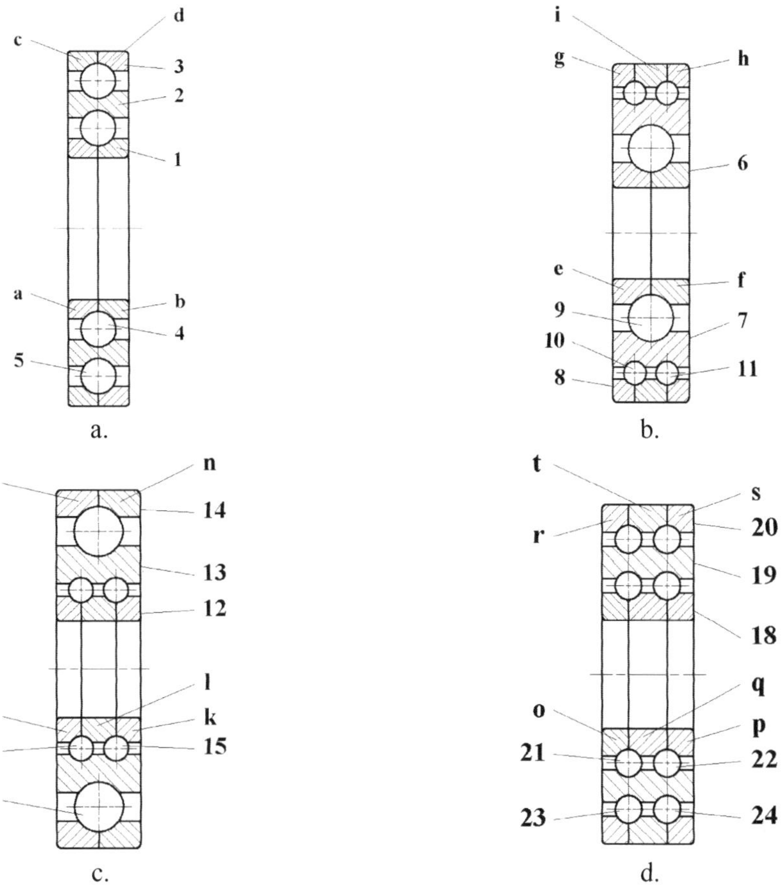

Figure 15. Radial-Axial bearings with three rings.

5. Conclusions

Several authors have proposed special bearings construction, designed to be used at high and very high speeds. Thus, the concepts of "stacked", "nested" and "planetary" bearings appeared. These concepts are principally identical, all identifying the presence of at least one intermediate ring and at least two layers of rolling bodies.

There are bearings available for which the value of the exterior diameter is equal to the interior diameter of other bearings. Two such bearings can be mounted one within the other, forming a stacked bearing obtained by assembling. The two bearings of a stacked bearing obtained by assembling can be of the same or of different types. The load capacity of this type of bearing is equal to the smallest load capacity of the component bearings, and the maximum limit speed is equal to the sum of the maximum limit speeds of the component bearings. This latter quality, the significant increase of the maximum limit speed of the bearings, is obtained by minimum constructive and technological modifications in relation to the existing practical solutions, and has represented another objective pursued by the authors.

In nested bearings, a specific machinery part can be identified, the intermediary race, which is not present in existing bearings, either radial or radial-axial. Nested bearings can be homogenous, if the two components are of the same type, other variants generating heterogeneous combination.

IOP Conf. Series: Materials Science and Engineering **399** (2018) 012043 doi:10.1088/1757-899X/399/1/012043

The diversity of bearing types creates the possibility to design, by different combinations, a high number of different nested bearings types. Their conception must also take into consideration the equalization of static and dynamic basic load rating of the two component bearings. This objective can be achieved by adequately establishing the radius value and the number of balls used in the two layers. An efficient solution is the utilisation of rolling bodies of different sizes in the two layers. Another solution highlighted by the authors is that of using two rows of rolling bodies in one of the layers of the bearing.

The main advantages of the nested bearing could be summarised as follows:

- strong increase in maximum limit speed;
- the eventual grip of rolling elements from one layer is not simultaneous with the other while the bearing still retains its functionality;
- there are no technological difficulties to manufacture such bearings since they are essentially the same as actual commercial products.

The disadvantages of these solutions are the following:

- larger radial dimensions;
- overall rigidity of the bearing is diminished;
- possible differences in static and dynamic loading capacity between the two layers of the bearing.

References

[1] Wang A, Yan S, Lin B, Zhang X and Zhou X 2017 Aqueous lubrication and surface microstructures of engineering polymer materials (PEEK and PI) when sliding against Si3N4. *Friction* , No. **5**, (4), p. 414-428, ISSN 2223-7690

[2] Zaretsky E V, Chiu Y P and Tallian T E 2013 Ceramic Bearings for Use in Gas Turbine Engines *J. of Materi Eng and Perform* **22** Issue 10, p. 2830-46

[3] *** www.engineersedge.com/bearing/ceramic_bearings.htm, Accessed: 2018-02-20

[4] Harris T A 1968 On the Effectivenss of Hollow Balls in High-Speed Thrust Bearings *ASLE Trans.* **11**, p. 290-294

[5] *** http://www.urb-s.ro/products/productAA1.html, Accessed: 2018-02-20

[6] Amann P 1924 *Dispositif de roulements à billes pour très grandes vitesses, à usure et étanchéité de joints normales*, Brevet d'invention FR no. 582661

[7] Maag A 1954 *Wälzlager,* Patent CH no. 300378

[8] *** www.koyo.com.br/upload/koyo/cat203ex%20-%20Ball%20&%20Roller%20 Bearings.pdf, p. 44-45, Accessed: 2018-02-20

[9] *** http://catalog.timken.com/guide/engineering-catalog/185.php, p 185, Accessed: 2018-02-21

[10] Cioară R, Mărăscu-Klein V, Pisarciuc C, Limbăşan I G and Cioară R M 2010 *Rulment radial-axial cu bile, cu trei inele (Radial-Axial Bearing with Three Rings)*, Romanian patent RO 123080

A study of a tensegrity structure for a footbridge

C Pleşcan[1*], M Conţiu[1] and A Dósa[1]

[1] Department of Civil Engineering, Transilvania University of Brasov, Romania

[*] E-mail: costel_plescan@yahoo.ro

Abstract. It is known that tensegrity systems are made of isolated bars subjected only to compression, connected by stretched cables. The use of these structures comes with advantages such as making large openings and low consumption of materials. This paper analyses the problems related to the building and calculation of a 36-meter-long pedestrian bridge. The structure of the bridge is made up of a double arch of compressed bars and tensioned cables, which supports the path by means of cables. The structure has a number of kinematics degrees of freedom and is statically indeterminate in the same time, which means that it can be pretensioned. Pretensioning introduces compression efforts in the bars and tension efforts in the cables, which fixes the kinematic degrees of freedom and confers rigidity to the system, thus resulting a structure which can support the external loads. The paper deals with aspects of structural analysis and static calculation of the structure subjected to loads.

1. Introduction

These structures were discovered in 1920 by Karl Ioganson and had different names until 1962 when Richard Buckminster Fuller called them "Tensile-integrity Structures" [1].

Due to the fact that these structures were in fact some mechanisms, they tended to attract engineers who gave them different uses, such as domes, towers, roofs for stadiums, temporary structures, exhibition pavilions and even aerospace structures [2].

Furthermore, in the field of bridges, many engineers and architects have conceived an impressive number of tensegrity bridges as follows: Kurilpa Bridge, Footbridge Proposal by Tor Vergata University, Footbridge Proposals by Wilkinson Eyre, Splash Bridge Proposal by Cullum, Bankside Bridge Proposal, Deployable Bridge Thesis Proposal, Bamboo Footbridge Proposal, Tube Bridge Proposal, Simplex Module Footbridge Proposal [3].

The first bridge of its kind is called Kurilpa Bridge and it was built in Brisbane, Australia on October 4, 2009. This bridge has a total length of 470m and a main opening of 120m; it weighs 540 tons and used about 6.8 km of cables [4].

While Kurilpa Bridge is integrally a tensegrity bridge, the other ones are artistic structural objects or theoretical proposals.

Even if the finite element method is successfully used in classical structures, the design of tensegrity structures is still a huge challenge for the designer from a structural and visual point of view [5]. Finding a form for a bridge structure is the most important step in the study; this is indicated also in literature [6]. Compared to the computerized modelling of tensegrity structures, the physical modelling of the model represents a second significant research step.

The structures analysed in this paper complied with the construction rules of tensegrity systems, therefore they are made from isolated bars subjected to compression only [7, 8], which are connected by stretched cables. By their mode of construction these systems are infinitesimal mechanisms with one or more degrees of kinematic freedom, they get geometrical deformability by tensioning the cables.

Since the rigid bars are not interconnected, the nodes are simplified. Rigidization of the structure by pretensioning leads to a small number of elements compared to the classical structures.

The structure is statically analysed, self-stress efforts are established and the stabilization effect of the pretension is emphasized.

2. The implementation methodology

2.1. Hypotheses

To facilitate the calculation, we apply a set of assumptions [9] like truss girder:

- the bars are straight and the bar axes are concurrent in nodes;
- external forces apply only to the nodes;
- the bars are perfectly articulated at the nodes.

Therefore the use of these assumptions, the systems' bars will have be subjected only to axial tension or compression efforts, and the resulting position of the deformed structure will be given only by the change in the bars' length.

2.2. Description of the pedestrian bridge structure

The main structural element is a double tensegrity arch, which has stretched bars on the outer side, and compressed bars, on the inside (figure 1). Figure 2 shows the transversal bars that keep the two arches in vertical position and, at the same time, give a lateral stiffness to the structure. Figure 3 shows the cross section of the pedestrian bridge at the middle of the opening, and a headroom inside a bridge, for pedestrians, with $h_g = 3.55$m and $b_g = 4$m.

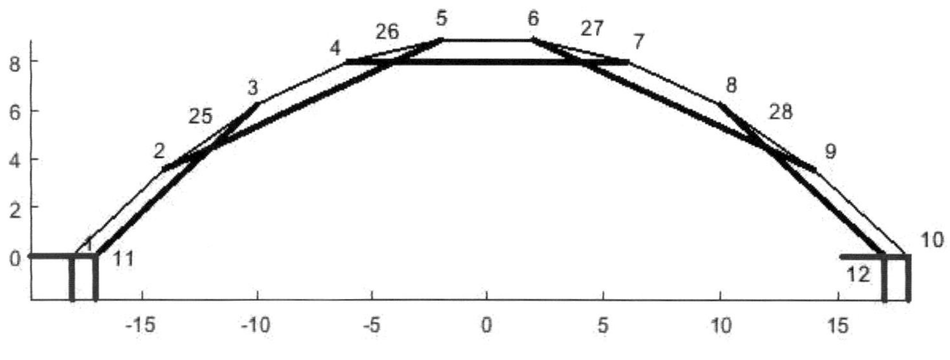

Figure 1. Side view of the pedestrian bridge.

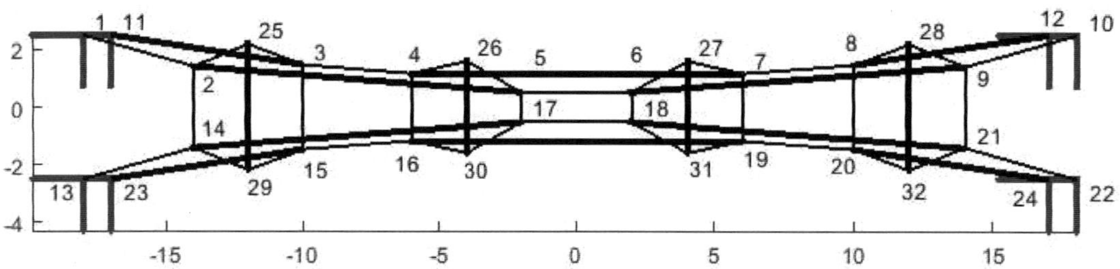

Figure 2. Plan view of the pedestrian bridge.

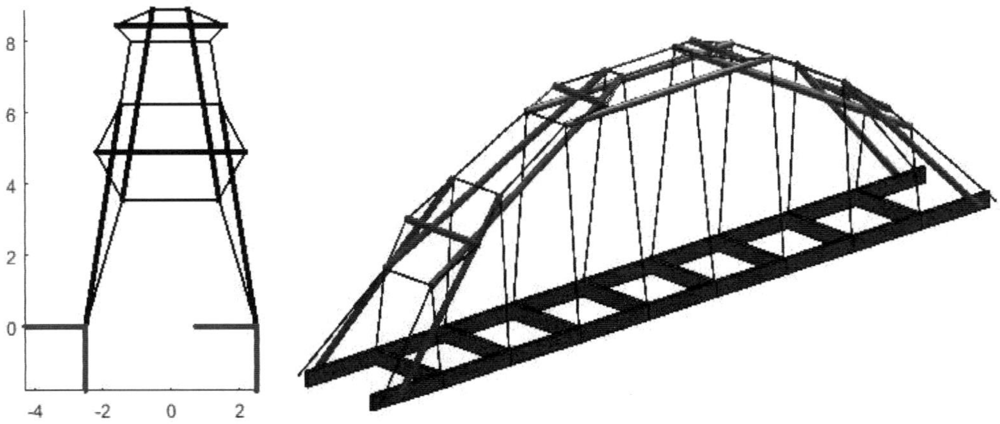

Figure 3. Cross section and 3D view of the pedestrian bridge.

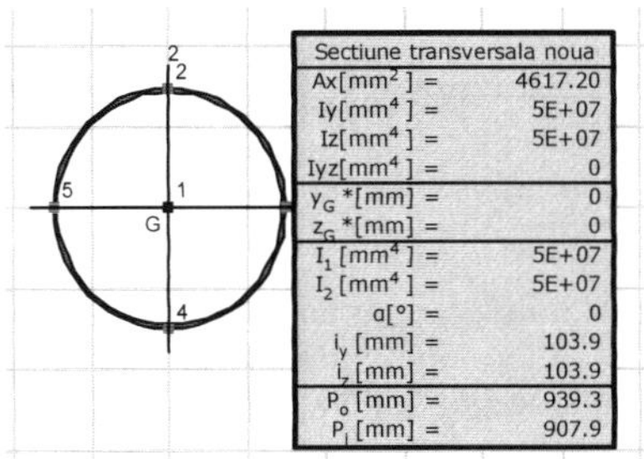

a. Section of the compressed bars;

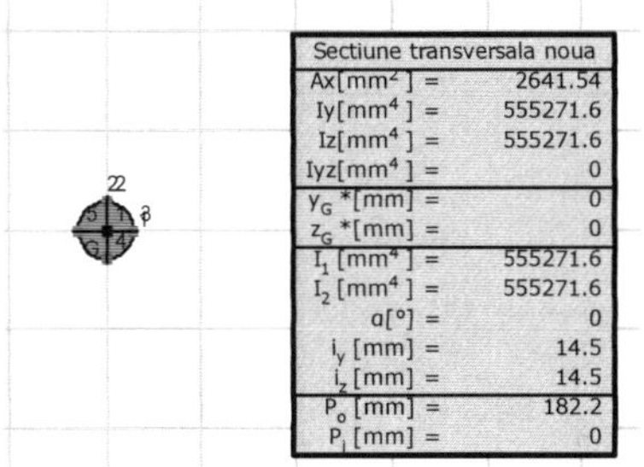

b. Section of the stretched bars;

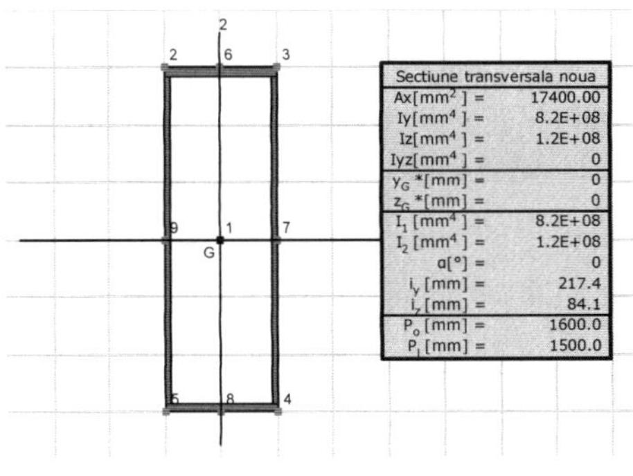

c. Section of the stiffening beam.

Figure 4. Characteristics of the structural elements.

The structure of the deck is suspended from the main structure by round steel bars. The nodes that secure the path suspension are located at 4 m horizontally. The solidarization of the two arches is accomplished by tensegrity polygons consisting of deviated arch cables, transverse cables and transversal bars. Figure 4 presents the sections used for making the spring bars and the stiffening beams for the deck.

The double arch is a truss system with the following features:

- no. bars b = 54;
- no. supports r = 24;
- no. nodes n = 32;
- no. of degrees of freedom m = 19;
- degrees of static indeterminacy s = 1.

The structure being statically undetermined, it can be pretensioned; cable pretensioning fixes the displacements of the m = 19 dofs, and ensures the rigidity of the structure.

2.3. Static analysis results

The structure was studied for two load cases:

1. dead load of 5.0 kN/m^2 and live load of 5.0 kN/m^2 on the whole span (P+Q);
2. dead load of 5.0 kN/m^2 and live load of 5.0 kN/m^2 on half of the span (P+Q on L/2).

The analysis of the structure shows that the arch alone is too flexible, especially in the case of the live load on half of the span, but the stiffening beam limits the deformations. The capable bending moment for the beam is Mc = 822 kN, the steel cable can support Ncap = 792 kN, the longest bar has a length of 13.13 m can take Ncap = 600 kN.

Table 1. Static calculation results.

	P+ Q		P+ Q on L/2	
	N	N/Ncap	N	N/Ncap
$N_{cable\ max}$	426.639	0.538	319.979	0.404
$N_{bars\ max}$	474.954	0.792	356.215	0.594
	M	M/Mcap	M	M/Mcap
M_{max}	44.578	0.054	227.848	0.277
W_{max} (m)	0.034		0.057	

Table 1 presents the results of the axial efforts in the bars and cables, the moment of the beam and the maximum vertical displacement of the stiffening beam, the effort recorded in the two load assumptions, and the ratio between the effective and the capable effort.

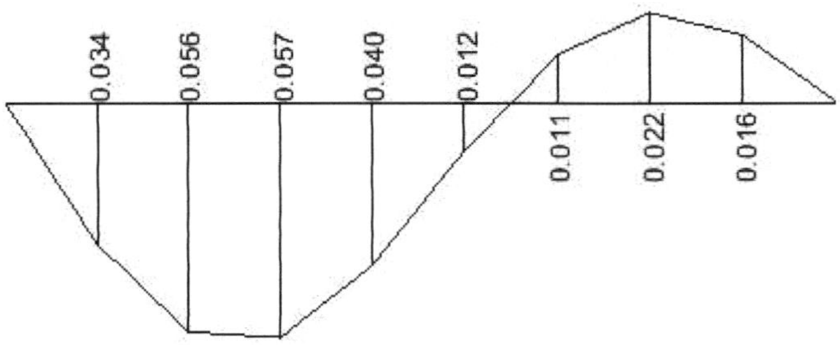

Figure 5. Vertical displacements in the rigidity stiffening beam of loads (p + u ½).

A short analysis of the report indicates that the structure has a significant reserve of load-bearing capacity. For the calculation of efforts and deformations, the pedestrian bridge was loaded with prestressing efforts, permanent loads from the weight of the deck and of the structure, and the loads of pedestrians which were applied in two load cases: load on the half span and load over the entire span.

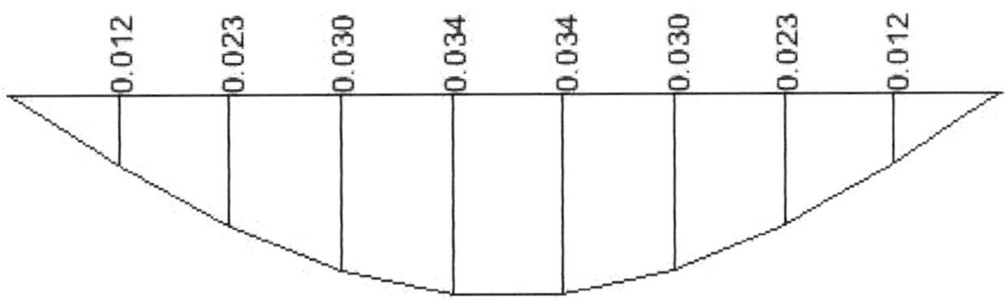

Figure 6. Vertical displacements in the stiffening beam of loads (p + u).

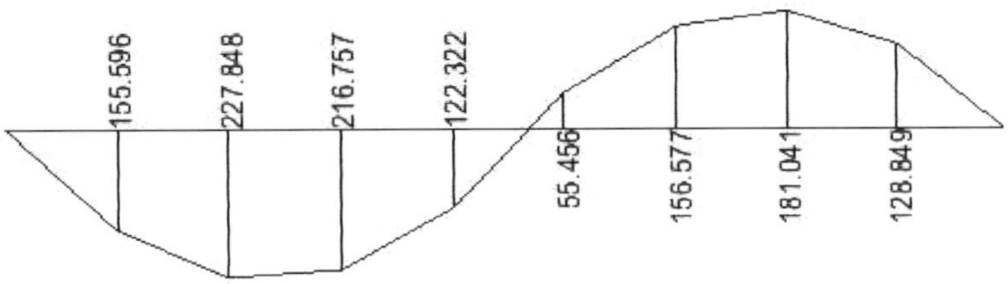

Figure 7. Bending moments in the stiffening beam from loads (p + u ½).

Figures 5, 6, 7 and 8 present the values of the vertical displacements and the bending moment diagrams in the stiffening beam corresponding to the load cases. It is obvious that when the beam is loaded at half the spam, we will have vertical and positive displacements (which can be neglected), but in the moment diagram, there is a negative moment with significant values.

The computations were solved by a SCILAB software of the authors. This software has program modules for the kinematic and geometric nonlinear analysis of truss systems. Also a special module was created to introduce the effect of the stiffening beams.

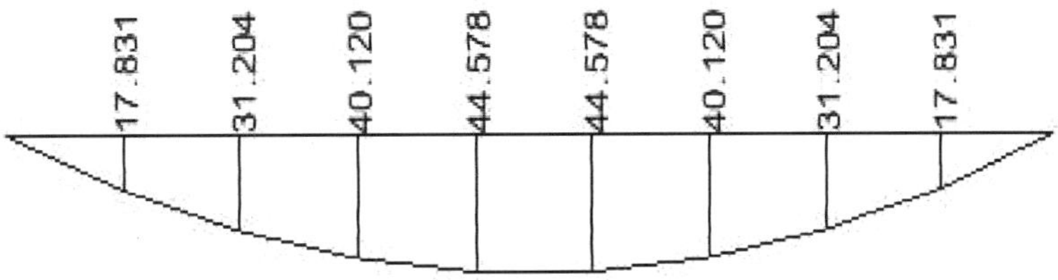

Figure 8. Bending moments in the stiffness beam of loads (p + u)

3. Conclusions

Tensegrity footbridges are lightweight and efficient structures with high strength to weight ratio compared with traditional ones. Using only a small number of compressed elements working in isolation one from another by a net of cable elements subjected to tension, less material and respectively less weight of material are used for supporting a given load. The stiffness of the structure is given by the effect of prestressing of the elements. The main disadvantages of tensegrity bridges consist in large deformations and vibrations. Tensegrity systems can withstand heavy loads, but their flexibility makes them to deflect considerably even under small loads. The stability, stiffness and load carrying capacity of the tensegrity structure studied in this paper were increased by using the supplemental stiffening effect of the longitudinal beams which support the deck.

As future research directions, we intend to study aspects like structure behaviour under static and dynamic loads in exploitation. We currently work to obtain a structure model to assess its behaviour by measurements.

References

[1] Fuller B 1962 Tensile-integrity structures U.S. Patent 3 063 521
[2] Sultan C 2009 Tensegrity: 60 Years of Art, Science, and Engineering *Adv. Appl. Mech.* Vol. 43
[3] Ilieva I M 2014 Potential applications of tensegrity structures to bridge construction *Int. Conf. on Traffic and Transport Engineering* (Belgrade)
[4] Micheletti A 2012 Modular tensegrity structures: the TorVergata footbridge *Mec., Mod. Meth. Civ. Eng.* LNACM 61 pp 375-84
[5] Burkhardt R W 2008 A Practical Guide to Tensegrity Design www.angelfire.com/ma4/bob_wb
[6] Yiyi Zhou 2007 General methods for creating tensegrity towers, arches, bridges and roofs, School of Civil, Environmental and Chemical Engineering RMIT University
[7] Skelton R E and de Oliveira M 2009 Tensegrity systems (Boston: Springer)
[8] Rhode-Barbarigos L, Bel Hadj Ali N, Motro R and Smith I F C 2010 Designing tensegrity modules for pedestrian bridges *Eng. Struct.* **32** (4) pp 1158-67
[9] PellegrinoSandCalladine CR 1986 Matrix Analysis of statically and kinematically indeterminate Frameworks, *Int. J. Solids Struct.* **22** (4) pp 409-28

Building information modelling design and analysis of glass structures

C Popa[1*]

[1] Technical University of Civil Engineering Bucharest, Romania

[*] E-mail: ciprian.popa@allbim.net

Abstract. The present article describes the importance of analysing, designing and maintenance of glass structures. Building Information Modelling is a work method that covers the entire lifecycle of a building by centralizing all the information inside a unique 3D model. Glass structures are a special case where precision and proper maintenance are vital for ensuring safe and economic exploitation. The article also illustrates a practical example of designing and maintaining a glass structure using different software solutions available on the market.

1. Introduction

Glass structures are special, and to most of the public they can generate fear and doubt regarding the safety they offer. There are many impressive glass structures in the world, having different sizes and functionalities. Most of them are designed in non-seismic areas, but in some areas such as Romania, an earthquake can be an additional source of fear when being inside a glass structure (figure 1), crossing a glass bridge (figure 2), or just swimming to an apartment situated across the street (figure 3). This article describes three different approaches when designing Glass Structures in Seismic Areas and also discusses a practical case study.

Figure 1. Apple Store, NY, USA [1].

Figure 2. Zhangjiajie Glass Bridge, China [2].

Figure 3. Swimming Pool, London, UK [3].

The case study presents the reactions and deflections a glass cube structure gets from horizontal forces, generated by wind loads in comparison to those that are generated by a typical Romanian Earthquake in Bucharest. The structure is the entrance to Apple Store, located in NY, USA, but the actions are evaluated as if this structure were located in Bucharest Centre. In figure 4 we can see a simulation in Google Earth of the Location of a Glass Cube Structure in Bucharest.

Figure 4. Glass Cube (Apple Store) analysed model, Bucharest, Romania.

2. Seismic design approaches for glass structures

In seismic areas, seismic actions have to be taken into account, as experience shows that the collapse of a glass element, whether it is a primary or secondary structural component, may result in a high number of deaths and injuries. Eurocode 1998-1 offers the combination rules for combining seismic loads with other permanent and live loads. ULS verification will include the effects of gravitational and climatic loads from combinations according to equation 6.10 and equation 6.12. [4].

When designing structural glass elements, three distinct situations can be identified:

- earthquake resisting elements made entirely of glass;
- structural elements made with glass and other materials which possess certain ductility;
- glass elements that are not part of the earthquake resisting structure, they are not loaded with horizontal force, but they are relevant for the inhabitants' safety (ex. Interior wall panels, edge curtain walls).

2.1. Earthquake resistant glass elements

The first category of elements will be designed according to 5.5.2.1 of EN 1990, to withstand the design seismic force without any damage (ex: cracking), for the EN Seismic Combination 6.4.3.4 of EN 1990). The seismic force will be evaluated according to EN 1998, assuming a behaviour factor of q=1. By using this elastic design approach, structural integrity is assured even after the seismic event.

According to Eurocode 8, the design seismic action for the no-collapse requirement is calibrated for a reference probability of more than 2% within the reference return period.

The structures for which earthquake resisting structure include glass elements and other materials will be designed so that the evaluated stresses will not cause cracking for the reference seismic action, in design combination with the other actions according to 6.4.3.4 of EN 1990.

The ULS check for the primary seismic-bearing elements must consider the contribution given by the glass structure to the seismic action resisting system considering a non-ductile behaviour. The glass structure may be part of the lateral and vertical force, but it cannot be part of the energy-dissipation systems.

The ULS checks will be based on linear analysis with energy behaviour factor q=1 (no energy dissipation and no ductility). A primary seismic member and its connections must be designed and detailed to support loads for the whole structure and from other elements (of superior order), in addition to its self-weight and out-of-plan load, when subjected to displacements caused by the most unfavourable seismic design scenario.

Also, glass structural members that do not pertain to an earthquake resisting structure, but are relevant for the inhabitants' safety, shall withstand without collapse the calculated stresses generated by the reference seismic action.

2.2 Structural elements made with glass and ductile materials

If the resisting structure for earthquakes includes a dissipative system (e.g. a hybrid seismic structure composed of the glass system and another material, like RC (reinforced concrete) or a steel frame), the design should follow the principles of Capacity Design (Hierarchy of Resistance). Failure of the glass system is only accepted for displacements greater than the ones produced by the seismic action correspondent to the no-collapse requirement. The hierarchy of resistance aims at ensuring an overall dissipative and ductile behaviour, as it is specific for structures made from different materials than glass. Additionally, the hierarchy of resistance aims at avoiding brittle failure and unstable mechanisms formed prematurely. To achieve this, the capacity design procedure will be adopted, which is used to obtain the hierarchy of resistance for the different structural elements and choosing failure modes that guarantee a favourable plastic mechanism and avoiding brittle failure modes.

The connection for the primary seismic loaded members shall be designed and checked with the seismic action associated with the no-collapse requirement. The check has to consider both relative displacements and internal stresses.

According to the French Technical Recommendation "Preposition de fische (CSTB et SNFA)" [5]: "Seismic action is divided in two types of solicitations: a dynamic solicitation due to ground movement and a static deformation induced by building floor drift. The amplitude of calculated action depends on building importance, type of ground and seismicity region. In France, the application of seismic regulations based on Eurocode 8 led to recommendations on façade conception and dimensioning. The validation criterion has been chosen as no elements fall, with performance conservation for important buildings (hospitals, firehouses…). The experimental tests carried out on glass façades (curtain walls and structural glazing kit) showed a large elastic deformability of the metallic frame under dynamic solicitation (succession of increasing accelerations until 16 m/s^2 at different frequencies between 1 and 15 Hz applied on a 3 m x 3 m mock-up) inducing few systems degradations. Degradations have been observed during floor drifting (static cyclic increasing displacement until 60 mm at the head of a 3 m x 3 m mock-up) with glass breakage. Recommendations concern calculation of anchoring to the structure, dimensioning of metallic frame, type of glass to use."

2.3 Glass elements that are not part of the earthquake resisting structure

When modelling an entire building, the strength and stiffness of the secondary elements is neglected, considering that in case of their failure, the ultimate resistance of the main structural system can guarantee the building's integrity and safety. Special considerations have to be made when dealing with lateral forces generated from seismic actions. In some situations, the in-plane stiffness provided by secondary elements can have a significant contribution to the inter-story lateral stiffness, even if, due to the lack of ductility, no allowance is actually given to including such stiffness in the calculation model [6].

Based on these premises, two requirements can be specified:

- under ULS (Ultimate Limit State), the strength and stiffness of the structural system, relying only on the primary elements, shall be verified and the secondary elements shall be checked for the lateral loads directly applied onto them; also, they will have imposed displacements which originate from the primary system deformations;
- under SLS (Serviceability Limit State), in case the contribution of secondary elements to the

overall system stiffness is relevant, the resultant stresses on the secondary elements will be added to the lateral loads directly applied onto them.

The inter-story drifts require being limited in accordance with the displacement compatibility in regard to glass elements. The secondary seismic member and its connections shall be designed and detailed to avoid cracking during the earthquake event associated with the no-collapse requirement. Moreover, these secondary elements and their connections must withstand their self-weight and also the out-of-plane loads when subjected to the displacements caused by the most unfavourable seismic design condition. Second Order effects ($P - \delta$) need to be taken into account when designing secondary seismic members.

Glass secondary structural elements in seismic areas should be executed after the hardening of the concrete structure or the assembly of the steel frame and after the displacement has settled. These elements are allowed to be in contact with the structures (ex: without separation joints) as long as they do not have a structural connection with them.

Beyond the safety margin of the compatibility check, appropriate measures should be taken into account for the secondary structural elements in order to avoid cracking, brittle failure or collapse of the lass during an earthquake event due to the drift of the structure. Partial or total out-of-plane collapse is not unlikely, thanks to the high strength-to-mass ratio of the glass elements.
This movement is allowed through the usage of a larger than typical clearance between the glass and the steel framing. A simple formula to calculate the required clearance for a flat rectangular panel is given in equation 1, which is taken from the American Standard ASCE 7-10 *Minimum design loads for buildings and other structures* [7].

$$\delta = 2c_1 (1 + hc_2 / bc_1) \quad (1)$$

where:
δ is the horizontal seismic drift over the height of the panel being considered,
c_1 is the average clearance on the vertical sides,
c_2 is the average clearance on the top and bottom,
h is the height of a rectangular panel,
b is the width of a rectangular panel.

This concept of movement and flexibility must be extended throughout the structure, especially in the point support areas.

3. Case study: seismic analysis of a glass cube structure in a seismic area

Figure 5. Real structure / Analytical model for Glass Cube Structure.

The Analysis Model in figure 5 is a perfect cube with the side of 9.75 m, enclosed with laminated glass panels with 2 intermediate laminated glass vertical fins per side. The roof is a laminated glass panel with

IOP Conf. Series: Materials Science and Engineering **399** (2018) 012045 doi:10.1088/1757-899X/399/1/012045

horizontal ribs also from laminated glass. The vertical and horizontal fins are connected via a hinged bolt connection. The structure has been described in more detail in a previous article in Bulletin of Transylvania University of Brasov [8].

The cube is the entrance to an Apple Store, and in this example only the superstructure has been modelled with a hinged support from the ground. The loads evaluated for this structure are specific for Bucharest, i.e. a snow load of 1.60 kN/m^2, a live load of 0.5 kN/m^2, a Temperature Variation of +25 °C, and wind load with a reference speed of 30 m/s. The Earthquake load has been modelled using the design spectrum method according to P100-1/2013 (figure 6).

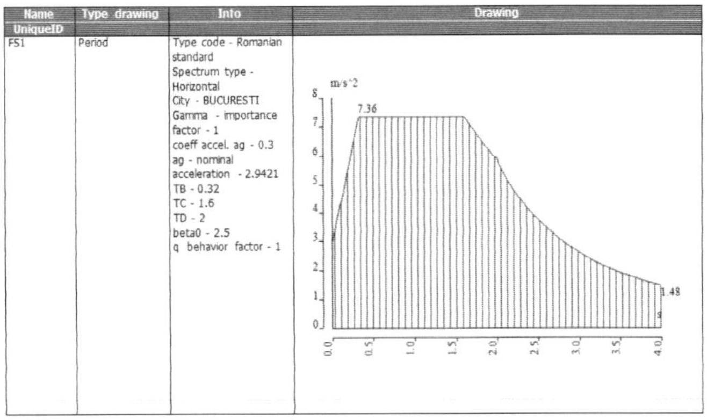

Figure 6. Design Seismic Spectrum according to P100-1/2013.

The results of the modal analysis can be seen in figure 7.

Calculation protocol

Solution of Free Vibration

Number of 2D elements	56373
Number of 1D elements	0
Number of mesh nodes	56596
Number of equations	339576
Combination of mass groups	MC 1 CM1
Number of frequencies	12
Method	Lanczos
Bending theory	Mindlin
Type of analysis model	Standard using improved reduced system (IRS)
Start of calculation	19.06.2018 13:41
End of calculation	19.06.2018 13:42

Sum of masses

[kg]	X	Y	Z
Combination of mass groups 1	28608.80	28608.80	28608.80

Modal participation factors

Mode	Omega [rad/s]	Period [s]	Freq. [Hz]	Wxi / Wxtot	Wyi / Wytot	Wzi / Wztot	Wxi_R / Wxtot_R	Wyi_R / Wytot_R	Wzi_R / Wztot_R
1	79.0897	0.0794	12.5875	0.0003	0.0000	0.5629	0.0000	0.0045	0.0000
2	119.5077	0.0526	19.0202	0.0027	0.4002	0.0000	0.2327	0.0003	0.0000
3	120.7057	0.0521	19.2109	0.7633	0.0010	0.0000	0.0011	0.0831	0.0000
4	132.0213	0.0476	21.0118	0.0001	0.5492	0.0000	0.0407	0.0000	0.0001
5	142.5300	0.0441	22.6844	0.1759	0.0000	0.0029	0.0000	0.1004	0.0000
6	153.9935	0.0408	24.5088	0.0050	0.0000	0.0043	0.0001	0.1219	0.0002
7	190.2137	0.0330	30.2735	0.0029	0.0000	0.0031	0.0002	0.1341	0.0004
8	250.0687	0.0251	39.7997	0.0000	0.0000	0.0000	0.0005	0.0009	0.6405
9	270.6709	0.0232	43.0786	0.0001	0.0071	0.0000	0.2121	0.0013	0.0002
10	305.2856	0.0206	48.5877	0.0000	0.0001	0.0000	0.0054	0.0006	0.1595
11	531.2685	0.0118	84.5540	0.0048	0.0000	0.0947	0.0002	0.0407	0.0000
12	593.9562	0.0106	94.5311	0.0000	0.0058	0.0000	0.0705	0.0000	0.0006
				0.9551	0.9634	0.6679	0.5635	0.4877	0.8016

Figure 7. Modal analysis results.

As can be seen, the structure is very stiff due to the strong ribs, having a dominant vibration period of just 0.05 seconds. The graphic display of the displacement can be seen in figure 8.

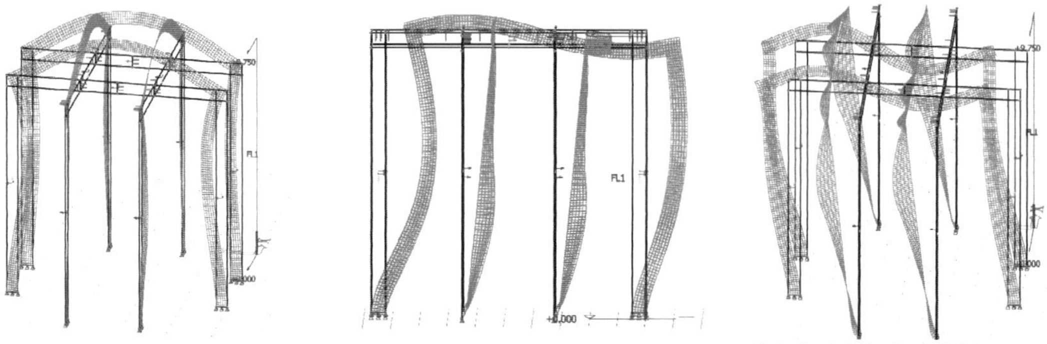

Mode 1: Vertical 56% mass Mode 3: Vertical 76% mass Mode 4: Vertical 55% mass

Figure 8. Rib vibration modes and mass participation factors.

The purpose of this analysis is to compare the maximum results in terms of stresses and deflections for a full complete glass structure in order to evaluate which scenario (Gravitational or Seismic) dominates the design. In figure 8 we can observe a comparison from the point of view of maximum 3D deflections. The iso-lines on the left side show a maximum value of 18.8 mm corresponding to the SLU Envelope of combinations generated according to equation 6.10 from EN 1990. On the right side, we can observe the maximum deflection of 4 mm from the SLU Envelope of combinations generated according to equation 6.12. In the first model, we can observe that the deflections are higher in the mid span due to the wind load which manifests a higher resultant force in the middle of the vertical panel, while in the second model (Seismic combinations), we see a smaller maximum deflection, situated on top of the structure. Thus, it may seem that the earthquake generates smaller deflections than wind.

Looking at the base shear resultant forces, the results are not consistent with the deflection ones. The maximum horizontal reactions from the envelope of combinations generated with equation 6.10 (Gravitational & Climatic loads) generate a maximum of 176.50 kN, while the maximum shear force generated with seismic combinations is 172.0 kN.

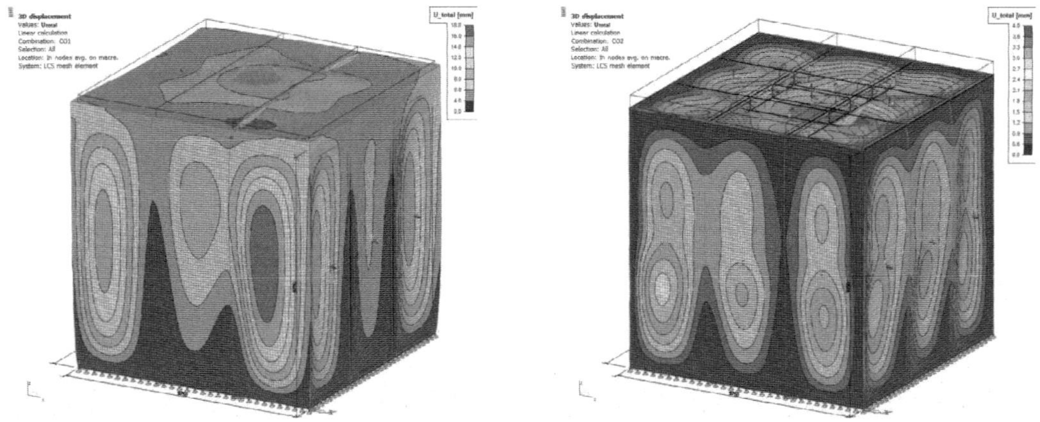

Figure 9. Maximum deflection from Ec. 6.10 combinations vs. maximum deflections from Ec. 6.12.

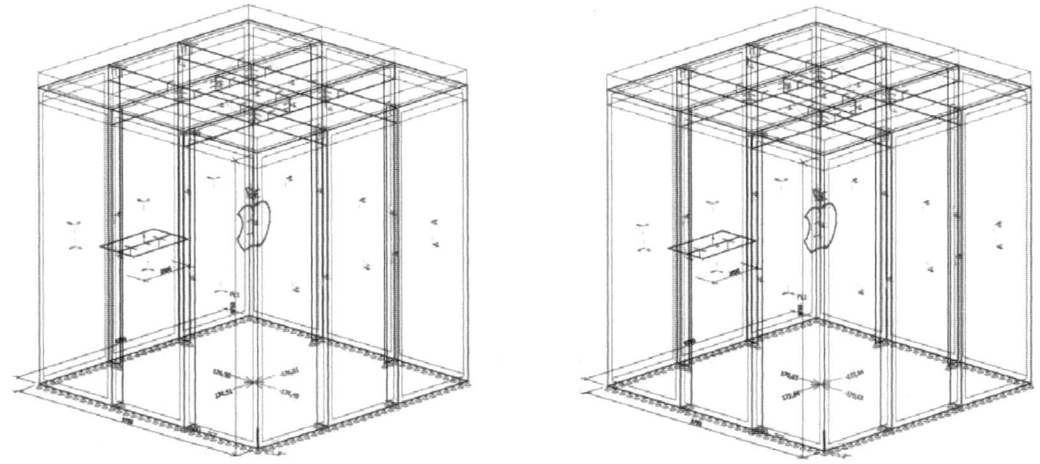

Figure 10. Maximum base shear force from Ec. 6.10 vs. maximum shear force from Ec. 6.12.

4. Conclusion

The case study presented above illustrates that the reactions from the base for a structure with a height/width ration of 1 and a rather tall height (~10 m) are almost identical to the horizontal forces resultant from Wind and Earthquake. Although the base shear forces are almost identical, the displacement distribution is very different. Maximum deflections from wind appear in the mid-section of the vertical panels, while maximum deflection from seismic forces appears on the upper part of the structure. Modal analysis also offers an insight into the way the structure behaves, and by studying the Eigen frequencies, the torsion sensitivity can also be taken into account. The modal analysis also helps determine the fundamental vibration period, which is essential in adopting or not a base isolation method.

In conclusion, it is important to consider seismic forces for any glass structure with primary elements made of glass, along with the other gravitational and climatic loads.

References

[1] www.apple.com
[2] http://www.meteoweb.eu/foto/cina-lo-spettacolare-ponte-vetro-sul-grand-canyon-gallery/id/707307/#13
[3] https://www.dezeen.com
[4] Feldmann, Kasper M 2014 Guidance for European Structural Design of Glass Components *JRC Scientific and Policy Reports* p. 68-70.
[5] Proposition de fiche (CSTB et SNFA): Disposition applicables aux facades légères en zones seismique
[6] O'Regan C 2015 Structural use of glass in buildings Volume 2, p.79
[7] ASCE/SEI7-10: Minimum design loads for buildings and other structures. Reston, VA: ASCE Press, 2010
[8] Popa C 2016 FEM analysis of glass structures using BIM software *Bulletin of the Transilvania University of Brasov* **9** (58) Proceeding of the International Scientific Conference CIBV 2016 p.109

IOP Conf. Series: Materials Science and Engineering **399** (2018) 012046 doi:10.1088/1757-899X/399/1/012046

Analysis on the setting method of tie member

G Qin[1*] and G Liu[1]

[1] School of Civil Engineering, Shenyang Jianzhu University, Shenyang, China, 110168

[*] E-mail: 1581011396@qq.com

Abstract. In order to provide a theoretical basis for the of construction scaffold design, the influence of the setting of tie member on the maximum bending moment of the scaffold under wind loads was studied. Based on the semi-rigid analysis of scaffold tube joints, the finite element ANSYS software is used to simulate and analyse the scaffold frame, and studying the change rule of maximum bending moment of scaffold upright tube when the setting of tie member is different. The maximum bending moment of the scaffold upright tube was obtained from simulation and analysis by using ANSYS. The comparison and analysis show that, when the tie member was deviated from the main node and only connected with the inner upright tube, the maximum bending moment of the upright tube was increased obviously, and the maximum bending moment of the inner upright tube increases significantly. Research has shown that the setting of the tie member should ensure the simultaneous connection with the inner and outer upright tube. According to the scaffold at different heights of building in the construction, the author proposed a value of the research for the reasonable position of the tie member and the maximum deviation from the main node.

1. Introduction

Scaffold is a working platform or protective frame, which is erected in engineering construction to meet the workers' operation needs. The temporary structure around the building in the construction, in order to facilitate the construction personnel up and down operation or external wall decoration work or stacking building materials is known as construction outside scaffold.

The setting of scaffold relates to the speed, quality and safety of engineering construction. Scaffold is an indispensable construction facility. With the increasing in the height of the building, the wind load role on the scaffold increases significantly. Due to structural forms or the size of the scaffold, the setting of the tie member will appear considerably different. In this paper, the key technical problems of such a as scaffold erection are analysed, when the scaffold tie member is deviated from the main node [1-3].

The research on scaffolding mainly focuses on two major aspects: a. node analysis model, b. load calculation method research. The analyses of representative works are mainly as follows:

In 2001, Professor Zongren Liu [4] the described calculation method in detail for the lower critical force limit of steel tubular scaffold with couplers. It is pointed out that in the actual construction, due to the insufficient torque of the coupler or its other factors affecting the force of the node of steel pipe joint, the "semi-rigid" node, can be simplified to the hinge contact.

In 2004, Professor Hongfei Ao and Professor Guoqiang Li [5-6] stated the fact that the ultimate bearing capacity of double row scaffolding is analysed by second-order finite element simulation.

Further considering the semi-rigid characteristics of the geometrically nonlinear beam column element and the coupler node, the rotational stiffness of the coupler is measured through experiments. Finally, the validity of the finite element analysis method is tested by two scaffold prototype tests.

In 2015, Guijuan Qin and Jiazhi Ren [7] analysed the wind load effect of scaffold tie member in Super high-rise building construction. It is concluded that the along-wind vibration factor should be considered in the calculation of wind load characteristic value for scaffolding construction in high-rise and Super high-rise buildings. The practical calculation method of the along-wind vibration factor in the calculation of the shape factor of wind load is given, and the specific calculation method of the along-wind vibration factor and the characteristic calculation formula for the horizontal wind load of a single tie member under different arrangement methods are given.

2. The setting of the tie member and the calculation of the standard value of the wind load

The longitudinal and transverse horizontal tubes are tightly fastened and tightly connected with the upright tube, which is called the main node. After the tie member is deviated from the main node, the right angle coupler is connected to the upright tube and the tie member. Their buckle contact point is not the main node (Deviating from the main node). The tie member is connected with the structure in construction. It is the most important force-carrying tube to transfer wind loads to the structure of the building. The reliable connection between the upright tube and the tie member, between the upright tube and the transverse tube, that can control the longitudinal and transverse deformation of the whole frame body. As the intermediate restraint of the frame body, the tie member can effectively reduce the calculation length of the upright tube and improve the bearing capacity of the frame body. The reasonable setting of the tie member can effectively increase the rigidity of the frame body, prevent the frame from overturning and ensure the lateral stability of the whole frame body [8-9].

2.1. Working condition analysis diagram

The different structure in the actual project, for example the door, window openings and other specific reasons, sometimes the tie member, cannot be connected to the upright tube and horizontal tube of the main node, but it can only deviate from the main node connection.

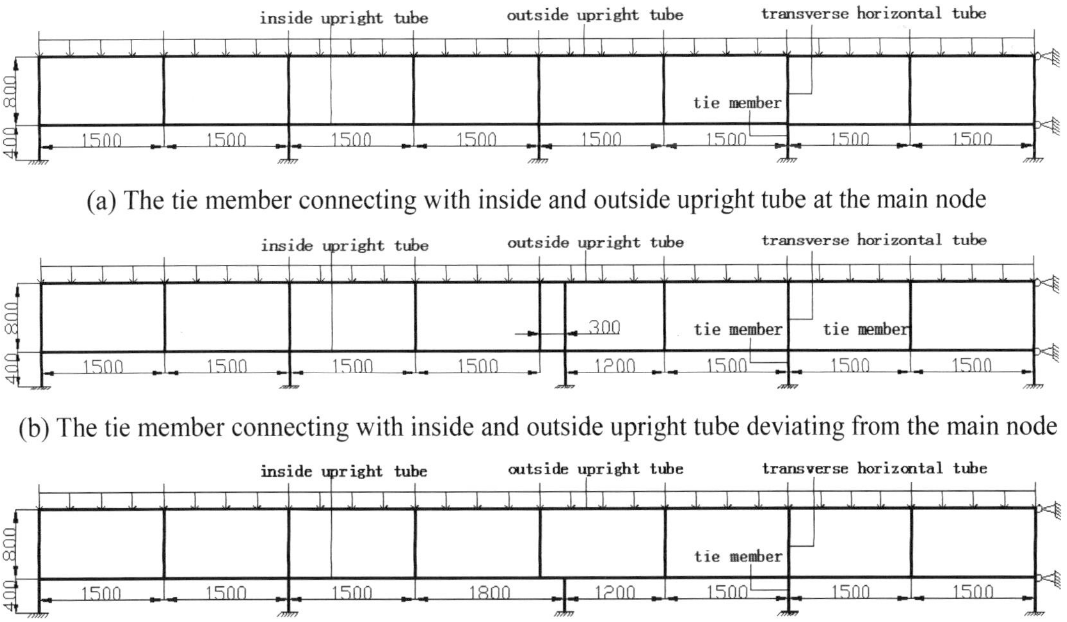

(a) The tie member connecting with inside and outside upright tube at the main node

(b) The tie member connecting with inside and outside upright tube deviating from the main node

(c) The tie member only connecting with the inside upright tube deviating from the main node

Figure 1. Working condition analysis diagram.

When the tie member is deviated from the main node connection, the tie member is divided into the tie member and the inside and outside upright tube simultaneously, and the tie member is only connected with the inside upright tube. In order to study the influence of a different setting on the frame design of the tie members of the main node under the wind load, this paper studies the influence of the bending moment of the tie member in the case of deviating from the 300mm of the main node. In this paper, the working conditions are analysed in figure 1.

2.2. Calculation method of standard value of wind load

The wind pressure is perpendicular to the air flow building plane on the pressure, and its expression is:

$$\omega(x,t) = \varpi(x) + \omega(t,x) \tag{1}$$

in the formula above:

$\varpi(x)$ - average wind speed static pressure,

$\omega(t,x)$ - wind fluctuating wind pressure.

Along-wind vibration is the dynamic effect caused the pulsating part of wind to the high structure.

In addition to the static effect, the high-rise building should consider the dynamic action. The dynamic action is related to the structural vibration, structural damping and structure height, and the pulsating wind pressure can be assumed to be a stochastic process of each state and deduced from the basic principle of stochastic vibration theory [10-12].

The along-wind vibration factor changes with the height of the frame body. The along-wind vibration factor is not only related to the building site, but also to the natural vibration characteristics of the structure. As for the construction scaffold of high-rise and super high-rise building, because the scaffold is connected with the structure through the tie member and based on the structure, we can only consider the effect of the first vibration mode on the structure when the structure is an equal-section structure with no change in shape and mass along the height [13, 14].

The value of the wind pressure on the building is related to the building's body size, aspect ratio and ground roughness. For the scaffold structure, the characteristic value of wind load can be simplified by static equivalent wind load. With the increasing of the structure height, the effect of the wind load is increasing, and the influence of along-wind vibration cannot be neglected. Therefore, in addition to considering the reference wind pressure value, shape factor of the wind load and exposure factor of wind pressure, the influence of along-wind vibration factor should be considered when calculating the characteristic value of horizontal wind load on scaffold. So the formula is as follows:

$$w_k = \beta_z \cdot \mu_z \cdot \mu_s \cdot w_0 \tag{2}$$

In the formula above : ω_k - characteristic value of wind load (kN/m^2);

μ_z - exposure factor of wind pressure;

μ_s - scaffold shape factor of wind load;

ω_0 - reference wind pressure value (kN/m^2).

The reference wind pressure value should select a 10-year recurrence period.

The rough category of the C ground:

$$\mu_z = 0.544 \times \left(\frac{z}{10}\right)^{0.44} \tag{3}$$

For the along-wind vibration factor β_z, the formula is :

$$\beta_z = 1 + 2g \cdot I_{10} \cdot B_z \cdot \sqrt{1+R^2} \tag{4}$$

B_z, R quoted from reference [7].

The rough category of the C ground: g=2.5， I_{10}=0.23.

The coefficients in formula (4) are strictly in accordance with the "Load code for the design of building structures" (GB50009-2012) [15]. For formula (2), the reference wind pressure value should select a 10-year recurrence period, ω_0=0.4kN/m^2 in the Shenyang area of China. According to the size of the analysis model frame (lift height and longitudinal spacing is 1.5m), the rough category of the C ground is taken for calculation. Taking Shenyang city of China building construction as example, ω_k value have been given in table1.

Table 1. ω_k value at different height of scaffold at the location.

	ω_k values at 50m-500m height above the ground									
Height (m)	50	100	150	200	250	300	350	400	450	500
ω_k Value (kN/m^2)	1.53	2.01	2.33	2.56	2.75	2.91	3.04	3.16	3.27	3.30

2.3. Calculation method of node semi-rigidity

The right angle coupler node is neither an ideal "rigid connection" nor an ideal "articulation", but a semi rigid link between the two.

This paper further proves that the properties and stiffness of semirigid and the experimental scheme diagram and the experimental device are shown in figure 2 and figure 3. When the material and component specifications of the scaffold selected are determined, rotational stiffness value of the coupler node is directly related to the size of the coupler tightening moment, the greater the tightening torque, the greater the stiffness of the node rotation, the rotation stiffness of the coupler nodes corresponding to different tightening torques is different.

The determination of the rotational stiffness can be measured by the test method, when the tightening moment requirements are met in construction, rotation stiffness can choose according to the specified indicators.

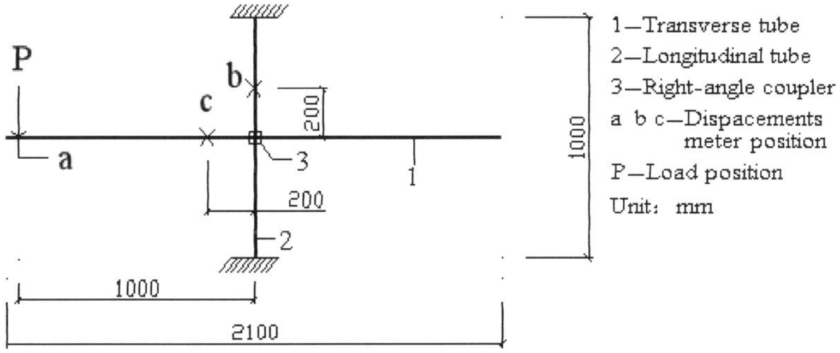

Figure 2. Simplified diagram of node rotation stiffness measurement scheme.

Figure 3. Test equipment.

IOP Conf. Series: Materials Science and Engineering **399** (2018) 012046 doi:10.1088/1757-899X/399/1/012046

3. ANSYS finite element analysis

(Take two-step of height and two-span of longitudinal as an example)

3.1 .Finite element analysis of at the main node

3.1.1. The establishment of finite element model. In this chapter, taking the analysis of tie member setting in two-step of height and two-span of longitudinal layout for the middle position as shown in figure 5, the tie member is set at the main node. Steel tube Specification Selection 48mm × 3mm, lift height and longitudinal spacing is 1.5m, distance between inside upright tube and building structure is 400mm, outside upright tube and building structure is 1200mm. Elastic modulus E=2.06 × 10^5MPa Poisson's ratio μ=0.3.The connection between the scaffold upright tube and steel I-Beam as articulation. The connection between tie member and the building structure is as rigid connection. In order to make the force of the fastener place better concord with the actual situation, the spring element Combin14 is introduced into the simulation process to simulate the coupler connection. The rotational stiffness between the tube and the tube through the coupler connection is 35kN·m/rad. The concrete setting of the steel tubular scaffold as shown in figure 4. The concrete arrangement of the tie member elevation as shown in figure 5. The symbol "× " represents the placement of the tie member. Number represents position number. ANSYS finite element model as shown in figure 6. Analysis of the position of the middle part of the whole frame body (figure 5 dashed section).

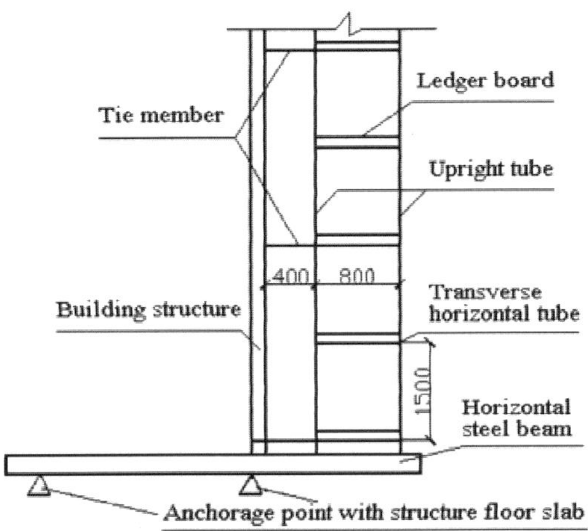

Figure 4. Cantilever steel I-beam tubular scaffold.

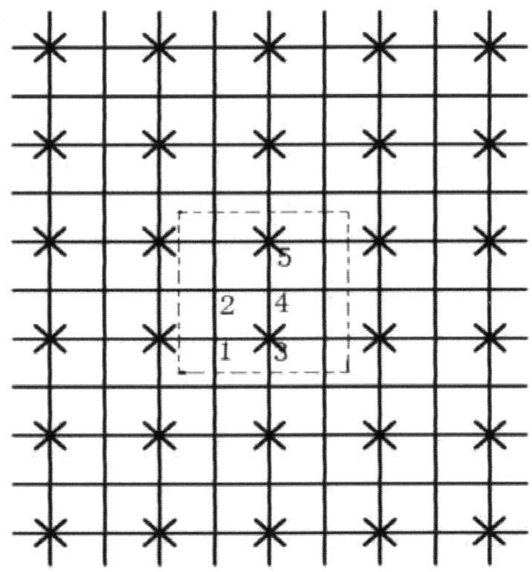

Figure 5. Elevation of tie member arrangements.

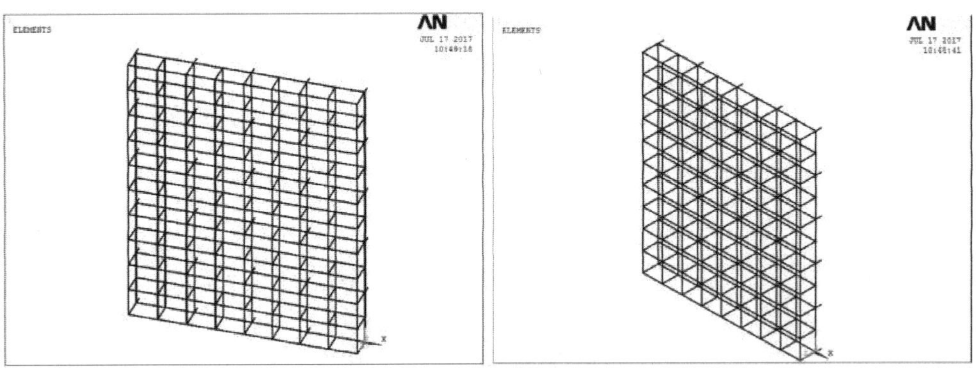

Figure 6. Finite element model.

IOP Conf. Series: Materials Science and Engineering **399** (2018) 012046 doi:10.1088/1757-899X/399/1/012046

3.1.2. Analysis of finite element result. By applying the equivalent characteristic of the value wind load of the scaffold at different heights of building, it is converted to the equivalent uniformly distributed line load and applied to the outside tube (facing wind side tube) of the frame body.

By extracting the middle part of the frame from the finite element data, the characteristic value of the moment of the scaffold at different heights is obtained. Results are shown in the table 2 and in the figures 10, 11, 12, 13, 14, 15.

Table 2. Bending moment characteristic value of node 1 to node 3 in figure 5 and figure 6.

Height above the ground (m)	Bending moment of upright tube（$\times10^5$N·mm）					
	position 1		position 2		position 3	
	Inside	Outside	Inside	Outside	Inside	Outside
50	3.36	1.80	-0.26	-1.81	4.3	2.92
100	4.76	2.70	-0.67	-2.73	8	6.2
150	5.67	3.29	-0.93	-3.31	10.5	8.6
200	6.34	3.72	-1.12	-3.74	13.4	10.55
250	6.88	4.07	-1.27	-4.10	13.4	13.4
300	7.33	4.36	-1.40	-4.38	13.4	13.4
350	7.71	4.60	-1.52	-4.63	13.4	13.4
400	8.05	4.82	-1.62	-4.85	13.4	13.4
450	8.36	5.02	-1.70	-5.04	13.4	13.4
500	8.45	5.08	-1.74	-5.11	13.4	13.4

Note: The corresponding bending moment value is positive when the stress at the upright tube of the frame body is tensile stress on the windward side.

3.2. Finite element analysis of tie member deviating 300mm from main node

3.2.1. The establishment of the finite element model. Finite element analysis of deviating from the main node connecting with inside and outside upright tube. In this chapter, taking analysis of tie member setting in two-step of height and two-span of longitudinal layout for the middle position as shown in figure 7(b), the tie member deviating from the main node.

Steel tube Specification Selection 48mm × 3mm, lift height and longitudinal spacing separately is 1.5m, distance between inside upright tube and building structure is 400mm, outside upright tube and building structure is 1200 mm. Elastic modulus E=2.06×10^5MPa Poisson's ratio μ=0.3. The connection between the scaffold upright tube and steel I-Beam as articulation. The finite element simulation analysis of the tie member deviating from the main node is basically the same as the simulation idea of the tie member at the main node. The only difference is that the middle position of the scaffold is set on the non-main node（the tie member deviates 300mm from the main node, for simulating the actual deviation of the tie member from the main node setting in the construction project. The node is got for arrangement of the tie member elevation shown as in figure 7. The tie member in figure 7 (a), which was originally set at main position 3, deviates from the height of the upright tube to position 6 as shown in figure 7(b). In figure 7, the symbol "×" represents the placement of the tie member. Analysis of the position of the middle part of the whole frame body, finite element model is as shown in figure 8.

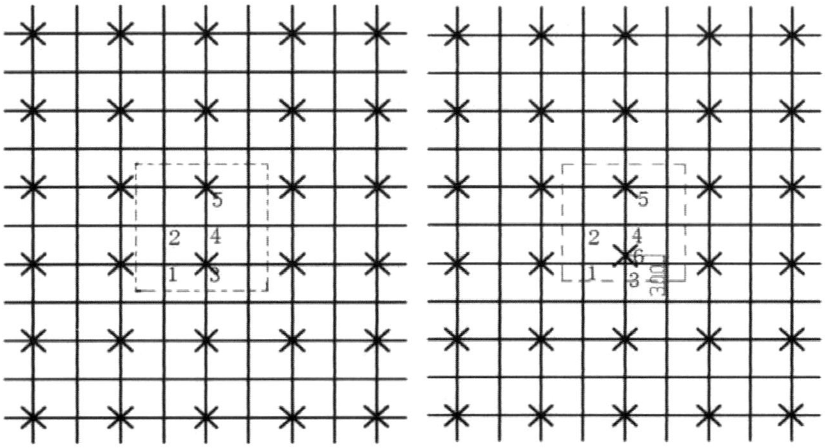

(a) At the main node (b) Deviating from the main node

Figure 7. Elevation of tie member arrangements.

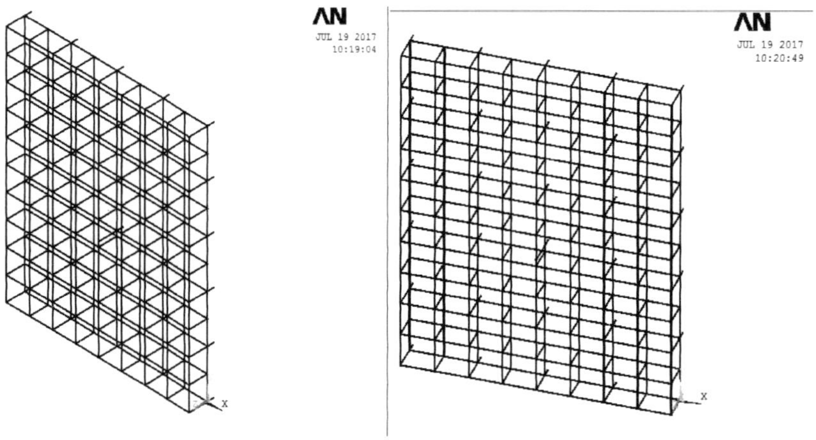

Figure 8. Finite element model.

3.2.2. Analysis of finite element result. By applying the equivalent characteristic of the value of wind load of the scaffold at different heights of building, it is converted to the equivalent uniformly distributed line load and applied to the outside tube (facing wind side tube) of the frame body.

By extracting the middle part of the frame from the finite element data, the characteristic value of the moment of the scaffold at different heights is obtained. Results are shown in the table 3 and in the figures 10, 11, 12, 13, 14, 15.

3.3. Finite element analysis of the tie member deviating from the main node only connecting with the inside upright tube

3.3.1. The establishment of finite element model. The finite element simulation analysis of the tie member deviating from the main node only with inside upright tube is basically the same as the simulation idea of the tie member deviating from the main node, the only difference is the removal of the small bar between the inside and outside of the transverse tube. This small tube cannot be set up for some reason in the actual construction project. The finite element model is shown in figure 9.

Table 3. Bending moment characteristic value of node 1 to node 3 in figure 7(b) and figure 8.

Height above the ground (m)	Bending moment of upright tube（×10^5 N·mm）					
	position 1		position 2		position 3	
	Inside	Outside	Inside	Outside	Inside	Outside
50	1.95	0.4	0.72	-0.82	1.61	0.39
100	2.9	0.85	0.96	-1.08	2.13	0.51
150	3.51	1.14	1.11	-1.25	2.46	0.6
200	3.97	1.36	1.22	-1.38	2.71	0.66
250	4.33	1.53	1.32	-1.48	2.91	0.7
300	4.64	1.68	1.39	-1.57	3.08	0.74
350	4.89	1.8	1.45	-1.67	3.22	0.78
400	5.13	1.91	1.51	-1.7	3.34	0.81
450	5.33	2.01	1.56	-1.76	3.46	0.83
500	5.4	2.04	1.58	-1.78	3.49	0.84

Note: The corresponding bending moment value is positive when the stress at the upright tube of the frame body is tensile stress on the windward side.

3.3.2. Analysis of finite element result. Based on the relevant data of the finite element model, the bending moment value of the joint of the tie member with the position of the inside upright tube. By extracting the middle part of the frame data, the characteristic value of the moment of the scaffold at different heights is obtained. Results are shown in the table 4 and in the figures 10, 11, 12, 13, 14, 15.

Table 4. Bending moment characteristic value of node 1 to node 3 in figure 7(b) and figure 9.

Height above the ground (m)	Bending moment of upright tube（×10^5 N·mm）					
	position 1		position 2		position 3	
	Inside	Outside	Inside	Outside	Inside	Outside
50	2.87	0.51	1.4	-0.68	2.68	1.21
100	4.28	0.91	1.82	-0.87	3.54	1.68
150	5.17	1.25	2.01	-1.02	4.21	2.23
200	5.84	1.56	2.31	-1.11	4.69	2.85
250	6.37	1.62	2.55	-1.23	4.86	3.36
300	6.83	1.66	2.76	-1.37	4.97	3.92
350	7.35	1.89	2.94	-1.45	5.12	4.44
400	7.55	2.01	3.21	-1.51	5.31	5.02
450	8.75	2.11	3.33	-1.59	5.56	5.51
500	9.96	2.23	3.51	-1.63	5.67	5.89

Note: The corresponding bending moment value is positive when the stress at the upright tube of the frame body is tensile stress on the windward side.

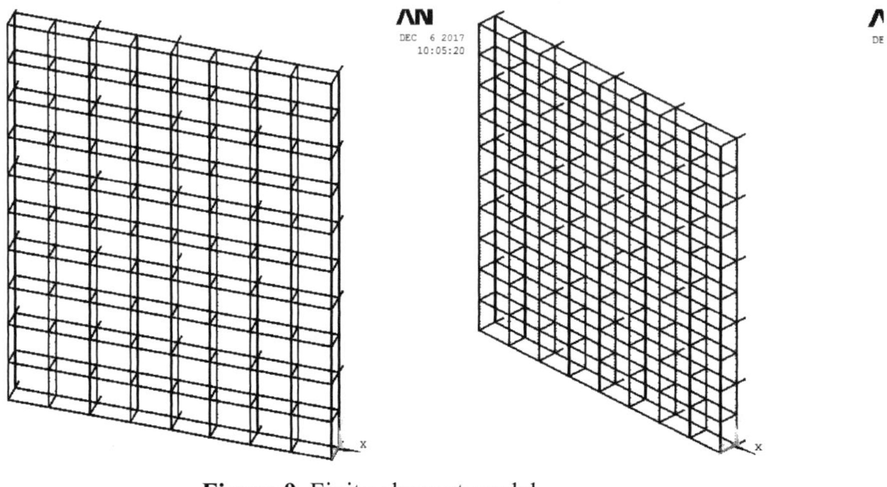

Figure 9. Finite element model.

Application of Originpro mapping and data analysis software, based on the above two groups of finite element analysis.

Taking the positions 1, 2, 3, 4, 5, 6 respectively to make a comparison, the bending moment characteristic changes the curve between the inside and outside upright tube, in these cases of the tie member at the main node, the tie member deviating from the main node, the tie member deviating from the main node only connected with the insider upright tube. The relation curve is as following :

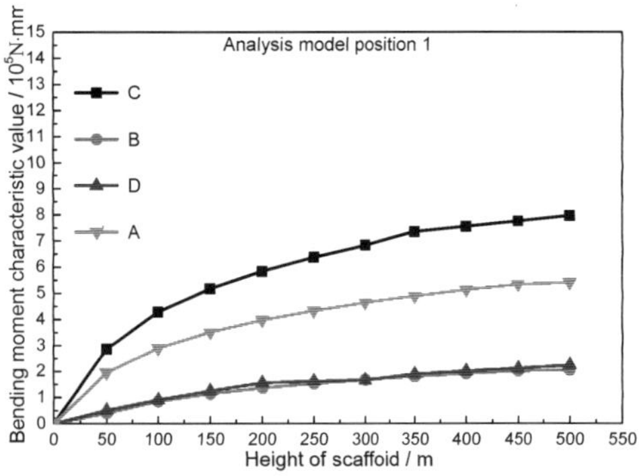

Figure 10. Curve of M-H at position 1.

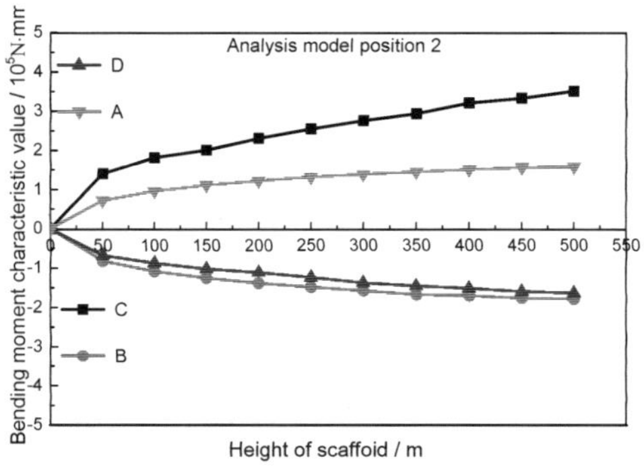

Figure 11. Curve of M-H at position 2.

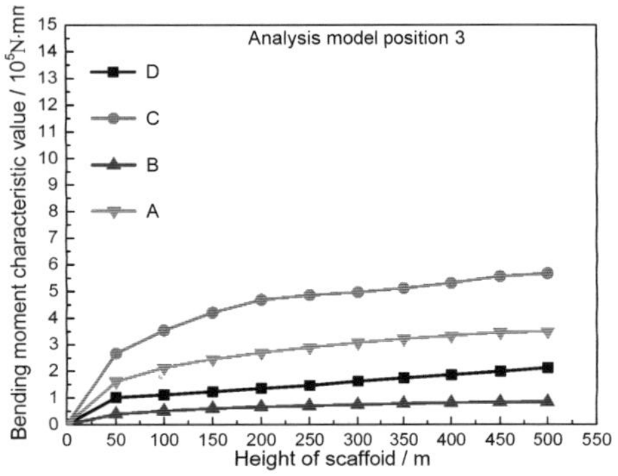

Figure 12. Curve of M-H at position 3.

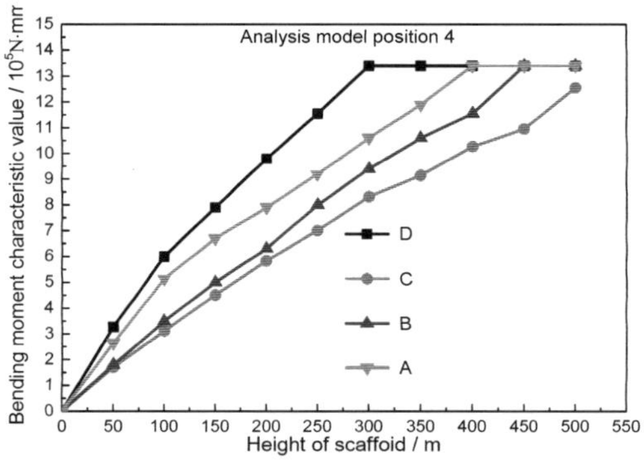

Figure 13. Curve of M-H at position 4.

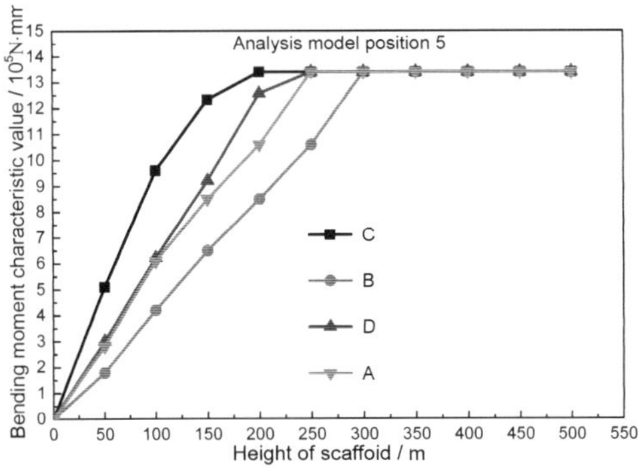

Figure 14. Curve of M-H at position 5.

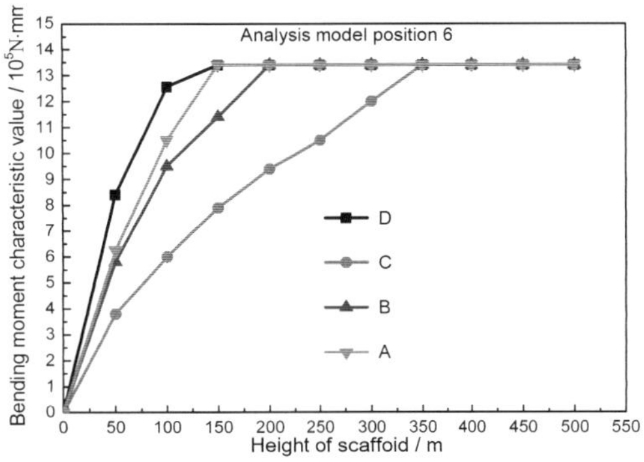

Figure 15. Curve of M-H at position 6.

Note: A-The bending moment value of outside upright tube, under condition of the tie member deviating from the main node (the tie member connecting with inside and outside upright tube).

B-The bending moment value of inside upright tube, under condition of the tie member deviating from the main node (the tie member connecting with inside and outside upright tube) .

C-The bending moment value of outside upright tube, under condition of the tie member deviating from the main node only connecting with the insider upright tube.

D-The bending moment value of inside upright tube, under condition of the tie member deviating from the main node only connecting with the insider upright tube.

4. Result analysis

For a single upright tube, when the frame height, step distance and span are the same, the curves of the finite element model are analysed as follows:

(1) In figure 10: the maximum bending moment of the upright tube at position 1 without tie member appears in the outside upright tube, The C curve is the bending moment value of the outside upright tube at the position 1 when the tie member deviates from the main node and is connected to the inside upright tube only. The C curve above the other A, B, D curve. In figure 10, the value of maximum bending moment C is 2.6×10^5N·mm more than A.

(2) In figure 11: the bending moment direction of the inside and outside upright tube is different. The C curve is the bending moment value of the outside upright tube at the position 2 when the tie member deviates from the main node and is connected to the inside upright tube only, it is an absolute value, much more than in other A B D situations. The absolute value of the maximum bending moment increases and it is bad for the stability of the upright tube.

(3) In figure 12: at position 3, the longitudinal horizontal tube intersects with the upright tube, The C curve is above others in spite of the tie member being set differently. The C curve is the bending moment value of the outside upright tube at the position 3, when the tie member deviates from the main node and the maximum moment value of the C curve is double than that of A curve.

(4) In figure 13: for the position 4 between the two tie members along the step direction, the maximum bending moment of the inside upright tuber increases rapidly. The maximum bending moment appears in the inside upright tube. The D curve is the bending moment value of inside upright tube under condition of the tie member deviating from the main node only connecting with the insider upright tube. Meanwhile, the maximum moment value of D curve is much larger than that in figures 10, 11, 12.

(5) In figure 14: for the position 5, the maximum moment of the frame body appears in the inside upright tube as C curve, and around 200m to reach the peak of the curve.

(6) In figure 15: for the position 6, after deviating from the main node, it can be seen from the figure that the maximum bending moment value of the outside upright tube is decreased as C curve, but the bending moment of the inner upright tube is increased obviously and around 120m to reach the peak

of the curve. The D curve is the bending moment value of the inside upright tube at the position 6, when the tie member deviates from the main node only connecting with the insider upright tube. Meanwhile, the maximum moment value of the D curve is much larger than that in figures 10, 11, 12, 13, 14.

Analysis of the figure 15: the tie member design arrangement deviating from the main node is 300mm, this design method being a feasible method. But over that size of 300 mm and the scaffold working height exceeding 100 m, that scaffold frame working is a hazardous situation. The proposed value of the maximum deviation from the main node is 200mm when the scaffold working exceeds 100m in height.

5. Conclusion

Based on the semi-rigid analysis of scaffold tube joints, the finite element ANSYS software is used to simulate and analyse the scaffold frame. Get the following important conclusions :

(1) The maximum moment value of the upright tube increases because of the tie member deviating from the main node, as compared with the tie member at the main node. Under construction, the tie member piece setting should be located on the main node, or around the main node, as near as possible.

(2) When the tie member deviates from the main node, the connecting with inside and outside upright tube at same time can reduce the maximum bending moment of the upright tube, which is beneficial to the stability of the upright tube. Because of the hole, etc., when the tie member could not be set at main node, the tie member should be connected with the inside and outside upright tube to reduce the maximum bending moment value of the upright tube.

(3) It is reasonable to deviate 300 mm from the main node when the scaffold using the height above the ground is less than 100m. When the using height exceeds 100m, the deviation size should be strictly limited to less than 200mm or the tie member quantity should be increased.

References

[1] Qin G and Zhang S 2016 Calculation and analysis of bending moment of scaffold upright tube in wind-load *Alternative translations* **S1**

[2] Qin G, Zhang S and Zhang X 2016 Analysis of maximum bendingmoment of upright tube wind load of construction scaffold of super high-risebuilding *Journal of Shenyang Jianzhu university (natural science)* **04**

[3] The Ministry of housing and urban-Rural Development of the People' s Republic China 2011 *Safety technical specification of construction fastener type steel pipe scaffold JGJ130-2011*(Beijing: China Architecture & Building Press)

[4] Liu Z and Ding Y 2001 Calculation method of lower limit of critical force for steel tubular scaffold with couplers *Architectural technology* **3 (8)** 541

[5] Ao H and Li G 2003 Analysis of ultimate bearing capacity of double polesteel tubular scaffold with couplers *Construction* **25 (3)** 214

[6] Ao H and Li G 2004 Research on ultimate stable bearing capacity of double pole steel tubular scaffold with couplers *Mechanics quarterly* **25 (2)** 213

[7] Qin G and Ren J 2013 Analysis of wind load acting on the tie member of sc-affold in super high-rise building construction *Journal of Shenyang Jianzhu University* **31(2)** 306

[8] Zhuo X and Zheng N 2005 Discussion on the Part of the Steel tubular scaffold with couplers specification *Construction technology* **34(3)** 78

[9] Chen X 2011 Development and application of new wall fasteners of external scaffold *Construction technology* **42(8)** 714

[10] Li Y, Xu W and Shen Q 2012 Analytical study on structural modeling for wall connector of steel pipe scaffold *Building construction* **34(11)** 1095

[11] Hadipriono F C and Wang H K 1986 Analysis of Causes of Formwork Failures in Concrete Structures *Const Engng Mgmt ASCE* **112(1)** 112

[12] Hadipriono F C and Wang H K 1987 Causes of Falsework Collapses during Contruc-Tion *Structure Safety* **4(3)**179
[13] Qin G and Xu W 2015 Study on the arrangement of the steel tubular scaffold with couplers in the construction of super high rise building *Journal of Shenyang jianzhu university(natural science)* **31(6)** 1023
[14] Siu C, Ho L and Ming G W 1994 Nonlinear Vibration analysis of steel frames with semirigid connections *Journal of Structural Engineering* **120(4)** 1075
[15] The Ministry of housing and urban-Rural Development of the Peoples Republic of China 2013 *Load code for the design of building structures GB50009-2012* (China Architecture & Building Press)

IOP Conf. Series: Materials Science and Engineering **399** (2018) 012047 doi:10.1088/1757-899X/399/1/012047

Steel antenna towers – from designing to manufacturing optimization

D Radu[1*] and A Feier[2]

[1] Faculty of Civil Engineering, Transilvania University of Brașov, 500152, 5 Turnului St., Brașov, Romania
[2] Department of Materials and Manufacturing Engineering, Politehnica University of Timișoara, Timișoara, Romania

[*] E-mail: dorin.radu@unitbv.ro

Abstract. The elaboration of a methodology for determining the acceptability of detected cracks/flaws in a structure has a major practical importance in the overall assessment and life integrity of a structure. The relation given by fracture mechanics links a parameter which describes the stress intensity at a crack tip to a material characteristic – fracture toughness. This relation provides the possibility of assessing the fracture conditions of the structural elements with defects (cracks). An Engineering Critical Assessment is an analysis based on fracture mechanics principles, of whether or not a given flaw is safe from brittle fracture, fatigue, creep or plastic collapse under specified loading conditions.

The paper presents some considerations on the designing and manufacturing process, taking into account structural assessment procedures using fracture mechanics. It also presents a study case – a steel monopole tower on which the method is applied, in order to assessing the possible discovered flaws in the designing phase. Ten types of flaws which can be discovered in the service time of the structure were assessed. Different types of locations were taken into account, thus resulting groups of flaws which were then assessed and compared.

1. Introduction

Due to the rapid growth of the telecommunication industry, the development of relevant infrastructure has been gaining importance. Modern telecommunication structures are essential to the present society. The emergence of new technologies creates a demand for additional facilities and for the introduction of new elements into our cities.

In fast developing countries, faced with high density of population in urban areas, great difficulties are experienced in finding land for the setting up of conventional lattice towers for communication purposes. A serious increase in land value has called for a suitable alternative to conventional lattice towers that should be friendly and unobtrusive to the environment. Environmental and economic pressures have require improved design approaches to make communication towers more environmentally acceptable and cost effective. Steel pole structures are used in different fields such as power transmission, communication, high way and stadium illumination. Steel tubular poles have a smaller plan dimension and are composed of fewer components compared to the lattice type towers.

When these poles are used as antennae supporting structures, they are more economical as far as the cost of land is concerned. These poles are either slip jointed or connected using bolted flange plates. The pole weight affects the overall cost of the system consisting of the overall cost of the material, manufacturing, transportation and erection. Following the need of upgrading to high standards of signal transmission, in Romania and not only, the telecommunication network is rapidly growing. There is an undeniable need for the design and erection of communication towers.

The aim of the present paper is to apply fracture mechanics in the assessment of the in structure possible discovered flaws. The assessment procedure, called ECA - Engineering Critical Assessment, can be applied starting with the design process, to elements' manufacturing or in service.

As a study case, the paper presents a steel monopole tower structure to which the assessment of the possible flaws in the design phase is applied.

2. Engineering critical assessment approach

Most welding fabrication codes specify maximum tolerable flaw sizes and minimum tolerable Charpy energy, based on good workmanship, i.e. what can reasonably be expected within normal working practices. These requirements tend to be somewhat arbitrary, and failure to achieve them does not necessarily mean that the structure is at risk of failure. An Engineering Critical Assessment (ECA) is an analysis based on fracture mechanics principles, of whether or not a given flaw is safe from brittle fracture, fatigue, creep or plastic collapse under specified loading conditions. An ECA can therefore be used: during design, to assist in the choice of welding procedure and/or inspection techniques; during fabrication, to assess the significance of known defects which are unacceptable to a given code [1], or during operation, to assess flaws found in service and to make decisions as to whether they can safely remain, or whether down-rating/repair are necessary.

The concept (also termed 'fitness-for-purpose analysis') is widely accepted by a range of engineering industries; however, in civil engineering it is not used very often.

For an analysis of a known flaw, the following information is needed: size, position and orientation of flaw, stresses acting on the region containing the flaw, toughness and tensile properties of the region containing the flaw. These information can be obtained partly as a result of a structural analysis made by a civil engineer, and partly by a materials testing and fracture mechanics analysis performed by a mechanical engineer.

The analysis is carried out in accordance with the British Standard procedure BS 7910 ('Guide to methods for assessing the acceptability of flaws in metallic structures') [2]. Although simplified analyses can be carried out based on code values of Charpy energy and maximum allowable stresses, it is usually necessary to carry out fracture-mechanics testing (critical K, CTOD or J) in order to obtain an accurate measurement of the material toughness. Additional stress analysis (e.g. by hand calculation or Finite Element Analysis) may also be required [3].

For design purposes, or for the analysis of weldments which fail to meet a toughness requirement, the ECA is based on a hypothetical 'reference flaw' which is highly unlikely to be missed during inspection.

An ECA can also be used to assess the significance of growing flaws, e.g. fatigue, creep or stress corrosion cracks, in order to make decisions on life extension and safe inspection intervals.

Considering the FM approach for a specific steel structure with welded joints, the with flaw elements can be divided into:

- safe in which the flaw/flaws will not grow to a critical size during the service life,
- safe if regularly inspected, the flaw/flaws is/are on the borderline of reaching a critical flaw size during the service life, and regular inspection is therefore necessary,
- repair is mandatory – the flaw/flaws is/are already so critical that there is every possibility of failure and they thus exceed the permissible risk in terms of safety.

ECA is characterized by applying FM in assessing specific flaws for steel elements joints, fatigue assessment of the steel elements with typical flaws, remaining lifetime calculation for steel structures containing flaws / cracks, determining the inspection interval for steel structures – risk analysis for different flaws [2, 4].

3. Case study – ECA applied for an antenna tower

The study is carried out following the need of the investor to assess the reliability in time of a number of 15 antenna towers. Given the number of sites, the ECA needed to be applied from the designing phase – for a given future discovered flaw to evaluate the implications in the behaviour of the structure.

The column is a 30 m height monopole type CHS profile (610 mm in diameter) column and it presents three segments – two segments of 12.00 m and one segment of 6.00 m. At the upper side (4.00 m from the top) a technical maintenance platform will be attached, as well as four antennas (with a total wind exposure area of 0.79 m^2) (figure 1).

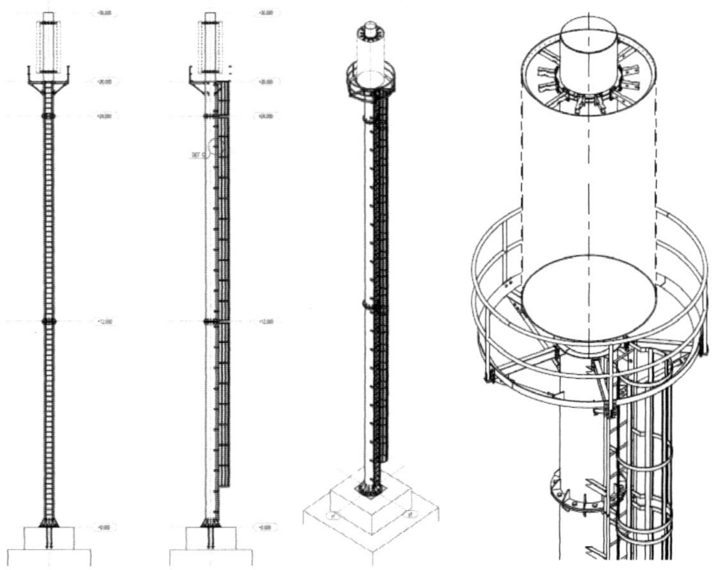

Figure 1. Geometry of the antenna tower – general views.

3.1. Loads evaluation and analysis methods

In order to apply the ECA from the design phase, four phases needed to be covered:

- loads evaluation with an in depth wind load assessment,
- global numerical analysis – linear elastic analysis,
- determining the critical stress in the given flaw,
- flaw geometry (assessment for a given flaw),
- material fracture toughness determination,
- applying failure assessment diagrams (level 1 and/or level 2) for a given flaw.

Regarding the wind, it was evaluated according to the Romanian normative CR1-1-4 [5], considering the structure in the area of q_b=0.6 kPa with the terrain category II. From the dynamic analysis what resulted was a frequency of the structure of 0.42 Hz, with coefficient $c_s c_d$ = 0.984. The wind on the tower was evaluated on the surface of the cylinder and on the top of the structure – onto the antennas. The total value of the wind load onto the antenna area was calculated $F_{w\ tot}$ = 22.11 kN according to [5]:

$$F_w = \gamma_{cf} \cdot c_d \cdot c_f \cdot q_p(z_e) \cdot A_{ref} \tag{1}$$

Following the linear analysis, a maximal stress of 311.12 N/mm^2 was determined, and it was identified in the area of the baseplate of the monopole. In order to assess a given flaw in a given area, multiple cases were detailed (table 1). An in-depth analysis (FEM type) was required in order to

IOP Conf. Series: Materials Science and Engineering **399** (2018) 012047 doi:10.1088/1757-899X/399/1/012047

evaluate the flaws in the joints. With Abaqus software [6], several analyses were ran, thus determining the stresses in the flaw risk areas (figure 2).

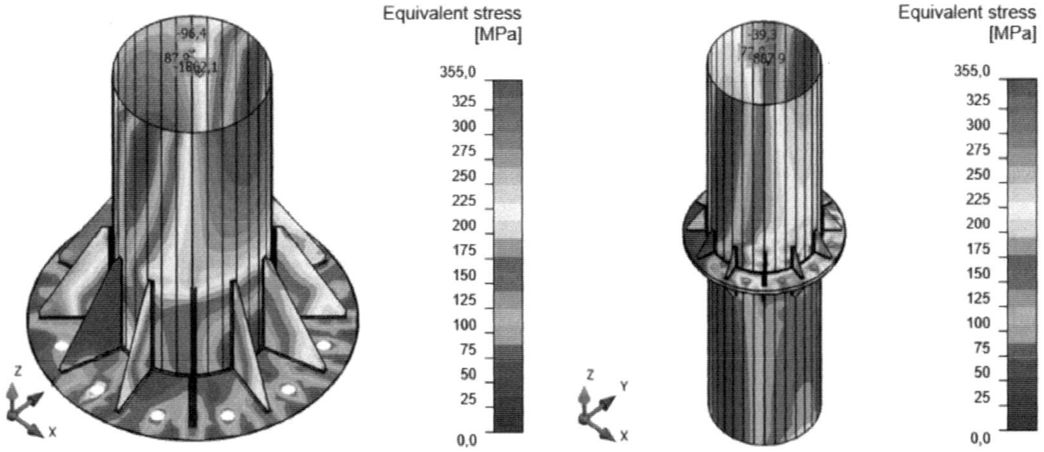

Figure 2. Equivalent stress values in the area of the (a) base plate joint; (b) segment joint.

Given the number of produced assemblies, the material fracture properties were determined prior to manufacture [7]. The experiments were conducted on the base material, thus the values of fracture mechanics properties were valid for the designed and future manufacture structure. The value of 78.92 $MPa \cdot m^{1/2}$ was determined on the specimens. This value will be used in further assessments.

Table 1. FEM analysis results - stresses in the flaw risk areas.

Flaw risk area	FEM analysis result (MPa)	Attached given flaw
Base plate - foundation connection	312.55	FP-TTF, FP-SF, FP-BF
Segment joint 1 (level of 12m)	202.45	FP-TTF, FP-SF, FP-BF, FP-EF
Segment joint 2 (level of 24m)	182.16	FP-TTF, FP-SF, FP-BF, FP-EF

A taxonomy was made of the flaws in order to have a complete behaviour view of the design structure with a clear description of each flaw (table 2).

Table 2. Description of each flaw.

Name	Flaw type	Description of the flaw
FP-TTF	Flat plate - through thickness flaw	Crack in the column base or in the segment endplate joint
FP-SF	Flat plate – surface flaw	Crack in the base material (tube wall)
FP-BF	Flat plate – buried flaw	Crack in the base material (tube wall)
FP-EF	Flat plate – edge flaw	Crack in the flange of the segment joint or column base in proximity of the welding longitudinal or transversal direction

IOP Conf. Series: Materials Science and Engineering **399** (2018) 012047 doi:10.1088/1757-899X/399/1/012047

3.2. Applying Failure Assessment Diagrams

The *Failure Assessment Diagram (FAD)* describes the interaction between the brittle fracture and plastic failure through the $F_f = f(S_r)$ function.

Structures using reasonably tough materials (high K_{Ic}) and having only small cracks (low K) will lie in the strength-of-materials regime. Conversely, if the material is brittle (low K_{Ic}) and strong S_r (high yield strength), the presence of even a small crack is likely to trigger fracture. Thus, the fracture mechanics assessment is crucial. The special circumstances that would be called into play in the upper right corner of figure 3 in this regime, a cracked structure would experience large-scale plastic deformation prior to crack extension [1].

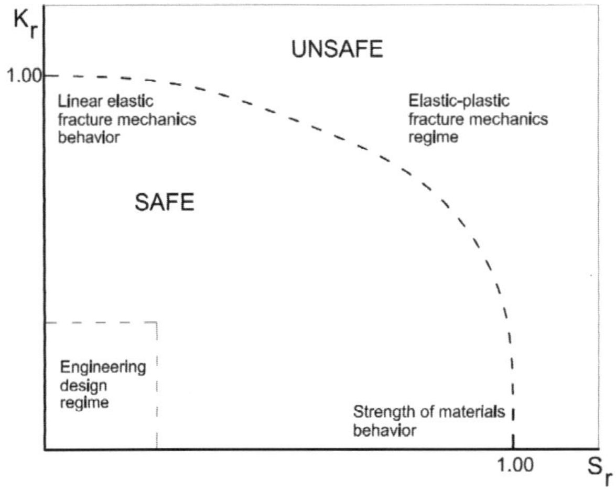

Figure 3. General plot of the ratios of the toughness and stress showing the relationship between linear elastic fracture mechanics and strength of materials as it relates to fracture and structural integrity [8]

The level 2 (FAD-2) assessment is the normal evaluation path for general application. The method presents an assessment line given by an equation of a curve and a cut-off line. If the assessment point is in the interior of the surface limited by the assessment line, the flaw is acceptable and if the assessment point is on the outside area, the flaw is considered unacceptable [9].

The equations describing the assessment line are:

$$\sqrt{\delta_r} \ \ or \ \ K_r = \left(1 - 0{,}14\, L_r^2\right)\left[0{,}30 + 0{,}70\, exp(\, -0.65\, L_r^6\,)\right] \text{ for } L_r \leq L_{rmax} \tag{2}$$

$$\sqrt{\delta_r} \ \ or \ \ K_r = 0 \ \text{ for } L_r > L_{rmax} \tag{3}$$

The cut-of line is fixed in point where $L_r = L_{rmax}$ where:

$$L_{r\,max} = (\sigma_Y + \sigma_u)/(2\sigma_Y) \tag{4}$$

in which:

σ_Y – the yielding resistance of the material

σ_u – the ultimate resistance of the material

For the assessment on level 2, a FAD is necessary to pass through the following phases [6]:

- determining the stresses – following a structural analysis. The assessments consider the real distribution of stresses in the proximity of the flaws – P_m, P_b, Q_m and Q_b.
- the fracture ratio K_r must be determined

$$K_r = K_I \,/\, K_{mat} \tag{5}$$

in which K_{mat} represents the fracture toughness of the assessed material, determined at the service temperature.

The stress intensity factor (SIF) – K_I is determined with the following relation:

$$K_I = (Y \cdot \sigma) \cdot (\pi \cdot a)^{1/2} \tag{6}$$

where

$$Y \cdot \sigma = (Y \cdot \sigma)_P + (Y \cdot \sigma)_S \tag{7}$$

in which:

$(Y \cdot \sigma)_P$ – contribution of the main stresses
$(Y \cdot \sigma)_S$ – contribution of the secondary stresses

$$(Y \cdot \sigma)_P = M \cdot f_w \cdot \{k_{tm} \cdot M_{km} \cdot M_m \cdot P_m + k_{tb} \cdot M_{kb} \cdot M_b \cdot [P_b + (k_m - 1)P_m]\} \tag{8}$$

$$(Y \cdot \sigma)_S = M_m \cdot Q_m + M_b \cdot Q_b \tag{9}$$

- Determining the ratio of stress L_r according with:

$$L_r = \sigma_{ref} / \sigma_Y \tag{10}$$

  in which σ_{ref} is obtain according with a relation specific with the flaw type. The point/points of assessment are represented graphically in (K_r, L_r) coordinates on the FAD level 2 [6].
- The evaluation of the position of the point is made according with the specifications.

For the in case structure assessment level 2 – FAD-2, assessments were made on different flaw types and flaw positons. The values of the input data are:

- σ_Y (yield strength)= 355 MPa; σ_T (ultimate strength)= 510 MPa; S355J2 steel type,
- K_{mat} = 78.92 MPa·m$^{1/2}$ was determined on the specimens (through testing),
- P_m = primary stress according to table 1 and table 2 (determined following a structural analysis),
- k_{tm} = 1; k_{tb} = 1 (stress concentrators factors),
- Q_{tm} = 0 (thermal membrane stress) and Q_{tb} = 0 (thermal bending stress),
- Q_m = 0 (residual membrane stress) and Q_b = 0 (residual bending stress).

3.3. Results

Considering the position of the flaw regarding the stress direction and position in the assembly of the steel shell element, 10 types of assessed flaws resulted, as presented in table 3. In figure 4 there is graphically presented the assessment results for the FP-TTF group.

Table 3. Results on the assessed flaws

Case	B mm	W mm	2a mm	a mm	2c mm	p mm	r_0 mm	h mm	t_w mm	L_r	K_r
FP-TTF-1	16	200	30							0.8318	0.6755
FP-TTF-2	32.63	200	30							0.8318	0.6755
FP-TTF-3	200	32.63	10							1.0195	0.4085
FP-SF-1	25	200	30							0.8318	0.6755
FP-SF-1	25	120	30							0.9427	0.6930
FP-SF-3	16	200								0.7644	0.7688
FP-EF-1	32.63	200		15						0.7644	0.7688
FP-EF-2	200	32.63		15						1.3086	1.6678
FP-EF-3	25	200		15						0.7644	0.7688
FP-BF-1	16	200	5		30	3		7		0.8080	0.8139

4. Conclusions

The engineering critical assessment procedure was applied from the design phase. Ten types of flaws which can be discovered in the service time of the structure were assessed. Different types of locations were taken into account, thus resulting groups of flaws which were assessed and compared.

The input data took into account the results from the FEM analysis of structure and the experimental results for material properties, all needed in the assessment procedures.

The comparison of the flaws' assessment with fracture mechanics procedures, revealed several problems:

- sensibility of the joints to the through thickness flaw in the endplate of the segment joint (FP-TTF-3 - crack in the column base or in the segment endplate joint) [10];

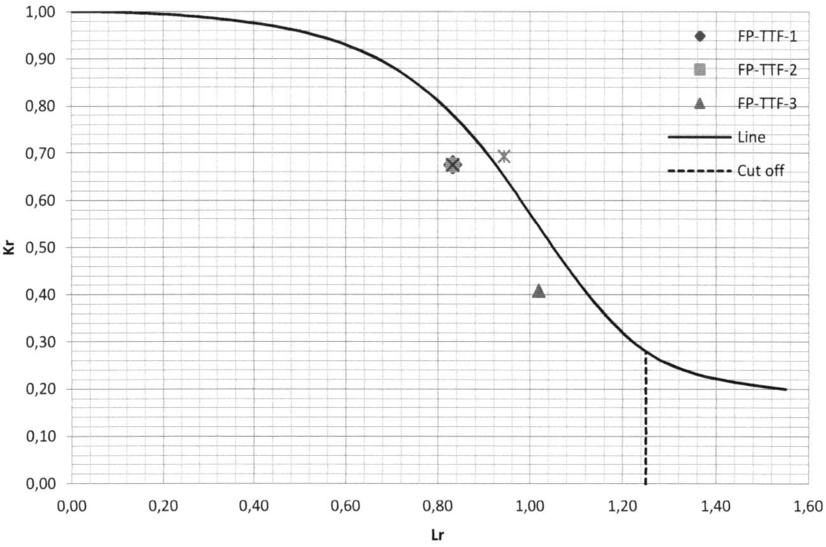

Figure 4. FP-TTF – Group of flaws – assessment.

- the edge flaw type – FP-EF-2 (Crack in the flange of the segment joint or column base in proximity of the welding longitudinal or transversal direction), is the most dangerous – a = 15 mm crack depth into welded joint is a critical flaw for which the joint is considered unsafe [11].

References

[1] Radu D and Gălățanu T 2017 Fracture mechanics critical assessment of the steel structures joints *Conference proceedings - 5th International Conference Contemporary Achievements in Civil Engineering* pp. 261-269

[2] ***, BS 7910/2013, "Guide to methods for assessing the acceptability of flaws in metallic structures", *BSI British Standards*

[3] Radu D, Sedmak A, Sedmak S and Dunjić M 2018 Stress analysis of a steel structure comprising cylindrical shell with billboard tower *Technical Gazette* **25** no.2/2018 pp. 212-217

[4] Milović Lj and Sedmak A 2007 Numerical and analytical modeling of elastic-plastic fracture mechanics parameters *Materials Science Forum* 555 pp. 565-570

[5] ***, CR1-1-4/20012 - "Cod de proiectare. Evaluarea actiunii vântului asupra constructiilor"

[6] ***, Abaqus software guide.

[7] Sedmak S and Sedmak A 2009 Fracture mechanics and non-destructive testing for structural integrity assessment *Key Engineering Materials* **399** pp. 27-36

[8] Kanninen M F and Popelar C H 1985 Advanced Fracture Mechanics *Oxford University Press*

[9] Petzek E 2006 Elaborarea de instructiuni pentru aplicarea principiilor mecanicii ruperii la stabilirea siguranței în exploatare și a duratei de viata rămase a podurilor metalice existente *Orizonturi Universitare Publishing house, Timișoara*

IOP Conf. Series: Materials Science and Engineering **399** (2018) 012047 doi:10.1088/1757-899X/399/1/012047

[10] Rakin M, Gubeljak N, Dobrojević M and Sedmak A 2008 Modelling of ductile fracture initiation in strength mismatched welded joint *Engineering Fracture Mechanics.* **75** (11) pp. 3499-3510
[11] Radu D, Sedmak A, Gălățanu T and Taus D 2017 Fracture Mechanics applied on investigation of the existing lattice structures *International Scientific Conference CiBV 2017 - Bulletin of the Transilvania University of Brașov* **10** (59)

IOP Conf. Series: Materials Science and Engineering **399** (2018) 012048 doi:10.1088/1757-899X/399/1/012048

The development status of construction equipment and technology in China

J Sun[1,2*], L Y Zhu[1] and J Wu[2]

[1] School of mechanical engineering, Shenyang Jianzhu University, Liaoning, China
[2] Shenxi Machinery Co., Ltd., Jiangsu, China

[*] E-mail: usunjia@163.com

Abstract. In recent years, with the continuous development of urbanization in China, construction equipment and technology in the field of building are developing rapidly. China has made much progress in this field, but there is also a gap in comparison. Rapid growth needs to follow the appropriate direction of development. In the future, China will further accelerate the speed of urban intensive construction, so it is of great significance to find a correct direction for the development of Chinese cities. Based on this purpose, this paper summarizes the current situation and rules of building equipment and technology in the view of the rapid development of modern architecture and related practical achievements. This article further analyses the building technology and equipment from the aspects of social culture, political economy, laws and regulations, and discipline development. The conclusion is that in the future development of buildings, the energy saving of construction equipment and technology will become an important symbol of construction technology progress. Finally, this paper puts forward the problems existing in the development of construction equipment and technology in China and the ways to solve these problems.

1. Introduction of Chinese architecture

From the background of modern Chinese architecture's development, Chinese architectural structure and quality mainly go through four stages.

The first stage was the period from the 1950s to the mid-1960s. The main purpose of the building during this period was to solve the residents' housing problems. The construction materials were relatively simple, the architectural styles were relatively common, and the quality of the buildings was not bad.

The second stage was the period from the subsequent 1970s to the 1990s, a period of reform and opening up for China. Especially during the 1990s, the rapid economic development led to large-scale construction, while, there was a shortage of technologies and resources. The architectural style of this stage began to be diverse, but due to the low level of construction technology, the quality of the building was unsatisfactory.

The third stage was from the 1990s until present. The nation's policies and regulations have gradually improved. The improvement of the construction supervision system, technical specifications, and improvement of corporate qualifications have greatly improved the overall construction technology level. The quality of construction has greatly improved, and the high-rise building construction equipment and technology have witnessed a rapid growth.

The fourth stage was the Chinese development into the "13th Five-Year" stage; Chinese architectural structure began to change gradually, in many fields in search of efficient, environmentally friendly and energy saving structure style. Construction equipment and technology also appeared and a new development direction was influenced by the construction structure.

(a) In the 1970s (b) In the 1990s (c) At present

Figure 1. Rural Architecture Style in Northern China.

(a) In the 1970s (b) In the 1990s (c) At present

Figure 2. Urban architectural style of China.

Figure 1a, b, c and figure 2a, b, c are miniatures of rural and urban architectural styles in China. After the development of building structure, such as brick-wood, brick, steel and reinforced concrete, modern architectural design, construction and maintenance determine the appearance, applicability and service life of the houses. Among them, construction as the core link of building formation is a complex process. It is necessary to apply materials theory, mechanics, structural science, technology and so on to solve various problems in construction, to ultimately ensure projects' quality and construction's safety [1].

The rapid development of the Chinese economy in the past decade has accelerated the pace of urbanization in China. The changes in the new countryside and new cities can be described as earth-shaking. Buildings with low security have experienced new generations of upgrades. Self-built houses in City Village and old houses, which called for hidden safety concerns, have all been greatly improved. Looking at the overall situation, the development of architecture meets law of human settlements. In the 19th century, it was the stage of bridges' and roads' construction. In the 20th century, it experienced the development of high-rise buildings and above-ground space. In the 21st century, it will enter the challenge of joint development of above-ground space and underground space [2].

2. Present situation in construction technology

The planning, reconnaissance, design and construction constitute the processes of project construction. Construction technology as its final link has important significance for the construction project. With the development of urban buildings to vertical heights and underground structures, the scale of

IOP Conf. Series: Materials Science and Engineering **399** (2018) 012048 doi:10.1088/1757-899X/399/1/012048

construction projects continues to expand, resulting in breakthroughs in engineering that rely more on improvements in construction processes and construction equipment [3].

Construction technology will be divided into foundation engineering, structural engineering, and decoration engineering. In this paper, the development of building construction technology will be described in terms of lines instead of surfaces.

2.1. Foundation engineering
The technology development focuses on pile forming technology, deep foundation pit supporting technology and waterproof technology. The traditional precast concrete piles have been replaced by concrete cast-in-place piles and immersed cast-in-place piles because of noise pollution. At present, muti-pile optimal construction mode has been formed in Chinese foundation engineering, according to the conditions, the different pile bodies and corresponding construction methods can be chosen. The technology of deep foundation pit support should deal with the choice of complex construction process and the implementation of waterproofing works. Waterproof construction technology is mainly applied to water conservancy projects, bridges, and underground constructions. Construct and material waterproofing technologies are gradually established. Waterproofing methods have developed to form a combination of "defense, row, cut-off, and plugging", which is suitable for the local conditions and comprehensive construction methods [4].

2.2. Structural engineering
The transportation of construction materials and the lifting of materials are the development directions of construction technology. The research of construction technology focused on noise control, identification of working conditions for power matching, reducing equipment wear, extending equipment life, reducing energy consumption, etc. While the construction equipment is designed and developed from a variety of equipment models according to the engineering form, the equipment type spectrum is also constructed.

The development of building structural materials is focused on reinforcing steel and concrete materials. For example, pre-stressed concrete technology has greatly improved the overall strength of the structure, making the reinforcement of concrete and concrete more stable, and the concrete mixing construction technology improves the performance of concrete as well. The traditional concrete that only meets the strength of the building is developed into a new type of concrete that can satisfy many performance criteria.

The development of concrete modification additives has also improved the anti-permeability, antifreeze, and anti-expansion saturation retardation properties of concrete construction. The formwork in structural construction has gone through the development of bamboo plywood, wood plywood, all steel formwork and all aluminium formwork system. In the future, the research of the concrete manufacturing process, the improvement of the forming precision of the formwork, the automatic processing and mechanized connection of the reinforced material in the concrete and the reduction the working strength of the employees will be the key issues. On the other hand, the progress of steel structure construction technology has solved the problems of long-span and earthquake resistance of buildings, greatly improving the development of assembly construction of building structures.

2.3. Decoration engineering
The standardization of curtain wall design and construction technology, vertical transportation technology and building external wall insulation technology have been formed. The development of various curtain walls has greatly improved the appearance, thermal insulation and heat protection of the exterior walls.

The progress of insulation materials and construction technology has also improved the overall technical level of decoration. The emergence and application of the automatic climbing frame constructional technology with mechanized greatly shortens the period of the high-rise building decoration engineering. The general use of the high suspended access platform reduces the working strength of the construction personnel, improves the construction safety, and can complete the construction from the bottom to the top. The construction quality is guaranteed effectively, the construction site is saved and the operation efficiency is improved, but better equipment is needed to ensure the safety and efficiency of the construction.

2.4. Construction industrialization

The industrialization of construction is being popularized in China and the rapid development in the construction technology is lack of construction equipment and technology [5].

From the view of building construction technology in China, the construction process is a variety of parallel operations in limited time and space, many kinds of materials are supplied at the same time, and various kinds of equipment are operated in synergy. Therefore, it is of great significance to seek scientific and advanced construction technology and construction equipment, and to reduce the consumption of a large number of human, material and financial resources in the construction [6]. This requires a reasonable organization and management measures.

The current stage of its development has experienced system-to-item, coarse-to-fine, continuous accumulation of construction experience, formation of construction regulations and improvement of construction processes. At the same time, the complex construction technology, such as the application of top-down methods in high-rise building construction, can effectively improve the quality of construction projects. At present, the focus of the construction process is mainly on the quality, safety, progress, and cost of the project construction. And its development trend is the development of environmental protection, ecological impact, green materials and green construction.

There is a certain contradiction between the transformation of construction technology and the development of market economy. In the contemporary Chinese construction, it is urgent to improve the system and concept of the construction process. In addition, BIM technology is a new technology that is being vigorously promoted in the construction industry [7]. It is a major change in engineering and can effectively coordinate multiple parts. It can play an active role in the entire life cycle of the project.

3. Comparison of the development of construction equipment

To realize the sustainable development of buildings, it is necessary to implement sustainable development ideas throughout the life cycle of construction decision-making, design, construction, and operation. The construction process is a stage of concentrated influence on the environment, it is also the stage of direct consumption of resources.

The technical level of the equipment will directly affect the construction level of the project, and it is the key factor for its evaluation. Therefore, the design and manufacture of energy-efficient, environmentally friendly equipment will be more conducive to promoting the development of green construction.

Here is only a comparison of typical equipment in construction. First of all, the development of lifting equipment, large scale, multi equipment co-operation, mobility, preparatory convenience and equipment specialization are its development direction. The development of hoisting equipment diversity in China is shown in figure 3a, b, c, d.

Second, the development of concrete equipment. Concrete preparation, concrete transportation, and concrete molding are the major tasks for such equipment. The preparation equipment is developing in the direction of automation, informatization, refinement, and uniform mixing. The transportation equipment is developing in the direction of high, far, and adaptability. The molding equipment is developing in the direction of mold releasing easier, higher molding accuracy, simple assembly of complex shapes, and high-efficiency vibration forming. At the same time, concrete injection equipment is rapidly developed.

With regard to the construction of reinforced concrete structures, steel processing equipment is also being developed toward large-scale, integrated, and automation. The emphasis is on improving equipment utilization, reducing the labour intensity of operators, and reducing material waste. Among them, the numerical control processing technology is better applied in the steel processing equipment. The development of piling equipment is based on the integration of mechanical and electrical fluid and information control technology, forming a series of piles, such as whole casing construction equipment, high end continuous wall grooving equipment, multi-function pile working equipment, water drilling equipment, large diameter DTH hammer, large vibration hammer and impact hammer. The last is the maintenance equipment outside the building, which is called the building maintenance units(BMU) in the European standard, which is often divided into two types, the permanent maintenance equipment and the temporary overhead operation platform, as shown in figure 4a, b, c.

(a) Large crane

(b) Machine cooperation

(c) Special lifting equipment

(d) Wheeled mobile crane produced by Xugong

Figure 3. Development of lifting equipment.

4. Problems in the development of construction technology

The rapid development of economy will inevitably lead to lax enforcement of the system. And the quality of construction is threatened. The problems presented in the development of construction technology can be analysed from the three levels of technology, systems, and concepts.

The main technical problem is that the rapidly expanding construction industry has caused the overall level of construction technology to be affected by the overall quality of the operators and the technical level of construction equipment.

It can be summarized as human factors and equipment factors. In the current rapid development of industrialized construction methods, it is obvious that a large number of qualified and qualified industrial operators and equipment are lacking.

(a) Temporary equipment (b) Industrial building equipment (c) Maintenance equipment

Figure 4. Construction maintenance equipment.

In institutional aspects, due to insufficient technical conditions or human negligence, the disordered competition and extensive management lead to construction quality defects and regional safety accidents. The vicious competition within the industry that has no technological breakthrough gradually leads to serious obstacles in the development of manufacturing enterprises, lease companies and construction companies of construction equipment. Sell shoddy product, cutting materials, counterfeiting, price war, qualification instead of supervision become high-frequency words. This problem will lead to safety-related industries in building construction and become a high-risk industry where the path is sought in the fog and walks on the steel wire.

As for the concept, it is influenced by the target of short-term interests, which leads to problems in the development of building construction, such as the lack of implementation of a series of norms and the neglect of the environment and ecology. Building construction becomes an important factor causing environmental damage. China lacks the professional team of construction engineering. The composition of construction personnel is complex, the overall quality is not high, and the phenomenon of aging is serious, which forms a disadvantage to the construction process and systems and technology promotion, and limits the scale development of advanced construction technology.

5. Green building and green construction

At present, to "establish a sound green and low carbon cycle of the economic system, form a green development mode and way of life", is the concept of scientific development in the new era of China. It is a rare historical opportunity to further promote and accelerate the development of green buildings and related technologies.

During the construction process, a lot of resources and energy are consumed, and the pollution of waste water, waste gas, solid waste, dust, noise and strong light will bring risks to the environment safety. The "sustainable development" and "green" concept promotes "green construction" and avoids safety and environmental accidents under the overall goal of energy saving, water saving, material saving, land saving, and environmental protection. And also reduces the impact on the lives of ordinary people. In the process of transformation from volume construction to concept construction, we need to carry out specific research and exploration on green construction theory, method, management and technology for green building projects. And to find out the standards and technical measures for the green construction management suitable for Chinese development. Researchers should give full play to the advantages of methods such as field investigation, literature review, quantitative analysis, repeated argumentation, and laboratory tests. Correctly understand the necessity and urgency of green construction. In practice, the management ideas of green value are highlighted, with green construction technical measures and technical points as the core, the problems existing in the construction are constantly exceeded, and the problems are solved in time and effectively and reasonably. Formulate long-term measures to prevent problems from expanding and happening again.

The construction design and construction should be standardized from the aspects of green materials, green machinery, green work method, green mold frame, construction industrialization,

IOP Conf. Series: Materials Science and Engineering **399** (2018) 012048 doi:10.1088/1757-899X/399/1/012048

information, resources and recycling of waste. Improving green construction theory and developing green construction technology. Construction of green building promote harmonious, healthy and sustainable development of man, man and nature, man and society. This is of great significance not only for China and the world, but also for building an ecological community.

Beset by environmental and ecological problems, traditional buildings will gradually be replaced by sustainable green buildings. The energy saving green construction technology not only conforms to the goal of the sustainable development of the construction industry, but also conforms to the trend of technical development. The construction technology of mechanical automation is gradually replacing the artificial construction which can not control the quality of the construction steadily. The fine construction will gradually replace the old and extensive construction. Green construction technology instead of high energy consumption and heavy pollution construction technology, and eventually form a green construction system.

The concept of new architecture, such as green building, is mainly to solve the problems of some traditional buildings, such as the waste of resources, environmental pollution, and ecological destruction. After the green building thinking has been put forward, more attention has been paid to green construction technology in construction. In the development of modern green management technology and the process of construction, designers should fully consider the idea of saving materials, water saving, energy saving, land conservation and environmental protection, and integrate the protection of the living environment into the construction technology. At the same time, the concept of green building and green construction has the characteristics of environmental protection, which can reduce pollution to the environment and improve the ecological balance. In green construction, more attention is paid to providing comfortable working environment and safety guarantee for operators, reducing noise and improving work efficiency. The development of modern BIM technology provides more detailed and visualized advantages for building construction. It can cope with modern complex architectural structure, optimize the overall design scheme and construction process, improve the rationality and operability of construction technology, and gradually form a green construction system.

New concepts, new historical periods and changes are bound to bring many new opportunities, accompanied by challenges. In the process of its advancement and development, people can refer to mature manufacturing and look for green development routes with manufacturing technology. Combined with the current situation and trend, green building and green construction technology are put forward to further enhance the overall operation capacity of green construction. The top-level collaborative design and development of construction processes and technological innovation, and the development of innovative equipment such as building machines, will form an innovative system of construction technology to achieve energy-saving and environmental protection throughout the entire life cycle of a building. Figure 5 shows the busy construction process at a modern construction site.

Figure 5. Busy construction site at a modern construction site.

6. Development and reflection

The role of government policy is still the main driving force for the development of construction equipment and technology in China. The improvement of market position also has a selection. In the future, building construction technology will move toward information, automation, artificial intelligence, and environmental protection and energy conservation. The development of information technology will experience the stage of data collection - optimization - standard contrast - guidance output - fully automated control. At present, the information technology is in the information collection stage to improve the construction site process monitoring information collection and processing. On the other hand, a large number of operations, such as simulation, evaluation, prediction and management, will greatly improve the construction process of construction projects by using computer technology. The development of equipment and technology follows mutual infiltration and mutual influence. The rapid development of information technology in the future will inevitably affect the innovation of construction equipment and the improvement of construction process, prompting the development of simultaneous development. At the same time, we need to pay more attention to the induction and summarization of technology, and to combine fast and slow as the development planning method. The minor problems in the process of construction should be paid attention to, and the professional problems such as terrain, surrounding environment, building structure and protection measures can be fully taken into account. Forming a whole cycle of sustainable development, green and healthy business philosophy-getting rid of interests is the only goal pursued by enterprises, and they dare to shoulder their due social responsibilities.

The construction company must clarify the market demand and reflect on the long-term relationship between the market and the technology level of the enterprise itself. The company must continuously improve its own construction technology, introduce new equipment and new technologies at a proper time to adapt to market development, and actively learn advanced construction technology. To improve its own defects and create a new pattern of continuous, stable and healthy development in the industry.

Development goals: The construction company must correctly grasp the diversity of the construction plan and select the best construction technology and construction plan after scientific demonstration. According to different projects, choose suitable methods according to local conditions, adopt scientific and effective management to implement the construction process.

Acknowledgments

Authors wishing to acknowledge assistance or encouragement from Shenxi Machinery Co., Ltd., at same time, this work was financially supported by the National Key Technologies R&D Program of the 12th Five- year Plan of China (Project No.2011BAJ02- B07).

References

[1] Editorial board 1999 *Chinese civil architecture encyclopedia dictionary* (China: Construction Industry Press) p 513
[2] Bo L, Kairong H and Qingguo L 2008 *China. J. Modern Tunnel Technol.* **45** L20
[3] Guohua Z 2017 *Jiangxi building materials* (Jiangxi: Materials) **21** 118-120
[4] Wei Y 2015 Research on cost control of construction projects *China: Value Engineering* **34** 16-18
[5] Xin M and Xiandong L 2017 Building construction branch of China Architectural Society *Proceedings of the 2017 annual conference of China building construction* (professional volume)
[6] Xiaosong X 2007 Key technology research and structural analysis of top-down construction method for large deep foundation pit (China: Diss., Tongji University)
[7] Jun W 2011 The application of BIM concept and BIM software in construction projects China: Diss., Southwest J T University)

IOP Conf. Series: Materials Science and Engineering **399** (2018) 012049 doi:10.1088/1757-899X/399/1/012049

Earth buildings in Romania. Tradition and perspectives

E Tugui[1], M Barnaure[2*] and M Coman[3]

[1] Faculty of Architecture and Urban Planning, Technical University of Cluj-Napoca, 28 Memorandumului st., 400114 Cluj-Napoca, Romania
[2] Faculty of Civil, Industrial and Agricultural Buildings, Technical University of Civil Engineering Bucharest, Lacul Tei 122-124, 38RO-020396, Bucharest, Romania
[3] Technical Sciences Department, Ion Mincu University of Architecture and Urbanism, 18-20 Academiei st., 010014, Bucharest, Romania

[*] E-mail: mircea.barnaure@utcb.ro

Abstract. Earth is one of the most commonly used traditional building materials. Recently its usage was reassessed, considering its qualities, with earth being viewed as environmentally friendly and sustainable. Extended research has been done to analyse material qualities and construction techniques. New buildings were developed, some of them with modern architecture and high aesthetic qualities. In this context, the paper proposes an interdisciplinary approach aiming to analyse earth buildings in Romania from an architectural and structural point of view. Furthermore, their potential use in contemporary architecture is evaluated. Currently, local regulations for the usage of earth as construction material are absent and there are only few examples of earth architecture in recent Romanian practice. We propose a necessary analysis for the process of retrieving a traditional building material, which has a high potential for contemporary architecture in our country. The study examines the historical usage of earth as a building material in Romania. The types of buildings and the construction techniques across various regions are identified. An analysis of the specific structural issues of earth buildings in Romania follows. Conclusions are drawn regarding the potential use of earth as a construction material in contemporary Romanian architecture.

1. Traditional earthen construction techniques

Earth as a building material in Romania has a long tradition. It was used for both rural and urban constructions, mainly for houses and auxiliary buildings, but sometimes also for churches and schools. From a geographical point of view, earth was commonly used across Romania, but building techniques vary from one region to another.

Earth was used for both structural and non-structural elements. Structural walls were made of rammed earth, adobe, or in combination with other materials (wood frames with earth infills - *paiantă*). Non-structural elements include interior walls, flooring and roofs (for earth lodges).

The methods of building with earth in different forms, in Romania, are very old. They were based on using local materials and craftsmen from the building site or neighbouring villages [1]. The raw material used is either clay in natural form (often moistened before being used), or in mixture with organic matter (chaff, straw, sawdust, etc.). The latter improve the mechanical properties by increasing the tensile strength of the mixture and reducing the deformations and cracks after drying [2].

Traditional methods include semi-buried lodges and surface dwellings. Earth lodges are nowadays only found in ethnographic museums. Surface dwellings have exterior walls made either out of earth (rammed earth, cob, adobe) or with timber framing and various earthen infills (clay mixed with straw or chuff, wattle-and-daub, sun-dried blocks).

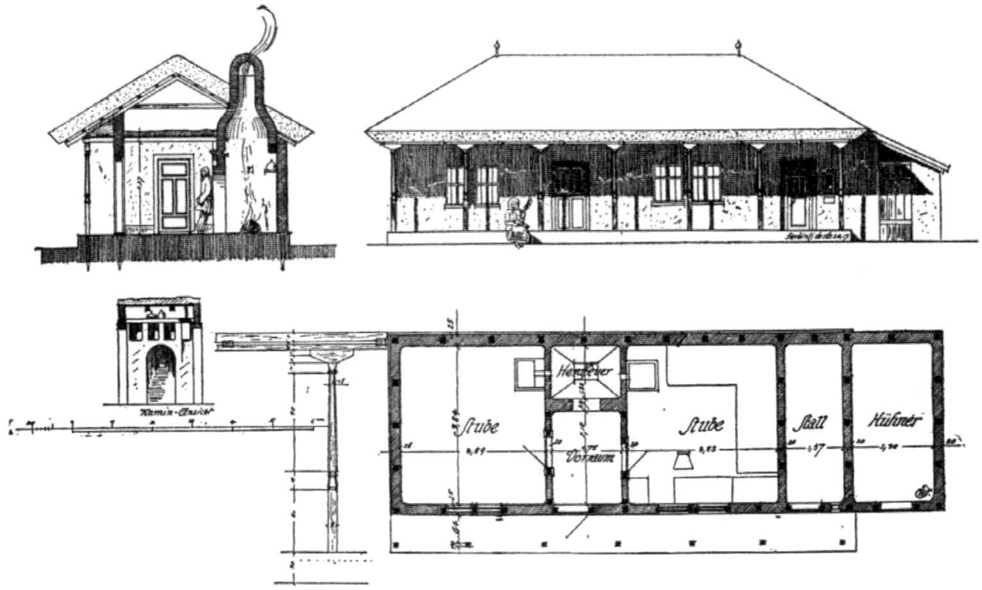

Figure 1. Traditional Romanian house made of timber framing with earth infills (*paiantă*) [3].

Old buildings did not generally have foundations (figure 1). The walls rested on massive horizontal round or cut wooden beams, directly on the ground or on large stones buried beneath the corners. More recent constructions had brick or even concrete foundations. The floor of the rooms was made of compacted earth, glued with clay smoothed by hand [1]. The hipped roof was made of round or carved wood, and the covering of wood shingle, reed or clay tiles. Walls were typically coated with loam plaster. The wall of the main façade was often decorated with specific elements embossed in the plaster [1].

1.1. Semi-buried buildings: earth lodges

Before mid 19th century, earth lodges, called *bordei,* were prevalent for housing in the plains of southern Romania [4, 5]. This construction type is present in Romania since the Neolithic period, when most of the buildings were oval or round-shaped [6, 7]. Archaeological research revealed 7-9th century and early medieval dugout shelters [6]. Around 1850, an important number of houses were semi-subterranean buildings, both in rural and urban areas [8]. In 1950, such buildings were still present only in some villages in the Oltenia region [6].

The construction of earth lodges in the southern regions of Romania has a long tradition, and their presence can be related not necessarily to the scarcity of other construction materials, but rather to a local archaic constructive tradition [4, 6]. Besides houses, semi-subterranean buildings sometimes had other destinations such as churches, grain/ice storage buildings, animal shelters, etc. [4, 6]. Earth lodges could have two or more rooms, and some configurations allowed for the construction of windows (figures 4 and 5) [4, 5, 9]. The tradition of semi-subterranean buildings is also found in northern Bulgaria [4] and other places in Europe [4, 6].

Earth lodges were constructed partially below and partially above ground level (figures 4 and 5). The building was covered with a thick layer of earth (figure 5) [4]. The interior floor level was generally at 0.6 – 1.2 m below ground [5].

Figure 2. Semi-buried house (*bordei*) from Muntenia (1860), engraving by Auguste Lancelot [10].

In the rudimentary buildings, the walls were covered only with a layer of clay, to create a flat surface. Predominantly, the walls were doubled by other materials, such as wood, wattle-and-daub, sometimes brick, and could be covered with lime plaster. Both the plan layout and the interior comfort of such constructions could be similar to the one of surface dwellings (figures 3 and 4). The amount of wood that could be used in semi-buried buildings was quite significant. Mainly oak was used, and the costs of such a building were significant [4, 5].

The pits where earth lodges were built were dug during summer, about 2 m deep. These pits were filled with wood, straw, corn cobs and then burned. Under the effect of the heat released by the fire in the pit, the clay in the earth solidifies and obtains properties similar to ceramics: low permeability and high strength [4].

The constructions had no attic. The ceiling was formed by tree trunks split in half, with the flat surface towards the interior, placed similarly to gable roofs. Split oak trunks with the flat side inwards were also placed on the outer walls. The wood was carefully sculpted on the visible side. The interior walls of the building were made of adobe or wattle-and-daub (figure 3) [6].

The plan was developed by adding side rooms, similar to the above-ground dwellings, and adding deep rooms (figure 4). The central room, where the earth was located, had a pyramid-shaped chimney made of wattle-and-daub or adobe, extending over the sloped roof (figures 2 and 5) [4, 6].

Figure 3. Interior image of a semi-buried house (*bordei*) [9].

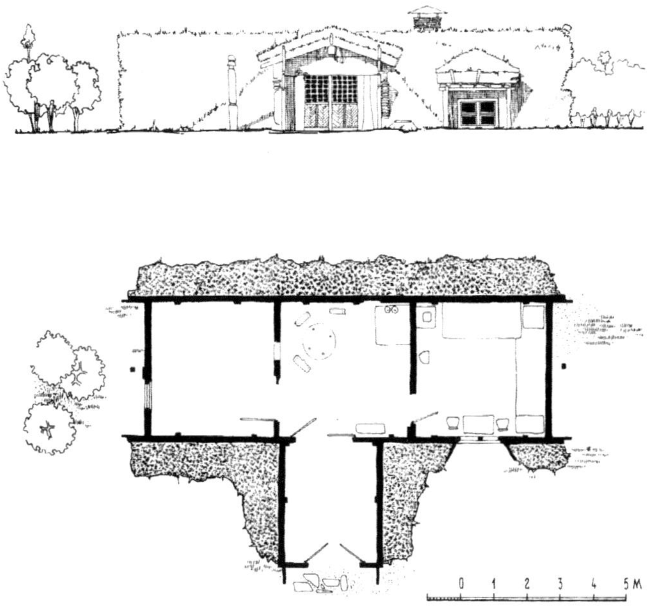

Figure 4. Semi-buried house (*bordei*) from Oltenia. Plan and South elevation [9].

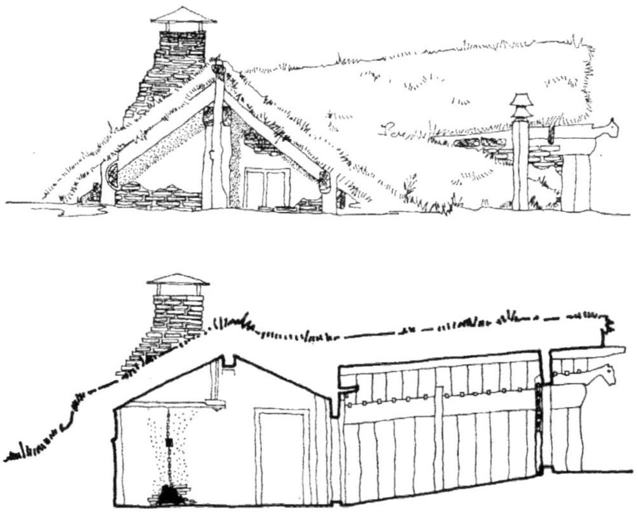

Figure 5. Semi-buried house (*bordei*) from Oltenia. Section and West elevation [4].

1.2. Buildings with earth walls

As in the case of earth lodges, the buildings with earth walls are concentrated especially in the plains of southern Romania [8]. In the 19th century, in the southern Romanian historical regions Muntenia and Oltenia, most of the surface constructions were made of wood (about 80% in the hill areas and about 53% in the plains). Buildings made of earth accounted for 20%, while brick and stone for 5%. Earth walls were also used, to a lesser extent, in Banat [11, 12, 13], northern Transylvania [13, 14] and in the central part of Moldova region [11]. Three building methods were used: rammed earth, cob and adobe.

1.2.1. Rammed earth ("pământ bătut cu maiul"). The mostly used method was similar to the technique of casting concrete into formwork. Earth without other mixture at its natural humidity [1] or moistened with water is placed in successive layers in a timber formwork. Each layer is then compacted to the maximum with wooden rams (figure 6). As the height of the wall increases, the wood planks are slid vertically and then the following layers are executed [5, 6]. The layers were at most 20 cm thick to allow a good compression of the earth. The final layer of compacted earth was 6-10 cm thick. In order to obtain homogeneous walls, the execution was to be carried out continuously, without interrupting the process until the completion of the construction [1]. In order to increase the resistance of the wall, vine chords were sometimes used [5]: every 3-4 layers, strings were placed horizontally, along the walls, primarily at the corners. Wood boards were placed on the top of the walls (sometimes the same planks used for formwork), so that the weight of the beams, the roof framing and the roof sheltering were evenly distributed on the walls [1]. The thickness of the walls varied between 50 and 80 cm, and in their thickness furniture elements (niches or cupboards) were sometimes integrated [1, 5]. The door and window openings were cut into the already built walls after these were finished [1]. Foundation thickness was between 70 and 100 cm. The foundation of the walls was made of compacted earth, burnt brick or, more recently, concrete. The foundations exceeded the thaw depth in the ground (80-100 cm) [1]. The rammed earth constructions could also have cellars [5]. Using the rammed earth techniques, solid constructions could be obtained. Their durability could reach up to 100-200 years under constant maintenance [1]. This construction technique was used in Romania until 1930s [4].

Figure 6. Rammed earth construction method (*pământ bătut cu maiul*) [13].

1.2.2. Cob. Another method of erecting earth walls was "fork building" - *pereți din pământ clădit cu furca* (figure 7), used in north Transylvania historical region [13, 14] and in the central part of Moldova region [11]. The clay, mixed with bigger chopped straw, was raised with forks by two men at the height of the wall, where a third one built it and stamped it with his feet. The walls were placed on a stone foundation or directly on the ground. The successive layers were flattened with a shovel and allowed to dry for a few days until a new layer was made [15].

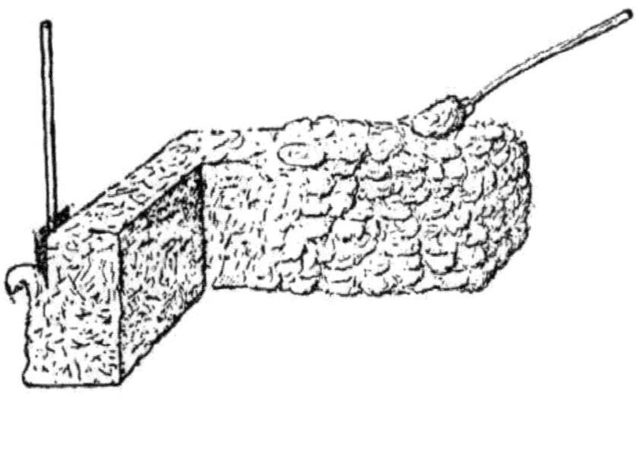

Figure 7. "Fork building" (*pereți din pământ clădit cu furca*) construction method [13].

1.2.3. Adobe. Adobe bricks (*chirpici*) is a construction material used for walls, made of unfired clay, formed in wooden moulds. The composition of the *chirpici* consists of a mixture of clay, with the addition of chopped straw, chaff, sawdust, etc. (*ceamur).* Pieces of dimensions ranging from 12x16x33 cm to 20x20x40 cm are not burnt, but sun-dried. The advantages of this construction material are the low price and good thermal insulation [2, 5]. The walls are made similar to masonry walls, using the adobe bricks and clay mortar [15].

1.3. Buildings with wood frame - earth infill walls ("paiantă")

The *paiantă* buildings date in Romania back from the end of the early Neolithic [16]. The walls of these constructions were made of wood beams, poles, braided rods / woven wickers, and roughcast from clay mixed usually with straw, and seldom with sand or leaves or of cane and thin rods, glued with adobe, supported on a wooden beam structure [17, 18]. Archaeological excavations have revealed the same constructive system dating from the Bronze Age until the Middle Ages, with stone foundations [6]. This constructive system was often used in the hill and mountain areas [18, 19]. In the rural areas, sometimes even schools [20] or churches [6] had the walls built with this technique.

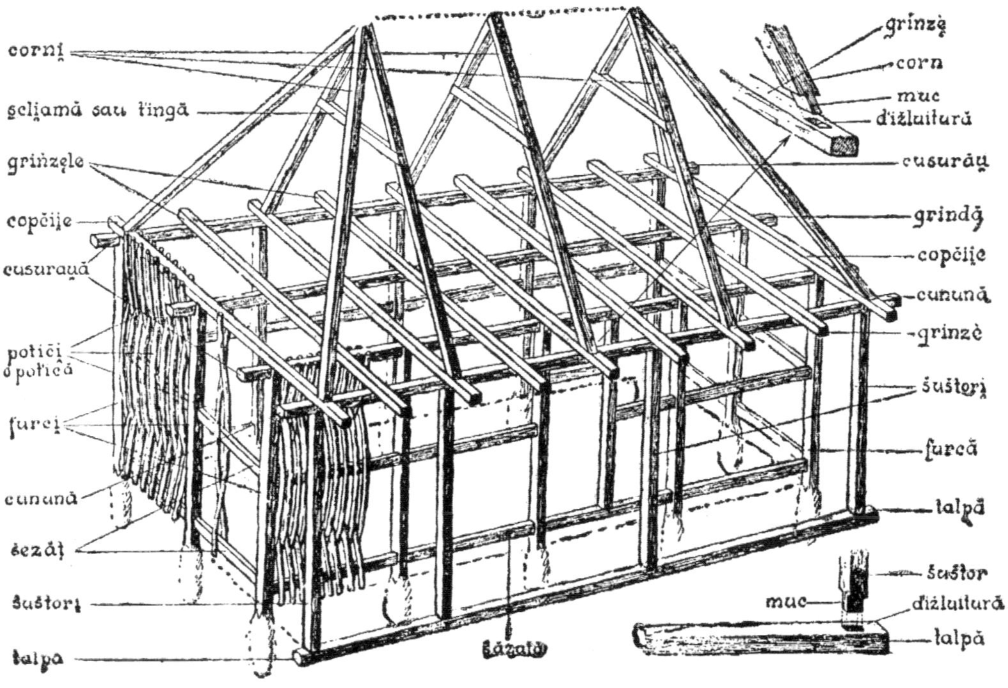

Figure 8. Wood structure of a building with wood frame and earth infill walls (*paiantă*) [13].

The wooded framing of *paiantă* walls had two structural typologies: vertical wood poles thrust in the ground ("fork system" / *în furci*, figures 8 and 10) or framing built on a base wooden beam (*pe tălpi*, figure 11) [15]. On the main frames, a wattle and daub wall was made (figures 8 and 11). Usually, rod wicker was made from the bottom to the top, then was covered on both sides with earth mixed with organic materials (straw or chuff) (figure 11) [5, 6, 15]. A variation of that system was to complete a vertical netting made of thin rods / wickerwork between the horizontal elements of the structure, or long pieces of broken trunks. Another alternative was interlacing of the vertically arranged beads with adobe "wefts" *(vălătuci)* arranged horizontally (about 7 cm thick). These were made of a layer of clay covering a layer of straws – rolled-up together [6, 15].

For wattle and daub walls, the first layer of daub was done on the inner side, and then, after it dried, the outside layer was made. In some cases, as to obtain a thicker wall, several successive layers of earth composed the wall. In the Danube river meadow and in the Delta, the rods were replaced with reeds, and for the filling was used the *ceamur* - a mixture of clay with straw, fibrous herbs, tow or chaff, squeezed using horses [15]. After the walls had dried out, they were covered on both sides with earth plaster. The surface of the walls was smoothed with a layer of clay and often whitewashed [20].

Figure 9. Buildings with wood frame and earth infill walls (*paiantă*). The drawing shows the building before the last layer of earth was added to cover the walls [9].

1.3.1. Wood poles structure ("în furci"). The system involved creating a vertical wood structure for the outer walls by thrusting into the ground 15-20 cm wide poles, spaced at around 1.20-1.50 m. The number of the vertical elements varied according to the dimension of the rooms and the desired structural strength (figure 10). The top of the vertical pillars was carved, for a better connection with the top girders, on which the beams of the ceiling were supported [1, 15, 20]. The depth at which the oak poles were buried was about 1 m, and their top reached 2 – 2.20 m above the ground [6, 15].

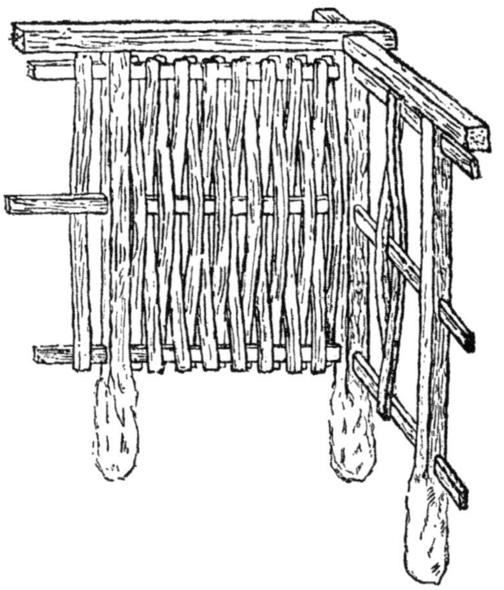

Figure 10. Construction system of a buildings with wood frame and earth infill walls (*paiantă*), with wood poles structure (*în furci*) [13].

The structure was strengthened by horizontal wood beams ("belts"). These were made by fixing, between the poles, oak or hornbeam horizontal wood joists, at about 70 cm from the ground and from the top (figure 8) [6, 13, 15]. The "forks" system was used in Transylvania, but also in Moldova, especially in the central and eastern part of Moldova [6, 15]. The disadvantage of this construction system is the contact of the wood elements with the soil moisture. To prevent decay, the tips of the poles were charred and soaked in oil. Another problem of the system is that, in time, under the weight of the construction, the poles were sinking in the ground, and it was necessary for the abode to be deepened [15].

1.3.2. Structures on wooden beams ("pe tălpi"). The most used solution was building the timber framing on a bottom wooden beam (figure 11). This beam was generally placed on a stone foundation, as to avoid moisture. The vertical columns (similar in size and distance to the forks) were fixed to the base beam and to the sill beam [15]. Sometimes, wooden diagonal braces were also used. In the western part of Moldova, the framing was reinforced by placing, on both sides, horizontal wood lath or beech rods. The space between these was filled with damp earth, mixed with chaff (figure 9). The walls were covered with a layer of earth mixed with chaff, then a thinner layer of earth mixed with cattle dung, then were lime-washed [1, 6].

Figure 11. Construction system of a buildings with wood frame and earth infill walls (*paiantă*) on wooden beams (*pe tălpi*) [13].

Buildings where earth is used as a construction material, in particular houses and auxiliary buildings made of wood and earth (*paiantă*), are still present in all the regions of Romania. Some existing buildings with high architectural value were moved from the place of origin to ethnographic museums, while other are listed as historic monuments. The need of preserving the existing buildings demands a modern approach towards the use of earth. Also, there is a rising interest for new green buildings made with traditional materials and construction methods.

2. Current issues related to the use of earth for buildings

In Romania, only standardized materials can be used in buildings. When national or European standards are not available (as it is the case for earthen materials), a technical agreement for the material must be issued prior to its use. This procedure involves costs that sometimes surpass the financial possibilities of potential clients. An alternative solution is to use already approved products. But this could lead to the impossibility of using local-available materials, while transport of materials to the site diminishes the eco-friendliness of the solution.

Regarding the structural solution, large areas of Romania are prone to high seismic hazard. Structural engineers, who are legally responsible for the entire lifespan of the buildings, are often reluctant to adopt solutions when lacking clear national or European code provisions. This is, to some extent, also the case for architects or engineers in charge of designing buildings for other essential criteria, such as thermal insulation, soundproofing, fire resistance, etc.

Romanian climatic conditions, notably rainfall and temperature, vary from one region to another in significant extent. For this reason, difficulties arise in implementing solutions that were validated in other countries, or even in other regions, without prior detailed studies.

Another difficulty arises from the loss of traditional building knowledge in the last 50 years. The introduction of national standards and regulations for construction was done, as in many other countries, in order to improve the quality of living and to minimize the risk of injury for people under extreme loads on buildings. While these regulations have their benefits, they also sometimes lead to the loss of traditional architecture. Some locally available natural materials and traditional construction techniques could no longer be used and had to be replaced with engineered materials that not only changed the architecture of the buildings, but sometimes even proved inadequate for some of the local conditions. Similar situations can be found in the literature for other countries. For instance, in [21] the case of Japanese Amami Island is discussed, where traditional hand-hewn timber buildings using local grown trees, clay and bamboo walls and thatch roofs were replaced with wooden stud walls with insulating core, steel fasteners and galvanized roof sheeting. This led to the loss of a sustainable and self-contained building system, of centuries-long carpentry traditions.

2.1. Experience of other countries

There are countries, some with high seismic demand (such as New Zealand, USA, India), where regulations have been developed and are currently in-force regarding the use of earth for buildings, either standards [22-24] or guidelines [25]. Standards allow for a rigorous design of engineered solutions. Guidelines are less detailed and strict, but address the general issues that can improve the structural behaviour of the buildings. The [22] and [23] standards do not only address the seismic design, but also the problems related to rain and wind-driven rain, water being, probably, the main cause of structural failure for earth buildings [26].

In other situations, research was conducted with the aim of evaluating the earthquake strength of existing vernacular buildings. These buildings do not comply with modern design principles and fall beyond the scope of in force technical regulations. Simplified evaluation procedures may highly underestimate the structural strength of the buildings. As an example, adobe buildings are generally assimilated to unreinforced masonry buildings and evaluated as such. Yet, a recent study [27] involving full scale laboratory testing on adobe walls had interesting results. Even if the typical failure modes are similar to those of masonry walls, the tested specimens exhibited a much larger ductility than expected, which leads to the conclusion that adobe buildings should not be simply designed or evaluated based on the methods developed for masonry.

A good practice example is shown in [28], where an extensive study was conducted for an area where earth buildings were traditionally built. The study led to the development of improved configurations for rammed earth buildings with modern living conditions. Over 400000 m^2 of buildings were constructed based on the research program.

Another example is shown in [29]. A prototype building with gypsum-adobe walls and wooden roofs was developed for usage in a highly seismic area (intensity 8). After detailed design and testing (including shaking table tests) over 1000 houses were built based on the prototype.

2.2. Recent developments in Romania

Recent projects and studies show that there is a rising interest in Romania related to earthen housing as a solution for green architecture and for the preservation of historical heritage.

2.2.1. Individual initiatives. There are rare examples of newly built engineered buildings using earth in Romania. One of these, a rammed earth structure was built in Lelese, Hunedoara county. Prior to construction, in 2014, extensive laboratory testing had to be undertaken in order to establish local soil composition, its mechanical properties and solutions for improving the reliability of the material [30].

2.2.2. Architectural monographies. Recently, architectural design guides were issued for some regions [31]. Over 140 guides are planned for release in order to cover all the Romanian historical regions. These guides recommend specific materials and construction techniques to be used in order to preserve the traditional housing styles, including earthen buildings. As only architects contribute to these guides, the conceptual structural solutions are not backed by engineering design and can not be directly put into use.

2.2.3. Engineering assessment. Structural engineering assessment of vernacular buildings is required by building owners or local authorities, in particular after extreme events (e.g. floods or landslides). The authors performed multiple evaluations of this type. As the buildings do not comply with current building codes, there is no clear assessment method. Evaluations based on engineering judgement can lead to different results, depending on the person conducting the analysis.

2.2.4. Engineering laboratory testing. Research based on laboratory testing is aimed at correctly assessing the structural vulnerabilities of non-engineered buildings and developing improved solutions. As an example, in [32], traditional buildings with timber frames and mud masonry infills are evaluated by in-plane static cyclic tests. Results are consistent with observations after major earthquakes, that traditional houses with infilled timber frames structure are often resilient enough so they protect the lives of their inhabitants.

2.2.5. Engineering technical methods proposals. Some engineering research goes beyond the structural problems of houses built with locally available natural materials and also addresses other problems such as thermal insulation and waterproofing. In [33], a modular construction method is proposed, based on wooden boxes with two compartments, filled with rammed earth and straw. According to the authors, this modular wall can provide structural strength and thermal insulation as required by current regulations, while offering an ecological and economical alternative to building with industrial prefabricated modules.

3. A vision towards the future

Existing earth buildings in Romania represent a part of our architectural heritage. Many of them might not comply with 21st century criteria regarding structural safety or quality of living. Yet, the solution of imposing only the use of engineered materials for their refurbishment will ultimately lead to their disappearance, along with centuries-old traditional craftsmanship. In light of the current trends towards sustainable buildings using local natural materials, there is a certain need of change in vision regarding earthen buildings.

Individual initiatives aimed at building a modern eco-friendly earthen house that complies with structural safety criteria, while also ensuring up-to-date conditions of living, must deal with high initial costs. These costs are related mainly to structural design that goes beyond the design codes and to material testing prior to obtaining a technical agreement for the product. People that lack the funds for preliminary studies are left with two solutions. The first one is to build a code-compliant masonry or concrete building (more costly and less ecological), which will further deteriorate the local existing architectural stock. The second one is building an earth building which, by lack of standards and detailed studies, might ultimately be improper or even unsafe.

Only initiatives by public authorities can overcome this problem. A feasible solution is to conduct integrated studies (architectural and structural) at regional level, funded either by local on national entities. These studies can analyse the technical local conditions (earthquake hazard, climate, soil types, etc.) in light of the cultural tradition and existing buildings. Based on these integrated studies, solutions can be developed, either as guidelines for the design of new buildings or the refurbishment of existing ones, or even as a catalogue of buildings that are adapted to the local conditions while ensuring modern living conditions.

References

[1] Stănculescu F, Gheorghiu A, Stahl P and Petrescu P 1958 *Arhitectura populară românească. Regiunea București* (Bucharest: Editura Tehnică)

[2] Bălan S *et al* 1985 *Lexicon de construcții și arhitectură* vol 1 (Bucharest: Editura Tehnică)

[3] Jänecke W 1918 *Das rumänische Bauern und Bojarenhaus* (Bucharest: König Carol-Varlag)

[4] Stahl H P 1972 L'Habitation enterrée dans la région orientale du Danube (XIXe et XXe siècles) *L'Homme* **12** n°4. pp. 37-61

[5] Enache S and Pleșa T 1982 *Zona etnografică Dolj* (Bucharest: Editura Sport Turism)

[6] Petrescu P, Florescu F B *et al* 1969 *Arta populară românească* (Bucharest: Editura Academiei Republicii Socialiste România)

[7] Mecu L 2013-2014 Considerații privind structurile de locuire și anexele lor în culturile Glina și Tei *Buridava. Studii și materiale* **XI** 359

[8] Bugă D and Oancea D 1986 Așezările omenești și tipurile de gospodării din Muntenia și Oltenia la jumătatea secolului XIX, *Litua. Studii și cercetări* **3** 542

[9] Ionescu G 1971 *Arhitectura populară în România* (Bucharest: Editura Tehnică)

[10] Oprescu G 1926 *Țările Române văzute de artiști francezi (sec. XVIII și XIX)* (Bucharest: Cultura Națională)

[11] Academia Română. Institutul de Etnografie și Folclor "Constantin Brăiloiu" 2003 *Atlas etnografic român. Volumul I. Habitatul* (Bucharest: Editura Academiei Române and Regia Autonomă Monitorul Oficial)

[12] Academia Română. Institutul de Etnografie și Folclor "Constantin Brăiloiu" 2010 *Habitatul. Volumul II. Banat, Crișana, Maramureș* (Bucharest: Editura Etnologică)

[13] Vuia R 1937 *Le village roumaine de Transylvanie et du Banat* (Bucharest) Academia Română. Institutul de Etnografie și Folclor "Constantin Brăiloiu" 2011 *Habitatul. Volumul III. Transilvania* (Bucharest: Editura Etnologică)

[14] Butura V 1978 *Etnografia poporului român. Cultura materială* (Cluj-Napoca: Dacia)

[15] Iordache G 1996 *Ocupații tradiționale pe teritoriul României* vol 4 (Craiova: Scrisul Românesc)

[16] Comșa E 1987 *Neoliticul pe teritoriul României. Considerații* (Bucharest: Editura Academiei)

[17] Florescu F B, Stahl P, Petrescu P 1967 *Arta populară din zonele Argeș și Muscel* (Bucharest: Editura Academiei Republicii Socialiste România)

[18] Bălan S *et al* 1988 *Lexicon de construcții și arhitectură* vol 3 (Bucharest: Editura Tehnică)

[19] Stoica G and Ilie R 1990 *Zona etnografică Câmpia Boianului* (Bucharest: Editura Sport Turism)

[20] Fukushima A 2012 Examination of the effect of building construction policies on the conservation of vernacular building techniques: Policies adopted after World War II in Japan in: *ICOMOS 17th General Assembly* 357-64

[21] Standards New Zealand 1998 NZS 4297, *Engineering Design of Earth Buildings*, Wellington

[22] Standards New Zealand 1998 NZS 4298, *Materials and workmanship for earth buildings*, Wellington

[23] ASTM International 2010 E2392 M-10, *Standard Guide for Design of Earthen Wall Building Systems*, Pennsylvania, United States

[24] Bureau of Indian Standards 1993 IS 13827, *Improving earthquake resistance of earthen buildings – Guidelines*

[25] Minke G 2012 *Building with earth: design and technology of a sustainable architecture* (Basel: Birkhäuser)

[26] Wu F, Wang H T, Li G, Jia J Q and Li H N 2017 Seismic performance of traditional adobe masonry walls subjected to in-plane cyclic loading *Materials and Structures* **50(1)** 69

[27] Liu J, Hu R, Wang R, and Yang L 2010 Regeneration of vernacular architecture: new rammed earth houses on the upper reaches of the Yangtze River. *Frontiers of Energy and Power Engineering in China* **4(1)** 93-9

[28] Zhou T G, Hu X, and Yu C X 2008 Shaking table model test and engineering practice of a new

gypsum-adobe walls dwelling in Xinjiang autonomous region, China. *Acta Seismologica Sinica* **21(3)** 319-24

[29] Silvestru M, Pauta R, Gruin A and Tirtea A 2014 Rammed Earth Houses As A Sustainable Construction System in *23rd IAPS Conference*, Timisoara, Romania

[30] Ordinului Arhitecților din România 2016 *Ghiduri de arhitectură pentru încadrarea în specificul local din mediul rural* www.oar.archi/despre-oar/publicatii-si-documente-oar

[31] Dutu A, Niste M, Spatarelu I, Dima D I and Kishiki S. 2018 Seismic evaluation of Romanian traditional buildings with timber frame and mud masonry infills by in-plane static cyclic tests. *Engineering Structures*, doi.org/10.1016/j.engstruct.2018.02.062

[32] Ciurileanu G T and Horvath I 2012 Modular Building Using Rammed Earth *Acta Technica Napocensis: Civil Engineering & Architecture* **55(2)** 173-81.

IOP Conf. Series: Materials Science and Engineering **399** (2018) 012050 doi:10.1088/1757-899X/399/1/012050

Modern technical solution of water tightness of reinforced concrete decanters. Case study

I Tuns[1*] and F L Tamas[1]

[1] Transilvania University of Brasov, Faculty of Civil Engineering, Brasov, 500152, Romania

[*] E-mail: ioan.tuns@unitbv.ro

Abstract. For reinforced special constructions with liquid storage roll, the performance exigencies concerning water tightness is essential. If the storage liquids represent potential sources of environmental pollution, the condition of removing external leaks is fundamental. This paper presents the case of the decanters of liquid residual products, resulting from the technological flow of processing animal proteins, employed in a wastewater treatment plant. The exigencies of water tightness performance of the reinforced concrete decanters are ensured by means of a modern, eco-friendly solution, based on the mass waterproofing of fresh concrete. This solution ensures water tightness without the need for any other protection works or external waterproofing, and any corrections can be applied even in the presence of water.

1. Introduction

1.1. General presentation elements

The modernization and the greening of the technological line related to a unit of processing / neutralization / incineration of by-products (not for consuming) and of animal waste in Brasov area led to the inclusion in the investment plan of the "Waste Water Treatment Plant" as in figure 1.

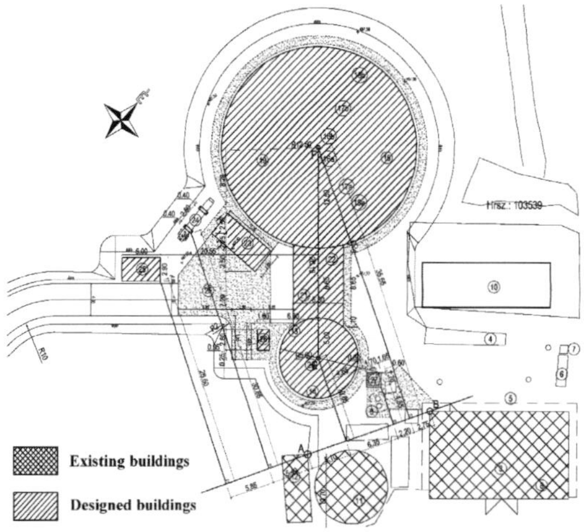

Figure 1. Wastewater treatment plants site plan.

IOP Conf. Series: Materials Science and Engineering **399** (2018) 012050 doi:10.1088/1757-899X/399/1/012050

The works related to the proposed investment of SC Protan SA Codlea refer to the sewage treatment line and consist, among other things, in the intercalation of two reinforced concrete collectors, one with the role of compensation, industrial accident, decantation, and the other with the role of biological neutralization of the compounds from the waste water.

The formwork plan and the corresponding section A-A of the biological decanter are presented in figure 2 and figure 3, respectively.

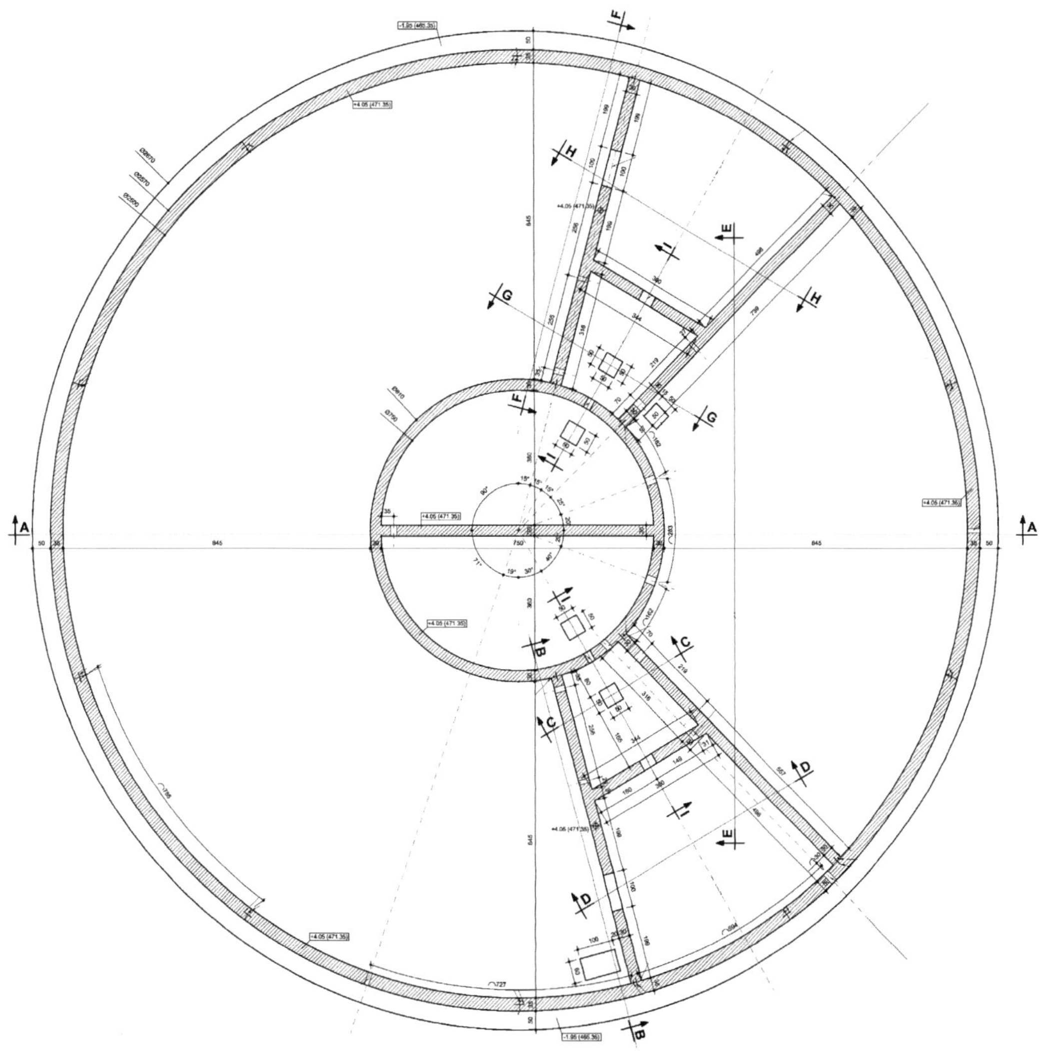

Figure 2. Formwork plan of the biological decanter.

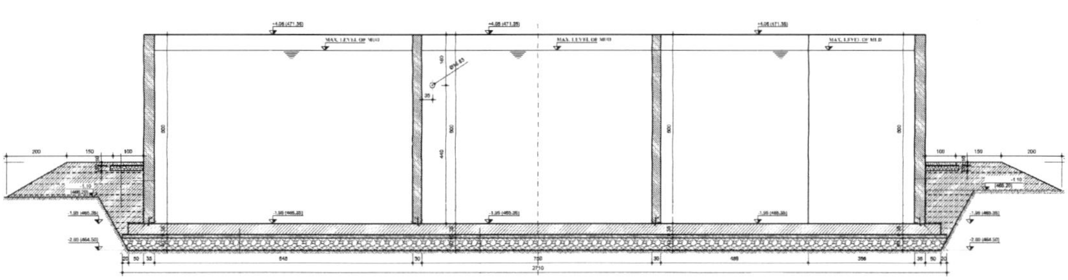

Figure 3. Section A-A corresponding to the biological decanter.

The geometrical and constructive components of the biological decanter are:

- circular shape in plan, having a diameter of 25 m, divided internally into five distinct functions;
- reinforced concrete raft foundation, thickness 35 cm;
- reinforced concrete exterior walls fixed in the raft foundation, 35 cm thick and 5.5 m high;
- reinforced concrete interior partition walls;
- the wall-raft foundation connection is insulated by a sealing profile and an expandable strip.

The decanter finishing is provided by apparent concrete with a high degree of water impermeability.

The main geometrical constructive characteristics of the sequential decanter of compensation, industrial accident, sedimentation are:

- circular shape in plan, having the outer diameter of 10.7 m, divided internally into three distinct functions;
- reinforced concrete raft foundation, thickness 35 cm;
- reinforced concrete exterior walls 35 cm thick and 5.5 m high;
- reinforced concrete interior partition walls;
- the wall-raft foundation connection is insulated by a sealing profile and an expandable strip.

The decanter finishing is also provided by the apparent concrete with a high degree of water impermeability.

The formwork plan and the corresponding section A-A of the compensation, industrial accident, decantation collector are according to figure 4, respectively figure 5.

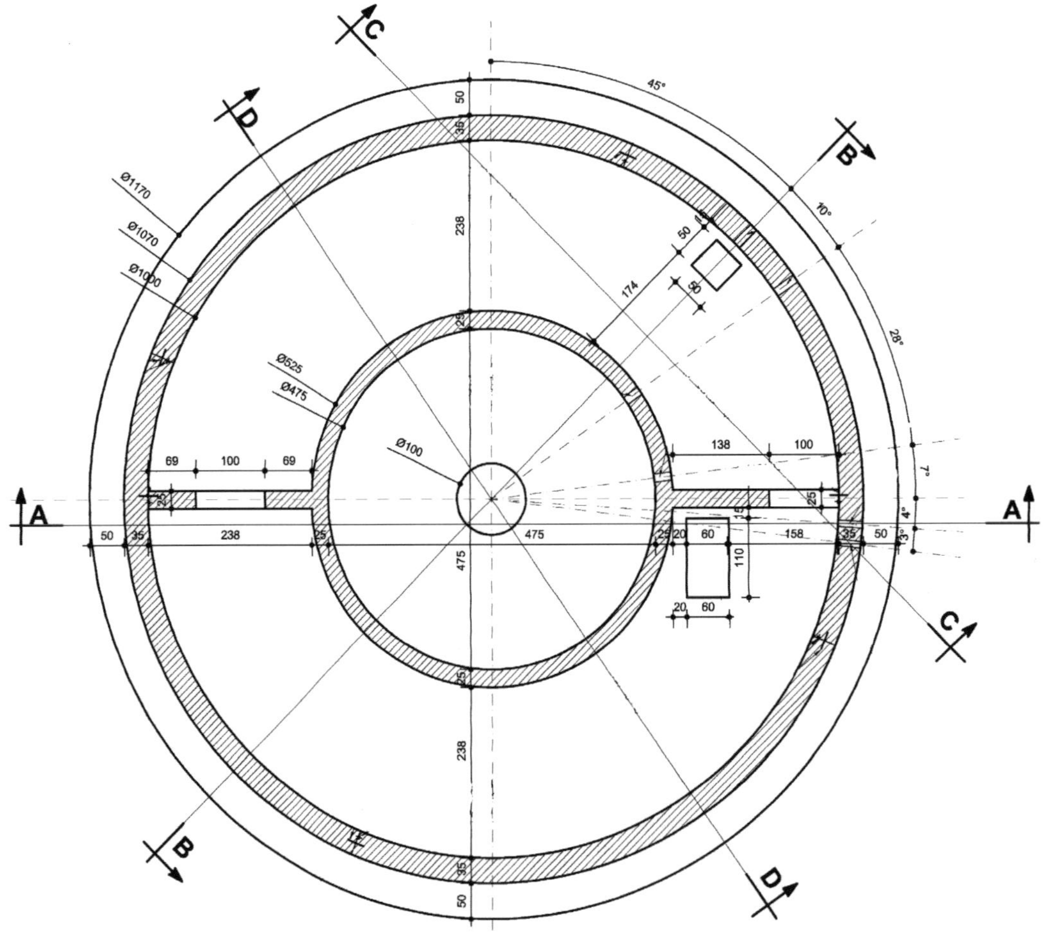

Figure 4. Formwork plan of the compensation decanter.

IOP Conf. Series: Materials Science and Engineering **399** (2018) 012050 doi:10.1088/1757-899X/399/1/012050

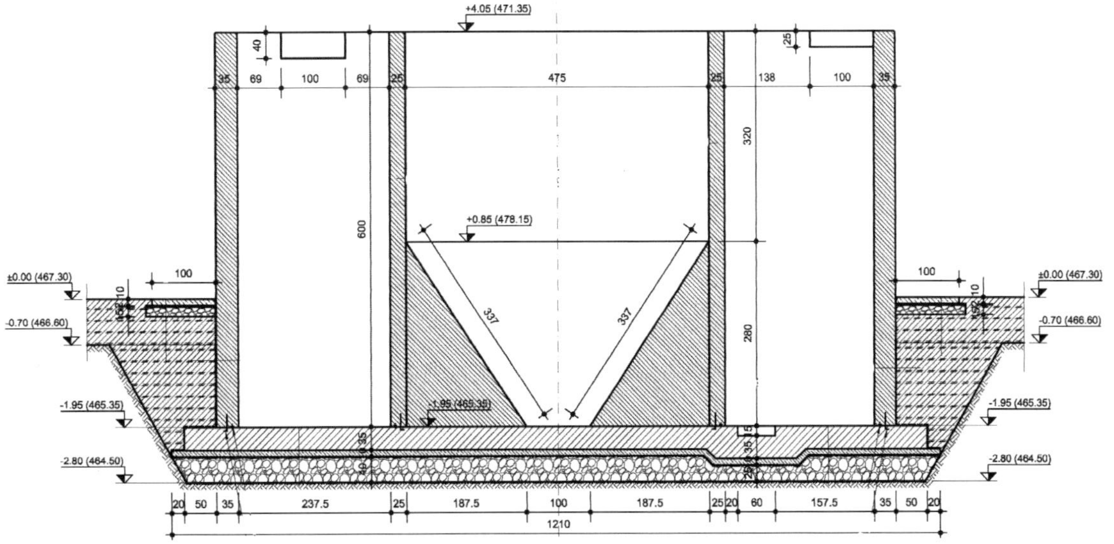

Figure 5. Section A-A corresponding to the compensation decanter.

1.2. Water tightness elements

According to the designed function, the two decanters need to ensure the storage and treatment of the wastewaters resulting from the technological process, in order to bring them to the quality parameters corresponding to the specific environmental requirements.

In this respect, all compartments for the two decanters must meet the water permeability performance requirements [1].

The design project contains specifications on the characteristics of the concrete used for the raft foundation and the walls, namely C35/45 XC4XD2XF1XA2 CEM IIA-S 42.5R Cl 0.2% D16 S4, degree of impermeability P12, waterproofing additive, G150 freeze-thaw degree [2].

Without specifying the type of waterproofing additive, as this falls into the contractor's responsibility, thorough investigations have been made in this regard, aiming at identifying a product compatible with the type of cement used. This product, recommended by the positive experiences of similar works, must ensure the possibility of repairing defects after pouring, even in wet conditions.

Having a practical experience of previously using Penetron products for waterproofing the basement of a building, under the hydrostatic pressure water conditions [3], this technology was chosen, supported by the following technical and economic factors:

- Penetron powdered products are part of the clinker mineralized binder, which is hydrated in the presence of water, with the abundant multiplication of the crystalline component, thus being compatible with the Portland cement used in the preparation of concrete;
- the hardening process occurs in the presence of moisture;
- crystals resulting from hydration reactions have high penetration capacity in the mass of concrete, thus ensuring pore filling;
- it is added into the concrete mixer with about 15 minutes before pouring;
- it does not require additional works to those normally provided for the pouring operation and follow-up protection of the concrete (wetting and sun protection);
- it provides relatively easy conditions for repair interventions of the defects, in the presence of moisture and water under hydrostatic pressure;
- it does not require distinctly working time for waterproofing decanters;
- it does not require additional or special protection works on the structural elements;
- it has no negative impact on the environment.

For the case study, the following products were selected:

- Penetron Admix – added to the fresh concrete;
- Penetron, Penecrete mortar, Peneplug and Penetron plus, to repair defects after concrete pouring.

2. Technological description of the applied method

2.1. Mass waterproofing of fresh concrete

Based on the concrete characteristics specified in the design project, which provides, in addition to the mechanical characteristics required by the construction importance and destination, a high level of water tightness under a severe exploitation regime in terms of aggressive factors for concrete, and taking into account the considerations outlined in chapter 1, paragraph1.2, the chosen method to waterproof fresh concrete was that by introducing Penetron Admix into its mass.

The technical instructions, the documentary material provided by the representative of the company selling Penetron products, displayed on the technical information networks were thus taken into account.

At the same time, in addition to the waterproofing requirement, we have also aimed at developing the process of evaluation / explanation the experimental data obtained following the application of the mentioned procedure, with the purpose of formulating valid conclusions for any further design and execution of similar works.

The Penetron Admix additive was inserted into the mass of fresh concrete directly into the concrete mixer, as shown in figure 6, followed by at least 15 minutes of mixing before pouring, with the following percentages of the amount of cement:

- 0.8% of the amount of cement for raft and two wall plots;
- 0.85% of the amount of cement for two wall plots;
- 0.90% of the amount of cement for two wall plots;
- 0.95% of the amount of cement for two wall plots;
- 1.0% of the amount of cement for two wall plots.

Figure 6. Inserting the Penetron Admix additive into the mass of fresh concrete with at least 15 minutes before pouring.

Pouring concrete into the decanters' walls, on predetermined design plots, was made with the purpose of limiting the opening of cracks from shrinkage and the amount of additive introduced into the fresh concrete complied with the recommendations contained in the technical instructions for use of the product. For each concrete pouring batch samples were taken at the frequency indicated in the norms provisions [4]. Pouring the concrete and its follow-up protection was done in accordance with the technical norms specific to these categories of works.

2.2. Waterproofing testing and correcting the noticed defects

28 days after pouring the last wall plot and sealing the technological pipe holes, the decanters were filled with water in order to perform the tightness test [5]. The water tightness test revealed the existence of some wall areas with a high level of moisture, generally located at the spacers used to ensure the thickness of the concrete cover layer of the reinforcements, at the joint with raft foundation, or in the pouring defects as presented in figures 7 and 8.

Figure 7. Areas of excessive humidity or leakage at the wall-raft foundation joint.

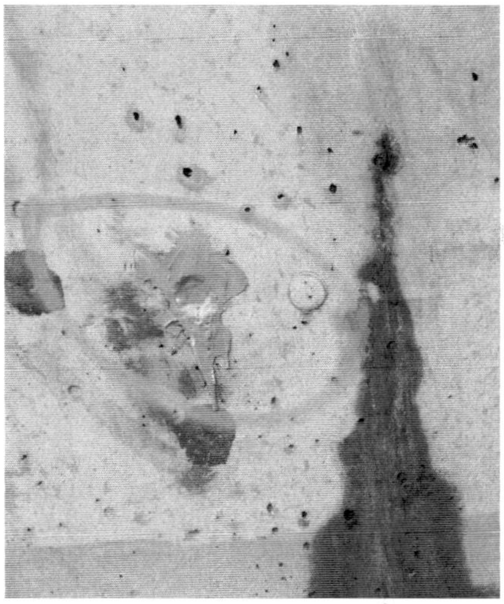

Figure 8. Areas of excessive humidity or leakage on wall.

Figure 9. Removing wall defects with Penecrete waterproofing additive.

Figure 10. Removing wall defects with Peneplug waterproofing additive.

The moisture spots on the wall, which were all due to the existence of pores interconnected from the concrete mass or to minor pouring defects, have disappeared within 3-14 days as a result of obstructing the formed spaces by depositing the crystals generated by the Penetron Admix additive in the presence of water. For the rest of the defects, they were opened, cleaned, deepened and repaired as appropriate:

- by brushing the emulsion of Penetron in water, ratio 5-2 and closing by pressing the mortar obtained from the mixture of Penetron powder with water, ratio 4-2, in the case of water without hydrostatic pressure, as indicated in figure 9;
- by vigorous pressing in the prepared hole of the mixture obtained from Peneplug powder and water, in the case of water with hydrostatic pressure, according to figure 10.

Following the remediation procedures listed, the water tightness test was successfully performed for both decanters, currently in operation, as can be observed in figure 11.

Figure 11. The wastewater treatment plant in-service.

2.3. Processing and interpretation of experimental tests

The conducted experimental tests aimed at highlighting the following aspects:

- optimizing the consumption of Penetron Admix waterproofing additive inserted in the mass of fresh concrete, subject to the dosage indicated in the technical instructions [6];
- the influence of waterproofing additive dosage on the concrete class;
- the closing time of the pore from hardened concrete due to the crystals generated by the Pennetron Admix additive, in the conditions of total immersion on one side of the wall.

Optimizing the consumption of Penetron Admix waterproofing additive inserted in the mass of fresh concrete. To optimize the dosage of a waterproofing additive, a multicriteria analysis was conducted based on:

- the percentage of wetted concrete surfaces after the water-tightness test, due to internal causes (interconnected capillary pores, entrapped air pores etc.), with the exception of execution defects;
- the percentage of self-repaired surfaces during water-tightness tests;
- the ability of crystals generated by the Penetron Admix additive to diffuse and penetrate in depth, highlighted by X-ray diffraction on concrete samples taken on site;
- the percentage of wetted surfaces remaining after self-repairing;
- the price/m^2 of the wall, including the costs of repairing water infiltrations.

Centralized data on the considered areas, the percentage of wet surfaces, repairs and costs are presented in table 1.

Table 1. Data of wet surfaces percentage / pouring plots / costs.

	Penetron Admix dosage[a]	Plots number	Plot surface [m^2]	Wetted surface [m^2]	Percentage of wetted surface [m^2]	Self-repaired wetted surface [m^2]	Repaired surface [m^2]	Price m^2/wall €
1	0.80% · C	2	130.3	2.30	1.76	1.70	0.06	29.10
2	0.85% · C	2	130.3	2.20	1.68	1.65	0.03	29.60
3	0.90% · C	2	130.3	2.15	1.65	1.63	0.02	30.20
4	0.95% · C	2	130.3	2.12	1.62	1.60	0.02	30.70
5	1.00% · C	2	130.3	2.12	1.62	1.60	0.02	31.20

[a] Percentage related to cement quantity, *C*.

Regarding repairing costs, the data reveal insignificant differences in the doses of analyzed waterproofing additive.

The multicriteria analysis based on the above-mentioned parameters recommends variant no. 2, which leads to a cost difference of 1.6 eur/sqm, compared to variant no. 5, and provides similar quality requirements, as shown in figure 12.

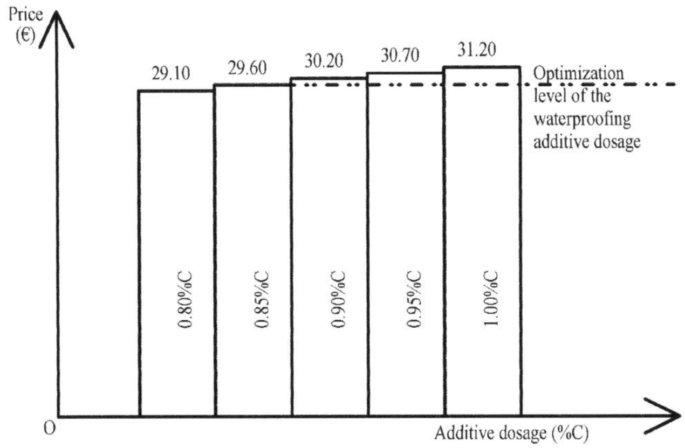

Figure 12. The level of optimization for the waterproofing additive dosage.

The influence of Penetron Admix waterproofing additive on the concrete class. The results presented in table 2 show that the Penetron Admix waterproofing additive introduced into the mass of fresh concrete does not influence the concrete class, determined on the basis of the compressive strength obtained on 150 mm side cubic samples, kept under standardized conditions and tested at 28 days.

Table 2. Results on concrete samples taken on site.

	Sample data	Concrete class	Object	Compressive strength, $f_{ck,cub}$ [N/mm^2]	Outside temperature [°C]
1	05.07.2017	C35/45	Compensation decanter. Raft	45.11	26
2	05.07.2017	C35/45	Compensation decanter. Raft	49.70	28
3	11.07.2017	C35/45	Biological decanter. Raft	45.24	29
4	11.07.2017	C35/45	Biological decanter. Raft	46.30	27
5	03.08.2017	C35/45	Compensation decanter. Walls	45.42	29
6	10.08.2017	C35/45	Compensation decanter. Walls	45.66	29
7	25.08.2017	C35/45	Biological decanter. Walls	50.58	26
8	14.09.2017	C35/45	Biological decanter. Walls	45.21	22
9	20.09.2017	C35/45	Biological decanter. Walls	45.24	22
10	26.09.2017	C35/45	Biological decanter. Walls	45.21	19
11	28.09.2017	C35/45	Biological decanter. Walls	45.33	16

The closing time of the pore from hardened concrete due to the crystals generated by the Pennetron Admix additive, in the conditions of total immersion on one side of the wall. The on-site investigation, conducted visually and by polls, has highlighted that areas where moisture spots were due to materials structure, except for execution mistakes, have self-repaired (disappeared) over a period of 3-14 days by diffusing and depositing the crystals generated by the Penetron Admix additive in the interior pores. This finding applies to all areas of the walls, regardless the dosage of the additive applied.

3. Conclusions

For the case study, the results of the interpretation and processing of the experimental data lead to the formulation of the following conclusions with validity for other municipal and public constructions, made of reinforced concrete, and meant for the storage of liquids:

- the mass concrete waterproofing solution with the Penetron Admix additive ensures that watertightness requirements are met without requiring additional protection or finishing work;
- Penetron Admix waterproof additive is compatible with Portland Cement;
- the optimization of the waterproofing additive dosage ranging whithin the manufacturer's recommended limits ensures a lower cost price with 1.6 eur per square foot compared to the upper limit of the range, under similar quality conditions;
- the insertion of Penetron Admix waterproofing additive in the fresh concrete mass does not influence the compressive strength;
- the closing time of all internal pores from the concrete mass after the total immersion of one face of the wall is two weeks.

References

[1] Design project *Waste water treatment plant – SC Protan SA Codlea*, 2017
[2] CP 012/1-2007 *Cod de practică pentru producerea betonului (Code of practice for the manufacturing of concrete)*, 2007
[3] Tuns I, Tamas F-L and Mantulescu M 2017 Waterproofing solution of an existing basement against water under hydrostatic pressure. Case study *Bulletin of the Transilvania University of Braşov* **10** (59) No. 1 - Series I: Engineering Sciences
[4] NE 012/2-2010 *Normativ pentru producerea şi executarea lucrărilor din beton, beton armat şi beton precomprimat – Partea 2: Executarea lucrărilor din beton (Norm for production and executio works of concrete, reinforced concrete and restressed concrete – Part 2> Execution of concrete works*, 2010
[5] STAS 4165-1988 *Alimentări cu apă. Rezervoare de beton armat şi beton precomprimat. Prescripţii generale (Water supply. Reinforced concrete and prestressed concrete decantors. General Prescriptions)*, 1988
[6] Penetron International documentation

Static linear analysis for trusses structure for supporting pipes

M Urdea[1*]

[1] Department of Manufacturing Engineering, Transilvania University of Braşov, Romania.

[*] E-mail: urdeam@unitbv.ro

Abstract: This paper is a study about truss deformations. This truss is supporting a pipe, it can be a bridge, or another structure and can be necessary for protecting rivers. The truss is a welded structure modelled with Weldment Structural Members, which is often found in industrial constructions and it can be constructed in different shapes. In order to sustain a pipeline over a damaged area, some trusses structures were modelled. The pipeline is supported by some support brackets mounted on the truss nodes. Finite Element Analysis is a process which can predict deflection and stress on a structure, here on a truss. A finite element model is a complete idealization of the entire structural part, including the node locations, the elements, material properties, loads and boundary conditions. The purpose of this paper is to find the best form for trusses using SolidWorks and to avoid oversizing these structures.

1. Introduction

Trusses are very common structural members; they can be used for sustaining a pipeline over a damaged area. The dictionary presents the truss as a framework, typically consisting of rafters, posts, and struts, supporting a roof, bridge, pipes or other structures. In engineering, a truss is a structure that consists of two-force members only, where the members are organized so that the assemblage as a whole behaves as a single object. A space frame truss is a three-dimensional framework of members pinned at their ends [1].

This paper presents several truss shapes. The aim of this study is to develop a truss modelling technique, to assign possible constraints and forces and to perform a linear static analysis of trusses.

The only inner force that develops in the member is the axial force. This is constant along its length and generates axial stress, uniform throughout the cross section. The 1D structural members are known as trusses. In finite element analysis truss members are modelled as truss elements. The truss is a special beam that can resist to only axial deformation [2, 3].

2. The flat truss study

The Warren, Howe and Pratt trusses have been made of wood, iron or steel since the19th century. The first truss modelled in SolidWorks is a flat truss design after Howe truss model with beam welded elements as in figure 1. For the Howe truss the diagonal members are in compression, the vertical members are in tension, except for the centre which is a null force [4].

After the truss sketch all truss tubes are generated in SolidWorks as Structural Members. For a truss the lines generated in sketch must exactly meet at common points named joints.

Due to the Structural Members generation in Features Manager Design Tree, a Cut list folder appears in which SolidWorks groups by shape and sizes all the profiles of a welded structure.

For this study the following truss characteristics were chosen: length, L=8m; the truss height, h=0,8m; the inside diameter of the truss elements, D1=70; the outside diameter of truss pipe, D2=75. It is made of steel OL37, equivalent with ANSI 1035, from the Library Material Catalogue, with following features: Young Modulus 2,049e+011N/m2 and Poisson Ratio 0.29. The admissible tension for OL37 (S 235 JR the new symbol, after SR EN 10025) is 90 - 120 N/mm^2 (MPa). The Mass Properties function informs us about the truss weight, namely 171048.44 grams.

In order to support some pipelines for crossing a river or over a rugged terrain it is necessary to use a structure like a truss. Figure 2 shows the scheme of the truss for supporting a pipe by some support brackets mounted on the truss joints. A static analysis for the truss is required to optimize its construction.

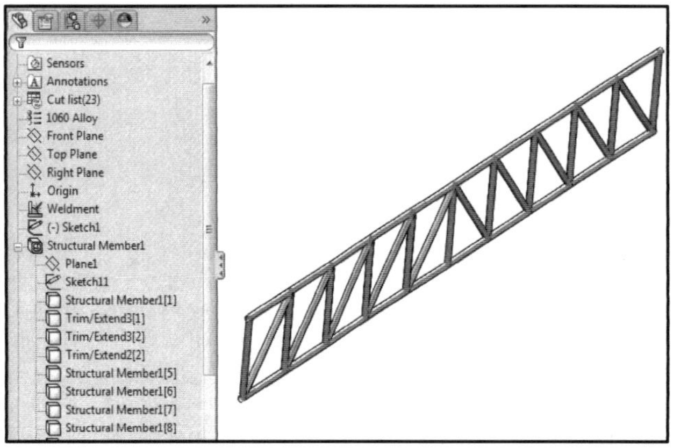

Figure 1. Flat trusses.

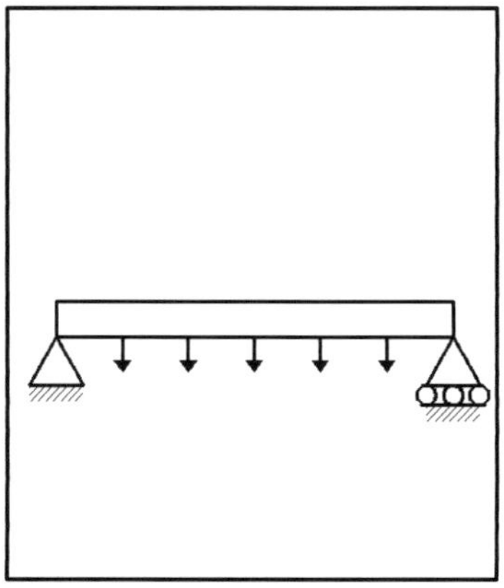

Figure 2. Truss scheme.

Figure 3 presents the Structural Member window, the form of the tubes was selected. In these studies: Standard - iso, Type - pipe and the pipe dimensions were chosen and this type will be modified. Four Structural Member groups were generated due to the truss shape. The way of generating the model may influence the development of the finite element analysis. In this first step the static simulation recognizes the weldment geometry and anticipates that they are beam elements. A Solid Body was created for each structural member and it was placed then in the Cut-List folder (figure 4) [5].

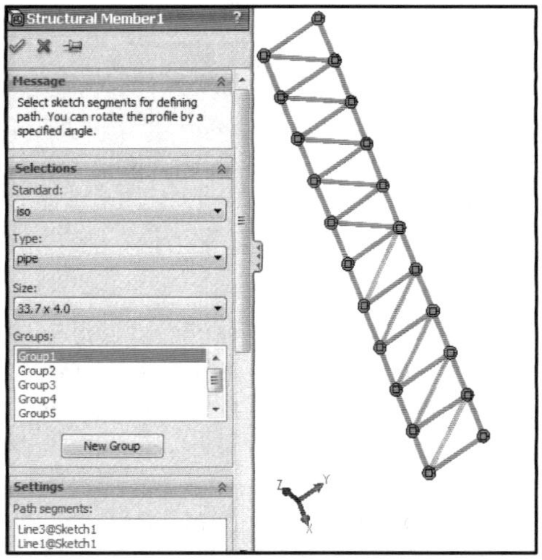

Figure 3. Structural member for flat trusses.

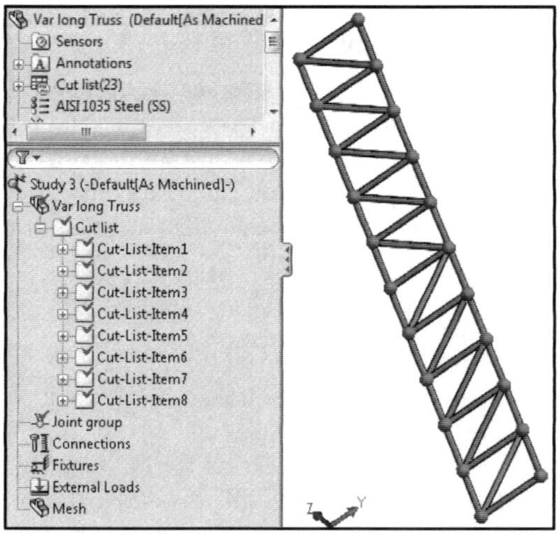

Figure 4. Cut-List folder.

In this Cut-List folder each element must be declared as a truss element, using the Apply /Edit Beam window. This activity is very important for the next step because, by default, this structure was declared with beam elements.

3. Static analysis for the flat truss

The static analysis assumes a complete idealization of the entire structural part (model), including the node truss locations, the elements, material properties, loads and boundary conditions. During this study, the following FEA steps were considered: Pre-Processing, Processing and Post-Processing. The paper includes diagrams and conclusions to illustrate the importance of this study. For this truss the external forces and the reactions are considered to act only at the nodes.

Edit Joints manages the joints with the Calculate button and all the joints are created automatically. The joints coincide with the pierce point of the weldment profile. The joints are presented in a Joint group folder (figure 4). It is noticeable that the joints transmit only forces. Moments are not present at the joints of a truss [1].

Due to the static analysis, the calculations of the forces on the every truss element can be checked by simulation, thus avoiding the complex calculation of the forces in each member of the truss and indicate whether the members are in tension or compression. It is known that the results obtained from the FEA analysis are less valid in nodal bonding areas (welds, riveting assemblies, screw assemblies) [2].

Due to this, the analysis model must have the same behaviour as the real model; the boundary conditions are defined for each of the extremes joints. To avoid any deformation of the structure due to temperature fluctuations, the truss was fixed using Fixed Geometry, namely at one end it was fouled constraints, that is Immovable (all three translations are zero on face, edge, or vertex). For the other end, using Use Reference Geometry, the translation of Oy and Oz axes was cancelled. The translation at this end can be achieved by using a translation guideline. Figure 5 and 6 present the fixtures at both ends. Every joint has six degrees of freedom per node; therefore distinguish between Fixed and Immovable restraint [6].

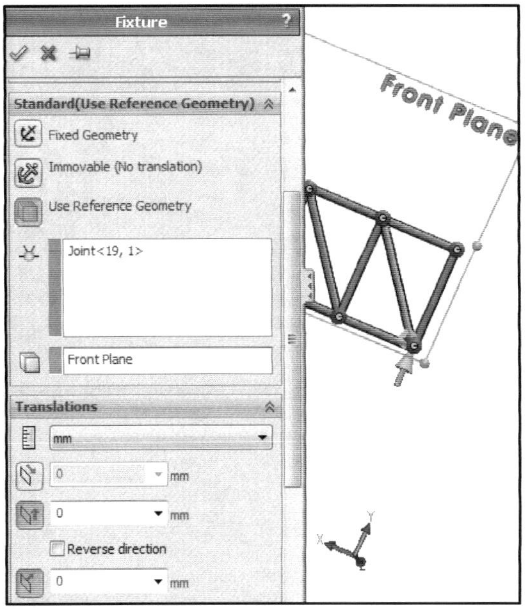

Figure 5. Fixture window - translation of Oy and Oz axes was cancelled.

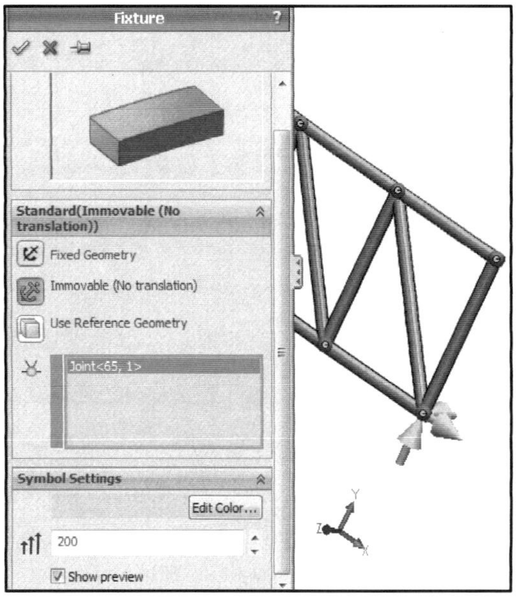

Figure 6. Fixture window - one end was Immovable.

On the other hand, all the joints are also constrained by using Use Reference Geometry. The constraints are in relation to the Front Plane and the displacement perpendicular to this plane is cancelled as in figure 7. These are restraints to prevent rigid body motion.

In this case the finite element analysis is based on a software system that assesses a material or an object by using the finite element method in order to find the applied stresses that have different effects on the design. A highly regarded advantage of the method for producers and engineers worldwide is that FEA can help determine a weakness in a design before the product is manufactured [7].

External Loads can be attached that is the Gravity and the Forces. The truss is loaded with its own forces. Its own weight is a consequence of the existence of inertial translational fields. The Forces, using External Loads, are applied to the five joints (figure 8), it is about concentrated forces at joints. The pipeline can be connected to these joints. In this study the forces are estimated to 11000 N for every 5 joints.

The next step is generating the Mesh, by launching Mesh function. It is important to note that a truss is a special beam element that can resist only axial deformation. As in this study the beam was considered as a truss there are no mesh control options for truss frames. A straight structural member identified as a truss is represented by one truss element [8].

The post-processing step and the computing part begin with Run action which starts the solver for the active study. The simulation calculates the space truss joint locations and their displacements.

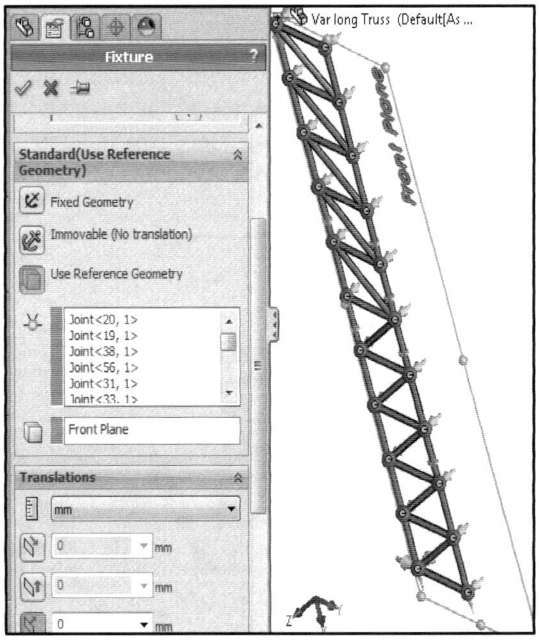

Figure 7. Reference Geometry for all the joints.

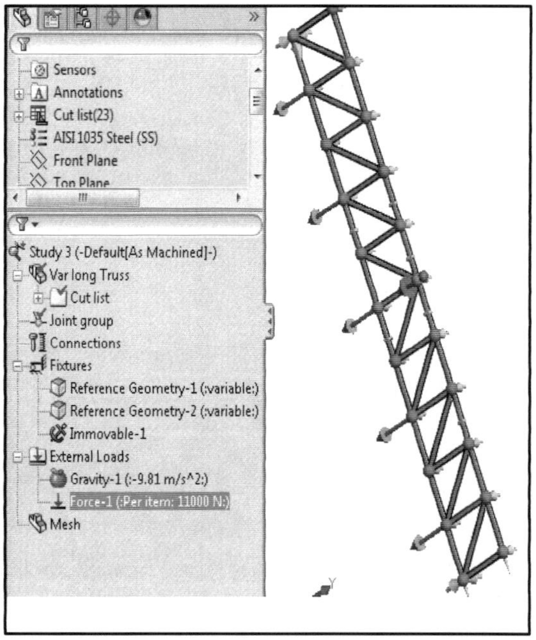

Figure 8. External loads.

The post-processing results are illustrated bellow, namely:

- upper bound axial and bending stresses (figure 9);
- displacement and deformation (figure 10).

The admissible tension of the material is 90-120 N /mm^2. The maximum stress is 104.535 N/mm^2 and the maximum displacement is 8.923mm. It is very important to define the results folder location for every FEA Study. After the post-processing the List Beam Forces and the Beam Diagrams also result.

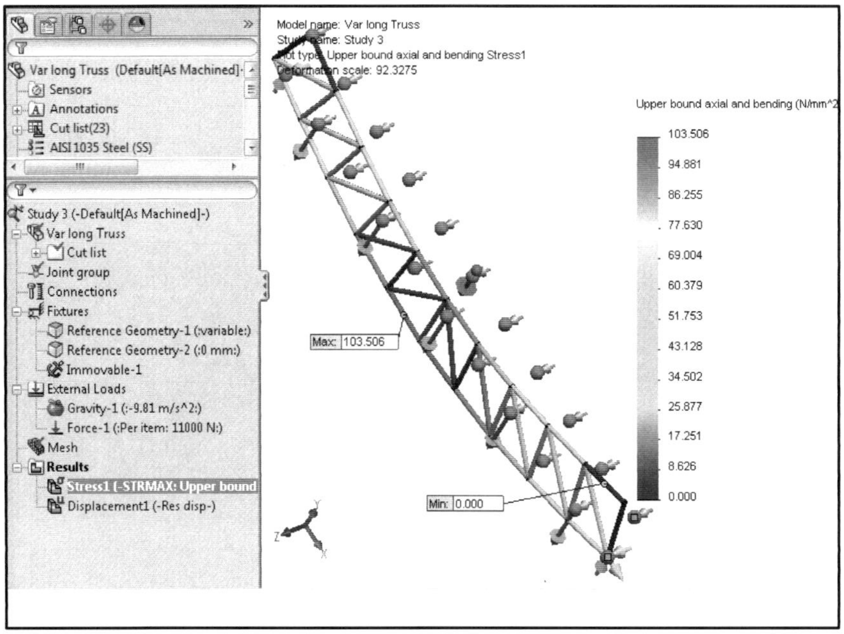

Figure 9. Upper bound axial and bending stresses for Howe truss.

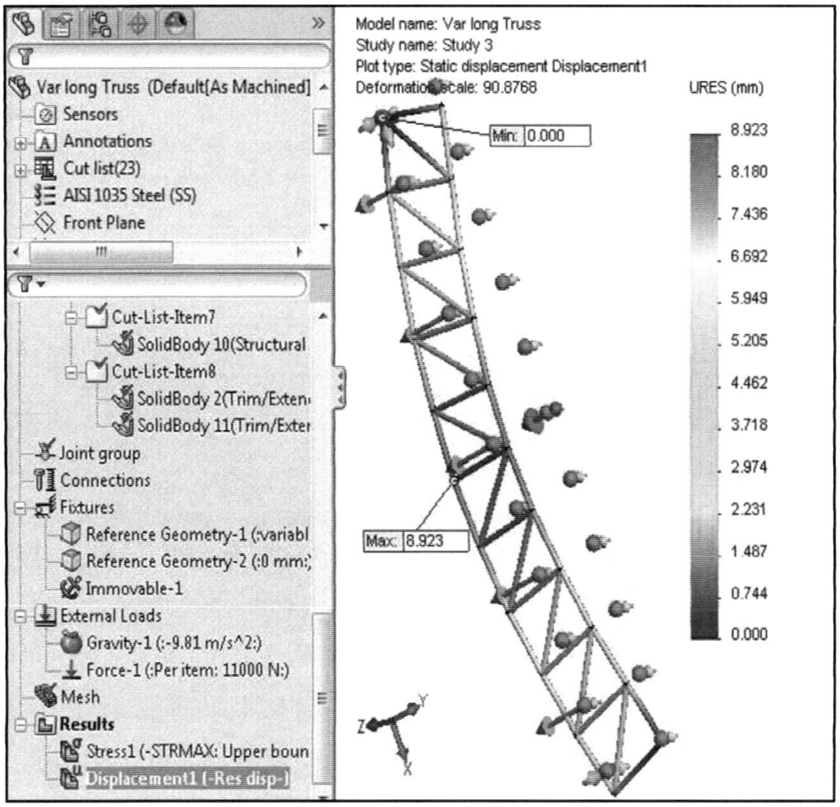

Figure 10. Displacement and deformation for Howe truss.

For the very popular Pratt truss the diagonal members are in tension, the vertical members are in compression, except the centre which is a null force. Figure 11 and figure 12 present the post-processing results for the Pratt truss. One can notice that the simulation results are very close for the Howe truss simulation results. The simulation conditions were similar.

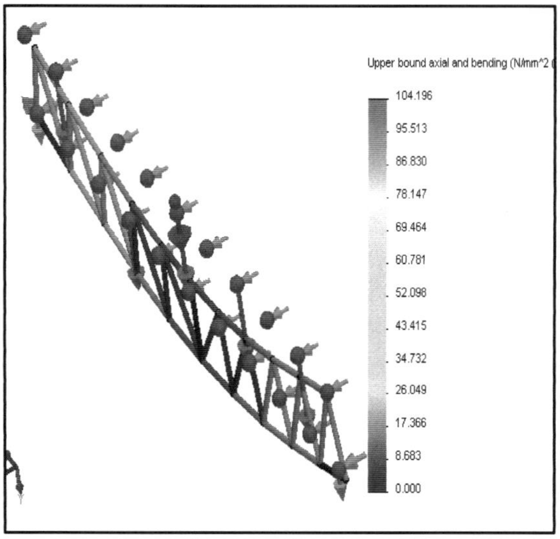

Figure 11. Upper bound axial and bending stresses for Pratt truss.

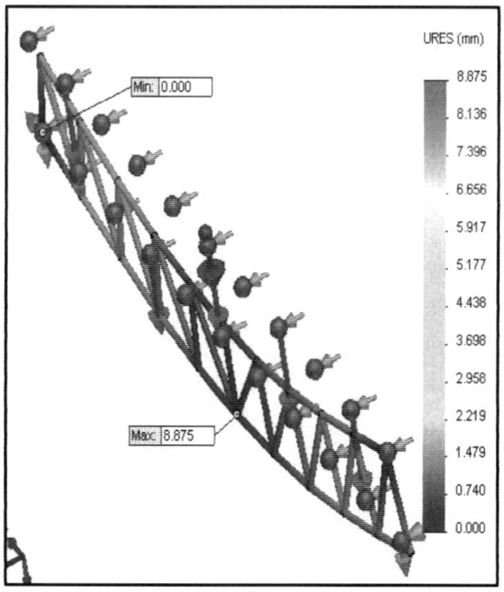

Figure 12. Displacement and deformation for Pratt truss.

4. Double flat truss study

The second truss modelled in SolidWorks is a double flat truss as in figure 13 and it develops from the Howe truss structure. In this case, the truss dimensions are: the truss length 8m; the truss height 0,8m, the truss width 0,6m; the diameters of the truss elements are the same. This study respects similar steps. Figure 14 presents the fixtures and the external loads. The pipeline can be connected to five joints as in figure 12. At the moment the forces are estimated at 11000 N for every 5 joints.

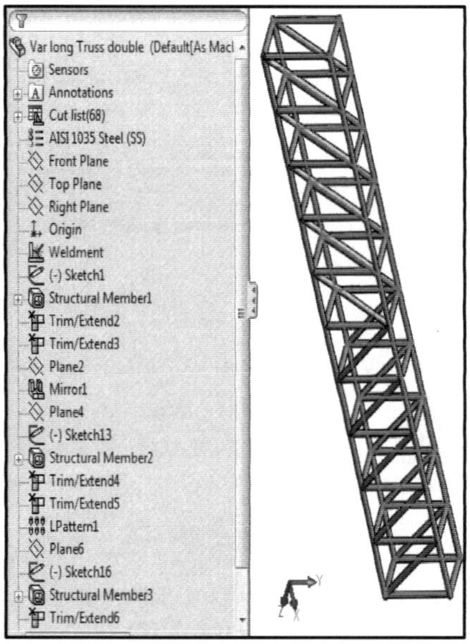

Figure 13. A double flat truss.

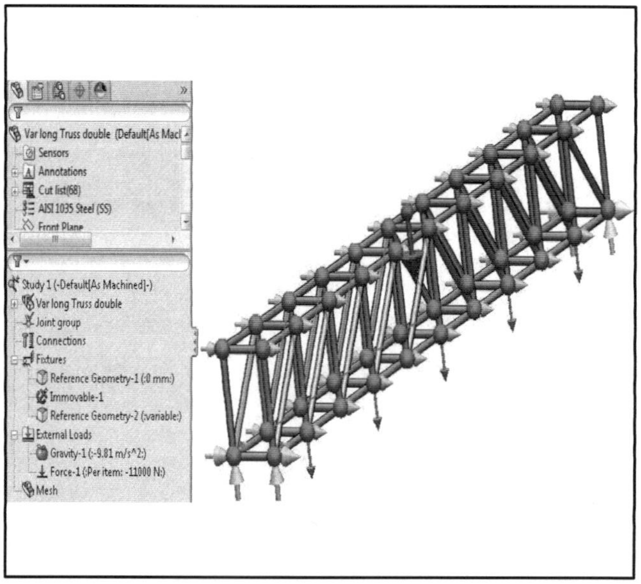

Figure 14. All fixtures and external loads.

The post-processing results are illustrated bellow, as follows:

IOP Conf. Series: Materials Science and Engineering **399** (2018) 012051 doi:10.1088/1757-899X/399/1/012051

- upper bound axial and bending stresses (figure 15);
- displacement and deformation (figure 16).

The admissible tension of the material is 90-120 N /mm^2. The maximum stress is 102.036 N/mm^2 and the maxim displacement is 8.726mm. There is a very small difference between the values of the two studies.

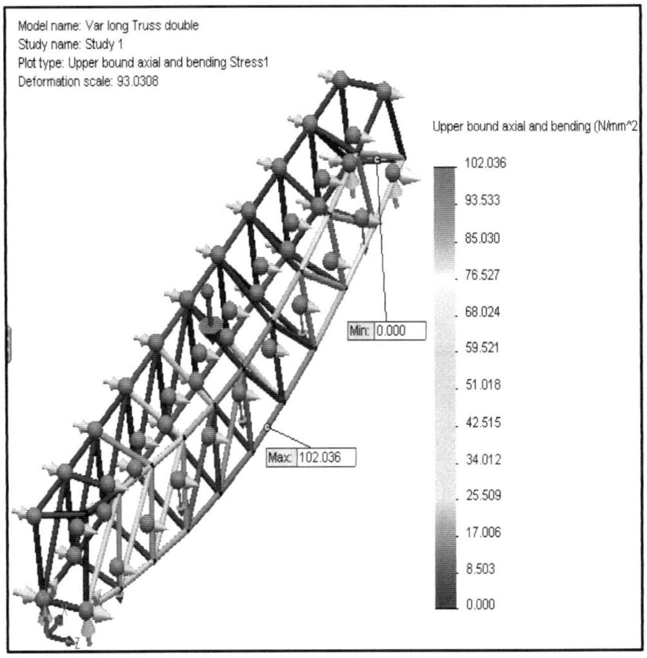

Figure 15. Upper bound axial and bending stresses for the double flat truss study.

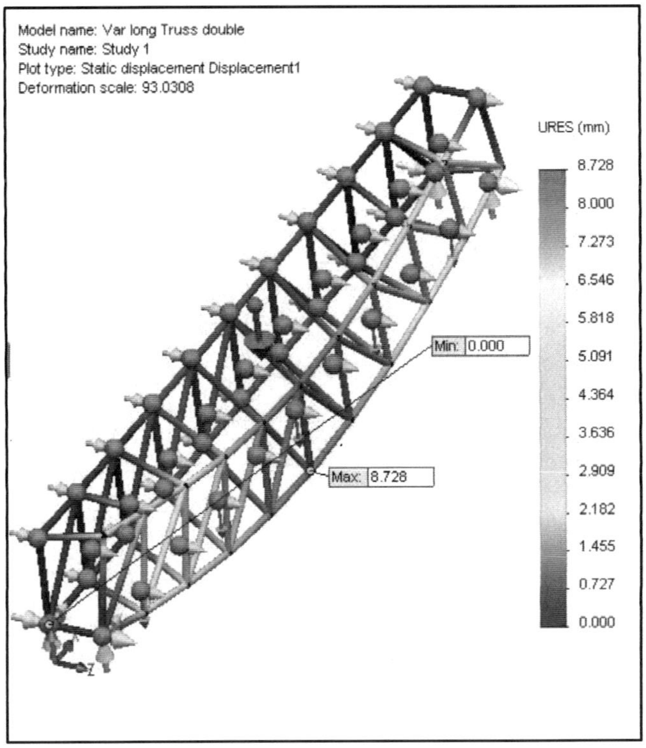

Figure 16. Displacement and deformation for the double flat truss study.

5. The third study

Considering the same second truss, a third study was conducted with another position of the pipeline; namely the possibility of introducing the pipeline into the truss was studied. In this situation the pipeline will be supported on five beam elements. The study will refer to a beam. In the Cut-List folders each element must be declared as beam element. The processing results with all the restrains and external loads are illustrated in figure 17.

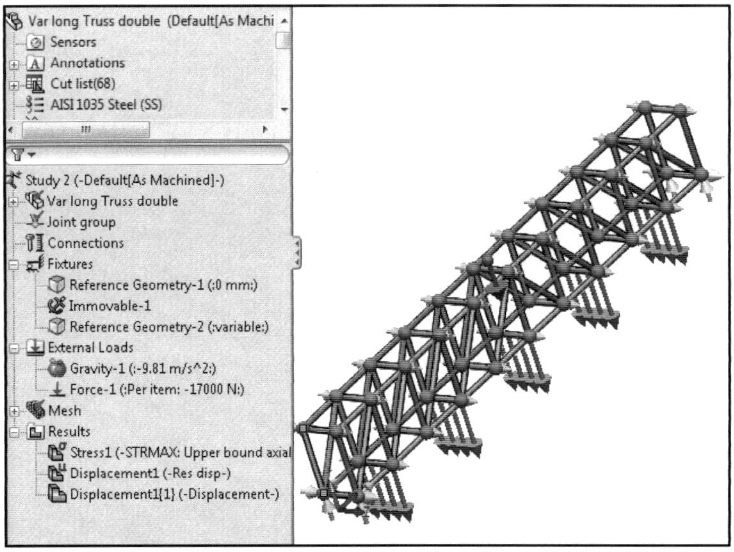

Figure 17. Results with all restrains and external loads.

As in this study the forces are estimated at 17000 N/mm^2 for every 5 beam elements, the post-processing results are illustrated in figures 18 and 19. The analysis of the fields of axial and bending stresses and displacements lead us to the result that the beam is a rigid structure. The maximum displacement (7.475mm) is in the central area where the maximum stress tension is 98.15 N/m2.

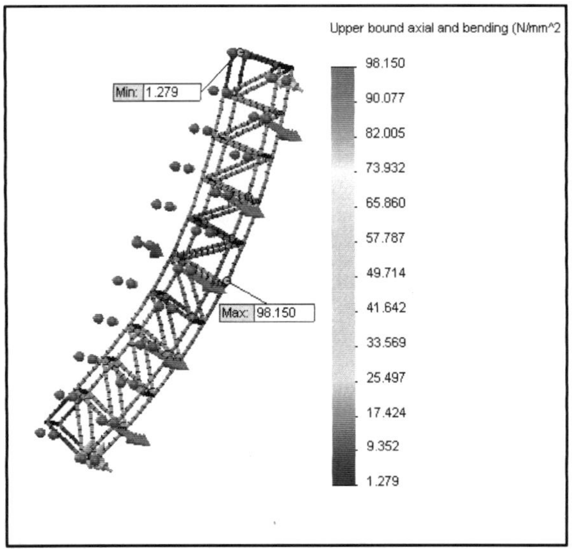

Figure 18. Upper bound axial and bending stresses for the third study.

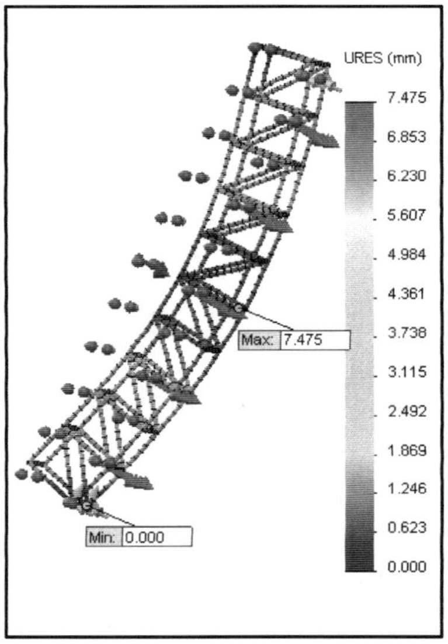

Figure 19. Displacement and deformation for the third study.

6. Conclusions

This paper is a study about the truss and beam deformations and the purpose is to find the best form for trusses using SolidWorks and to avoid oversizing these structures. The truss can be modelled by using solid elements, thin-shell elements, or beam elements, depending on the desired result. In this situation, for good results, the weldments structure with Structural Members was used.

The studded trusses are the Pratt truss and the Howe truss. The verticals elements which are included in these trusses are for raising the span length of the truss bridge.

Three different trusses designs were considered. After the first studies, Axial stresses, Displacement and Deformation in the same simulation conditions resulted with similar values namely, there is a very small difference between the values of the two studies. For these trusses, loads can be applied only to the joints. In the second study the double truss has a more complex structure and yet the results are almost similar.

In this situation, the third study refers to a beam and in the simulation the truss elements into beam elements ware changed. By analyzing the fields of displacements and tensions the result is that the truss is a rigid structure. For the third study, the use of the beam simulation method and higher forces led to the maximum displacement (7.475 mm) and the maximum axial stress tension (98.15 N/m^2) which are lower than in the first variants.

References

[1] SolidWorks Help http://help.solidworks.com/2017/english/solidworks/cworks/c_Beams_Trusses.htm

[2] Mogan G L and Butnariu S L 2007 *Finite element analysis in engineering* (Braşov: Transilvania University of Braşov) p 197

[3] Radostina V P 2014 *Introduction to Static Analysis Using SolidWorks Simulation* (Boca Raton: CRC Press) p 353

[4] Garrett Bridges https://www.garrettsbridges.com/design/

[5] Akin J Ed 2009 *Finite Element Analysis Concepts via SolidWorks* (Houston: Rice University) p 303

[6] Urdea M 2017 *Int. Conf. on Engineering Graphics and Design ICEGD* vol 1 (Constanţa: JIDEG) pp 155-158

[7] Croitoru E I, Magurian D and Oancea G 2016 Finite Element Analysis of Heating-Cooling Systems Used for Laminating Tools *Transactions of FAMENA* **40(1)** 111

[8] Kurowski P M 2011 *Engineering Analysis with SolidWorks Simulation 2011* (Mission KS: SDC publication) p 455

IOP Conf. Series: Materials Science and Engineering **399** (2018) 012052 doi:10.1088/1757-899X/399/1/012052

The inspirational value of unintentional design occurrences: freeing aesthetic

IS Vuscan[1*] and S Feng[1]

[1] College of Arts, China University of Petroleum, No. 66 Changjiang Road, Huangdao District, Qingdao City, 266580, East China

[*] E-mail: sonia.vuscan@upc.edu.cn

Abstract. The paper stresses the importance of taking into consideration those unintentional design occurrences that leave imprints on the object-oriented dimension of the environment. The visual traces registered on surfaces, represent a trigger for creativity. The information captured undergoes a metamorphosis in the perception of the viewer who intuitively puts it to use. The value of these unintended occurrences is given by the incipient quality of further becoming. To this, the perceiver assigns potential renditions in terms of aesthetics or functionality. The unintended occurrences, as mediums for innovation acquires inspirational value. The environment and personal space are open fields of research and a metaphorical bank of data cumulating information on human behaviour and the results of human-object, inter-*objectual* or object-environment interactions. The paper aims at stressing the importance of decoding visual markers to benefit creativity and aesthetics.

1. Introduction

The enquiries based on the results of which this article is grounded followed a thought with roots in intuition. It is the effect of observation, registration, and analysis of particular visual contexts which actually have, if any, poor relation to design. The article intends to raise awareness of those unintentional design occurrences with embedded inspirational value. In the design scenario in which race in finding inspiration is a given fact, it is so often found in creations already established. Usually the object triggering inspiration is so highly and exhaustively developed that we could only be content with pastiche with a personal twist.

In order to create chances for < the new > in design:

- questioning the establishment is mandatory;
- analyzing design antecedents as means of understanding design now;
- looking for the drivers behind < the face > of human habitat;
- finding reasons for contemporary < societal scenic requisites >;
- defining < the new > in design based on observation and critique of the material world and that within oneself, the ambient and the personal perception of it.

The article signals the benefit of commuting between past and present, observing and studying preferences and usages, analyzing object-oriented circumstances, human conditions and impressed surroundings, the whole plethora of environmental contexts, filtered through a specific scrutiny, and implying a particular eye.

Following the outlined method, the paper aims at answering questions in relation to the coined concept of unintentional design occurrences:

- defining the unintentional design occurrences.
- which are the circumstances conducive to their appearance?
- what sustains their prolonged existence?
- how they connect to aesthetics?

2. Defining the unintended design occurrences

The following circumstances could be nominated as *unintentional design occurrences*:

- incidents caused by choices regarding: shapes, colors, materials, associations of the three, which are found under the influences of various forces that impose upon them a change in their initial status.
- visual consequences of actions performed on materials and products, shapes and surfaces, and reactions to uninformed association of the above mentioned properties.
- patterns on materials, imprinted as a result of mutual influences or frictional results.
- aesthetic valences encountered within compositional results of random objects' association.

Figure 1. Private parking space signals.

Figure 2. Metal barrels as flower pots. Where being old, rusted and distorted add value to the aesthetic image of the ensemble.

This article is focused on determining the inspirational and aesthetic value of those unintended design situations as induced by objects which, through their visual appearance, and peculiar context, raised awareness and triggered an interest in their status. The analytical process has as starting point tracing visual manifestations of peculiar inter-objectual relations, or object – space associations, that induce focus and aesthetic review, and suggest latent potential for generating inspiration in thinking innovation. Intuition and chance are intrinsic factors of data registration in a borderless inspirational area. The transcendence of the ordinary occurring within an object while taken outside its initial context, acquires a halo of extra-ordinary value (figure 1).

The aesthetic appeal of what resides in such unintentional occurrences refers to the objects themselves. It is predetermined by a whole sequential imagery composing a certain scene, the scenic decorum in situ. The scrutinized object-oriented composition is a result of actions dictated by fulfillment of functionalities with the particular objects as prime characters. In such a case, the aesthetic value is unaccounted for, the sole purpose being that of strict functionality. The objects occupy a strategic place in order to perform a certain utilitarian function, in the case illustrated in figure 1 that of signaling private parking space. Improvisation with objects and intuitive usage of objects is a subject acknowledged, dissected and naturalized in the domain of design [1]. The objects are readapted to perform previously unassigned tasks, and by this, the user restores usefulness (figures 2 and 3).

Figure 3. Plastic barrels as pigeon shelters.

Figure 4. Motorcycle helmets as pigeon shelters.

This seems like a migration of function, taking us a step further into discussing function as self-dependent, freed from the objects which mediate its fulfilment. Such frame of judgment is somehow ancestral [2]. An object at the end of its functional life is reinstated into a utilitarian circuit if rehabilitation prevails in the detriment of discarding (figure 4). Through postponed discarding, the object is kept outside the landfill, therefore determining the delay in the purchase of a new object.

On a personal level, through this act of postponed discarding, one submits to the tenets of sustainable practice [3]. Emotional liaisons are the drivers behind postponed discarding [4]. If the object resonates with the user on an emotional level, this increases the chances of updating, repairing, readapting or preserving the object. Emotional liaisons are the drivers behind postponed discarding [4].

If the object resonates with the user on an emotional level, this increases the chances of updating, repairing, readapting or preserving the object. Discarding is postponed and the object is preserved for fulfilling uncertain tasks. Even if uninstalled in a utility post, they follow other usage patterns, inchoative of improvisation, where creativity is expected to interfere.

The quality of the unintentional design occurrences lies in details and random associations. Within the contexts most likely to accommodate unintentional design occurrences, cheapness and improvisation is raised to higher spheres. The rectification of poor design is already at hand. People mediate the existence of particular spaces through guiding intervention that best suits their, if solely, practical purpose. In the process, they are creating chances for aesthetic appeal. Its existence, if occurring, is involuntary. The motive - long lived objects exhibiting patterns dictated by time and usage. The objects (acted upon), materials, were chosen following no other criteria but their handiness, lowest price and strict functionality.

The intention is to highlight the existence of unintentional design occurrences that could be found with predominance in un-designed scenarios, or within a limited range, in case of alteration imposed on designed scenarios. They are the fortuitous outcomes of submitted improvisation, or adaptation of existing design scenarios to better fit a personalized space. Unintentional design occurrences often escape categorization and are discarded as incapable of sustaining an aesthetic scrutiny. The unintentional design occurrences are, at their inception, often unfamiliar with the notion of aesthetics, or aesthetic value, but are linked to functionality and economy. Derived from these characteristics springs what in my perception adheres to beauty.

2.1. The inspirational value of the unintended occurrences

The article stresses the inspirational value of the unintended design occurrences. It is intended as a theoretical approach to what is un-designed, for faulted, improvised, ignored, left to chance – unintentional design occurrences - considered in terms of inspirational value for freeing aesthetic. Each one of them is a manifesto against uniformity, and pleading for diversity, amalgam and boldness in deciding forms. They advocate against prolonging the ubiquity of dullness as a weary presence in created environments, and common dismissal of improvisation for being gestures of incompetence.

Figure 5. Aesthetic markers. Traces on soft soil.

Figure 6. Aesthetic markers. Pipe network.

The article draws attention to the unlimited possibilities that these unintentional occurrences could bring to the field of design, discussing unintentional design occurrences not solely from the perspective of what they can offer to endeavors concerning optimization of functional properties, accommodating consumer's needs, but mainly freeing aesthetics (figure 5). Usage of cut from context objects, bear the outcome of materials reaction to outer stimuli. Opting for a loose control over green spaces is a form of rebellion, pushing the boundaries of permissiveness as regards established urban aesthetics.

As if following an organic development pattern, the abstract graphics given by intermingling and juxtaposing of cable networks or pipe networks thrive in a setting abandoned to improvisation (figures 6 and 7). Constituting annexes out of salvaged, disparate materials with no attention given to integration into décor, they become in time, through the evenness patented by passing of time, a pastiche of design.

The article advances the validity of the assumption that, by selecting, registering and understanding certain design situations as being consequences of prior states, this would conduct the process toward advancing design remedies. Training into discovering incipient design corrections in unintended design situations, may ease coupling to an inexhaustible inspirational network.

The method implied for gathering information has as a starting point in which a certain design situation is perceived as aesthetically appealing, it's followed by photographic capturing, then further inquiry and repartition, all evolving in prospectuses for further designs. The fact of strictly observing, analyzing, acknowledging the power of human creativity, intuition, skill is in itself a finite action. While in search for inspiration, one is predisposed to observe intentionally and to discover randomly. Such is the case of encountering deficiencies in present design, signaled by counter reactions from the users.

Unintended design occurrences are recipients of inspiration and guides for correcting design. This article attempts to instill a method for sharpening attention. Users point to the faults, signaling *where* and *what*, they are in a rudimentary sort of way showing *–how* [5]. One could take this in and work out solutions or appreciate the ingenuity in itself.

Figure 7. Aesthetic marker. Cable network.

Figure 8. Wind shield. Correction marker.

Within a contemporary contextual development, the uttered need of perpetuating traditional elements considered of perennial validity is to be repudiated, as elements of composition. Unintentional design occurrences have a saying in confectioning forms and structures, materializing concepts, shaping contemporaneity through incipient inspirational value. The melancholy for a reverberating past will impede thinking innovation.

This article pleads in favor of creativity coming in new and experimental samples. The direction can be intuited in composing chameleonic structures changing and growing from within, developing into something of continuous becoming [6]. The unintended design occurrences could be considered as fortuitous happenings with inspirational value and guides for improving design (figure 8).

2.2. *Human imprints on the ambient*

The unintended design occurrences are based on human imprints on the ambient, the result of which developed into new distorted environmental situations. The ones considered incubating an intrinsic quality that could be further dissected with gains of any kind was subjected to photographic registration. We based our analysis upon this selective collected material. Conclusions were based on categorizing and analyzing the body of information which was perceived as valuable for our research from the distraction that the rest of it represented. The difficulty raised was choosing correctly and establishing the usefulness of the elements retained. The action resides in recognizing the elements of interest and the relation they cultivate with objects or materials composing their vicinity.

The consideration that there is aesthetic appeal and precepts of inspiration in the abject, improvisation, amalgam and accumulation, determined our decision to bring under discussion the possibilities this research could reveal: There is also a nostalgic quality of feeling that could affect human thinking and behavior.

For further analyzing human imprints on the ambient one must:

- scrutinize the object-oriented environment (under particular conditions);
- perceive personal space as research filed in decoding human behavior;
- understand the involvement of the inhabitant in private habitat - gradually manifesting a personal design strategy in relation to space (according to one's values, needs, possibilities);
- improvisation or applied capability in accommodating personalized needs that vary in tandem with changing persona;
- determine the importance of personal imprints on the perpetual change of ambient;
- question the habit of distorting existing design situations;
- register the urge for manual activity;
- reestablish self-credibility in handicraft skills.

An accelerated intervention on an object or space manifests as a form of assimilation. The result is a humanization of the subjected object or space, the finality of which would be the creation of a person with increased individualism evolving within the range of an increased individualized, humanized space.

3. Human alienation from objects and object oriented environment

People should not disconnect themselves emotionally from the material environment. With design ubiquity, human alienation from the environment occurs, as well as receding involvement with the environment. One does not undertake the task of creating the object, therefore one is subjected to reduced emotional attachment. Enhancing emotional attachment is transferred to selecting the body of objects that sum up one's requisite. Actuating on one's tools shifts from creating to selecting. This prerogative complicates due to multiplied choices. The immense variety of objects with similar functionality conducts to confusion, visual oversaturation, purchasing dissatisfaction or shopping frenzy. Possible reactions to product oversaturation:

- accumulation (due to indecisiveness);
- restrain (overcame by a feeling of pointlessness);
- improvisation (procuring function outside purchasing).

There is an increased tendency toward improvisation, preservation and upgrade. The article intends to place under discussion personal space and perceptions on public space, under the presumption that they facilitate better understanding on user's needs, usage and behavior.

3.1. Behavior patterns imprinted on the object-oriented environment

Users' behavior imposed on material possessions is a trigger for deforming, transforming, adapting, upgrading, collecting, discarding. Perpetuating certain actions upon objects and surfaces determine consequential changes in their aesthetics, functionality and context. We direct our enquiries towards the visual consequences. Accumulation of objects, taken outside a utilitarian circuit, is a tenet of pauperism. The complexity resides in properly applying methods to critique the ensemble of contextual juxtaposition of objects, and giving a purposeful orientation to the information collected in the process. Under scrutiny is the un-designed body of objects, composing the requisite and ambient of pauperism. In conditions of deprivation human creativity invests in creating techniques for reaching dead ends.

Existential conditions impose on our conduit, exercising creativity while recalibrating the scale of values. Under specific conditions one might witness exceeding behavioral patterns, marked by: hyperactivity (prolific manifestations in a restricted space with the purpose of space optimization); improvisation (transform to reuse); safeguarding (accumulate for future use); preserving (careful maintenance); adapting (multiple functionality); prolonging (extend the useful life of a product to the limit); postponed discarding of the product.

Figure 9. Appendix assembly.

Figure 10. Metal buckets as private parking signals.

Negative behavior is to be counted among contextual states: apathy (consequence of a perceived feeling of futility); indifference (due to a feeling of hopelessness); disregard (ignoring alternatives to the status quo as pointless, minimizing expectations, contemplating life at the lower end); deficiency (motive for adopting survival techniques in a context dictated by privations); accumulation (compulsory tendency to heap objects to which questionable valor has been attached); disorder (subjecting space to a conscious disorder as part of slipping into indifference); misery (existing state from which the aforementioned states reverberate).

4. The versatile value of improvisation

Improvising spaces to fulfill personalized needs gives the initiator a satisfying feeling of gain and bargain (figure 9). Matching salvaged objects to functions also fit in the same description (figure 10). So is the case of vegetable gardens in improvised recipients, or prepared soil beds in arbitrary places (figure 11). People make use of their ambient intuitively, adjusting it to fulfill their needs (figure 12). The aesthetic appeal resides in the unexpected. The environment ability of accommodating improvisations stirs amazement.

4.1. Improvise for fitting

Improvisation abhors uniformity, embraces accidents and handicrafts imperfections.

No matter how aesthetically perfected and well-intended, unidirectional aesthetic nurtures incipient dullness. Correcting means enhancing creativity by avoiding rules and allegiance to styles which impose rules, demand obedience, limit growth and freedom.

If design means fulfilling needs and we accept changeability in people, their ideas, moods, needs, desires, appetites, given the fact that everything is perpetually changing, in the field of design this translates in embedded capability of following change.

Improvisation is welcomed hazard. It opposes pastiche, a technique that could metaphorically be described as running for the new blindfolded. It is tangential with creativity. The hazardous outcomes of improvisation are equivalent with the outcomes of performing various tactics for enhancing creativity: experimenting with materials, shapes, colors, becoming absurd, bombastic, becoming artificial.

Figure 11. Ceramic sinks as seedlings soil beds.

Figure 12. Branches as support for clothes.

In search for inspiration one must turn away from what is organized, overdesigned, or subjected to inhibitory rules. Creativity is the equivalence of rebellion in thought. In creation, one must opt for personal appeals and drives, while pursuing outside influences in designing specifics. One must get infused with the outside through impressions, influences and suggestions while designing based on inspiration. Other means of enhancing creativity:

- disobedience;
- improvisation;
- observation;
- sounding potentialities.

5. Objects as Analysis Factors

The body of objects, composing our surroundings and requisites, are analysis factors, evidences of who we are as individuals. Objects are reminders of our desires, tastes, activities, pleasures and investments but also recklessness, waste, greed, oscillations, search of identity. Objects are a proof of personal history. Object selection, preservation, collection is an important stage in exhibiting or manifesting identity, empowering what is *objectual* with the status of embodied identity - things do not create one, they reflect one.

The material possessions add to our societal persona. Collecting and treasuring objects are tracks to reach individual or collective past - marks in tracing our life course. Personal space and requisite stands for what we were, are and planning to become. The body of tools which are used in organizing and completing actions are the incipient of future experiences. Given the incessancy of use and ubiquity, one witnesses a peculiar case of valor minimization in perceiving the material world. One witnesses a decreasing interest in the possessed object and an increasing interest in the desired object. Accelerated tiredness of the object in one's possession and the safety of assured superabundance of objects are at the basis of frequent detachment, followed by discarding and replacing.

Opting for repairing objects as opposed to discarding objects gives the possessor a feeling of activated responsibility. Further investment in the defected object mediates a bond between the possessor and the object possessed. Objects should be provided with built-in quality of bearing unlicensed user intervention in case of repairing, replacing damaged parts, etc.

5.1. *The ritual and the objects*

Objects tend to change properties when taken outside their contextual accolade or predetermined task. In the designing process think of objects as future vestiges: a bottle, a comb, a computer case, a spray can, etc., how they would advocate for our culture through their material, shape, color and texture.

Imagine objects as cult objects subjecting usages as in performing a desecrated ritual. This would bring something new for one deceived in its expectations regarding following behavioral etiquette. Considering nostalgia for obsolete elegance and an impulse to revitalize a conduit, one could enquire if objects could determine a behavioral trajectory through suggestive formal features. In the theatrical manifestation of conduit witnessed in today's social scenario, objects should ennoble user behavior.

5.2. *Object-oriented congestion*

The object-oriented congestion is an agglomeration of objects with emotional charge and inspirational value. They could be perceived as emotion recordings, objects as mediums to access past memories. They could undertake the task of concomitantly initiating the new and securing the past. People are fallible, show gaps in performance, determination, stability, constancy – therefore, there is a need for objects as attention signals.

As spiritual creatures, humans develop spaces of desire - shrines under disguise - the symbolic of which welcome inquiry. One must acknowledge the inclination toward mystifying objects, empowering them with immaterial attributes, resulting in collecting, exhibiting and manifesting.

5.3. *The kaleidoscopic satire of matter agglomeration*

It is the unexpected and unrepeated in the formation of matter conglomerates. It is a guided act, subjected to the logic and whims of the performer. Unintentional design occurrences come into existence as a result of the fulfilled act, registered at a propitious moment of its completion.

The conglomerates are under a state of uncertain composition, under the imminence of further addition. Added components have decisive share in the perpetual transformation with repercussions in the aesthetic of the whole. Upon perceiving the abject in its *objectual* manifestation, one is subjected to reactions of uneasiness and curiosity, vexation, forced acceptance and understanding of the mechanism of this grotesque machinery [7]. While placing matter agglomeration under the auspices of perpetual evolving, one should enquire upon the laws dictating changeability, the forces that act upon it.

If perceived as a staged scene, the agglomeration of objects is taking a stance against further acquiring, affording a feeling of matter satiety. The refuse, scattered in afforded premises, substitute the thought of purchasing with that of disembarrassing. Accommodating refuse demands space, which is a scarcity. Under normal circumstances, the next logic act would be selecting and disposing of unwanted matter, however, the context described assimilates refuse, becoming a constructive element of composition.

We count on a busy mind, finding convenient ways of disposing refuse. Possibly trigger an urge to sort and dispose of the objectionable. The intriguing persistence of the augmented agglomeration of matter requires an unconventional method of analysis. Agglomeration adds organically, under an inner law, which demands questioning the laws which govern its prolonged toleration by both the provocative factor and those who relate to the state of fact. The state of affair is an imposed situation dictated by evolved tolerance to such scenarios; an absurd opacity, selective observation and acknowledgment of elements comprised in a given space. It imposes searching into a stratum of infantile logic. In a customary scenario, the habitat is subjected to laws dictated by societal assessment, governed by order and logic. The scenes are seen as in a reversed enhancing lens. The receiver is overwhelmed by such an uncomfortable feeling of suffocation and satiety; perceiving an end to such state of kaleidoscopic use of space - objects arranged as in a game of derailed kaleidoscopic structure.

One tends to place an adverse judgment on empty space and surface. Void is considered an irritating remembering of an avoided *something*; it creates anguish and a feeling of unfinished and provisional state; it stirs self-awareness and gambling on spirituality. At the opposite pole, heaping is in relation to an infantile drive of shifting and distorting *objectual* identities. Matter agglomeration is an exacerbating presence of humanity, creating a den that secures one from exposure, intrusion.

6. Freed aesthetics

There is no purposeful aesthetic dimension of these unintentional design occurrences. It is intriguing that these occurrences are developed through processes with negative connotations. Other laws dictate their occurrence: shortages, habits, accessing the vernacular by acting based on tips acquired in an over-the-fence kind of verbal transmission. Improvisation stands for the need of keeping an appearance of mobility. It implies emotional involvement. The overall impression is that of stringed chaos. It touches the realm of the grotesque. Its outcomes could provoke contradictory reactions and critiques. They are what might be called abject manifestations in design.

Were these, accidentally achieved, aesthetically appealing ambiences, intended to emit aesthetic stimuli? Could it be that they have strived to inoculate aesthetic appeal since their inception? The idea of unintended aesthetic is based on our preconception that these occurrences are the consequences of strictly functional manifestations. They are developed in time, with an obliterated inception momentum, perpetually adjusting to accommodate spontaneous adjustments. As the results of improvisation, they present deficiencies and malfunctions. The objects lack consideration for concepts as: aesthetics, safety, ergonomics and ecology. They follow the sole purpose of fulfilling a certain task.

6.1. The aesthetic of decadence

The unintended design occurrences are channeled by contextualized chances and faults of many kinds (deficiencies in quality of: design, engineering, installation, materials, use, maintenance, etc.) (figure 13).

Figure 13. Ferruginous leaks. Expression of material deficiency.

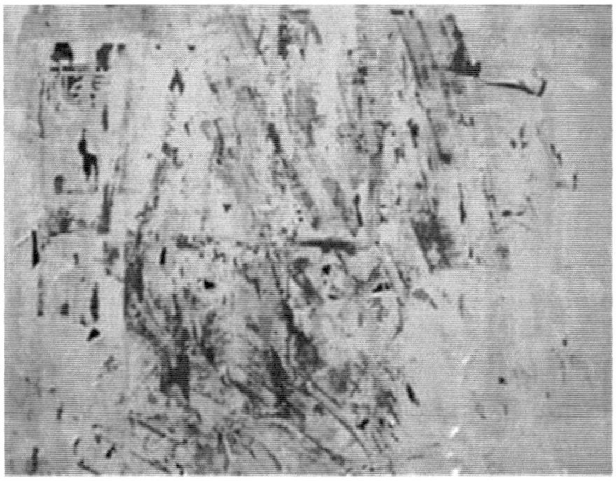

Figure 14. Aesthetic marker. Billboard.

They can be traced to many reasons but our attention is toward results. Shapes, colors, textures – marks signalizing the aesthetic of decadence. Collages of negligence, indifference, and the dictum getting it done no matter how – a caricature of efficiency. Advertisements in approximate thick juxtaposing layers recall bas-reliefs, with the scrapes of desuetude bursting at the edges, on the conquered exhibiting surface of walls, poles and what not (figure 14).

Peeling paints of fickle color choosing, corroding metals and carcasses with fading logos report on actions and reactions (figure 15).

Human-object interaction, the friction with and within matter, lay at the surface and is subjected to interpretation. These incidents which evoke response should be registered, categorized, kept as incipient of inspirational value in developing aesthetics.

Figure 15. Rusty metal. Expression of material deficiency.

They contain endurance evolved in time, they are as many reaped results of prolonged repeatability with presumed variations, waiting for a receptive eye to access their hermeneutics. One has to substitute for an explorer of human condition at the edge of society but a pulsating presence with the pride of outliving change. Their value resides in the contrast and the absurd. While establishing a future direction in design, one has to probe both poles. Appreciate what was previous in design, for the data recorded and forwarded in time, helping in making proper adjustments to contemporary designs, and launching prognosis for future ones. Obsolete products have a value dictated by traces left through usage, marks left by incidents, reparability, replacing of parts, all adding to their particularity. These traces are records of bearing characteristics, durability or functionality's history, etc. new has to evolve out of something ticking history. Products which occupied designated places and performed specific tasks in their functional time pile up in unintelligible conglomerates and amalgamation, rousing melancholy and a sense of waste. While *making future* for a product one must start acknowledging the remnants of a product's past. One must take responsibility for this long and careless past based on a false presumption of perpetual abundance, tolerability and assimilation. The trash has the shape of former shelf products pointing to their "last generation" tag of fame and pride. The "old" could testify against errors and deficiencies - helpful clues for future directions in designing tomorrow.

Acknowledge and interpret the changes suffered at the surface of objects. They are alterations suffered in time, proofs of defaults. During their functioning period objects suffer alterations and through this processes accumulate information. In a group of identical products their scars indicate appurtenance. If not the quality of their material, one has to appreciate their servitude.

6.2. The aesthetic appeal of unintended design occurrences

The article's enquiry is in relation to those unintended design occurrences to be found if searched for with an open mind and an eye trained to discover. The element of surprise is conspicuous. It is about a scenario in which one might incidentally register a glimpse of beauty revealing itself with humility in raw utterances of creativity. Beauty manifests itself in ample variations. The aesthetics of niche endeavors in the field of object manipulation and installation within a certain space.

What seems of questionable aesthetic valor could build up into solid compositions by permutation in an optimal space, and in relation with other objects, which bring mutual value through fortuitous association: surrogates of gardens grown in peculiar formats and sites, artisanal window frames in color outside chromatic harmonies and juxtapositions, the droppings from the ventilation systems, leaving vertical and ferruginous traces in gradient values on the wall, etc.

These occurrences happen accidentally, following not many laws. They perpetuate their existence in a continuous state because of negligence, omission, or hazard. Unintentional design situations could occur most unexpectedly in propitious circumstances.

The questions that arise while acknowledging these facts are:

- what induces visual appealing?
- which are the circumstances that foster such occurrences?
- how could they benefit design? Suppositions on design potentialities conditioned by aforementioned unintended occurrences.

People adjust space, enhance the possibilities of space and through this aesthetic phenomena could occur. The viewer reactions range between confusion, bewilderment, frustration - the result of contemplating the ephemeral character of the aforementioned occurrences. The need of solving what puzzles about this visual appeal emanating from such unintended occurrences is incessant. The article intends to lay the foundation for further research to elucidate this phenomenon.

Unintended design occurrences are the visual result of constant interactions evolved in time. The effect tends to become more acute if the cause of the aesthetic incident is not removed or attenuated. The circumstances under which these occurrences thrive are with preponderance antiquated spaces with finessed renovations, which have stagnated or undergone an organic pattern of development. In the case of urban spaces, they imply law's indulgence, a blind eye from the citizens, abstraction from demolition, reconstruction, re-design, from all that could restrict this peculiar assimilation of space. Its existence is determined by a process of alteration of an initial status quo.

The foothold for the occurrence of aesthetic design incidents are in relation to that which no longer corresponds to the stage of an existing time, which is obsolete and under permanent threat of extinction.

Perpetual remodeling of urban space and accelerating rate of architectural renewal is the end of these plagued design occurrences. It is speed turned into shape. Vacant land is incorporated and old neighborhoods are being reformed with the ease of a child knocking down a house of blocks and feverishly starting anew. These hazardous staged spaces under scrutiny are outlived. The new is overwhelming and seldom exhibits that which could pass as something of melancholic beauty.

The inquiries are in relation to those unintended design occurrences taken place in such hazardous staged compositions. They result as accidents and become integrated into the scene while concomitantly transforming the scene by altering it. For example, a broken glass filtrates the sun beams and becomes an object of contemplation. It transcends its purely utilitarian status and through hazardous alteration accumulates aesthetic value. Compared to its vicinity (the un-deteriorated panels of glass) its valor shifts owing to its newly acquired quality of absorbing and reflecting light thus creating an ever-changing guise. The history of this glass panel is brought to the attention of the viewer (initial state, alteration factor, visual result– are all circulating phases of the object provoking judgments).

The becoming is transient in its temporary existence. The intriguing character, the disruption of harmony and symmetry and the feeling that something was given for free, all collaborate to the exchange that occurs between the viewer and that which is viewed.

7. Conclusions

The unintended occurrences placed under scrutiny are as many aspects of drifting design placed at the border of permissiveness and acceptance. The characteristics which propelled them into exile, the quilts attached to them: alienation, repudiation, misery augmentation, propensity to propagate, resilience to order, endurance.

The article's interest was in analyzing what is of a corrupted, deluded, nonetheless conceptual quality in such ambient and environmental contexts, highlighting that in the aforementioned contexts, there is valid information at disposal with repercussions in design and architecture. There is an unexplored dimension of the abject under whose denominator all these conceptual values reside, values that could be transplanted into the domain of design, inspirational values with reverberation in design.

The article is a recording of consecutive stages of scrutinizing unintentional occurrences and their imprints on the surface of objects and ambient. The practice traces a starting line with no predetermined finish. A disposable cup at the toe cap of one's shoe imprinted with a discernable logo is as much litter as it is free ad. By permuting the discussion into another assessing register, litter becomes disposed matter and data, simultaneously. Humans leave behavior markers on disposed products. There are imprints, traces, proofs at the length of a step. To the common tendency when assessing heaping: intrigue, indifference, discarding or criticism, one could add the potential for extracting facts of pragmatic and inspirational valor. What one may see as faults and clutter, an exuberant manifestation of the abject in design, another may see the logic result of enumerable causes, and indices for better design and innovation. The unintended design occurrences are the results of a successful pursuit.

References

[1] Norman D A 1988 *The Psychology of Everyday Things* (New York: Basic Books Inc)
[2] Gibson J 1979 *The Ecological Approach to Visual Perception* (Boston: Houghton Mifflin Company)
[3] Barr S 2003 *Strategies for Sustainability: Citizens and Responsible Environmental Behavior*. Vol. 35, No. 3 (Sep., 2003), pp. 227-240
[4] Norman D A 2004 *Emotional Design: Why We Love (or Hate) Everyday Things* (New York: Basic Books)
[5] Fulton S J and IDEO 2005 *Thoughtless Acts?: Observations on Intuitive Design* (San Francisco: Chronical Books)
[6] Benyus, J M 2002 *Biomimicry: Innovation Inspired by Nature* (New York: Perenial)
[7] Kristeva J 1982 *Powers of Horror: An Essay on Abjection* Trans. Leon S. Roudiez (New York: Columbia UP)

IOP Conf. Series: Materials Science and Engineering **399** (2018) 012053 doi:10.1088/1757-899X/399/1/012053

Civilized enclaves of wilderness: substitutes for an alienated urban nature

I S Vuscan[1*] **and S Feng**[1]

[1] College of Arts, China University of Petroleum, No. 66 Changjiang Road, Huangdao District, Qingdao City, 266580, East China

[*] E-mail: sonia.vuscan@upc.edu.cn

Abstract: The article stresses the inspirational capacity of spontaneous vegetation condensed in spaces defined by human oblivion, such as derelict structures and untamed lots. According to the holstering capacity of the medium, isolated cases (weeds sprouting on crumbling walls), or emulating natural habitats (wasteland flora) developed into a wasteland nature propagating as a consequence of dysfunctional urbanization. Because of increased infrastructure development and urbanization, spaces passable of accommodating such unintended design occurrences recede. Such crude source of inspiration for inoculating nature into architecture and design is to be found within the parameters of enduring, unintended, free form compositions of opportunistic vegetation. The article aims at hosting an alternative to current urban order, in what concerns a contextualized human-nature interaction, launching a thinking scheme for rewinding and sustaining the urge toward soil and nature.

1. Introduction

This article wishes to outline manifesting outcomes of nature, with a melancholic call for freedom and wilderness. Within the length of its pages, the article attempts to thrust forward means of restoring nature's wild predisposition within the urban context. Because of decreasing space for oases of green and repose, improvisation and adaptation to the new possibilities offered by a throughout built environment are to be perceived as a solving solution. Urban expansion does not occur in tandem with that of greenspaces, on the contrary, due to increasing demand but limited land availability, greenspaces are stagnating or, more often than not, receding [1]. While scrutinizing spaces at a venture in search for inspiration, a passible eye (designer or architect) might discovered the enduring and obstinate character of nature in unintended occurrences of spontaneous bursts of vegetation, emerging in spaces affording their self-propagation. The observation of those obvious clues of nature's way (spontaneous outbursts of vegetation), is to be followed with receptive designs, materializing in greenspaces and built structures that seem to genuinely evolve into what might become the utopian juxtaposition of the material and natural dimensions of urbanity. In order for this to concretize, the creator (designer or architect) must shift practice from imposing it, to submitting to the way of nature.

The article stresses the aesthetic possibilities offered by spontaneous burst of vegetation found randomly in peculiar spaces (vacant lots, decaying surfaces, repose surfaces, ruinous structures, erroneous structures, etc.). It brings forward the incipient aesthetic qualities of unintended occurrences, in vegetal compositions that have not undergone human intervention but evolve and propagate within the limits of human omission.

The article underlines the existence of such unintended aesthetic occurrences while scrutinizing the unaccounted for presence of heathers, that sprouted opportunistically in spaces that were not intended for their propagation. The intention is to highlight the aesthetic value of spontaneous vegetation. Acknowledge the inspirational possibilities offered by heaths, which could provoke a change in human perception and appreciation, if subjected to a process of observation, selection, restrained intervention and replication in spaces where highlighted and framed by subtle design could acquire status recognition. The dictums under which the ideas were constructed and enquires have been conducted, the key to read the symbolism of photo-registered information are as follow:

- let nature be natural by restraining from over-manipulating greenspaces;
- design should be literally green;
- make chances for nature in design and architecture;
- acclimatize wilderness into a concrete oriented urbanity.

The mechanics of change has at its basis gaining raw aesthetic and ecological knowledge, finding inspiration in the crude and primal, replicate cropped compositions from aleatory incipient habitats in intended spaces of thought and structure.

2. The natural element and urban nature

The natural element is inclusive to the design of the environment, being counted among its building materials. It follows a certain pattern in a repetitive way, given to an established design scheme. One could identify a certain urban area according to its natural décor pattern, the preponderance concerning plant species, comprised vegetation, objectifying aesthetically secure designs. Greenspaces are subjected to design, the natural being distorted through a process of adjustments and fittings with the purpose of creating a sense of controlled space. At the opposed pole there is the alternative of freed vegetation with particular attention given to the enclosure. Design should be the prerogative of a material environment with nature being allowed a lent process of de-civilization.

Vegetal decoration is not to be considered equivalent to decorating with nature. While the former involves vegetal constituent elements of urban flora manipulation for reaching a level of accepted and established aesthetic, the latter implies accepting the unpredictability of nature's behavior in designated spaces (figure 1).

The bizarre beauty of such unintended aesthetic occurrence, as that attached to the nature engulfed built structure, is something that escapes the appreciation and comprehension of neighboring inhabitants and passersby. It is reputed for being disordered, peculiar, reverberating into an oasis of welcomed and intriguing wilderness among arid compositional creativity. The free form composition dares to move away from the accepted and expected form of the material environment. For neophytes it is an upsetting and discordant presence, for a specific viewer a source of aesthetic revelation, calling for a reevaluation of its unexpected, unrepeated intrinsic compositional possibilities.

Urban nature shares its scarcity among conformist greenspaces (subdued through over-manipulating processes of maintenance and design) and transient greenspaces (circumstantial vegetal compositions in a state of dereliction, underrated heaths that could be valorized with minimum intervention, seeding and trimming). One must acknowledge the ingenuity with which spontaneous vegetation reclaims hostile space, its exuberant development, spreading and resilience to detrimental conditions.

Urban nature lacks a defined status and aesthetic, with processed vegetation and spontaneous vegetation competing for dominium. The presence of spontaneous vegetation was mentioned by Tallmadge (2004) with pejorative connotation.

Figure 1. Example of nature engulfed built structure.

Figure 2. Nomadic vegetable garden.

Tallmadge (2004) denounces the duality of a sterile material component (such as roofs and pavement) of urbanity and too much wildness (such as weeds) with regard to urban vegetation [2]. He translates these visual signals as waist and a call for action. While lacking appurtenance, purpose and quantifiable attributes, they are not short in incipient inspirational value, creating, through their flexible rebellion, spaces of wonder.

Interaction with greenspaces is restricted to visual consuming. With everything already established, the inhabitants are educated to submit to the establishment. In today's over-manipulated greenspaces, nature is being amended from being itself. This highly contradicts Thoreau's reflection on nature's obstinate inclination toward a wild state [3]. Our acceptance of what is commonly considered urban nature: preserves, parks, and gardens, embody what Thoreau (2010) mentions as nature imported at any price. That which is acknowledged as nature in an urban framework is shaped by culturally dominant metaphors of nature. These metaphors validate that nature which is accommodated in parks and gardens or protected as reminiscences of native landscapes effaced by urban development [2]. Moralizing dictums and the augmentation of the fragile character of urban nature, reprimands and penalties, conduct to citizenry's estrangement and sole participation at a visual level. While viewers are being designated to coercive pedestrian areas, urban nature is reformed according to oppressive designs into decorative, non-interactive patterns. Both humans and nature are subjected to permutations from the natural, a restructuring with repercussions in the way humans relate to nature and the natural.

3. The soil at the block's foot

In their study Ryan and Simson (2002) underline the necessity to reconsider urban woodlands as both economic value and improvement stimuli for urban life and environment. The acknowledgeable gains are reason enough for undertaking conservation of existing woodlands and development of further ones. They discuss the propagation of urban woodlands not only at the outskirts but in the urban environment, and furthermore on 'people's doorsteps' [1].

Rebellious spaces, coming in green of many hues and textures from inceptive indulgent intervention, create precedents of urban palliation for nostalgia, in shapes of origin and brutal beauty. The article advocates for a change in human conduct toward permitted nature in a space of concrete and soil-phobia. It also states the need to revise laws and restrictions regarding unpermitted intervention in greenspaces and bordering soil bands. It solicits the right of gaining responsibility over the greenspace in the immediate vicinity of one's own residence - the soil at the block's foot, as exemplified in figure 2, dormant soil revitalized into improvised patches of vegetables which add value to the overall appearance of the accommodating space. The mutable character of the nomadic vegetable gardens is demonstrated by absence of continuity.

The disposable surface of vertical walls could be considered as a lucrative surface for exhibiting intended vegetal compositions. Figure 3 illustrates an unintended encounter with vegetal elements, an example of nature's stance in creating chances for its accommodation, in a scenario advancing the motto – design and architecture should be literally green. Vegetal elements sprout spontaneous in a passive revolt against the concrete corset.

Figure 3. An example of unintended design occurrence – creating chances for nature in design and architecture.

Figure 4. A living roof composed by vegetation which was not caused by human intervention.

Nature spouts in a tenacious effort to proliferate where workmanship and materials are defective. Rooftop surfaces are not to be considered as sterile material components in the framework of green design and architecture. Figure 4 illustrates a living roof – status acquired for its existing vegetation, not caused by human intervention, but self-developed as a consequence of other exterior random agents. Due to a series of fortuitous events, an unintended design occurrence with highly aesthetic value emerged.

Aiming at gaining space for green in architecture, one must consider built-in terraces (figure 5). Within the context of receding the horizontal manageable space, the issue of gaining space rises. Creating opportunities for green patches, be it ornamental or vegetable/herbs patches, is in increased demand. Built-in terraced surfaces could be adapted to host such endeavors while balancing built structures with the idea of suspended gardens.

Expect amalgamation of derailed patterns in composition, confusion in style and tilt toward humanity in its faults and graces. If the enabling environment emerges, one can expect to encounter the unexpected under the form of unintended aesthetic occurrences considered as mediums for astonishment, and resuscitation of curiosity. The unexpected element is the one accounting for dismissing repeatability and predictability, resulting in gradually omitting the visual friction with an urbanized nature.

Spontaneous vegetation should suffer a shift in human perception from a reproachful to an accepting attitude. Spontaneous vegetal elements are to be seen not as derogative but a provided presence in the developing process toward completion of intended design. New schemes in design should embed nature in its structures as implied constructive elements. With variable location and composition, such aesthetic phenomena prove a contingent happening, freed from space pertaining. They are fortuitous encountering, in relation with abandoned or partially abandoned structures. A proliferation of green roofs could be a consequence of acknowledging and fancying such aesthetic encounters. Letting nature coat the material environment will come as an ecological solution for the aesthetic and environmental problem - vegetation advances, where human intervention recedes.

Figure 5. Spontaneous vegetation captured to exemplify an unintended design occurrence.

Figure 6. Nomadic vegetable gardens, improvised in soil gained in an urban rivulet's valley.

The attribute of citizenship in man became antagonistic with nature cohabitation. It has to be the path of concrete to safely guide you through synthetic jungles of limited and repetitive patterns. The danger is opportunistic and nature wild and spreading, therefore one was trained by words and symbols. The hermeneutics of human and nature behavior was inseminated in our urban minds since infancy. Restriction come in words and geometric plates inoculating decency and clinic relation with nature: *Don't step on the grass*. Awakening and waiting are the drives to experience nature. The action of turning toward the petty land at hand would have as a consequence increased empathy with the environment.

Citizenry should have a stance in shaping urban environment, thus being able to develop particular patches of green and repose, intuitively guiding exuberances of vegetation (figure 6). Allowance accredited to citizenry into shaping environment will result in objectifying spaces to meet their needs and preferences. So is the case exemplified in figure 6, where dwellers' perception of the land at hand, inspired a change in its purpose and therefore its appearance.

The way in which humans perceive their environment is in relation to what it affords, becoming the medium for attaining human purpose. Purpose, which might be influenced and triggered by the order of the environment's constituting elements and the characteristic of these elements [4].

Nomadic vegetable gardens, improvised in soil gained by circling trunks of decorative trees in an urban rivulet's valley. These diminutive gardens are proofs of un-receded agricultural impetus inherent in the urban transplanted rural man, in need of soil and cropping food. The peculiarity of the spontaneous outbursts of green and food reside in their placement in inadequate space and shaping to adapt to the peculiarity of space. Interdictions are implemented sympathetically due to the fact that the clandestine gardeners are superficially urbanized rural individuals. The space becomes a medium for healing melancholies, and offering comfort by accommodating a continuation of disrupted practices. Space is thus seen as soothing grounds for alienated spirits. Periods of intensive agricultural activity is followed by periods of abstinence (owing to applied interdiction) equivalent with a period of vegetal pauperization and bareness of the space.

Drawing back on an emphatic relationship with nature by launching a tendency toward human – environment interaction on a level of:

- acknowledging – the visual interconnection with the green environment would be of enhanced effect if that which is visualized breaks with convention and follows the pattern of nature as encountered in a wild environment, therefore appealing to memory and emotions. Professional intervention should be discreet and minimal;
- using – within the urban environment, the human and natural element should interconnect on an experiencing level. This would request rethinking green space design in a way that could embrace spontaneous human intervention without creating prejudice to the design project;
- experiencing – human's drive toward a tactile or physical relation with nature must be considered. To limit the interaction to a strictly contemplative experience is coercive;
- Imprinting personality marks – humans leave marks on the environment in an attempt to signalize an existing individuality. While doing this, they attach emotion to the place thus marked. It is a feeling of attachment toward that surface which suffered personal intervention;
- rehabilitating nature - a liberating gesture of letting nature be natural. Equivalent to McHarg's (1900) 'redemptive urban nature,' rehabilitated nature has as rationale conserving remnants of native habitats and species, as well as restoring degraded ecosystems. Throughout McHarg's (1900) narrative of restoring urban nature, urban habitats must recall former landscapes previous to the city's inception. Within McHarg's (1900) discourse the controversial issue of 'invasive species' is introduced, identified as non-native species transplanted and acclimatized through human intervention. McHarg (1900) instigates an ideationally struggle with invasive, non-native vegetation stirred by a vision of urban landscapes as weedlands standing for urban ecosystems. The driving force is the belief in the malignant influence of predatory weeds acting as impediments in restoring the sourcing vegetal ecosystem [5].

4. A neophyte stamp on greenspace development

Old communities maintain customary habits and transpose practices from rural to urban habitats, skipping laws or regulations. Annexes to annexes, interwoven roofs, unintentional associations of colors and materials, mixed styles, cohabitation of old and new, everything unintentional. The explanation could reside in the association of poor financial conditions with the need of expanding space, gaining space, absence of restrictions, or unauthorized, unimpeded intervention on the architecture of built structures, etc.

Nomadic, new settlements, coagulating in the vicinity of the construction sites, dwelling cells of transit and toil, are as many examples of alleviating improvisations of high efficiency but low comfort. The enclosing walls become substitute means for cultivating spontaneous vegetable gardens (figure 7).

The article's intention is not only to ovate the inspirational power of improvisation (in spite of exhibiting confusion in style and maneuvering), but also to amend the opposing overdesigned landscaping. There is limited place for fortuitous accidents given the universality of forms and patterns, repeatability, tamed creativity, dullness and dehumanization, absence or limited unauthorized intervention.

The inducted indolence of the citizenry turned into spectators, in case of greenspace design, should be converted into communion and participation. Citizens should be encouraged to get involved with the surrounding environment. Involvement conducts to a spirit of appropriation. This will concretize the relationship between the inhabitant and its habitat.

At an incipient phase in reshaping urban greenspaces lies researching and experiencing in situ the predetermined site. Ideas should be the embodiment of resulted information and citizenry's response to the forwarded design solutions.

The article insists on the aesthetic, ecological and sociologic fortuitous consequences through the design process of replicating wilderness, the creation of spaces of interaction, self-development and permitted neophytes' involvement.

Figure 7. Improvise for full efficiency. Adapt any medium to satisfy needs, as supplying food and provide a substitute for melancholic imagery.

Failing to provide involving opportunities for citizenry would degenerate overtime into a declining empathy for nature and an increasing alienation from nature:

- receding inclination toward involvement with the natural environment;
- submitting to the mainstream's restrained behavior in relation to the environment;
- accepting design scenarios without questioning;
- restraining from expecting and stimulating change.

In order to reverse the process of alienation from nature and the material environment designers should:

- stimulate curiosity towards the natural and the object oriented environment;
- encourage a critical attitude and induce a continuous rise in demands;
- create chances for human - nature interaction within the urban landscape;
- guide attitudes toward human - nature harmonious cohabitation;
- acclimatize wilderness to the urban context.

4.1. Nomadic vegetable gardens

Enquiries have been made with regard to acknowledged improvised vegetable gardens in opportunistic but not intruding places (customs sustained and propagated by old folks transplanted from rural to urban habitats). The young generation lost inclination toward gardening due to an interruption in the custom flow. Practicing gardening becomes more a matter of fulfilling a spiritual rather than a practical need in a scenario in which all can be conveniently purchased, and, moreover, the costs often surpass the harvest's value [6].

The existence of nomadic vegetable gardens is being sustained by the still persistent habit within the new wave of urbanized citizens. With the old generation gone, the new generation is still to unveil its orientation but it is already there so real and touchable, it's so foreseeable (figure 8). The mutable character of the nomadic vegetable gardens is demonstrated by absence of continuity. Adaptable and improvised vegetable beds – substitute decorative plants in case of melancholic urban expression caused as a result of a longing for the origins and a feeling of fallow acutely felt. Developing the habit of gardening could be thought over as either triggered and maintained by the perspective of an activity generating income (producing and delivering food, designing greenspaces), or recreation and hobby [6].

Figure 8. Nomadic vegetable gardens – dormant soil revitalized into improvised patches of vegetables which add to the overall appearance of the accommodating space.

In the framework of urban greenspaces or green bands, these nomadic vegetable gardens become productive mediums through their inceptive property of providing food [6]. These green patches of soil will build on the appearance of the pertaining site, substituting for a stroke of human touch, while ensuring seasonal products, wellbeing, and creating opportunities for learning and socializing [7].

Nomadic vegetable gardens stand for whatever soil surface is able to provide proper medium for seeding and growing, unconditioned by productivity. It is considered proper for hosting adaptable vegetable beds any vacant patch of soil, liable to suffer a distortion in availability and affordance toward cultivation and domestication. The low productivity level makes me conclude that the proliferation of these adaptable vegetable beds are caused by motives other than material, profitable, but rather oscillating in the psychological, affective domain.

While signalizing opportunistic vegetable gardens, springing as saluted oases of green spaces improvised and unfitted for the task, one turns toward Olmsted's aim to provide a natural counterpoint to the increasingly dehumanized and abstract urban conditions. His ideas emphasized the possibility of acclimating countryside to the urban environment [8].

5. Ovate nature and the natural

Through design, the receptors could be persuaded to accept an unpredictable aesthetic as dictated by nature in a wild state of development. The designer's task is to instigate the receptors to accept wilderness and natural caption and reception of nature through evocative examples and experiences. Taste and preferences should undertake a shift from overdesigned to continent design, or humbly guiding and trimming naturally developed greenspaces. The appetite for an aesthetic of the natural should be induced indirectly by engaging citizens in admiring, appreciating, experiencing naturally grown nature, designing around nature and creating chances for self-developing nature in urban green spaces and architectural projects.

In the absence of a genuine wilderness as accessible source of inspiration, one might shift attention from long praised and inspirationally exhausted examples, to what Lynch (2001) mentions as 'ruinous attractions', vegetation engulfing derelicts, deteriorating structures, vacant lots found in a transient state of defining and a source of intriguing active and perceptive experiences. While under an uncertain formative state, these marginal, uncontrolled spaces abound in affordances guiding spontaneous behavior in an otherwise restrict environment. Consigned to oblivion by municipality, these spaces are motive for uninhibited imagination and action. While considered accidental, they form as residue manifestations of urban flora. Design and planning embody our "expectations of nature" within the urban context, through choices made regarding vegetal components [9].

Unintentional proliferation of spontaneous vegetation within the framework of unmanaged vacant lots or ruins is rejected by human planning and design. While vacant land is perceived as triggering negative economic, environmental and social influences, urban planning is intended to embody positive opportunities by filling vacant land with infrastructure or opting for parks and gardens, nonetheless this concretizes in spontaneous vegetation substitution.

Resorting to natural conglomerates for inspiration while designing greenspaces would trigger diversity and highly inspirational creations. This would result in emotional attachment, respect for, and protection of thus created counterpart spaces. Users intuitively apprehend involvement while experiencing imagery and material surroundings. As a reaction to designers' involvement citizens might readjust attitudes toward greenspace as encountered in urban design compositions, take pride in spaces of natural beauty, appropriate and return involvement. Indifference and mediocrity would trigger adverse stances. This *don't like, don't look, don't care* attitude has to recede.

One tends to opt for perfunctory alternatives: don't grow, don't repair and don't care since everything is replaceable, conveniently purchasable. The tendency moves toward a previously documented opting, or exercising preferences according to pertinent advising at a choosing stage, therefore I appeal to the opinion shapers who could curve tendencies toward nature and the natural. Society propagates an already existing niche adviser to guide opinions, teaching how to behave for fashion. Likes and dislikes should be the result of a responsible thinking process and should not be reduced to an impulsive concluding act effectuated as a result of facile persuasion by media stimuli, or through perpetual encountering of typical visual expressions. The plethora of choices are perceived through ubiquitous visual stimuli. Decisions in relation to taste are often taken according to what is commonly perceived as beautiful and right.

5.1. The Natural: embedding nature in design

The incentive the article is making to the environment shapers is striving to achieve the natural as opposed to striving to achieve beauty, since achieving the former means achieving the latter. Human habitat is a composition of man-created objects, assembled with preponderance from man-created materials through the mediation of man-created technology. The exit from that which is logical and oppressive within human habitat is expected to be provided by nature which is to be permitted under censorship as man created the natural environment. In such case nature is denaturalized and forced into predetermined shapes. They are the result of inducted order into nature's own order, one which doesn't satisfy our sense of order. In their struggle to reach an unnatural exterior condition of their environment, humans suffer an inner distortion of their conduit and perceptions.

In order to educate the viewer into accepting nature's exuberances, the designer or architect must:

- make nature the prime décor element;
- emphasize the quality of nature which transcends time and aesthetic values;
- suggest to the viewer the presence of nature;
- introduce the viewer to the complexity and diversity of nature.

The shrinkage to which design and architecture subject nature is a consequence of a competing urbanity. Instead of elevating our condition to a state of enhanced comprehension, the shape givers of the urban environment and educators of our perception of it, are taking great efforts in leveling nature to fit our preconceptions. Conventional design and architectural practice propose a restricted nature instead of educating citizens of the transformative implications of achieving nature ubiquity.

Urban nature should dissipate almost imperceptibly in its surroundings, engulfing, framing infrastructure, thus easing its bearing upon human perception. Nature is no longer natural and the surroundings are other human distortions of it. Humans lost balance by underestimating nature's natural value, trying to subdue nature by shrinking it into dictated shapes.

When designing and planning, one must persist in remembering the fact that nature is a source of inspiration and a bank of answers for existing and arising problems - we are from nature, and should follow similar laws. Thoreau (2010) instigated toward regarding man firstly as part of Nature, and secondly as appertaining to society [3]. The need for nature and the natural has been distorted through the medium of design in restrictive outcomes, as for the other "genuine needs" they are being predominantly neglected by the designer [10]. Design should embrace Commoner's dictum that 'nature knows best' [11]. Design should be dictated by feelings of humility, decency, respect and appreciation of nature. Following natural laws in designing green spaces will bring benefit to both humans and the environment, while elevating urban aesthetic.

By embracing a principle of respect while designing the shape and composition of urban nature in greenspaces, aesthetics would have to undergo a change. Aesthetic now has short circuited contact with nature. Aesthetic is a coined term that does not apply to nature but to human expressions in relation to nature. Human developed structures should be minimal, while nature should be maximal, a perpetually perfecting design, while guiding engulfing nature. Nature is value in itself. Adorn design with nature while welcome nature saturation in the environment.

Aiming to restore nature in an urban environment fits within McHarg's (1900) strives to initiate a counteraction to the accelerating degradation of the original ecosystem upon which a city was built. Within the frame of his ideas, this indigenous ecosystem is restored as greenspace amidst urban space. In the designs influenced by his ideas, the restoration privileges native flora, as authentic nature, to restore within the urban habitat [5].

5.2. A living façade

To take the generous stance toward nature to an extreme would be to assign subtle designed frames for unleashed nature in a highly coercive society. Within this context, the rebellious manifestations of nature would be a relief valve, a welcomed breach toward envisioning an alternative in conducting existence, stirring an appetite for behavioral change.

Pointing toward the objective visual expository of green facades, one has but to question the scarceness of their happening in an undisputed concrete glassed environment. The partially or totally covered façade is an architectural personality index, testifying for human aptness towards nature and aesthetic attitude. The existence of such casual occurrences provides as many clues of the feasibility of such objectively denominated green design.

Decorating with nature could mean, as in figure 9, capitulating to the siege of vines which adds chameleonic attributes while saving repetitive renovation of the host built structure's facade. The whole built structure gains in aesthetic valor, changing appearance, while moving toward the peculiar style. Not many prefer such alternative of letting and guiding nature. More often the superannuated is talked about and is seems to be in dissonance to that which is progressive and mainstream. A minimal design and a maximal burst of nature, giving free hand to envisioning patterns and a mixture of hues and colors offered by leafs, flowers and fruits are a boundless source of compositional possibilities.

The core idea is to seek inspiration in visually experienced unintentional aesthetic occurrences, imprinted in the perceiver's aesthetic reader. Through a process of observation, registration and evaluation, perceived images will be stored as data, signalizing similar previous encounters.

The rationality is in relation to the concept of living, green walls, partially or totally embedded in an entanglement of vines. They add with minimal demands, permitted being the only repressed action needed for their cyclicality. The expression is that of an unarticulated intention - wall surface monopolization. Under discussion are the extreme cases of abandonment and neglect, having as a consequence the aforementioned unintentional aesthetic occurrences.

Separating cause and effect – the latter being one of aesthetic impact, the discussion could be turned in the direction of unintentional aesthetic occurrences, within the limits of which the intercepted visual experience becomes one with embedded inspirational value. It is a matter of taking inspiration from those unintentional aesthetic occurrences which are to be defined under the given circumstances as:

- aesthetic accidents which imply and are caused by randomly associated natural elements, arranged in visually appealing compositions;

- the interconnection of natural elements and material elements within the range of a given space or context that are solely subjected to the laws of natural development.

One must admit the contradictory impression around a site similar to the one discussed. There is the characterized peculiarity that stirs opposed feelings, possibly based on:

- personal taste – the perceiver's consideration regarding the visual outcome of an engulfed structure by spontaneous vegetation, it is a matter of subjectivity; it is also a matter of social image since perceivers tend to consider such visual expressions as malign proofs of decadence;
- awareness of negative effects – it is an alternative which requires in addition expenses and uncommon interventions in preparing the targeted surface to avoid imminent decay;
- predisposition to nature and the natural – one might opt for a nature engulfing built structures as opposed to a pristine alternative and unabated conflict to rebellious vegetation.

Despite contradictory opinions on the subject, it is an undisputed fact that sites hosting such aesthetic occurrences, are to be considered under a beauty denominator. There is the existing peculiarity given by the spreading of nature in form of isolating layers of green on surfaces that will demarcate them from bare homologues (figure 10.). Weeds and vines are the appanage of ruins and vacancy, of that what is disabled and abandoned. Wherever humans step out there is an exuberance of wilderness ready to step in. Ceased activity and human withdrawal have as consequence nature monopolization.

Figure 9. Decorating with nature.

Figure 10. An example of nature monopolization.

Human perception is trained to see urban wilderness as pejorative. The article advocates for a shift in the citizens' prejudgment. While design should be minimal and perfected, nature should be welcomed in its intended concrete/soil connections. The continuance and resiliency of wild vegetation should be coopted and adapted to complete design into a revitalized urban environment.

There is the habit of subjecting human deeds to interpretation and critic in terms of their aesthetic valor, as for the nature and its myriads extensions one cannot overdo, exaggerate in adding too much of it while beautifying.

One must acknowledge the incubation of that "something" of special valor which circumscribes these hosting buildings from others. There are the so perceived negative effects of the intrinsic quality of nature to accommodate life: parasites, dampness, leaks, a surplus of shadow, sedimentation of dead organic matter, etc. This might conduct the reasoning toward unfavorable conclusions and words would be uttered such as: stagnation, abandonment, or excessive, peculiar (figure 11.).

Life with its cyclicality is to be experienced more acute within the premises of such environment. In relation to experience: the light and shadow, sounds, colors, smells would provide stimuli in a perpetual changeability and at various intensities. The hypothetic inhabitant would substitute for a witness of movement and perpetual transformation.

IOP Conf. Series: Materials Science and Engineering **399** (2018) 012053 doi:10.1088/1757-899X/399/1/012053

Figure 11. Inspiration for alternative roofing. Let nature sprout and mature in places it fancies as propitious. Allowing is the only requirement for escaping architectural anonymity.

Man transcends its pettiness entering in a relation of concentricity with nature and the natural, recalibrating values, purposes. Man reconnects with humanity through a process of naturalization. Justify a directed engulfment of the urban habitat with rebellious vegetation, through the increased demand for green, based on an awakened urge for living, feeling, trying.

5.3. *Keep of the tender grass!*

This is a warning that forms generations of men estranged from nature, alerting to keep nature in a register of contemplation rather than experimentation. People feel the need for communion which is more probable to occur with nature as scenery. The signal is breaching the impulse of interpersonal communion. It results in increased frustration, detachment, the interdiction being perceived as a form of assault to personal freedom. As opposed to the over-exercised control within the green space premises, offering nature and natural alternatives are to be considered. Citizens possess an unconscious oversaturated perception of restrictions, concrete and repetitious patterns. While travelling the way of asphalt, one might encounter breaches in the otherwise flat surface (figure 12 and 13).

Figure 12. Beautified by nature – moss coating sidewalk creates an exquisite example of spontaneous vegetation as décor element.

Figure 13. Beautified by nature – grass invading fissures in the asphalt on a road with infrequent traffic.

This obsessing concrete coating is an outcome of disguised phobia of soil and nature's unpredictability. One must embrace errors that result in unintended aesthetic occurrences, such as the cracks in the asphalt which accommodate grass and weeds in becoming opportunistic green nettings.

In the breach between the street and the side-lane, the errored joints leave space for stray plants needing but ignoring and letting be, in what might be considered the "green stiches" of an injured environment. Such insertion is perceived as the hallmark of inadvertence in attendance. In nature, sustainability is to be found as embedded quality.

Humans acquire sustainability through a cognitive process and implementation developed in time. In succeeding, we need to rely on learning and acquiring skills. It is but logical to follow proved capability, as the one employed by nature. Humans strive to humanize nature. If initiated and attended, supported in functioning to full capacity, green spaces could become genuine urban oases of wilderness, offering citizenry a chance to experience a quality substitute for a nature denied on a daily basis.

5.4. *Nature and the natural perceived as the apanage of rurality*

Nature in free form can be accommodated in urban spaces. Nature is not to be restrained to rural spaces and receding wilderness. If one is to consider that rural spaces lack the benefits of urbanity concerning enhancement in life quality, chances of affirmation, cultural and social experiences, the urbanity lacks environmentally friendly surroundings, peace and a slower rhythm of life. The urban and the rural with the acceleration of urbanization will soon collide and further intermingle. It is a chance in creating an idyllic concentric urban/rural space for the benefit of the inhabitants of both administrative organizations.

One cannot perceive genuine nature in the urban space, nature experience is labelled as a rural component. Nature, in its natural state within an urban environment, is considered unrefined, unsophisticated, un-urban, henceforth the need to civilize it.

Where nature is untrimmed a feeling of abandonment and disregard persists, while littering and improper behavior are more probable to occur. Nature must be subjected to subtle design, and landscape designers must practice restrain while taking genuine natural spaces as inspiration.

The article addresses the need for creating natural green space compositions, recalling the aesthetic and biological composition of the initial landscape (using indigenous plants, implying nonaggressive, ecologically sound techniques), subtly framed by design, framing, ads to the natural composition the status of under signature project. Viewers will attach to the perceived site respect for an authoritative validation of an existing design project.

6. Decorating with nature

There is a cohabitation of past and present design patters inspired by nature. The selection of inspirational vegetal patters, their stylization conducted to reach a symbolic state recall pattern's origin. The selection, synthetization and coding is implicitly subjected to a decanting process developed in time. Some patterns survive and transcend time, spreading an aesthetical orientation.

The article advocates for nature as décor. Inspiration from nature could come uncut, not complying to the human aesthetic filter. In natural state, it comes in opposition to a designer's artistic individuality of which residues due to author's perception, processing visual information and expressing personal interpretations, tend to distort the motive of inspiration.

This article tends to signalize the beauty of nature's rebellious manifestations, spontaneous vegetation, valorized through restrained design. Not only processed urban landscapes functioning under rules and restrictions are subjects of appreciation or evaluated as being worthy of protection, but heaths defined as beauty in crude state, spaces of escape, unfortunately surviving only due to calculated disregard or negligence, perpetual delay in space rehabilitation. Such spaces are to be encountered at the city's outskirts, purlieus.

Tallmadge (2004) found them in vacant lots with unfulfilled capabilities. He uses the term *wilderness*, trying a description of these landscapes, with reserve. His reserve toward pushing defining terms is justified by the impossibility of justly allotting urban heaths to what one was trained to categorize as wilderness. This urban made wilderness suffer, according to his statement, too much human impact, and the amalgamation of alien species together teaming with the absence of specific fauna, add up to conclude on an unfit use of the term [2].

Where trees raise undressed and grass sprouts untrimmed and following a syncopate rhythm, the human impact is more acute owing to a circulating impression of unsanctioned intervention. One tends to allocate valor to spaces in obvious custody. This article exhibits unintended aesthetic occurrences of uncut beauty found accidentally in unattended spaces of spontaneous nature. Urban landscapes as such, because of their blended composition and lack of compositional norms and structure, do not fit with established notions of beauty and value in nature [2].

Those compositions are to be taken as sources of inspiration, overtaken and subjected to subtle interventions for reaching their full potential. One must search, select and make the unintended aesthetic occurrences something repeatable. Sometimes nature aligns in certain ways appealing to our aesthetic perception, all due to nature and hazard.

Tallmadge refuses implying qualifying notions of *nature* or *ecosystem* with regard to opportunistic vegetation and wastelands due to the poor quality of the vegetation and unfortunate human marks in form of waste amalgamation [2]. Spontaneous vegetation is a product of human oblivion and nature's opportunism, the closest urban nature gets to its wild counterpart. Spontaneous vegetation, heaths, wastelands comprises elements of a rebellious kind of nature, whose ecological and aesthetic valor has still to be identified.

Greenspaces counting as décor elements: bare trees, undressed bushes and uncut grass, all un-cosmeticized, are not enjoying visual appreciation, being estimated as heaths and abstaining from further enquiries into their ecological or aesthetic status. With minimal design, they can be upgraded to spaces resulted as objectified design scheme even if abstinent in revelation. In order to be acknowledged as valuable spaces, they must signalize the embodiment of a design project, otherwise risking citizen's belief of entering a space of forgetfulness and omission.

The natural element, if permeated, spreads through itself, beautify pro bono. One must stop seeing negligence in nature's rebellion. In every crack there sprang nature. Offered a chance, nature would coat and permeate our world of concrete and lifelessness, shaping it in something alive and ever changing. The question that arises is if, through a process of intentional omission, green spaces could exhibit unintentional aesthetic occurrences, further developing into natural aesthetic organisms.

Within the urban context, the observer's preconceived opinions regarding the ecological and aesthetic value of nature are based on traditional concepts, as acquired through past encounters with promoted green space designs. Persisting in preconceptions in relation with nature's behavior and exhibits results in lack of awareness, understanding and acceptance of new manifestations of nature within the urban environment. This spontaneous vegetation, self-propagating and flourishing in neglected urban wastelands and margins like vacant lots and other urban waste spaces could, with minimum design interventions, become a valid representation of a substitute nature.

6.1. Create opportunities for nature: Returning a gesture of hospitality

Within the brackets of urban design, green spaces have to be approached from the perspective of interpolating a missing element – nature in genuine form. Nature's intrinsic character of propagation could be exploited with gains in the endeavor of satisfying a rising demand for the natural element while decreasing the efforts of imagining and maintaining. With minimal intervention but proper initiating and maintaining, the expected result would be one approaching genuine cuts out from a remote natural environment, opposing a meticulously designed urbanity.

The visual effect would be one of high contrast, speculating the novel element, and targeting transformations in human perception and involvement, via human's contained capability of bewilderment. Where expectations are toward encountering a unitary pattern concerning urban nature, these oases of untamed nature would instigate to reaction and cogitation.

The driver behind such endeavors of symbolically transplanting (through registration and inspired design schemes) genuine natural composition into clearly delimited urban spaces is the human instinct for nature and the natural. Irrespective of the deviation degree, the natural instinct is under a latent state. It needs a trigger for accessing those distorted altered liaisons. Embracing the tendency in urban landscaping of emulating natural habitats in artificial surroundings would benefit both the urban environment and citizenry.

Schmitt (1990) discusses the subject of 'nature in need of redemption', fulfilling a decided need for green as a beneficial factor within the urban industrial environment. Urbanized nature follow the trajectory of traditional green space design compositions concretized into recreational spaces in which human agency holster accredited nature [12].

7. Conflictual aesthetic

Marsh (2003) raised awareness of the ill-effects of human agency, punctuating the 'unforeseen and unintended consequences', as well as the wrong application of technological enterprise. Man as a 'disturbing agent' and its predisposition to irreversibly mingle inorganic matter and organic life finds itself in antagonistic relation with nature [13].

Through a nostalgic recall, one must admit to a reversing tactic: while at the beginning nature dictated the behavioral patterns in humans, now we are facing the effects of an Anthropocene Age. Gobster *et al.* (2007) signals anthropogenic risks in relation to contradictory ecological and aesthetic values of landscapes, in that the aesthetic qualities could conflict ecological processes while developing landscape appearance. As prevention method against anthropogenic landscape change due to fulfilling aesthetic preferences Gobster *et al.* (2007) bring forward the notion of ecological aesthetic. According to Gobster *et al.* (2007), it is advisable for humans to resonate to the aesthetic stimuli of landscapes that comprise beneficial ecological functions. Therefore, consuming landscape aesthetic will intrinsically contribute to healthier ecosystems, and through extension, human health and welfare. A controversial enquiry is launched whether, in an antagonistic scenario in which ecological goals and aesthetic preferences are contradictory, the aesthetic should take the form of practicality and ethical consideration toward the environment [14].

8. The crave for wilderness

Fulfil citizen's tamed appetite for wilderness by supplementing that which is provided in urban context by our contemporaneity. Welcome a symbiosis of the two, human and nature cohabitation, conducting to the completion of the contemporary citizen. Design projects should involve a natural nature into the urban decorum by creating chances for its spreading into ubiquity. Goods are protected by laws under the menace of applied penalties. Nature/wilderness should be protected, while placing it at the core of citizens thinking and living. Having something under permanent scrutiny will behave as a commercial. Wide exposure will add familiarity. Reinstall familiarity with a natural nature by adding it in the center of our urban existence.

McHarg (1900) points toward the need of redefining urban nature, which could emerge by embedding nature in urban design. The expected result is that of restoring the ecological function to the damaged urban ecosystem, while adding a new aesthetic to the urban environment [5].

There is a rising request for green as shown by the emphasis placed in advertisement upon forwarding utopic shades of green and blue. It is not ignorance but indifference. Seize the proper moment based on the mechanics of request and demand, while competing with the artificial plants market. Where there is no space, improvise space for the green and natural.

Greenspace design projects lost the unexpected element. Greenspaces are as predictable as the building structures composing their vicinity. To overcome visual apathy, the designer must take as inspiration genuine manifestations of nature as found in a wild state, replicate wild scenarios and learn during the process.

On a brick fence, among broken glass and barbed wire there spring a tree sprout. The nature doesn't care for trans-passing. It doesn't know better but to spread and sustain cyclicality following nature's pattern. The law of decency collapses under extreme circumstances. Nature is at war, making the best of what is left to it.

Weeds sprout out of concrete cavities, being as many examples of working methods for growing plants on walls and fences (figure 14). Spontaneous vegetation could be envisioned as an active agent for developing green design and architecture projects, aiming at reaching an urbanity engulfed in nature.

8.1. Creating a context for nature

While imitating nature's ways, create a designed alternative with an open result of green walls, paths, fences, spaces that seemed easily adaptable to their natural surroundings (intended cracks, drawing with nature, tracing frail contours of seeded plants mingling with weeds).

The moss on the shop arcade, the dripping plant taking its downside course and dead vegetation deposited in eaves go unnoticed (figure 15).

Figure 14. Self-initiated and propagating green wall – visual example of spontaneous vegetation.

Figure 15. Unnoticed presence of nature - example of unintended green design occurrence.

In search for new material for proving the inspirational validity of *unintended design occurrences,* one must be prepared to observe freed from preconceptions and seek beauty in the unconventional, ignored, and underrated. Such is the case illustrated in figure 15, where the vegetal elements self - developed indiscriminately to the ugliness and dullness of their circumstantial habitat. One has but to hope for a continuation of a lax maintenance practice.

Spontaneous vegetation could be accommodated: intended cracks with incipient nature, walls and structures, roofs, sidewalks, etc. A crack in the wall, pavement, a moment of oblivion and you witness a proof of nature's survival (figure 16).

Figure 16. Unintended integration of nature in design and architecture.

Accepting, releasing, admiring, placing the emphasis on integrating with subtlety in design, architecture, and planning the effusions of nature. The designed environment must create the context for nature, provide a scene on which it can perform in plenitude. If vegetation finds its way in adverse conditions, one could just imagine its magnitude in specially designed ones. Opting for subtlety, naturalness, and raw materials, minimal intervention in form and almost imperceptible designs, expect a subjective compositional structuring. Choosing and collating, keeping and weeding while exercising restrain from creative effusions and trusting the medium (soil) for its latent qualities and possibilities. Simplify as a form of efficiency, minimize as a form of beauty. Metaphorically speaking, within a green space context, design is the frame for exhibiting nature thus analyzed and accepted as valid.

The imperative question is if we are prepared to accept nature's insubordination, within an urban environment. Change perspective on permissiveness, on that which is commonly accepted in an urban environment (in the crevices created by the defective joining of dales composing the sidewalk pavement, plants strive for existence; weeds sprout in thin mater sedimentation in the defective walls). Delaying prompt intervention in removing unexpected vividness of stray vegetation results in what might be considered an abandoned space, the dominion of negligence and low interest for the place found under described auspices. This article advocates against a hasty and doubtless intervention (removal of the unaccounted for vegetation), and considers such occurrences as fortuitous encounters with inspiration. Could it be that the presence of stray vegetation piercing its way in between asphalted streets and concrete curbs mediate the chance for experiencing an unintended aesthetic phenomenon?

9. Guidelines for urban regularization of human-nature relationship

One witnesses in the development of normalized green space designs the traces of ecological damages caused by:

- the futility of an unnatural drive for manipulating vegetation into forms and patterns;
- abuse in submitting nature to aesthetic rules for the sake of visual gimmicks.

There is a need for changing rules, approaches and ideas based on respect and amiable communion with nature in a form and composition concordant to its natural state. Nature is uncivilized by fact, in this residing its power and beauty. In natural state, there is a receding aesthetic characteristic. A restructuring in green space design is imperative. Initiating a redesigning campaign could be interpreted in terms of embracing a revival and integration of native plants and natural aesthetic into urban spaces. Nature in a natural state accommodated into an urban context is also a psychological support for the citizenry. One must admit the psychological dimension of spaces turned wild, enclosed in the framing of minimal design.

Conducted enquiries were in relation to signalizing and structuring inspirational venues into a working methodology to benefit landscape, architecture and urban design.

In order to initiate a change in regularizing human-nature relationship and concentric ties within the urban environment, few guidelines are as follows:

- finding inspiration from nature and wilderness;
- keep choices locally. Reintroduce plants from local flora and reconnect space with local ecosystems;
- design subtle frames while using with preponderance natural materials;
- think the whole cycle of a design project (changes of seasons, sprout, blooming, withering, etc.). Take into account the way in which these metamorphoses transform design;
- think human interaction – allow and encourage human touch;
- consider the human element as an implicit part in your designs;
- maintain, while guiding without enforced interfering in nature's self-development;
- think to create and propagate diversity;
- create contrasts and accents (always one center of interest within the composition);
- keep in mind that nature is not decoration – nature *is*, and must be emphasized by subtle framing;
- add to nature's encouraged invasiveness, the need of a strategic maintenance in order to preserve biodiversity (selective weeding, trimming and planting);
- accept, encourage, guide and appreciate nature's exuberance.

10. Conclusions

The article stresses the importance of rethinking laws on human and greenspace interaction in the urban context. With citizens restrained in asphalted areas and nature distorted by invasive projects, the result is the alienation of man from nature. While man is not at ease in a natural environment or when confronted with that which is in a natural state, nature is distorted by excessive rationality.

The proposed design scenario is one related to urban green spaces being dictated by abandonment to chance and a subtle orchestrated wilderness. It is a matter of nurturing a visually effective contrast between a subtle design intervention (subtle and facile assimilation of the material components in a natural context) and the chaotic but plainly perceptible invasion of the natural.

Based on the information received while examining the revelatory examples offered by nomadic vegetation found in certain aesthetic relationship with a hosting material environment, the designer or the architect, could find as practical and inspirational the ways of opportunistic nature, and orient design-related decisions toward accommodating vegetation in projects with ecological and sustainable agendas.

Following the principle of *green literality*, nature coating as opposed to concrete coating could be perceived as an ecologically sound design solution for both aesthetic and environmental issues within the urban context.

References

[1] Ryan J, Simson A 2002 'Neighbourwoods': identifying good practice in the design of urban woodlands *Arboricultural Journal* **26:4** 309-31.

[2] Tallmadge J 2004 *The Cincinnati Arch: Learning form Nature in the City* (Georgia: University of Georgia Press)

[3] Thoreau H D Walking *The Atlantic Monthly, A Magazine of Literature, Art, and Politics. Boston: Ticknor and Fields.* ***IX*** *(LVI): 657 – 674. June 1862*

[4] Kaplan, S 1988 Perception and landscape: conceptions and misconceptions In: *Environmental Aesthetics: Theory, Research and Applications* (Cambridge: Cambridge University Press)
[5] McHarg I L 1900 *Design with Nature* (New York: Wiley)
[6] van Leeuwen E, Nijkamp P and de Noronha Vaz T 2010 The multifunctional use of urban greenspace *Int. J. Agr. Sustain.* **8: 1-2** 20-5
[7] Armstrong, D 2000 A survey of community gardens in upstate New-York: implications for health promotion and community development *Health & Place* **6** 319-27
[8] Martin J 2012 *Genius of Place: The Life of Frederick Law Olmsted* (Boston: Da Capo Press)
[9] Lynch K 2001 *Wasting Away - An Exploration of Waste: What It Is, How It Happens, Why We Fear It, How to Do It Well* (New York: Random House)
[10] Papanek V 2005 *Design for the Real World: Human Ecology and Social Change* (Chicago: Chicago Review Press)
[11] Commoner B 1971 *The Closing Circle: Nature, Man, and Technology* (New York: Random House)
[12] Schmitt P J 1990 *Back to Nature: The Arcadian Myth in Urban America* (London: Johns Hopkins University Press)
[13] Marsh G P 2003 *Man and Nature: Or, Physical Geography as Modified by Human Action* (Seattle: University of Washington Press)
[14] Gobster P H, Nassauer J I, Daniel T C and Fry G 2007 The shared landscape: what does aesthetics have to do with ecology? *Landscape Ecol* **22** 959–72

IOP Conf. Series: Materials Science and Engineering **399** (2018) 012054 doi:10.1088/1757-899X/399/1/012054

The study of steel transfer parts in assembled building connection

H Wang[1], M Li[1*] and N Qiao[1]

[1] School of Civil Engineering, Shenyang Jianzhu University, Shenyang, 110168, China

[*] E-mail: 297531635@qq.com

Abstract. At present, most prefabricated components in the assembled buildings could be assembled at the construction site. But the connection method is still of traditional nature, such as steel connection and seam filled with cast-in-place concrete. So, our research group proposed a series of different new types of connection schemes based on the idea of connecting steel transfer parts. According to the different parts of the connection, including how to connect the assembled shear wall and the coupling beam by the U type bar connector, how to connect the shear walls by the method of mortise joint in the horizontal direction and how to connect the shear wall with the columns and beams in the frame-shear wall structure by the keyways. As compared with the traditional way, the following observations can be made: By using the steel transfer parts to connect prefabricated components, the internal force can be transferred effectively and the connection reliability is ensured. The design requirement of "strong connection" is easy to be achieved, which can increase the seismic performance of the structure. The field wet operation can be reduced and connection parts no longer need to be welded. The assembly rate is raised greatly in the whole structure. In the meanwhile, the proposed schemes still exhibit some drawbacks which need to be improved by finite element simulations and experimental studies.

1. Introduction

In recent ten years, with the improvement of science and the development of assembled building technology in overseas countries, the development of assembled concrete structure has been paid attention to in our country. Rapid progress has been made in scientific research and practical projecting. Under this background, more and more research into assembled concrete structure connection technology has been carried out in our research group, including the connection technology between shear wall and frame column, the connection technology between shear wall and frame beam, the connection technology between coupling beam and shear wall, etc. [1-4]. From the research results at home and abroad, we can infer that the connection technology of assembled concrete structure components still focuses mainly on steel connection. In this method, the steel load-transferred path has hardly changed and the calculation of the whole structure is the same as the cast-in-place structure. However, the construction technology is relatively complicated, which leads to many constraints.

The use of steel transfer parts to connect concrete components can simplify the construction technology efficiently and also the connection mode flexibly and changeably. By using the steel transfer parts to connect prefabricated components, issues, such as load-transferred path transformation and uncoordinated deformation may occur. But, at the present scientific and technological level, these problems could be solved. Even based on such a connection method, the structural system which has better seismic performance may be found improve the industrialization of assembled buildings in future and promote the development of assembled buildings. Accordingly, our research group put forward a variety of connection schemes based on the idea of connecting steel transfer parts. Some of the following schemes will be introduced to expatiate the function of steel transfer parts in assembled buildings.

2. The assembled concrete connection schemes based on the ideal of connecting steel transfer parts

2.1. The connection scheme between precast shear wall and coupling beam

In the current assembled concrete structure, the cast-in-place method is mainly adopted as the connection method between shear wall and coupling beam. By using this method, the concrete pouring quality is difficult to be effectively guaranteed because the amount of concrete is relatively low. Methods, such as steel connected by sleeve and then poured concrete is not available to connect shear wall and coupling beam. To solve these problems, our research group proposed a connection scheme by using the U type steel bar connector to connect precast shear wall and coupling beam, which is based on the idea of connecting steel transfer parts. This scheme mainly includes the precast coupling beam with connector, the precast shear wall with connector, the U type steel bar and the steel hoop, which is shown in figure 1(a), figure 1(b), figure 1(c) and figure 1(e), respectively. At the same time, the connector is composed of embedded steel, steel plate and U type plate with connecting hole, which is shown in figure 1(d).

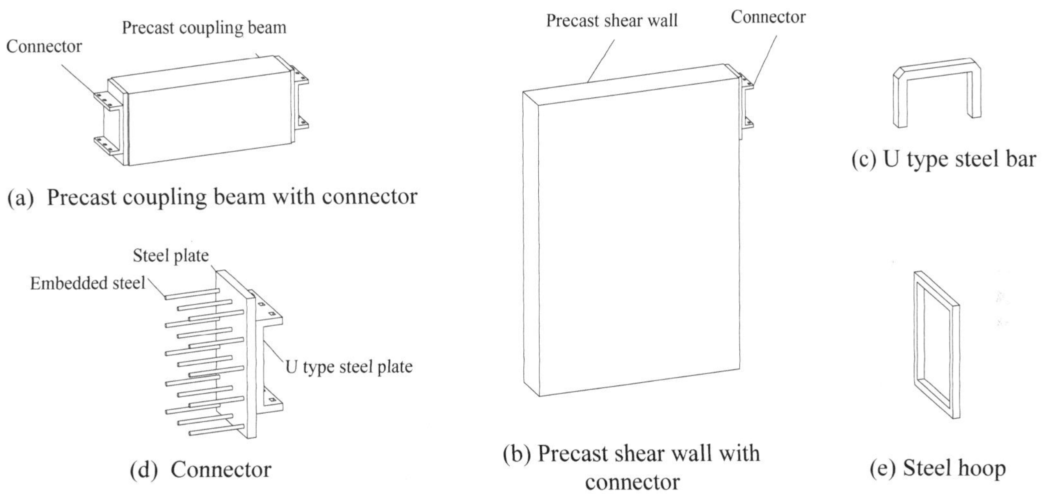

Figure 1. Components of assembled shear wall and coupling beam connected by U type steel bar.

The implementation process of this scheme mainly includes as follows: first, the precast shear wall is fixed in the reserved position and the steel hoop round the connector of shear wall. Second, the precast beam is hoisted in the reserved position, the U type steel bar is inserted into the reserved hole of two connectors to ensure close contact. Then, the steel hoop skids off and round on the outside of U type steel bar, as shown in figure 2(a). Last, the U type framework is supported on the outside of connection location, the grouting material is poured into the framework. After grouping solidified material, the U type framework is dismantled to finish the connection between the coupling beam and the shear wall, as shown in figure 2(b).

In this scheme, the tension tress is caused by the bending moment between coupling beam and precast shear wall, which is transferred by the U type steel bar. The shear force is transferred by grouting material and steel hoop, the pressure is transferred by the U type steel bar and grouting material. In theory, through this method, the bending moment and shear force between precast components can be transferred effectively to satisfy the design requirement of "strong connection". The connection reliability can be ensured in a suitable way for the coupling beam with any size or reinforcement. The connection process has the advantages of easy operation, small seam and no need for welding.

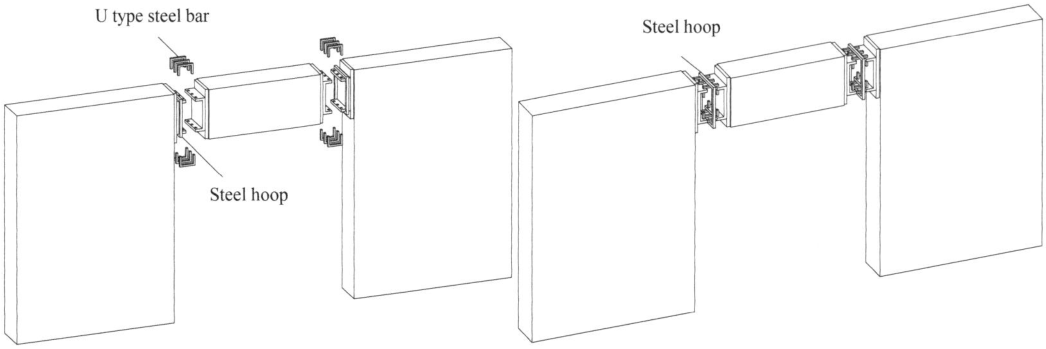

(a) The 3D graph of connection process (b) The 3D graph of connection completion

Figure 2. Assembly process of frame shear structure and coupling beam connected by U type steel bar connector.

2.2. The connection scheme of precast shear wall structure

In the assembled concrete structure, a shear wall needs to be dismantled into 2 or more shear walls if the length is large. Then, the shear walls will have a horizontal connection in the construction site after dismantling. In the current practical project, the shear wall connection in the horizontal direction occurs mainly by inserting longitudinal steel into outstretched horizontal steel. The stirrup set around longitudinal steel to form the embedded column [5, 6]. Based on this, our research group proposed a connection scheme by using the mortise joint to connect shear walls in the horizontal direction, which is based on the idea of connecting steel transfer parts. This scheme mainly includes the lock catch connector and shear wall with C type card slot steel plate connector, which is shown in figure 3(a) and figure 3(b). The C type cart slot steel plate is shown in figure 3(c), which is inserted through the steel plate and welded in the perforated steel plate. There are two grooves in the C type card slot front. The holes on the steel plate are meant to weaken stiffness, which can save steel. The horizontal steel of shear walls goes through the perforated steel plate and welded with perforated steel plate after bending 90 degree.

The implementation process of this scheme mainly includes as follows: First, one precast shear wall is hoisted and fixed in the reserved position. Second, another shear wall is hoisted on the side of first shear wall to ensure that the C type card slot is fitted closely. Then, the lock catch connector is locked in the grooves of the C type card slot, as shown in figure 3(d). Last, the steel framework is supported between two shear walls. The grouting material is poured into the framework to finish the horizontal connection between two shear walls.

In this scheme, the horizontal shear force between two shear walls is transferred to the C type card slot margins of the lateral steel plate, then the shear force transition is completed by the lock catch connector. In the meanwhile, the grouting material between shear walls can also transfer partial shear force. So, this method can achieve the reliability of connection. The connection process also has the advantages of easy operation, high assembly rate, fast construction speed, small seam and no need for welding.

IOP Conf. Series: Materials Science and Engineering **399** (2018) 012054 doi:10.1088/1757-899X/399/1/012054

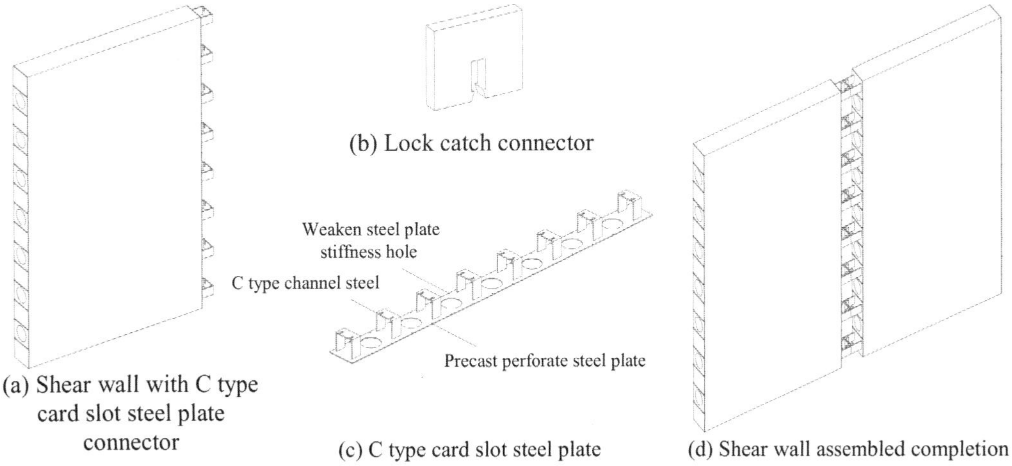

Figure 3. Components of assembly shear wall and shear walls connected by tenon and mortise in horizontal direction.

2.3. The shear wall and column, shear wall and beam connection scheme of precast frame-shear structure

The popular technologies of assembled concrete structures mostly confined to pure frame structure or pure shear wall structure. There is little research done to investigate the assembled concrete frame-shear wall structure. Until now, the concrete frame-shear wall structure has been usually constructed by means of the cast-in-place method [7]. The conventional practice in assembled frame structure is the beam and column precast, the joint casting. The corresponding problems are steel binding difficulty and the joint construction quality is hard to be guaranteed. So our research group proposed a connection scheme by using the keyways to connect the prefabricated components of assembled frame-shear wall structure, which is based on the idea of connecting steel transfer parts. The scheme mainly includes: The precast column with column-wall embedded parts, which is shown in figure 4(a), the cast-in-place foundation beam with beam-wall embedded parts, which is shown in figure 4(b), the precast frame beam with beam-wall embedded parts, which is shown in figure 4(c), the precast shear wall with column-wall embedded parts and beam-wall embedded parts, which is shown in figure 4(d). The I-section column-wall connect key, the groove type beam-wall connect key and the cross type beam-wall connect key are shown in figure 4(e), figure 4(f) and figure 4(g), respectively. The column-wall embedded part and the beam-wall embedded part are shown in figure 4(h) and figure 4(j), respectively.

The implementation process of this scheme is shown in Figure 5 and includes the following steps: First, the precast column is hoisted above the foundation beam to ensure that the steel stretched at the bottom of foundation beam is inserted into the sleeve at the bottom of the precast column. The grouting material is poured into the sleeve, as shown in figure 5(a). Second, the precast shear wall is hoisted above the foundation beam to ensure that the beam-wall embedded parts on the lower side of the shear wall, the beam-wall embedded parts on the upper side of the foundation beam and the column-wall embedded parts on the side of column are aligned. Third, the I-section column-wall connectors, the groove type beam-wall connectors and the cross type beam-wall connectors between the foundation beam, the precast column and the precast shear wall are fixed. The groove type beam-wall connectors are inserted into the holes of the beam-wall embedded part webs, the cross type beam-wall connectors are fixed on the outside of the beam-wall embedded parts. Then, the precast frame beam is hoisted above the precast beam to ensure that the longitudinal steel is inserted into the polyvinyl chloride pipe of core node area. The grouting material is poured into the polyvinyl chloride pipe. Last, the I-section column-wall connectors and the groove type beam-wall connectors between the precast frame beam and the precast shear wall are fixed, as shown in figure 5(b).

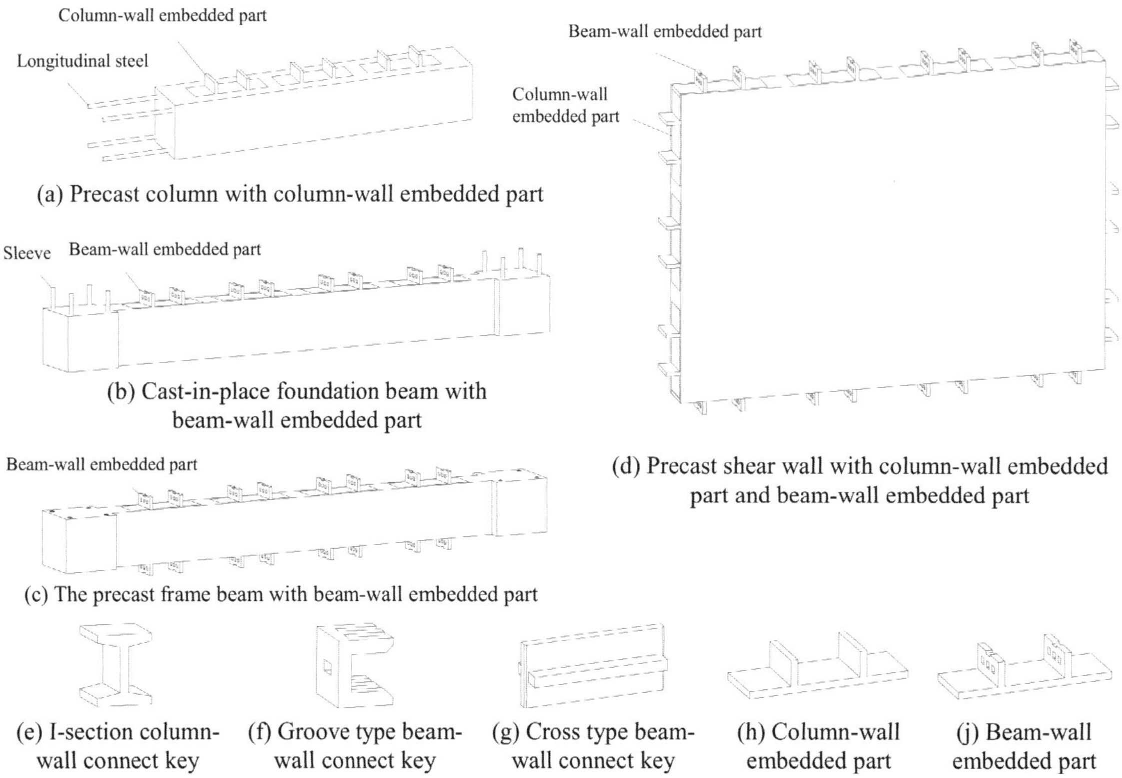

(a) Precast column with column-wall embedded part

(b) Cast-in-place foundation beam with beam-wall embedded part

(c) The precast frame beam with beam-wall embedded part

(d) Precast shear wall with column-wall embedded part and beam-wall embedded part

(e) I-section column-wall connect key (f) Groove type beam-wall connect key (g) Cross type beam-wall connect key (h) Column-wall embedded part (j) Beam-wall embedded part

Figure 4. Components of assembly frame shear structure connected by key-groove method.

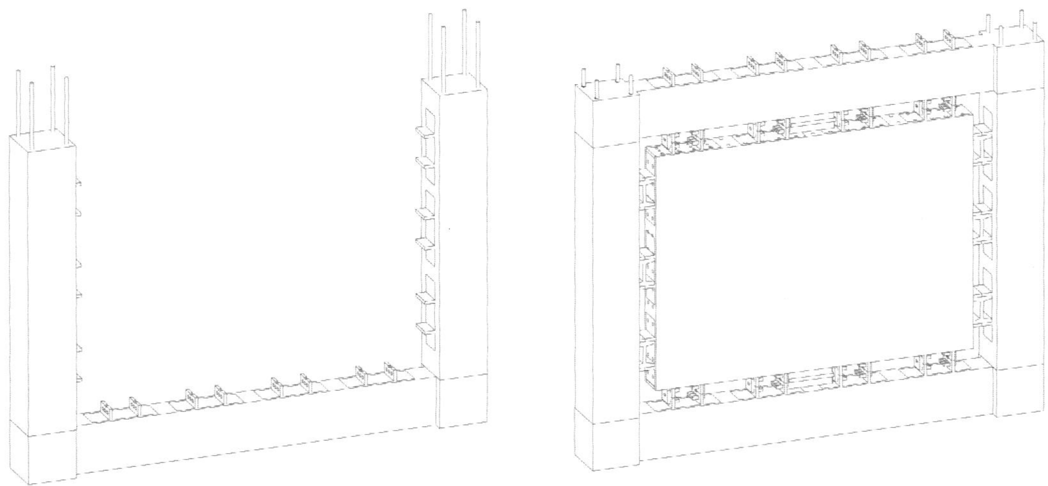

(a) Assembly of precast column and foundation beam (b) Middle frame beam and precast shear wall connected completion

Figure 5. Assembly process of frame shear structure connected by key-groove method.

In this scheme, the pulling force between the frame beam and the shear wall is transferred by the groove type beam-wall connectors locked in the beam-wall embedded part web holes. The shear force between the two is transferred by the grouting material at gaps and the webs of cross type beam-wall connectors. The pressure between the two is transferred by the grouting material, the webs of beam-wall embedded parts and the groove type beam-wall connectors. The reciprocal shear between the frame column and the shear wall is mainly transferred by the I-section column-wall connectors and the grouting material. In theory, the force between the prefabricated components can be transferred effectively and ensure the reliability of the connector. The connection process also has the advantages of easy operation, high assembly rate, fast construction speed, small seam and no need for welding. It can also reduce the field wet operation. Besides, the deformation of the frame-shear wall structure along the horizontal direction can be controlled through adjusting the size of the embedded part and the connector, which makes it obtain better seismic performance.

3. Conclusion

Through introducing the connection schemes of precast concrete structure components based on the idea of connecting steel transfer parts, it can be observed that:

(1) By setting the steel transfer parts, the force between the precast shear wall and the coupling beam, the force along the horizontal direction of the structure and the force between the prefabricated components of the frame-shear structure can be transferred to satisfy the design requirement of "strong connection" in the structure. So, the schemes proposed by our research group have certain feasibility.

(2) In the connection schemes based on the idea of connecting steel transfer parts, the internal force is transferred by steel transfer parts, which makes the reliability of connection easily ensured. The connection process has the advantages of easy operation, high assembly rate, fast construction speed, small seam and no need for welding. The adjustment of the size and shape of steel transfer parts can make the structures have better seismic performance. So, this kind of method has good engineering prospects.

(3) In the connection schemes based on the idea of connecting steel transfer parts, there are still some problems, such as load-transferred path transformation and un-coordination deformation. In addition, the schemes themselves also have many problems that need to be improved, which require plenty of finite element simulations and experimental studies. Our research group will also carry out relevant work to provide technical support for the development of precast concrete technology.

Acknowledgement

This work is financially supported by the Ministry of housing and urban construction projects (2016-K5-009) and projects of Shenyang plan of science and technology (18-013-0-29).

References

[1] Li M, Wang L, Wang Y, Tao W, Shan R and Chen J 2015 Force behavior of parallel double coupling beams with different width *J. Industrial construction.* **45** 239-423

[2] Zhao W and Sun C 2012 Precast concrete frame-shear wall assembling connection structure and assembling connection method *P.* China patent ZL201210281096.4

[3] Zhao W, Tong J and Yuan S 2015 Research on the connection method of column-wall member in precast fram-shear wall structure *J. Journal of Shenyang Jianzhu University* **31(3)** 408-417

[4] Zhao W, Ren G, Jin Q and Lv A 2017 Finite element analysis of assembled shear wall structure by steel cage vertical connection *J. Engineering and Technology Research* **3** 1-2

[5] Zhao W, Guo W, Jin Q and Dui X 2014 State of art research on connection type of vertical components for precast concrete shear wall systems *J. Industrial construction* **44(4)** 115-121

[6] Zhi Q, Guo Z, Xiao Q, Yuan F and Song J 2017 Quasi-static test and strut-and-tie modeling of precast concrete shear walls with grouted lap-spliced connections *J. Construction and Building Materials* **150** 190-203

[7] Guo L, Rong Q, Qu B and Liu J 2017 Testing of steel plate shear walls with composite columns and infill plates connected to beams only *J. Engineering Structures* **136** 165-179

IOP Conf. Series: Materials Science and Engineering **399** (2018) 012055 doi:10.1088/1757-899X/399/1/012055

Study on vibration extraction and control characteristics of the high speed spindle online dynamic

Z Wang[1,2*], W Tu[1,2], J Li[1,2], M Liu[1,2], F Zhu[1,2] and H Chen[1]

[1] School of Mechanical Engineering, Shenyang Jianzhu University, 9 Hunnan East Road, Shenyang 110168, China

[2] National-local joint Engineering Laboratory of NC Machining Equipment and Technology of High-Grade Stone, Shenyang Jianzhu University, 9 Hunnan East Road, Shenyang, 110168, China.

[*] E-mail: juven1126@163.com

Abstract. In order to solve the mechanical spindle low processing accuracy and precision elements problems which were caused by high speed spindle noise and vibration, a robust and reliable vibration extraction and control scheme of mechanical spindle was proposed. Based on the electromagnetic balancing system and by applying LabV1EW visual instrument, high speed spindle noise reduction and vibration signals filtering were pre- processed and vibration signals sampling, quantization, interception were post-processed. Base frequency vibration signals amplitude and phase were extracted by using different signal extraction methods. Through experiment verification and comparative analysis, the optimal vibration signal amplitude and phase extraction method was selected. By using the influence coefficient method, the influential factors including spindle speed, trial weights magnitude and phase were analysed. The high speed spindle unbalance was calculated and compensated. The high speed spindle electromagnetic balancing device block was controlled accurately. Experimental results showed that the vibration amplitude of the mechanical spindle was obviously reduced and the balance efficiency was greatly improved. The influence coefficient method and the electromagnetic balancing system took an active role in the process of dynamic balancing regulation. Such achievements enriched online dynamic balancing theory and promoted the relevant control technique developments.

1. Introduction

The continuous development of high-speed spindles in recent years and the research and application of rotary rotors represented by high-speed spindles, have become more numerous, so it has put forward higher requirements for the rotation precision and stability during the operation of the spindle [1]. The vibration generated by the operation of spindle causes the failure rate of the high-speed spindle and other components to reach 60% to 70% [2]. For the spindle of the machine tool, although the balance had been corrected and the rotation accuracy had met the requirements, there will still be imbalance of spindle vibration and reduction of rotary precision caused by some factors such as design, manufacture, fatigue wear, load shock in the long run process. Therefore, the unbalanced vibration characteristics of the spindle and the regulation characteristics of the on-line dynamic balance have a great influence on the precision of the machine tool. It is of great significance that the operating characteristics of the spindle system are improved based on research of characteristics of spindle vibration signal and control characteristics of dynamic balance.

At present, scholars at home and abroad mainly focus on the development and innovation of online dynamic balance methods and their balancing devices. K. Fedem [3] proposed the criteria for discriminating the flexible rotor. For the flexible rotor, the vibration of the spindle caused by the imbalance is related to many factors: magnitude of imbalance, phase angle, stiffness, damping, structure, supporting conditions, and rotation speed. Therefore, the main spindle dynamic balance control methods can be summarized as [4]: modal balance method, no trial weight balance method and influence coefficient method [5]. Liu Rongqiang and Xia Songbo [6] proposed dynamic balance control methods of residual influence coefficients to solve the problem of too large correction weights; Kang Kaining, Wang Peijun et al. [7] proposed a balanced method minimizing the maximum residual vibration, MINMAX method, which could calculate well to eliminate vibration; Zhu Xiangyang, Zhong Binglin [8] studied the method of least squares influence coefficient balance algorithm using the minimized iterative weighting method; Zhang Xiaoxuan, Tang Yunbing et al. [9] proposed a method that could optimize test mass block and residual vibration of spindle based on the least square method of improved genetic algorithm;Field dynamic balancing experiment of grinding wheel shaft finished by College of Mechanical Engineering, Xi'an Jiao Tong University proved that grinding wheel shaft balancing device could control vibration of the spindle accurately and efficiently, it was obvious that the vibration of the spindle was reduced [10]; the non unloading dynamic balance method based on the influence coefficient method proposed by the researcher of the National Key Laboratory of Mechanical Transmission of Chongqing University [11] had the advantages of high balance precision, simple operation and high balance efficiency; the researchers coming from College of Mechanical Engineering and Applied Electronics Technology, Beijing University of Technology had achieved the precise balance of key parts of machine tool spindle [12]. Meltal had arrived at Modal Balancing Technique through main vibration type of rotor type spindle system at each critical speed; Bishop and Saito, based on the conclusion of Meltal, had solved some problems such as how to choose the mode balance correction and deal with the gravitational drooping in the balance process after a long study [13]. Xu Bingang and Qu Liang Sheng [14] in Xi'an Jiao Tong University realized the combination of modal balance method and hologram technique. Under the condition of a non critical speed, the method could achieve simultaneous equilibrium in a two order mode for rotary rotor with asymmetrical structure; the researchers of the Institute of vibration engineering of Northwestern Polytechnical University have improved the mode balance technology. Using instantaneous amplitude information of spindle raising speed response during the starting process of flexible spindle, flexible rotor was two-plane balanced. The method could quickly and effectively reduce transient residual vibration of rotary spindle without obtaining vibration phase data, and greatly improve the accuracy and efficiency of balancing [15]. Institute of Aeronautical Power Machinery of Aviation Industry Corporation of China and School of Power and Energy Engineering of Northwest University proposed a Balancing method of flexible rotor across second order without trial weights, which solved some problems such as the size of the mass block and multiple shutdown in the process of spindle balance [16]. The vibration amplitude of the rotor was effectively reduced, and numbers of shutdown were reduced in the process of spindle balance. So, accuracy and efficiency of in the balance of the spindle were greatly improved. In summary, some achievements have been made in characteristics of spindle vibration signal and control characteristics of dynamic balance, which has laid a certain theoretical foundation for this study, however, the research results are still lacking in unbalanced vibration signal extraction and control characteristics of dynamic balance of high speed spindle, and vibration signal characteristics and dynamic balancing control characteristics also play an important role in the dynamic balancing efficiency of spindle. This paper mainly studies the extraction and analysis of unbalanced vibration signals of high speed spindle and analysis of control characteristics of online dynamic balance.

2. High-speed spindle unbalanced vibration signal feature extraction method

2.1. High-speed spindle unbalanced vibration signal feature extraction simulation and experiment
The correlation filtering method can strictly control the DC component and noise interference in the signal and achieve accurate extraction of the unbalanced signal. Therefore, the correlation filtering method is applied to extract unbalanced signals. The principle of correlation filtering is shown in figure 1.

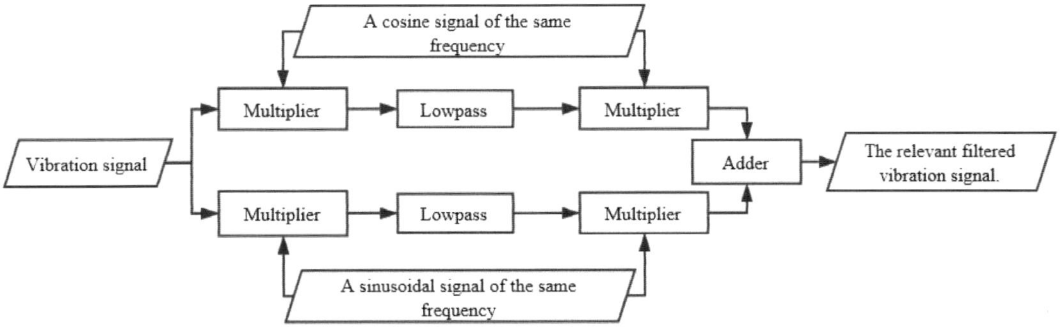

Figure 1. Correlation filtering principle.

The correlation analysis method issued to extract the amplitude and phase of the vibration signal. This method has the characteristics of fast operation, high accuracy and it is also not easily disturbed by other noise and other signals. The on-line dynamic balance of the spindle is established and completed from the signal pre-processing, post-processing, the extraction of the characteristic value of the fundamental frequency vibration signal, and the on-line dynamic balance using the influence coefficient method as the control method. The flow chart is shown in figure 2.

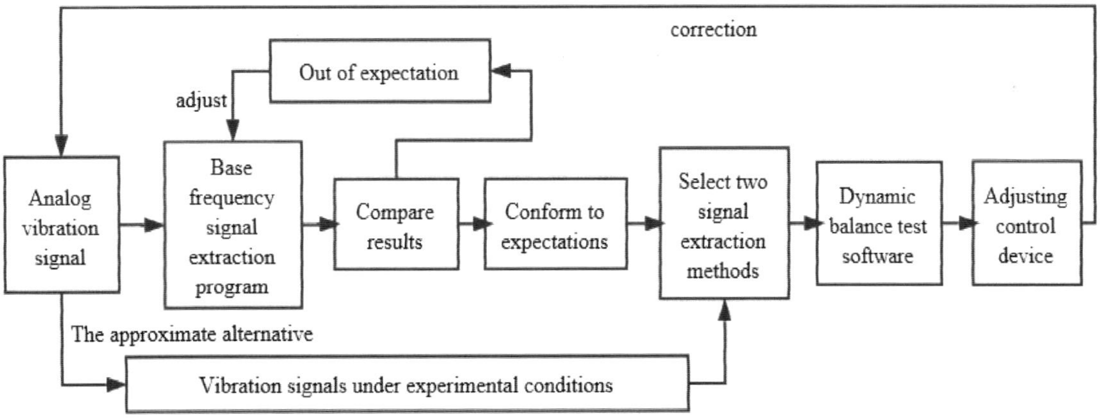

Figure 2. Flow chart of regulation of online dynamic balance vibration signal.

2.1.1. Analysis of simulated vibration signal results. The vibration signal expression is as follows:

$$y(t)=a_0+A\sin(\omega t+\delta)+\sum_{i=1}^{n}a_i\sin(i\omega t+\varphi_i)+s(t). \quad (1)$$

In the formula, a_0 is the DC component; $i\omega$ represents different signal frequencies; φ_i is the phase of the vibration signal at each frequency; A is the fundamental frequency amplitude of the vibration signal; $s(t)$ is noise and other interference signals; $a\sin(\omega t+\phi)$ is a fundamental component signal.

To obtain a more realistic spindle vibration signal, set the following analog simulation signal. The expression of the interference signal of uniform white noise and white noise of Gaussian white noise is expressed as:

$$x(t) = 2+6\sin(\omega t + 30)+3\sin(2\omega t + 20)+2\sin(3\omega t + 10) + s_1(t)+s_2(t) \quad (2)$$

The information of the time waveform of the simulated vibration signal is: uniform white noise, standard deviation of a single Gaussian white noise signal is 1, DC component is 2μm, fundamental frequency is 10 Hz, fundamental frequency vibration signal is 6μm, frequency double vibration signal is 3μm, triple frequency vibration signal is 2μm and the corresponding initial phase is respectively 30°, 20°and 10°. The number of sampling points is 1000 and the sampling frequency is 1000Hz.For the simulation of the time-domain waveform of the vibration signal, the harmonic wave signal mixed with white-noise and white Gaussian noise is used as the analog vibration signal in the time domain waveform in the program. The front panel of the program is shown in figure 3.

In this paper, five basic frequency vibration signal extraction methods are selected. In the whole cycle of 10Hz, the vibration signal amplitude and phase extraction operations are respectively performed, and the first 50 sets of calculation results are averaged. The results are shown in table 1.

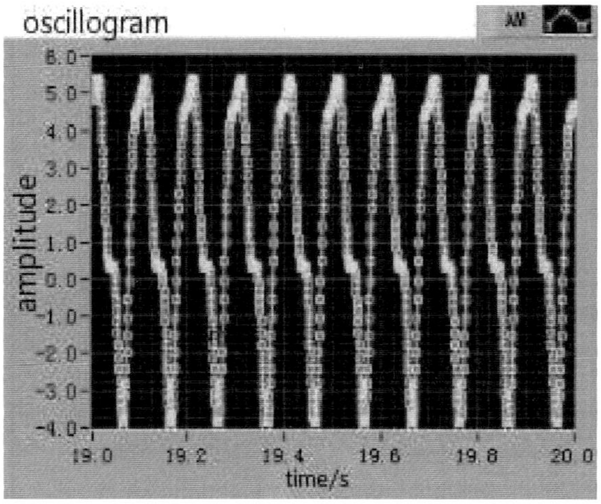

Figure 3. DC component, uniform white noise, Gaussian white noise mixed harmonic signal.

Table 1. Comparison of the extraction results of the baseband signal under simulation.

Extraction Method / Vibration parameters	Correlation	Full cycle interception	Traditional FFT	Mutual power	Sine approximation
Average amplitude/μm	6.01	5.98	5.97	6.02	6.05
Average phase/°	30.12	29.97	30.19	30.22	29.88

2.1.2. Experimental analysis of signal extraction methods. This research is based on the built-in electromagnetic slip ring type dynamic balancing device. The hardware part is composed of motor, eddy current displacement sensor, photoelectric sensor, preamplifier, data acquisition device, system controller and so on. In the dynamic balance measurement, the control command is transmitted to the control device by the computer, so that the control device can accurately control the speed of the motor. The function of the preamplifier is to amplify the vibration signal detected by the eddy current displacement sensor. The photoelectric sensor detects the speed pulse signal of spindle speed; Eddy current displacement sensor is used to detect vibration signal of vibration point; the data acquisition device adopts the PCI-9234 product from National Instruments in the United States. The function of the product is the A/D conversion function, analog signals detected by sensors are converted into digital signals, and then the digital signals are transmitted to the LabVIEW software to data analysis and processing. The experimental platform is shown in figure 4. The internal structure of the electromagnetic ring is shown in figure 5.

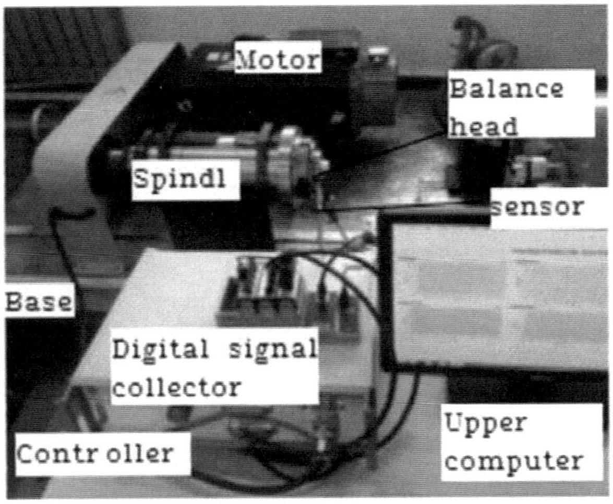

Figure 4. Experimental platform.

Figure 5. The internal structure of the electromagnetic ring.

Under the experimental conditions, the sampling frequency of the vibration signal is set to 1000Hz, the number of sampling points is 1000, the spindle working speed is set to 1500r/min, and the theoretical spindle frequency is 25Hz, so as to better observe the vibration of the spindle working speed, in the shaft the end of the test surface placed mass 10g trial weight. The results of the extraction are shown in table 2.

From table 2, we can see that the theoretical spindle frequency is 25Hz, but during the experimental operation, the spindle frequency detected by the Hall element is not 25Hz. At this time, the analog signal does not fully represent the spindle under actual operating conditions signal.

Using the various vibration signal extraction methods compiled by LabVIEW software, vibration signals with the same frequency as the spindle vibration are extracted, and 50 amplitude and phase data of the rear panel of the program are selected, as shown in figures 6 and 7.

IOP Conf. Series: Materials Science and Engineering **399** (2018) 012055 doi:10.1088/1757-899X/399/1/012055

Table 2. Comparison of the results of the fundamental frequency signal extraction under the experiment.

Extraction Method / Signal parameters	Correlation	Full cycle interception	Traditional FFT	Mutual power	Sine approximation
Amplitude/μm	5.52	5.48	5.42	5.35	5.31
Phase maximum/°	33.21	33.18	34.25	34.18	35.34
Phase minimum/°	30.68	31.04	29.97	29.32	30.33
Phase difference/°	2.53	2.14	4.28	4.86	5.01

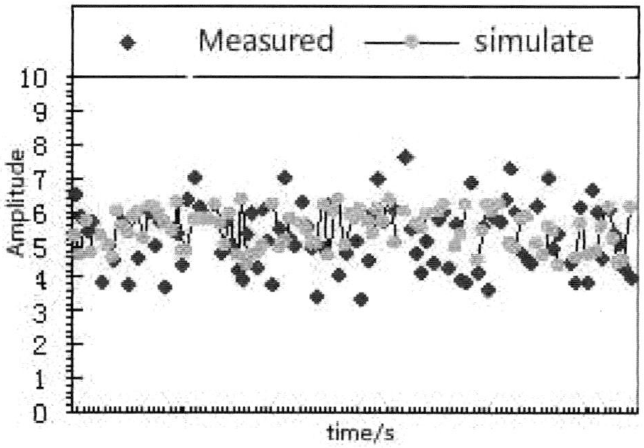

Figure 6. Vibration amplitude comparison.

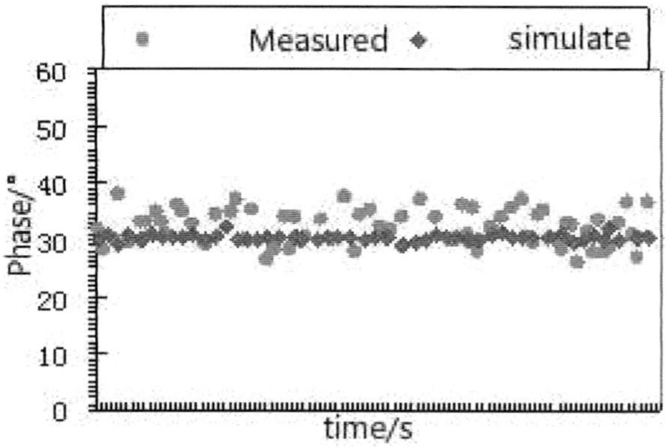

Figure 7. Vibration phase comparison.

The results obtained under the experimental conditions show that the unbalanced vibration caused by the spindle plus the appropriate test weight has a similarity of 90.78% between the amplitude of the fundamental frequency vibration and the vibration amplitude of the fundamental frequency under the simulated conditions. The vibration phase fluctuates less, the phase average absolute error is 2.15 degrees, and the accuracy rate is 92.53%. The harmonic signals including the fundamental frequency, double frequency, and triple frequency harmonic signals, mixed noise, and direct current components are approximately used to characterize the vibration of the actual working condition of the spindle. After completing the experiment for the verification of different vibration signal extraction methods and the similarity of 90.78% between the vibration amplitudes of the fundamental frequency under the experimental and simulation conditions, the experimental results can be used as feedback values to correct and modify the spindle of SYL04H-1 lathe based on the experimental platform. Spindles of lathes put forward more realistic expressions of vibration signals, paving the way for follow-up research and constructive reference.

The spindle working condition of this experiment is performed under the condition that the working speed is set to 1500r/min. Therefore, under the condition that the spindle frequency is 25Hz, the vibration signal under the simulation condition is used to replace the spindle vibration signal under real conditions. In the pre-processing of the signal, noise and DC components at different frequencies are subjected to wavelet de-noising, filtering, and other operations. Therefore, in the section of the method for extracting the amplitude and phase of the vibration signal, there is no need to address the type of noise, the size, DC component, etc. as they are studied precisely. The expression of the vibration signal will ignore this part. Only the single-term formula can be used as a symbol substitution. The spindle vibration can be caused by the interference of the mixed harmonic signals of uniform white noise, Gaussian white noise and DC component. The signal expression is:

$$x(t) = e + 5.5\sin(\omega t + 32.5) + 3.2\sin(2\omega t + 20) + 2.4\sin(3\omega t + 10) + s(t) \, . \tag{3}$$

In the formula, the fundamental frequency of the spindle is 24.6Hz, the vibration signal of the fundamental frequency is 5.5μm, the frequency of the double frequency vibration signal is 3.2μm, and the frequency of the triple frequency vibration signal is 2.4μm. The corresponding initial phase is 30°, 20° and 10°, respectively. The number of sampling points is 1000 and the sampling frequency is 1000Hz. DC component and noise are expressed separately e , $s(t)$.

2.2. The effects of high-speed spindle unbalanced vibration signal extraction on dynamic balance and experiments

The cross-correlation algorithm and the DFT signal extraction method applied to the whole cycle were applied to the dynamic balance software test system, respectively. The vibration signal extracted by the cross-correlation algorithm and the DFT extracted whole-period vibration signal were used as the influence coefficient method for on-line dynamic balance. For the input of the item parameters, the parameters such as the influence coefficient, calibration quality and residual unbalance are obtained. The comparison of the spindle dynamic test balance effect and the experimental data are shown in figure 8 and table 3.

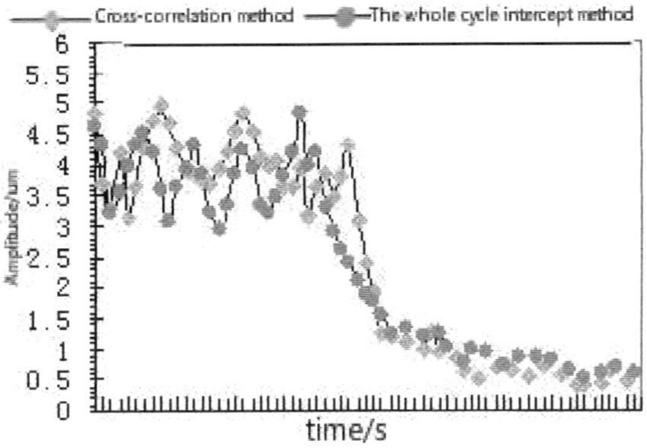

Figure 8. Comparison of amplitude before and after balance.

Table 3. Spindle dynamic balance experimental data.

Balancing parameters	Cross correlation method	Full cycle interception
Initial imbalance	4.75μm∠30°	4.96μm∠85°
Try to add weight	10g∠210°	10g∠265°
Unbalance after aggravation	5.36μm∠85°	5.84μm∠110°
Correction weight	11g∠260°	19g∠115°
Residual imbalance	0.52μm∠130°	0.67μm∠70°
Influence factor /μm/(g·mm)	0.0078∠290°	0.0042∠260°

Influencing coefficient method online dynamic balance many experiments show that the vibration signal extracted by the cross-correlation algorithm is used as the input of the influence coefficient method, the amplitude of vibration is significantly reduced, the balance accuracy is 89.78%, and the residual balance is lower than the value obtained by the whole cycle interception DFT method. Unbalanced vibration is effectively suppressed. It can be seen that the embedded cross-correlation algorithm program is applied to the dynamic balance software. Each performance parameter meets the design requirements. The effect of the single-plane dynamic balance process is ideal. This test system can be applied to the dynamic balance of other similar working conditions. Promote high-speed spindle double-sided balancing test.

The results show that using the cross-correlation algorithm, the spindle balance efficiency and accuracy are higher, and the spindle vibration is more effectively suppressed. The balance effect is better than the whole cycle interception DFT method, which ensures that the influence coefficient method is the balance regulation strategy accuracy.

3. Experimental research and analysis of the characteristics of on-line dynamic balance of high-speed spindle

3.1. Effect of rotation speed on experimental platform vibration

In order to study the vibration of the spindle rear end after the spindle rotates at higher speed, the vibration displacement sensor is first placed on the side of the end face of the spindle rear end. Recording is started from 1200 r/min without changing the spindle support structure. Every time the 200r/min increase, the amplitude of vibration is continuously increased to 3400r/min, and the vibration intensity of the spindle end is increased.

Change the spindle support method, and place rubbers with light elasticity at the front and rear ends of the spindle. At the rear end of the spindle, the displacement sensor is again placed at the same position as the original one, and the spindle end face vibration detection is performed. After the same experimental measurement method and stepping speed as the previous step, the following experimental results are obtained, as shown in table 4.

During the experimental research, it was found that the vibration of the main shaft and the motor housing was small at the other speeds of the main shaft system. However, when the rotation speed of the rotor system exceeded 2400 r/min, the rear end vibration of the main shaft began to have greater amplitude fluctuations. From the perspective of installation, the rear end of the spindle is relatively weak, and the motor and the belt are connected. The detected roundness of the spindle end face is also large, because the belt drive system affects the vibration of the rear end of the spindle.

Table 4. Vibration amplitudes across the experimental platform.

Rotating speed r/min	Spindle front end vibration amplitude μm	Base vibration amplitude μm	Motor vibration amplitude μm	Rear spindle vibration amplitude μm
300	0.261	0.04	0.551	0.461
600	0.244	0.063	0.328	0.430
900	0.249	0.111	0.279	0.397
1200	0.467	0.189	0.405	0.325
1500	2.717	0.291	1.033	5.894
1800	3.45	5.148	22.522	8.190
2100	2.637	3.438	10.895	15.572
2400	1.302	1.397	2.478	2.650
2700	2.131	0.599	0.809	0.814
3000	3.228	0.695	1.731	1.421
3300	4.926	0.66	1.496	1.511
3600	5.724	0.539	0.717	3.660

3.2. Effect of rotation speed on spindle dynamic balance

When the mechanical spindle runs at a low speed, the spindle can be regarded as a rigid system. After the dynamic balance, the balance effect is relatively reasonable and the effect is relatively stable. However, at high speeds, the spindle system may enter a flexible state. If the spindle is still considered to be a rigid system, the system will have an effect on the dynamic balancing effect after entering a certain speed. Measuring the spindle speed at 3400r/min, 3600r/min, 3800r/min, 4000r/min four different speeds, dynamic balance of the vibration response measurement. The vibration amplitude at different speeds is shown in figure 9.

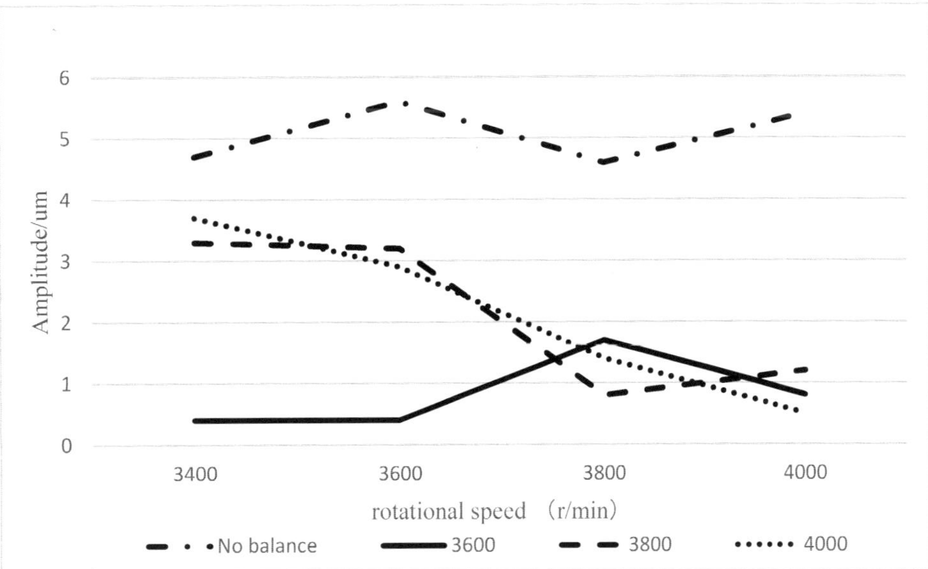

Figure 9. Vibration amplitude at different speeds.

It can be seen from the experimental results that, firstly, when the main shaft is not balanced, the vibration is greater at each rotating speed. After 3600r/min dynamic balance, the vibration amplitude is basically unchanged from 3400r/min to 3800r/min and 4000r/min, and the amplitude is increased; after 3800r/min dynamic balance, the speed is reduced or increased, and the amplitude is increased. There are different degrees of improvement; the same, after the 4000 r/ min dynamic balance and then the speed is reduced, the amplitude also has varying degrees of improvement. Changing the rotational speed has an effect on the dynamic balance effect. Changing the spindle rotational speed mode also has a certain influence on the vibration amplitude change of the main shaft. The change rules of the system at 3400r/min and 3600r/min are similar, and the change rules at 3800r/min and 4000r/min are similar.

Changing the rotational speed has an effect on the dynamic balance of the experimental platform in this paper. The reason is that the entire system has crossed the critical rotational speed and is working in a flexible state. It cannot be simply based on the rigid shaft dynamic balance method. The double-sided influence coefficient dynamic balance method of the flexible shaft should be used.

3.3. Effect of test block plus angle on spindle dynamic balance

In the experiment, try to add a weight of 10.0g of the same mass at different angles on the spindle end face. The trial weight angle is added once every 30°, and 12 times during the test. The vibration of the spindle before and after adding the trial weight is measured and the vibration amplitude is obtained. The difference between the difference and the trial plus the mass point angle correspond one to one. The results are shown in figure 10 and figure 11.

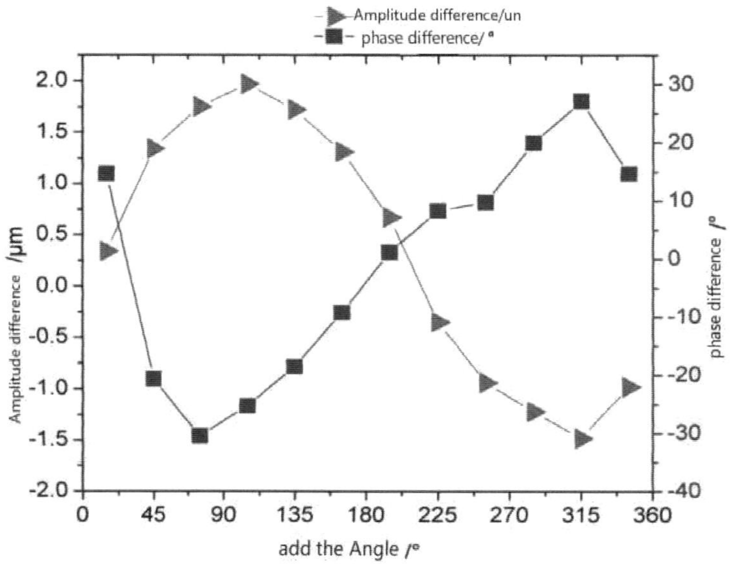

Figure 10. Relationship between amplitude difference and phase difference and test weight angle.

As can be seen from figure 10, the amplitude difference measurement changes show a "sinusoidal" change, and the phase difference shows a "cosine curve" change. At a test angle between 45° and 135°, the difference between the amplitude difference and the phase difference varies a lot and then begins to narrow. When the trial angle is between 225° and 315° the amplitude difference, the change in the difference between the value and the phase difference reappears and shows a gradual increase.

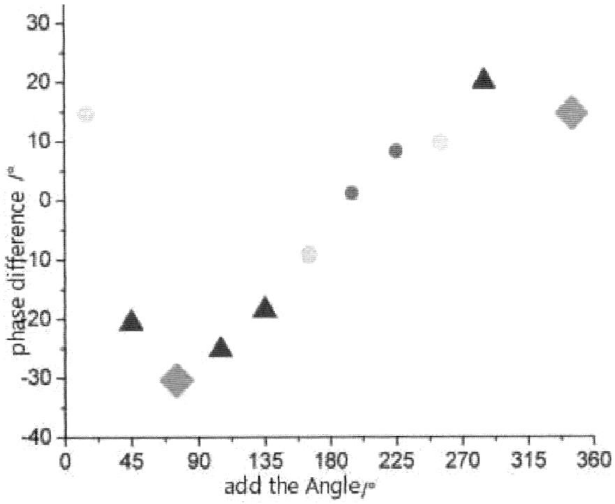

Figure 11. Influence factor size vs. angle.

From figure 11, we can see that the size of the influence coefficient changes with different trial angles, and the size of the influence coefficient is reduced to three sizes with different shapes. Circles represent smaller influence coefficients, and quadrilaterals represent larger influence coefficients. The distribution area of the coefficient of influence represented by different shapes has a certain degree of regularity: the distribution of influence coefficient is larger in the area where the difference between the amplitude difference and the phase difference is larger, and the distribution of smaller influence coefficient is the area where the difference between the difference in amplitude and the difference in phase is small. After this, the relationship between the size of the influence coefficient and the difference in the difference between the difference between the amplitude and the phase is explored. According to figure 10 , a total of four angles, ie, 75°, 315°, 195°, and 255°, in which the difference between the difference between the amplitude difference and the phase difference change greatly and are smaller and selected. The test of heavy-moving balance was performed again. Comparison of the vibration before and after the dynamic balance and the balance efficiency are shown in table 5.

Table 5. Effect of balancing after adding test weight in different angles.

Angle	Balanced vibration amplitude/μm	Balanced vibration amplitude/μm	Decrease percentage
75	5.1	1.1	78.43%
195	5.8	1.8	68.98%
255	5.3	1.5	71.69%
315	5.4	1.3	76.92%

The experimental results show that when the test weight is added to two positions of 75° and 315°, the amplitude ratio of vibration is reduced obviously, and the dynamic balance effect is better.

By comparing the final balance effect of the experiment, the following conclusions are obtained: firstly, the larger the difference between the amplitude difference and the phase, the larger the influence coefficient obtained, indicating that the unit mass is most sensitive to the amount of imbalance compensation, and the amplitude difference and phase the smaller the difference is, the smaller the influence coefficient obtained is, indicating that the unit mass is the least sensitive to the amount of imbalance compensation; secondly, the difference between the amplitude difference and the phase difference directly determines the size of the influence coefficient, and indirectly affecting the on-line dynamic balance of the spindle, the greater the influence coefficient, the more effective the amplitude reduction effect.

4. Conclusion

Experimental verification and analysis results show that cross correlation method has the best effect on amplitude and phase extraction of vibration signals after eliminating noise and filtering preprocessing during the operation of the spindle. The vibration signal obtained by the cross correlation algorithm is used as the input of the influence coefficient method. The vibration amplitude is obviously reduced, the balance precision is 89.78%. The method has great speed and accuracy in calculation.

Fundamental theory of regulation of online dynamic balance based on influence coefficient method is studied, and vibration parameters are detected from multi-directions. Spindle amplitude with control method of single plane influence coefficient was reduced to 87.26%, compared with single plane dynamic balance in balancing efficiency and accuracy, the method avoids the greater error caused by the original error accumulation, it improves the efficiency and accuracy and realizes more precise measurement.

After the critical speed of 1800r/min, the spindle is in a flexible working state, bearings supporting stiffness will be reduced to cut down on external influence. With the increase of spindle rotation rate, there will be a downward trend in the balance efficiency of the spindle, but it still maintains more than 80%. The experimental results show that the two parameters of the amplitude and the angle of test block directly determine the influence coefficient and indirectly affect the effect of online dynamic balance. Through the experiment, if the mass block formula for the first trial addition is improved, spindle response will be obvious, it is helpful for on-line control and restraining spindle vibration. The research results of moving strategy of mechanical double balance block balancing device show that the movement path of mass blocks do not cross when moving at the same time and the mobile strategy that mass blocks successively arrive is the most reasonable. The feasibility and accuracy of the move strategy are verified, some problems such as the error and the long balance time in the mass movement process are solved.

Acknowledgments

This project is supported by the National Natural Science Foundation of China (No.51675353), Shenyang science and technology project (No.F16-096-1-00), Natural Science Foundation of Liaoning Provincial (No.201602638; No.20170540746), Education Research Project of Liaoning Provincial Department (No.LJZ2016013; No. LJZ2016007), key Laboratory Open Foundation of Shenyang Jianzhu University (No.SJSC-2015-14) and Science and Research Foundation Project of Shenyang Jianzhu University (No. 2014007).

References

[1] Xuesen Z, Long C and Zengqiang L 2014 Status Research of Active Balancing Machine for Spindle of Ultra-precision Machine Tool *J. Aviation Precision Manufacturing Technology* **50** *6*

[2] Zhan W, Fenglong Z, Wei T and Chiyu Z 2017 The High-speed Spindle Online Dynamic Balance Experimental Regulation and Characteristics Analysis *J. Modular Machine Tool & Automatic Manufacturing Technique* **2017** 18

[3] Vegte J V 1964 Continuous automatic balancing of rotating systems *J. Journal of Mechanical Engineering Science* **6** 264

[4] Mingquan Y, Minqing J, Heng L and Weihong F 2011 Online automatic balancing control method and simulation for high-speed grinding process *J. Modular Machine Tool & Automatic Manufacturing Technology* **2011** 44

[5] Saito S, Azuma T 1983 Balancing of flexible rotors by the complex modal method *J. Journal of Vibration Acoustics Stress and Reliability in Design* **105** 94

[6] Rongqiang L and Xinhua W 1994 Influence factor balance method and its application in initial bending rotor balancing *J. Turbine Technology* **1994** 200

[7] Kaining K and Peijun W 1996 Minimization study on maximum residual amplitude of flexible rotors when balance *J. Journal of Mechanical Strength* **18**1

[8] Xiangyang Z and Binglin Z 1999 Calculation of the optimal balance mass of flexible rotor with upper bound constraints *J. Journal of Mechanical Engineering* **35** 34
[9] Yixuan Z, Yunbing T and Guihuo L 2005 Improved least square influence coefficient methods *J. Journal of Nanjing University of Aeronautics & Astronautics* **37** 110
[10] Bingang X, Liangsheng Q and Ruixiang S 2000 Balancing of flexible rotors without test weights *J. Journal of Xi'an Jiaotong University* **34** 63
[11] Zhijiang X, Yike T and Yuanyou L 2002 Double-face dynamic balance method of spot rotator without removing the test weights *J. Journal of Chongqing University (Natural Science Edition)* **25** 101
[12] Shihai Z, Liangsheng W and Dashuai Z 2009 Design of double-plane online balancing system for machine tool spindle *J. Modular Machine Tool & Automatic Manufacturing Technology* **2009** 79
[13] Yun Z, Xuesong M, Donglin Z， Gedong J and Dongsheng Z 2012 Modal analysis and Fourier transform method for flexible rotor-free test-weight balance *J. Journal of Vibration and Shock* **31** 7
[14] Borghesani P, Pennacchi P, Randall R B and Ricci R 2012 Order tracking for discrete-lane separation in variable speed conditions *J. Mechanical Systems and Signal Processing* **30** 1
[15] Xiaofeng L, Longxi Z and Zhenxia L 2013 Theoretical and experimental research on balancing of flexible rotors without trial weight *J. Journal of Vibration, Measurement & Diagnosis* **33** 565
[16] Gangqi L, Longxi Z, Qing M and Jingjing H 2014 Balancing method of flexible rotor across second order without trial weights *J. Acta Aeronautica ET Astronautica Sinica* **35** 1019

A slate grinding machine based on green design

C Wei[1], Y Cai[1*], Q Zhang[1] and Q Jia[1]

[1] School of Mechanical Engineering, Shenyang Jianzhu University, 9 Hunnan Road, Shenyang 110168, China

[*] E-mail: weiyu001@163.com

Abstract. In this paper, a new slate grinding machine based on green design is investigated. To save a lot of cast iron, large natural stones are used in the structure of the machine. Three key components of the body frame, the crossbeam and the power head are introduced in detail. In view of the body frame structure, the process of combination of stone and steel is given. The design method of the important parameters of the power polishing head is proposed. The deformation of the crossbeam under the gravity action of the 12 power polishing heads is analysed. Due to the crossbeam moving laterally with all the power heads, they have great inertia. Therefore, it is difficult to control the motion. In this case, a new buffer brake mechanism is proposed and the dynamic process of the crossbeam reversing is analysed. Finally, the calculation method of the required output torque of the driving motor is proposed. The results show that the functions of the slate grinding machine meet the requirements of use, the proposed method of design and analysis is correct and the concept of green manufacturing is realized.

1. Introduction

Slate is a common building material and it is widely used in indoor and outdoor surfaces, walls, etc. Surface polishing is an important process of slate processing. The slate grinding machine is a special equipment for slate surface polishing and it plays a key role in the quality of stone processing, whether it is the production line of stone sheet processing or the production line of large slate.

With the rapid development of the construction industry, the demand for slate is increasing more and more. The slate grinding machine production enterprise has got a good opportunity for development. The quality and performance of the slate grinding machine have also been greatly improved. Some of the most advanced slate grinders are produced in Italy. The main companies are SIMEC, MORDENTI, GAS-PARIMENOTTI, BRA, TE-MAFRU-GOLI, PEDRINI S.P.A, BREETN, GREGORI, BARSANTI, BRETON, PROTEC, ACIMM, COGE-MAR, LOFFLER, FIGKRT & WINTERLING, SCHULZE DR GMBH, BAVELLONI, TESMEC, etc. [1]. China's slate grinding machines production enterprises have also achieved great development. A new type of continuous grinding machine with resin grinding disc is developed by HUAXING machinery company limited [2]. The PLC control system has been successfully applied in slate grinding machine [3, 4]. PLC automatically collects the shape of the stones by collecting data information from all parts of the field, and completes automatic control of more than ten valves, grinding heads and swing arms on site by means of high-speed counting of the encoder and the HT7A00T touch screen makes the maneuverability of the stone equipment control system more convenient and flexible [4]. To improve the precision and the life of the stone mill measurement system, an online detection system of continuous rock grinder is proposed based on the machine vision [5]. Achievements have been obtained in mainly three aspects, driving power grinding head with two stage planetary reducer, using multifunction grinding disc and adding a grinding head pressure regulating device [6].

From the trend of the development of slate grinding machine, high automation and new material and new structure, green and environmental protection are the main directions. In this paper, a new slate grinding machine based on green design is introduced and studied, whose main feature is the use of a mixed structure of stone and steel. China's Xinjiang, Kuandian and Xiuyan in Liaoning are the main producing areas of large granites. This provides natural raw materials for the new slate grinding machine studied in this paper.

2. Structure design

A new slate grinding machine is designed, which is composed of the body frame, the power polishing head, the crossbeam, etc. The whole work system is shown in figure 1 and the structure of the machine is shown in figure 2. The main parameters are listed in table 1.

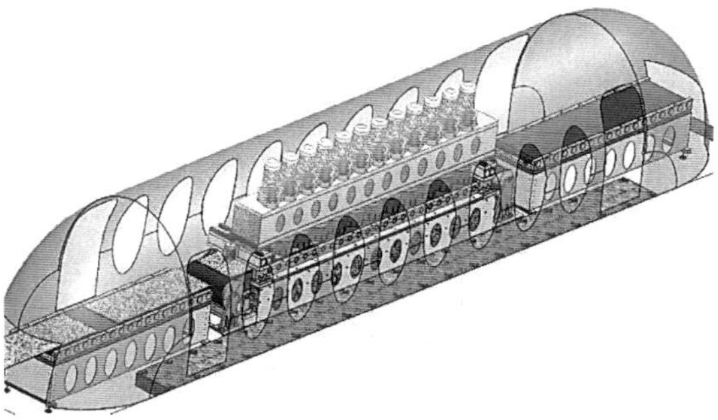

Figure 1. Whole work system of the new machine.

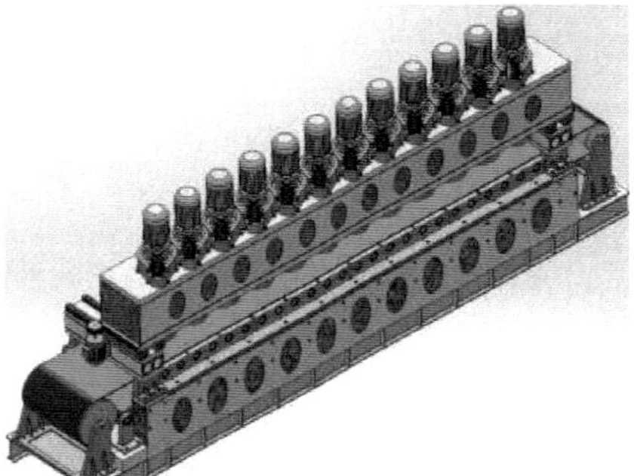

Figure 2. Structure of the machine.

Table 1. Main parameters.

Parameters	Values
Working thickness	8-50 mm
Working width	⩽800 mm
Conveyor belt speed	0.2-5.0 m/min
Crossbeam swinging speed	1-60 m/min
Total power	100 Kw
Grinding head power	7.5 Kw
The number of grinding head	12
Machining precision	⩽0.20 mm

2.1. Body frame

The body frame is the foundation of the whole slate grinding machine and supports other parts and slates. It is an important component of the processing precision. In this paper, a mixed structure of stone and steel has been adopted, which is shown in figure 3. The processing of large special-shaped stones can be completed by many companies in China, such as Zheng Shu Technology Co., Ltd. in the Zhejiang province. Since stone and steel have different physical properties, their combination process is very special. Stone connection with left and right steel plates is taken as an example, which is shown in figure 4. The epoxy resin adhesive is the connection medium of the steel and the stone and plays an important role in vibration absorption.

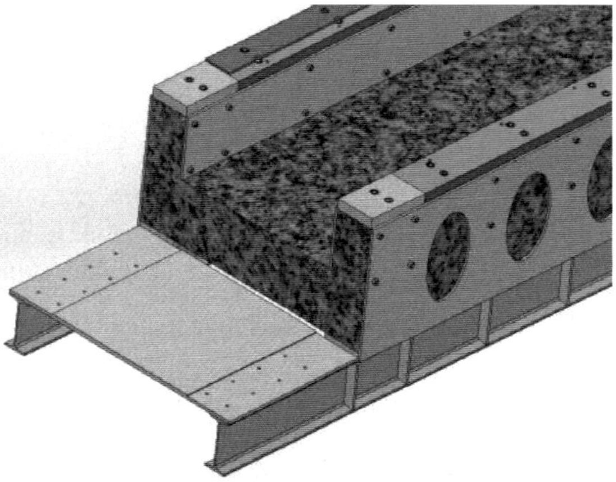

Figure 3. Body frame of the machine.

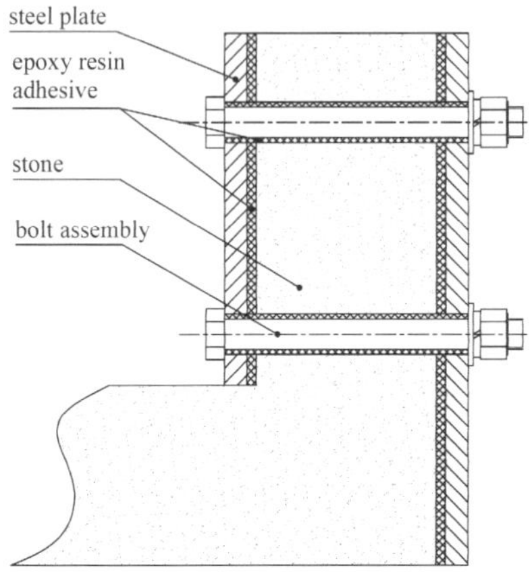

Figure 4. Connection of stone and steel.

2.2. Power polishing head

The power polishing head is the most important component in the slate grinding machine. In this paper, a new power polishing head is designed, which has 2 degrees of freedom and with automatic locking device. The overall structure of the power polishing head is shown in figure 5. The motor I drives the principal axis to rotate, so that the main motion of the grinding process is realized. The gear I is meshed with the gear II, and the gear II and the long shaft form a thread pair. Hence, the motor II can make the polishing head with principal axis and motor I move vertically by the gear pair and the thread pair. The gravity of the polishing head, the principal axis, the long shaft, the motor I and other auxiliary parts, acts on the thread pair. Therefore, in order to reduce the force acting on the thread pair, the spring plays a key role. The motor III can drive the locking mechanism, which includes gear meshing and thread meshing, as shown in figure 6. The gear III is meshed with the gear IV, and the gear IV is fixed with the rotating sleeve with internal thread. The rotating sleeve and the upper and lower wedge-shaped sleeves with different screw threads form two thread pairs, respectively. The upper and lower wedge-shaped sleeves move vertically, so the holding block can be pushed to lock or release the long shaft by them.

The initial pretension and stiffness of the spring are designed according to the following equations:

$$\begin{cases} k\Delta x - Mg = F_s \\ k\left(\Delta x - \dfrac{H}{2}\right) - Mg = 0 \end{cases}, \quad \min(F_s) \tag{1}$$

subject to

$$\begin{cases} \Delta x > H \\ \left|k\left(\Delta x - H\right) - Mg\right| \le F_s \\ \left|k\left(\Delta x - H\right) - Mg + pS_h\right| \le F_s \end{cases} \tag{2}$$

where: k is the spring stiffness, Δx is the initial maximum compression of the spring, M is the total mass of the polishing head, the principal axis, the long shaft, the motor I and other auxiliary parts, F_s is the force acting on the screw pair between the gear II and the long shaft, H is the total length of the thread of the long shaft, p and S_h are, respectively, the pressure and action area of the polishing head acting on the slate, g is the acceleration of gravity.

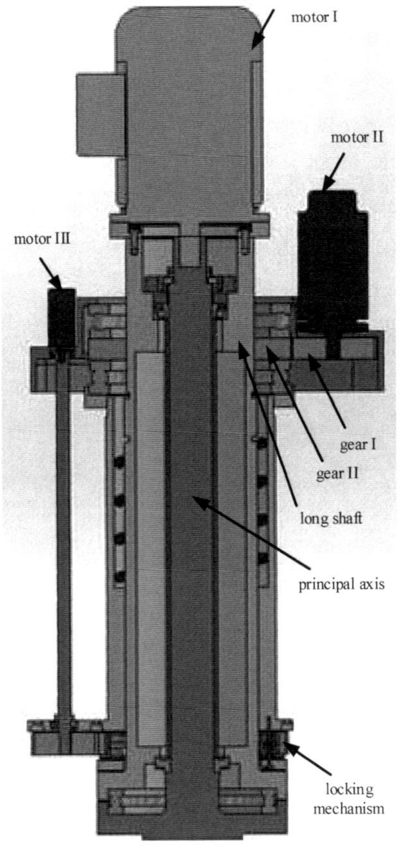

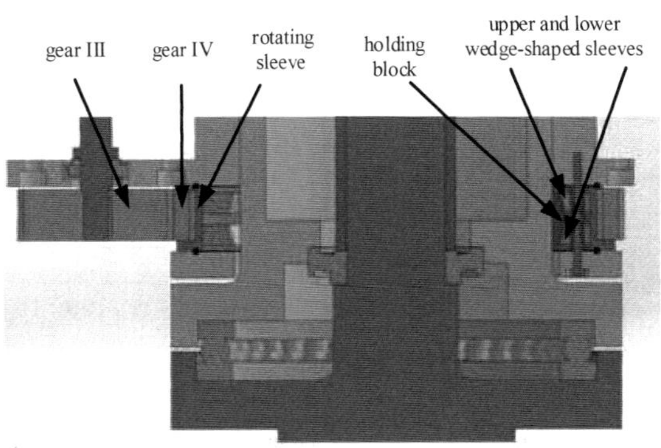

Figure 5. Cross-section of the power polishing head. **Figure 6.** Locking mechanism details.

2.3. Crossbeam

The crossbeam is also a part of a mixed structure of stone and steel, shown as in figure 7. The epoxy resin adhesive is also the connection medium of the steel and the stone. Welding is used between the upper and lower steel plate and the side plate. Under the gravity of its own and total 12 power polishing heads, the maximum deformation of the crossbeam is about 0.115 mm, as shown in figure 8.

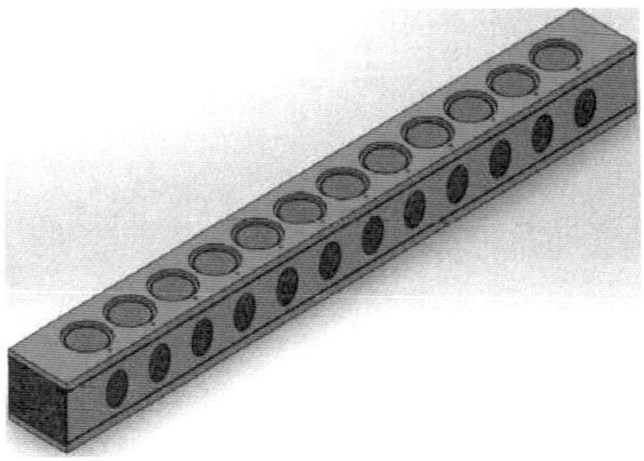

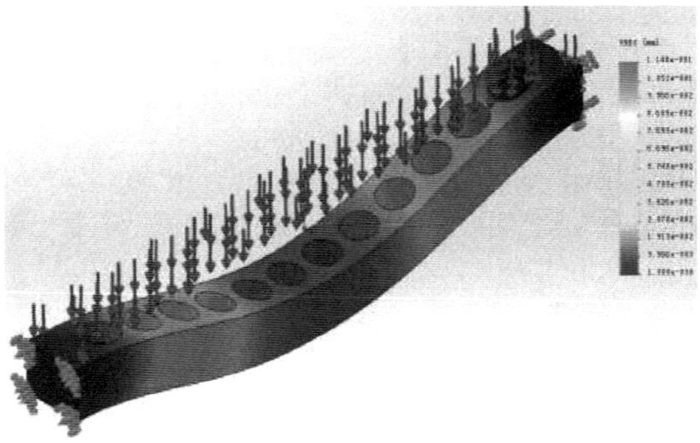

Figure 7. Structure of crossbeam. **Figure 8.** Maximum deformation of crossbeam.

3. Dynamic simulation of crossbeam movement

As all power polishing heads are mounted on the crossbeam, they are swinging laterally during grinding. In the process of the reciprocating movement, the acceleration will change greatly. This is an important basis for calculating the parameters of the driving mechanism. Because the inertia of the crossbeam is very large, the barrier brake mechanism is added on the basis of the original structure of the crossbeam, as shown in figure 9(a).

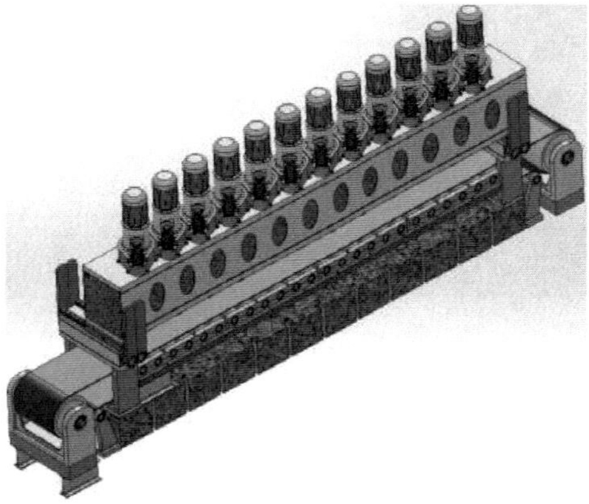

(a)

(b)

Figure 9. Barrier brake mechanism.

In this paper, in order to obtain the output torque required for the driving mechanism, a computational simulation model is established according to the actual structure. The crossbeam begins to decelerate after contact with the baffle and its speed drops to 0 when the location is shown in figure 9(b), and then they start the reverse acceleration. For this deceleration and reverse acceleration process, the velocity of the crossbeam is designed as cosine curve and can be expressed by:

$$v(t) = v_{\max}\cos\omega t,\quad t\in[0,T] \tag{3}$$

where: $v_{\max}$ is the main parameter of the crossbeam, that is, the maximum motion velocity, ω is the angular frequency to be determined, T is the total time of the deceleration and reverse acceleration process, $T = 2\pi/\omega$.

The maximum compression of the spring of the barrier brake mechanism is the braking distance of the crossbeam, which depends on the distance that the centre of the polishing head is beyond the slate. We can get the equation

$$d_{\mathrm{bk}} = \int_0^{T/4} v_{\max}\cos\omega t dt \tag{4}$$

where d_{bk} is the braking distance of the crossbeam.

In this paper, $d_{\mathrm{bk}} = 0.07$ m and $v_{\max} = 1$ m/s. Solving the Eq. (4), $\omega = 14.2857$. Hence, the crossbeam swings from the middle position with $v_{\max}$ as the initial velocity, the velocity curve of a complete cycle can be obtained, as shown in figure 10.

When the crossbeam contacts with the baffle, the output torque of the drive motor changes the direction in a moment. For the process of the deceleration and reverse acceleration, neglecting all frictions, the following equation can be obtained:

$$\begin{cases} T_q = F_t r \\ 2F_r + 4k_{bk}s + f = M\ddot{s}\ ,\ t\in\left[0,\frac{T}{2}\right] \\ s = \int_0^t v_{\max\cos\omega t dt} \end{cases} \tag{5}$$

where: T_q is the output torque of the drive motor, F_t is the driving force acting on the crossbeam, r is the radius of the gear in the gear and rack meshing pair, k_{bk} is spring stiffness of the barrier brake mechanism, s is the displacement from the beginning to the stop, f is the friction force of slide rail.

MATLAB software is often used in the process analysis by using motion equations [7-12]. According to the eq. (5), F_t and T_q can be obtained by using MATLAB/SIMULINK software. The known parameters are shown in table 2. The curve of T_q is shown in figure 11, for the process of the deceleration and reverse acceleration, which is an important basis for selecting the drive motor.

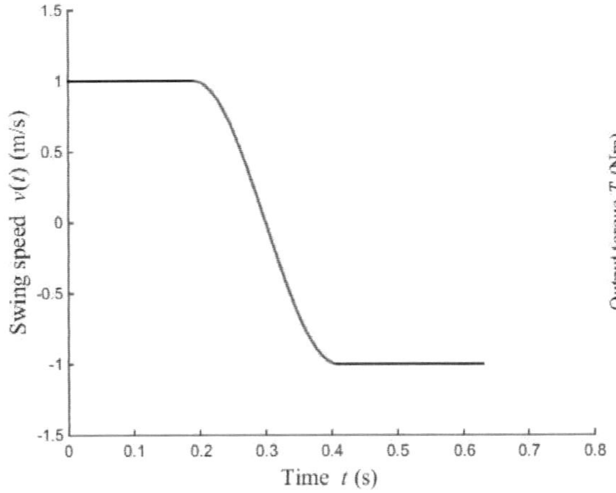

Figure 10. Swing speed of crossbeam.

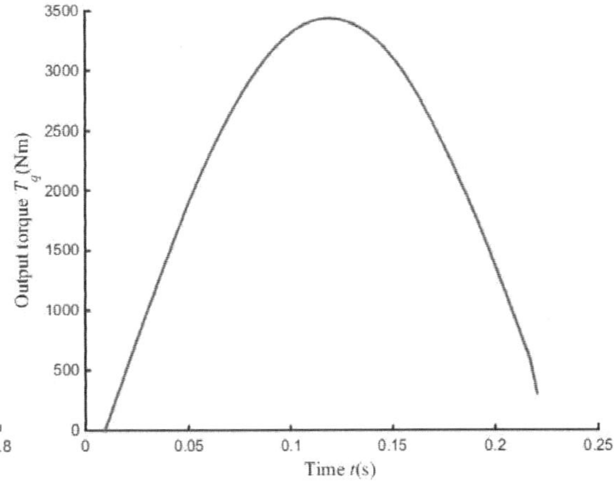

Figure 11. Output torque of the drive motor.

Table 2. Known parameters.

Parameters	Values
r	150 mm
k_{bk}	500,000 N/m
f	130 N
M	13,000 kg

4. Conclusions

A new slate grinding machine based on green design is described and studied. China is rich in natural stone resources, especially large stones, and China also has advanced stone processing technologies. All these provide basic conditions for the design of slate grinding machine with mixed structure of stone and steel. The three key structures of the machine are introduced and analyzed. Due to the large inertia of the crossbeam, the method of adopting the buffer brake mechanism is given in this paper, and the movement process of the crossbeam is analyzed. For a given ideal crossbeam swing speed, the required output torque of the driving motor is obtained, which is the most important parameter of the dynamic device of the crossbeam movement.

References

[1] Jinhua T 2004 Development trend of stone continuous mill at home and abroad *J. Stone* **9** 16
[2] Haizhou M, Changqing G, Yaoguo Z et al 2012 Development of a new type of continuous grinding machine with resin grinding disc *J. Stone* **2** 9
[3] Peimin C 2005 The control system of PLC in the stone grinding machine *Microcomputer Inf.* **21(2)** 42
[4] Feng S 2014 Application of LE PLC in stone industry *Int.J.Mechatr. Techn.* **1** 36
[5] Zhe Y, Hanchi Z and Zifang Q 2014 Study on the Slab Width Detection System of Continuous Rock Grinder Based on the Linear CCD *J. of Shenyang Jianzhu University* (Natural Science) **30(6)** 1113
[6] Qiang F and Changting Y 2016 Full automatic floor polishing lapping machine *J. Stone* **9** 45
[7] Tahmasebi M and Esmailzadeh S M 2018 Modeling and co-simulating of a large flexible

satellites with three reaction wheels in ADAMS and MATLAB *Int. J. of Dyn. Contr.* **6** 79

[8] Li H S and Cao Z J 2016 Matlab codes of Subset Simulation for reliability analysis and structural optimization *Structural and Multidisciplinary Optimization* **54** 391

[9] Fang J, Fengyu X, Hang Z et al 2017 Optimization and simulation of the operational motion of a pantograph: Uplift and retraction *J. of Mech. Sci. and Techn.* **31(1)** 41

[10] Karakaya S, Kucukyildiz G and Ocak H 2017 A New Mobile Robot Toolbox for Matlab *J. of Intelligent & Robotic Syst.* **87** 125

[11] Monteiro T P and Canale A C 2017 Development of a direct-dynamic model for a passenger vehicle on MATLAB Simulink *J. of the Brazilian Society of Mechanical Sciences and Engineering* **39** 385

[12] Rangaswamy T, Vidhyashankar S, Madhusudan M et al 2015 Performance Parameters Analysis of an XD3P Peugeot Engine Using Artificial Neural Networks (ANN) Concept in MATLAB *J. of The Institution of Engineers* (India) Series C **96(2)** 175

IOP Conf. Series: Materials Science and Engineering **399** (2018) 012057 doi:10.1088/1757-899X/399/1/012057

Ceramic motorized spindle for NC machine tool

Y Wu[1*] and S Li[1,2]

[1] National-Local Joint Engineering Laboratory of NC Machining Equipment and Technology of High-grade Stone, Shenyang Jianzhu University, No.9, Hunnan East Road, Hunnan District, Shenyang, 110168, China
[2] School of Mechanical Engineering, Shenyang Jianzhu University, No.9, Hunnan East Road, Hunnan District, Shenyang, 110168, China

[*] E-mail: wuyh@sjzu.edu.cn

Abstract. In order to reduce the inertia force and centrifugal force generated when the spindle bearing unit rotates at a high speed and to increase its limit rotation speed, precision, rigidity, service life and reliability, it can meet the high speed and high precision requirements of the spindle system. The innovative design of the ceramic motorized spindle unit, testing and analysis of the comprehensive performance of ceramic motorized spindle and the related research on the intelligentization of the ceramic motorized spindle were carried out. Finally, the current limitations and challenges are discussed, and the future development trend of the numerically-controlled machine tool ceramic motorized spindle is forecasted.

1. Introduction

With the requirement of high-speed precision manufacturing technology and the development of machine tool technology, motorized spindle technology develops toward high speed, high power, high precision, high stiffness, high reliability and long lifetime. The ordinary motorized spindle has limited rotational speed, stiffness, precision, and relatively short lifetime, which are due to the limitations in bearing, lubrication and cooling technologies. This has also seriously affected the development and application of high-performance NC machine tool. Prof. M Weck in the Laboratory of Machine Tools and Production Engineering (RWTH Aachen University), who plays an international leading role in the research on motorized spindle of high-speed NC machine tool, pointed out: “The static and dynamic properties of spindle system are mainly affected by the spindle's bending strength and the moment of inertia. The flexural deformation of spindle is the main factor influencing the static and dynamic deflection of its front end. The moment of inertia of spindle directly influences the acceleration time of spindle. High-speed high-precision spindle system must meet the requirements of high stiffness and fast acceleration. The application of new structural materials such as engineering ceramics can help solve above problems” [1, 2].

Therefore, high-performance engineering ceramics such as hot isostatically pressed silicon nitride and zirconia toughened ceramic are considered most suitable for manufacturing high-speed high-precision parts like ceramic bearings, spindle and so on since they have excellent properties including low density, high strength, high stiffness, high temperature resistance, wear resistance, being non-magnetic and being insulated [3, 4]. When fabricating motorized spindle of NC machine tool, the use of high-performance structural ceramics aims to reduce the inertia force and centrifugal force generated by spindle-bearing unit during high-speed rotation and improve its extreme speed, precision, stiffness, lifetime and reliability, thus meeting the high speed and high precision requirements of the spindle system.

2. Innovative design of ceramic motorized spindle unit

2.1. Structural design of ceramic motorized spindle without inner ring

Engineering ceramics have properties of high stiffness and high brittleness, which are completely different from those of steel materials. This causes a series of technical problems for the structural design, manufacturing and control of engineering ceramics. For example, the problems of installation and positioning of full ceramic bearings on the axis of a ceramic rotor have to be solved if the traditional design method for motorized spindle is adopted. However, since ceramic materials have high stiffness, extremely small thermal expansion coefficient, extremely high compressive strength and relatively low tensile strength, it is impossible ensure the precision assembly of ceramic inner ring and ceramic rotor. In addition, thread needs to be drilled into the rotor axis in order to achieve the axial positioning of ceramic bearings. However, because of the high stiffness and brittleness of ceramic materials, it is hard to ensure the precise machining of threads into the axis of ceramic motor. Moreover, there may exist unpredictable machining cracks, which will affect the strength of the rotor axis. Therefore, in order to solve these problems, an inner ring-free all-ceramic ball bearing was creatively designed as shown in figure 1 [5, 6].

After many considerations, it was decided to use a Y_2O_3-partially stabilized ZrO_2 (Y-PSZ) ceramic as the main spindle, a hot isostatic pressing silicon nitride (HIPSN) ceramic for the ball, and Y-PSZ and HIPSN for the bearing outer ring, respectively. In order to adapt to the high-speed rotation of the ceramic motorized spindle, an outer race guided cage was adopted which can help improve the lubrication of the outer ring raceway of ceramic bearing. Polyetheretherkerone (PEEK) was used for the ceramic bearing cage. The coefficient of the friction between Si_3N_4 ceramic and PEEK is only 0.028. Clearly, PEEK is very suitable for the high-speed ceramic ball bearing.

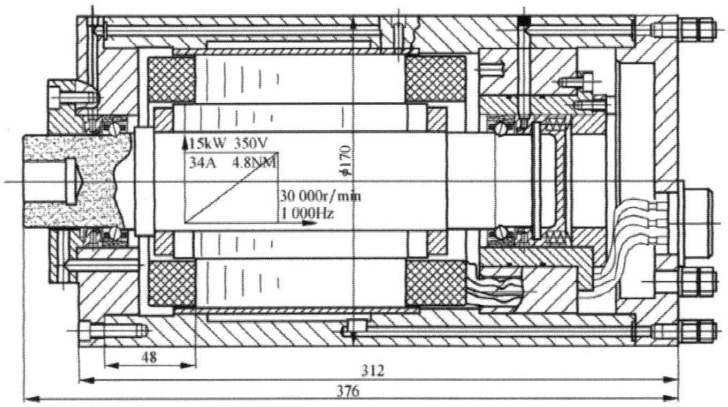

Figure 1. Optimization design of ceramic motorized spindle.

2.2. Ceramic ball bearing design

The high speed, high-precision bearing technology is one of the key technologies of the high-speed spindle system. For the first time, the design and development of all-ceramic ball bearings without inner rings have been proposed. Because ceramic materials have different properties from metal materials, the structural parameters of the ceramic ball bearing have to be designed and optimized according to the properties of ceramic materials. Moreover, the mechanical analysis, failure mechanism and life prediction of the inner ring-free all-ceramic ball bearings are discussed.

(1) Mechanical analysis of ceramic ball bearings. In general, angular contact ball bearings for high-speed spindles only work properly with axial preload. Preloading not only eliminates the axial clearance of the bearing, but also improves the stiffness of the bearing and the accuracy of the rotation

of the spindle while suppressing the vibration and slippage of the ceramic ball during rotation. By mechanically analyzing the bearing, we can obtain the minimum preload of the bearing that prevents the gyro from slipping by the following formula:

$$F_{\mathrm{a}} \geq \frac{M_{\mathrm{g}} Z \sin\alpha}{\mu D_{\mathrm{b}}} + 1.9 F_{\mathrm{r}} \tan\alpha , \tag{1}$$

where J is the moment of inertia of ceramic ball, $J = \frac{1}{60}\rho\pi D_{\mathrm{b}}^{5}$ (kg·m^2); ω is the absolute angular speed of ring (rad/s); ω_m is the angular speed of revolution (rad/s).

(2) The failure mechanism and life prediction theory of the inner ring type full ceramic ball bearings. The failure modes of the fully ceramic ball bearing are fatigue spalling, wear and fragmentation. And under light-load working condition the fatigue spalling of ceramic materials should be the main failure mode. Micro-pores or impurities on the surface of ceramic raceway are the main reason causing fatigue spalling. Figure 2 shows the scratch marks on the raceway of ZrO_2 ceramic outer ring caused by the sudden fracture of Si_3N_4 ceramic balls. Figure 3 shows a pit caused by contact fatigue spalling on the raceway of ZrO_2 ceramic outer ring.

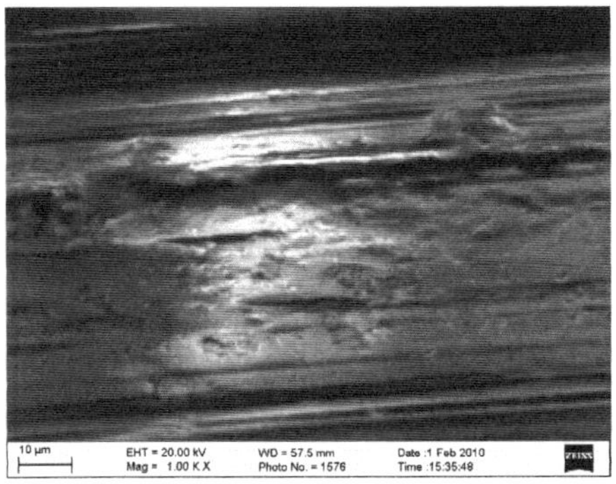

Figure 2. Scratches on the raceway of ceramic bearing ring.

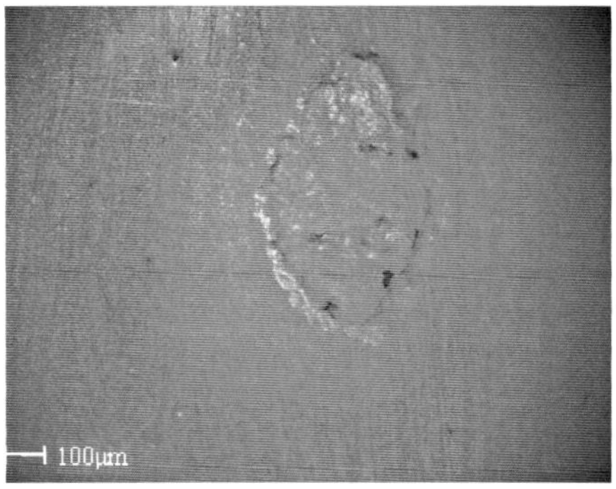

Figure 3. Fatigue spalling of the inner raceway of ceramic bearing.

In addition, the failure mode of ceramic ball bearings is mainly manifested as the fatigue failure of the ceramic bearing rings (mainly the bearing outer ring). However, the contact fatigue spalling failure mechanism and life estimation model of all-ceramic ball bearings lacks systematic theoretical basis. In this paper, the life prediction a model of ceramic bearings is established by modifying the life prediction model of steel bearings as shown in equation 2:

$$L_{\mathrm{hn}} = \left(\frac{10^{6}}{60n}\right)\left(\frac{C}{f_{\mathrm{p}} P}\right)^{\varepsilon} f_{\mathrm{R}} f_{\mathrm{m}} f_{\mathrm{T}} f_{\mathrm{u}} f_{\mathrm{E}} , \tag{2}$$

where: L_{hn} is the life of the ceramic ball bearing when the reliability is (1-h)%; f_p is the load factor; f_R, f_m, f_T, f_u, f_E are the lifetime corrections for reliability, material, temperature, clearance, axial deflection, respectively.

(3) Optimization design of the structural parameters of ceramic ball bearing. We considered the failure mechanism and lifespan prediction of ceramic ball bearings to optimize the design parameters of ceramic ball bearings. The failure mode of ceramic ball bearing is mainly shown as the fatigue failure of ceramic bearing rings (mainly the outer ring of bearing). Therefore, this study designed and optimized the internal structural parameters of ceramic ball bearing with the aim to prolong fatigue life.

The ceramic motorized spindle was designed on the basis of 7008C bearing. The main internal structural parameters of the final ceramic ball bearing were determined according to the optimization

IOP Conf. Series: Materials Science and Engineering **399** (2018) 012057 doi:10.1088/1757-899X/399/1/012057

design results shown in table 1. The curvature radius coefficients of the inner and outer raceways of the full ceramic ball bearing (f_i and f_e) should be smaller than those of steel bearing or even hybrid ceramic ball bearing. This can avoid the case where the contact (Hertz) stress is too high and affects the lifetime of the ceramic bearing.

Table 1. The optimization design results of ceramic ball bearings.

Bearing type	Size (mm)	Contact angle	D_b (mm)	Z	D_m (mm)	f_i	f_e
7008C	40×68×15	15°	7.144	18	54.005	0.505	0.510

2.3. Tool interface design of ceramic motorized spindle

We have optimized the overall structure of the ceramic spindle, developed a variety of ceramic shaft tip tool interface technology and successfully developed the front threaded hole ceramic shaft and external cone ceramic shaft.

For ceramic motorized spindle with built-in connecting rod, the internal thread in the front end of the shaft requires high precision. However, it is difficult to ensure the precision of the internal thread during the sintering process and precision machining of ceramic motorized spindle. Moreover, the rate of end product is very low. Thus, we proposed an alternative design plan such as in figure 4. Internal thread would not be directly sintered in the front end of shaft. Instead, a steel threaded insert was bonded to corresponding site by using high-strength adhesive. In this way, the manufacturing cost is reduced and the precision of the assembly of connecting rods is ensured. At the same time, the use of a steel threaded insert has a "buffering effect". Specifically, it can avoid the load's direct action on the ceramic motorized spindle and ceramic bearing under overloading and causes destruction of the whole structure.

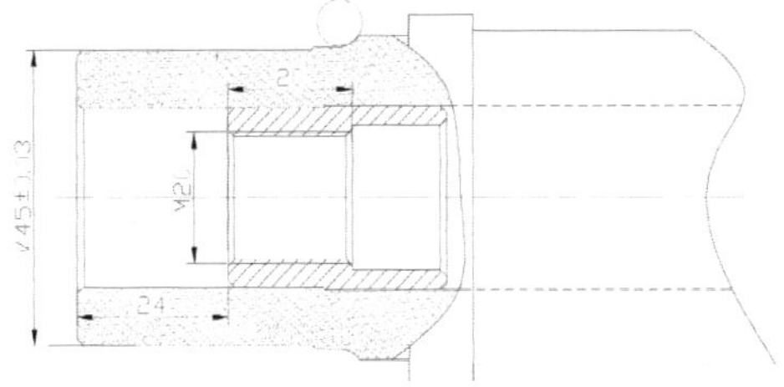

Figure 4. Design of the ceramic motorized spindle with built-in steel thread insert.

2.4. Precision assembly of the ceramic motorized spindle

In order to meet the design requirements including high rotational speed, high rotation precision, high stiffness and little vibration for the ceramic motorized spindle, high-speed dynamic balancing of spindle-bearing rotor system must be performed. In addition, the precision assembly of the main parts of the ceramic motorized spindle system should be also ensured. The spindle torque output is achieved through the interference fit between the ceramic motorized spindle and the motor rotor. Ceramic materials have different characteristics from traditional steel materials, and different ceramic materials have different thermal expansion coefficients.

$$\Delta = \frac{k_c(1-v^2)M_t}{\pi\mu EBa}\left(\frac{1+c_e^2}{1-c_e^2}+\frac{1+c_i^2}{1-c_i^2}\right)+\frac{\rho\omega^2(1+v)(3-2v)(1-c_e^2c_i^2)a^3}{2Ec_e^2} \tag{3}$$

where k_c is safety factor; M_t is the transmission torque of shaft (N·m); ω is the angular velocity of motor rotor (rad/s); ρ is the average material density of motor rotor (kg/m^3); *B* is effective contact length of matching surface (m); μ is the coefficient of friction between matching surfaces.

After calculation, the amount of interference fit between ceramic shaft and motor rotor in the spindle system designed in this research reaches 0.08–0.10 mm. The rotor is loaded into the spindle after the rotor is heated to about 200°C for 6-8 hours. The motor rotor is then subjected to precision turning and grinding so that its outer diameter can meet the design requirement. Subsequently, the weight balance method is used to achieve the dynamic balance of the motor rotor. The precision of dynamic balance can reach G0.4 level. Figure 5 shows the main components of the ceramic motorized spindle and the assembled ceramic motorized spindle prototype.

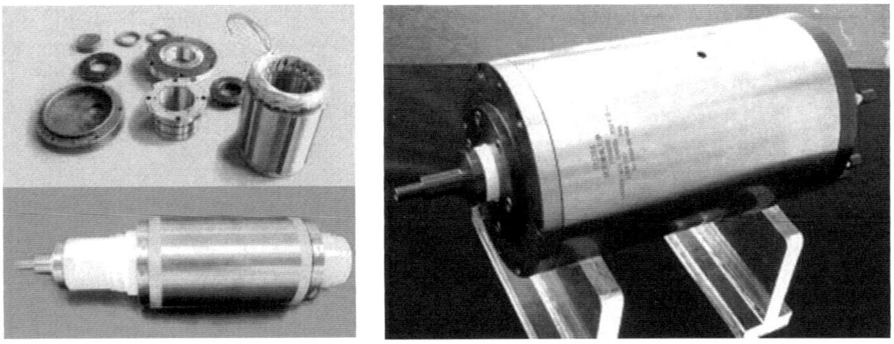

Figure 5. Ceramic motorized spindle prototype.

3. Testing and analysis of the comprehensive performance of the ceramic motorized spindle

Testing and evaluating the comprehensive performance of the ceramic motorized spindle is one of the most important steps in the design and manufacture of high-quality motorized spindles. Because of the high hardness and high brittleness of ceramic materials, evaluating and analyzing the comprehensive performance of the ceramic motorized spindle is particularly important. In this study, a prototype of the ceramic motorized spindle for NC machine tool was successfully fabricated and a platform for testing and evaluating its comprehensive performance was also built. In order to further improve the quality and performance of the ceramic motorized spindle for product upgrading, the ceramic motorized spindle was tested for its load characteristics, temperature rise, vibration, noise, precision, stiffness under different working conditions [7-15].

3.1. Test conditions

The testing object was SJD170SD30 ceramic motorized spindle without an inner ring designed and fabricated in this research. The main specifications of the ceramic motorized spindle tested are listed in table 2. It was tested in laboratory with constant temperature and humidity. The temperature in the laboratory was 20 ℃, and the environmental noise was not higher than 35 dB. The platform for testing the comprehensive performance of the ceramic motorized spindle is shown in figure 6.

Figure 6. Platform for testing the comprehensive performance of the ceramic motorized spindle.

IOP Conf. Series: Materials Science and Engineering **399** (2018) 012057 doi:10.1088/1757-899X/399/1/012057

Table 2. Specifications of the ceramic motorized spindle.

Rated power (kW)	Rated torque (N·m)	Maximum rotational speed (r/min)	Rated current (A)	Rated voltage (V)	Rated frequency (Hz)
15	4.8	30000	34	350	1000

3.2. Load characteristics test of the ceramic motorized spindle

Figure 7 and figure 8 show the changes of the input power, output power and efficiency of the ceramic motorized spindle as the load increases from 0 to 4.8 N·m at 6000 r/min and 8000 r/min, respectively. It can be seen that the efficiency of the ceramic motorized spindle under full load operation can reach 85.64%, which is quite close to the designed efficiency of ceramic shaft motor.

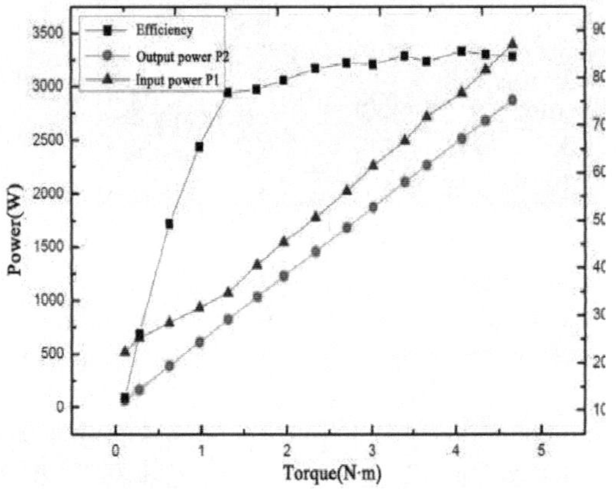

Figure 7. The output power of ceramic motorized spindle at 6000 r/min in loading process.

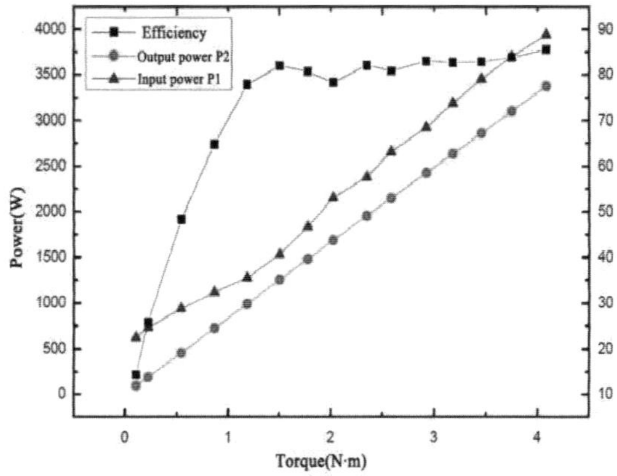

Figure 8. The output power of ceramic motorized spindle at 8000 r/min in loading process.

3.3. Temperature rise test of the ceramic motorized spindle

(1) The influence of the lubrication system on the temperature rise of the ceramic motorized spindle. In this research, an orthogonal experiment was performed to investigate the influence of the main design parameters of the oil/air lubrication system on the temperature rise of spindle bearings. We designed an orthogonal experiment with four factors and three levels as shown in table 3.

Table 3. Factors and levels for temperature rise experiments of ceramic motorized spindle.

Factors	Amount of oil supplied each time A (mL)	Oil supply interval B (min)	Pressure of oil supplied C (MPa)	Viscosity of lubricating oil D (cSt)
Level 1	0.02	2	0.3	15
Level 2	0.04	5	0.4	32
Level 3	0.08	10	0.5	68

The results of the temperature rise experiment for the ceramic motorized spindle under different parameter settings of lubrication system are shown in figure 9.

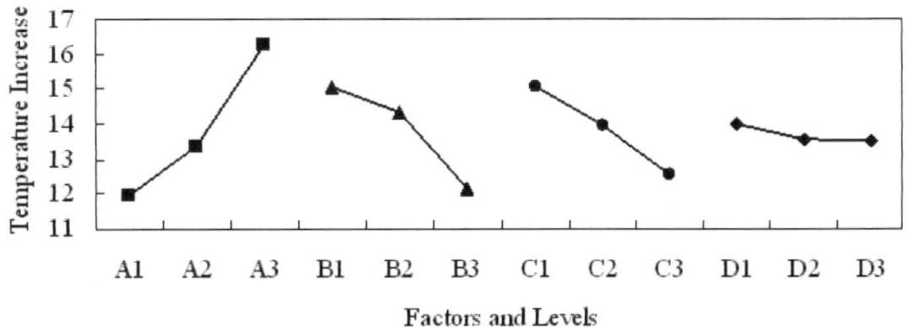

Figure 9. Effects of the parameters of lubrication system on the temperature rise of ceramic motorized spindle.

The influences of the four factors of the lubrication system on the temperature rise of the ceramic motorized spindle are shown in figure 10. In order to achieve the minimum temperature rise, experimental results demonstrate that the amount of oil supplied each time by the oil/air lubrication system should be set at 0.02 mL .The oil supply interval of the oil/air lubrication system should be set at 10 min. The pressure of the oil supplied should be set at 0.5 Mpa. The viscosity of the lubricating oil should be 68 cSt for the full ceramic spindle-bearing system.

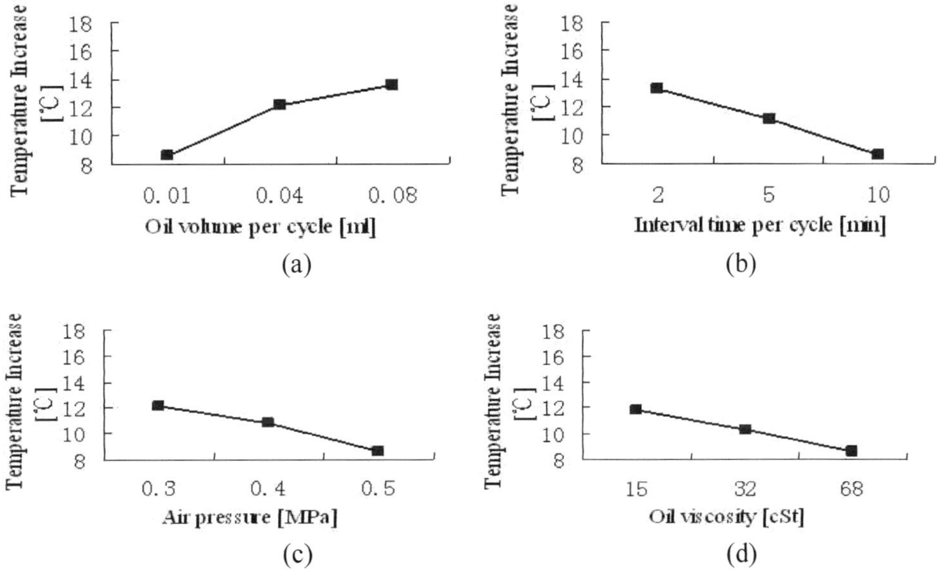

Figure 10. Influences of the four factors on the temperature rise of the ceramic motorized spindle.

(2) The influence of the preloading force on the temperature rise of ceramic motorized spindle. As shown in figure 11, we analyzed the temperature rise of the front and rear bearing of the ceramic motorized spindle under optimum lubricating conditions and under preloading forces of 400 N, 600 N and 800 N. Experimental results suggest that the temperatures of the front and rear ceramic bearings increase slowly when the preloading force is 400 N. When the preloading force of ceramic bearings reaches 800 N, the temperatures of the front and rear ceramic bearings increase rapidly. When the rotational speed of spindle reaches 20,000 r/min, the temperature rise of the front bearing exceeds the set highest temperature rise 35 ℃. Thus, the preloading force of ceramic bearings should be within 400–600 N so as to ensure the high-speed rotation of the ceramic motorized spindle.

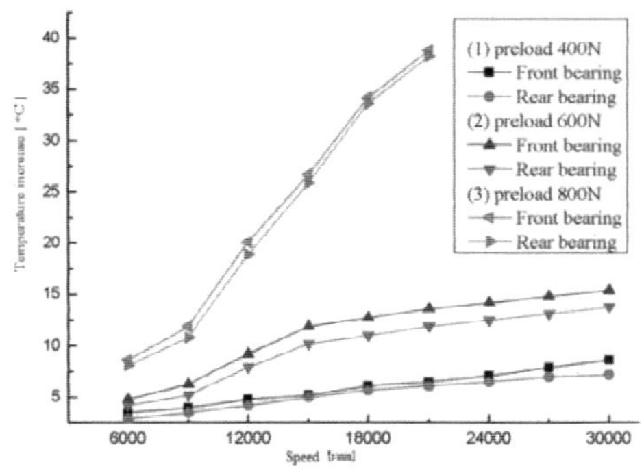

Figure 11. Temperature rise under different preloads and rotational speeds of ceramic motorized spindle.

3.4 Vibration and Noise test of the ceramic motorized spindle

The experimental results show that the noise of the ceramic motorized spindle reaches 95 dB at a speed of 30,000rpm, which exceeds the traditional motor spindle noise control. This high-frequency noise generated by a ceramic electric spindle is considered to be related to the material properties. In addition, many factors affect the noise generated by the ceramic spindle. So far, the theories of vibration and noise of spindles and bearings are still not mature. Through high-speed rotation experiments, we discovered and proposed the vibration and noise problems of the ceramic electric spindles. However, we only proposed a simple discussion and analysis. The in-depth analysis and control of high-frequency noise of the ceramic motorized spindle should become the key research topic in the future [17].

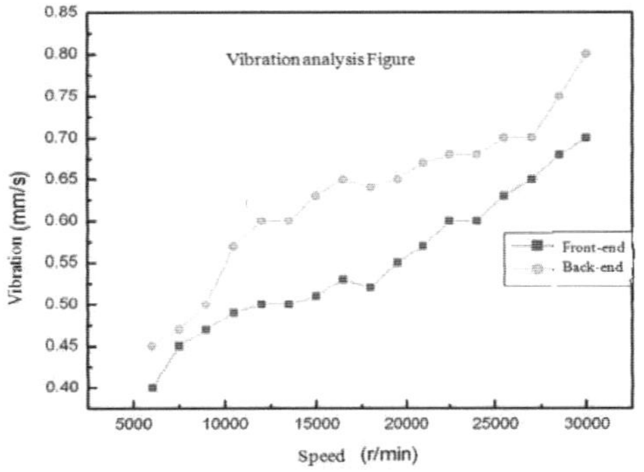

Figure 12. Vibration performance of the ceramic motorized spindle.

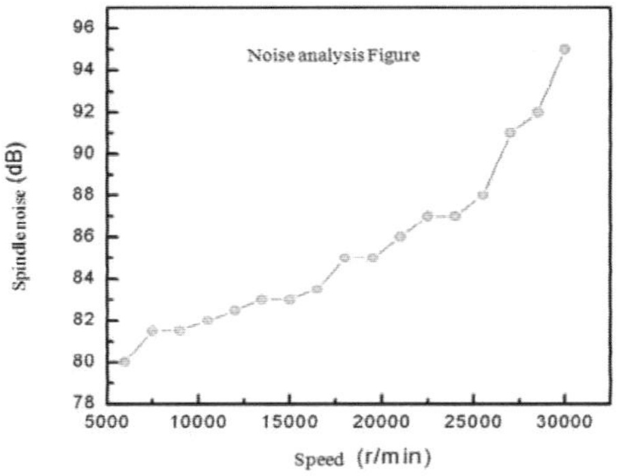

Figure 13. Noise of the ceramic motorized spindle.

4. Intelligence of the ceramic motorized spindle

Intelligent prediction of the temperature rise of the motorized spindle and the intelligent control technology of the spindle bearing preload are studied.

4.1. Intelligent prediction of the temperature rise of the ceramic motorized spindle

The prediction of the temperature field in the interior of the motorized spindle is of great significance to the construction and optimization design of early warning systems for motorized spindles. The combination of the measured temperature of the spindle surface and the finite element model is conducive to the improvement of the prediction accuracy of the whole temperature field in motorized spindles. In other words, the heat transfer coefficient can be optimized based on the measurement data of the surface temperature and then this coefficient can be taken as the boundary condition for the finite element model. Subsequently, the temperature field can be calculated. In order to reduce calculation amount and time, as well as to quickly obtain the heat transfer coefficient, a model for the intelligent and accurate prediction of the temperature field in motorized spindles can be constructed on the basis of the optimization of the heat transfer coefficient by a genetic algorithm[16].

Let *fi* be the objective function of the optimization problem and *fit* be the fitness function.

$$f_i = \frac{1}{m}\sum_{i=1}^{m}\sqrt{(T_{ei} - T_{si})} \tag{4}$$

$$fit = \frac{1}{1+f_i} \tag{5}$$

In equations (4) and (5), m is the number of the temperature measuring points, T_{ei} is the steady-state temperature of motorized spindle monitored in the experiment, and T_{si} is the steady-state temperature of motorized spindle obtained by finite element simulation. The fitness value was set as $f_i \leq 0.67$. When $f_i \leq 0.5$, iteration ended. Then, the optimal heat transfer coefficient and the optimal temperature field of motorized spindle were output.

As shown in table 4, the initial value of the heat transfer coefficient is input into the finite element model of the motorized spindle. Then the initial temperature field before the optimization of the heat transfer coefficient is calculated by simulation. The initial steady-state temperature field obtained by simulation was extracted. Combined with the steady-state temperature monitored by experiments, a genetic algorithm was used to optimize the 100-generation iteration to obtain the best heat transfer coefficient and an accurate temperature field prediction model.

Table 4. The initial values of heat transfer coefficients.

Parameter name	Heat transfer coefficient between motorized spindle and external air h_1 (W/m^2·℃)	Heat transfer coefficient between rotor and stator h_2 (W/m^2·℃)	Heat transfer coefficient of shaft end h_3 (W/m^2·℃)	Heat transfer coefficient between bearing and compressed air h_4 (W/m^2·℃)	Heat transfer coefficient between stator and cooling water h_5 (W/m^2·℃)
Initial values	9.7	146.81	121.35	71.42	190.12
Optimization values	19.99	188.42	188.20	127.71	500.29

The experimentally monitored temperatures at certain points of shell, front bearing, stator, rear bearing were compared with the simulated steady-state temperatures before and after optimization as in figure 14. Before optimization, the average error reached 8 ℃. After optimization, the average error decreased to 0.81 ℃.

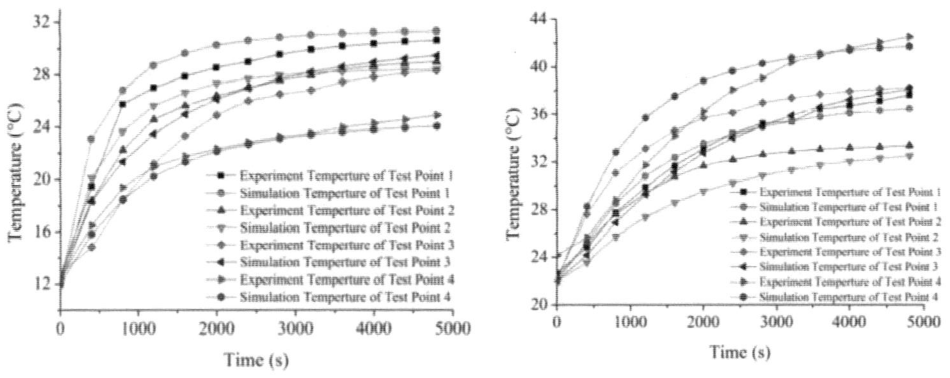

Figure 14. The comparison between the experimentally monitored temperatures of various parts of the motorized spindle and the simulated steady-state temperatures before and after optimization.

IOP Conf. Series: Materials Science and Engineering **399** (2018) 012057 doi:10.1088/1757-899X/399/1/012057

4.2. Research on the intelligent control method of preload of the ceramic motorized spindle

A variable preload control platform for a ceramic motorized spindle based on a piezoelectric actuator was designed. The aim is to determine the required preload of the machine tool spindle under different working conditions, and apply pre-tightening force to the bearing through the piezoelectric actuator to meet the full-speed performance requirements of the high-speed machine tool spindle, thereby realizing a highly efficient spindle system. In order to realize the control platform, some related technologies need to be studied. The output force characteristics of the piezoelectric actuator is studied in this article.

(1) As the inner ring-free all-ceramic motorized spindle shown in figure 1 is modified. Eight piezoelectric ceramic actuators are uniformly arranged at the rear end of the motorized spindle to preload the bearings to ensure that each piezoelectric ceramic actuator is evenly applied. Tightening and maintaining of good contact with the bearing outer ring determine its axial position with bolts.

The front and rear bearings are respectively provided with a temperature sensor on the outer ring and a vibration sensor is arranged above the corresponding housing of the bearing to monitor the working state of the main shaft. By adjusting the driving voltage of the piezoelectric ceramic driving power source, the magnitude of the output force is changed and the control of the bearing preload during the working process of the motorized spindle. This device scheme provides the corresponding bearing preload under the premise of different working conditions of the motorized spindle, and can also compensate for the thermal expansion of the spindle. The final design of the preload control device based on piezoelectric ceramics is shown in figure 15.

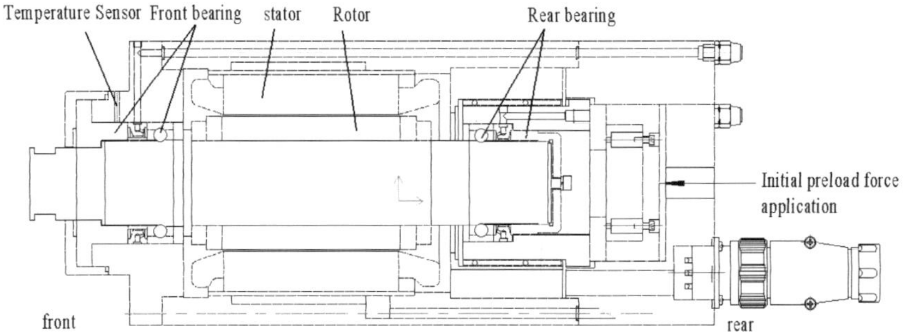

Figure 15. Variable preload control platform structure of the ceramic motorized spindle.

(2) Research on the force output characteristics of piezoelectric ceramic driver. A device for testing the output characteristics of piezoelectric actuators based on 9257B plane dynamometer and HVA-150.A1 piezoelectric actuator driving power was designed. The force output characteristics of PSt 150/4/7 VS9 piezoelectric actuators at different initial loads and voltage frequencies were analyzed by test. The test device is shown in figure 16.

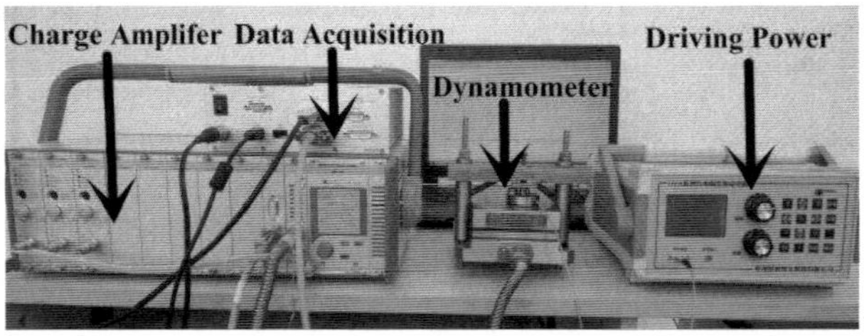

Figure 16. Test device of piezoelectric actuators.

The piezoelectric actuator was tested at different initial voltages of 150 N and the driving voltage of 0-150 V. The test result is shown in figure 17.We can see that the maximum output force of the piezoelectric actuator is maximum at 150N initial preload.

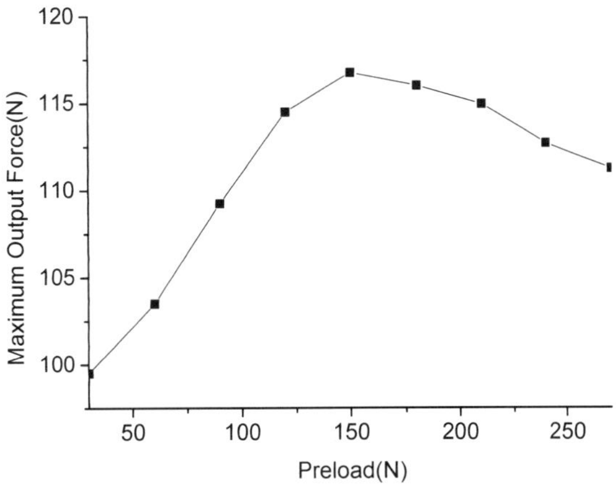

Figure 17. Maximum output force under different preload.

(3) Modeling and prediction of hysteresis main ring of piezoelectric actuator. In this paper, we model and analyze the hysteresis at 150N initial pressure and 1Hz driving frequency. The first 4 working cycles are used as a training set to predict the output power of the 5th duty cycle. The input layer voltage is determined using the principle of the Preisach model, and the BP neural network is used to complete the identification of the density function [18].

As can be seen from figure 18, the prediction results of the model based on the time-series input voltage are linear, and the prediction result based on the input voltage of the Preisach model is close to the expected value. This kind of hysteresis prediction model can better predict the output force of the piezoelectric actuator in different working cycles. The maximum prediction error is 2.9252N, the minimum prediction error is -3.3327N, and the average prediction error is 1.54%. The prediction effect of this model is good and the accuracy of the prediction model can basically meet the control requirements.

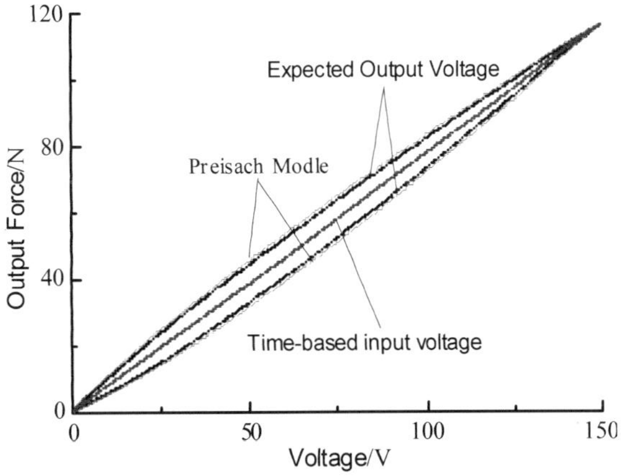

Figure 18. Contrast of different models.

5. Conclusion and Outlook

5.1 Conclusion

The innovative design of the ceramic motorized spindle unit including the structural design of the ceramic motorized spindle without inner ring, ceramic ball bearing design, tool interface design of ceramic motorized spindle and the precision assembly of the ceramic motorized spindle have been studied. What is more, testing and analysis of the comprehensive performance of the ceramic motorized spindle including load characteristics test of temperature rise test and vibration and noise test have been done. In addition, intelligence of the ceramic motorized spindle including the intelligent prediction of the temperature rise and intelligent control method of preload of ceramic motorized spindle have been studied. A variable preload control platform for a ceramic motorized spindle based on a piezoelectric actuator has been designed and a model for predicting the hysteresis main ring of piezoelectric actuator has been built.

5.2. Outlook

(1) In the innovative design of ceramic electric spindle unit, the failure mechanism and life of all ceramic ball bearings are worthy of further study and discussion. (2) In testing and analysis of the comprehensive performance of the ceramic motorized spindle, due to the large number of factors affecting the noise of ceramic motorized spindles, the theories regarding spindle and bearing vibration and noise are recognized as immature. The in-depth analysis and control of the high-frequency noise of ceramic motorized spindles will be worth focusing on in the future. (3) In intelligence of the ceramic motorized spindle, one of the future trends is likely to move toward prognostics-centered control and maintenance [19]: Techniques that support condition-based control and maintenance can be divided into two main categories: diagnostics and prognostics [20]. Diagnostics, as a posterior event analysis, deals with fault detection, isolation and identification when an abnormality occurs. Prognostics, as a prior event analysis, deals with fault and degradation prediction before they occur. Prognostics-centered control and maintenance is much more efficient than the diagnostics-based strategy in achieving zero-downtime performance of intelligent spindles.

Acknowledgments

This work was supported by NSFC Grant Numbers 51675353，Liaoning Natural Science Foundation 2015020149, Shenyang Science and Technology Project F16-205-1-15 and RC170216. The Project is also sponsored by “Liaoning BaiQianWan Talents Program”.

References

[1] Weck M and Koch A 1993 Spindle-Bearing Systems for High-Speed Applications in Machine Tools *J. Annals of the CIRP* **42**(1) 445-448
[2] Weck M 1998 Machine Tools for High Speed Machining Proceedings of the Improving Machine Tool Performance J. **7** 27-41
[3] Wang C, Zhang W, Liu Z et al. 2003 Present Status of Full - Ceramic Bearing and Its Application Prospects *J. Modern technology ceramics* **3** 30-32
[4] Zhang K, Wu Y, Wei F et al. 2002 Research on experiment of ceramic ball bearings in CNC speed spindle *J. Machinery manufacturing* **4** 33-35
[5] Boglietti A 2003 Predicting iron losses in soft magnetic materials with arbitrary voltage supply: an engineering approach *J. IEEE Trans, Magnetics* **39**(2) 981-989
[6] Depenbrock M. 1988 Direct Self-Control (DSC) of inverterfed induction machine *J. IEEE Transactions on Power Electronics* **3**(4) 420-429
[7] Li S, Li X, Wu Y, Zhang K 2014 A Research on Comprehensive Performance Test System of High-Speed Motorized Spindle *International Conference on Instrumentation* 613-617
[8] Zhang L, Wang L, Xu M et al. 2012 Analysis and Improvement Measures of Motorized spindle Deflection Caused by Action of Unbalanced Magnetic Pull Force *J. Micromotors* **45** (11) 13-15
[9] Li S and Wu Y A 2010 study on oil/air lubrication and preload of a high frequency fully ceramic motor spindle *IEEE International Workshop on Automatic Control (IWAC2010)* November 4-6, Zhengzhou, Henan, China. Vols.6 3611-3614
[10] Zhang K and Wu Y Research on Precision Grinding System and Key Technology Based on PMAC-PC *Advanced Materials Research* Vols. 97-101 (2010) pp 1942-1946

[11] Tang Y, Wu Y, Zhang K et al. Atomistic simulation of the bending deformation of a single-crystalline nano-beam 2010 *International Conference on Mechanic Automation and Control Engineering*, MACE2010 404-407
[12] Wen H and Su J 2012 Experimental research on thermal and vibration performance of high-speed spindle *J. Advanced Materials Research* **411** 767-771
[13] Zhou S, Wu Y, Wang Q, Du J, Zhao D, Lu F 2014 Study on Nonlinear Dynamic Characteristics of the Double Worm Gear Anti-backlash Drive Mechanism *Machinery Transmission J.* **38**(10) 109-113
[14] Zhang L, Wu Y, Wang L et al. 2012 The Effect of the Air-gap Eccentricity on the Rotor System of the Ceramic Motorized Spindle *J. Mechanical Science and Technology for Aerospace Engineering* **31**(9) 112-1521
[15] Cho J K and Lee D G 1998Thermal characteristics of the spindle bearing system with a gear located on the bearing span *J. International Journal of Machine Tools and Manufacture* **38** 1017-1030
[16] Bossmanns B and Tu J F 1999 A thermal model for high speed motorized spindles J. *International Journal of Machine Tools & Manufacture* **391** 345-1366
[17] Vafaei S, Rahnejat H and Aini R 2002 Vibration monitoring of high speed spindles using spectral analysis techniques *J. International Journal of Machine Tools and Manufacture* **42** 1222-1234
[18] Zhu Y and Xu H 2014 Dynamic Preisach Model of Giant Magnetostrictive Actuator Based on Hysteresis Function of Hyperbolic Tangent Function *J. Journal of Mechanical Engineering* **50**(3) 165-170
[19] Cao H, Zhang X and Chen X 2017 The concept and progress of intelligent spindles: A review *International Journal of Machine Tools & Manufacture* **112** 21–52
[20] Jardine A K S, Lin D and Banjevic D 2006 A review on machinery diagnostics and prognostics implementing condition-based maintenance *Mech. Syst. Signal Process.* **20** 1483–1510

Research on surface quality of Si_3N_4 ceramics by internal grinding

H Yan[1,2*], Y Wu[1,2] and H Wang[1,2]

[1] School of Mechanical Engineering, Shenyang Jianzhu University, Shenyang 110168, China.
[2] National-Local Joint Engineering Laboratory of NC Machining Equipment and Technology of High-Grade Stone, Shenyang Jianzhu University, Shenyang 110168, China.

[*] E-mail: sjzuwuyh@163.com

Abstract. The machined surface quality of hard and brittle materials was studied based on the analysis of the material removal mechanism of ceramics. According to the geometrical relationship of the internal grinding, the equivalent diameter of grinding wheel was deduced. In the internal grinding, the models of undeformed chip thickness, peak-valley surface roughness and arithmetic mean surface roughness were established with the assistance of planar grinding. The proposed models were verified utilizing the orthogonal test of internal grinding Si_3N_4 ceramics. The trend of experimental results was consistent with the theoretical model. The scientific validity of the proposed models was supported by the experimental results, and the investigation provides a theoretical reference and experimental basis for further study on the grinding of ceramic materials.

1. Introduction

Ceramic materials such as Si_3N_4 ceramic have advantages of high strength, high hardness, oxidation resistance, corrosion resistance, wear resistance and ablation resistance. Ceramic products have been used in military engineering, aerospace engineering and other high-tech fields [1]. With the development of Si_3N_4 ceramics, the requirements of machining quality, precision, shape, dimension precision and surface complexity are improved gradually. Especially in recent years, with the advent and development of ceramic motorized spindle, the machining accuracy requirement of ceramic bearing rings is extremely demanding. However, ceramic materials are difficult to process because of their hard and brittle properties [2, 3]. The efficient and precise machining has become a focus recently. Nowadays, the common method of precision machining for ceramic materials is grinding with diamond wheel [4-6].

Grinding is generally applied in the final stage of processing mechanical parts to ensure the shape accuracy and diminish surface roughness [7]. The formation of grinding surface is a constant interference between abrasive grains and workpiece surface. Due to the complex process of grinding, numerous factors can interfere in the surface roughness, thus the relationship between them is difficult to calculate accurately [8]. Generally, the proposed theoretical models about grinding surface roughness were based on some assumptions to better simulate the mechanism of workpiece surface formation. The theoretical formula showed high calculation accuracy, whereas the calculation results were often smaller than the actual values of surface roughness [9, 10]. As the empirical formula is suitable for specific grinding conditions, it has a narrow scope for analysing and forecasting workpiece surface roughness values, which limits its application [4]. When the grinding condition changes greatly, the calculation accuracy decreases. To choose a proper grinding wheel is also crucial [11]. Reducing the grain size, which can improve the machining surface morphology, determines a reduction of the material removal rate and prolongs the processing cycle. In addition, Liu et al. [12] investigated the influence of planar grinding on Si_3N_4 ceramics, which proved that the grinding parameters regulated the surface roughness. Jing et al. [13] established a model between the surface morphology and process parameters based on BP neural network, providing the support for surface morphology optimization. Mochida et al. [14] pointed out that small grinding depth was necessary for low damage grinding by increasing the wheel speed at an even high material removal rate. Rodriguez et al. [15] studied the influence of grinding depth and workpiece velocity on Si_3N_4 ceramics surface damages. The experimental results showed that the surface damage decreased under "hard" grinding conditions.

In sum, the existing literature focused on the prediction for surface roughness of the hard and brittle materials after grinding. As the mechanism model of grinding surface roughness is limited to planar grinding, the mechanism of internal grinding is reported rarely. In this paper, taking Si_3N_4 ceramics as an example, the mechanism of internal grinding ceramic materials was investigated by means of the orthogonal test. Peak-valley surface roughness, arithmetic mean surface roughness and surface morphology were utilized to characterize the machined surface quality. Theoretical models of undeformed chip thickness and surface roughness for internal grinding were established. Furthermore, the reliability and accuracy of models were verified by means of the orthogonal test to provide a theoretical basis for surface roughness prediction in actual internal grinding.

2. Removal mechanism of ceramic materials in internal grinding

2.1. Equivalent diameter of grinding wheel

As shown in figure 1, similar to surface grinding, internal grinding can be considered a material removal process, in which the abrasive grains were utilized as cutting tools. Under the same grinding condition, the contact arc length of internal grinding is larger than that of planar grinding, thus the chip is not easy to eliminate. Due to the poor heat dissipation of internal grinding, the workpieces are prone to be burnt by high temperature [16].

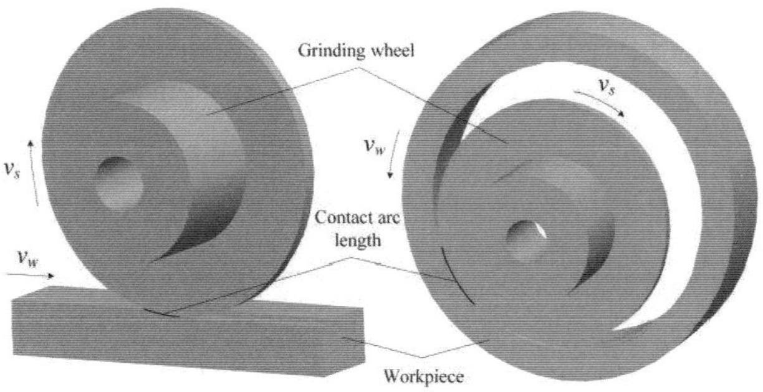

Figure 1. Sketch of planar grinding and internal grinding.

Figure 2 shows the process of brittle removal and surface formation of internal grinding for ceramic materials. Because of the character of hardness and brittleness, the ceramic workpiece easily produces cracks on the surface even in subsurface during the procedure of machining [17]. Also, in the grinding process, the abrasive grains cross the workpiece and remove the material directly, forming the chip and separating it from the workpiece. In addition, the material removal by internal grinding is in form of tiny debris, forming certain scratches on the machined surface and even visible microcracks.

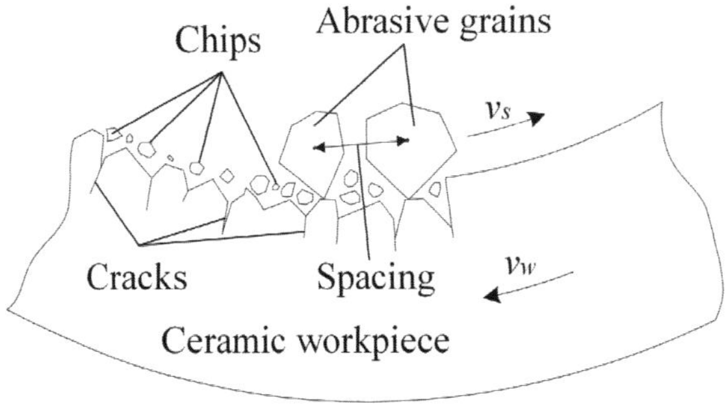

Figure 2. Material removal of ceramics by internal grinding.

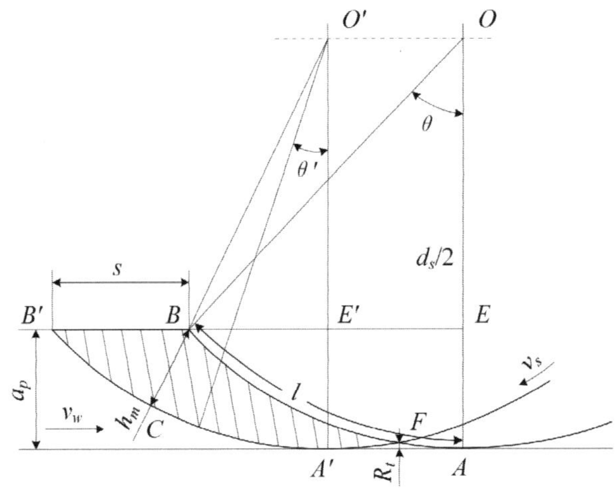

Figure 3. Undeformed cutting thickness and surface roughness of planar grinding.

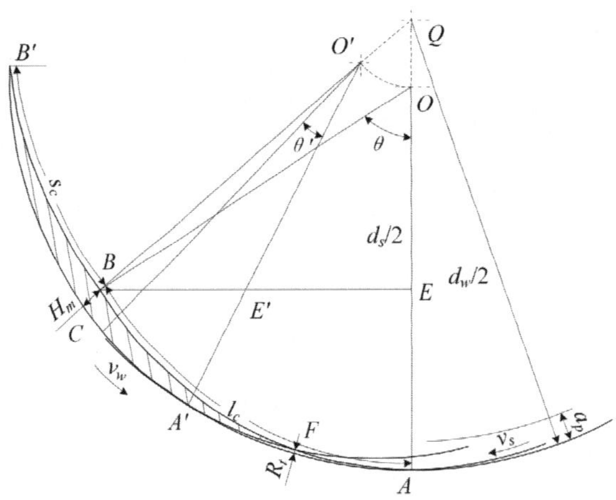

Figure 4. Undeformed cutting thickness and surface roughness of internal grinding.

In the process of high speed finish grinding, the contact arc length was also tiny due to the minute grinding depth. In this paper, plunge grinding was applied as an example to study the mechanism of internal grinding of ceramic materials.

As shown in figure 3, in the planar grinding, the contact arc length l is considered the chord length AB because the grinding depth is quite tiny. So,

$$l \approx AB = (a_p \cdot d_s)^{\frac{1}{2}} \tag{1}$$

where a_p is the grinding depth, d_s is the external diameter of the grinding wheel.

As shown in figure 4, similar to planar grinding, the contact arc length l_c of internal grinding [18] can be expressed as:

$$l_c = (AE \cdot d_s)^{\frac{1}{2}} \tag{2}$$

where AE is equivalent to grinding depth of planar grinding. In the process of internal grinding chip formation, AE is solved by geometric relation as equation (3):

$$AE = \frac{a_p}{1 - d_s/d_w} \tag{3}$$

where d_w is the internal diameter of workpiece.

The contact arc length l_c of internal grinding is finally denoted as:

$$l_c = (a_p \cdot d_e)^{\frac{1}{2}} \tag{4}$$

where d_e is defined as equivalent diameter of wheel.

The d_e of internal grinding is calculated by:

$$d_e = \frac{d_s}{1 - d_s/d_w} \tag{5}$$

2.2. Undeformed chip thickness of internal grinding

Undeformed chip thickness is the maximum cutting depth of a cutting edge, which is often referred to as "abrasive cutting depth". In planar surface grinding, the workpiece is stationary when the cutting edge is working, and a distance of s, i.e., the feed per cutting edge, is moved instantaneously before the next cutting edge enters the cutting process. As shown in figure 3, the undeformed cutting thickness of the planar surface grinding [18] can be described as:

$$h_m = 2s\left(\frac{a_p}{d_s}\right)^{\frac{1}{2}} - \frac{s^2}{d_s} \tag{6}$$

where s is:

$$s = \frac{Lv_w}{v_s} \tag{7}$$

where: L is grain spacing, which indicates the distance between two adjacent grains, v_s is grinding wheel speed, v_w is workpiece velocity.

As shown in figure 4, in internal grinding, s_c is the moved arc length of the center of the grinding wheel, which correspondsto the feed per cutting edge of the planar surface grinding. s_c is defined as follows:

$$s_c = s(1 - d_s/d_w) \tag{8}$$

The undeformed chip thickness of internal grinding for ceramic materials can be obtained using s_c in equation (8) instead of s in equation (6). This is actually the same as the result that d_e in equation (5) substitutes for d_s of equation (6). H_m can be represented as:

$$H_m = 2s\left(\frac{a_p}{d_e}\right)^{\frac{1}{2}} - \frac{s^2}{d_e} \tag{9}$$

The equations (5) and (7) are plugged into equation (9) to obtain the final expression of H_m, shown as equation (10).

$$H_m = 2L\left(\frac{v_w}{v_s}\right)a_p^{\frac{1}{2}}\left(\frac{1-d_s/d_w}{d_s}\right)^{\frac{1}{2}} - L^2\left(\frac{v_w}{v_s}\right)^2\left(\frac{1-d_s/d_w}{d_s}\right) \tag{10}$$

2.3. *Surface roughness model of internal grinding*

As shown in figure 3, the theoretical formulas of peak-valley surface roughness R_t and arithmetic mean surface roughness R_a in the planar surface grinding are given [18], respectively, by:

$$R_t = \frac{1}{4}\left(\frac{v_w L}{v_s d_s^{\frac{1}{2}}}\right)^2 \tag{11}$$

$$R_a = \frac{1}{9\sqrt{3}}\left(\frac{v_w L}{v_s d_s^{\frac{1}{2}}}\right)^2 \tag{12}$$

When the d_s in equation (11) is regarded as the equivalent diameter of d_e which is replaced by equation (5), the peak-valley surface roughness in internal grinding ceramic materials can be rewritten as:

$$R_t = \frac{1}{4}\left(\frac{v_w L}{v_s}\right)^2 \frac{1-d_s/d_w}{d_s} \tag{13}$$

According to the equation (13) and figure 4, when d_w tends to infinity, i.e. $d_s/d_w = 0$, here equation (13) is the same as equation (11), which is equivalent to the peak-valley surface roughness of the planar surface grinding.

Similarly, the arithmetic mean surface roughness of internal grinding ceramic materials is determined as:

$$R_a = \frac{1}{9\sqrt{3}} \left(\frac{v_w L}{v_s} \right)^2 \frac{1 - d_s / d_w}{d_s} \tag{14}$$

According to the model established above, the undeformed cutting thickness and surface roughness are not only related to the geometrical size of grinding wheel and workpiece, but also to the cutting parameters and the grain spacing. Average spacing of grains L_{avg} can be indicated as equation (15) [19].

$$L_{avg} = 137.9 M^{-1.4} \sqrt[3]{\frac{\pi}{32 - S}} \tag{15}$$

where: M is the grain size number of abrasives, S is the structure number of the grinding wheel. When the grain size of the wheel is settled, the average grain spacing is determined.

3. Test procedures

3.1. Materials and equipment

As shown in figure 5a, the test was carried out in high precision MK2710 NC internal and cylindrical modular grinding machine tool manufactured by Wuxi Machine Tool Group of China. The four axes of the machine tool are composed of a servo motor and a ball screw through a flexible coupling. It can realize no gap and high sensitivity movement, and the motion resolution of each axis can both reach 1 μm.

Figure 5. Experimental machine tools and grinding test.
a) MK2710 grinding machine and b) Internal grinding.

Table 1. Performance of grinding wheel.

Wheel attribute	Values
External diameter (mm)	55
Thickness (mm)	23.5
Aperture (mm)	10
Ring width (mm)	3
Ring height (mm)	5
Abrasive	RVD
Concentration (%)	100
Average grain spacing (μm)	108

Table 2. Properties of Si_3N_4 ceramic.

Material attribute	Values
Density ($g \cdot cm^{-3}$)	3.25
Coefficient of thermal expansion ($10^{-6} \cdot K^{-1}$)	3.2
Elastic modulus (GPa)	310
Poisson ratio	0.26
Fracture toughness ($MPa \cdot m^{1/2}$)	6.0
Bending strength (MPa)	900
HRC (GPa)	80
Inner diameter of material ring (mm)	65

As shown in figure 5b, the grinding test of Si_3N_4 ceramic inner ring was carried out by using a diamond grinding wheel with metal bonded and grain size of 120#. The grinding wheel parameters and material properties are shown in tables 1 and 2, respectively. The grinding fluid was water-based emulsified oil (1 : 20).

3.2. Test scheme

To investigate the parameters which affect the surface formation mechanism of Si_3N_4 ceramic via internal grinding, an orthogonal test was carried out, which included wheel speed, workpiece velocity and grinding depth. The orthogonal test design is shown in table 3.

Table 3. Orthogonal test design.

Experiment number	v_s (m/s)	v_w (mm/min)	a_p (μm)
1	34	1200	4
2	34	6000	12
3	34	12000	20
4	51	1200	20
5	51	6000	4
6	51	12000	12
7	17	1200	12
8	17	6000	20
9	17	12000	4

4. Results and discussion

4.1. Surface morphology

As shown in figure 6, the surface morphology of workpiece after grinding was observed under the scanning electron microscope (S-4800). In the internal grinding, the influence of grinding parameters on the surface quality of Si_3N_4 ceramic was analysed.

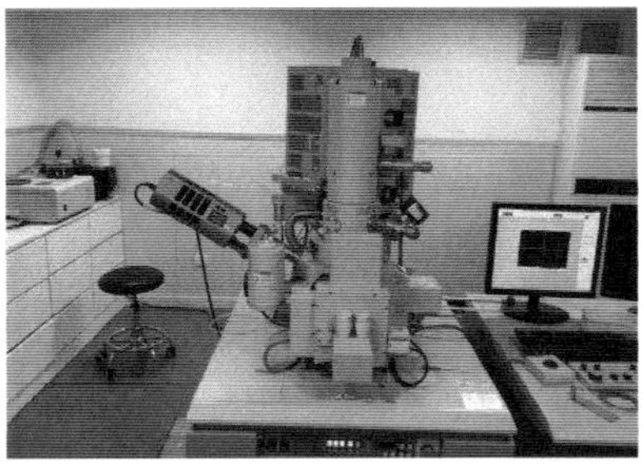

Figure 6. Scanning electron microscope.

Figure 7 shows the surface morphology of Si_3N_4 ceramics by internal grinding. When the grinding wheel speed was low, the gully regions were wide and the bulge parts were high on the surface (figure 7a and b). Along with the increase of the wheel speed, the gully was obviously narrowed (figure 7c). As the wheel speed continues to increase, the bulge on the surface may be removed (figure 7d). From all the pictures in figure 7, we can find that the surface scratch caused by the plastic removal and fracture pit by brittle removal were obvious. The increase of the wheel speed reduced the maximum undeformed chip thickness, which resulted in a shallow action depth and a narrow action width of the abrasive grain. As a result, the gully became shallow and narrow. Furthermore, the increase of the wheel speed led to an increase of the number of abrasive grains participating grinding per unit time, so that the bulges were removed in time or squeezed into the gully, so the increase of the wheel speed was beneficial to improve the machined surface. Therefore, in the internal grinding, the wheel speed is an important factor which affects the surface morphology of Si_3N_4 ceramics.

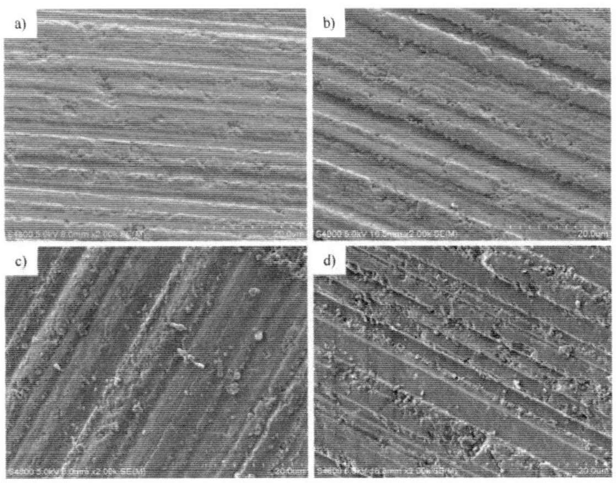

Figure 7. Surface morphology of Si_3N_4 ceramics by internal grinding. a) Experiment number 1, b) Experiment number 3, c) Experiment number 6 and d) Experiment number 8.

In addition, according to figure 7a and b, when the grinding wheel speed was the same, the increase of grinding depth would widen the gully and slightly heighten the bulge in the machined surface. From figure 7b and d, a better surface quality could be obtained using a smaller workpiece velocity. With the increase of grinding depth, the removal of material is mainly in a brittle manner [20]. The increase of the workpiece velocity reduced the time of abrasive grain cutting in the same area, thus the gully and bulge of the machined surface will be increased. Consequently, it is not conducive to achieving a better surface quality through increasing the grinding depth and the workpiece velocity.

4.2. Surface roughness

The roughness of machined surface was measured by Taylor Hobson roughness measuring instrument, and the results are shown in table 4. According to the results of orthogonal test, the relation curve of the surface roughness with the cutting parameters is drawn as shown in figure 8.

Table 4. Experimental results.

Experiment number	R_{tE} (μm)	R_{aE} (μm)
1	1.4763	0.3957
2	1.5861	0.4134
3	1.7903	0.4762
4	1.5160	0.3635
5	1.4358	0.3793
6	1.6842	0.4216
7	1.7907	0.4598
8	1.8723	0.4874
9	1.9212	0.5424

As shown in figure 8, the surface roughness decreased and the surface quality was improved by the gradual increase of the grinding wheel speed. Increasing the workpiece velocity led to an increase of surface roughness, which resulted in the deterioration of the surface. The increase of the grinding depth also made the surface roughness an increase trend, but the change was small. In this experiment, when the wheel speed increased from 17 m/s to 51 m/s, the surface roughness value R_{tE} decreased from 1.8614 μm to 1.5453 μm, down by 16.98% and the surface roughness value R_{aE} decreased from 0.4965 μm to 0.3881 μm, reduced by 21.83%. When the workpiece velocity increased from 1200 mm/min to 12000 mm/min, the R_{tE} increased from 1.5943 μm to 1.7986 μm, an 12.81% upwards and the R_{aE} increases from 0.4063 μm to 0.4801 μm, an increase of 18.16%. The effect of the grinding depth on the surface roughness was that the grinding depth increased from 4 μm to 20 μm, the R_{tE} increased from 1.6111 μm to 1.7262 μm, which increased by up to 7.15%. The R_{aE} decreased slightly from 0.4391 μm to 0.4316 μm when the grinding depth increased from 4 μm to 12 μm, while increasing the grinding depth to 20 μm, the R_{aE} increased to 0.4424 μm. In the process, the increase of the grinding depth led to an increase of 0.75% only for the R_{aE} value.

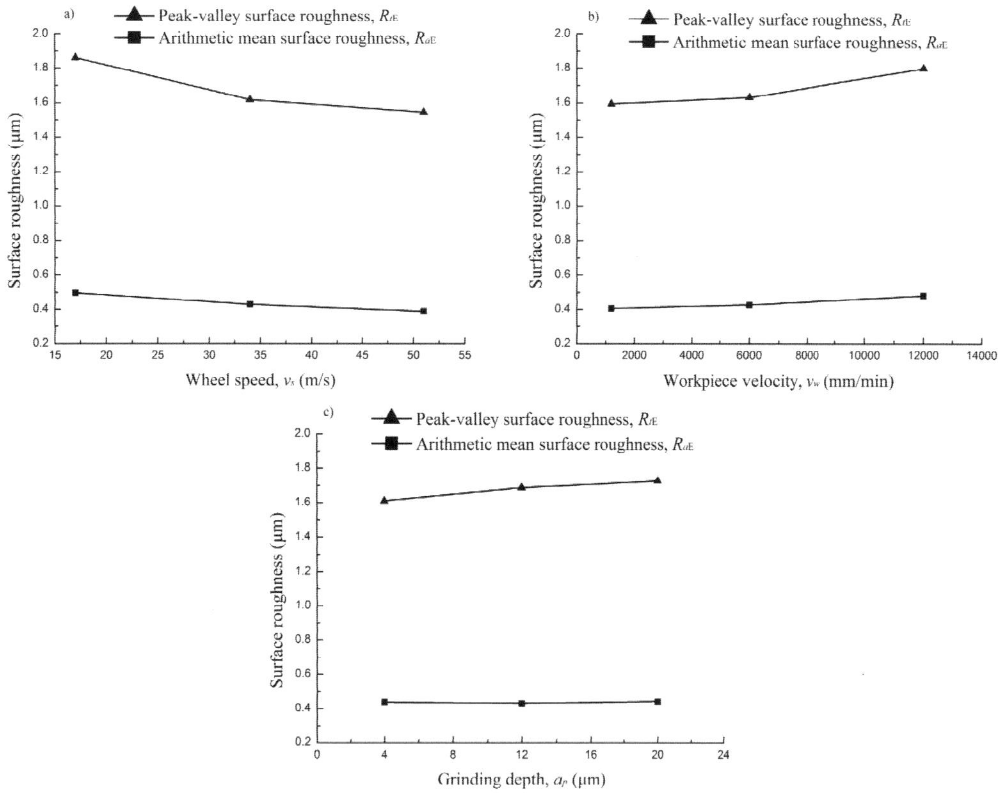

Figure 8. Effect of grinding parameters on surface roughness in the internal grinding process of Si_3N_4 ceramics. a) Wheel speed, b) Workpiece velocity and c) Grinding depth.

By increasing the grinding wheel speed, the number of effective abrasive grain raises in per unit time, which can improve the machined surface quality. Reducing the workpiece velocity can diminish the height of the residual area, which leads to the decrease of the surface roughness value. For the grinding depth, the increase of the grinding depth can cause the increase of the surface roughness, but the change is not obvious. In the actual grinding process, an increase in cutting depth will increase the number of grains that participate in cutting simultaneously, bringing a tremendous friction extrusion in the contact area. In the meantime, the heat produced by the process is not easy to dissipate in internal grinding, leading to a rise of cutting temperature, which may also burn the workpiece surface and reduce the process quality.

Tables 3 and 4 show that the minimum value of R_{tE} was obtained (1.4358 μm) under the following conditions: wheel speed v_s at 51 m/s, workpiece velocity v_w at 6000 mm/min and grinding depth a_p at 4 μm. Whereas the surface roughness value R_{aE}, 0.3793 μm, under the same conditions, was not its minimum value. The best condition for R_{aE}, 0.3635 μm, was when the wheel speed v_s was 51 m/s, the workpiece velocity v_w was 1200 mm/min and the grinding depth a_p was 20 μm. Correspondingly, the value R_{tE} was 1.5160 μm, also not the minimum.

From the analysis presented above, it results that, in this experiment, the wheel speed has the main influence on the surface roughness, followed by workpiece velocity, with the minimum effect of grinding depth. Under the test parameters of the No. 5, that was $v_s = 51$ m/s, $v_w = 6000$ mm/min and $a_p = 4$ μm, the optimal surface quality was obtained by comprehensive analysis of R_{tE} and R_{aE}.

5. Comparison between theoretical and experimental results

5.1. Theoretical values

The measured values of surface roughness after processing were larger than the theoretical values, due to the machine vibration, size error, wheel dressing or the measurement accuracy, which would affect the experimental results. In this paper, all these factors can be attributed to the size effect M_r, so the formulas (13) and (14) respectively are corrected as:

$$R_t = \frac{M_r}{4}\left(\frac{v_w L}{v_s}\right)^2 \frac{1 - d_s/d_w}{d_s} \tag{16}$$

$$R_a = \frac{M_r}{9\sqrt{3}}\left(\frac{v_w L}{v_s}\right)^2 \frac{1 - d_s/d_w}{d_s} \tag{17}$$

In order to compare the change of undeformed chip thickness and grinding depth conveniently, an amplification factor K_H is added into formula (10), as shown in equation (18).

$$H_m = 2K_H L\left(\frac{v_w}{v_s}\right) a_p^{\frac{1}{2}} \left(\frac{1 - d_s/d_w}{d_s}\right)^{\frac{1}{2}} - K_H L^2 \left(\frac{v_w}{v_s}\right)^2 \left(\frac{1 - d_s/d_w}{d_s}\right) \tag{18}$$

The size effect M_r and the coefficient K_H are calculated by the latest square method. The values of M_r and K_H are 4×10^5 and 700, respectively. Table 5 presents the theoretical and experimental values of the modified model.

Table 5. Theoretical and experimental values.

Experiment number	a_p (μm)	H_m (μm)	R_{tE} (μm)	R_{tT} (μm)	R_{aE} (μm)	R_{aT} (μm)
1	4	0.7658	1.4763	0.0075	0.3957	0.0019
2	12	6.6321	1.5861	0.1870	0.4134	0.0480
3	20	17.1238	1.7903	0.7480	0.4762	0.1920
4	20	1.1416	1.5160	0.0033	0.3635	0.0008
5	4	2.5526	1.4358	0.0831	0.3793	0.0213
6	12	8.8427	1.6842	0.3324	0.4216	0.0853
7	12	2.6528	1.7907	0.0299	0.4598	0.0077
8	20	17.1238	1.8723	0.7480	0.4874	0.1920
9	4	15.3160	1.9212	2.9921	0.5424	0.7678

5.2. Undeformed chip thickness and grinding depth

Figure 9 shows the calculated undeformed chip thickness and the experimental grinding depth. From the data of view, the overall variation trend of undeformed chip thickness and grinding depth was similar. It proved that the theoretical model of undeformed chip thickness established in this paper can reflect the effect of cutting parameters on the undeformed chip thickness. Thus, a brief description of the material removal and surface formation by internal grinding on the hard brittle materials was obtained.

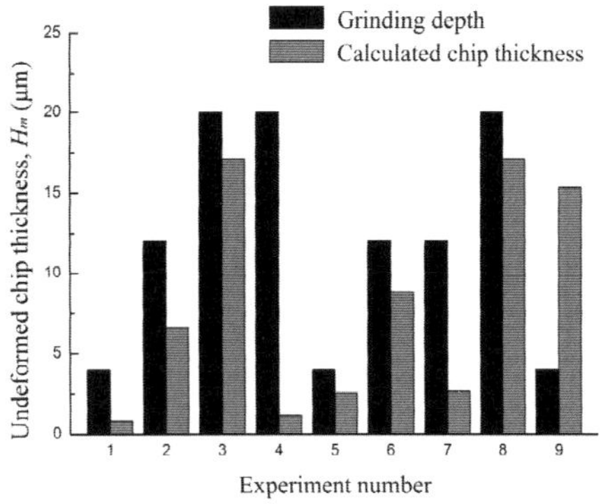

Figure 9. Comparison between undeformed chip thickness and grinding depth.

5.3. Measured values and theoretical values of surface roughness

Figure 10 shows the relationship between peak-valley surface roughness and arithmetic mean surface roughness, in which the fitting curve is obtained by the least square method. The coefficient of the fitted curve (k_F = 3.8124) is very close to that of the theoretical curve (k_T = 3.8971), and the error is only 2.17%. The relationship between peak-valley surface roughness and arithmetic mean surface roughness obtained by the experiment is consistent with the theoretical prediction and validates the correctness of the theoretical model.

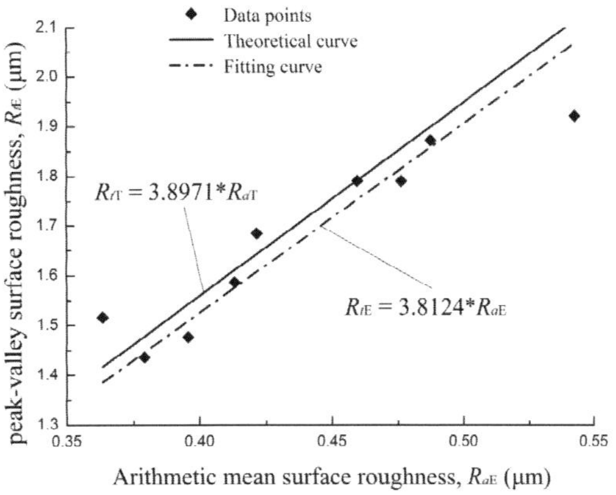

Figure 10. Relationship between peak-valley surface roughness and arithmetic mean surface roughness.

The surface roughness values calculated by the theoretical model and obtained by the experimental measurement are presented in figure 11. In figures 11a and b, both the calculated values of test No. 9 are greater than the measured values. The others show an opposite trend, where the actual measured values of surface roughness are greater than the theoretical values. According to the variation trend of peak-valley surface roughness, the experimental measurement results had a parallel trend with the theoretical values. The same results were observed in arithmetic mean surface roughness.

The theoretical model was built based on pure cutting state, which did not consider the effects of axial feed on the undeformed chip thickness and the surface roughness, assuming that the grinding wheel abrasive cutting edge was evenly distributed and had the same radial bump height. Furthermore, several factors that affect the undeformed chip thickness and surface roughness were not included in the modelling process, such as wheel wear, scratch overlap, cutting temperature and vibration of machine tool and grinding wheel. Moreover, the surface morphology obtained completely according to the theoretical cutting path is unsatisfactory in actual grinding. Comprehensively analyzing the above factors, it is not difficult to understand that the actual measured surface roughness values are much larger than the theoretical values, while in a similar trend. Therefore, the theoretical model for the surface roughness of the internal grinding is identified by the experimental results. The model can provide a theoretical reference for the investigation of surface formation mechanism of hard and brittle materials by the internal grinding, and also a valuable application for grinding internal surfaces of ceramic bearings.

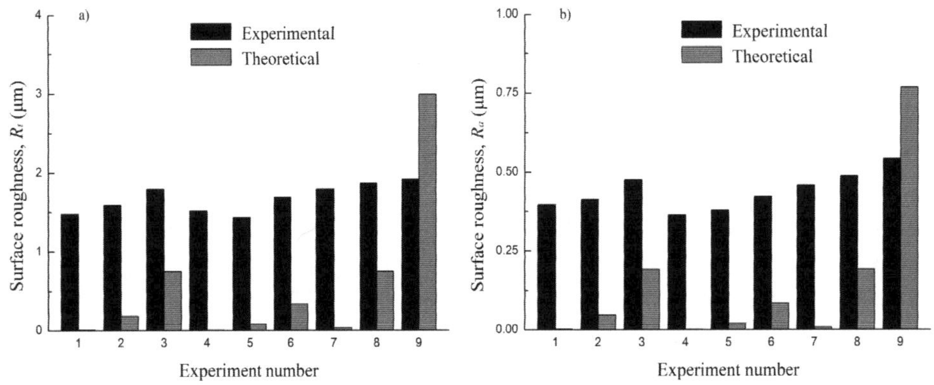

Figure 11. Comparison between theoretical and experimental values of surface roughness. a) Peak-valley surface roughness, R_t and b) Arithmetic mean surface roughness, R_a.

In addition, the influence of grinding depth on the surface roughness is not considered in the theoretical model. As the experimental results showed, the grinding depth had little effect on the surface roughness. From this point of view, the theoretical model is reliable.

6. Conclusions

This paper investigated the Si_3N_4 ceramics internal grinding material removal mechanism. By analysing the internal grinding process, the wheel equivalent diameter of internal grinding was deduced. Furthermore, a theoretical model for the internal grinding undeformed chip thickness and surface roughness of Si_3N_4 ceramics was developed based on the planar grinding model. Additionally, the size effect was utilized to correct the theoretical model to better simulate the machined surface quality. Internal grinding test of Si_3N_4 ceramics was carried out using the orthogonal test scheme., Then, the influence law of wheel speed, workpiece velocity and grinding depth on the Si_3N_4 ceramics grinding surface roughness were investigated and discussed according to the experimental results. And through the analysis of the theoretical model and the experimental results, the conclusions are described as follows:

1. Surface morphology of Si_3N_4 ceramics showed that the grinding wheel speed was an important factor which affected the surface formation in internal grinding. By increasing the grinding wheel

speed and reducing the workpiece velocity, the quality of the machined surface, on which the gully and bulge tend to shrink can be improved.

2. According to the experimental results, the surface roughness will be alleviated by increasing grinding wheel speed and reducing workpiece velocity. Analysed comprehensively, the increase of the grinding depth accelerated the surface roughness a little. The experimental results were consistent with the trend of the theory model. It can be also obtained from the test results, according to which the grinding wheel speed has the greatest influence on the surface roughness, followed by the workpiece velocity, and the influence of the grinding depth is last.

3. Through the comparison of undeformed chip thickness and actual grinding depth, experimental and theoretical values of peak-valley surface roughness and arithmetic mean surface roughness, the experimental results have a similar trend with the theoretical values. The results showed that the theoretical model can reflect the influence of cutting parameters on surface formation and surface roughness. It provides a theoretical and experimental basis for the study of the formation mechanism, which can predict the surface roughness of the hard and brittle materials such as ceramics.

Acknowledgments
The work was supported by the National Natural Science Foundation of China (Grant Numbers 51675353 and 51375317).

References
[1] Bengisu M 2001 Applications of ceramic materials *Engineering Ceramics* (*Engineering Materials*) (Berlin Heidelberg: Springer) chapter 6 pp 407-445
[2] Foschini R C, Souza E A C, Borges A F S and Pintão C A F 2016 A new approach to measure the elasticity modulus for ceramics using the deformation energy method *J. Mech. Sci. Technol.* **30** 3585
[3] Ren J, Kang R and Wang X 2011 *Grinding technology of difficult-to-machine materials* (Beijing: Publishing House of Electronics Industry)
[4] Wu Y, Zhu Z and Wang H 2014 Analysis on the factors that influences on the internal grinded surface of silicon nitride ceramic *J. Shenyang Jianzhu Univ.* (*Nat. Sci.*) **30** 312
[5] Zhang B and Fu Y 2013 Grinding of brittle materials with brazed diamond grinding wheel *Int. J. Adv. Manuf. Tech.* **67** 2845
[6] Malkin S and Hwang T W 1996 Grinding mechanisms for ceramics *CIRP Ann-Manuf. Techn.* **45** 569
[7] Zhang B, Zheng X, Tokura H and Yoshikawa M 2003 Grinding induced damage in ceramics *J. Mater. Process. Tech.* **132** 353
[8] Bigerlle M, Najjar D and Iost A 2003 Relevance of roughness parameters for describing and modeling *machined* surfaces *J. Mater. Sci.* **38** 2525
[9] *Agarwal* S and Rao P V 2005 A probabilistic approach to predict surface roughness in ceramic grinding *Int. J Mach. Tool. Manu.* **45** 609
[10] *Agarwal* S and Rao P V 2010 Modeling and prediction of surface roughness in ceramic grinding *Int. J Mach. Tool. Manu.* **50** 1065
[11] Cai M, Li X *and* Rahman M 2007 Study of the mechanism of nanoscale ductile mode cutting of silicon using molecular dynamics simulation *Int. J Mach. Tool. Manu.* **47** 75
[12] Liu W, Deng Z, Shang Y and Wan L 2017 Effects of grinding parameters on surface quality in silicon nitride grinding *Ceram. Int.* **43** 1571
[13] Jing J, Feng P, Wei S and Zhao H 2017 Investigation on surface morphology model of Si_3N_4 ceramics for rotary ultrasonic grinding machining based on the neural network *Appl. Surf. Sci.* **396** 85
[14] Mochida Y, Nishioka T, Yamakawa A, Tanaka Y, Ogura Y, Nakao H and Miyake M 1997 Ultra-high-speed grinding of Si_3N_4 ceramics *J. Ceram. Soc. Jpn.* **105** 784

[15] Rodriguez R, Leriche A and Carrerot H 2004 Influence of grinding parameters on Si_3N_4 ceramic mechanical properties *Silic. Ind.* (*4th International Symposium on Nitrides*) **69** 325
[16] Biermann D, Holtermann R, Menzel A and Schumann S 2016 Modelling and simulation of thermal effects in internal traverse grinding of hardened bearing steel *CIRP Ann-Manuf. Techn.* **65** 321
[17] Thomas R S 2013 Grinding brittle materials *Ceram. Ind.* **163** 18
[18] Malkin S, Cai G, Gong Y and Song G 2002 *Grinding technology theory and applications of* machining *with abrasives* (Shenyang: Northeastern University Press)
[19] Cheng J and Gong Y 2013 Experimental study on ductile-regime micro-grinding character of soda-lime glass with diamond tool *Int. J. Adv. Manuf. Tech.* **69** 147
[20] Bifano T G, Dow T A and Scattergood R O 1991 Ductile-rgime grinding, a new technology for machining brittle materials *ASME J. Eng. Ind.* **11** 184

Packing circle items in an arbitrary marble slab

Z Yuan[1], Y Zhang[1], M V Dragoi[2*] and X T Bai[1]

[1] School of Mechanical Engineering, Shenyang Jianzhu University, Shenyang 110168, China
[2] Manufacturing Engineering Department, Str. Mihai Viteazu nr. 5, Braşov, 500174, Romania

[*] E-mail: dragoi.m@unitbv.ro

Abstract. Nature marble is widely used in architecture, but generally it is irregular in shape. It is necessary to pack regular shapes such as circular and rectangular into irregular marble slabs, with an optimistic way to save material. However, when the arbitrary region is a concave set, it is difficult to recognize its characteristics if the circular item is in the region. An angle bisector method was proposed to avoid overlapping inside circles and ensure the circle is contained in the marble slab. Meanwhile, the mathematical model for packing circle items in an arbitrary region was established based on the heuristic algorithm. The experimental results show that the approach is effective and it can be used not only in a convex region, but also in a concave one.

1. Introduction

Packing questions are a class of optimization problems in mathematics that involve attempting to pack objects together into containers. Many of these problems can be related to real life packaging, storage and transportation issues, for example, packing circle items in an arbitrary marble slab.

Many scholars have studied packing circular items, Szabó, P. G [1] outlined the history and wide application of the circular layout problem. George, J. A [2] used a method that combines quasi-random and genetic algorithms to solve the problem of arranging different size circles in a rectangle. López, C. O [3] presented a heuristic for the problem based upon formulation space search, to solve the problem of packing different size circles in a circular container. According to Newton–Raphson, to get nonlinear equations, Birgin E. G [4] solved the problem of arranging circles of the same size in circles, triangles, and rectangles. Galiev S. I [5] et al. transformed the problem of packaging circles of the same size into a 0-1 linear programming problem, and proposed a heuristic algorithm based on a linear model to solve the problem. In the optimal layout of circular parts, X X Song [6] et al. proposed the placement algorithm Arc Search Algorithm (ASA), and used the hybrid inheritance algorithm as a search strategy which reduces the calculation time and increases the material utilization rate to be arranged effectively. The maximum hole degree strategy was proposed by Huang et al [7], to solve the problem of arranging circles of different sizes in the garden. Akeb et al. [8] proposed an algorithm that embeds the beam search in a binary local search with a minimum container radius. The basis of the algorithm search is that the largest hole metric branches out of a node. Hifi M [9] et al. proposed a heuristic algorithm based on simulated annealing for both constrained and unconstrained circular cutting problems. They defined an energy function, the minimum value of this function representing the good results. In Grosso et al. [10] an algorithm was proposed based on monotonic basin hopping [11], which solved the problem of round and round circles of equal packing in a round container with the smallest radius. An algorithm that combines local search and nested partitioning of feasible space was proposed [12]. The search process of this algorithm is tabu search, and the tabu list is used to find the best of the arrangement of circles; then, the feasible set is explored by dividing it. A solution method based on a combination of the branch-and-bound algorithm and the reduced gradient method were proposed by Stoyan Y. G [13] ,it arranged different size circles in the rectangle. Mladenović N [14] et al. compared some of the round layout issues and developed a general reformulation descent heuristic algorithm to solve the problem of the arrangement of circles of the same size. Castillo I [15] et al. presented illustrative numerical results using generic global optimization software packages, Birgin E. G [16] confirmed the relevance of global optimization in solving circle packing problems.

The above algorithm is mainly aimed at studying and solving the optimization of the circle layout in the regular graphics, but the study of packing circles in an irregular shape is less represented.

This paper focuses on the problem of packing circular elements in an irregular shape. The circles cannot be overlapped on each other, and the circles must be contained in the region as a constraint condition. A mathematical model was proposed based on the angle bisector method which involved the constraint condition and overlapped condition. Meanwhile, the limited branch tree search algorithm, as one of the heuristic algorithms, was used to solve the mathematical model.

2. Build mathematical model

P_i, P_j are defined as circular items ($i \neq j$) which will lay out in the region P. Then the mathematical description of the constraints of packing circular items in an irregular region is as follows $P_i \cap P_j = \Phi$, P_i, $P_j \in P$

2.1. Non-overlapping

There are n pieces of circle items with the same size which will be put into the irregular region. Set r as the radius of the circle, $O(x_o, y_o)$ is the centre of the circle. The distance between any two centres is greater or equal to double radius to ensure there is no overlap.

$$\left((x_{oi}-x_{oj})^2+(y_{oi}-y_{oj})^2\right)^{1/2} \geq 2r \tag{1}$$

Formula (1) can be expressed as:

$$\min\left[0,(x_{oi}-x_{oj})^2+(y_{oi}-y_{oj})^2-4r^2\right]^2=0 \tag{2}$$

So the non-overlapping formula of all the circle items is:

$$\sum_{i=1}^{n}\sum_{j=1}^{n}\min\left[0,(x_{oi}-x_{oj})^2+(y_{oi}-y_{oj})^2-4r^2\right]^2=0 \tag{3}$$

2.2. Inside the region

2.2.1. Angle bisector method. The article proposes a new angle bisector method to determine the condition of circular items inside the irregular region. Because the irregular region can be divided into convex and concave, the condition can be described in two cases.

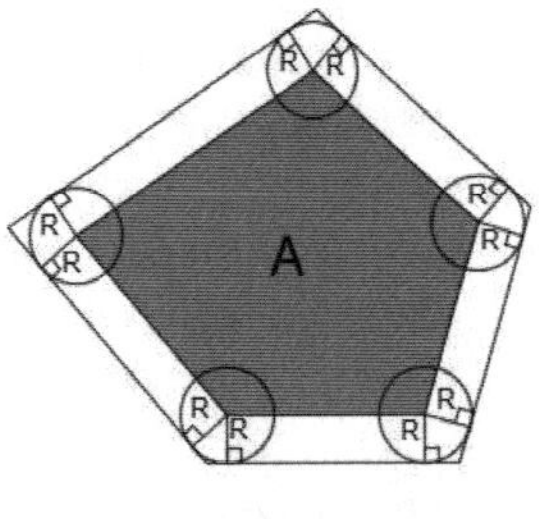

Figure 1. The feasible region.

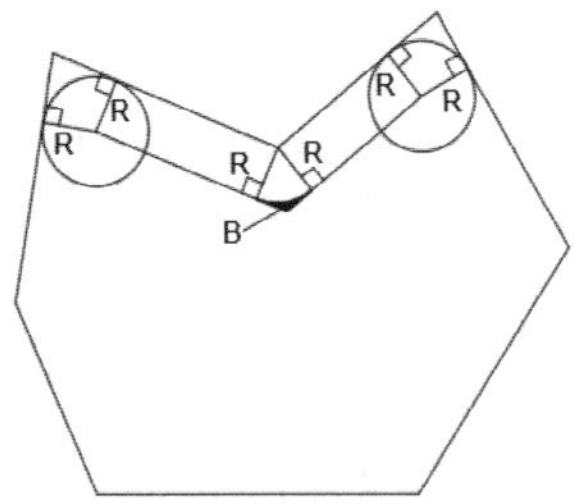

Figure 2. The ignored area B.

(1) The initial plate is a convex polygon: When the circle is tangent to the two sides of the polygon, the centre of the circle is the point where r is from both sides and on the angle bisector of the top corner. These circle centre points are connected in turn to form the feasible region A. As shown in figure 1, the A region represents the centre of the circle inside the region. It can be used to test whether the circle is inside the region or not.

(2) The initial plate is a concave polygon: Compared to the convex polygon, an additional condition needs to be considered at the concave vertex. The path of the circle centre is the fan shape in the concave vertex so the feasible side involves straight lines and curves. The area of shaded B is very small compared to the circular items and the irregular region, as shown in figure 2, so the feasible region will be linked with the straight line, and the shaded area B was ignored to simplify the calculation process.

2.2.2. Formulation of circular parts in sheet interior conditions. Setting any coordinate of the vertex of the irregular region is (x_m, y_m), the corresponding centre point on the vertex is $q(x_{qm}, y_{qm})$. The distance from this centre point to the side line is equal to radius.

$$\begin{cases} \dfrac{\left|(y_m - y_{m-1})x_{qm} - (x_m - x_{m-1})y_{qm} + (x_m - x_{m-1})y_m - (y_m - y_{m-1})x_m\right|}{\left[(y_m - y_{m-1})^2 + (-(x_m - x_{m-1}))^2\right]^{1/2}} = r \\ \dfrac{\left|(y_m - y_{m+1})x_{qm} - (x_m - x_{m+1})y_{qm} + (x_m - x_{m+1})y_m - (y_m - y_{m+1})x_m\right|}{\left[(y_m - y_{m+1})^2 + (-(x_m - x_{m+1}))^2\right]^{1/2}} = r \end{cases} \tag{4}$$

There are many algorithms for judging the position between the points and region A, a simple and effective algorithm is the Angle summation method. Suppose the polygon region A has M edges, and the vertices are $P_1, \ldots, P_M$. The coordinate of the point that will be tested if it is inside $P(x, y)$. There is a function

$$f(x, y) = \sum_{k=1}^{M-1} \angle P_k P P_{k+1} + \angle P_M P P_1 \tag{5}$$

If $f(x, y) = 2\pi$, P is the point inside the polygon, if $f(x, y) = \pi$, P is the point on the edge of the polygon, if $f(x, y) = 0$, P is the point outside the polygon. The□$P_i P P_{i+1}$ can be deduced as follows:

$$\begin{aligned} \angle P_i P P_{i+1} = \arccos & \frac{(x_i - x)\times(x_{i+1} - x) + (y_i - y)\times(y_{i+1} - y)}{\left[(x_i - x)^2 + (y_i - y)^2\right]^{1/2} \times \left[(x_{i+1} - x)^2 + (y_{i+1} - y)^2\right]^{1/2}} \\ & \times \operatorname{sgn}[(x_i - x)\times(y_{i+1} - y) - (x_{i+1} - x)\times(y_i - y)] \end{aligned} \tag{6}$$

Plug formula (6) into (5)

$$f(x, y) = \sum_{k=1}^{M} \begin{Bmatrix} \arccos \dfrac{(x_k - x)\times(x_{k+1} - x) + (y_k - y)\times(y_{k+1} - y)}{\sqrt{(x_k - x)^2 + (y_k - y)^2} \times \sqrt{(x_{k+1} - x)^2 + (y_{k+1} - y)^2}} \\ \times \operatorname{sgn}[(x_k - x)\times(y_{k+1} - y) - (x_{k+1} - x)\times(y_k - y)] \end{Bmatrix} \tag{7}$$

According to whether the centre is in the feasible region, we determine whether the circle is in the irregular area. We substitute the centre coordinates $O(x_o, y_o)$ into the formula (7), and the mathematical formula for determining whether a single circle item is inside the irregular region can be express as

$$\min\left\{ \sum_{m=1}^{M} \begin{Bmatrix} \arccos \dfrac{(x_{qm} - x_o)\times(x_{q(m+1)} - x_o) + (y_{qm} - y_o)\times(y_{q(m+1)} - y_o)}{\left[(x_{qm} - x_o)^2 + (y_{qm} - y_o)^2\right]^{1/2} \times \left[(x_{q(m+1)} - x_o)^2 + (y_{q(m+1)} - y_o)^2\right]^{1/2}} \\ \times \operatorname{sgn}[(x_{qm} - x_o)\times(y_{q(m+1)} - y_o) - (x_{q(m+1)} - x_o)\times(y_{qm} - y_o)] \end{Bmatrix} - \pi, 0 \right\}^2 = 0 \tag{8}$$

When the number of circles is n, the formula (8) can be express as:

$$\sum_{i=1}^{n}\min\left\{\sum_{m=1}^{M}\left\{\begin{array}{l}\arccos\dfrac{\left(x_{qm}-x_{oi}\right)\times\left(x_{q(m+1)}-x_{oi}\right)+\left(y_{qm}-y_{oi}\right)\times\left(y_{q(m+1)}-y_{oi}\right)}{\left[\left(x_{qm}-x_{oi}\right)^2+\left(y_{qm}-y_{oi}\right)^2\right]^{1/2}\times\left[\left(x_{q(m+1)}-x_{oi}\right)^2+\left(y_{q(m+1)}-y_{oi}\right)^2\right]^{1/2}}\\ \times\operatorname{sgn}[\left(x_{qm}-x_{oi}\right)\times\left(y_{q(m+1)}-y_{oi}\right)-\left(x_{q(m+1)}-x_{oi}\right)\times\left(y_{qm}-y_{oi}\right)]\end{array}\right\}-\pi,0\right\}^2=0 \quad (9)$$

2.3. Mathematical model

Set $O_1,\dots,O_n$ is the coordinate of the centre point of the circular items. From the above two constraints (3) and (9), the mathematical model for arranging circular elements of the same size in irregular plates can be express as:

$$f(O_1,\cdots O_n)=\sum_{i=1}^{n}\sum_{j=1}^{n}\min\left\{0,[(x_{oi}-x_{oj})^2+(y_{oi}-y_{oj})^2-4r^2]\right\}$$

$$+\sum_{i=1}^{n}\min\left\{\sum_{m=1}^{M}\left\{\begin{array}{l}\arccos\dfrac{\left(x_{qm}-x_{oi}\right)\times\left(x_{q(m+1)}-x_{oi}\right)+\left(y_{qm}-y_{oi}\right)\times\left(y_{q(m+1)}-y_{oi}\right)}{\left[\left(x_{qm}-x_{oi}\right)^2+\left(y_{qm}-y_{oi}\right)^2\right]^{\frac{1}{2}}\times\left[\left(x_{q(m+1)}-x_{oi}\right)^2+\left(y_{q(m+1)}-y_{oi}\right)^2\right]^{\frac{1}{2}}}\\ \times\operatorname{sgn}[\left(x_{qm}-x_{oi}\right)\times\left(y_{q(m+1)}-y_{oi}\right)-\left(x_{q(m+1)}-x_{oi}\right)\times\left(y_{qm}-y_{oi}\right)]\end{array}\right\}-\pi,0\right\}^2=0 \quad (10)$$

3. The solving algorithm

The search tree algorithm, one of the important heuristic algorithms, is widely used in the packing problem. In this paper, the principle of the algorithm is to arrange the next circular item on the left and bottom on the basis of the arranged circular items.

3.1. Layout strategy

The distance between the two circular parts when they do not overlap is at least 2R. The critical point of the circle item is shown in figure 3. The small circle in the middle of the figure is the circle P_j that has been arranged, the circle C is the trajectory of the centre of the circle P_i ready to be placed.

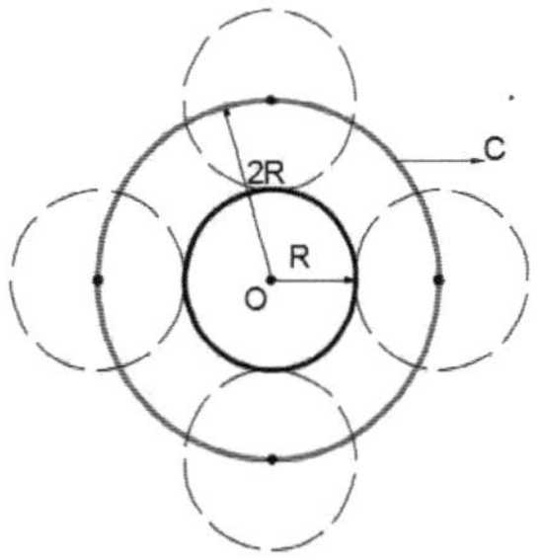

Figure 3. The critical point of the circle item.

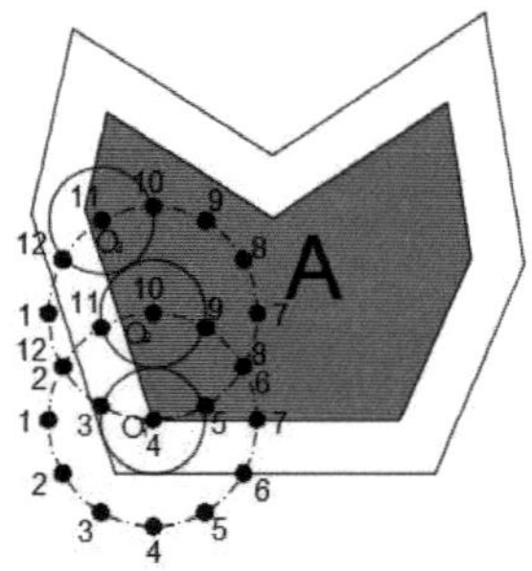

Figure 4. The layout strategy of the circle.

In order to calculate faster, path C should be separated into discrete points. The greater the number of points, the more accurately the result is. This article sets the number of equally divided into 12, the final experimental results prove that the value is more appropriate. In order to reduce the complexity of the calculation, the final optimization result is not affected. This article uses the bottom-left placement strategy to determine the unique position of a newly-arranged circle part. The first centre point is placed at the bottom left position of the polygon, and each new centre point that is laid out takes precedence over the leftmost position, only in the same left situation, the lower position is given priority. Figure 4 shows how the circle is laid out in a polygon. The polygonal shaded area in the figure is the feasible area of the centre of circle. Firstly, the first centre point O_1 is selected in the bottom leftmost position of the centre of the feasible region A, the centre O_2 is pushed out by the centre O_1, the centre O_1, O_2 launches O_3 position. This process is repeated until the feasible region A cannot arrange a new centre point.

3.2. Finite branch tree heuristic algorithm
This paper uses the branch tree structure algorithm. In the search tree, the corresponding coordinate of the root node is the lowest leftmost points of the feasible region, the parent node represents the best position of the circle centre, the child nodes represent all centre positions of the next circular part to be laid out. Each parent node can generate 12 child nodes. The number of layers in the tree represents the number of circle parts that the polygon can arrange. Figure 5 describes the growth of the branch tree.

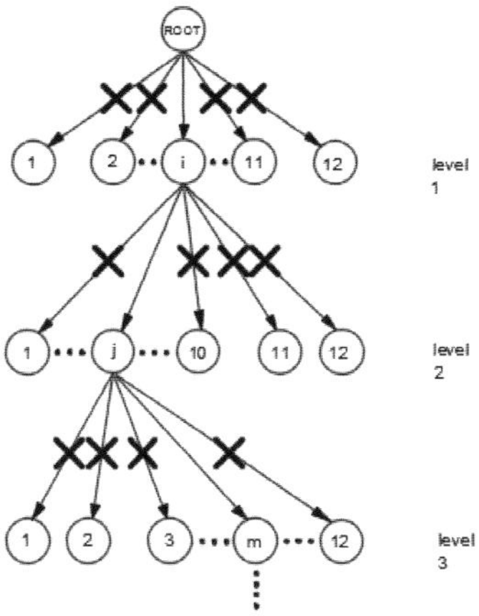

Figure 5. The growth process of the finite branch tree.

4. Numerical simulation
Example data is shown in table 1.

Table 1. Experimental date

Question	Vertex coordinates	Polygonal area (mm^2)	Round radius (mm)	Part area (mm^2)	Utilization rate
1	{(0,0)(19,0)(10,19)}	180.5	1	135.1	74.8%
2	{(0,0)(23,0)(23,13)(5,14)}	278	1.5	204.9	73.7%
3	{(0,0)(26,0)(12,8)(8,16)}	168	1.5	127.2	75.7%
4	{(0,0)(14,0)(18,5)(11,19)(−10,13)	345	2	251.3	72.8%
5	{(0,0)(18,0)(20,18)(8,13)(−1,19)}	302	2	213.6	70.7%

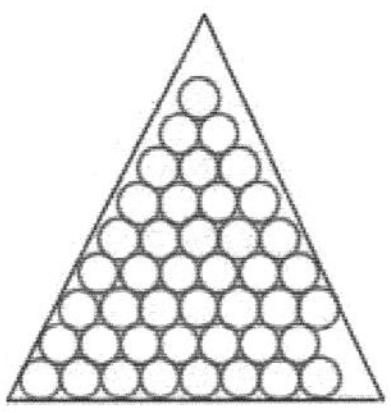

Figure 6. Problem 1 layout result.

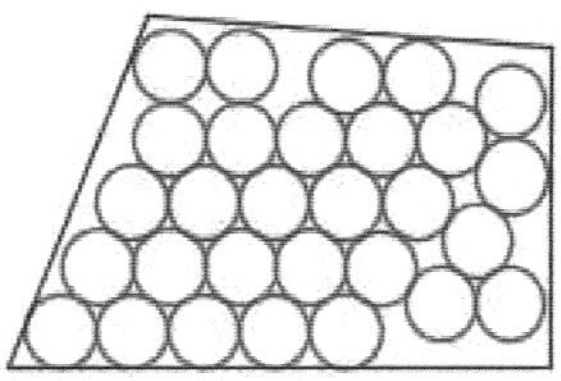

Figure 7. Problem 2 layout result.

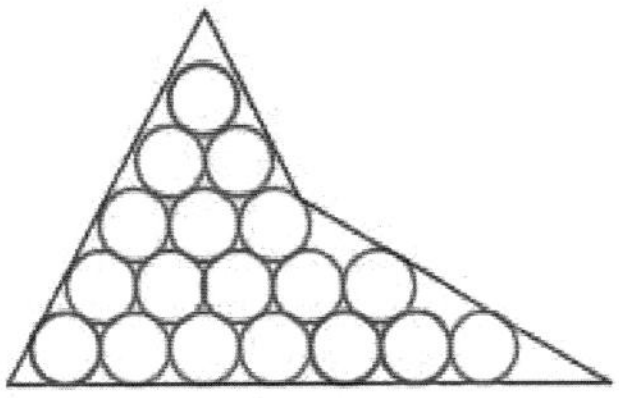

Figure 8. Problem 3 layout result.

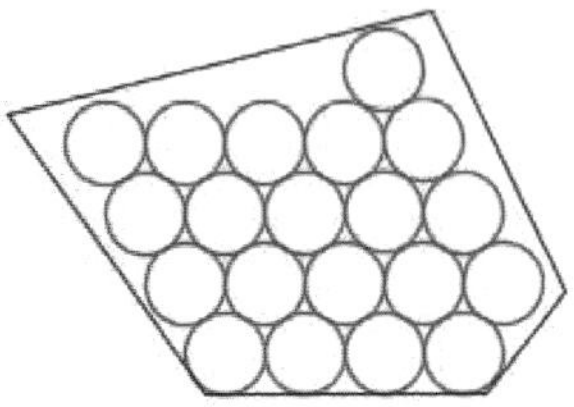

Figure 9. Problem 4 layout result.

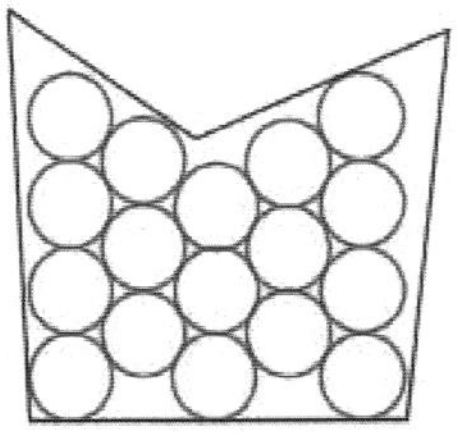

Figure 10. Problem 5 layout result.

As shown in the data of the example in table 1, the utilization of the plates is effectively improved, as displayed in figures 6 to 10.

5. Conclusion

(1) This paper proposes the angle bisector heuristic algorithm for the irregular graphics layout and solves the problem of packing circular items in irregular regions.
(2) The paper proves that the algorithm can be used successfully not only for convex set graphics, but also for concave set graphics.
(3) The numerical simulation proves that the approach can pack circle items in irregular marble slab effectively.

Acknowledgements

This research was supported by the National Natural Science Foundation of China (Grant No. 51705342) and Supported by the Natural Science Foundation of Liaoning province (Grant No. 20170540754).

References

[1] Szabó P G, Markót M C, Csendes T, Specht E, Casado L G and García I 2007 New approaches to circle packing in a square: with program codes *Springer Science & Business Media* **6** 44
[2] George J A, George J M and Lamar B W 1995 Packing different-sized circles into a rectangular container *European Journal of Operational Research* **84** 693
[3] López C O and Beasley J E 2016 A formulation space search heuristic for packing unequal circles in a fixed size circular container *European Journal of Operational Research* **251** 64
[4] Birgin E G and Gentil J M 2010 New and improved results for packing identical unitary radius circles within triangles, rectangles and strips *Computers & Operations Research* **37** 1318
[5] Galiev S I and Lisafina M S 2013 Linear models for the approximate solution of the problem of packing equal circles into a given domain *European Journal of Operational Research* **230** 505
[6] Song X and Li Y 2006 Application of hybrid genetic algorithm in optimal layout of circular parts *Microcomputer Information* **22** 170

[7] Huang W. Q, Li Y, Li C M and Xu R C 2006 New heuristics for packing unequal circles into a circular container *Computers & Operations Research* **33** 2125

[8] Akeb H, Hifi M and M'Hallah R 2009 A beam search algorithm for the circular packing problem *Computers & Operations Research* **36** 1513

[9] Hifi M, Paschos V T and Zissimopoulos V 2004 A simulated annealing approach for the circular cutting problem *European Journal of Operational Research* **159** 430

[10] Grosso A, Jamali A R M U, Locatelli M and Schoen F 2010 Solving the problem of packing equal and unequal circles in a circular container *Journal of Global Optimization* **47** 63

[11] Leary R H 2000 Global optimization on funneling landscapes *Journal of Global Optimization* **18** 367

[12] Al-Mudahka I, Hifi M and M'Hallah R 2011 Packing circles in the smallest circle: an adaptive hybrid algorithm *Journal of the Operational Research Society* **62** 1917

[13] Stoyan Y G and Yas' kov G 2004 A mathematical model and a solution method for the problem of placing various-sized circles into a strip *European Journal of Operational Research* **156** 590

[14] Mladenović N, Plastria F and Urošević D 2005 Reformulation descent applied to circle packing problems *Computers & Operations Research* **32** 2419

[15] Castillo I, Kampas F J and Pintér J D 2008 Solving circle packing problems by global optimization: numerical results and industrial applications *European Journal of Operational Research* **191** 786

[16] Birgin E G, Martínez J M, Nishihara F H and Ronconi D P 2006 Orthogonal packing of rectangular items within arbitrary convex regions by nonlinear optimization *Computers & Operations Research* **33** 3535

Design and analysis of acoustic reforms of studio

J Zhang[1*], Y Zhang[1], Z Zhou[1] and M Fang[1]

[1] School of Architecture and Urban Planning Shenyang Jianzhu University in China

[*] E-mail: hongmamm@sina.com

Abstract. After evaluating the studio acoustics environment quality and finding problems, related reforms about acoustic were performed, such as consulting the acoustic design code related studio, through testing on background noise, reverberation time, evenness of acoustic field and other acoustic indeces at the coalface, making antithesis and analysis. Catching the defects of studio acoustic design implied that: electronic equipment running lead to the background noise in the studio is too large; the interface material in the room sounds absorption is too little, bringing the reverberation time is too long; the wall sounds absorbing material and reflector position is unreasonable begets the lack of evenness of the acoustic field. According to the problems above, a theoretical analysis and calculation of the overall acoustic of the studio was conducted, finally advancing the reforms plan. The reformation plan consists of processing door and window insulate sounds, increasing the sounds absorption of the material in the studio, and arranging a reflector on the side of the wall evenly. Thus, the plan will cater for the requirements of acoustics ultimately and provide technical reference for the design of an acoustic studio in the future.

1. Overview

In recent years, with the continuous development and progress of the society, people have become increasingly demanding of the acoustic environment. In pursuit of better sound quality, more stringent requirements have been set on the design of the recording studio. This project is mainly aimed at acoustic acceptance of established recording studios and found that acoustic requirements cannot be fully met. For this purpose, we carried out corresponding acoustic modifications to meet the functional requirements.

The recording studio is also called the studio. It is a special recording site built by people to create a specific recording environment and acoustic conditions. The main function is to record movies, songs, music, etc. The acoustic characteristics of the recording studio are used for the quality of the recording production and its products. They play a very important role.

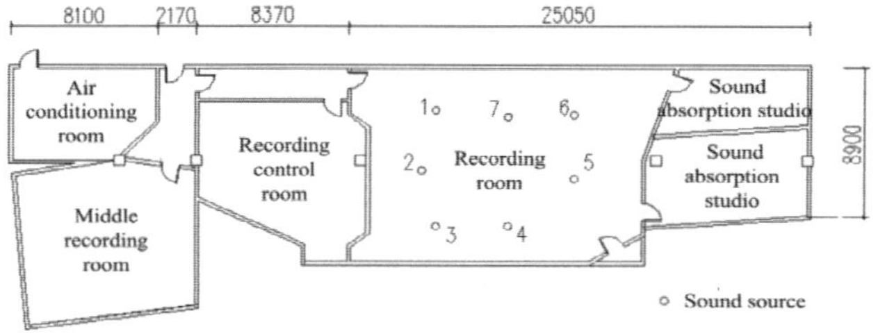

Figure 1. Recording studio floor plan.

Figure 2. Recording studio interior.

Figure 1 and figure 2 is a Shenyang art school recording studio. In the process of acceptance testing, it was found that there were certain problems in the aspects of background noise, reverberation time, sound field uniformity, etc., which did not achieve the desired acoustic effect and required acoustic modification.

2. Recording studio acoustic design requirements

2.1. Acoustic specifications

- Choosing the appropriate reverberation time based on the indoor area. The recording studio uses short reverberation as much as possible, and the frequency characteristics of the reverberation time require straightness.
- Sound quality requirements are clear and intelligibility is high, followed by good fullness.
- Under use conditions, suppress sound quality defects such as echo, tremor echo, etc.

The specific acoustic technical indicators are as shown in table 1.

Table 1. Acoustic specifications.

Room name	Reverberation time T60 (s)	Noise evaluation curve (NR)	Soundproof door insulation (dB)	Soundproof window insulation (dB)
Recording studio	0.4+0.1S	NR-30	>35	>50

2.2. Recording studio acoustic processing

It is mainly divided into two aspects, one is the control of noise, and the other is the control of reverberation time.

2.2.1. Noise control. The noise of the studio is partly from outside the building; on the other hand, it comes from inside the building, including the noise outside the studio and the noise inside the studio.

There are mainly three ways to spread noise: one is that noise generates vibrations through walls, floors, and ceilings, and transmits sound energy into the recording studio; the other is through the gaps or holes, etc., and enters the recording studio through airborne sound; the third is through the recording studio; the rigid connection of the wall, ceiling or floor with the outside world, through the solid sound into the recording studio.

Treatment methods can be soundproof windows, soundproof doors, composite soundproof walls, and attention should be paid to reducing acoustic bridges, which can reduce the impact of external noise on the recording studio.

2.2.2. Reverberation time control. The reverberation time is the time after the sound field in the room is stable, the sound source suddenly stops sounding, and the sound self-stated sound pressure level is attenuated by 60 dB. In the design of the sound-collecting space, proper reverberation time makes the music full and contagious. Too short reverberation time makes the sound dry and ineffective; long reverberation time makes the speech intelligibility lower, and the sound quality lacks rhythm and intensity, so we must choose an optimal reverberation time.

2.3. Material selection instructions

The acoustic design of the recording studio adopts green, flame-retardant materials and should meet the requirements of the domestic industry. As the studio is a television technology room, in order to ensure quality, especially to ensure the acoustic index, the selected materials must have good decorative and acoustic properties.

3. Recording studio acoustic test

3.1. Background noise

In the recording studio, four points were evenly distributed to test the noise level at each point. Compared with the acoustic standards, it was found that the air-conditioning room was not opened and the background noise met the requirements; the air-conditioning room was turned on, and the background noise did not meet the requirements. The measured data are shown in table 2.

Table 2. Recording studio background noise.

Position	Position 1	Position 2	Position 3	Position4
No noise in the air-conditioning room LA	32.8dB	29.2dB	33.3dB	30.6dB
Air-conditioning room noise LA	42.8dB	39.2dB	39.3dB	36.6dB

It can be seen from the data in the table that, when the recording studio air conditioner is not on, the background noise in the recording studio can basically meet the requirements of the specification. However, when the air conditioner is turned on, the background noise in the studio increases by 2-3 times, and it is completely unable to reach the recording studio. The background noise is standard.

3.2. Background noise

At a wall angle of 1.5 meters from both sides of the wall, a 16-sided non-directional sound source is placed, 7 points are selected in the entire room, and the reverberation time of each point is tested, and it is found that the high frequency reverberation time of the recording studio is also problematic. The data is shown in table 3.

Table 3. Studio reverberation time.

Frequency	Position 1	Position 2	Position 3	Position 4	Position 5	Position 6	Position 7
250Hz	0.48s	0.56s	0.42s	0.40s	0.37s	0.51s	0.42s
500Hz	0.42s	0.40s	0.42s	0.37s	0.50s	0.58s	0.40s
1kHz	0.40s	0.35s	0.42s	0.45s	0.53s	0.82s	0.40s
2kHz	0.54s	0.44s	0.51s	0.54s	0.64s	0.55s	0.44s
4kHz	0.52s	0.45s	0.46s	0.47s	0.48s	0.50s	0.48s

According to the measured data, the reverberation time of each frequency in the recording studio is longer, and the reverberation time in the middle and high frequencies is too long. Obviously, this will lead to a more severe impact on the luster effect.

3.3. Acoustic field uniformity

In the recording studio, 4 points were evenly distributed, and the sound pressure level at each frequency was tested. The measured data are shown in table 3.

Table 4. Studio reverberation time.

Frequency	Position 1	Position 2	Position 3	Position 4
250Hz	26.0 dB	25.8 dB	27.9 dB	27.0 dB
500Hz	25.8 dB	22.5 dB	26.5 dB	25.1 dB
1kHz	22.2 dB	17.6 dB	19.8 dB	19.5 dB
2kHz	18.7 dB	13.8 dB	20.4 dB	14.7 dB
4kHz	14.9 dB	12.0 dB	18.0 dB	9.5 dB

As shown in the sound pressure levels at various points in the table, the high-frequency sound pressure level difference at some points exceeds 3 dB, which means that the sound field uniformity of the recording studio is not ideal, and the uniformity of the high-frequency sound field performance is lower. Although the wall surface in figure 3 was also diffused, the low frequency sound field uniformity is good, but the high frequency effect is not obvious.

Figure 3. Diffusion

4. Recording studio reforms plan

4.1. Background noise

The effect of the background noise on the recording studio is very important. The main task is to do sound insulation. In particular, sound insulation of doors and windows is the key to noise reduction. The soundproof door is an important part of the recording studio to prevent the influence of outdoor noise. After the plane analysis research and field test, the noise of the recording studio mainly comes from the air conditioning room system, the sound insulation of the wall meets the requirements, and the door cannot meet the sound insulation requirements. Soundproof doors can reduce the noise impact on the interior of the studio.

- The quality of the main body of the soundproof door in the recording studio is increased. According to the quality law, the sound insulation performance is improved; the frequency of the materials in the inner layers is staggered to avoid the “blind effect”.
- Between the door panels filled with ultra-fine glass wool, to ensure the integrity, cannot be broken cotton, improve sound absorption properties.

- Interlayer connections add damping material to suppress vibration transmission and avoid acoustic bridges.
- Closed treatment is required between the door frame and the wall surface, as well as between the door leaf and the door frame, to ensure closeness.
- The original design used a soundproof window, as shown in figure 4. Four layers of glass are used, and the layers are not parallel and the thickness of the glass is not the same. However, sound absorption materials are not well arranged on the window sill between the glass, resulting in poor sound insulation effect. The modification scheme is as shown in figure 5, and the sound absorption material between the window frames is increased.

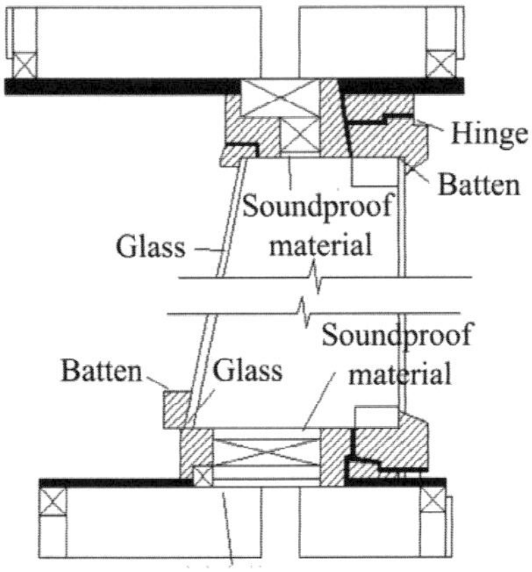

Figure 4. Soundproof window. **Figure 5.** Soundproof window reforms plan.

After the reform, soundproof doors and soundproof windows are shown in Figure 6.

Figure 6. Soundproof door and window.

4.2. Reverberation time

As shown in figure 2, the top of the recording studio is made of mineral wool board, and the walls are pasted with thin sheets of outer cover fiber fabrics. As shown in table 2, indoor high-frequency reverberation takes a long time, and extra porous glass wool felt, mineral wool felt, and other porous sound-absorbing materials need to be added, including glass fiber cloth and other breathable properties, but also a certain strength of the material for the surface layer, through the high frequency sound reverberation time calculation, reducing the indoor high frequency reverberation time. After the reform， the indoor effect is shown in figure 7.

Figure 7. Recording studio interior after reform.

4.3. Acoustic field uniformity

Reproduce part of the walls of the recording studio: The diffuser and the sound-absorbing material are evenly arranged on the two side walls so as to avoid large areas of parallel and opposite acoustic reflection surfaces, so that the uniformity of the indoor sound field can be optimized. After the reform, the indoor effects shown in figure 8.

Figure 8. Recording studio interior after reform.

IOP Conf. Series: Materials Science and Engineering **399** (2018) 012060 doi:10.1088/1757-899X/399/1/012060

5. Conclusion

By processing the sound insulation of doors and windows, increasing the sound absorption performance of sound-absorbing materials, and evenly spreading diffusers on the side walls, the existing problems in the recording studio are solved, the background noise is reduced, the reverberation time is reduced, and the sound field uniformity is increased, with the degrees ultimately meeting studio acoustics requirements. The retrofit program provides technical reference for the acoustic design of the recording studio in the future.

Acknowledgments

The study has been supported by the China National Key R&D Program (Grant No. 2016YFC0700700). Supported by Program for Liaoning Innovative Research Team in University (LNIRT).

References

[1] Zhang J, Fu Q and Xu S.2014 Acoustic Reconstruction Design and Analysis of the Audiovisual Classroom:Taking Audio-visual Classroom in Shenyang Jianzhu University for Example *Huazhong Architecture* **4** 16

[2] Zhang J, Fu Q and Xu S 2014 Classroom acoustic environment test analysis and improvement in Shenyang Jianzhu University for Example*J. of Shenyang Jianzhu University(Social Science)* **1** 4

[3] Zhong X 2012 Building sound-absorbing materials and sound insulation materials Chemical Industry Press

[4] Xiang R 1992 Practical architectural acousticsChina Building Industry Press

[5] Liu X 2010 Architecture Physics (3rd Edition) China Building Industry Press

[6] Liu J 2011 Architecture Physics (4th Edition) China Building Industry Press

IOP Conf. Series: Materials Science and Engineering **399** (2018) 012061 doi:10.1088/1757-899X/399/1/012061

Control method of shaft and hole mating based on convolution neural network in assembly building prefabricated components

K Zhang[1], S Tong[1*], J Zhao[1] and Q Wang[1]

[1] College of mechanical engineering, Shenyang Jianzhu University, China.

[*] E-mail: tongshenghao@sjzu.edu.cn

Abstract. In order to further improve the automation degree of assembly building construction, realize prefabricated component intelligentized positioning, this paper applies the convolutional neural network in positioning control in the field of prefabricated component of assembly building, and puts forward a probability search into target region to establish a new method for loss equation algorithm, the method based on multi label classification loss equation modelling with probability search algorithm, in the training process to continuously improve the positive incentive probability score. The simulation results show that the designed convolution neural network can achieve a correct prediction of the assembly location control instruction, and the proposed method can improve the prediction accuracy and convergence of the network model.

1. Introduction

In prefabricated building construction, after the prefabricated component is hoisted to a designated area, the alignment of the connection hole at the bottom of the prefabricated component and the embedded reinforcement is a special “shaft-hole” assembly problem. When assembling and positioning, the multiple connection holes at the bottom of the prefabricated wall panel must be aligned with the steel bars at the same time, which makes it more difficult. Therefore, it has become the most important issue that restricts the intelligent construction of prefabricated building construction. In recent years, scholars have proposed a variety of theories and methods for automated assembly technology, mainly including the Cartesian impedance control theory, the force/position integrated control assembly theory, the digital assembly technology based on laser tracking, the vision-based servo control theory, and assembly movement. Jokesch [1] proposed a general algorithm for axle hole assembly based on Cartesian impedance control, which was applied to the automatic connection and insertion of an electric vehicle charging device and charging pile, and completed seven pins and holes with asymmetrical positions aligned at the same time, and evaluated the algorithm from different starting positions. Touati [2] proposed an automatic assembly method for force/position comprehensive control based on artificial neural network. First, the neural observer used the data obtained during the contact process of the assembly as an input for off-line training; then, the training parameters were applied. In the neural network with online learning function, the assembly closed-loop control system has good performance. Song [3] proposed a complex shape part assembly control strategy. It controls the forces during assembly movements based on visual geometry information and

CAD models, using cameras to track the position and orientation of the parts. In the assembly process, the impedance control method is used to control the contact force and avoid the occurrence of excessive contact force.

The above control method uses a large number of sensors and requires accurate modelling of the space position of the shaft. Therefore, the amount of calculation is very large, and requires a higher assembly environment, and system parameters need to be readjusted when the assembly environment changes [4]. Obviously, the above control method is not applicable to construction sites with complex construction environment and changeable weather conditions. Therefore, a prefabricated component assembly method, based on the convolution neural network, is proposed in this paper. The relative position image of the bottom connection hole of the prefabricated component and the top of the pre embedded steel bar is used as the input, and the control instruction of the prefabricated component six degrees of freedom hoisting platform is used as the output, and the neural network is built and trained. This method can reduce the degree of dependence on sensors, reduce control variables and increase the intelligence of assembly of the prefabricated component. Convolution neural network has been applied in various fields in recent years [5], and has made breakthroughs in speech recognition, image recognition, general object recognition, motion analysis, Natural Language Processing and even brain wave analysis. For the first time, this paper combines the big data of construction site with the predictive classification advantages of convolutional neural networks, and proposes a prediction method of prefabricated building in-situ control instructions, based on convolutional neural networks. It can be stated that the deep learning was applied to an exploratory research in the field of building construction. The specific operation is shown in figure 1.

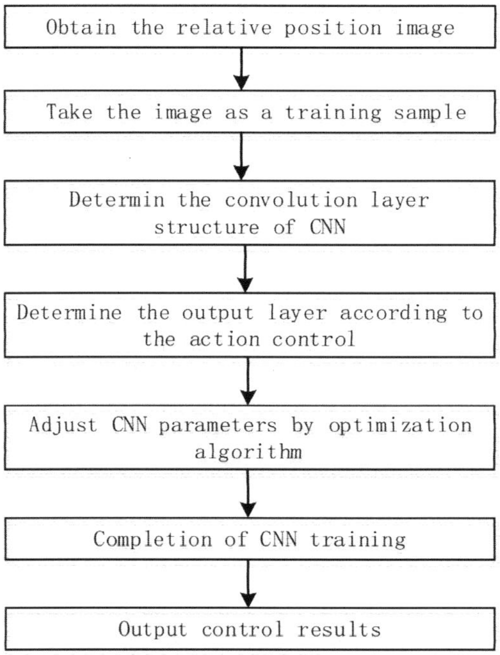

Figure 1. Flow chart.

2. Algorithm and structure

2.1. Algorithm

Because neural network is only interested in some areas of input image, how to search and locate the target area and give feedback in time is an important factor of the neural network. The existing method of visual target recognition usually relies on training data with a large number of annotated images [6].

The image labeling method based on the bounding box is complex and involves many human activities, which takes time and reduces the accuracy of identification. This paper presents a probabilistic search into the target region to establish a new method for loss equation algorithm, this method in multi label classification loss equation, based on adding probability search algorithm, the network training process to continuously improve the positive incentive probability score, can further improve the model prediction accuracy and convergence. This algorithm adds a gradient factor on the basis of the original algorithm, reduces the search time of the significant area of the image, and makes the network model converge quickly. The specific algorithms are as follows:

If target region output $f_k^{(t-1)}(x)$ at $t-1$ time, corresponding probability model is $P[f_k^{(t-1)}(x)]$, the increment is $\Delta f_k^{(t)}(x)$ at time of t, the Image features can be represented as:

$$\phi_i[f_k^{(t)}(x)] = \phi_i[f_k^{(t-1)}(x) + \Delta f_k^t(x)] \quad (1)$$

The probability model of the corresponding time target region can be expressed as:

$$\begin{aligned} P[f_k^{(t)}(x)] &= P[f_k^{(t-1)}(x) + \Delta f_k^t(x)] \\ &= \sum_{i=1}^{k} \omega_i \phi_i [f_k^{(t-1)}(x) + \Delta f_k^t(x)] + b_i \\ &= P[f_k^{(t-1)}(x)] + P[\Delta f_k^t(x)] \end{aligned} \quad (2)$$

ω_i the formula comprises: the corresponding weight value, b_i the bias, ϕ_i the i eigenvalue of the pixel point.

The loss equation is the sum of all two logic regression loss functions [7], from which we could get:

$$L[f_k^{(t)}(x), y_k] = \sum_k \log(1 + e^{-y_k f_k^{(t)}(x)}) \quad (3)$$

$f_k^{(t)}(x)$ is the output of the neural network to the input x at time t, $y_k \in (-1,1)$ represents the output target category k of input x.

2.2. Network Structure

Different from the traditional convolutional neural network structure, in order to adapt to the algorithm idea of this paper, we have redesigned the next full-connected layer of the single-channel convolutional neural network according to the training characteristics and training objectives adopted. Most DCNNs use a softmax layer to calculate the classification results after they have fully connected layers [8], and this paper calculates the target output directly through the fully connected layer. Deep network structure and more nodes or filters can improve the performance of neural networks, but it also increases the complexity and training difficulty of the network [9]. Based on the data size of the training set, this paper weighs the relationship between network performance and training difficulty. We use a network structure that consists of 3 convolution layers and 2 full connection layers. After the full connection layer, each active action corresponds to a Separate output layer.

Input layer: 128 × 128 RGB image as input to the network.

Convolutional layer: The network consists of three convolutional layers. The first layer uses a convolution filter size of 7×7, and the other two layers have a filter size of 3×3. The step size is set to 1×1.

Pooling layer: 3 activation function layers followed by 3 pooling layers, all using maxpooling, filter size 3×3, step size 2×2.

Fully connected layer: 3 convolutional layers followed by 2 fully connected layers and 2 active function layers.

Output layer: After two fully connected layers are output layers for each valid action corresponding to a single output.

3. Experimental analysis

3.1. Experimental data

The experimental data and images are from the V101150.5 industrial building component lifting installation automatic control digital simulation system. The system has a complete prefabricated library including PC board, PCF board, composite floor, and more than 10 thousand assembly building construction method pieces of information [10]. The sample image is the relative position image of the embedded hole at the bottom of the precast member and the embedded steel bar joint on the floor, which is used to simulate the real lifting condition and enlarge the sample capacity at the same time, adjust the brightness and contrast of the image, as shown in figure 2. 5000 sample images are used for training, and 1000 sample images are used for testing. The 5000 samples were divided into 50 batches, each containing 100 samples.

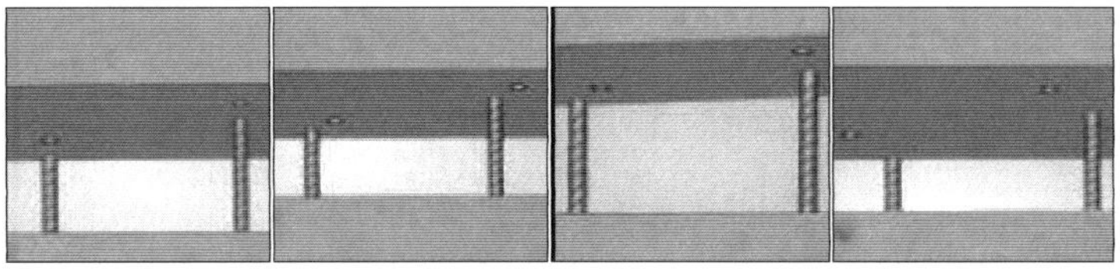

(a)Original image

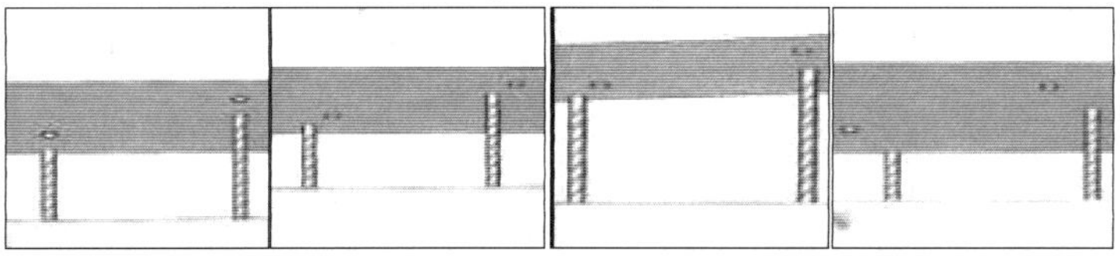

(b)High brightness and high contrast

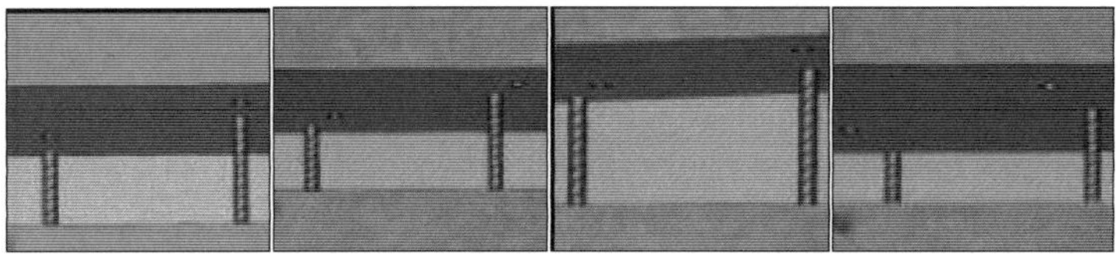

(c) Low brightness and low contrast

Figure 2. Sample image processing.

3.2. Analysis of test results

After the training is completed, the convolution neural network is used to classify the test set [11], and there are 12 action instructions in the output terminal, as follows: movement of the X axis positive direction, movement of the X axis negative direction, movement of the Y axis positive direction, movement of the Y axis negative direction, movement of the Z axis positive direction, movement of the Z axis negative direction, clockwise rotation of X axis, counter clockwise rotation of X axis, clockwise rotation of Y axis, counter clockwise rotation of Y axis, clockwise rotation of Z axis, counter clockwise rotation of Z axis. The reference coordinate system is built on the basis of the building XY plane. The output of the neural network is grouped by the direction of the control instruction [12], and compares the output of the neural network with the expert instruction after grouping. The same decision is correct. The correct output instruction number is divided by the total number of instructions of the group to determine the correct rate of the control instruction of the group. The correct rate of each group is calculated, and then the average correct rate of average output is determined. The result of the experiment is compared with the result of expert labelling, the average correct rate is 90.1%, and the specific results are shown in table 1. The table illustrates that the prediction accuracy of movement of the Y axis negative direction is the highest, and the prediction accuracy of clockwise rotation of X axis and counter clockwise rotation of the Y axis is lowest. The prediction accuracy of the mobile output along three axes is higher than that of its revolving around the axis.

Table 1. Prediction results.

Action command	Correct rate
Positive direction of X axis	91.5%
Negative direction of X axis	93.6%
Positive direction of Y axis	93.1%
Negative direction of Y axis	93.2%
Positive direction of Z axis	94.8%
Negative direction of Z axis	94.9%
Clockwise rotation around the X axis	84.5%
Reverse clockwise rotation around the X axis	84.6%
Clockwise rotation around the Y axis	85.2%
Reverse clockwise rotation around the Y axis	84.5%
Clockwise rotation around the Z axis	90.2%
Reverse clockwise rotation around the Z axis	90.6%
Average value	90.1%

In order to explore the effects of different distances of the prefabricated component bottom and the top of rebar on the output performance of the network model, and then to do the model simulation experiment, grouped by distance of the prefabricated component bottom and the top of rebar from the packet input samples to test the network, there is need to obtain the network prediction accuracy with distance of prefabricated component bottom and the top of rebar. As shown in figure 3, the correct rate of prediction increases with the distance of the prefabricated component bottom, and the top of rebar increased, when the distance of prefabricated component bottom and the top of rebar is up to 7.8cm, the network model to predict the average highest correct rate is 96.5%, then the correct rate gradually decreases with the increase of distance.

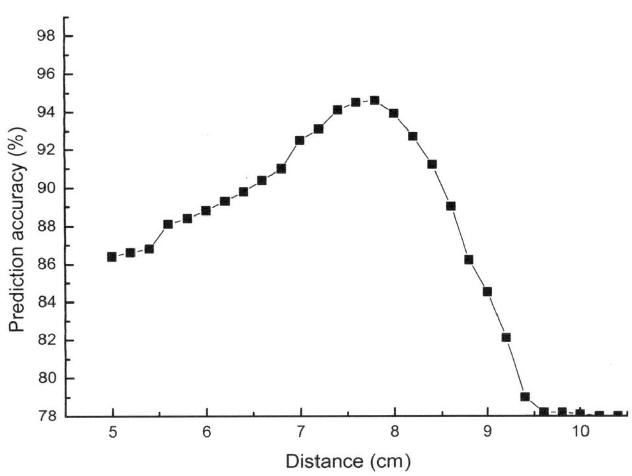

Figure 3. Influence of the distance between the bottom of the prefabricated component and the top of the rebar on the predictive performance of the network in the sample.

4. Conclusion

Considering the difficulties raised by the position control of the assembly building prefabricated component, we have designed the multi-layer convolution neural network, which proposes a probability search into the target region to establish a new method for loss equation algorithm. Our experiment shows that:

(1) the convolution neural network designed in this paper can realize the correct prediction of the position control instruction of the prefabricated component of the assembly building;

(2) this method improves the prediction accuracy and convergence of the network;

(3) the influence of the distance between the bottom of the prefabricated component and the top of the rebar on the prediction performance of the network is studied, which lays the foundation for the follow-up research.

Acknowledgments

1. National Key R&D Program of China "Research on automatic control technology and system for installation and installation of industrial building components" 2017YFC0703903.
2. Shenyang science and technology program project F16-096-1-00.
3. National Natural Science Fund, 51675353.

Reference

[1] Jokesch M, Suchý J, Winkler A, et al.2016 Generic Algorithm for Peg-In-Hole Assembly Tasks for Pin Alignments with Impedance Controlled Robots *Robot 2015: Second Iberian Robotics Conference* (Springer International Publishing) p 1949

[2] Touati Y, Amirat Y and Saadia N 2007 Artificial Neural Network-based Hybrid Force/Position Control of an Assembly Task *International IEEE Conference on Intelligent Systems* (IEEE)

[3] Song H C, Kim Y L and Song J B 2014 Automated guidance of peg-in-hole assembly tasks for complex-shaped parts *Ieee/rsj International Conference on Intelligent Robots and Systems* (IEEE) p 4517

[4] Tompson J, Stein M, Lecun Y, et al. 2014 Real-Time Continuous Pose Recovery of Human Hands Using Convolutional Networks *Acm Transactions on Graphics* **33 (5)** 1

[5] Mnih V, Kavukcuoglu K, Silver D, et al. 2015 Human-level control through deep reinforcement learning *Nature* **518 (7540)** 529

[6] Cireşan D, Meier U, Masci J, et al. 2012 Multi-column deep neural network for traffic sign classification *Neural Networks - Official J. of the Int. Neural Network Society* **32 (1)** 333

[7] Wei Y, Liang X, Chen Y, et al. 2016 STC: A Simple to Complex Framework for Weakly-supervised Semantic Segmentation IEEE Trans. on Pattern Analysis & Machine Intelligence **39 (11)** 2314

[8] Vallance R R, Morgan C and Slocum A H 2004 Precisely positioning pallets in multi-station assembly systems *Precision Eng.* **28 (2)** 218

[9] Wang X G, Sun Y and Tang XO 2015 From unified subspace analysis to joint deep learning: progress of face recognition in the last decade *Com. of the CCF* **11 (4)** p 8

[10] Oquab M, Bottou L, Laptev I, et al.2015 Is object localization for free? - Weakly-supervised learning with convolutional neural networks *Computer Vision and Pattern Recognition* (IEEE) p 685

[11] Dauphin Y N, De Vries H and Bengio Y 2015 Equilibrated adaptive learning rates for non-convex optimization *J* **35 (3)** 285

[12] Bellemare M G, Veness J and Bowling M 2012 Investigating contingency awareness using Atari 2600 games *Twenty-Sixth AAAI Conference on Artificial Intelligence* (AAAI Press) p 864

Technique and outfit of large-scale intelligent PC external wall panel mixing production line

D F Zou[1*], S Y Wen[1], W D Yu[1], S Y Zhang[1] and Y Sun[1]

[1] School of Mechanical Engineering, Shenyang Jianzhu University, SY 110168, China

[*] E-mail: 59408598@qq.com

Abstract. The production of concrete products is the key problem of our intrastructure construction, and the automation degree of this production equipment shows the level of science and technology in our construction industry. To promote the development of concrete production technology and increase the degree of its equipment, the production technology and main parts of PC external wall panel mixing production line are analysed and annotated in this paper. This production line consists of a moving device and carriers, cleaning spraying and marking system, placing concrete and vibrating system, surface treatment system and curing System, and so on. This production line is observed, checked and controlled by a cluster control system consisted of center control sub-system, cycle control sub-system of concrete tray, and curing sub-system of concrete product. The study of the large-scale intelligent PC external wall panel mixing production line helps promote our production technology of the concrete products and drive the development of our construction industry.

1. Foreword

The complete set of equipment technology for PC component production is one of the key technologies of construction industrialization in the world at present. It takes prefabricated component production, prefabricated construction as the mode of production, integrated design, production and construction as the whole industrial chain. As compared with traditional architecture, it improves the quality of building products, and achieves the maximum value of building energy efficiency, environmental protection and life cycle. It is an internationally recognized technology industry of sustainable development [1, 2].

In China, Vanke is one of the earliest real estate companies which took housing industrialization as an important strategic goal and put industrial technology into commercial development. In 2008, Shenyang Vanke cooperated with Shenyang Modern Construction Industry Co., Ltd of Yatai Group. A three-story industrial test building was built in Vanke Charming City. After the successful cooperation, Yatai Group introduced a set of automatic production lines for PC components, the main equipment of which was provided by Hebei New Earth Machinery and Electrical Manufacturing Co., Ltd. [5]. China construction companies, Shenyang Wanrong Modern Construction Enterprise Co., Ltd., Northern Heavy Industry Group Co., Ltd., and some large domestic enterprises and real estate companies also build and use the concrete factory production line. After nearly ten years of development, both component users, component production units and component production equipment suppliers have accumulated rich experience in the field of housing industrialization. However, as compared with the developed countries such as Japan, there is still a big gap [6-10] in our country. As far as China is concerned, there are still some problems, such as low prefabrication rate, poor production quality, simple production equipment and low degree of intelligence, etc. High intelligence complete set technology and equipment of the production line still rely on import [11-14]. Therefore, we are still facing problems of digestion and absorption, and then independently develop the large-scale intelligent PC component production line that is suitable for China's national conditions.

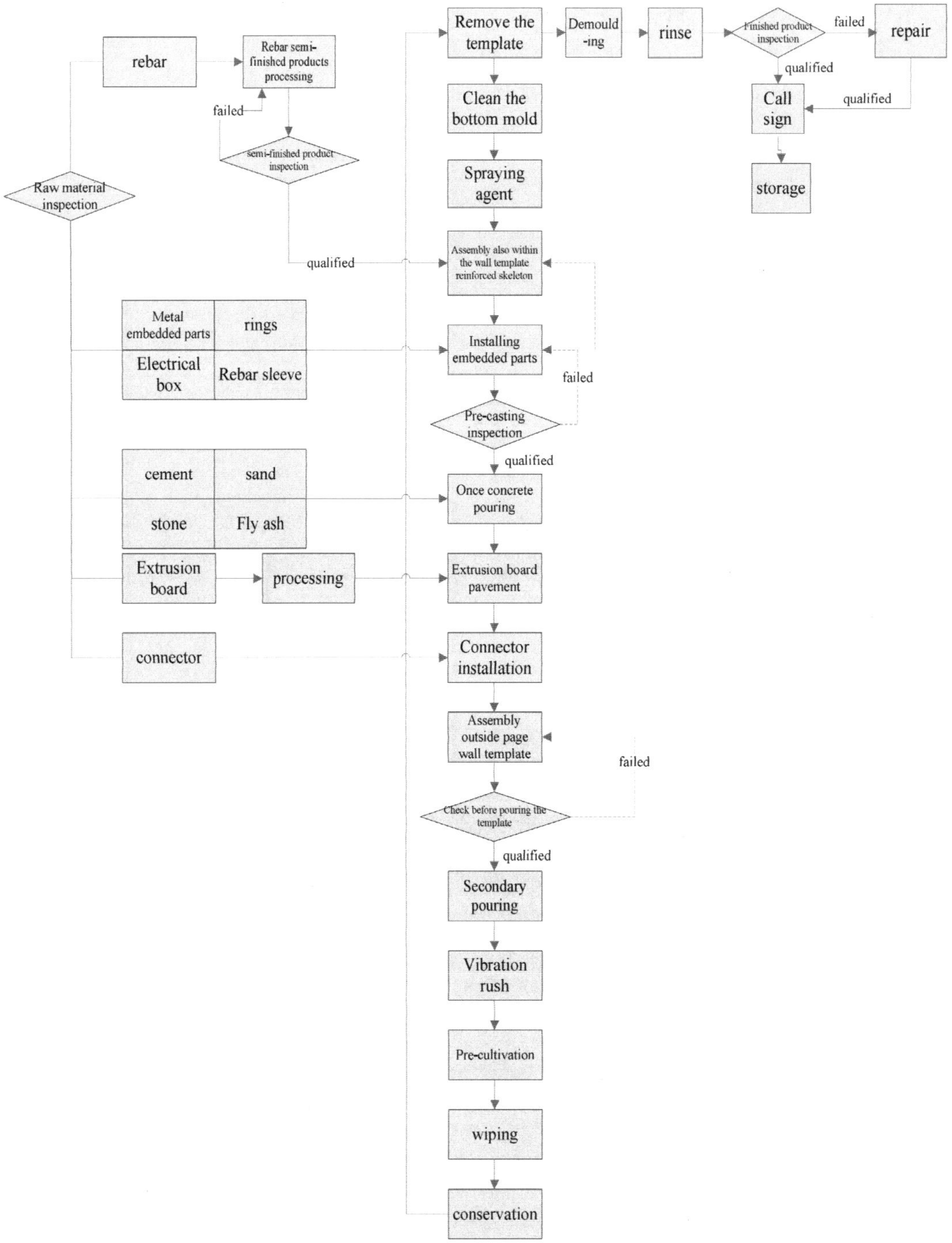

Figure 1. Flow chart of PC external wall panel mixing production line.

On the basis of participating in the production of large-scale intelligent PC external wall panel mixing production line, this paper analyses and interprets the complete process, key components system and working principle of the production line, which is used in the production of large-scale intelligent PC external wall panel mixing production line by the Northern heavy Industry Group Co., Ltd. The research in this paper not only provides the basis for the domestic design and manufacture of large-scale intelligent PC external wall panel production line, but also provides practical guidance for the further study of the key parts of the production line.

2. Technology of PC external wall panel mixing production line

The main products of domestic assembled integral shear wall structure system are external wallboard, internal wallboard, composite floor slab, balcony board, air conditioning board, staircase, and so on. External wallboard, internal wallboard, composite floor slab can be mass produced on the mixed automation production line. The process technology of PC external wall panel mixing production line mainly includes: product line design, layout of process equipment configuration and drawing of process flow chart; determine the composition of production equipment, main parameters, workshop layout and water, electricity, gas supply requirements; according to the layout of equipment in the workshop, plant design requirements, civil construction requirements, etc. Figure 1 is the flow chart of PC external wall panel mixing production line.

The PC external wall panel mixing production line is a comprehensive multifunctional production line, which can produce different sizes and different types of board components. The main production parts of PC external wall panel mixing production line include: moving device and carriers, cleaning spraying and marking system, placing concrete and vibrating system, surface treatment system and curing System, as shown in figure 2.

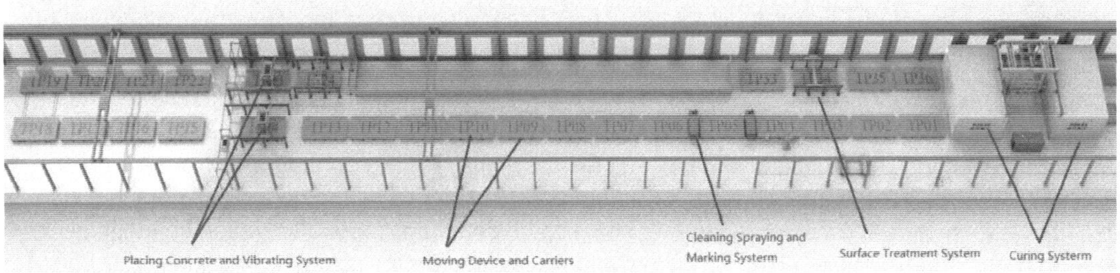

Figure 2. Process layout of PC external wall panel mixing production line.

2.1. Moving device and carriers

PC external wall panel mixing production line is a continuous annular production line, which is a set of complete logistics system, in which the moving device of the production line includes moving wheels, driving wheels and transversing devices, and the carriers of the production line is the mould platform.

The moving wheels and the driving wheels are fixed in the production line. The moving wheels are mainly used to support the mould table. In general, there are more than five pairs of moving wheels under each die stand at the same time (depending on the size of the drive wheel, the load of the mould table). The driving wheel is used to drive the mould table, so in the production line, one drive wheel is placed every few moving wheels. Ensure that at least one pair of driving wheels are driven under the mould table. The moving wheel and the driving wheel can only cause the mould table to move in one direction, that is, the longitudinal direction. In order to make the mould table move horizontally, a ferry vehicle must be used. The transversing device is the equipment that carries the platform transverse and realizes the switching line track. The transversing device is used in pairs, and it is shown in figure 3.

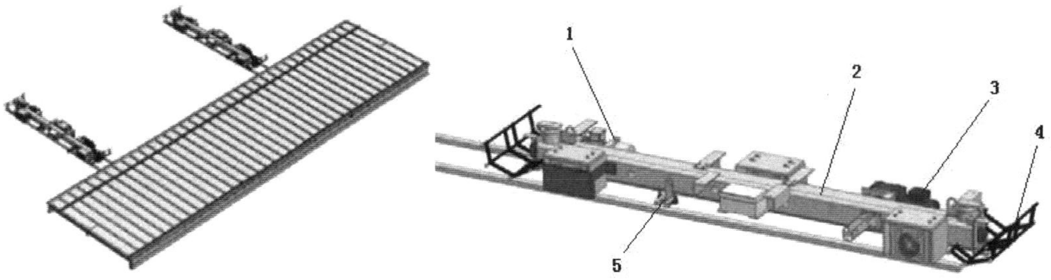

Figure 3. Transversing device.
1-Hydraulic system; 2-Frame; 3-Moving drive system; 4-bull bar; 5-Close switch

The mould table is the carrier of the PC external wall panel. The whole process of placing, vibrating, shaping and curing the concrete members of the slab type is carried out in the mould table. A production line needs dozens or even hundreds of mould tables according to its production capacity. The mould table has a tool type. Universal, fluidity features, can produce different sizes of wallboards and floors, and can produce a variety of building component products.

2.2. Cleaning spraying and marking system
Cleaning spraying and marking system is the production preparation system, which prepares for the production of PC components, including: cleaning machine, spray machine and marking machine.

Cleaning machine is fixed in the cleaning station, used for cleaning surface of the mould table. The front end of the cleaning machine is equipped with scraper. The function of the scraper is to remove the large concrete attachment. The height and angle of the scraper can be adjusted to avoid the damage caused by the scraper to the surface of the mould table during the cleaning process. A rolling brush is arranged at the back of the mould table sweeper for cleaning local attachments. The rolling brush is driven by the motor and rotates around. The concrete residue is removed by the tangential force formed by rotation. The height of the rolling brush is adjustable. According to the use of rolling brush, it needs to be replaced regularly. A mobile crap box is arranged at the bottom of the mould cleaning machine for waste collection and disposal. The crap box can move along the track below it to facilitate the rapid cleaning of waste. The cleaning room is equipped with negative pressure suction and dust removal device to control the discharge of fly dust and ensure no dust in the work station.

After the marking machine is arranged on the mould table cleaning machine, most of the marking machines used in the current production line of prefabricated components are numerical control marking machines, which can quickly and accurately draw the position of the edge die embedded parts on the bottom die and improve the placement of the edge die. It is a matter of accuracy and speed of embedded parts. NC marking machine directly inputs information from U disk or production data line, automatically draws points and lines on mould table as required, draws the external outline and characteristics of components to mould table according to drawing requirements. Ink jet printing is used in the marking mode, and the side die mounting position will install the mould strip according to the drawing line datum to prepare the concrete around construction. The marking machine moving system is controlled by servo controller, the line is controlled by a servo controller,

The spraying machine is arranged after the marking machine, spraying the mould release agent on the mould table, so as to facilitate the smooth release of the maintained components. When the mould table moves to the spraying machine station, the spraying machine starts, and the nozzle and brush roller of the spraying machine are adjusted to an appropriate height through a lifting device. According to the instruction, the nozzle corresponding to the area where the prefabricated components are placed is opened, the brush roller starts to rotate, the nozzle is sprayed, the brush is driven by the motor, and the spray area of the mould table is forced to be coated evenly. According to the movement of the spray gun, the Spraying machine is mainly of two types: Rotary spraying and reciprocating spraying.

Rotary spraying is mostly used for special-shaped components or small area spraying, while reciprocating spraying is mainly for prefabricated buildings. Board, prefabricated wall panels and other large area spray. The spraying machine is shown in figure 4.

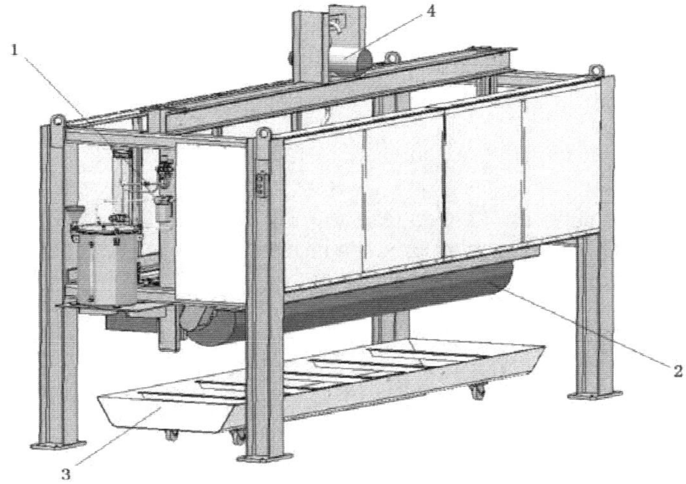

Figure 4. Structure of spraying machine.
1-Spray device; 2-;Brush roller device; 3-Waste liquid recovery unit; 4-Lifting device

2.3. Placing concrete and vibrating system

Placing concrete and vibrating system are the key links of the production line. The main equipment includes concrete placing machine and vibrating machine. Whether the concrete placing machine is uniform or not, whether the vibration is dense or not, it determines the quality of the components. In the production line, the concrete placing machine station coincides with the dense vibrating station, and the efficiency of the production line is improved by using the method of vibrating while placing.

Figure 5. Concrete placing machine and simulated placing concrete.

The concrete placing machine is used to place concrete evenly and quantitatively into the mould of concrete members. In the production line, the concrete placing machine is required to have plane two coordinate motion control and vertical hopper lifting function. Due to the need of the production line, it is required to have a high degree of automation of the feeder. According to the drawing size and the design thickness, the uniform concrete placing machine can be controlled by the program, and the control system has a computer interface. The control system has a computer interface, which is easy to realize the function of reading drawings and data directly from the central control room computer system. The concrete placing machine requires fast concrete placing machine speed and easy operation, and its moving speed can be adjusted without electrode. The concrete placing machine can be manually controlled and automatically controlled. The concrete placing machine and simulated placing are shown in figure 5.

The main function of the vibrating machine is to remove the air and excess moisture from the concrete and to make the edges or edges of the concrete to be formed accurately, and then to obtain a high quality prefabricated component product, as shown in figure 6. There are two forms of vibration table commonly used in precast component production line: horizontal vibration, including horizontal ring and X, Y directional vibration, combined vibration, including horizontal ring and X, Y horizontal vibration, and Z direction vertical vibration. The exciting force in the process of vibration is mainly obtained by the moment of inertia produced by the vibrating motor or the eccentric body driven by the motor, in which the frequency and amplitude of the vibrating motor and the magnitude of the moment of inertia can be set according to the requirements of PC component products.

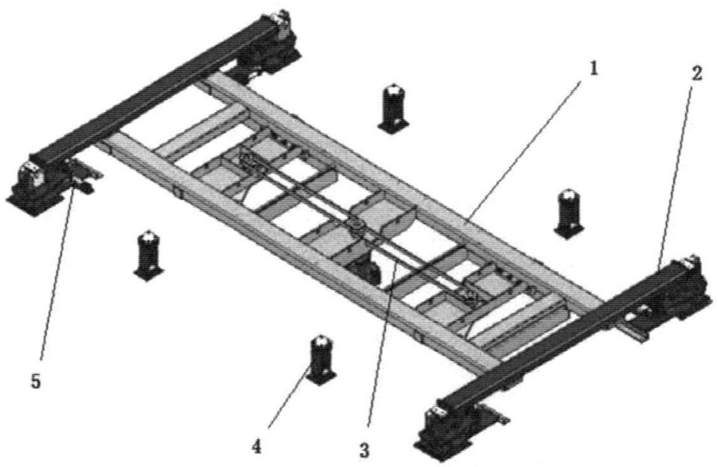

Figure 6. Vibrating machine.
1-Vibrating frame; 2-Double fixed bearing; 3-Synchro belt system;
4-Moving wheel; 5-Hydraulic system

2.4. Surface treatment system

Surface treatment equipment includes a vibration smoothing machine, polishing machine and napping machine.

The effect of the vibration smoothing machine is to smooth the concrete surface in the template, and its structure is shown in figure 7. The vibration smoothing machine can flatten the plate components before and after, at the same time, the trolley moving device is mounted on the cloth machine or its own cross beam to move longitudinally. By moving in the horizontal direction and the longitudinal direction, the complete covering of the components on the mould table can be realized. In order to reduce the harm of vibration to the equipment, the leveling device should be used to reduce the damage to the equipment. Generally, the two-stage damping method, such as the joint damping of rubber shock absorber and rubber pad, effectively solves the vibration problem between the vibrating rod and the vibration frame.

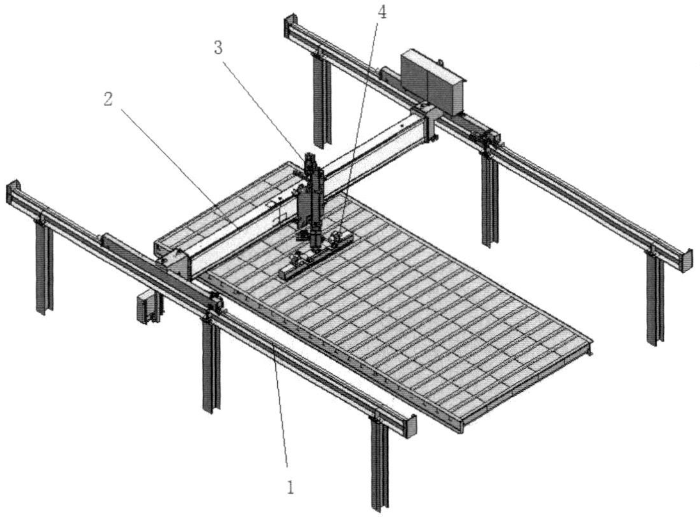

Figure 7. Vibration smoothing machine.
1-Support structure; 2-Single beam; 3-Vibrator; 4-Leveler

The napping machine is mainly used in the production of superimposed floor. After the vibrating process, when the preform of the superimposed floor is not solidified, an isometric shallow ditch is drawn, which makes the surface of the superimposed floor rough and increases the roughness of the surface. When the laminated floor plate is loaded by the laminated floor through the drawing machine, the unsolidified preform is contacted at the bottom of the isometric drawing plate of the drawing plate group, and the preform moves forward with the mould table, and the resistance to the drawing group and the gravity of the drawing group itself make the drawing sheet form a certain swing angle. The bottom side of the right angle on the surface of the prefabricated parts with equal spacing groove. The napping machine is simple in structure and easy to maintain. It can save energy by using gravity and mould table.

Figure 8. Polishing machine.

The polishing machine is placed in front of curing kiln, and its function is to achieve the highest degree of smoothness of visual facet of precast concrete components. It consists of a leveling disc for rough leveling and an electric mud knife for fine leveling. The mud knife directly touches the surface of the prefabricated member, and the head of the mud knife is connected to the adjustable guide rail side plate through a threaded rod to form a moving track up and down. The clay knife head is fixed on the wing arm and polished according to the surface quality of concrete. The head of the mud knife can be adjusted up and down, and the angle of the floating blade can be adjusted mechanically. At the same time, the speed of the rotary machine of the mud knife can also be adjusted. The polishing machine is shown in figure 8.

2.5. Curing System

The curing System is another key link in production line, which is mainly composed of curing kiln and stacker.

The curing kiln is a vertical three-dimensional silo structure, each containing a temperature control device and a humidity control device, which can automatically control temperature and humidity, as shown in figure 9. The temperature curve of heating-constant temperature-cooling is used to avoid cracks or other defects affecting the strength of the members. The curing kiln is designed for stacking, the lowest layer height is realized, the kiln body is small, the building elevation is greatly reduced, the building construction cost is saved, the maintenance volume is reduced, and the maintenance energy consumption is saved. Kiln body insulation board and door plate high temperature resistance, fire resistance, flame retardant, good insulation effect. Kiln with steam heating pipe, humidification pipe, and ventilation piping.

Figure 9. Curing kiln.

Figure 10. Stacker.

Stacker is a lifting transportation equipment which can store the cast concrete components in the corresponding hole of steam curing kiln. Stacker is shown in figure 10. The stacker is a highly automatic equipment. The position of kiln door is automatically located, the time of entering and leaving the kiln is recorded automatically, and the kiln door is opened efficiently and smoothly, and the position of mould table is automatically fixed and locked before entering and leaving the kiln. In safety protection, there are marching, lifting, opening and closing, in and out of the kiln must have a safe interlocking function, the operation of sound and light alarm device. The winch mechanism is equipped with anti-skipping detection to ensure the safety and reliability of the mechanism. The quality of PC components and the production capacity of the whole production line are directly affected by the performance of the stacker and the level of technical level. The structure of the stacker can be divided into the upper track hanging type and the bottom track moving type, which can travel along the track direction. The hoisting device can make the platform body and the roller trolley move along the vertical direction, so that the stacker can push the components of the different openings of the stacker.

3. Control system of mixed production line for PC exterior wall panel

According to the site layout and functional area classification of the PC outer wall panel hybrid production line, the control system of PC outer wall board hybrid production line is divided into three basic subsystems: central control subsystem, mould table circulation subsystem, component maintenance subsystem, and so on. The production line with high degree of automation uses a manipulator to carry on the side touch cleaning and the setting, the reinforcement net placement and the lattice beam placement. Therefore, the control system also includes the edge die cleaning and placing subsystem, Steel mesh placement subsystem and lattice beam placement subsystem [15].

The central control subsystem mainly used for the management and control of the production process of precast concrete components. The specific functions are as follows: effective arrangement and control of the motion process of the mould table, optimization of all operation data and transportation process. At the same time, the automatic detection and transmission of fault information can be realized. The initial data is transmitted to the monitoring system for preprocessing, and the processed data is transmitted to each independent device in a precise time. The equipment is produced on the basis of received data. Throughout the process, production-related data are stored and analyzed for use in other production plans and ERP System control.

Mould table circulation subsystem is mainly used to control the movement of each equipment on the production line. The control subsystem consists of one master station, one input and output module slave station. Several frequency conversion drive slave stations (corresponding to each equipment frequency conversion debugging requirement) and one pneumatic valve slave station. The system controller runs three PLC units, which are respectively used to control transporting and fixing mould table, moving transvers device, dumping component, vibrating concrete, smoothing component, opening and closing curing kiln door, and placing concrete, and so on.

The central controller of maintenance subsystem adopts industrial computer. Its operating interface is very friendly, which is convenient for man-machine interaction and suitable for field use. The operating system has a perfect function and can set the parameters of temperature and humidity. It can record the temperature curve or report in real time. It has the function of printing the data report and the playback of the record temperature. The control cabinet is composed of PLC system, industrial temperature controller, temperature sensor, humidity sensor, etc. Multi-channel digital and analog signal input module. After receiving the process parameters of the upper computer, it can form a closed-loop control system, according to the arrangement in the maintenance kiln multi-point. According to the temperature sensors arranged in the kiln, the temperature signals of different positions are collected and the steam valves are automatically adjusted to form a temperature environment in which the temperature gradient is required in the curing kiln. At the same time, the spray valve can be sprayed or ventilated automatically according to the sensor signal.

The edge mould cleaning and placement subsystem, the steel mesh placing subsystem, the lattice beam placing subsystem. Each of the three subsystem controllers runs one PLC. The edge die cleaning and placing subsystem PLC unit mainly completes the cleaning of the edge die and the bottom die of the die. The placement of edge molds, And put the cleaned edge die into the edge die library and take out. The steel mesh placement subsystem PLC unit completes the finished production and stores the steel mesh into the mould table. The lattice beam placement subsystem mainly completes cutting the finished lattice beam according to the requirements. And put into the mould table.

4. Comparison and discussion

At present, China's domestic PC component production equipment is generally not automated, with excessive manual participation and serious concrete waste. Foreign imported equipment has a high degree of automation and less manual intervention. However, due to the different design concepts of PC board components at home and abroad, most of the foreign PC components are of standardized design, and the steel bars do not leak out. This makes automation easier. The domestic PC components are mostly personalized design, and the leakage of steel bars leads to the inability of mold design to be standardized. The molds can only be placed by hand. At the same time, the mold life cycle is short and waste is serious; due to leakage of component steel bars and embedded parts, surface treatment equipment cannot work normally, and manual intervention is often required, or manual treatment is not required by surface treatment equipment. The concrete placing machine does not reach the actual automatic fabric, the placing concrete accuracy is not high, often requires secondary placing concrete, resulting in decreased placing concrete efficiency. Therefore, to make the PC board production line process and complete sets of equipment automation intelligent, and in line with China's national conditions, but also need to invest considerable energy.

5. Conclusion

(1) The PC external wall panel mixing production line can be systematized, intelligent and automated. The work links of each component part have very strong coupling in mould table carrier moving, mould table cleaning, marking, placing and vibrating concrete, smoothing, polishing, napping and the whole process of the curing and storage. Under the action of cluster control system, all key parts have good coordination, and the high quality large concrete prefabricated parts can be produced efficiently.

(2) As compared with the traditional concrete processing technology, large-scale intelligent PC external wall panel mixing production line has the following characteristics: high level of process equipment, complete semi-automatic or automatic control, less operation workers, small error caused by human factors, high processing efficiency. Furthermore, its subsequent scalability is strong, and can be changed and expanded according to requirements.

(3) The research and development of large-scale intelligent PC external wall panel mixing production line break the dependence of our country on foreign concrete production technology and equipment and inject continuous power into the domestic concrete factory production. It is of great significance to improve the production efficiency and resource utilization ratio.

Acknowledgments

Project fund: National Key R&D Program of China (2017YFC0704003); Shenyang Jianzhu university scientific research innovation cultivation plan project (CXPY2017005).

References

Sun Q X, Pan Z Q and Han X H 2012 Investigation on production equipment and process of cement concrete prefabricated components in Germany *Concrete world* **7** 92.

Jiang Q J, Liu H and Zhong Z Q 2011 Thoughts on the development and orientation of concrete prefabricated components industry *Concrete world* **4** 20

Sun Q and Zheng Z H 2012 Talking about the manufacture of concrete prefabricated members *Commercial concrete* **9** 99

Jiang Q J and Zhong Z Q 2012 Development of precast concrete components in China in 2011*Concrete world* **1** 94

Yu M, Ren X, Liu X G and Li C Z 2015 Research on key technology of concrete prefabricated components and development of complete sets of equipment *Scientific and technological innovation guide* **12 (30)** 67

Vale C P and Abrantes V T 2014 An earlier example of portuguese concrete prefabrication, the "torre da boavista" condominium in porto *Sustainable Housing Construction* **12** 16

Sun H, Sun J, Wu Y H and Zhao K J 2015 Brief introduction of large Intelligent PC component Automation production Line *Concrete and cement products* **3** 35

Chen Y, Okudan G E and Riley D R 2010 Decision support for construction method selection in concrete buildings: Prefabrication adoption and optimization *Automation in Construction* **6(19)** 665

Jiang Y 2015 Research and Application of numerical Control production Line for Small-Scale concrete Prefabricated components *Logistics engineering and management* **37 (8)** 116

Mai Y F 2016 Key Technology of making concrete components in NC moving Mould table production Line *Building operations* **38 (6)** 767

Peter W 2010 Post-treatment of concrete Improvement of the visual quality of fair-faced concrete surfaces *Betonwerk und Fertigteil-Technik/Concrete Plant and Precast Technology* **76 (7)** 36

Choi H K, Choi Y C and Choi C S 2013 Development and testing of precast concrete beam-to-column connections *Engineering Structures* **11 (56)** 1820

Sun H, Sun M, Hou X L, Zhao K J and Wang R T 2014 Dynamic characteristics of concrete Mixer during Start-up *Journal of Shenyang Construction University (Natural Science Edition)* **4 (30)** 720

Liu H Q, Quan Z G and Guo Q 2014 Process Design and equipment selection of PC production Line in Xi'an Jiangong Weibei Industrial Park *Tile* **4** 34

Lu Y 2015 PC Application of Automation Technology in production Line of prefabricated components *Electric drive* **45** 69